Introduction to
Solid State Physics

Introduction to
Solid State
Physics

Amnon Aharony
Ora Entin-Wohlman

Ben Gurion University of the Negev, Israel
& Tel Aviv University, Israel

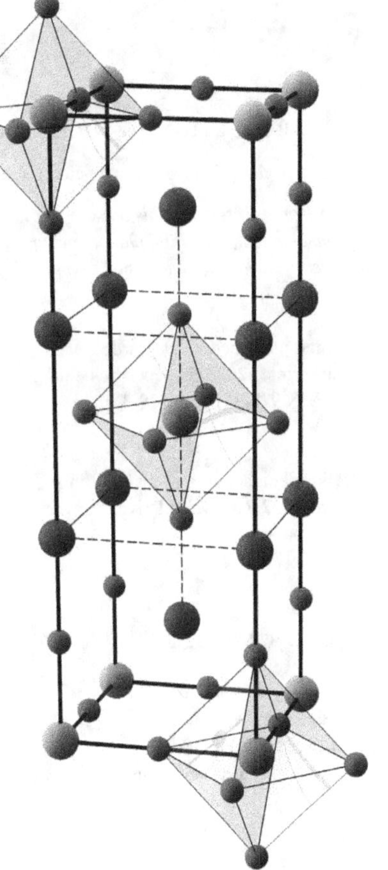

World Scientific

NEW JERSEY · LONDON · SINGAPORE · BEIJING · SHANGHAI · HONG KONG · TAIPEI · CHENNAI · TOKYO

Published by

World Scientific Publishing Co. Pte. Ltd.

5 Toh Tuck Link, Singapore 596224

USA office: 27 Warren Street, Suite 401-402, Hackensack, NJ 07601

UK office: 57 Shelton Street, Covent Garden, London WC2H 9HE

British Library Cataloguing-in-Publication Data
A catalogue record for this book is available from the British Library.

INTRODUCTION TO SOLID STATE PHYSICS

First published 2019 (Hardcover)
Reprinted 2020 (in paperback edition)
ISBN 978-981-122-129-3 (pbk)

ISBN 978-981-3272-24-8

For any available supplementary material, please visit
https://www.worldscientific.com/worldscibooks/10.1142/11041#t=suppl

Desk Editor: Rhaimie Wahap

Preface

Almost all natural materials solidify at a sufficiently low temperature. At equilibrium, these solids are crystals built of elementary units that repeat themselves periodically. This book introduces the reader to the physics of the solid state, and is focused on the properties of periodic crystals. It presents just a part of the much broader field of condensed matter physics, which encompasses also the study of noncrystalline materials, e.g glasses and soft materials, as liquids, liquid crystals, polymers, and membranes; the book sets a base for further studies of these more advanced topics.

The authors are indebted to many friends and colleagues whose comments and observations improved considerably the representation of the various topics. Special thanks are due to professors Yoseph Verbin and Yoram Kirsh, who followed closely our progress, commenting on each step. We thank professors Assa Auerbach, Dror Orgad, and Yaacov Kantor who reviewed critically our initial proposal for the contents of the book, and professors Alexander Palevski and Efrat Shimshoni who read each of the chapters and improved the contents and the representation in so many ways. Shlomi Matityahu and especially Yaacov Kleeorin are acknowledged for their invaluable help in the preparation of the exercises and many of the illustrations. Professors Ron Lifshitz, Yuri Rosenberg, and Hagai Shaked shared with us their expertise in central topics (quasicrystals, X-ray scattering and neutron scattering). We are grateful to professors Elyahu Eisenberg, Ady Stern and Moshe Schechter for their useful comments. Last but not least, the staff of the Open University of Israel, under whose auspices the Hebrew version of this book was prepared, is gratefully acknowledged.

A. Aharony and O. Entin-Wohlman

Contents

Chapter 1

Introduction

This chapter begins, in Sec. 1.1, with a survey of the different phases of matter. The particles in a gas are located at random positions; in a liquid (or in a glass) there is a larger statistical probability to find pairs of particles for which the two members are at a certain distance from one another, and in the crystalline solids the particles are arranged on a periodic lattice. The information on the particles' locations is encapsulated in the correlation function that quantifies the probability to find pairs of particles at a given distance in-between them. This book explains the fundamental concepts needed to describe the physics of such many-body systems, with examples focusing mainly on periodic solids (i.e., crystals), that are the energetically-favored structures at low temperatures. The understanding of crystalline structures explains the wide range of solid materials' properties, and paves the way to the development of advanced technologies.

Carbon appears in Nature in many versatile periodic structures, that exist in three spatial dimensions, but also in two dimensions (planar formations), in one dimension (chain-like networks) and also in zero dimensions (large molecules). These structures are described in Sec. 1.2, exemplifying many of the structures encountered in the subsequent Chapters.

Section 1.3 outlines the contents of the book's chapters: the explanation of crystalline structures, the description of experimental methods for identifying them, the analysis of the binding energies determining those structures, the discussion of the atoms' vibrations in the crystal, and the derivation of the electronic properties of solids.

1.1 What is the nature of the solid state?

The physics of the solid state is concerned with systems comprising a very large number of atoms or molecules; attractive forces cause them to occupy a finite volume in space, in a rigid formation. The physical laws that govern these forces are based on the theories of **mechanics**, **electricity**, and **quantum mechanics**. The large number of atoms requires also the application of the laws of **statistical mechanics**.

The same microscopic laws lead to the formation of various arrangements of atoms (or molecules) in the (macroscopic) solid, and thus to a variety of physical properties. The theory of the solid state aims to explain the way in which the same physical laws give rise to widely-disparate physical properties. For instance, why are certain solids good electrical conductors (i.e., metals), while others are electrically insulating, or inversely, superconductors? Why are there transparent materials that allow the passage of light (or sound), and materials that reflect the light (or sound) waves impinging on them? Why are certain materials rigid, and others are soft? How does it happen that certain solids possess magnetic properties and others possess ferroelectric (i.e., comprising electric dipole moments) ones?

The theory of solid-state physics not only explains materials found in Nature, but also enables the development of **novel materials** that can be used in **new technologies**. Conspicuous examples from the previous century include the advent of transistors, of integrated circuits and microelectronics, of magnetic memories, and more. The history of solid-state physics begins with the discoveries of von Laue and Bragg (the father and the son) related to the periodic diffraction patterns created by X-ray scattering off solid crystals (Chapter 3). Other important milestones are the explanation of the properties of electrons moving in the periodic potential of crystals, given by Bloch (Chapter 6), the discovery of the transistor by Bardeen, Brattain, and Shockley, the understanding of magnetism by Néel, Anderson, and van Vleck (Chapters 2, 4, and 6), the discovery of superconductivity by Kammerlingh-Onnes and its explanation by Bardeen, Cooper and Schrieffer, the theory of Josephson, and the experiments of Giaver *et al.* on tunneling between superconductors. More recent developments, taking place in the last 40 years, include the invention of the scanning tunneling microscope, the discovery of superconductivity at high temperatures, of systems possessing giant magneto-resistances, of quasicrystals, of the graphene films, of topological phases, and of the quantum Hall effect. Several of these important topics are surveyed in this book. Many of the physical examples given in the following Chapters are related to materials and properties that have been discovered only recently.

Phases of matter. Very crudely speaking, matter can assume one of the three **states** or **phases**: at high temperatures materials are in the **gaseous** state, in which the atoms or the molecules are moving freely in the vessel containing them, filling it randomly and homogeneously. The knowledge of the location of a certain atom conveys no information about the locations of the others; the **average density** of the gas is determined by the ratio of the number of atoms and the vessel's volume. (At much higher temperatures there appears the **plasma** state, in which atoms are disintegrated into nuclei and electrons.) As the temperature is reduced, it eventually reaches the **boiling temperature**, where the material separates into a denser portion, called **liquid**, and another portion, the vapor of that material, which behaves as a **gas**. This is the **gas-liquid phase transition**. Due to gravity, the denser fluid fills the lower part of the vessel. The surface tension of the liquid causes the formation of a film separating the fluid from the gas above it. The liquid is not rigid, and adjusts itself to the shape of the vessel; it is easily deformed. As in the gaseous phase, the location of a certain atom in the liquid state has no one-to-one correspondence with the other atoms' locations; however, there exists a probabilistic relation among these locations. As the temperature is decreased further it reaches the **melting temperature**, at which the **liquid-solid phase transition** takes place. At low temperatures almost all materials become solids. The **solid state** is a rigid configuration of the atoms or molecules, where their motion is restricted; at most they vibrate slightly around their equilibrium positions. A well-known example for transitions among the various phases is provided by **water**: at ambient pressure, water (which is a liquid) freezes and solidifies into ice below $0°$ Celsius, and turns into aqueous vapour above $100°$ Celsius. (The two temperatures are defined as the melting and boiling temperatures, and build the temperature scale of Celsius). These temperatures vary with pressure; e.g., under higher pressures it is harder for the molecules to turn into a gas, causing the boiling temperature to rise. Below we mention more phases of matter.

Forces among atoms (or molecules). The various forces operating among the atoms (or molecules) in the solid state are discussed in Chapter 4. Ubiquitously, the potential energy, V, that gives rise to the force between a pair of particles, depends on the distance between the particles , r, as illustrated in Fig. 1.1. At short distances, the derivative (with respect to r) of the potential energy is negative and hence the force is repulsive, preventing the particles from approaching too closely one another. At larger distances the derivative is positive and the force is attractive. The force vanishes at the minimum of the potential energy, $r = r_m$. Classically, (i.e., ignoring possible oscillations around the minimum) the point $r = r_m$ marks the **equilibrium** configuration of the pair.

A simple description of a large ensemble of particles is based on the assumption that the force operating between each pair of them is derived form the potential energy displayed in Fig. 1.1. Initially the particles are located at random positions, keeping the density as dictated by the number of particles divided by the volume of the container. An initial velocity, of a random magnitude and a random direction, is assigned to each particle; these velocities are chosen from a probability distribution that depends on the temperature and is found according to the laws of statistical mechanics. One then solves the Newtonian equations of motion to obtain (numerically) the particles' locations at later times. This method is named **"molecular dynamics"**. Allowing the computer code to run a long enough time, the system presumably achieves its equilibrium configuration. Figure 1.2 shows three instantaneous snapshots obtained by this procedure, pertaining to three different densities. Each particle is represented by a sphere of a finite radius, to emphasize the repulsion among the particles that prevents them from approaching each other (the spheres cannot penetrate one into the other). At equilibrium the instantaneous snapshots taken at different times are qualitatively the same; they are statistically similar.

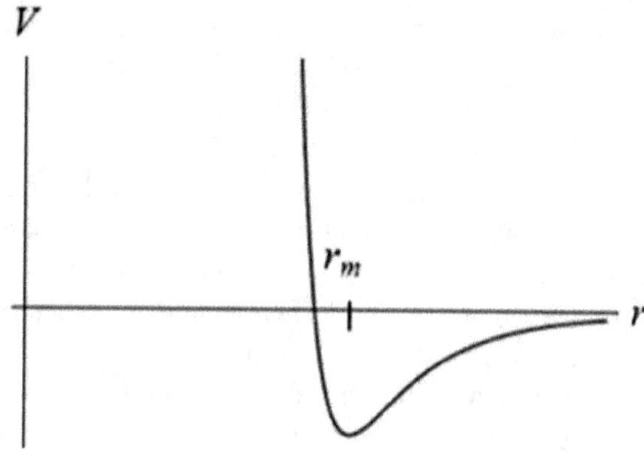

Fig. 1.1: Schematic plot of the potential energy of a pair of particles as a function of the distance in-between them (details in Chapter 4).

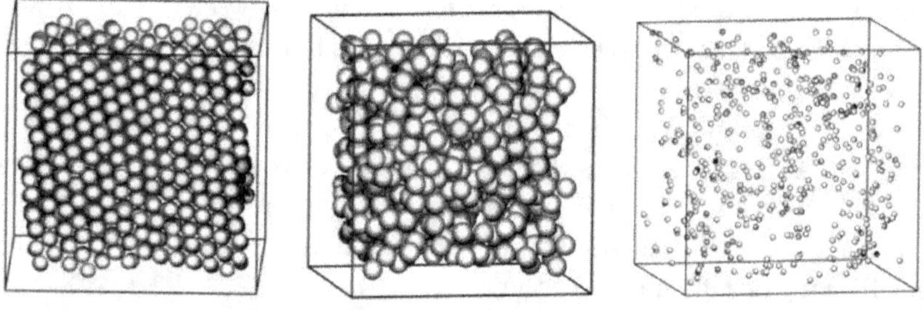

Fig. 1.2: Instantaneous snapshots taken from molecular-dynamics simulations of 500 particles, whose pair potential-energy is that of Fig. 1.1. The panels differ by the density of the particles: 0.01 (right panel), 0.5 (mid panel), and 1 (left panel). The density varies following variations of the vessel volume. Calculations by C. R. Iacovella and S. C. Glotzer. Taken from http://web.archive.org/web/20071007200029/http://matdl.org/matdlwiki/index.php/softmatter:Radial_Distribution_Function, by permission of Prof. S. C. Glotzer, University of Michigan.

The correlation function. The different panels in Fig. 1.2 correspond (from right to left) to the phases of gas, liquid, and solid. A quantitative way to distinguish among these phases is based on the **radial correlation function**. This function is computed as follows. Locating the origin at the center of a certain particle in each instantaneous snapshot, one finds the number of all other particles whose centers are within a sphere of a radius r around the origin. (The entire sphere is located within the vessel containing the particles.) Then one finds the number of particles in a slightly larger sphere, whose radius is $r + \Delta r$ (also contained in the vessel). The difference between the two numbers, ΔN, is the number of particles in the spherical shell of volume $4\pi r^2 \Delta r$, and the density in that shell is $\Delta N/(4\pi r^2 \Delta r)$. This density is **the probability density to find two particles separated by a distance** r. This procedure is repeated, applied on spherical shells centered around other particles, on various instantaneous snapshots, and for different initial velocities (but with the same spatial density and the same velocity distribution). One then averages the results over all snapshots, to obtain the radial correlation function

$$g(r) = \left\langle \frac{\Delta N}{4\pi r^2 \Delta r} \right\rangle , \tag{1.1}$$

where the angular brackets denote the average. In a similar fashion, the correlation function of a two-dimensional system is derived from the particles' density in a circular layer around a center of a certain particle in the plane, $g(r) = \langle \Delta N/(2\pi r \Delta r) \rangle$. Chapter 3 describes **radiation scattering** off material samples; it is shown there that the intensity of the scattered radiation is directly related to the Fourier transform of the correlation function.

Gas. Figure 1.3 displays the correlation functions computed for the snapshots shown in Fig. 1.2. In all cases the correlation function is rather small at short distances, as the particles are prohibited from approaching each other too closely (due to the repulsion among them). The correlation function in the left panel of Fig. 1.3 has a single peak at a distance that approximately corresponds to the minimum of the potential energy of pairs of particles (see Fig. 1.1). At larger distances, the probability to find another particle does not vary with r and remains constant. This is typical for a gas. The peak indicates that momentarily, the probability to find pairs of particles separated by the optimal distance is higher, owing to the fact that this yields a lower potential energy. Nonetheless, the correlation function does not convey any information about the relative locations of all other particles. As mentioned, one cannot predict the relative locations of all other particles (situated at distances far larger than r_m) from the location of a certain particle.

Liquid. In the mid panel of Fig. 1.2 the particles are much denser (as compared to the gaseous phase) and their motion is rather confined in space, though it is not completely blocked. The correlation function in the mid panel of Fig. 1.3 displays several broad peaks: these imply that around each particle there are several spherical layers in which the density of atoms exceeds its average value. Such a correlation function corresponds to a **liquid**. As opposed to the gaseous state, for which there are no correlations among the particles' locations, in the liquid state there are different probabilities for various separations between pairs of particles. The leftmost peak in the correlation function of the liquid resembles that of the gas; the next one represents particles that are neighbors to those featuring in the first peak, and so on.

Glass. When the motion of particles as those exhibited in the mid panel of Fig. 1.2 is stopped abruptly (by reducing precipitously their velocities, an operation equivalent to a sudden reduction of the temperature), the particles "freeze" in their instantaneous positions, and attain a random pattern, typical to **glasses** and to **amorphous materials**.

Ordered solid. As opposed to the mid panel of Fig. 1.2, most of the particles in the right panel there are **arranged in a periodic pattern:** the surroundings of all particles are identical and the entire structure repeats itself when translated from one particle to the other. Ordered rows of spheres touching one another can be discerned in the right panel in Fig. 1.2; generally, each sphere (particle) is surrounded by six others, located at equal distances away from it. The mathematical definitions of **periodic lattices** (also called **crystals**) are discussed in Chapter 2. It is shown there that the periodic arrangement displayed in the right panel of Fig. 1.2 is the densest one possible for adjacent spheres arranged on a plane (see Fig. 2.3), and the densest spatial ordering is built of such planes (the upper row in Fig. 2.24). As each neighbor of a certain sphere in the plane is also surrounded by six others, and so on, the entire planar structure is uniquely defined; the locations of all spheres can be predicted given the positions of a few of them around a central one (ignoring the small oscillations of each sphere around its equilibrium position). The same is true in three dimensions, where each particle is surrounded by 12 neighbor particles. Consequently, the correlation function in the right panel of Fig. 1.3 contains many peaks; the intervals in-between them correspond to the spherical shells in the ordered structure. The peaks become narrower as the sample volume is increased and the statistics is improved. This is the description of a **crystalline solid**, the main issue of the present textbook. In particular, it is found in Chapter 3 that each periodic structure has its own correlation function, that can be uniquely identified from the data collected by **scattering radiation** off the material; such experiments measure the Fourier transform of this correlation function.

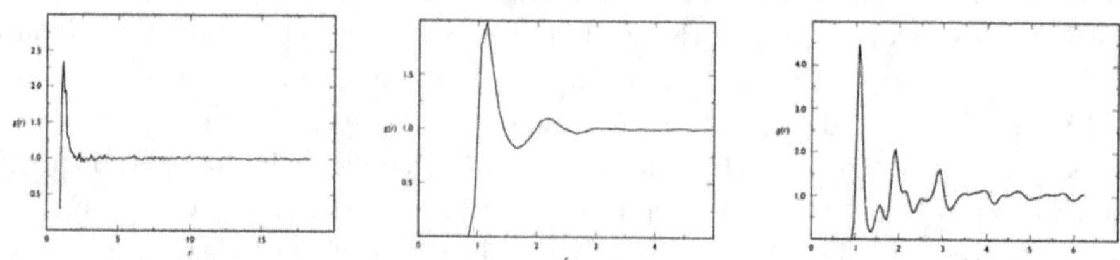

Fig. 1.3: The radial correlation function (normalized such that it approaches 1 at large distances; the radial distance is 1 for $r = r_m$), for the instantaneous snapshots depicted in Fig. 1.2. Taken from the same source as in Fig. 1.2.

Order and disorder. The right panel in Fig. 1.2 displays a well-organized pattern, in which the particles are located on a periodic crystal. In contrast, there is no definite order of the particles' locations shown in the other two panels there. At low temperatures the total energy, E, of the periodically-ordered structure is the lowest. Any deviation away from this order increases the energy. As the temperature is raised, the particles move faster, and more arrangements of them become possible. This, in turn, gives rise to an increase in the **entropy**, S; the entropy is related to the number of possible configurations of the system at a given energy, and it increases as that number grows. At any temperature T the system is in the state which has the minimal **free energy**, $F = E - TS$. As the temperature is raised, the entropy term becomes more dominant and can reduce the free energy, even though the energy is raised. Therefore, following the increase in the entropy, the amount of **disorder** in the system increases as well. In this way one may understand the transitions from solid to liquid, and from liquid to gas. At times these transitions are referred to as **order-disorder phase transitions**.

Quantum **fluctuations** (i.e., random changes or oscillations), like those of a harmonic oscillator even in its quantum ground state, can also lead to disorder. These fluctuations are quite significant in helium (because of the low mass of the helium atom), and consequently at atmospheric pressure helium remains in the liquid state (and even in the viscosity-free **superfluid** state, a unique phase by itself) down to very low temperatures. Hydrogen is lighter than helium, but since the bonds among hydrogen atoms are stronger than those in helium (whose electronic "shells" are full, forbidding the passage of electrons from one atom to the other), hydrogen does solidify at about $13°$ K, or $-260°$ C. The various bonds among atoms and the way they affect the specific structure of the material are discussed in Chapter 4. Interestingly enough, in two dimensions the oscillations of the atoms around their equilibrium locations destroy the long-range periodic structure; the periodic crystalline structure is then replaced by **topological phases** (see Sec. 5.5; these unique phases are currently at the forefront of the research in condensed matter).

Periodic solids *vs.* **glasses.** When the motion of all particles is halted, the resulting pattern is that of a **glass**; it resembles the one in the mid panel of Fig. 1.2. This is an extreme example of a **non-periodic solid**. Window glass is created upon suddenly cooling liquid silicone oxide, SiO_2. Cooling abruptly this material causes the particles (atoms or molecules) to freeze in their random positions, acquiring the same amorphous formation they had in the liquid state. At low temperatures, the free energy of this state is higher than that of the periodically-ordered solid. However, in order to reach the ideal periodic pattern corresponding to the lower temperature, the particles have to re-arrange themselves. Numerical molecular-dynamics studies indicate that this is not easily accomplished: each particle is captured at a local minimum of the free energy, and needs to overcome a potential barrier in order to move away from it. The material therefore remains in a state of a free energy higher than the "correct" (i.e., thermodynamic) one of that temperature; this is a **meta-stable state**. The lifetime of this state, that is, the duration of time until the true thermodynamic state is reached, is extremely long at low temperatures (for instance, that of window glass is hundreds of years), and therefore this state appears as a stable one during typical experimentally-accessible times. Figure 1.4(a) exhibits the amorphous arrangement of silicone and oxygen ions in a two-dimensional sample of glass. This structure differs qualitatively from the periodic structure displayed in Fig. 1.4(b). Periodic crystals of SiO_2 are obtained by melting the glass, and then cooling it slowly enough to allow the particles ample time to reach their optimal arrangement. Certain ceramic materials, as well as alloys of metals (e.g., steel) also attain amorphous patterns.

Quasicrystals. There are alloys (e.g., stainless steel) that arrange themselves in a periodic pattern of their constituents; others freeze in a non-periodic structure. But surprisingly enough, when the cooling rate is neither too slow nor too abrupt, the resulting solid is found in an intermediate phase, named **quasicrystalline**. The order of the ions in this phase is not periodic; however, their pattern does possess unique symmetries – quasicrystal show certain phenomena also found in periodic crystals (see Secs. 2.8 and 3.11).

Condensed matter. Solids can be crystalline, as in the left panel of Fig. 1.2 and in Fig. 1.4(b), or amorphous, as in the mid panel of Fig. 1.2 and in Fig. 1.4(a). In addition to the liquid and solid states, materials can assume other phases, for instance that of **liquid crystals**, which are built of two-dimensional parallel layers, with freely-moving molecules on each of them; **polymers**, i.e., very long, practically one-dimensional molecules, moving freely in space; **membranes**, that are two-dimensional surfaces oscillating along the normal to their plane, and more. These phases that comprise properties of liquids and solids, are included in the (relatively novel) scientific discipline called **condensed-matter physics**. The present book discusses mainly **periodic solids**, like the one shown in the left panel of Fig. 1.2; this is part of condensed-matter physics (see also Sec. 7.2).

The periodic table. Many of the specific examples discussed in this book refer to crystals based on atoms, i.e., on elements. The disparities and similarities among the various elements follow from their respective locations in the **periodic table**, displayed in Fig. 1.5. The entries in this table contain the atomic number of the element (below the letter indicating its name) and its atomic weight (above that letter). The different hues indicate the melting temperature of the crystal built of each element. As seen, the melting temperatures of the noble gases (in the right column of the table) are rather low. In contrast, the melting temperature of diamond, built of carbon atoms, is very high. The melting temperature is higher when the **binding energy** of the crystal (the energy released as the material is transformed from the gaseous state to the liquid one) is larger. Binding energies are calculated in Chapter 4; the different values obtained for them explain the disparity in the melting temperatures of various elemental crystals. The periodic table reappears also in other contexts: Fig. 2.22 of Chapter 2 displays the lattice structures of the crystalline elements; Fig. 6.24 of Chapter 6 identifies their electrical properties (i.e., metals, insulators, *etc.*). Beside atomic crystals, there exist more

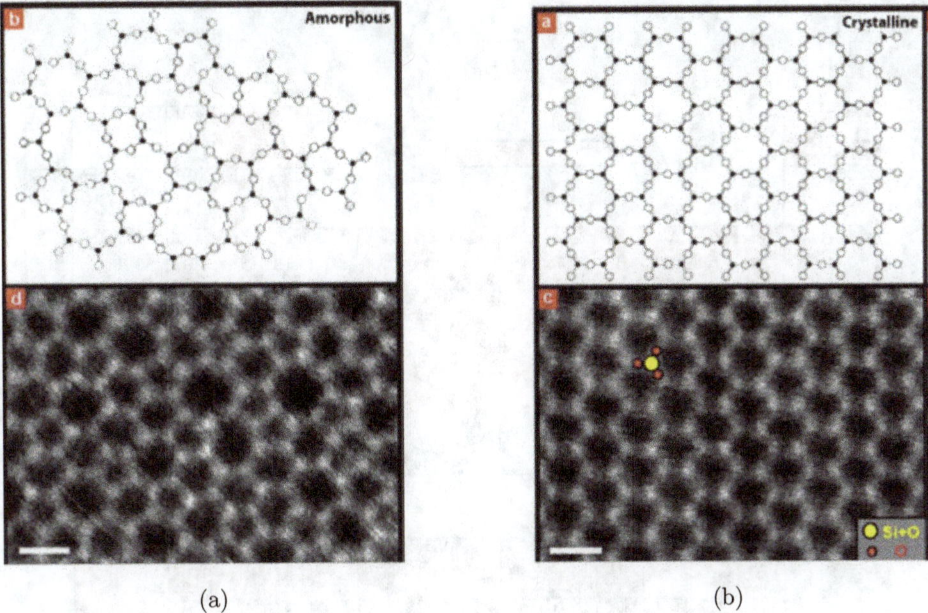

(a) (b)

Fig. 1.4: Amorphous (a) and periodic (b) arrangements of silicone ions (full circles) and oxygen ones (empty circles) in two-dimensional SiO_2. Top panels: theoretical models; lower panels: scanning electron microscope (see Sec. 3.1) measurements of a single layer absorbed on a graphite substrate (see the next section). Each silicone atom is connected to an additional oxygen atom, located below the plane of the picture, so that each Si is coupled to two O's. [P. Y. Huang, S. Kurasch, A. Srivastava, V. Skakalova, J. Kotakoski, A. V. Krasheninnkov, R. Hovden, Q. Mao, J. C. Meyer, J. Smet, and U. Keiser, *Direct imaging of a two-dimensional silica glass on graphene*, Nano Lett. **12**, 1081 (2012).]

complex crystals composed of several atoms; an example is table salt, comprising ions of sodium and ions of chlorine, depicted in the right panel of Fig. 2.17. The structures of such crystals are related to the effective radii of the constituent ions (and these in turn are determined by the wave functions of the electrons in each ion); these are shown in Fig. 4.6 of Chapter 4.

1.2 The spatial dimensions

Solid phases. The periodic crystalline phase is quite versatile: certain materials may appear in various **structural formations**, that have divers properties. These are called **polymorphs**, and are **different phases of the same solid material**. For instance, to date there are sixteen(!) disparate periodic structures of **ice**, some of them discovered only recently. The beautiful patterns of snow flakes are related

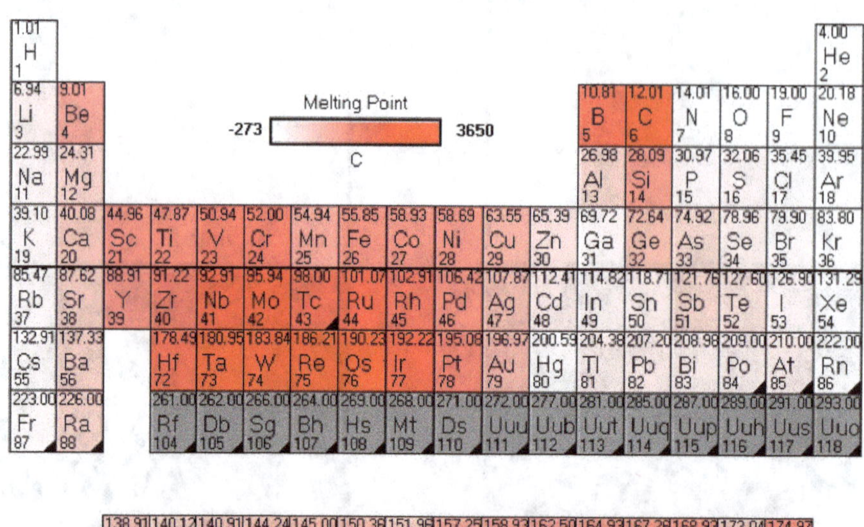
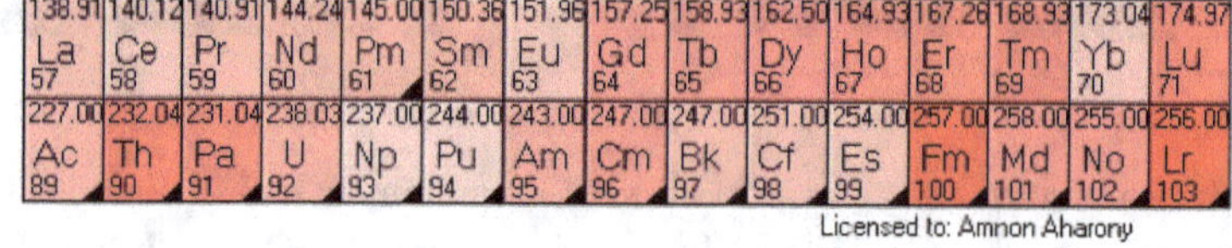

Fig. 1.5: The periodic table. The various hues indicate the melting temperatures of the elements, according to the scale above (darker hues correspond to higher melting temperatures). Created by Atomic PC, see http:/www.blackcatsystems.com/software/periodic-table-of-the-elements-software.html.

to several of these phases (see Fig. 4.29 of Chapter 4). Temperature or pressure variations, (i.e., changes in the energy and in the entropy) lead to transitions among the various phases, and hence to modifications in the material properties; e.g., from an insulator to a conductor and even to a superconductor (whose electrical resistance vanishes at low enough temperatures), from a non magnetic substance to a **ferromagnet** that has a finite magnetic moment (like that of the compass needle), from a dielectric material to a **ferroelectric** one, which possesses an electric dipole moment, and so on.

Carbon. Carbon appears in Nature in more than forty different structures; the polymorphs of elements (like carbon) are called **allotropes**. Several of these are surveyed in this section. In **three dimensions**, as explained in Chapter 2, there are mainly two carbon allotropes: (1) the **graphite** (see the right panel of Fig. 2.9). This structure comprises parallel planes, each contains a periodic array of hexagons built of the carbon atoms (see Sec. 2.3). (2) When prepared under high pressures, carbon has the **diamond** structure, shown in the left panel of Fig. 2.20. In this configuration, each carbon atom is surrounded by four others, as displayed in Fig.

4.14. Graphite is soft, and is a good electrical conductor, while diamond is hard and is an insulator. Carbon attains more conformations, that can be characterized by a smaller number of spatial dimensions. Chapter 4 details the covalent bonds among the carbon atoms, which explain these various forms.

Graphene. In **two dimensions**, the carbon atoms form the structure shown in Fig. 1.6 [see also Figs. 2.8(a) and 2.10]. This planar structure is identical to one of the planes of graphite (shown in the right panel of Fig. 2.9). Graphene was originally produced by peeling single-atom layer off graphite, using a sticky tape; its unique properties are currently subjected to intensive research.

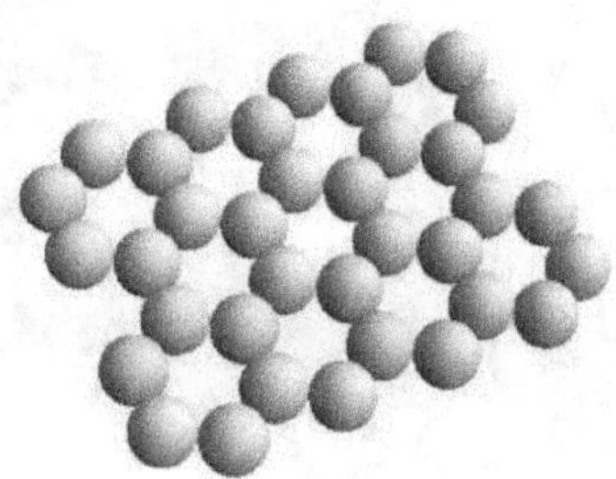

Fig. 1.6: Schematic illustration of the carbons in graphene.

One-dimensional allotropes of carbon. Carbon possesses several **one-dimensional** periodic structures. Figures 1.7(a) and (b) depict two long molecules of carbon. The **cumulene**, Fig. 1.7(a), is built of a periodic arrangement of carbon atoms, with the same spacing between each two neighboring atoms. In the **polyyne** [Fig. 1.7(b)] on the other hand, there are two types of bonds, marked by a single line and by a triple line. These bonds alternate along the molecule, and therefore the structure repeats itself after each second carbon. The formal definitions and the descriptions of these arrangements are detailed in Chapter 2, and the electronic configurations leading to them are analyzed in Chapter 4.

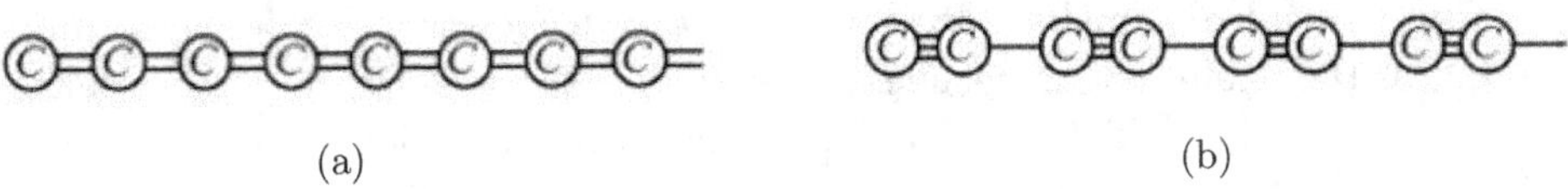

(a)   (b)

Fig. 1.7: One-dimensional chains of carbon. (a) Cumulene; (b) polyyne.

One-dimensional allotropes of carbon can be also prepared by cutting graphene into **narrow ribbons** or **nano ribbons**, of widths that comprise only a small number of atoms. Figures 1.8(a) and (b) display two options for this cutting. The structure on the cut (on the left side) in Fig. 1.8(a) is called **armchair**, and that in Fig. 1.8(b) (again on the left hand-side) is called **zigzag**. Each carbon on the cut has a "superfluous" electron, that can bind chemically a hydrogen atom.

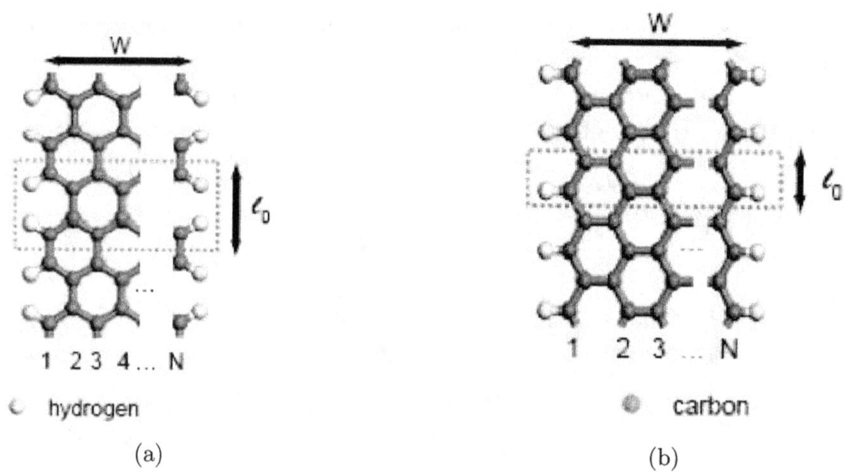

Fig. 1.8: Carbon nano ribbons. [P. S. E. Yeo, K. P. Loh, and C. K. Gan, *Strain dependence of the heat transport properties of graphene nano ribbons*, Nanotechnology **23**, 495702 (2012).]

Carbon nanotubes. When a ribbon of graphene (see Fig. 1.8) is folded, and the two opposite edges are attached together (instead of coupling there hydrogen atoms), a nanotube, Fig. 1.9(a), is created. This tube can be quite long. Nanotubes draw currently vast interest due to their very unique transport properties.

Fullerene. Other big carbon molecules may assume a spherical shape, like the molecule C_{60} shown in Fig. 1.9(b). It is called **Buckyball, Buckminster-fullerene**, or **fullerene**, owing to its similarity to the spherical dome designed by Buckminster-Fuller for the 1967 exhibition in Montreal. There exist also "half" molecules of this type, connected at the end of a nanotube.

The spatial dimension. It is customary to define the spatial dimension of a sample as the number of independent directions in space along which it is spanned over long distances. The ideal periodic solid is infinitely long along each of its dimensions. According to this definition, graphite and diamond are (spatially) three dimensional, graphene is two dimensional, while the cumulene, polyyne, nano ribbons and nanotubes are one-dimensional; the fullerene is just zero dimensional. The properties of a solid are significantly affected by its spatial dimensions.

Nanotechnology. The present day highly-advanced facilities enable the production of zero-, one-, and two-dimensional samples in the lab, and to exploit them for various applications. At ultra-low temperatures and extremely small length

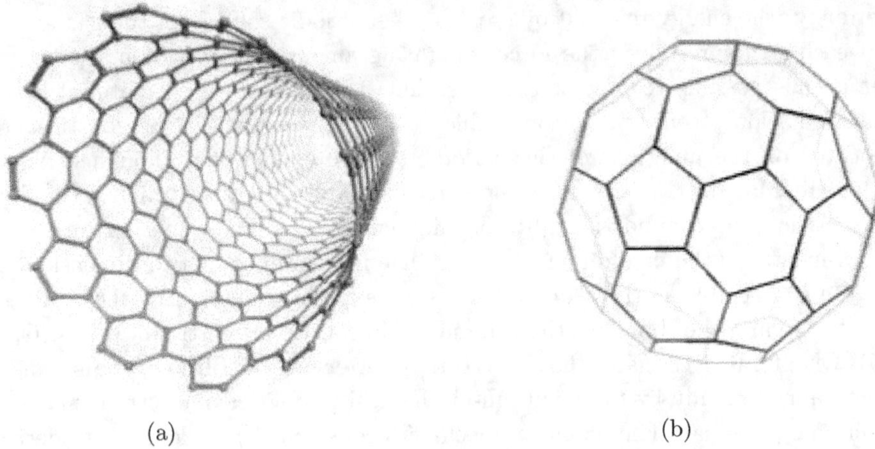

(a)   (b)

Fig. 1.9: (a) A carbon nanotube; (b) the Buckminsterfullerene molecule, or Bucky-ball. The carbon atoms are situated at the junctions, so that each has three carbon neighbors like in graphene.

scales, the electronic properties of such systems are dominated by quantum mechanics. Those samples are therefore termed **quantum dots, quantum wires**, and **two-dimensional electron gas, 2DEG**, respectively. The branch of condensed-matter physics centered on materials of low dimensions, in particular on small samples (of length scale in-between the macroscopic and the microscopic ones), is termed **"mesoscopic physics"**. This field encompasses both basic research, focusing on the physical properties (which are determined by quantum mechanics), and applied research, aiming to develop new technologies. As the size of these systems is typically about nanometers (1 nanometer= 10^{-9} meter) this particular branch of condensed-matter physics is called **nanotechnology**.

1.3 The present book

As mentioned, this book is centered primarily on the properties of **periodic crystals**. The (almost) perfectly-periodic crystal is the structure that most materials assume at very low temperatures. The discussion of its properties introduces the basic concepts, required to grasp and appreciate more advanced topics. Here is a terse survey of the following Chapters. Chapter 2 finds that only **a finite number of crystalline formations is possible**. Their comprehension allows for the analysis of **experiments** in which X-rays, electrons or neutrons **are scattered off** the solid material; in this way various crystalline patterns are uniquely identified (see Chapter 3). Moreover, such experiments convey valuable information on the **internal structure** of each unit of the crystal. The fact that there is just a finite number of crystalline conformations enables the calculation of the **binding energy** (the energy released by creating the periodic solid from the gaseous state) of each structure, given the forces operating among its constituents (Chapter 4). It is thus possible to predict the crystalline formation of each material by calculating its lowest energy, the one achieved in the state where the constituents are located on the junctions of the periodic structure. One has though to allow for the **thermal motion** (at finite temperatures) and the **quantum fluctuations** of the basic units around this state of minimal energy. These motions are at times detrimental to the crystalline structure (like in helium or at two dimensions), and cause eventually (as the temperature is raised) the transition to the liquid state. They are also the source of the thermal and acoustic properties of solids (Chapter 5). Chapter 6 investigates the **electronic properties of solid crystals**. These are determined mainly by the quantum-mechanical properties of electrons moving in a periodic potential. The quantum-mechanics considerations allow one to derive the electrical properties of various solids, and to distinguish among metals (conductors), insulators, and semiconductors. Chapter 7 surveys certain advanced topics.

Chapter 2

The crystalline structure of solids

The focal point of this chapter is the **geometry of periodic crystals**. The first two sections present one- and two- dimensional periodic crystals, with evaluations of their packing densities. Section 2.3 brings the formal definitions of Bravais lattices and crystals comprising a lattice and a base, and lists several examples, mainly two-dimensional ones (for instance, graphene). The formal description of periodic lattices is continued in Sec. 2.4 with the definitions of lattice vectors and unit cells, e.g., the primitive and the Wigner-Seitz unit cells. The next two sections survey the ubiquitous three-dimensional periodic lattices: the cubic crystals and the hexagonal-closed-packed ones, and Sec. 2.7 explains the classification of the lattices according to their symmetry properties. It is proven that there is a finite number of possible Bravais lattices, and hence each crystal can be assigned to a certain Bravais lattice. The remaining sections are devoted to several advanced topics, which are still actively investigated. Quasicrystals are described in Sec. 2.8, epitaxial adsorption, thin films, superlattices and multi-layered structures, are reviewed in Sec. 2.9; Sec. 2.10 surveys tersely magnetic crystals. Appendix A lists the five platonic solids; these bodies are referred to throughout the main text, serving as demonstrations of various symmetries of solids.

2.1   Lattices and crystals in one dimension

Periodic solids. The main part of this chapter is focused on the **symmetry properties of perfectly-periodic solids**. Metallic alloys that occasionally appear in non-periodic **quasicrystalline** structures are briefly discussed in Sec. 2.8. The discussion ignores the **vibrations of the atoms in the crystal** (those exist even at zero temperature due to the quantum uncertainty principle) and is centered on the geometric pattern formed by the average locations of the atoms (or ions) in the material. Lattice vibrations are studied in Chapter 5.

Ideally, the atoms (or the molecules) are arranged in a **periodic structure**, termed a **"crystal"**. The basic unit of the crystal is the **"base"**. The base may consist of a single atom (or an ion) or of a group of atoms (ions). The entire

15

material is a periodic array of such basic units. A basic unit volume, called the
"unit cell", is assigned to each base. These unit cells are adjoint to each other and
fill periodically the entire space. Representing each base by a single point (e.g., the
center of mass of its constituents) produces an ensemble of an infinite number of
points, termed a **"lattice"**. Examples of lattices and crystals are presented below,
followed by formal definitions of the concepts introduced above. The last three
sections survey several advanced topics, based on these concepts.

One-dimensional crystals. The simplest example is that of **one-dimensional lattices.** As mentioned, there exists nowadays a vast experimental
endeavor to study one-dimensional systems, in the context of **mesoscopic physics
and nanotechnology.** The upper panel in Fig. 2.1 displays a periodic array of
identical points which form the most elementary one-dimensional periodic lattice.
Such a system can be derived for instance from Fig. 1.7 displaying the **cumulene,**
when each carbon atom there is replaced by a single point located at its center.
Formally speaking, a **lattice is an infinite collection of points** and the periodic
lattice is a special case of such an ensemble. In the case of Fig. 2.1, the "surroundings" of each site in the lattice is identical to that of any other site. Alternatively,
when one "resides" on any of the lattice sites and "looks" around, one observes
exactly the same view. **A lattice in which all points are surrounded by
identical environments is a Bravais lattice.**

Lattice vectors and lattice constants. The second row in the upper panel of
Fig. 2.1 repeats the very same lattice, but with certain additions. The arrow there
represents a **vector a**, that connects two neighboring lattice sites. Such a vector is
a **lattice vector**. The length of the lattice vector, denoted a, is termed the **lattice
constant**. It is identical to the distance in-between two nearest-neighbor sites. An
important property of the one-dimensional Bravais lattice is that **each point on it
can be reached from any other by an integral number of lattice vectors.**
In other words, the nth site on the lattice is located at $\mathbf{R}_n = n\mathbf{a}$ (relative to an origin
assigned to an arbitrary lattice site), or alternatively, the original lattice cannot be
distinguished from the one obtained upon displacing all lattice sites by the lattice
vector **a** (assuming an infinitely long lattice). This property identifies the lattice as
a **periodic one.** It can also serve as a definition of a Bravais lattice.

The unit cell. In that second row of upper panel of Fig. 2.1 there appear
also two rectangles, of length a and of an arbitrary width. (In the one-dimensional
case discussed here, each "rectangle" is in fact a straight segment of zero width;
the finite widths are guides to the eye.) One may assign a lattice point to each
"rectangle" (e.g., the point located at the center of the right rectangle, or the point
located at the left end of the left rectangle). When each of the lattice sites is
covered by one and only one of these rectangles, then the collection of rectangles
covers the entire axis. Each of these rectangles (or segments) may be declared as
the **primitive unit cell** of the lattice. The **primitive unit cell is the shortest
segment (or the smallest area in two dimensions, or the smallest volume**

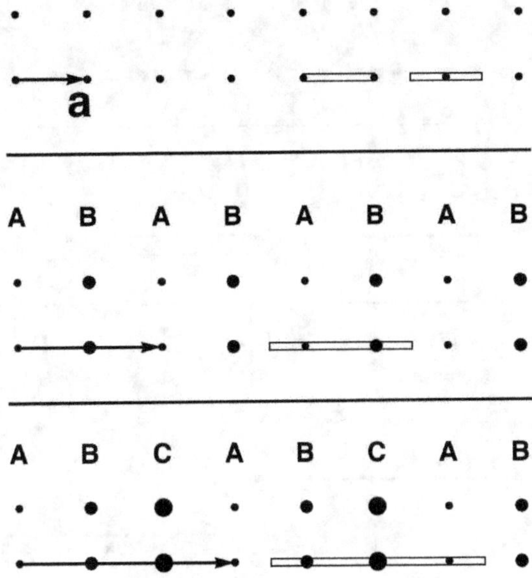

Fig. 2.1: Examples of one-dimensional crystals. The arrows represent the lattice vector that transforms the crystal to itself. The points reached by the arrows form the Bravais lattice. The rectangles represent possible unit cells.

in three dimensions) that contains the base of the crystal (a single point in the example above). Displacing the primitive unit cell by all integral multiples of the lattice vector covers the entire axis (or plane, or space). The primitive unit cell is not uniquely defined: any segment ("rectangle") of length a that covers a single lattice site can serve as the primitive unit cell.

Lattices with a base. All the structures displayed in Figs. 2.1 and 2.2, except the one in the first panel of 2.1, are **not** Bravais lattices, since the surroundings of the sites in them are not identical. For example, the lattice featuring in the second panel of Fig. 2.1 contains "small" points (A), placed in-between "large" ones (B). These points may represent, e.g., two types of atoms of different radii. The surrounding of the A atoms differs from that of the B ones (the nearest neighbors of the A's are B's, and *vice versa*). On the other hand, the surroundings of the small atoms are all identical, and so are also those of the large atoms. Hence, this crystal can be described by a basic unit containing a pair of points, like those within the rectangle. In that case the system is a **crystal**, the pair of atoms is its **base**, and the rectangle containing them serves as the **primitive unit cell** of the crystal. (That is, it is the smallest possible segment which includes solely a single base, and upon being displaced by the lattice vector –shown by the arrow–covers the entire axis.) In summary: the base is the cluster of sites **within** each unit cell, and the cell

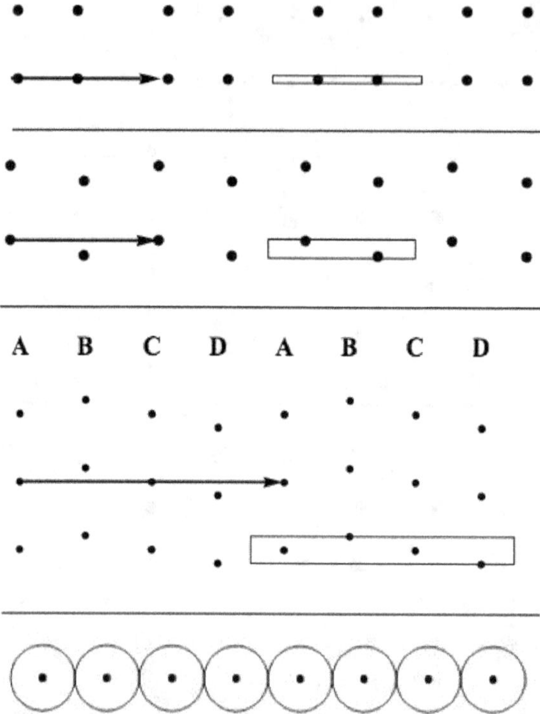

Fig. 2.2: One-dimensional lattices with larger unit cells, and a lattice built of circles (or spheres) arranged densely on a straight line.

itself (a segment in the one-dimensional crystal) is the volume including the base. The crystal consists of periodically-repeated bases. The length of the primitive unit cell in this example is twice that of the previous one shown in the upper panel of Fig. 2.1; so is also the length of the lattice vector (the arrow there).

When the two points (atoms) become identical, the crystal illustrated in the second panel of Fig. 2.1 is the same as the first panel there. The unit cell containing two atoms, whose length is $2a$, may still faithfully describe the lattice. However, that cell is **no longer primitive**, as there exists a smaller cell (like the one in the upper panel of Fig. 2.1) that contains a single atom, and that covers the entire axis when displaced by multiples of the lattice vector. It is occasionally convenient to use unit cells that are not primitive; the physical properties are independent of the choice of the unit cell (as long as its displacements by multiples of the lattice vector covers the entire axis).

Similarly, the lowest panel in Fig. 2.1 displays a crystal whose base includes three different atoms, with a unit cell and lattice vector of length $3a$. In the first three panels of Fig. 2.2, the lattice sites are shifted as compared to their locations in the "simple" case in the first panel of Fig. 2.1. For instance, the lattice exhibited in the first panel in Fig. 2.2 represents the one-dimensional crystal of carbon, the polyyne [Fig. 1.7(b)], for which the distances among neighboring carbon atoms assume two distinct values. Again, the surrounding viewed from each lattice site is not identical to that seen from all others, resulting in a crystal that contains bases of two atoms (first two panels) or four (third panel in Fig. 2.2).

2.2 Two-dimensional lattices

Dense packing. Section 2.1 describes a lattice as an ensemble of points. Alternatively, one may consider the lattice in the first panel of Fig. 2.1 as a collection of identical segments ("rectangles", as those shown there) contiguous to one another. Replacing each lattice site by a rectangle results in the **densest possible packing** for these rectangles. A similar conclusion is reached when each lattice site is enclosed in a sphere: the densest packing of spheres whose centers are confined to lie on a straight line is obtained for a periodic array, such that each sphere touches its two neighbors, see the last row of Fig. 2.2. In this example, the lattice built of the spheres' centers is one-dimensional, but the base assigned to each lattice point is a three-dimensional sphere. There is only a single dense-packed array in one dimension.

(a)   (b)

Fig. 2.3: The densest possible packing of identical circles (a) and of identical spheres (b) lying on the plane.

In higher dimensions there are more ways to densely-pack spheres. Figure 2.3(a) displays a dense arrangement of identical circles (e.g., coins) on the plane. Encompass the dark circle at the center by identical circles, each being tangential to the dark circle and to the circle introduced before; the dark circle is then surrounded by exactly six circles (the dashed-grey ones in the figure). Adding more circles such that each is tangential to as many circles as possible results in the array displayed in Fig. 2.3(a). This structure is repeated until filling the entire plane. **The packing ratio of this structure is defined as the ratio of the total area of the circles to the total area of the plane.** The structure in Fig. 2.3(a) forms the

densest packing that can be attained by identical circles on the plane, i.e., it is
the array with the highest packing ratio. The proof is outlined below: one shows
that there is only a finite number of periodic lattices in two dimensions and then
calculates the packing ratios of them all, to find that the pattern in Fig. 2.3(a) is
the densest. **The number of circles that are nearest-neighbors to any given
circle, usually denoted by** z**, is termed the "coordination number"**; it is 6
for the structures of Fig. 2.3.

Another example for dense packing is realized in the following experiment: ping-
pong balls are thrown on a swimming pool. As long as the number of balls is small,
they land randomly on the water surface. However, as their number increases, the
balls come closer to each other. Eventually they arrange themselves in the pattern
displayed in Fig. 2.3(b) (observed from above). Such a layer of ping-pong balls
is exploited in the northern countries in out-of-doors swimming pools to prevent
the water there from freezing in winter. When a person jumps into such a covered
pool (the northern people do so, after taking a hot sauna followed by a splash in
the snow), the balls are scattered at all directions; but once the swimmer has gone
somewhere else the balls re-arrange themselves in the dense-packing pattern. This
indicates that this is their optimal arrangement. Though each unit in this "crystal"
is a three-dimensional sphere, their centers all lie on a plane, and therefore the two
lattices displayed in Fig. 2.3 are considered as two-dimensional.

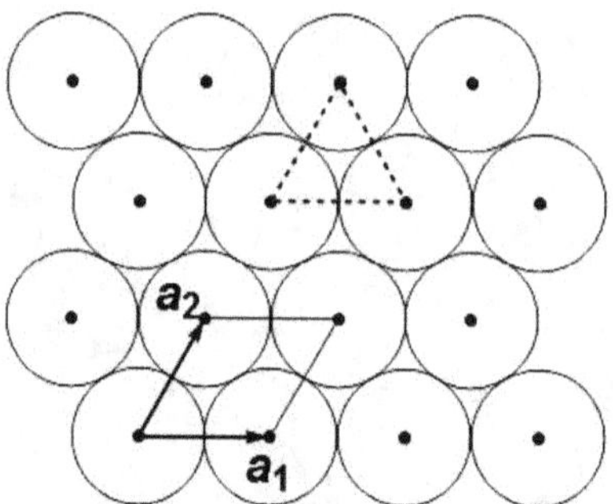

Fig. 2.4: Close-packed identical circles on the plane. The points at the centers of
the circles form the triangular lattice. The arrows represent the two lattice vectors,
$\mathbf{a}_1$ and $\mathbf{a}_2$. Displacing all points by each of the lattice vectors transforms the entire
lattice back to itself. The length of each lattice vector equals the lattice constant:
it is the distance between neighboring lattice sites, a. In the example of Fig. 2.3(a),
this distance is the diameter of each circle. The rhombus spanned by the two lattice
vectors represents a possible choice of a primitive unit cell.

The triangular lattice. Leaving the center of each circle (or sphere) in Fig.
2.3 as is, and decreasing the radii to an infinitesimal value, results in the points

depicted in Fig. 2.4. **This collection of points** is an example of a **periodic lattice in the plane.** The lattice portrayed in Fig. 2.4 is termed the **"triangular lattice"**: connecting together three nearest-neighbor sites gives an equilateral triangle (indicated by the dashed lines); the entire lattice is built from equilateral rhombi (each comprising two equilateral triangles) contiguous to each other and repeating themselves periodically. Such a rhombus (see Fig. 2.4) can serve as a primitive unit cell. At times this lattice is called the **"hexagonal lattice"** for reasons clarified below.

Crystal growth. A physical way to build the triangular lattice is based on a collection of particles, with a central force operating between each pair of them. The force is repulsive when the two particles are close together, and is attractive when they are farther away from one another. Such a force results from the potential energy plotted in Fig. 1.1. This potential energy characterizes various types of crystal bonds (see Chapter 4). The total potential energy of the particles forming the crystal is often dominated by the energies related to the forces between nearest neighbors; the forces exerted by farther-away particles are less effective. At equilibrium, the distance between the centers of neighboring particles is roughly the one that corresponds to the minimum of this two-particle potential energy. Intuitively, it is plausible that when many such particles are distributed on the plane, the maximal gain in the potential energy of the whole system is reached when each particle has as many nearest neighbors as possible (i.e., the largest feasible coordination number), each residing at a distance corresponding to the minimal value of the pair-wise potential energy. This gain is realized for the structure shown in Fig. 2.4, where the center of each particle is located at a vertex of the triangular lattice and has six nearest neighbors. It is impossible to construct a two-dimensional periodic lattice made of identical particles with a coordination number larger than 6.

The process outlined in Fig. 2.3(a), that begins with a single particle to which more and more particles are added, imitates a method for **crystal-growing based on a gradual adsorption of identical particles on a planar substrate.** Once the first particle (the dark circle) is attached to the substrate, a second one is added. The two particles move on the plane, attracted to one another by the force operating in-between them, until they form a little cluster in which they are separated by the distance corresponding to their minimal potential energy. A third particle arriving on the plane moves towards this cluster, to form with its two predecessors an equilateral triangular cluster (the dashed triangle in Fig. 2.4), where each pair of particles is separated by the distance at which the potential energy is minimal. More particles join in a similar manner, thus building the triangular lattice. The triangular lattice is formed in the plane by identical particles exerting on each other short-range forces such that each one adheres to its neighbors. Crystals comprising more types of atoms (for instance, table salt, built of positive sodium and negative chloride ions) for which long-range forces are also effective (i.e., the Coulomb force among the ions) grow in a different pattern.

In reality, the situation is more complicated (see below), as clusters of various perimeters can grow around different initial locations, and do not necessarily join together smoothly. This may result in an aperiodic metastable structure; an example is illustrated by the atoms B and C in Fig. 2.23. **A metastable state occurs when the system is at a local minimum of its energy, but not in the global minimum.** To reach the latter, the system must overcome potential barriers, a process not easily accomplished at low temperatures. Warming up the substrate on which the crystal is growing helps to overcome these barriers, and allows small clusters to merge into larger ones. Ideally, the particles reach the structure of a single triangular lattice.

There are other methods to grow crystals. Crystals grown under controlled conditions are needed for certain precise measurements or for building up artificial compounds which do not exist in Nature. Other ways to grow crystals are not detailed here (but see Sec. 2.9).

The unit cell. Figure 2.5 illustrates the concept of a **unit cell**, mentioned in the context of one-dimensional crystals. Filling the area around each circle in Fig. 2.3(a) up to the tangents common to that circle and all of its nearest neighbors, results in a periodic collection of hexagons that tiles the entire plane. Each hexagon represents a unit cell "belonging" to the circle inside it (or to the lattice site at the center of that circle). Each of the hexagons in Fig. 2.5 **contains all the points in the plane that are closer to the lattice site at its center than to any other center.** This unit cell is termed the **Wigner-Seitz unit cell**. Its hexagonal shape explains why the triangular lattice is occasionally referred to as the "hexagonal lattice". The Wigner-Seitz cell is a primitive unit cell; it contains a single lattice site (which is the base of this crystal); its displacements by the lattice vectors (displayed in Fig. 2.5) tile the entire plane. The Wigner-Seitz unit cell is not the only possible primitive cell: for instance, the plane can be covered entirely by identical rhombi as the one displayed in Fig. 2.4. Since both the rhombi and the hexagons tile the whole plane, and each contains a single lattice point, their areas must be equal (see problem 2.5). The rhombus can be displaced so that the lattice site lies at its center (rather than at its left lower corner); this gives rise to yet another optional primitive unit cell.

Problem 2.1.

a. *What is the packing ratio of the planar structure in Fig. 2.3(a)?*

b. *Each circle there is replaced by a sphere, and the spheres are densely arranged on a plane as in Fig. 2.3(b). Another plane, tangential to the spheres, is added from above. What is the resulting Wigner-Seitz cell? What is the ratio of the total volume of the spheres to that enclosed between the two planes? (this is the ratio of a sphere's volume to the volume of the unit cell). This ratio is termed the* **volume packing-ratio** *of the system.*

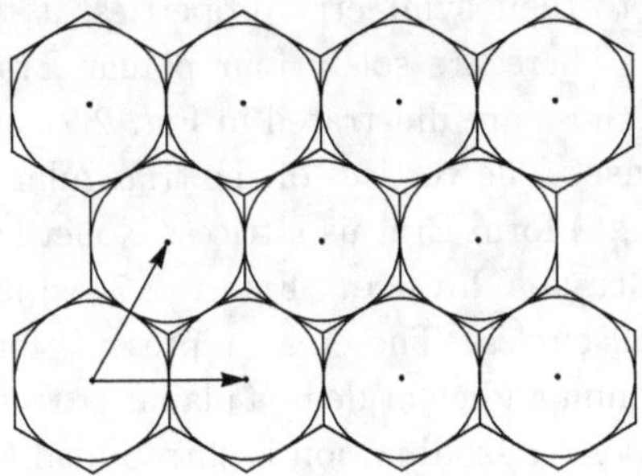

Fig. 2.5: Hexagonal primitive unit cells (Wigner-Seitz cells) representing the triangular lattice of Fig. 2.4. The lattice points of Fig. 2.4, that are also the centers of the circles in Fig. 2.3(a), are located at the centers of the hexagons. The arrows are the lattice vectors (as in Fig. 2.4).

Problem 2.2.
Figure 2.6(a) portrays an array of mutually-tangential circles, each touching solely four others at the end points of two perpendicular diameters. The lattice formed by the circles' centers (marked by dots) is called the **"square lattice"**.
a. *Enclose each circle in a Wigner-Seitz cell, so that the cells completely cover the plane. What is the shape of this cell?*
b. *What is the packing ratio of this structure?*
c. *What is the volume packing-ratio for spheres arranged on a square lattice [Fig. 2.6(b), when the spheres are caged between planes from below and from above]?*

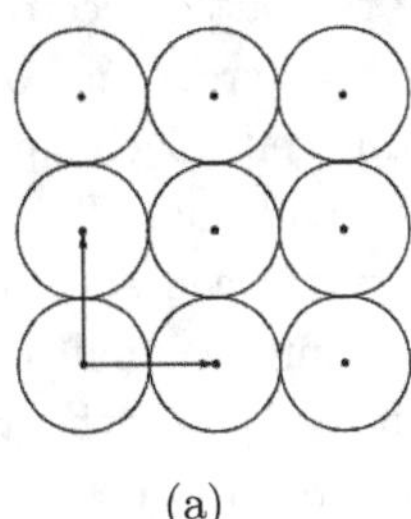

(a)

(b)

Fig. 2.6: The square lattice, built of adjacent circles (a) or contiguous spheres (b). The arrows represent the lattice vectors.

2.3 Planar Bravais lattices with a base

Bravais lattices. The triangular and the square lattices are two members of the family of **Bravais lattices**. As mentioned, a **Bravais lattice is an array of sites, each of them surrounded by identical environments.** Bravais lattices are classified according to their symmetry properties; it turns out that there is a finite number of them. There are solely four planar Bravais lattices, and seven three-dimensional ones; those are illustrated in Fig. 2.27 and Fig. 2.28.

Crystals with a base. The vertices of the triangular lattice (Fig. 2.4) and of the square lattice (Fig. 2.6) form Bravais lattices. Nonetheless, **not any periodic array of sites is a Bravais lattice.** Figure 2.7 exhibits two periodic arrays featuring large and small circles. These can represent, for instance, the (positive and negative) ions of sodium and chloride in a planar cross section of the **table-salt crystal**, in which each positive sodium ion is surrounded by four negative chloride ones, and *vice versa.* (Since the chloride ion is negatively charged and the sodium one is positively charged, and as positive and negative charges attract one another, this arrangement has the lowest possible energy, see Chapter 4.) Thus, the crystal looks differently when "viewed" from the center of the large circle or from that of the small one. If the center of each circle is considered as a site on a square lattice, the emerging lattice is **not** a Bravais lattice according the definition given above. It is still possible, though, to describe the structures in Fig. 2.7 in terms of arrays of sites from which one does "observe" the same surrounding. For example, the environments of each of the centers of the large circles are identical. One may therefore join together these centers, as shown by the dashed lines, and obtain a Bravais lattice. This is a square lattice, rotated by 45° relative to the original one and its lattice constant is $\sqrt{2}$ larger. Alternatively, the surroundings viewed from each of the vertices of the square bounded by solid lines are also identical to each other, and therefore these vertices form a Bravais lattice as well, similar to the previous one but displaced relative to it.

Any of the squares in Fig. 2.7 can serve as the primitive unit cell of the Bravais lattice corresponding to this crystal, again indicating the redundancy in the choice of the unit cell. Repeated Translations of each square by the lattice vectors (marked in the figure) tile the whole plane by such cells. The square that connects the centers of the small circles is also a proper primitive unit cell. Each square in this Bravais lattice contains two circles, a large one and a small one. This is obvious regarding the square bounded by solid lines which contains the centers of two circles, large and small. The one marked by dashed lines contains a small circle at its center, and four quarters of the large circle, and hence it also packs the areas of the large and small circles. In contrast with the planar structures discussed in Sec. 2.2, in which the unit cell represents a single lattice point, the unit cell of the lattices here contains two lattice points (the centers of the small and the large circles). As mentioned, a unit comprising the constituents contained in the primitive unit cell (e.g., a small circle and a large one) is termed the **base** of the periodic array. The base consists of the constituents that repeat themselves, while the primitive unit cell is the area that is repeated *ad infinitum* and covers the whole plane. There is one-to-one correspondence between a certain choice of the primitive unit cell and the base contained in it.

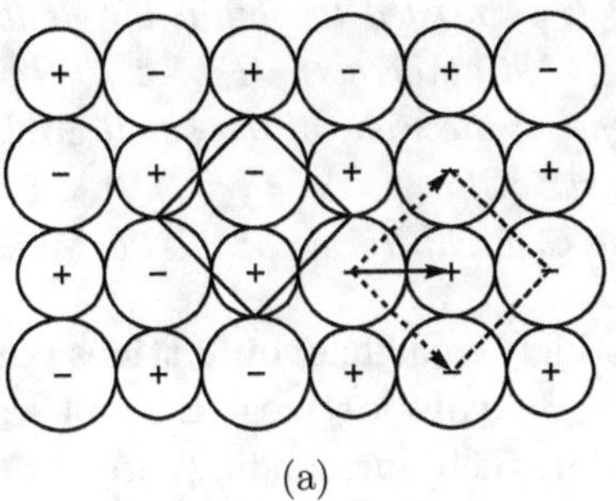

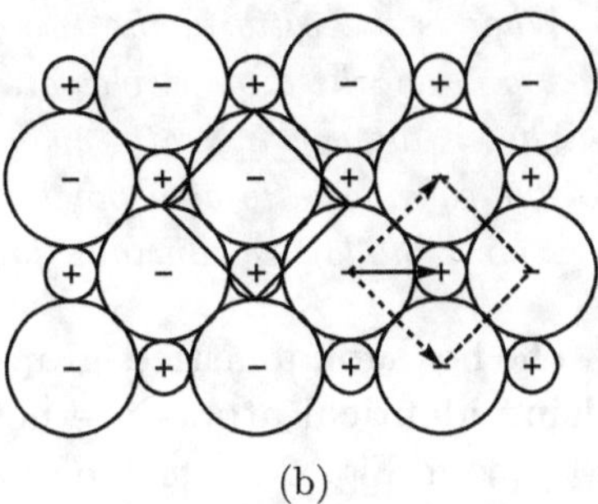

Fig. 2.7: Examples of a planar crystal built of two types of circles. The crystal may represent a planar cross section of table salt, NaCl. The two panels differ by the ratio of the circles' radii, and consequently also by their packing ratios, but they do share certain symmetry properties. The big circles in (a) are not tangential to each other while in (b) they are; therefore the packing of (b) is denser. The squares represent possible unit cells of the corresponding Bravais lattice: the surroundings of each vertex of the squares are identical. The unit cell contains a single base comprising a big circle and a small one. The dashed arrows are lattice vectors and the solid vector is $\mathbf{r}_2$, connecting two circles in the base that belongs to the dashed unit cell.

Any periodic material comprises a Bravais lattice of points; each of them represents a base of the material. This collection of bases is called a "crystal". The crystal is built of a lattice with a base.

A practical way to consider the lattice displayed in Fig. 2.7 is to note that the large circles alone (deleting the small ones), form a square lattice with a unit cell given by the dashed square; accounting only for the small circles (and obliterating the large ones) again yields a square lattice, identical to the previous one. In the crystal portrayed in Fig. 2.7, the lattice of the small circles is shifted relative to that of the large ones by the vector $\mathbf{r}_2$, the solid arrow in the figure. Hence, the crystal consists of two square sublattices, displaced relative to one another. A similar description pertains to any crystal with a base: the crystal always consists of several identical sublattices, each corresponding to one of the elements contained in its base. Each of these sublattices is a Bravais lattice, identical to the Bravais lattice of the entire crystal.

Problem 2.3.
The small circles in Fig. 2.7(a) are tangential to the large ones, but the latter are not tangential to each other.
a. *Find the ratio of the total area covered by the two types of circles and the area of the whole plane, i.e., the planar packing-ratio ρ of this structure, denoting the radii of the small circles and those of the large ones by $r_<$ and $r_>$, respectively.*

b. *What is the ratio of the two radii, $x = r_</r_>$, needed to obtain the densest packing [when the large circles are also touching each other, as in Fig. 2.7(b)]?*
c. *What is the packing ratio upon further decreasing the radii of the small circles? Plot the packing ratio as a function of the radii ratio, for $0 < x = (r_</r_>) < 1$.*
d. *Express the lattice constant, for each of those cases, in terms of the two radii.*

The hexagonal lattice–graphene. Interestingly enough, **even lattices comprising identical atoms are not necessarily Bravais lattices.** Examples are portrayed in Fig. 2.2: the atoms are identical, but their surroundings are not. A two-dimensional realization is shown in Fig. 2.8(a), which exhibits the **hexagonal lattice**, also known as "honeycomb" for obvious reasons. As mentioned in Chapter 1, this lattice is currently intensively studied, since it describes the structure of **graphene** for which each point on the lattice represents a **carbon** atom (Fig. 1.6).

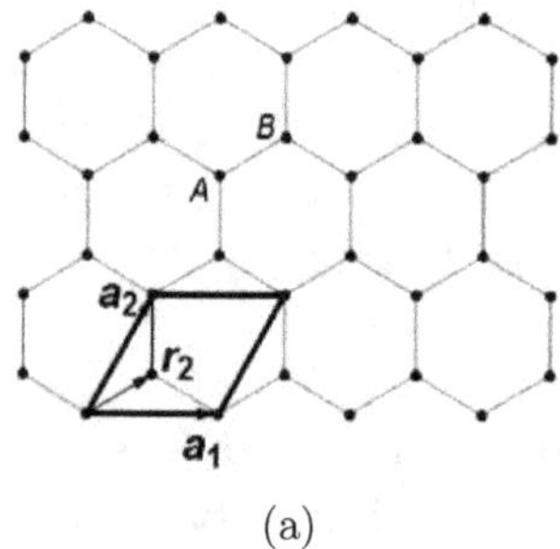
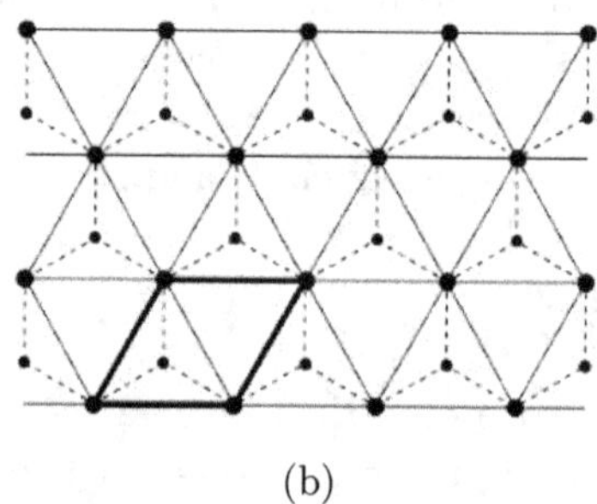

(a)　　　　　　　　　　　　　　　　　(b)

Fig. 2.8: The hexagonal lattice. (a) The lattice sites at the corners of adjacent hexagons. The figure displays two types of lattice sites, A and B, whose surroundings are different. The rhombus represents an optional unit cell, containing a base of two sites connected by the vector $\mathbf{r}_2$. (b) The A–type points are drawn as larger dots and the thin solid lines demonstrate that these points build a triangular lattice, with the same unit cell as in the other panel. The smaller dots are the B–type lattice sites. These too are located on a triangular lattice shifted relative to the former one.

Graphite. As mentioned in Chapter 1, carbon appears in numerous crystal structures (or **polymorphs**) due to the multitude of possibilities for chemical bonds among its atoms. The most famous is the **diamond**; under common circumstances the stablest three-dimensional structure of carbon is **graphite** (diamond is formed in Nature only under strict pressure and temperature conditions, and hence is rather rare). The graphite is made of parallel planar layers inside each the atoms are ordered in an hexagonal array as in Fig. 2.8(a), see also the right panel in Fig. 2.9. This three-dimensional structure is also called the **hexagonal lattice**. Graphite can be found in pencils; the layers are only loosely coupled, and the process of writing "peels" two-dimensional layers off the black material onto the paper. In

2004 Geim and Novoselov succeeded to peel two-dimensional layers of graphite and thus opened the frontal research field of single layers, i.e., of graphene.

Figure 2.10 displays a graphene specimen, hanging freely in space. Though topologically a plane, the graphene may attain a wavy form or fold itself in various shapes. As mentioned, the two-dimensional graphene possesses certain spectacular physical properties, e.g., high rigidity of structure and high electrical conductivity. These unique features, and the rare possibility to easily produce usable samples, have invoked the extensive interest in this material.

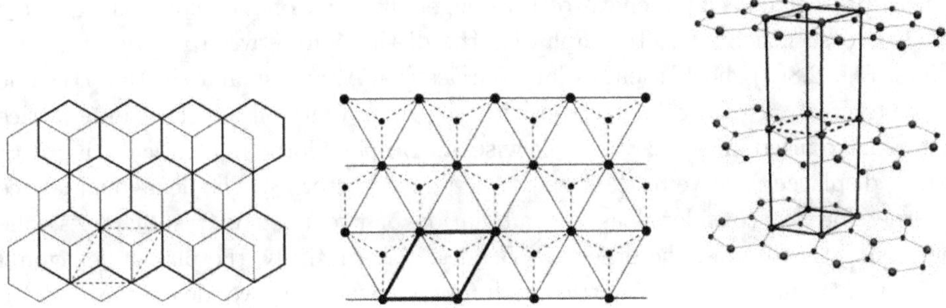

Fig. 2.9: Right panel: the hexagonal lattice of graphite. Each layer in the lattice is an hexagonal lattice, identical to that displayed in Fig. 2.8. The atoms in the lower and upper planes lie precisely one on top of the other, while the hexagonal lattice forming the layer in-between is displaced with respect to those layers. The thick lines show the unit cell. The central panel displays the arrangement of the atoms in the middle layer [note that Fig. 2.8(b) shows the atoms in the lower layer]. The $A-$type atoms occupy the same sites in the two planes, but the $B-$type ones are displaced so that they are located above the centers of the hexagons in the layer below (or below the centers in the layer above). Left panel: the lattices of the two planes are shown together (viewed from above). The thick lines pertain to the lower layer and the thin ones refer to the middle plane.

Graphene as a crystal with a base. The planar hexagonal lattice, Fig. 2.8(a), can be characterized by the two points A and B. The surrounding observed from A differs from that seen from B. For instance, "glancing" to the right of A reveals a $B-$point at a higher location (on the perpendicular axis), and a vacancy at the mid point of the hexagon below. As opposed, on the right of B there is a lattice point at a lower location (on the perpendicular axis) and a vacancy at the center of the hexagon above it. In other words, A resides at the lower edge of the hexagon, and hence the empty site at the center of this hexagon is above it, while B is found at the upper edge of the hexagon, and the empty site of its center is below it. The environment viewed from A is the same as seen from all the lower corners of the hexagons, while all the upper corners share the same surroundings as point B. Since these two environments are not identical, the hexagonal lattice is **definitely not** a Bravais lattice, but it can be described as a crystal, i.e., as a **Bravais lattice with a base that contains two sites.** Figure 2.8(a) displays a rhombic unit cell which consists of A points at its corners and a B point inside it. Thus, A and B form together the base of this crystal. Denoting the lattice vectors by $\mathbf{a}_1$ and $\mathbf{a}_2$ as in the figure, the vector joining the two base sites is $\mathbf{r}_2 = (\mathbf{a}_1 + \mathbf{a}_2)/3$. (The length of this vector is 2/3 of the height of the equilateral triangle spanned by the two lattice vectors, since the second point is located at the middle of this triangle–the locus of its heights. The length of the longer diagonal of the rhombus, $\mathbf{a}_1 + \mathbf{a}_2$, equals two such heights.) To emphasize the distinction between the two types of points, Fig. 2.8(b) shows them as large circles (for A) and small ones (for B). The A points create a triangular lattice, with the same unit cell and the same lattice vectors introduced in Fig. 2.8(a). Likewise, all B points form an identical triangular lattice, displaced relative to the former one by the vector $\mathbf{r}_2$. The hexagonal lattice can therefore be considered as a combination of two triangular sublattices, one comprising the A sites, the other the B ones. Alternatively, the planar hexagonal lattice is a triangular Bravais lattice with a base containing two identical atoms (to avoid confusion: the lattice is triangular but it can be described by a rhombic unit, as in Fig. 2.4).

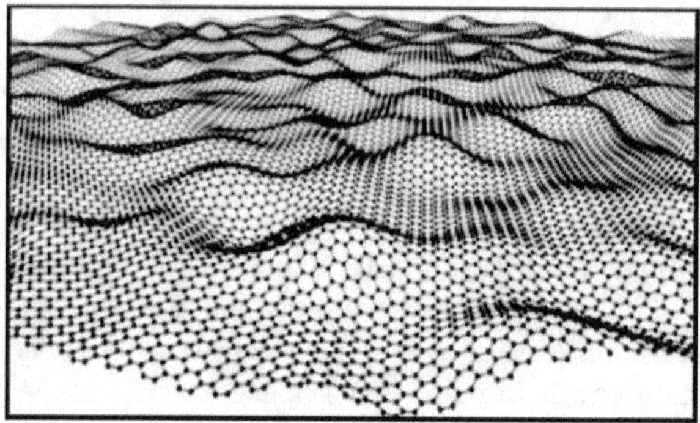

Fig. 2.10: A graphene sample, hanging freely in space. Courtesy of A. Geim.

The lattice structure of graphite. The three-dimensional lattice of graphite is illustrated in the right panel of Fig. 2.9. The atoms in the lower and in the higher layers are exactly above one another; in the mid layer atoms A are above the atoms A in the layer below, but atoms B are displaced compared to their locations in the lower plane, and are each above a vacancy at the center of an hexagon in that plane. In this configuration, the B atoms are located at the same distance from the six atoms forming the hexagon in the lower layer and from the similar six atoms in the plane above, instead of being close to only a single atom in each such plane (had the two hexagonal lattices been placed precisely one above the other). Though the distance from each of the 12 neighbors is longer than in the other option, their large number compensates for this difference and yields a lower potential energy. The central panel of Fig. 2.9 exhibits the atomic array in the mid layer. The large circles, representing the A atoms, are still located as in Fig. 2.8(b) (which describes the lower layer of the graphite), but the B atoms are displaced by the vector $\mathbf{r}_2$ relative to their locations there. Glancing at the rhombus that represents a unit cell in Fig. 2.8(b) reveals that the rhombus above it [in the mid layer, shown by the thick lines in the central panel of Fig. 2.9 and by the dashed lines in the right panel there] also contains two atoms, A and B, but the latter, instead of being at the mid point of the left triangle of the rhombus as before, is now at the mid point of the triangle on the right side.

Other materials described by the hexagonal lattice. Two comments are called for. First, the lattice displayed in Fig. 2.8(b) describes as well crystals comprising two different atoms, e.g., boron and nitrogen (BN), situated at the locations of the large and small circles in the figure. This material is also at the forefront of contemporary research. A somewhat simpler example are the crystals of silicene and germanene, in which silicon or germanium atoms replace the carbon ones forming the graphene. Much like the graphene, these materials also possess intriguing physical properties, nowadays under active investigations. In those, the two atoms of the unit cell are located in two different parallel layers (Fig. 2.11). The dissimilarities between these crystals and graphene is discussed in Chapter 4.

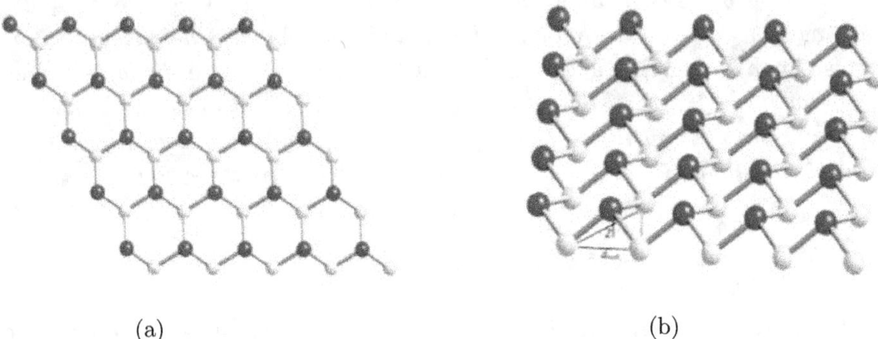

(a)   (b)

Fig. 2.11: Silicene lattice; (a) viewed from above, (b) a side view, both showing the different heights of the two sublattices [A. O'Hare, F. V. Kusmartsev, and K. I. Kugel, *A Stable "Flat" Form of Two-Dimensional Crystals: Could Graphene, Silicene, Germanene Be Minigap Semiconductors?,* Nano Lett. **12**, 1045 (2012).]

Second, both graphene and BN are examples of the more general case, shown in Fig. 2.12, of a triangular lattice with **three types of sites,** marked A, B, and C.

The square lattice featuring in Fig. 2.7 consists of **two identical sublattices** displaced relative to one another; the triangular lattice can be described by **three sublattices**. The unit cell of the lattice of Fig. 2.12 is still identical to those presented in Fig. 2.8 and the mid panel of Fig. 2.9, but its base contains three types of sites. The lattices portrayed in Fig. 2.8(b) and 2.9 can be derived from the one in Fig. 2.12 provided that all C or all B points are deleted, respectively. Hence, the two different layers in the graphite lattice are only particular cases of Fig. 2.12. Discarding all A points results in a hexagonal lattice identical to the formers, but shifted relative to them. It is possible to build three-dimensional hexagonal lattices by arranging differently these three planar structures, one on the top of the other.

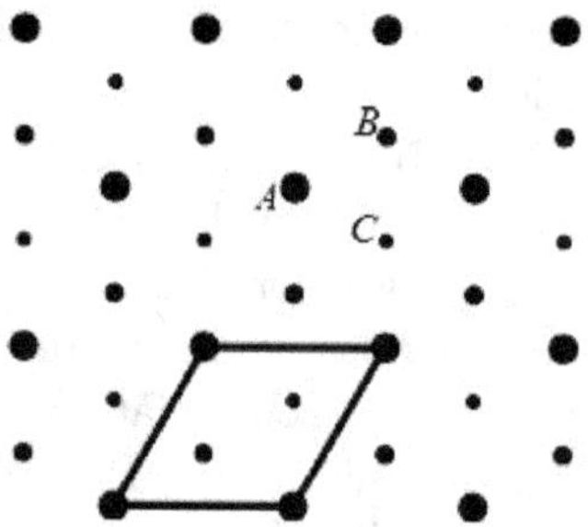

Fig. 2.12: The triangular lattice with three types of sites.

Problem 2.4.

A circle of radius $r_>$ is drawn around each "large" point in Fig. 2.8(b), and a circle of radius $r_<$ is drawn around each "small" one. Derive the packing ratio of the structure as a function of $x = r_</r_>$, in two cases: (i) when the large circles touch one another; (ii) when the large circles touch the small ones. For which x is the packing ratio maximal? Compare it with the one of the square lattice. Obtain the lattice constant as a function of the two radii. What is the result for graphene?

2.4   The lattice vectors and the unit cell

Translations in the lattice. The periodicity of the triangular lattice can be expressed in terms of the lattice vectors, marked by arrows in Figs. 2.4 and 2.5. Each vector connects two neighboring lattice sites. In the infinite lattice, any translation by one of these vectors (or by an integral multiple of it) transforms the lattice into another one, completely identical to the original lattice. In formal terms, a **three-dimensional Bravais lattice is the ensemble of points in which each pair can be connected by the vectors**

$$\mathbf{R} = n_1\mathbf{a}_1 + n_2\mathbf{a}_2 + n_3\mathbf{a}_3 \ , \tag{2.1}$$

where n_1, n_2, and n_3 are arbitrary integers (positive, negative, or zero), and $\mathbf{a}_1$, $\mathbf{a}_2$, and $\mathbf{a}_3$ are three independent displacement vectors (i.e., they are not parallel to each other and are not in the same plane), which are called **lattice vectors.** Alternatively, any of the lattice sites may be chosen as the origin, and then any other lattice point is given by Eq. (2.1), with the proper integer coefficients. The Bravais lattice is defined in Secs. 2.1 and 2.3 as the set of **points with identical environments.** This indeed pertains for all the points connected by the vectors given in Eq. (2.1). Hence, **the two definitions are equivalent.** Lattice vectors of planar two-dimensional lattices are represented by the pairs of vectors emerging from the same site in Fig. 2.4 or 2.5. A three-dimensional example is portrayed in Fig. 2.13: the three lattice vectors $\mathbf{a}_1$, $\mathbf{a}_2$, and $\mathbf{a}_3$ form a parallelogram. When each of the vertices of the parallelogram is a lattice site, and when the entire lattice is built of such identical parallelograms adjacent to each other (along the common faces), then each lattice point can be reached from the origin by a displacement vector of the form Eq. (2.1). The lattice so created is a Bravais one, as the surroundings of all the points are identical.

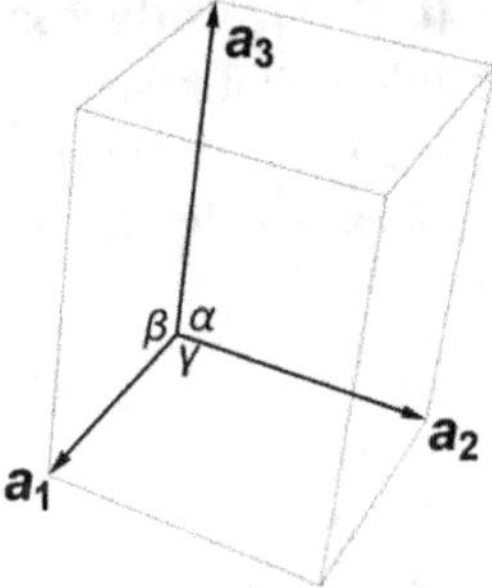

Fig. 2.13: A unit cell built of the three lattice vectors $\mathbf{a}_1$, $\mathbf{a}_2$, and $\mathbf{a}_3$, with a lattice point located at each vertex. The angle α is between $\mathbf{a}_2$ and $\mathbf{a}_3$, β is the angle between $\mathbf{a}_3$ and $\mathbf{a}_1$, and γ is the one between $\mathbf{a}_1$ and $\mathbf{a}_2$.

The unit cell and the primitive unit cell. Since the lattice consists of the collection of parallelograms, the one in Fig. 2.13 can be declared as the **unit cell.** Each parallelogram can be identified with a single lattice site, situated, e.g., at the origin (from which the three vectors emerge). In other words: **the density** of the material described by this lattice is the ratio of the mass of the base represented by the lattice site, and the **volume of the primitive unit cell,** given by

$$V = |\mathbf{a}_1 \cdot [\mathbf{a}_2 \times \mathbf{a}_3]| \tag{2.2}$$

(check!). The three lattice vectors cannot be coplanar, lest the volume vanishes. The area of the primitive unit cell in a two-dimensional lattice, described by two lattice vectors, is

$$S = |\mathbf{a}_1 \times \mathbf{a}_2| \tag{2.3}$$

(check!). The choice of the lattice vectors is not unique. Figure 2.14 displays four examples of pairs of lattice vectors in the plane. Linear combinations of the pair $\mathbf{a}_1$, $\mathbf{a}_2$, or of the pair $\mathbf{a}_5$, $\mathbf{a}_6$, or of the pair $\mathbf{a}_7$, $\mathbf{a}_8$, with integer coefficients [according to Eq. (2.1)] displaces any point on the lattice to another one on it. In contrast, linear combinations of the pair $\mathbf{a}_3$, $\mathbf{a}_4$, with integer coefficients, **cannot** reach **all** lattice sites. For instance, a linear combination with **non-integer** coefficients is needed to reach the point in the center of the parallelogram built of these two vectors. The parallelograms built on the former pairs have the same area, which is the specific area corresponding to a single lattice point (check!); the area of the parallelogram built on the fourth pair is twice that area, and hence this parallelogram contains **two** lattice sites (e.g., the one at the lower left corner and the one at the mid point). There are other pairs of lattice vectors that can span the whole planar lattice; there are also many possibilities to choose the lattice vectors of a three-dimensional lattice. It is therefore customary to define **primitive lattice vectors: for the vector $\mathbf{a}_1$ one picks up the shortest vector connecting two lattice sites. The vector $\mathbf{a}_2$ is then chosen as the shortest one which is not parallel to the former, and $\mathbf{a}_3$ is the shortest vector between nearest neighbors which does not lie in the plane spanned by its two predecessors.** When the vectors $\mathbf{a}_1$ and $\mathbf{a}_2$ have equal lengths it is possible to interchange their notations to $\mathbf{a}_2$ and $\mathbf{a}_1$, as happened in all the examples presented above. In the case of Fig. 2.14, the primitive lattice vectors are $\mathbf{a}_1$ and $\mathbf{a}_2$, which are normal to one another and have the same length.

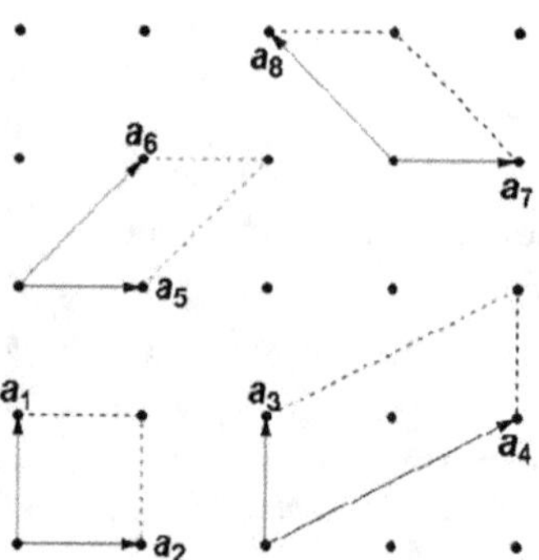

Fig. 2.14: Possible pairs of lattice vectors (and unit cells) in the plane.

The choice of the primitive unit cell is not unique. In Fig. 2.5 the unit cell is chosen as the hexagon encompassing a single circle, i.e., as the Wigner-Seitz cell. This hexagon includes all the points in the plane which are closer to the center of the hexagon than to any other lattice site; it is encased by the mid orthogonals of the segments connecting the central lattice site with its six neighbors. The extension to three dimensions is clearcut: one chooses a lattice site and connects it with other lattice sites. One then bisects each segment by a perpendicular plane. **The three-dimensional Wigner-Seitz cell is the volume confined by these planes, that contains the initial lattice site**. Figure 2.13 offers a different definition of the unit cell: the primitive unit cell is a cuboid, or a parallelepiped, built of the **three primitive lattice vectors** of the lattice. In Fig. 2.4, this unit cell is a rhombus, with edges of length equal to the distance between nearest-neighbor lattice points, like the one shown at the lower left corner of the figure. The whole plane can be described as a collection of such rhombi, or as a collection of hexagons (those are shown in Fig. 2.5). Since each such cell, be the rhombus or the hexagon, represents a single lattice site, and their lot tessellates the plane, one expects their areas to be identical (problem 2.5). Generally, the three lattice vectors can attain different lengths, as in Fig. 2.13. These are termed **"lattice constants"** (extending the definition given for the one-dimensional case, Sec. 2.1). For the triangular lattice, the lengths are equal and there is a single lattice constant. Whether to use the primitive unit cell or the Wigner-Seitz one is a matter of convenience. Both represent faithfully the lattice (and they are not the only possible unit cells). When non-primitive unit cells are exploited, the definitions of the lattice constants have to be readjusted.

Problem 2.5.
a. *Calculate the area of the Wigner-Seitz unit cell of the triangular lattice (Fig. 2.5) and that of the rhombus in Fig. 2.4, in terms of the lattice constant a. Are they indeed equal?*
b. *Determine the corresponding unit cells (the Wigner-Seitz one and the one built on the lattice vectors) of the square lattice, discussed in problem 2.2. Find their areas.*
c. *Given the lattice vectors of the square lattice, whose sites are occupied by two types of circles, what are the lattice vectors of the crystal in Fig. 2.7? What is the area of the primitive unit cell spanned by them?*

    Crystal with a base. When the base of a crystal includes more than one lattice site, the choice of the origin relative to the locations of the base points is not unique. Figure 2.7 illustrates two choices of a unit cell, each containing the base points at different locations; one choice is mapped on the other by a displacement of the entire lattice. The same is true for the unit cell shown in Fig. 2.8(b) and the mid panel in Fig. 2.9. Quite generally, locating the lattice site that represents a certain unit cell at the axes' origin, the **locations of the base points** relative to the origin (i.e., within the unit cell) are denoted by

$$\mathbf{r}_i = x_1^i \mathbf{a}_1 + x_2^i \mathbf{a}_2 + x_3^i \mathbf{a}_3 \; , \tag{2.4}$$

where $i = 1, \ldots, n_B$ enumerates the n_B base sites, with $|x_\alpha^i| < 1$. In the unit cells marked by dashed lines in Fig. 2.7, the two base points are at $\mathbf{r}_1 = 0$ and

$\mathbf{r}_2 = (\mathbf{a}_1 + \mathbf{a}_2)/2$, where the lattice vectors $\mathbf{a}_1$ and $\mathbf{a}_2$ are marked by the dashed arrows there.

Problem 2.6.

Prove that the unit cell spanned by the vectors $\mathbf{a}_3$ and $\mathbf{a}_4$ in Fig. 2.14 describes a square lattice with a base comprising two sites. Determine the locations of the base points, in terms of these vectors. Find the area of the unit cell.

Problem 2.7.

Find the lattice vectors of the unit cell marked by solid lines in Fig. 2.8, in cartesian coordinates (expressed in terms of the length in-between nearest-neighbors on the original hexagonal lattice). Give the locations of the base points in terms of these vectors. What is the answer for the unit cell in the mid panel of Fig. 2.9?

2.5 The cubic lattices

The simple-cubic lattice. Most of the crystals found in Nature are **three-dimensional**. The simplest way to construct three-dimensional lattices is to begin with a two-dimensional one, and to lay above it an identical two-dimensional lattice, such that each lattice point in the second layer is located precisely above a lattice point in the lower plane. An example is the square lattice of spheres shown in Fig. 2.6; one places successive layers identical to the initial one, such that the spheres in each layer are located above those in the layer below. The resulting structure is displayed in Fig. 2.15(a). The lattice built of the centers of all these spheres has orthogonal lattice vectors of identical length; they form the edges of a cube. In cartesian coordinates these lattice vectors are $\mathbf{a}_1 = a\hat{\mathbf{x}}$, $\mathbf{a}_2 = a\hat{\mathbf{y}}$, and $\mathbf{a}_3 = a\hat{\mathbf{z}}$, where a is the lattice constant (whose length is the cube's edge and the diameter of the spheres), and the unit vectors $\hat{\mathbf{x}}$, $\hat{\mathbf{y}}$, and $\hat{\mathbf{z}}$ point along the cartesian axes. This cube may be chosen as the primitive unit cell of the lattice [see Fig. 2.15(b)], which is duly called **the simple-cubic lattice**, abbreviated as **SC**.

Problem 2.8.

a. *Find the volume of the (primitive) unit cell of a **SC** lattice built of spheres whose diameter is a. Find the volume packing-ratio and the coordination number.*
b. *What is the shape of the Wigner-Seitz cell of this lattice?*
c. *Propose two sets of lattice vectors that span all lattice points. What is the volume of the unit cell constructed of these lattice vectors?*

Cesium-chloride type crystals. As found in problem 2.2, the packing ratio of circles arranged on a square lattice, 0.5236, is not the optimal one; the densest packing of circles on the plane, 0.6046, is reached for circles on a triangular lattice. Likewise, the packing ratio of the simple-cubic lattice (see Fig. 2.15) is not the optimal one. (The reader is encouraged to try and identify, for instance by exper-

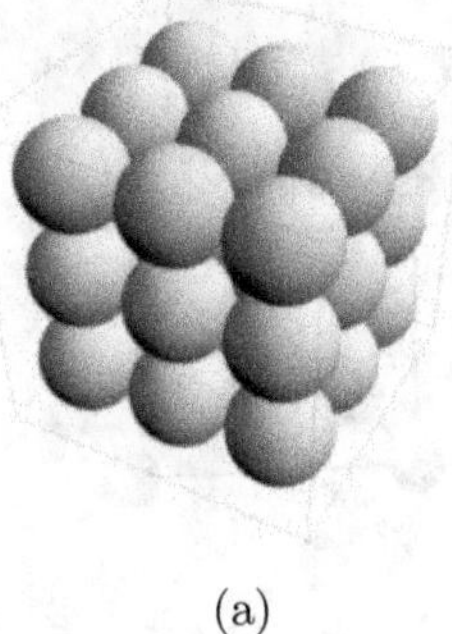

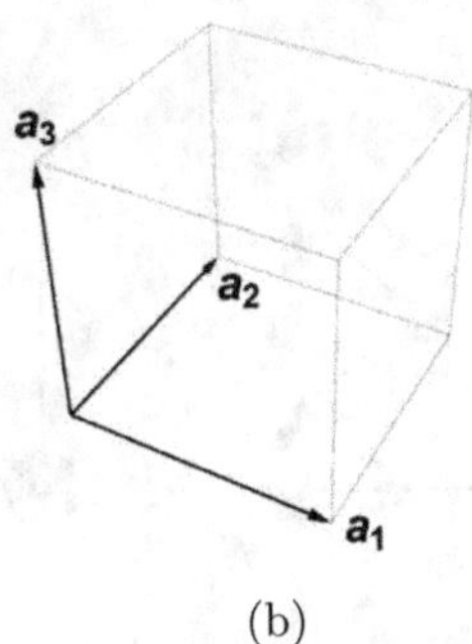

(a) (b)

Fig. 2.15: (a) A sample built of eight such unit cells, each hosting a sphere. (b) The unit cell of the simple cubic lattice.

imenting with balls, the lattice which possesses the densest packing of spheres in space; the answer is below.) Around the center of the unit cell of the **SC** there is an empty zone, that is responsible for the low packing ratio. This ratio can be increased by, e.g., adding a small sphere, of radius $r_<$, at the center of each cube, i.e., around the point $\mathbf{r}_2 = (a/2)(\hat{\mathbf{x}} + \hat{\mathbf{y}} + \hat{\mathbf{z}}) = (\mathbf{a}_1 + \mathbf{a}_2 + \mathbf{a}_3)/2$. This is similar to its planar analog, depicted in Fig. 2.7. This small sphere can be enlarged until it touches the eight surrounding large spheres. Since the length of the cube's diagonal is $a\sqrt{3}$, and the radius of the large sphere, $r_>$, is $a/2$ (the spheres on each edge of the cube are contiguous), the large spheres touch each other as long as $2(r_> + r_<) \leq a\sqrt{3}$. (The diagonal has to host two radii of the large spheres, and a diameter of the small one.) Hence, $r_< \leq (\sqrt{3} - 1)r_> = 0.732r_>$. As the radii of the large and small spheres are different, the lattice so formed in not a Bravais lattice, but **a simple-cubic lattice with a base**, that contains one large sphere (e.g., at the corner of the cube) and one small sphere (e.g., at the center of the cube). Exploiting the original lattice vectors, the vector connecting the two spheres in the unit cell is $\mathbf{r}_2 = (\mathbf{a}_1 + \mathbf{a}_2 + \mathbf{a}_3)/2$. This unit cell is shown in the right panel of Fig. 2.16. Disregarding the radii ratio of the spheres there, this unit cell belongs to the **cesium chloride** (CsCl) crystal; all other crystals sharing this unit cell are hence termed **cesium-chloride type crystals**. The bigger negative Cl ions occupy the corners of the cell, while the small positive Cs ion resides at its center.

Occasionally it is expedient to describe such a unit cell by its projection on the lower planar face, as done in the central panel of Fig. 2.16. Except for the ions residing on this plane (marked by the dark circles), all other ions are represented by circles which contain their height above the plane, in units of the lattice constant along the normal to the plane (in the present case, $1/2$). The cesium-chloride crystal is in fact built of two simple-cubic sublattices, shifted relative to one another by $\mathbf{r}_2$. As mentioned, **a crystal with a base may be described by sublattices, each containing different atoms or ions of the base.**

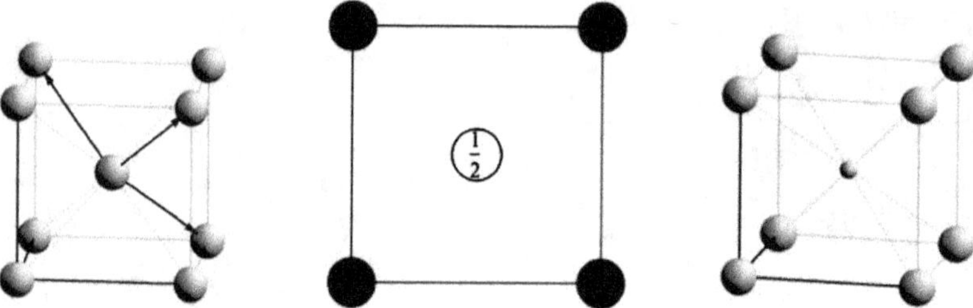

Fig. 2.16: Right panel: the unit cell of a crystal of the cesium-chloride (CsCl) type. (The radii of the spheres are not in scale; they are reduced for clarity.) Central panel: the projection of the unit cell on the lower planar face. The dark circles at the corners represent the four chloride ions on that face (on the XY plane) and the empty circle is the cesium ion, located at height $a/2$ above (the height, $1/2$ in units of the lattice constant, is marked within the circle). Left panel: the unit cell of the body-centered cubic lattice (**BCC**). The lattice vectors [Eq. (2.5)] are marked by the arrows.

The body-centered cubic lattice (BCC). A special case of the crystal displayed in the right panel of Fig. 2.16 is realized when the ion (or atom) located at the center of the cube is identical to those occupying the corners; this is shown in the left panel of Fig. 2.16. The lattice there is the **body-centered cubic lattice,** abbreviated as **BCC**. The surrounding of the central site in this structure is identical to those of the sites at the corners, and therefore the **BCC** lattice is a Bravais lattice; the cubic unit cell in the figure **is not** the primitive unit cell of this lattice (as it contains two lattice sites, e.g., at the center of the cube and in one of its corners). The lattice vectors of this Bravais lattice are the three vectors connecting a point in the lattice with three of its nearest-neighbors. A lattice point placed at the origin is surrounded by eight nearest neighbors, at $(a/2)(\pm\hat{\mathbf{x}} \pm \hat{\mathbf{y}} \pm \hat{\mathbf{z}})$ (in cartesian coordinates). A possible choice for the primitive lattice vectors is hence

$$\mathbf{a}_1 = (a/2)(-\hat{\mathbf{x}} + \hat{\mathbf{y}} + \hat{\mathbf{z}}) \,, \quad \mathbf{a}_2 = (a/2)(\hat{\mathbf{x}} - \hat{\mathbf{y}} + \hat{\mathbf{z}}) \,, \quad \mathbf{a}_3 = (a/2)(\hat{\mathbf{x}} + \hat{\mathbf{y}} - \hat{\mathbf{z}}) \,.$$

$$(2.5)$$

These are portrayed by the arrows in the left panel of Fig. 2.16. Though the cubic unit is not the primitive unit cell of the **BCC** lattice, it is often convenient to use it (with a base including two identical atoms) rather than the primitive one built of the three vectors in Eq. (2.5). This is because the cubic unit cell reflects the symmetry under rotations by $90°$, which are hard to observe from the primitive cell. It is also more convenient for carrying out various calculations due to the efficiency of the cartesian coordinates.

    The preceding discussion implies that had identical spheres been placed at each point of the **BCC** lattice, and increased until each one touches its eight neighbors, then the spheres at the corners do not touch one another (because the radius of each sphere, r, equals a quarter of the diagonal: $r = a\sqrt{3}/4$, and thus is smaller than $a/2$). The volume of the cube is a^3 and it contains two spheres (one at the center, and $1/8$ of a sphere at each corner); the total occupied volume is $2 \times 4\pi r^3/3$. Hence, the volume packing-ratio is $\pi\sqrt{3}/8 \approx 0.680$. This packing is denser than that of the **SC** lattice.

Problem 2.9.

What is the packing ratio of the cesium-chloride type crystals as a function of the radii ratio $x = r_</r_>$? What is the maximal packing ratio of this structure?

Problem 2.10.

a. *Find the volume of the cell built on the primitive lattice vectors, Eq. (2.5). What is its relation to the afore-mentioned volume of the cubic unit cell?*

b. *Another optional unit cell for the **BCC** lattice consists of the vectors $\mathbf{a}_1 = a\hat{\mathbf{x}}$, $\mathbf{a}_2 = a\hat{\mathbf{y}}$, $\mathbf{a}_3 = \mathbf{r}_2 = a(\hat{\mathbf{x}} + \hat{\mathbf{y}} + \hat{\mathbf{z}})/2$. Find the number of atoms in this unit cell, and its volume. Are these primitive lattice vectors? Is it a primitive cell?*

Table salt and the face-centered cubic lattice (FCC). When the sites in the **SC** lattice are occupied alternately by two types of atoms (or ions), as done in the planar analog, Fig. 2.7, the outcome is the three-dimensional structure of **table salt, sodium chloride** (NaCl), shown in the right panel of Fig. 2.17. The same structure pertains (in Nature) to crystals in which the sodium is replaced by some other alkali atoms from the first column of the periodic table, for instance lithium (Li) or potassium (K), and also when the negative halogenic ion is replaced by others from the column before the last one in the same table, e.g., fluorine (F) or iodine (I). Quite generally, when a lattice site of the cubic lattice accommodates a small ion (for example, sodium), then all its six nearest neighbor sites (two along the positive and negative directions of each cartesian axis) contain the large ion, (e.g., chloride), and *vice versa*. This structure has the minimal potential energy for these ions, for a certain range of the radii ratios (see Chapter 4). The crystal displayed in Fig. 2.7 is a planar cross section of the structure in the right panel of Fig. 2.17, for instance, the upper plane in that figure. Note though that neighboring planes are not placed precisely one on top of the other; rather, above each sodium ion there is a chloride one, and *vice versa*. The neighboring planes are shifted relative to each other. As in Fig. 2.7, the ions in the right panel of Fig. 2.17 do not share the same environments, and consequently this structure is not a Bravais lattice. The base of this crystal contains two neighboring ions, represented by a small sphere and a large one; see the atoms in the ellipse on the right panel. Alternatively, the crystal consists of two identical simple cubic sublattices displaced relative to one another along one of the cartesian axes, e.g., by the vector $\mathbf{r}_2 = (a/2)\hat{\mathbf{z}}$ connecting the two ions in the base. As found in the two-dimensional example, each such base can be represented by a single lattice point. For instance, the crystal can be described by the lattice built solely of the large spheres, with each lattice site (at the center of the sphere) representing the entire base. The resulting lattice is shown in the central and left panels of Fig. 2.17 (projected on the plane of the lower face).

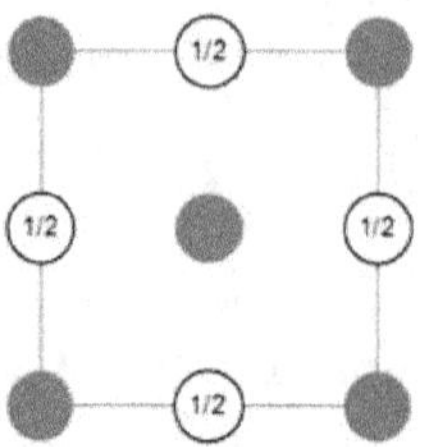 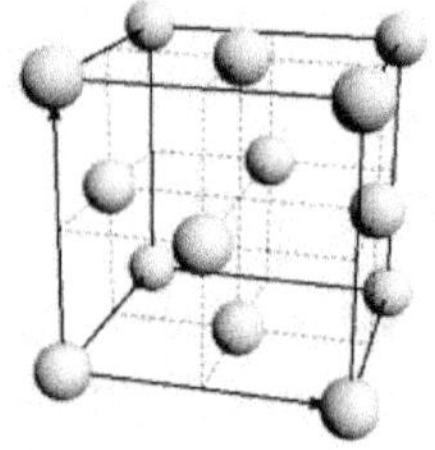 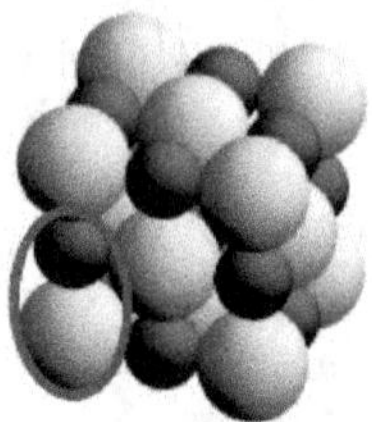

Fig. 2.17: Right panel: the table-salt crystal. The ellipse encompasses the base (comprising a chloride ion–the larger sphere–and a sodium one–the smaller sphere). Central panel: the face-centered cubic lattice (**FCC**), the Bravais lattice of table salt. Each lattice site represents the base (see the right panel). The arrows show the cubic lattice vectors. Left panel: a projection of the cubic unit cell of the **FCC** lattice on the plane of the lower face.

The lattice shown in the central panel of Fig. 2.17 can be described by using the cube there as the unit cell that repeats itself. Though this is not the primitive unit cell of this lattice, the description is most convenient because the cubic unit cell reflects the symmetry of the lattice under rotations by $90°$, for instance around each of the cartesian axes. When the cube is the unit cell, the lattice vectors, each of length a, are directed along the cartesian axes and are normal to each other precisely as for the cubic lattice in Fig. 2.15. As opposed to the unit cell of the **SC** lattice, though, the present unit cell contains four lattice sites. Each of the square faces has a point at its center, leading to the term **"face-centered cubic lattice"**, abbreviated as **FCC**. To see that each cubic cell contains four lattice points, one may slightly shift the cube away from the origin towards the negative sides of all three axes, without moving the lattice points themselves. The displaced cube contains the point at the origin, and its three nearest neighbors located at

$$\mathbf{r}_1 = (a/2)(\hat{\mathbf{x}} + \hat{\mathbf{y}}) \ , \quad \mathbf{r}_2 = (a/2)(\hat{\mathbf{z}} + \hat{\mathbf{x}}) \ , \quad \mathbf{r}_3 = (a/2)(\hat{\mathbf{y}} + \hat{\mathbf{z}}) \ . \tag{2.6}$$

Another method to obtain the very same result leaves the cube untouched, like in the figure, but divides the "mass" of the ion at each lattice site among the unit cells sharing that site. A lattice point at the corner of the cube is common to eight cubes, while a lattice point at the center of each face resides in two cubes (at the two sides of this face). Hence each corner "contributes" a "mass" of $1/8$, and each lattice point on the face has a "mass" of $1/2$. The cube possesses eight corners and six faces; hence the total "mass" in the cell is $8 \times (1/8) + 6 \times (1/2) = 4$. The volume of the cube is a^3, and the specific volume (per unit base) is $a^3/4$. Recalling that each lattice point represents the base of table salt, which consists of two ions, one concludes that the base of the cubic unit cell of table salt includes eight ions, four sodiums and four chlorides.

The **FCC lattice is a Bravais lattice**. Each **FCC** lattice site is surrounded by the same environment of 12 nearest neighbors. For instance, the lattice site at the origin of the cartesian coordinate system (say the one at the lowest left corner in the mid panel of Fig. 2.17) has four nearest neighbors on each of the planes XY, ZX, and YZ, leading to a total of 12 nearest neighbors. The same is true for any other lattice site, e.g., the one at the center of the face. This large number of nearest neighbors implies that the packing of spheres in the **FCC** lattice is very dense. Indeed, the densest packing of identical spheres in three dimensions, as shown in the following, is on an **FCC** lattice.

The primitive unit cell of the FCC lattice. As mentioned, the cubic unit cell is not the primitive unit cell of the **FCC** lattice. Primitive lattice vectors are the shortest lattice vectors, those that connect an arbitrary lattice site with three of its nearest neighbors. The twelve nearest neighbors in the **FCC** lattice are all at the same distance (i.e., half of the face diagonal, $a/\sqrt{2}$). Hence, the lattice vectors may be chosen as any triplet of non-planar vectors that connect the origin with its nearest neighbors. An example for such a choice and the primitive cell built on them, is given in Eq. (2.6) and shown in Fig. 2.18.

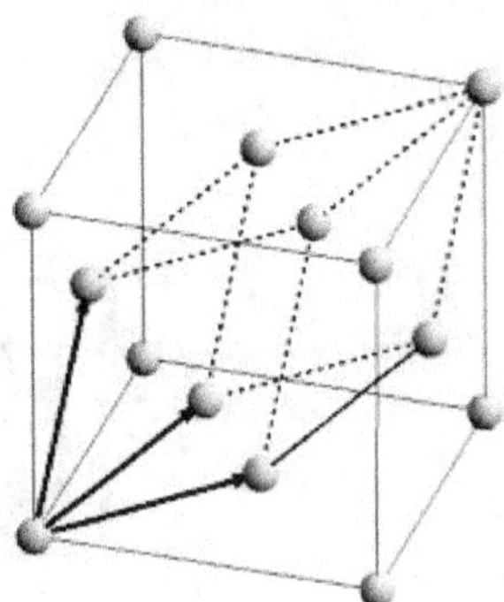

Fig. 2.18: Thin lines: the cubic unit cell of the **FCC** lattice, comprising four lattice sites. The arrows are the primitive lattice vectors, which together with the dashed lines form the primitive unit cell of this Bravais lattice.

Problem 2.11.
a. *Find the volume of the primitive unit cell in Fig. 2.18.*
b. *Around each point in the **FCC** lattice a ball is blown up until all balls are touching each other and cannot be grown any further. This is the densest packing of this lattice. What is the packing ratio?*

Wigner-Seitz cells. As explained, the Wigner-Seitz unit cell contains all points in space which are closer to a given lattice point than to any other lattice point. Figure 2.19 displays the Wigner-Seitz cells of the **FCC** and the **BCC** lattices. In the **FCC** lattice the 12 nearest neighbors are located at $(a/2)(\pm\hat{\mathbf{y}}\pm\hat{\mathbf{z}})$, $(a/2)(\pm\hat{\mathbf{x}}\pm\hat{\mathbf{y}})$, and $(a/2)(\pm\hat{\mathbf{z}}\pm\hat{\mathbf{x}})$. The Wigner-Seitz cell is confined by planes perpendicular to each of these vectors, passing through their mid points. Consider for instance the neighbor marked by a black sphere, Fig. 2.19(a). The plane normal to the line connecting it with the central site is marked by the square within the thick lines. Next consider the four neighbors connected by a rectangle (joined together by the dash thick lines). The planes related to those intersect the former one on the thick lines, which by symmetry, create a square. (Symmetry dictates that all edges and all angles of the area so created should be identical.) Since this argument holds for all other neighbors, it follows that the Wigner-Seitz cell of the **FCC** lattice is a polyhedron built of 12 identical square faces. This body is termed **rhombic dodecahedron**. Albeit being built of 12 identical squares, it is not one of the Plato solids, since its vertices are not identical (see Appendix A). The Wigner-Seitz cell of the **BCC** lattice consists of two types of faces, eight are equilateral hexagons (separating the central lattice site from its 8 nearest neighbors at the corners of the cube) and six are squares (located on the faces of the cubic unit cell), which separate the central lattice site from its next six neighbors. By symmetry, the edges of the hexagons and of the squares are all equal.

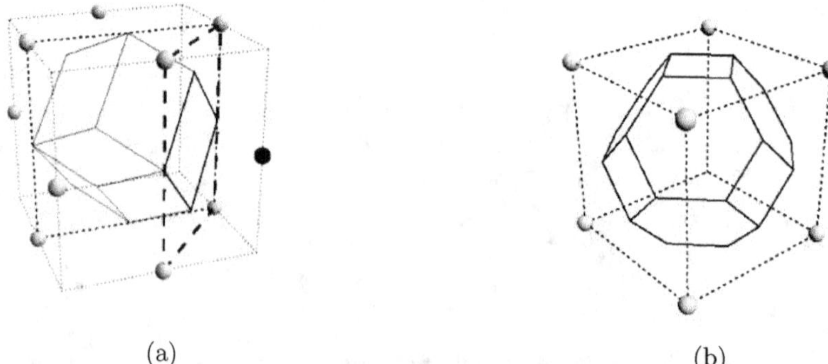

(a)   (b)

Fig. 2.19: Wigner-Seitz unit cells of the **FCC** (a) and the **BCC** (b) lattices. Each face of the cell is a plane normal to the line joining the site at the centre of the cube with one of its nearest neighbors (marked by spheres), and cutting it at its center.

Problem 2.12.

*Find the length of the edge of the Wigner-Seitz cell of the **FCC** lattice.*

The zinc-blendecrystal. In a certain similarity to the cesium-chloride crystal, the crystal of table salt may be constructed from an **FCC** lattice of chloride ions, to which the (smaller) sodium ones are added at points $\mathbf{r}_2 = (a/2)\hat{\mathbf{z}}$ away from each Cl. In the case of table salt, the sodium ion "enters" into the cavity formed in-between the six chloride ions "surrounding" it, precisely as the planar structure of table salt is described in conjunction with Fig. 2.7 (where each sodium ion is situated in the cavity created among four chloride ions). It is possible to compute the packing ratio for crystals of the NaCl type, for various values of the radii ratio [see problem **s.2.7**). Quite a number of crystals in Nature are described by sublattices of the **FCC** structure, with various fillings of the cavities of this structure. The right panel of Fig. 2.20 shows one of the structures of zinc sulphide sublattice, ZnS, whose German name is **zinc-blende**. The larger spheres represent the sulphur ions, and the smaller ones describe the zinc ones, each belonging to an **FCC** sublattice. The two sublattices are displaced relative to one another, by the vector $\mathbf{r}_2 = (a/4)(\hat{\mathbf{x}} + \hat{\mathbf{y}} + \hat{\mathbf{z}})$ (a quarter of the cube's diagonal). The projected cubic unit cell on the lower-face plane appears in the central panel of Fig. 2.20. Note that the zinc ion is situated in the space between four sulphur ions, but these are not located on the same plane; e.g., the ions at the origin and at the three points given by Eq. (2.6). Connecting all of these four lattice points to each other (the thick dashed lines) creates a **tetrahedron**, whose center (the locus of its three heights) is occupied by the zinc ion. (The tetrahedron is the simplest of Platonic solids; it consists of four equilateral triangular faces, see Appendix A.) The zinc-blende structure is of particular importance for **semiconductors**: it describes gallium arsenide (GaAs), aluminium arsenide (AlAs), and indium arsenide (InAs), all used in many electronic devices.

The diamond. As in the example of Fig. 2.16, the crystal of the zinc-blende type attains a unique configuration when the two sublattices contain the same atom. The lattice so formed is exhibited in the left panel of Fig. 2.20; it is the lattice of **diamond**. In this structure each carbon atom is surrounded by four nearest neighbors (i.e., each carbon atom resides at the mid point of a tetrahedron built of four other carbon atoms), precisely like in the zinc-blende structure. These structures are explained in Chapter 4. The spatial orientations of the various tetrahedra differ for different carbon atoms in the unit cell. This can be observed on the right panel of the figure: the environment viewed from a "dark" lattice point is not the same as the one observed from a "bright" lattice site; the tetrahedra are oriented differently. Nonetheless, all environments of the darker points are identical, and so are also those of the brighter ones. The lattice of the diamond, therefore, must be described by **two** sublattices, i.e., by an **FCC** lattice with a base that contains two atoms, much like zinc-blende.

Perovskites. The three structures of ionic crystals examined so far comprise solely two types of ions: the cesium-chloride-type, the table-salt-type, and the zinc-sulphide-type. The number of positive nearest neighbors of each negative ion (and *vice versa*) in these structures decreases from eight to six to four. This disparity is explained in Chapter 4 in terms of the different ratios of the ions' radii. The unit cells become increasingly complex when the crystal is made of more than

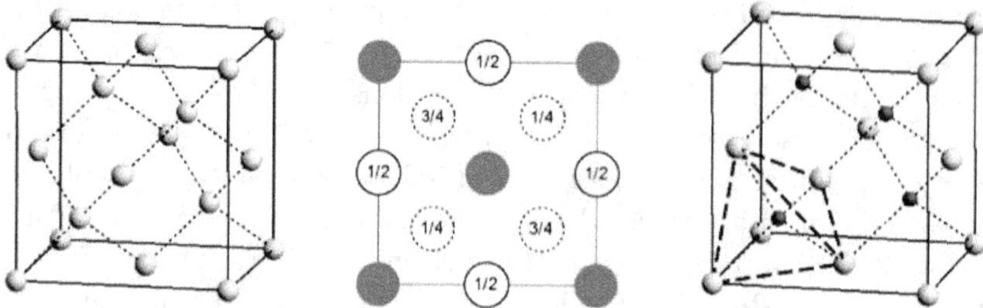

Fig. 2.20: Right panel: the structure of zinc-blende. The larger (lighter) spheres represent sulphur, and the smaller (darker) ones represent the zinc. Each zinc ion has four sulphur ions for nearest neighbors (connected to it by the dashed lines), and *vice versa*. The thick dashed lines create the tetrahedron caging a zinc ion. Central panel: the projection of the cubic unit cell on the lower-face plane (the dashed circles represent the zinc ions). Left panel: the cubic unit cell of diamond.

two kinds of ions. Examples of such crystals, important from both basic-research and technological points of view, are the perovskites. Figure 2.21(a) portrays the perovskite structure of the chemical formula ABO_3 (where A and B stand for certain positive ions), for instance, $SrTiO_3$. The strontium ions reside in the corners of the cubes and the titanium ones are located at the centers, precisely as in the structure of cesium chloride. The difference is that these two ions are positive, Ti^{4+} and Sr^{2+}; the negative oxygen ions, O^{2-}, are located at the centers of the faces, and form an octahedron shown by the thick lines in Fig. 2.21(a); this octahedron cages the B ion (or Ti) situated in the center of the cell. [The octahedron, built of eight identical equilateral triangles, is one of the Platonic solids, see Appendix A.] Another example of a perovskite crystal is $BaTiO_3$, much explored because it possesses a ferroelectric phase transition: at low temperatures the positive ions move relative to the negative ones, thus creating an electric dipole (see the end of Sec. 2.7). Perovskites are generally cubic at high temperatures, as in Fig. 2.21(a), but transform into a tetragonal or an orthorhombic structure at low temperatures, see Fig. 2.28.

Generalized perovskites: Figure 2.21(b) displays the structure of lanthanum cuprate, La_2CuO_4; it is an example of the A_2BO_4 structure. This is not a cubic crystal, since the vertical edge of the unit cell is longer than the other two that form the basis (which are equal). The lattice vectors are orthogonal to each other. A lattice with mutually orthogonal lattice vectors, of which only two are of the same length, is termed **tetragonal**. A certain similarity does exist between the lanthanum cuprate and the perovskite structures: in both there are planes of AO or of BO_2. However, in the structure shown in Fig. 2.21(b) there are two planes of the first type in-between two planes of the second, as opposed to a single such

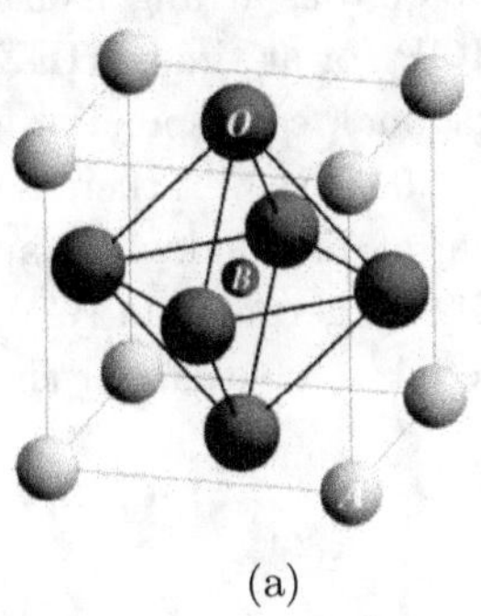
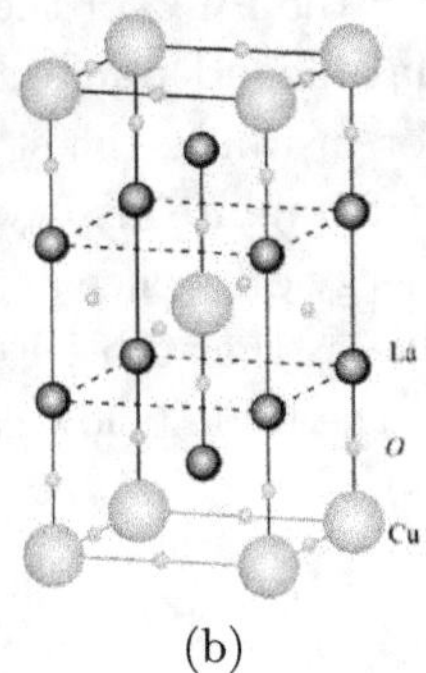

(a)   (b)

Fig. 2.21: (a) The perovskite structure of ABO_3, with, e.g., A=Sr, La, Ba, and B=Ti, Al, Mn; (b) the tetragonal structure of lanthanum cuprate, La_2CuO_4. The smaller spheres represent the oxygen ions, and the others are lanthanum or copper ions, as marked in the figure. The lattice constants are $a_1 = a_2 = a$ and $a_3 = c$.

plane in the structure illustrated in Fig. 2.21(a). Each copper ion in the lanthanum cuprate is surrounded by an octahedron of oxygens. The structure displayed in Fig. 2.21(b) is known as a **"generalized perovskite"**. The lanthanum-cuprate compound is famous as being the first "high-transition-temperature-superconductor"; it has been found that by replacing a fraction of the lanthanum ions by barium or strontium ones the material becomes a superconductor below $35°K$. At the time of this discovery, this transition temperature was considerably higher than the superconducting transition temperatures known till then. It was later found that one may add more planes in-between the copper ones, and so increase the transition temperature further, even up to $138°K$. The mechanism responsible for these high transition temperatures is still a topic of intensive research.

Problem 2.13.
Determine the Bravais lattice of the strontium titanate, Fig. 2.21(a); find the primitive unit cell and the base of this crystal. What is the relation of the base to the chemical formula of the compound?

Problem 2.14.
Is the tetragonal cell of the lanthanum cuprate shown in Fig. 2.21(b) the primitive one? If not, what is the primitive cell and what is the base? Find the lattice vectors.

2.6 Dense packing in space

Lattice structures of chemical elements. Figure 2.22 displays the periodic table of the elements in Nature. The drawing in each entry indicates the **lattice structure of the crystal** built on the corresponding element (in the solid phase of the material, at low enough temperatures). It is seen that many of the metals are arranged in the **FCC** or the **BCC** lattices. Another ubiquitous arrangement is the **hexagonal closed-packed** crystal, abbreviated as **HCP** (or at times, **HEX**); see below. For instance, the stable form of carbon is hexagonal, corresponding to the structure of graphite displayed in the right panel of Fig. 2.9. However, there is only a minute energy difference between this structure and that of diamond which appears in the table as the stable **FCC** structure of silicon (Si) and germanium (Ge), both located in the table below carbon (and therefore expected to have similar chemical properties).

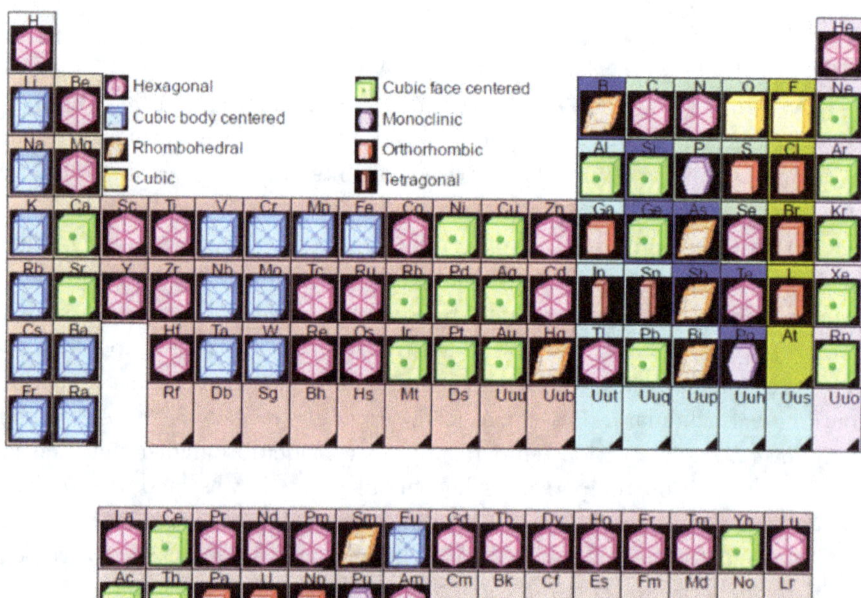

Fig. 2.22: The lattices formed by the elements in the periodic table: the figure in each square indicates the structure (face-centered cubic **FCC**, body-centered cubic **BCC**, hexagonal **HEX**, and more).

Close-packing in space. As discussed in Sec. 2.2, the triangular lattice corresponds to a dense packing of circles on a plane, e.g., the ping-pong balls in a swimming pool, see Fig. 2.3. The volume packing-ratio of this planar layer of spheres is 0.6046, as found in problem 2.1(b). The **simple hexagonal lattice** is formed when such triangular-lattice layers are placed one above the other, with straight lines of lattice points along the direction normal to the layers. When the spheres are also adjacent to those in the neighboring layers, the resulting volume packing-ratio is identical to that of a single layer. This ratio decreases when the spacing in-between the layers is extended (and the spheres in neighboring layers are not touching each other).

The packing ratio of the simple hexagonal lattice is not the optimal one. A larger ratio can be obtained by the following procedure. One begins with the dense packing on the plane, i.e., from the infinite planar triangular arrangement of spheres as in Fig. 2.3, and adds gradually more spheres. The first sphere to be added has to initiate a new layer. This sphere is equally attracted to each of the spheres in the first layer (as they are all identical), and therefore it "aims" to be close to as many spheres as possible. It therefore lands in the valley created at the middle of the triangle formed by neighboring spheres in the first layer (and not precisely above a single sphere there, as happens in the simple hexagonal structure). Joining the center of the newcomer with the centers of those three spheres creates a **tetrahedron**. The tetrahedra repeat themselves all over the second layer; they form a three-dimensional structure of close-packed spheres, much as the equilateral triangles are a basis for the close-packing in two dimensions.

Each sphere in the first layer is surrounded by six valleys. However, once one of the valleys is occupied, the two next to it remain empty (as all spheres are of equal size, if two of them reside in neighboring valleys they have to partially "penetrate" into one another). As a result, only three valleys around each sphere of the first layer are occupied, for example in the arrangement of the dashed circles in Fig. 2.23 (marked by B). In this way a triangular lattice of dashed circles is formed. On the other hand, when the valley next to the one exploited before is occupied, the triangular lattice that emerges is the one marked by C in the figure (thin circles). When the lattice is grown by adding atoms one by one, domains of the type B and domains of the type C can be formed in the second layer, and they cannot be joined together. These metastable states can be removed by carrying out carefully the growth procedure (see Sec. 2.2); at times several cycles of heating and cooling are needed in order to reach an arrangement based on one of these two options.

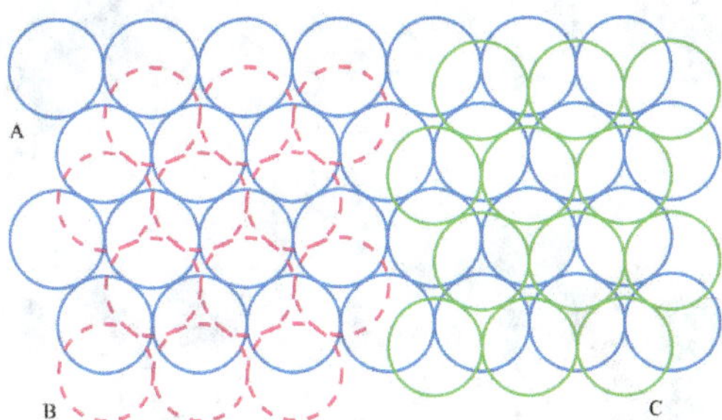

Fig. 2.23: Possible locations of spheres in the second layer above a triangular lattice. The letter A represents the first layer (thick blue circles), and the letters B (dashed red circles) and C (thin green circles) represent the two possible arrangements of the second layer.

There are also two options in constructing the third layer. Denoting the first layer by A and the second by B, the spheres in the third layer can locate themselves exactly above those in the first layer, as in Fig. 2.24(b), and then the third layer is A. The periodicity of the structure so formed, ABABAB..., [depicted also in the lower part of Fig. 2.24(b)] is of two layers in the direction normal to the plane. This structure, called **hexagonal close-packed** and abbreviated **HCP**, has the densest packing ratio of spheres in space. Alternatively, the centers of the spheres in the third layer can place themselves above the empty valleys in the two preceding layers as in Fig. 2.24(a), and then the third layer is marked by C, and the structure is ABCABCABC... This structure is identical to the **FCC** lattice, and it also appears under the name **cubic close-packed, CCP**. The lower panel of Fig. 2.24(a) displays the cubic unit cell of the **FCC** lattice, as does also the mid panel in Fig. 2.17. Note that the green spheres are all lying on the plane spanned by three corners of the cube (two opposite ones in the base of the cube and an opposite one in the layer above, connected by thin dashed lines). These green balls are part of a triangular lattice. Likewise, the red balls, again lying on a plane passing through three corners of the cube, are also part of a triangular lattice (parallel to the previous one), and each of them resides in a valley formed in the green plane. The blue sites on each side of these two planes create the third layer in the ABCABC... arrangement.

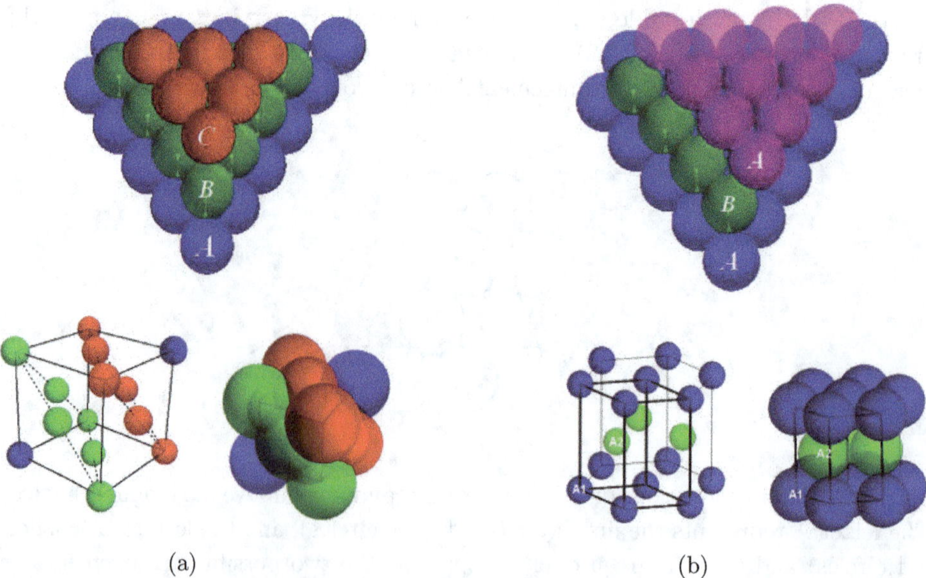

(a)   (b)

Fig. 2.24: (a) The **FCC≡CCP** structure; (b) the **HCP** structure. Each color marks a dense planar layer.

The hexagonal close-packed lattice. The hexagonal prism in the lower part of Fig. 2.24(b) can serve as the unit cell of the **HCP** lattice, but it is not the primitive unit cell. The primitive unit cell of the triangular lattice, portrayed in Fig. 2.4, is a rhombus; in the **HCP** structure, the layer above is also a triangular lattice and consequently the primitive unit cell of this lattice is the prism with a rhombohedral base, indicated in the figure by thick lines. The two planar lattice vectors in Figs. 2.4 and 2.5, i.e., $\mathbf{a}_1 = (a,0,0)$ and $\mathbf{a}_2 = (a/2, a\sqrt{3}/2, 0)$ (both are of length a, with an angle of $60°$ in-between them) are reinforced by a third vector normal to the plane, $\mathbf{a}_3 = (0,0,c)$, of length c. The prism contains two lattice sites, marked in the figure by $A1$ and $A2$. The base points in the unit cell are located at the origin and at $\mathbf{r}_2 = (a/2, a\sqrt{3}/6, c/2)$ (just above the mid point of the equilateral triangle below). In the dense hexagonal structure, the spheres in the mid layer touch the spheres in the lower and upper layers. Adding the constraint $|\mathbf{r}_2| = 2r = a$ (where $2r = a$ is the diameter of each sphere) yields $c = 2a\sqrt{2/3} \approx 1.633a$ ($c/2 = a\sqrt{2/3}$ is the height of the tetrahedron formed by the sphere inside the unit cell and the three others in the basis touching it; the edge of this tetrahedron is of length a. Check!) Nonetheless, in many hexagonal materials $c > 1.633a$. The notation **HCP** is kept solely for the **dense structure** for which $c = 2a\sqrt{2/3} \approx 1.633a$.

Problem 2.15.
a. *Calculate the packing ratio of the* **HCP** *lattice.*

b. *Compare with that of the* **FCC** *lattice (problem 2.11). Are you surprised?*

Problem 2.16.
Triangular lattices of spheres are arranged in layers, forming the structure ABCBABCBA.... Determine the primitive unit of this lattice and calculate its packing ratio.

The two lattices, **HCP** and **FCC**, share the same density, 0.74, which is the largest packing ratio of identical spheres in space (see problem 2.15). They also share the same potential energy if only that of nearest-neighbor pairs is accounted for. As can be seen in the periodic table, Fig. 2.22, these two structures are rather ubiquitous. The factors that determine which of them is more stable hinge on delicate considerations, like the interaction between farther-away neighbors (Chapter 4). The subtle disparity in the arrangements of successive planes leads at times to more complex structures, e.g., ABCBABCBABCBA... (problem 2.16), ABCACBABCACB... (with periodicity of six planes), and more. All these structures share the same densest packing ratio. Nature produces at times also random arrays, with no periodicity along the normal to the planes.

Kepler's conjecture. The question "what is the largest packing ratio of identical spheres in space?" had been already asked by the astronomer Johannes Kepler in 1611 (following the one raised a century before regarding the best arrangement of cannon balls in the ships sailing to America). Kepler conjectured that the maximal packing ratio is 0.74, but was not able to prove it mathematically. In 1831 Gauss succeeded to show that when the spheres are arranged periodically then indeed the densest structures are the **HCP** and **FCC** lattices. All the periodic structures which are possible in space are sorted out below; a simple computation of the corresponding packing ratios reproduces the afore-mentioned result. However, the question whether there exist denser non-periodic structures remained open; it has been even included by Hilbert in his 1900 list of unsolved mathematical problems. This question can be explored by exploiting formidable computer codes that pursue all possibilities. In 1998 Thomas Hales argued that he had proven the Kepler's conjecture a formal proof was published in 2014 by him and his collaborators. Non-periodic structures of identical spheres or other bodies are quite essential for explaining the physics of **granular materials**. Granular systems are created when grains are poured at random into a vessel (e.g., wheat grains poured into a silo, or medication tablets into their container). The research of granular material is going on for quite a number of years, but there are still many open questions, challenging physicists and engineers.

Hexagonal crystals with a base – wurtzite. Examples of crystals built of **two types of ions** have been encountered in the discussion of the cubic lattices. For instance, the structure of table salt or of zinc-blende (Figs. 2.17 and 2.20) contains two interwoven **FCC** lattices, of the sodium (zinc) ions, and of the chloride (sulphur) ones. It turns out that ZnS assumes also the structure called **wurtzite**, exhibited in Fig. 2.25. As for the zinc-blende, this structure is constructed such that the zinc ions fill the tetrahedral cavities among the sulphur ions; in the case of wurtzite, though, the sulphur ions occupy the sites of an **HCP** lattice. This structure comprises two **HCP** lattices, one of the zinc ions and the other of the sulphur ones, displaced relative to one another along the $c-$axis of the unit cell by 5/8 the height of the unit cell. Due to the above-mentioned minute differences between the **CCP≡FCC** and the **HCP** lattices, ZnS appears in Nature in the two polymorphic structures, the particular one is chosen by subtle details, like long-range interactions, the growth method, and the ambience conditions (temperature and pressure).

Problem 2.17.
Find the primitive unit cell of the wurtzite structure. What is the base of this crystal?

2.7 Classification of periodic lattices according to their symmetries

The symmetry groups of crystals. Each lattice (or crystal) possesses certain operations that transform it to itself. Such an operation is called a **"symmetry operation"**. For instance, a translation by a vector of the type given in Eq. (2.1) leaves the lattice as is, and therefore it is a **symmetry operation of the lattice**.

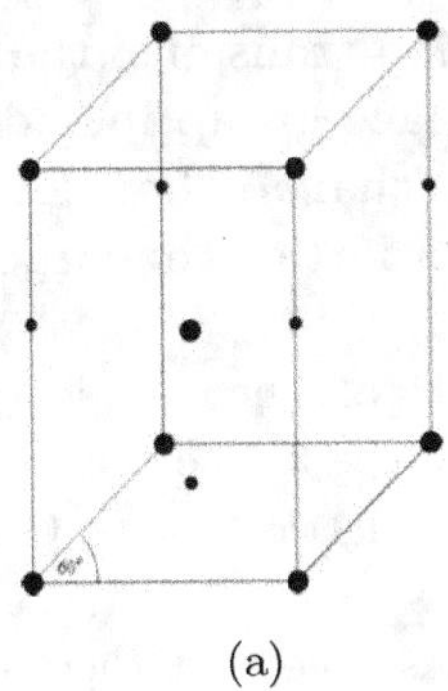
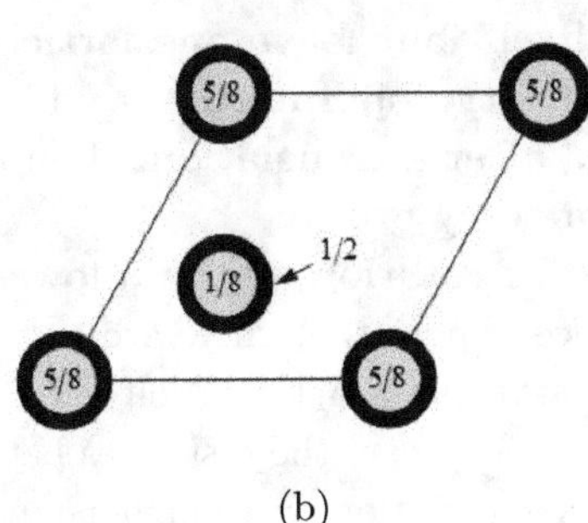

(a)   (b)

Fig. 2.25: (a) The structure of wurtzite. The larger points indicate the sulphur, and the small ones the zinc. Each type of ions resides on an **HCP** lattice. (b) The projection on the base plane of the hexagonal unit cell (the sulphur ions–large dark circles–are at height 0 or 1/2, and the zinc ions–bright small circles–are at 1/8 height and 5/8 height of the unit cell).

The collection of the Translations by all linear combinations with integer coefficients of the lattice vectors forms an algebraic **group**: the sum of any two such combinations, or their product by an integer, are also vectors of the same character, and hence the symmetry operations they represent belong to the same group as well. (In fact, these properties are the formal definition of a group.) This group is called **the translation group of the lattice**.

Likewise, most of the lattices reproduce themselves after being **rotated**. For example, the planar triangular lattice is not changed upon being rotated by 60° around an axis normal to the plane and passing through one of the lattice sites. Each of the cubic lattices is invariant (remains itself) when rotated by 90° around an axis that coincides with one of the lattice vectors of its cubic unit cell, or by 120° around an axis which lies on a diagonal of the cubic unit cell. There are other symmetry operations as well (see below). In this section it is established that all the Bravais lattices can be classified according to their symmetries. There is only a finite number of classes of lattices, where the lattices in each class share the same symmetry operations that transform them back to themselves. In Chapter 3 it is found that X-ray diffraction patterns of each lattice are invariant under the same symmetry operations as those of the lattice itself. As the number of lattices' classes is finite, their structure can be identified by inspecting the symmetry operations of the diffraction patterns.

Since a lattice is invariant under a certain symmetry operation, it follows that applying two symmetry operations leaves it unchanged as well, and hence the combined operation is also a symmetry one. This property implies that the ensemble of all the symmetry operations of a certain lattice is an algebraic **group**, termed the **"symmetry group of the lattice"**. It is customary to distinguish between two types of symmetry operations and symmetry groups: **a point group of symme-try operations is the collection of operations which transform the lattice to itself, but leave one lattice point intact. A space group includes also symmetry operations in which all lattice points change their locations, e.g., after a translation. The more symmetric is a lattice the larger is its symmetry group.**

The rotation angles allowed for a periodic lattice. The periodicity of the lattice implies that there are only a few angles by which the lattice can be rotated and still come back to itself. Consider the straight line of lattice points, CAA'D in Fig. 2.26, where the distances between any neighboring sites is the lattice constant of that direction, a (and hence they are all equal). Assume that there exists a rotation axis passing through the point A which is normal to the plane of the page. The lattice is rotated around this axis by an angle ϕ and comes back to itself. As a result of the rotation, the point C moves to the point B (the points B and C are both located at a distance a from A, AB=AC=a). The existence of this rotation axis implies that the lattice is invariant also under the inverse rotation, by the angle $-\phi$. Such a rotation around an axis passing through the point A' and parallel to its predecessor, brings the point D to the point B', with A'D=A'B'=a. The points B and B' create a straight line, parallel to the original one, CD. Since both A and A' are on the lattice, the distance between them is necessarily an integer product, m, of the lattice constant (along this direction), i.e., AA'=ma ($m = 1$ in the figure). As the line BB' is parallel to the line CD, the distance between the points B and B' must be an integer product of a as well, with a different integer coefficient m', i.e., b =BB'= $m'a$ ($m' = 2$ in the figure). The length difference between AA' and BB' is twice the projection of BA on CA; a simple trigonometric calculation shows that difference to be $(m' - m)a = 2a\cos\phi$. Hence,

$$\cos\phi = (m' - m)/2 \ . \tag{2.7}$$

As m and m' are integers and $|\cos\phi| \leq 1$, the allowed values for the angle ϕ are obtained for $m' - m = -2, -1, 0, 1, 2$, leading to $\phi = \pi$, $2\pi/3$, $\pi/2$, $\pi/3$, 0 (or $\phi = 180°$, $120°$, $90°$, $60°$, 0).

Usually the allowed rotation angles are denoted as $\phi = 2\pi/n$, and then the integer n can be solely 1, 2, 3, 4, or 6. In this notation the rotation axis is termed **"rotation axis of order n"**. ($n = 1$ represents a rotation by 360°, which is identical to no rotation at all.) This proof **forbids** any other rotations, e.g., of order 5 or 10.

Additional symmetry operations. In addition to Translations and rota-tions, a Bravais lattice may be mapped back on itself by the following symmetry operations: (i) **reflection by a planar mirror**; for instance, when the mirror is normal to the $\hat{x}$−axis at the origin, the reflection transforms the point (x, y, z) into the point $(-x, y, z)$; (ii) **inversion through a point**; each vector **r** originating at that point is transformed into the vector $-\mathbf{r}$; (iii) **complex rotation**; interweav-ing a rotation and a reflection, or a rotation and a translation (like the motion of

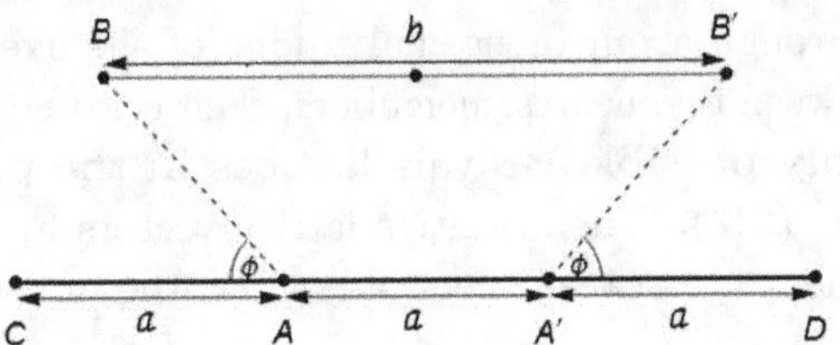

Fig. 2.26: Rotations of a lattice. The rotation axes, passing through the points A and A', are normal to the plane of the page. The lattice constant along the lines CD or BB' is a.

a screw). Some of these operations belong to the point group (e.g., the inversion through a lattice point, or the reflection through a mirror passing through a lattice site), and other belong only to the space group (e.g., the inversionthrough a point which is not a lattice site, or a complex rotation). Appendix A describes the Platonic bodies, whose symmetry groups include numerous elements. As mentioned there, several of these symmetry groups are shared by periodic lattices. An example are the symmetry groups of the cube and of the octahedron which in many aspects are largely identical to those of the cubic lattices.

Classification of one-dimensional lattices. In one dimension there is only a single Bravais lattice, built of a periodic array of points located at $\mathbf{r} = n\mathbf{a}$, where n is an integer and $\mathbf{a}$ is the lattice vector [see Fig. 2.1]. This lattice is mapped upon itself by rotations of order 1 or 2 around an axis normal to the one-dimensional lattice in two- or three-dimensional spaces, passing through a lattice site or through the mid point between two lattice sites. The symmetry group of this lattice also includes a reflection by a mirror perpendicular to the lattice at a lattice site or at half way between two nearest neighbors, and is also invariant to inversions via these locations. (These two operations are identical in one dimension.)

Classification of two-dimensional lattices. The **square lattice** in Fig. 2.6 is invariant under rotations of orders 2 and 4 (e.g., around an axis perpendicular to the plane and passing through a lattice site or through the center of each square. Each rotation axis of order 4 is a rotation axis of order 2 as well, but not the other way around: an axis normal to the plane at the mid point between two nearest neighbors is only a rotation axis of order 2). The **triangular lattice** (or the hexagonal lattice) in Fig. 2.4 comes back to itself after rotations of order 2, 3, or 6. These symmetry properties are independent of the **length** of the lattice constants, and therefore they characterize **all** lattices of this structure. The **rectangular lattice** for which the lattice vectors are normal to one another but are of disparate lengths has only rotation axes of order 2. The most general planar lattice, the **oblique lattice** shown in Fig. 2.27, has two lattice vectors of different lengths, and the angle between them is not a right one. This lattice has also only rotation axes of order 2. The rectangular lattice is more symmetric and indeed it is also invariant under a reflection through a mirror normal to one of the axes at a lattice site or at the mid-point between two nearest neighbors. These **four** lattices, illustrated in Fig. 2.27, are the **only possible Bravais lattices in the plane, according to the constraint Eq. (2.7).** The primitive lattice vectors $\mathbf{a}_1$ and $\mathbf{a}_2$ belonging to each lattice, the angle γ in-between them, as well as the unit cell, are indicated in the figure.

The rectangular-centered lattice. As mentioned in Sec. 2.5, it is at times expedient to describe a lattice exploiting a non-primitive unit cell. The planar example is the **rectangular-centered lattice**, for which each rectangle's center is also occupied by a lattice site. The right lower panel of Fig. 2.27 depicts both the rectangular unit cell (with a base comprising two points, marked by dashed lines) and the primitive one, which resembles that of the oblique lattice. With the former cell, one may take advantage of the simpler cartesian coordinate system. Figure 2.27 lists separately the rectangular-centered lattice, since it is more symmetric than the oblique one, and shares certain symmetry operations with the rectangular lattice: e.g., its symmetry group includes reflections through mirrors normal to the plane and passing through one of the edges (in contrast with the oblique lattice). For this reason, the rectangular-centered lattice is occasionally presented as yet another planar Bravais lattice, increasing the total number of planar Bravais lattices to **five.**

The point groups of the lattices in Fig. 2.27 are as follows (in all cases, the rotation axes and the mirror planes are perpendicular to the lattice plane). The oblique lattice possesses solely rotations of order 2 around axes passing through lattice sites, through the mid point of each edge, and through the center of each cell. The rectangular lattice has in addition reflection planes normal to each of the axes, at each lattice point, at the mid point of each edge, and at the center of each rectangle. Increasing further the symmetry, the square lattice possesses also rotation axes of order 4 through each lattice site and through the center of each square, and mirror planes passing through the diagonals of each square. Finally, the most symmetric triangular lattice is characterized by rotations of order 2 around the mid point of each edge, rotations of order 3 around the center of each triangle, and rotations of orders 2, 3, and 6 around each lattice point. The symmetry group of the triangular lattice also contains mirror planes through each edge, normal to edge at its mid point.

Problem 2.18.

Why is the rectangular-centered lattice not mentioned in the preceding paragraph?

Classification of three-dimensional lattices. Figure 2.28 displays the allowed Bravais lattices in space. The lengths of the lattice vectors and the angles in-between them are defined in Fig. 2.13. **Seven** of the lattices have a primitive unit cell that contains a single lattice site. In three of them the lattice vectors are normal to each other: in the **cubic lattice** (all lattice vectors are of the same

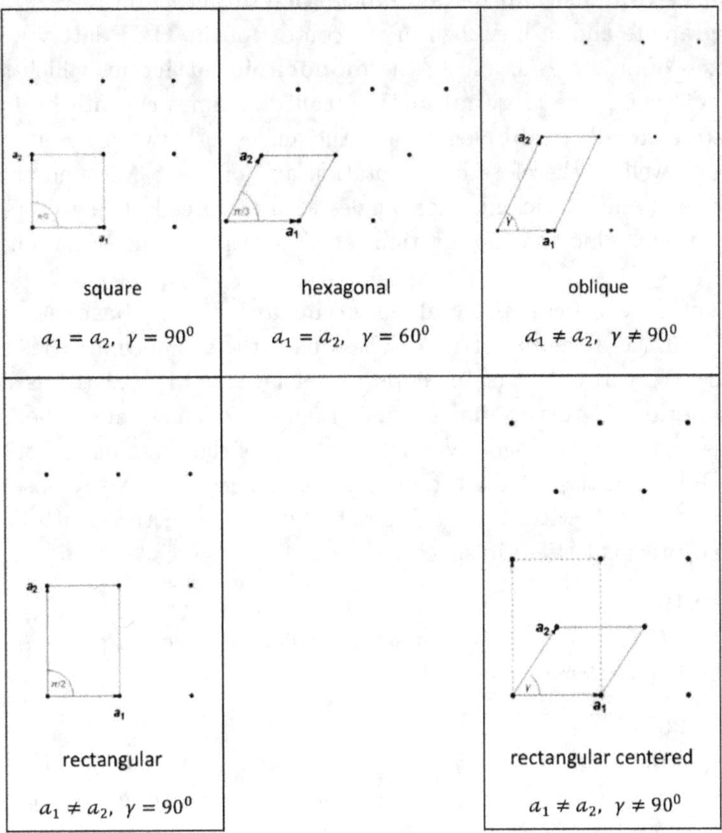

Fig. 2.27: The four Bravais lattices in the plane: the square, the triangular, the oblique, and the rectangular. Also are shown the lattice vectors, the angle in-between them, and the primitive unit cell. The primitive lattice vectors of the fifth lattice, the rectangular centered, are like those of the oblique lattice, with an arbitrary angle γ. The lengths of this lattice are such that $\cos\gamma = a_1/(2a_2)$, and the vectors $\mathbf{a}_1$ and $2\mathbf{a}_2 - \mathbf{a}_1$ form a right angle. Indeed, these two vectors create a rectangle, shown by the dashed lines. This rectangle represents the unit cell of a rectangular lattice, with a base containing one point at the corner and one at the center.

length), in the **tetragonal lattice** (two of the vectors have the same length) and in the **orthorhombic lattice** (all lengths are different). These three types possess rotation axes of order 2, and the first two have in addition axes of order 4. The symmetry group of the cubic lattice also includes axes of order 3 (around the cube's diagonal). Note that all the symmetry operations of the cubic lattice are included in the point group of the rhombic dodecahedron, shown in Fig. 2.19(a). There is also the **three-dimensional hexagonal lattice** (displayed in Fig. 2.9 in connection with graphite and in Fig. 2.24 in the context of the **HCP** lattice), which has rotation axes of orders 2, 3, and 6; the **monoclinic lattice** for which there exist only axes of order 2; the **trigonal lattice** (called at times **rhombohedral**) whose lattice vectors are all equal in length and the angles in-between them are equal to each other as well; it therefore has a rotation axis of order 3 around the diagonal of the unit cell (which forms identical angles with all three lattice vectors); and the **triclinic lattice**, which has no rotations at all, except for the trivial one, of order 1.

Several of these lattices appear also as **structures with a base**, e.g., the **BCC**, the **FCC**, and the diamond lattices. These have the same symmetries as the **SC** lattice. The structure of lanthanum cuprate, displayed in Fig. 2.21(b), is an example of a **body-centered tetragonal lattice**. Figure 2.28 shows also other structures with a base, that have the same symmetries as one of the seven basic lattices. All of them are Bravais lattices, though their symmetries are more easily observed when considered as crystals with a base. Together with the structures with bases, there are fourteen Bravais lattices in space.

Problem 2.19.

The list of spatial Bravais lattices does not include the base-centered tetragonal lattice and the face-centered tetragonal lattice. Why?

Problem 2.20.

*Show that the lattices formed by the primitive unit cells of the **BCC** and **FCC** lattices, depicted in Fig. 2.16 and 2.18, are particular cases of the tetragonal lattice. Determine the angles in-between the lattice vectors in each of them.*

Problem 2.21.

Prove that any Bravais lattice is mapped back upon itself under an inversion through a lattice point.

Problem 2.22.

a. *List the symmetry operations of the tetragonal lattice.*
b. *Which members of this group disappear once the lattice is distorted into a monoclinic structure?*
c. *Which operations are added to the group when the tetragonal lattice is distorted to be a cubic one?*

Problem 2.23.

A circle is placed at each lattice point of the (planar) oblique lattice, such that all circles are touching each other. Find the planar packing ratio of this arrangement, and show that (in the appropriate limits) the result reproduces the packing ratios of all other lattices in Fig. 2.27. Compute the maximal packing ratio. To which lattice does it belong? [Hint: analyze first the angle in the range $60° \leq \gamma \leq 120°$ and then show that other angles can be mapped into this range.]

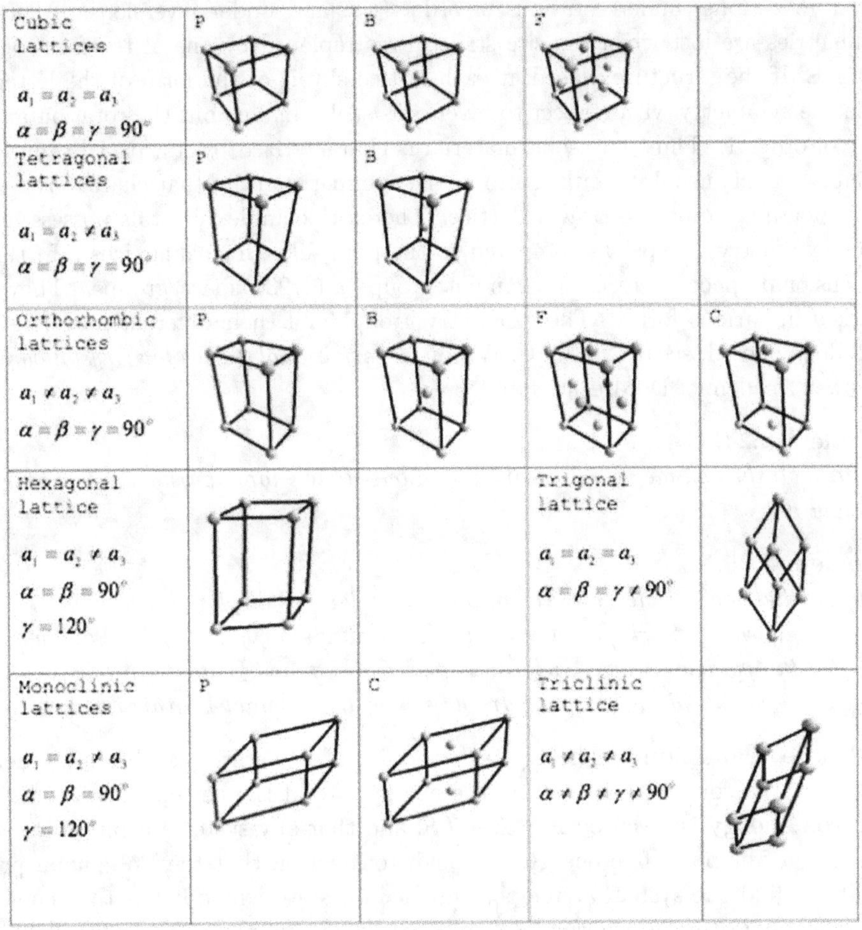

	P	B	F	C
Cubic lattices $a_1 = a_2 = a_3$ $\alpha = \beta = \gamma = 90°$				
Tetragonal lattices $a_1 = a_2 \neq a_3$ $\alpha = \beta = \gamma = 90°$				
Orthorhombic lattices $a_1 \neq a_2 \neq a_3$ $\alpha = \beta = \gamma = 90°$				
Hexagonal lattice $a_1 = a_2 \neq a_3$ $\alpha = \beta = 90°$ $\gamma = 120°$		**Trigonal lattice** $a_1 = a_2 = a_3$ $\alpha = \beta = \gamma \neq 90°$		
Monoclinic lattices $a_1 = a_2 \neq a_3$ $\alpha = \beta = 90°$ $\gamma = 120°$	P	C	**Triclinic lattice** $a_1 \neq a_2 \neq a_3$ $\alpha \neq \beta \neq \gamma \neq 90°$	

Fig. 2.28: The seven spatial Bravais lattices (with a single lattice site in the primitive unit cell) and the seven Bravais lattices with a base (each sharing the same symmetries with one of the formers). The letter P indicates "primitive", B is "body-centered", F is "face-centered", and C is "based-centered". The lengths of the three lattice vectors are denoted by a_1, a_2, and a_3, and the angles among them are defined in Fig. 2.13.

Subgroups of the symmetry group. The symmetry of a crystal with a base, or a lattice in which each site is occupied by an atom or a molecule with a non-spherical internal structure, is lower than the symmetry of the point-like lattice. For example, when each lattice site of the one-dimensional lattice [Fig. 2.1] is occupied by a structure of the form of the digit 0, this lattice still returns to itself under reflections, under rotations of order two, and under inversions. All these symmetries are lost when the site structure is replaced by one of the form of the letter E. If the structure placed at each lattice site is of the form of the letter W then the symmetry with respect to inversions still remains, but the rotation axis is only of order 1. Thus, any solid material is characterized by its spatial symmetry group, which is usually a **subgroup** of the symmetry group that characterizes the corresponding point-like Bravais lattice. Therefore complex crystals possess many more symmetry groups (as compared to the point-like Bravais lattices). In three-dimensional space there exist 32 point groups and 230 space groups. They are marked in various forms. The symmetry groups and their different notations are well-documented, see e.g., A. J. C. Wilson, *International Tables for Crystallography* (Kluwer Academic, Dordrecht, 1995).

Problem 2.24.
Identify all the symmetry operations of the one-dimensional crystals shown in Figs. 2.1 and 2.2.

Problem 2.25.
Suppose that each of the unit cells of the planar lattices displayed in Fig. 2.27 is occupied by a base comprising an atom at the coordinate origin of the cell and another atom at the point $\mathbf{r}_2 = (\mathbf{a}_1 + \mathbf{a}_2)/3$. *Determine the modifications in the symmetry groups, as compared to the symmetry groups of the original Bravais lattices.*

Phase transitions. The crystal structure of a material is ultimately determined by the competition between the energy E and the entropy S. Recall that the **free energy** is given by $F = E - TS$, and that any system "aspires" to be in the state of minimal free energy. [Enough to mention that the Boltzmann probability to find the system at temperature T in a state where it has the energy E is proportional (up to a normalization given by the sum over all probabilities) to $g(E)\exp[-E/(k_\mathrm{B}T)]$, where $g(E)$ is the degeneracy of the energy E (the number of different states with the same energy), and k_B is the Boltzmann constant. When the entropy is written in the form $g(E) = \exp[S/k_\mathrm{B}]$, this probability is proportional to $\exp[-(E - TS)/(k_\mathrm{B}T)] = \exp[-F/(k_\mathrm{B}T)]$, and consequently there is a higher probability for the system to be found in a state where the free energy F is lower.] The second term in the free energy, $-ST$, is dominant at high temperatures, and then states with higher entropy, i.e., with higher symmetry, are more favorable. As the temperature T is reduced, the significance of this second term is diminished; consequently the symmetry can be reduced and the system may attain more ordered states and thus gain energy (the energy is usually smaller as the symmetry is reduced). Therefore, many materials undergo **phase transitions** among various crystalline structures as the temperature is varied. These transitions occur at temperatures termed **"transition temperatures"** and may be either of **first order**, when the material "jumps" discontinuously from one structure to another, or of **second order**, when the structure changes continuously as the temperature drops below the transition temperature. It is common to refer to such transitions as **symmetry-breaking transitions**. Thus, e.g., it is possible that a material crosses over (as the temperature is reduced) from its gaseous state to a liquid one, then to a solid phase with cubic symmetry, from it to a tetragonal structure, which is less symmetric, and then to an orthorhombic structure, which is even less symmetric. This is the case with the two perovskite structures described in Fig. 2.21. An hexagonal lattice may reduce its symmetry by transforming into a trigonal or an orthorhombic lattice.

Ferroelectricity. A conspicuous example for a symmetry-breaking phase transition is found in **ferroelectric materials**, e.g., barium titanate, $BaTiO_3$. At high temperatures this material has the perovskite structure, see Fig. 2.21(a). The barium ions are located at the corners of the cube, marked in the figure by A. At high temperatures the structure is cubic, like the one shown in the right panel of the figure, and therefore the centers of mass of the positive and the negative ions (in each unit cell) are found exactly at the same point: for the cubic cell in Fig. 2.29, there are $+4$ charges in the sites at the center of the cube, eight $1/8$'s of $+2$ charges in the sites at the corners, and six $1/2$'s of -2 charges in the sites at the centers of the faces. Altogether, there are $+6$ charges in the unit cell, and -6 charges that can be assigned to the center of the cube. Because the centers of mass coincide, this lattice lacks an electric dipole [which exists only when the center of mass of the positive charges and the one of negative charges are separated by a finite distance; the moment is equal to the product of either the total positive charge or the total negative one (equal up to a sign) with that distance]. Below the transition temperature, the titanium and the barium ions move along one of the axes (chosen at random), whereas all other ions remain in their locations. As a result of this displacement, the symmetry is reduced from cubic to tetragonal, and an electric dipole moment is created in the displacement direction, see the left panel of Fig. 2.29. In this new state the center of mass of the positive charges is shifted relative to the center of the cube, along one of the cartesian axes. If the crystal is placed in an electric field parallel to one of the axes during the cooling process, the ions move so that the dipole moment created below the transition temperature is parallel to the electric field; it is so oriented also after this field is switched off. Reversing the electric field direction inverts the direction of the displacement, and thus also the direction of the dipole moment.

This new phase is called **"ferroelectric"**. The "ferro" part of this term originates from the similarity between this phenomenon and ferromagnetism (Sec. 2.10); it indicates that each unit cell carries the same dipole moment, all dipoles pointing at the **same** direction (like the magnetic dipole moments in a ferromagnet). The "electric" part of the name characterizes the response to an electric field; the dipole moment of the crystal follows the direction of an applied electric field. This property of ferroelectric crystals enables them to keep and process information in

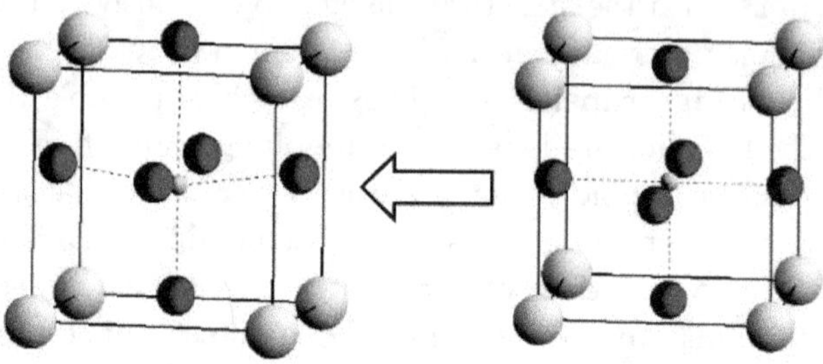

Fig. 2.29: The phase transition of barium titanate from the cubic structure (on the right) to the ferroelectric tetragonal structure (on the left). The barium and the titanium ions move slightly away from their locations in the perovskite lattice shown in Fig. 2.21(a).

small samples. For this reason ferroelectric materials are very useful for various technological applications.

Problem 2.26.
Based on Figs. 2.1 and 2.2, propose a one-dimensional model for the ferroelectric phase transition.

2.8   Quasicrystals

The discovery of Shechtman –quasicrystals. The allowed rotation axes in a Bravais lattice are proven to be solely of order 1, 2, 3, 4, or 6, and therefore the lattice can be mapped back on itself only after rotations by 360°, by 180°, by 120°, by 90°, or by 60°. Chapter 3 discusses the experimental techniques to identify the crystal structure of a given material. A particularly useful method is based on the diffraction picture created by the interference of radiation waves scattered from the periodic array of the crystal. The diffraction pattern reflects the symmetry properties of the crystal; specifically, it contains the same rotation axes that exist in the crystal. While investigating the diffraction patterns of alloys grown from a mixture of aluminium and manganese, Shechtman had observed in 1982 "circles" around the center of the picture, comprising **ten** points, and invariant under rotations by 36°. Such a symmetry operation, of a rotation axis of order 10, contradicts Eq. (2.7), and is therefore forbidden according to the analysis above. Indeed, the scientific community had doubted this result, hindering its publication till 1984. The discovery was fully recognized in 2011, when Shechtman was awarded the Nobel prize in chemistry. By now, many more materials of such symmetries have been discovered, as well as crystals possessing other "forbidden" rotation axes, like 8 or 12, and it is completely accepted nowadays that Shechtman discovered **a new phase in condensed matter**. It became clear that the "new" symmetries are possible since the atoms in such materials are not arranged periodically; these are **not Bravais lattices**. They are called **"quasicrystals"**, and their intriguing properties are currently being investigated by many research groups.

The diffraction pattern conveys information on the **Fourier transform** of the density probability in the sample, see Chapter 3. The density of the material itself is obtained by computing the inverse transform of the diffraction pattern. This technique, applied to the data of Shechtman, produces complex structures in which the atoms are arranged according to certain rules, but **are not periodically ordered**. Alternatively, one may explore the spatial structure of the atoms in quasicrystals by measuring the properties of their surfaces. It is possible to observe the surface directly by using a scanning electron microscope (see Chapter 3), and it is also feasible to adsorb certain atoms on the surface and then monitor their locations. The atoms place themselves at minima of the potential energy of the surface; measuring these locations enables one to deduce the structure of the surface itself. The right panel of Fig. 2.30 shows a picture, taken by a microscope, of silver atoms adsorbed on the surface of a Al-Pd-Mn quasicrystal (made of aluminium, manganese, and palladium). The left panel of Fig. 2.30 portrays the map of the potential energy of the surface, as deduced from the microscope picture in the right panel. Darker areas represent lower potential energy. The figure emphasizes a place of strong attraction, which indeed adsorbed a silver atom. The left panel contains many points whose local surrounding is invariant under rotations of order 5.

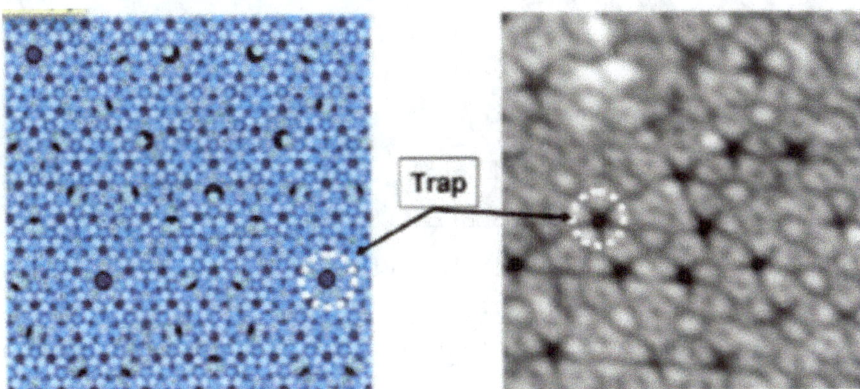

Fig. 2.30: Right panel: a picture taken by a scanning electron microscope of silver atoms adsorbed on the surface of the quasicrystal Al-Pd-Mn. The left panel shows the corresponding potential-energy map. Taken from: B. Unal, V. Fournée, K. J. Schnitzenbaumer, C. Ghosh, C. J. Jenks, A. R. Ross, T. A. Lograsso, J. W. Evans, and P. A. Thiel, Phys. Rev. B **75**, 064205 (2007).

A new definition of a "crystal". Following Shechtman's discovery, it became clear that quasicrystals **are not periodic** and hence cannot be described by

Bravais lattices, as this description requires a **single** unit cell that repeats itself. Nonetheless, these structures do possess certain symmetry properties similar, in a sense, to those of lattices, and do have a certain order. In addition, the diffraction patterns of quasicrystals comprise discrete radiation points – like those of a periodic lattice (Chapter 3). The discovery of Shechtman caused the International Union of Crystallography to modify the definition of a "crystal", omitting the word "periodic". As of 1992, a **crystal is a material whose diffraction pattern comprises discrete points.**

Fibonacci sequence. Perhaps the simplest example of a quasicrystals is provided by the **Fibonacci sequence and the related one-dimensional lattice.** The sequence was proposed in 1202 by Leonardo da Pisa, alias Fibonacci, as a model for the growth of hare populations. In this model, L (for long) represents a couple of mature hares, and the letter S (for short) is a pair of babies. Each month a mature couple gives birth to a young pair, and therefore each L is replaced by LS (this notation indicates an ordering in which S appears to the right of L). During that period a young pair grows up, and therefore S is replaced by L. The rows in Fig. 2.31 show the time evolution of the population according to these rules; each row represents a generation. The row reached after an infinite number of generations is the **Fibonacci sequence.**

S

L

LS

LSL

LSLLS

LSLLSLSL

LSLLSLSLLSLLS

LSLLSLSLLSLLSLSLLSLSL

Fig. 2.31: Successive generations of hare populations. (The order within each row matters!)

Fibonacci series. Each row in Fig. 2.31 is constructed by merging together the preceding two rows, placing the previous row to the left of the row before. Hence, the number of letters (hares) in the nth row obeys

$$F_n = F_{n-1} + F_{n-2} \, . \tag{2.8}$$

Beginning with S, like in the figure, then $F_0 = 1$ and $F_1 = 1$; as a result $F_n = 1, 1, 2, 3, 5, 8, 13, 21, 34, \ldots$. This sequence of numbers is the **"Fibonacci series"**, or Fibonacci numbers. For very large n it can be shown that the ratio of numbers in two successive generations, F_n/F_{n-1}, approaches a constant, τ; this constant is the positive root of the quadratic equation $\tau = 1 + 1/\tau$. It is called the **golden ratio**, $\tau = (\sqrt{5}+1)/2 = 1 + 2\cos(2\pi/5) = 1.618\ldots$ (see problem 2.27). This irrational number appears in numerous contexts in physics and mathematics.

Fibonacci lattice. The letters in each row of Fig. 2.31 can be replaced by linear segments: the letter L is replaced by a segment of length τ, and the letter S by a shorter segment, of length 1. The row before the last in Fig. 2.31 is then replaced by the one-dimensional lattice shown in the lower part of Fig. 2.32, and the line before it by the lattice in the upper part. (The lengths are different in the two parts of Fig. 2.32; for reasons to be clarified below they are chosen such that the two lattices have the same length.)

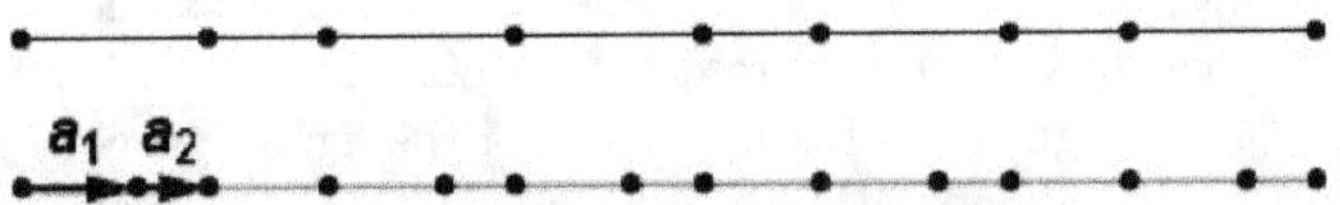

Fig. 2.32: Fibonacci lattice, built of linear segments according to the Fibonacci rules. The long segment is denoted $\mathbf{a}_1$, and the shorter one is $\mathbf{a}_2$.

By inspection, the sites in Fig. 2.32 do **not** form a Bravais lattice. It is impossible to reach them all by integer multiples of a single lattice vector. On the other hand, each point can be reached after an integer number of L steps and another integer number of S steps. The location of a point on the lattice (relative to the first one) can be written as $\mathbf{x}_n = m_1 \mathbf{a}_1 + m_2 \mathbf{a}_2$, where m_1 and m_2 are certain integers, and the vectors $\mathbf{a}_1$ and $\mathbf{a}_2$ (see Fig. 2.32) represent the steps, L and S, respectively. Although the "lattice" is one dimensional (the points are located on a line), it is described by **two different lattice vectors**, whose lengths' ratio, τ, is irrational. It is paramount to note that, contrary to the aforementioned lattices, the coefficients m_1 and m_2 **cannot** assume any integer value! These values are determined by the Fibonacci rule, which limits their choice. For instance, the combinations SS or LLL never appear in the rows in Fig. 2.31. The distance from the origin (chosen to be one of the sites) of the point $\mathbf{x}_n$ is $|x_n| = m_1\tau + m_2$. Were m_1 and m_2 allowed to be any integer, then $|x_n|$ would have approached arbitrarily close any numerical value. For example, $518\tau - 835 \approx 3.14161$ is equal to π up to the fourth digit (the reader is encouraged to approximate another arbitrary real number in a similar way). It follows that by allowing the coefficients to be arbitrary integers, one could have obtained lattice sites as close as one wishes to any point on the line, leading to a very dense arrangement, in contrast with the lattices in Fig. 2.32. There, the distance between nearest neighbors is never less than 1. Although the two lattices in Fig. 2.32 are not periodic, the locations of the sites are determined by the Fibonacci rule, and so are inter-related. This regularity causes the diffraction patterns of the Fibonacci lattice to contain only discrete points. This lattice is thus a one-dimensional quasicrystals.

Problem 2.27.
a. *Prove that the ratio $t_n = F_n/F_{n-1}$ approaches the golden ratio for large n.*
b. *Find the ratio between the number of longer segments to the number of shorter ones in an infinite Fibonacci lattice.*

Inflation symmetry. As mentioned, the Fibonacci lattice is not periodic. However, the symmetry with respect to discrete Translations is replaced by another symmetry: this lattice is self-similar under **inflation transformations**. Stated differently, when lengths are measured by rulers with different unit lengths, the lattice looks the same, independent of the units' lengths. This property can be grasped by mapping the lattice in the lower line of Fig. 2.32 back onto the previous generation. Each pair of segments represented by LS, whose total length is $\tau + 1$, is replaced by a single segment in the upper line of the figure, which is termed L'. Similarly, each L which is not followed by S, is copied to the upper line with no change in length, and is denoted S'. It is then seen that the upper row in Fig. 2.32 represents faithfully the previous generation (the third from the end line in Fig. 2.31) provided that the new long segments are L' and new short ones are S'. This mapping procedure yields that the length of L' is $1 + \tau$, and that of S' is τ. The ratio, $L'/S' = (\tau + 1)/\tau$ is τ, as dictated by the quadratic equation determining the golden ratio. Thus the lattice in the upper line of Fig. 2.32 is identical to the one in the lower row, provided that the unit of length is multiplied by τ! Such a symmetry under inflation characterized many fractal structures, see Chapter 7.

Fibonacci lattice as a projection of a square lattice. The description of the one-dimensional Fibonacci lattice by two lattice vectors is not accidental. This lattice can be built by a **projection** of a square lattice. Figure 2.33 portrays the projection of a square lattice on a straight line, which represents the physical space. One draws a straight line tilted by the angle α relative the $\hat{\mathbf{x}}-$ axis in a fictitious square lattice, with $\tan(\alpha) = 1/\tau$. All lattice points located within a finite-width band that encompasses the straight line and is parallel to it (shown in the figure by the two dashed lines) are projected on the straight line. One may choose the band width, w, such that $3a\sin(\alpha) < w < 2a\cos(\alpha)$. For instance, $w = a[\cos(\alpha) + \sin(\alpha)]$ (the final result is independent of this arbitrary choice). In this case the collection of lattice points projected on the straight line forms a staircase structure, as seen in the figure. The left inequality ensures that no successive three horizontal segments appear on the same stair, while the right inequality ensures that the height of each stair does not exceed the length of a single step. Denoting the coordinates of these lattice sites by $\mathbf{r}_n = m_1\mathbf{a}_1' + m_2\mathbf{a}_2'$, where $\mathbf{a}_1'$ and $\mathbf{a}_2'$ are the lattice vectors of the square lattice, the projected points on the straight line are located at $x_n = a[m_1\cos(\alpha) + m_2\sin(\alpha)]$. The distance between two successive points on the straight line assumes only one of two values, corresponding to the projections of the two lattice vectors, the vertical and the horizontal, of the square lattice. For example, the distance between the points x_1 and x_2 is L= $a\cos(\alpha)$, and the distance between the points x_2 and x_3 is S= $a\sin(\alpha)$. The ratio of these is τ. The conditions on w eliminate configurations of the types SS or LLL, and therefore it can be proven that the order in which the segments L and S appear on the thick line is identical to the order in the Fibonacci sequence. In particular, the arrangement of the segments in Fig. 2.33 is identical to the one shown in the last row of Fig. 2.31, beginning at the fourth letter there (the starting point is arbitrary and is related to the choice of the width w of the band).

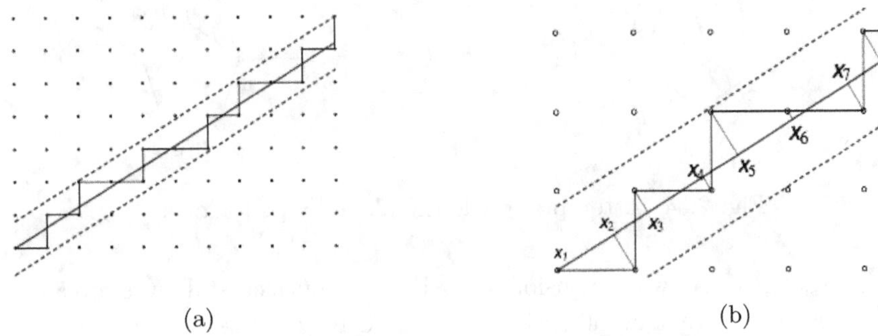

(a)   (b)

Fig. 2.33: Projecting a square lattice on a one-dimensional quasicrystal. (a) The lattice points in-between the two dashed lines are projected on the thick line. (b) A magnification showing the points x_n on the thick line.

Problem 2.28.
Prove that the distances of the points on the thick line in Fig. 2.33 from the origin are given by

$$x_n = S[n + ||n/\tau||/\tau] \,, \tag{2.9}$$

where $||y||$ is the integer closest to y. This expression enables the extension of a Fibonacci lattice to negative n's. How does one prove that this is indeed a Fibonacci lattice?

Problem 2.29.
a. *Suppose that the thick line in Fig. 2.33 passes through one of the square-lattice sites. How many more lattice sites are then on this line?*
b. *Suppose that $\tan(\alpha)$ in Fig. 2.33 is a rational number, i.e., $\tan(\alpha) = p/q$ where p and q are integers with no common factor. Describe the lattice that will appear on the straight line.*
c. *It is possible to approximate the irrational number $\tau \approx 1.61803$ by a sequence of rational numbers, 1.6, 1.61, 1.618. Which lattice is obtained by projecting from two dimensions onto a straight line for each of these approximations?*

The one-dimensional quasicrystals is described by **two** lattice vectors, whose lengths' ratio is irrational. This is a **general property of quasicrystals: they can be described by several lattice vectors whose number is larger than the spatial dimension; the corresponding lattices can be obtained from a projection of a periodic lattice of a higher dimension.**

(a)   (b)

Fig. 2.34: Attempts to tile the plane by pentagons.

Penrose tiles. A two-dimensional version of a quasicrystals was already analyzed by Kepler in the seventeenth century, and by Penrose in the last century. Penrose explored the question of **covering the plane by tiles, i.e., tiling the plane**. As seen in Fig. 2.27, Bravais lattices can completely tesselate the plane, by using identical oblique, square, triangular, rectangular, or hexagonal tiles. In contrast, this cannot be accomplished by identical pentagons: attempting to encircle a point by identical pentagons, with the corner angle $3\pi/5 = 108°$ (as the pentagon can be divided into three triangles, the sum of the five corner angles is 3π) leaves a space of $36°$, see Fig. 2.34(a). However, Fig. 2.35(b) illustrates that the plane can be covered by pentagons, provided that rhombus-shaped tiles of opening angle $36°$ are added. That is, the **plane can be tiled by two types of tiles.** Penrose demonstrated that the plane can be covered by two types of rhombi with identical edges but different opening angles, $36°$ and $72°$, displayed in Fig. 2.35(a). As shown there, at certain points there appears symmetry with respect to rotations of order 5 (e.g., where 5 "big" rhombi meet at a point). Assigning a unit length to the edge of the rhombi, the length of the long diagonal of the wide rhombus is τ, and that of the short diagonal of the narrow rhombus is $1/\tau$ (check!). Interestingly enough, there are cuts of the structure of Penrose, along special directions, whose distances create the Fibonacci lattice, see Fig. 2.35(b). It can be also proven that the ratio of the numbers of wide and narrow rhombi equals $1/\tau$, as in the one-dimensional Fibonacci lattice. Though the tiles of Penrose cover the plane, and albeit the fact that rotating by $72°$ the structure in Fig. 2.35(a) brings it to a qualitatively similar one, i.e, **it is statistically symmetric**, the structure is not periodic. It does not come precisely back to itself after the rotation, and it is not described by a single unit cell. However, the **two-dimensional structure of Penrose can be obtained**

by a two-dimensional projection of a five-dimensional cubic lattice. The quasicrystalline materials discovered so far can all be described by two or three different unit cells, which fill completely the space. As shown by Steinhardt and Levin in 1984, an extension of the Penrose structure to three dimensions, with two different unit cells (containing the ratio τ as well) reproduces the diffraction pattern found by Shechtman. This requires **six basic translation vectors** (in place of the three of a Bravais lattice), indicating a projection from a six-dimensional periodic lattice.

(a)

(b)

Fig. 2.35: (a) Penrose tiling. (b) Parallel lines placed at distances that form the Fibonacci sequence. Taken from `http://www.physics.princeton.edu/~steinh/` `QuasiIntro.ppt`, courtesy of P. Steinhardt.

Problem 2.30.
a. *Propose five vectors in the plane, such that each point in the lattice of Penrose can be described as a linear combination of them, with integer coefficients.*
b. *Show that these five vectors can be obtained from a projection of the lattice vectors of a five-dimensional cubic lattice on a plane normal to one of the (five) diagonals.*

On the symmetries of quasicrystals and Platonic bodies. Appendix A lists the five Platonic bodies, symmetric polyhedra with equilateral faces. The tetrahedron, cube, and octahedron have already been mentioned in the context of periodic lattices. The bigger Platonic bodies, **the icosahedron and the dodecahedron**, are not found in periodic lattices, but they do appear in quasicrystals. The original findings of Shechtman were identified as representing icosahedral quasicrystals: the diffraction patterns display an icosahedral symmetry point, and they can be described in terms of six lattice vectors pointing at the vertices of an icosahedron. There exist also descriptions of the Al-Mn alloys as packings of icosahedra, with Al and Mn atoms occupying their corners.

2.9 Epitaxial growth, thin films, superlattices and multi-layered structures

Crystal growth. Certain materials are found in Nature in their crystalline forms; crystals are also created by slowly cooling liquids, or upon crystalization of salts from a solution. Frequently, however, it is required to grow crystals under controllable conditions in the laboratory. As mentioned in Secs. 2.2 and 2.6, crystals can be grown by adsorption. Another ubiquitous technique is the **epitaxial growth from molecular beams**, abbreviated as **MBE (molecular beam epitaxy)**. The name refers to a regulated growth procedure of atomic (or molecular) slabs, where the layers adjust themselves to the periodicity of the substrate on which they are being deposited. Another method is the deposition off the vapor of a solid material (vaporized upon heating by a laser beam in a vacuum chamber); this procedure is called **PDE (pulsed laser deposition)**.

Epitaxial growth on a crystalline substrate. As different materials possess different lattice symmetries and disparate lattice constants, there are numerous options for choosing the adequate substrate to grow upon a certain material. Cutting a crystal along a specific direction creates planar surfaces with symmetry properties and lattice constants that are determined by the cut direction. For instance, different cuts of the **FCC** lattice (Figs. 2.17 and 2.24) yield planes with sites that form a triangular or a square lattice, with lattice constants dictated by the direction of the cut (check!). Thus, one has to pre-design not only the substrate material, but also the direction of the cut on which the crystal is to be grown. Modern technologies enable the growth of **thin films**, comprising a single layer of the adsorbed atoms, or a small number of such layers. These technologies allow for the investigation of two-dimensional slabs (recall that these are lying on a substrate and hence are not exactly two dimensional, in contrast with graphene, Fig. 2.10). It is also possible to grow one-dimensional systems, e.g., by an adsorption on a substrate crossed by a dent, followed by a removal of all deposited material except that sunk in the dent.

One-dimensional epitaxial growth. When the growing crystal comprises planes with symmetry identical to that of the substrate, and when the lattice constants match, the growth is accomplished smoothly. However, such a perfect harmony is rarely encountered. Consider, e.g., the process described in Sec. 2.6, in the simpler version of a one-dimensional substrate. It begins with a straight line of circles, on top of which additional circles, of a different radius, are to be adsorbed, as shown in Fig. 2.36. In the right and mid panels of Fig. 2.36 the radius of the new circles is $\sqrt{2}$ times larger than the radius of the substrate circles. When the attraction between the new circles and those in the substrate is larger than the attraction in-between themselves, they "prefer" to land in the "valleys" of the substrate, as in the mid panel of Fig. 2.36. The result is a periodic lattice with a lattice constant twice that of the substrate. This new lattice constant is also that of the joint lattice of the two layers together. This is a **commensurate** arrangement.

A similar periodic lattice is obtained also when the radii ratio is a rational number. For instance, in the left panel of Fig. 2.36 the radii ratio is 3/2. The adsorbed circles can still touch one another and yet sink into one of three "valleys". The constant of this periodic lattice is thrice that of the substrate.

On the other hand, when the attraction among the new circles is stronger than the one between them and the circles on the substrate they "prefer" to touch each other, even at the price of distancing themselves from the substrate. The result is presented in the right panel of Fig. 2.36. The requirement that the circles touch each other causes all new circles (except the first one) not to be **exactly** inside a dent; i.e., the length of the new unit cell of the joint lattice is infinite: the distance of any point from another cannot be covered by an integer number of finite-length lattice vectors. This outcome is expected whenever the radii ratio is irrational and the adsorbed circles are strongly coupled; the resulting structure is **incommensurate**. Recall that this feature appears also in quasicrystals: there again the ratio of length scales is irrational, the lattice periodicity is lost, and the crystal is described by an infinite unit cell. Indeed, the atoms' distribution on any symmetry axis of the quasicrystals is determined by a finite number of lattice vectors (two or more), with lattice constants whose ratio is irrational. A similar description pertains to the incommensurate arrangement: each layer is characterized by its own lattice vector and hence each lattice point can be reached by a linear combination of the two lattice vectors, with arbitrary integer coefficients.

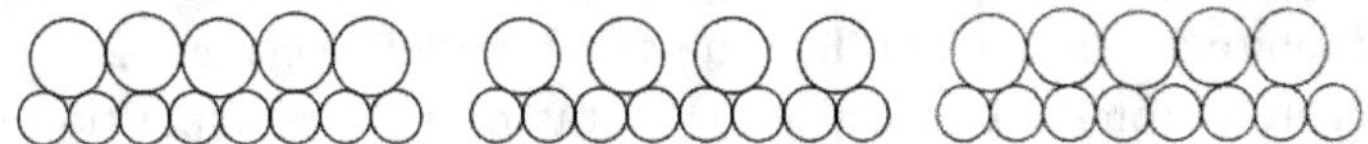

Fig. 2.36: Adsorption on a one-dimensional substrate. Right panel: Incommensurate adsorption. Mid and left panels: commensurate adsorption, with a double and a triple step of that of the substrate lattice.

Problem 2.31.
Determine the length of the unit cell of the joint two-layer lattice, when the ratio between the radius of the circles in the upper layer and the radius of those in the substrate is $r_2/r_1 = p/q$, where p and q are integers with no common factor. What is the base?

Lock-in of commensurate arrangements. Figure 2.37(a) displays an incommensurate arrangement of circles of radius $\sqrt{3}$ on a one-dimensional substrate made of circles of radius 1. Like in the right panel of Fig. 2.36, the unit cell of the joint structure is infinite. Nonetheless, note that the fifth circle from the left end in the second layer is very close to the eighth valley in the substrate. It then may happen that the system gains energy if that circle moves slightly to the right, into the valley, even though it leaves a narrow space relative to its neighbor. This configuration is shown in Fig. 2.37(b). Once this happens, the surrounding of the fifth circle in the second layer coincides with that of the first circle there, and the structure is commensurate, i.e., it is a Bravais lattice with a unit cell of length 7 (in units of the diameter of the circles in the substrate). Such a transformation, from an incommensurate pattern to a commensurate one, is called **lock-in**, and generally occurs when there is a good rational approximation to the irrational ratio of the circles' radii. In the example discussed here, the distance between the centers of the first and fifth circles, $4\sqrt{3} \approx 6.9282$, is replaced from by 7. This replacement amounts to the approximation of $\sqrt{3}$ by the rational number $7/4 = 1.75$. A similar lock-in transformation takes place when $\sqrt{3}$ is approximated by any other rational number, e.g., $173/100$.

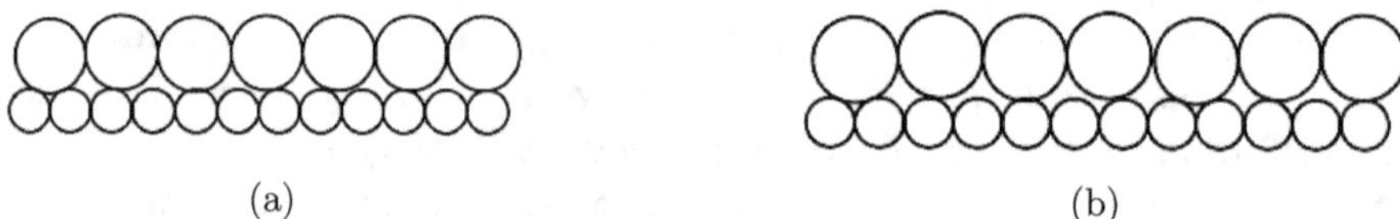

(a)   (b)

Fig. 2.37: (a) Incommensurate adsorption. (b) Commensurate "lock-in" of the adsorption in (a).

Two-dimensional epitaxial growth. In two dimensions there are many more options for commensurate and incommensurate arrangements. An example is the growth of a noble gas, like krypton, on a graphite substrate. Recall that the planes building graphite are hexagonal, like the graphene lattice (Figs. 2.8 and 2.9). The way the krypton atoms are organized is dictated by the competition between two interactions. Each krypton atom is attracted to the minimum of the potential energy, located at the center of each hexagon. Even when this interaction is the only one operating, the krypton atoms still face a problem, as their radius is larger than the distance between the centers of neighboring hexagons. Therefore, these atoms cannot occupy **all** hexagons. In this scenario, the adsorbed atoms create a new triangular lattice, that covers only one third of the hexagons (marked in red and by the letter α in Fig. 2.38). As seen in the upper part of the figure, there are three different configurations to cover the substrate, giving rise to regions with different arrangements of the krypton layer (over hexagons of different colors); this is similar to the two possibilities to arrange the second layer of spheres in the close-pack configuration shown in Fig. 2.23.

So far the interaction between pairs of krypton atoms, due to which they would have preferred to arrange themselves in a close-pack configuration, has been ignored. With this interaction alone, the lattice constant of that configuration is determined by the minimum of the potential energy. Once the attraction in-between the krypton atoms is considerably stronger than that between them and the substrate, they tend to arrange themselves in the pattern displayed in the lower part of Fig. 2.38. There, the krypton structure is indeed incommensurate with the substrate: each layer has

Fig. 2.38: Adsorption of krypton atoms on a graphite substrate. The three colors represent three optional locations of the adsorbed atoms. Top: commensurate adsorption; bottom: incommensurate adsorption. [R. J. Birgeneau and P. M. Horn, Science **232**, 329 (1986).]

its own lattice constant, and the ratio between those constants is irrational (the krypton atoms are not exactly spheres, and the radii in the figure are adjusted to the various possible structures). Intermediate configurations are also possible: when the two interactions are comparable in magnitude, it may happen that after several cycles of the upper layer, one of the krypton atoms reaches close to the center of a certain hexagon; it then can gain energy by sliding into it, [as in Fig. 2.37(b)] on the expense of the energy it loses by increasing the distances to its neighbors, or on the expense of creating a small space in-between two dense regions (i.e., it creates a **"domain wall"**). Such a lock-in transformation of the new layer produces a joint lattice of the two layers, with a larger unit cell. The lattice constant of this new lattice is the distance between these spaces.

Problem 2.32.
Determine the joint unit cell of the two layers in the commensurate lattice shown in the upper part of Fig. 2.38. What is the base?

Superlattices. It is possible to grow "sandwiches" comprising several layers in a controllable manner. Figure 2.39 exhibits several layers of lanthanum manganate and strontium manganate, in a periodic array; each period contains a single layer of one of these perovskites; along the normal axis the structure is built of alternating layers of the two materials. Their lattice constants are close to one another, and therefore it is relatively easy to build a periodic array. It is an example of a **"superlattice". Superlattices are structures in which different crystalline**

layers of finite thicknesses repeat themselves periodically. The material in the figure is $La_{1-x}Sr_xMnO_3$, with $x = 1/2$; it is an anti ferromagnetic insulator (see Sec. 2.10). At lower strontium concentrations, $x < 1/2$, the material is a metallic ferromagnet. Due to the disparate properties (i.e., elastic, electric, or magnetic properties) of the parent compounds, it is feasible to engineer novel materials with pre-designed characteristics, for instance, high rigidity, or an ability to store magnetic information.

Fig. 2.39: A superlattice. Layers of lanthanum manganate and strontium manganate; the oxygens are located on the corners of octahedra, each containing a manganese ion [like in Fig. 2.21(a)]. The purple and green spheres represent La and Sr ions, respectively [T. S. Santos, S. J. May, J. L. Robertson, and A. Bhattacharya, *Tuning between the metallic antiferromagnetic and ferromagnetic phases of $La_{1-x}Sr_xMnO_3$ near $x = 0.5$ by digital synthesis*, Phys. Rev. B **80**, 155114 (2009)].

Problem 2.33.
Determine the unit cell of the structure in Fig. 2.39. How is the answer modified for the periodic structure with $x = 1/3$?

Planar surfaces between two three-dimensional lattices. At times, the contact layer in-between the substrate and the growing crystal above it has unique properties by itself. A conspicuous recent example is the contact surface between the two perovskites [Fig. 2.21(a)], strontium titanate and lanthanum aluminate, as shown in Fig. 2.40. The lattice constants of these two materials are very close, and consequently they are smoothly coupled. Since each layer of aluminium and oxygen is negatively charged (as deduced from a summation over the various charges), a surplus of electrons is accumulated on the contact layer, giving rise to unique electric and magnetic properties (even superconductivity!), which can be modified by varying the number of layers of $LaAlO_3$. This surface layer is currently under active research in many laboratories.

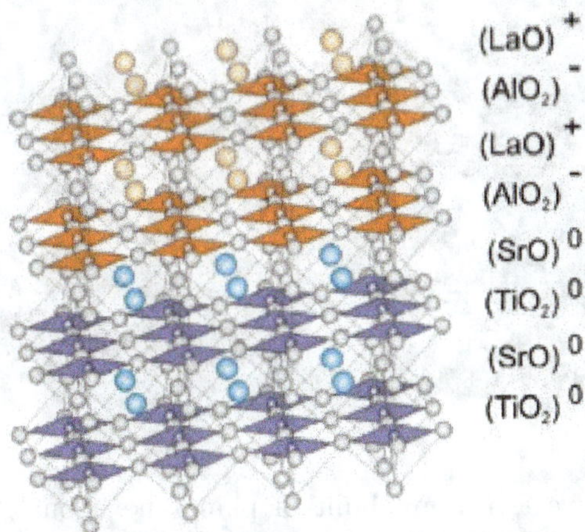

Fig. 2.40: Growth of LaAlO$_3$ on a SrTiO$_3$ substrate [M. Huijben, A. Brinkman, G. Koster, G. Rijnders, H. Hilgenkamp, and D. H. A. Blank, *Structure–Property Relation of SrTiO$_3$/LaAlO$_3$ Interfaces*, Adv. Materials **21**, 1665 (2009)].

Heterostructures. Another type of pre-designed materials are heterostructures. As mentioned, graphite is built of successive layers of graphene. Other planar materials of identical structure are boron nitride and silicene, as well as MoS$_2$ (in which the sulphur ions are located above and below the lattice points of one of the sublattices in the hexagonal lattice), WSe$_2$, and more. These are all characterized by strong bonds within each plane, while the coupling between the planes is significantly weaker (see Chapter 4). Figure 2.41 is taken from a paper that surveys the various possibilities for growing multi-layer structures, which consist of successive layers of two-dimensional lattices. The growth process is compared to a game with lego blocks–each layer is represented by lego blocks of a certain color. Of particular potential importance are structures comprising hexagonal lattices like graphene.

2.10 Magnetism

Magnetic phase transitions. As mentioned, crystalline materials can undergo **phase transitions**, such that below a certain transition temperature they acquire new physical properties. Those are often related to a symmetry change of the crystal (e.g., the ferroelectric phase mentioned at the end of Sec. 2.7). Here are discussed phase transitions in the context of **magnetism**.

Atoms and ions often possess **magnetic moments** stemming from the **angular momentum** associated with the electrons' motion, and/or that of the particles

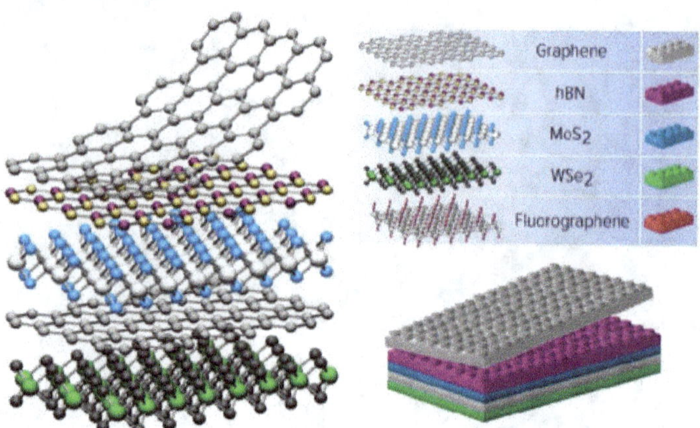

Fig. 2.41: A multi-layer structure built of planar hexagonal lattices of different materials [A. K. Geim and I. V. Griegorieva, *Van der Waals heterostructures*, Nature **499**, 419 (2013)].

forming the nuclei, and from the **spin** of all these particles. The electronic magnetic moment is measured in units of the **Bohr magneton**, $\mu_{\rm B} = e\hbar/(2m_0c) = 9.247 \times 10^{-21}$ erg/Gauss, where e and m_0 are the charge and the mass of the electron, respectively, $\hbar$ is the Planck constant, and c is the speed of light. In an elementary picture, non-metallic magnetic materials are described by assigning to each lattice site i a magnetic moment represented by the vector $\boldsymbol{\mu}_i$. (This picture does not pertain to metals, in which the conduction electrons possess magnetic moments as well, and these magnetic moments are not localized.) At temperatures low enough as compared to the transition temperature into the magnetic phase, the magnitude of these moments is generally of the order of a Bohr magneton for each magnetic atom (or ion) in the crystal. Several forces operate among the moments. The well-known one is the electromagnetic force between dipole moments, that decays slowly with the distance between them. In addition, there are also **exchange forces**, originating from the quantum-mechanical treatment (see Chapter 4). The interplay among these forces may result in a geometrical arrangement of the magnetic moments, when the temperature drops below a certain transition temperature.

Magnetic patterns. In the simplest case, the interactions among the magnetic moments cause them to coordinate themselves in parallel with each other, in a **ferromagnetic** arrangement (line B in Fig. 2.42). Consequently, the material acquires a **macroscopic magnetic dipole moment**. However, at times the interactions between neighboring moments cause them to be antiparallel, and the result is the **antiferromagnetic** order shown in row C of Fig. 2.42. Interactions among farther away neighbors give rise to more complicated patterns, e.g., those portrayed in the other rows of the figure. Such interactions can also lead to **incommensurate arrangements** of the magnetic moments (relative to the structural lattice). For instance, the moments may rotate periodically along a certain direction in the lattice, with a periodicity that is incommensurate with the lattice constants.

Spin glasses. The random directions of the magnetic moments in row A of Fig. 2.42 represent two possible configurations. At higher temperatures the system passes through states in which the moments are oriented at all possible directions, similar to atoms in a gas which may pass through all points in space; then row A reflects an instantaneous pattern of these moments. The temporal average of each moment vanishes, and there is no global magnetic order. The only order in the system is that of the lattice of atoms (or ions) building the material, with a unit cell that contains a single atom (or ion). Upon decreasing the temperature abruptly to a very low one, it may happen that the moments "freeze" in the random arrangement they had at the higher temperature, and then row A describes a steady state of the system, which does not correspond to a minimum of the free energy, but is **metastable**. This configuration is called **"spin glass"**, and is analogous to the "mundane" glasses, in which the atoms freeze in non-periodic arrangements (see Chapter 1). At times, a spin glass is created when the magnetic interactions compete with each other: a fraction of them prefers parallel arrangements of the moments, while other attempt to make them antiparallel. The competition gives rise to a **frustration** of the moments, and results in moments oriented at random (see below). This unique situation is an example of a **random system**, a field which is still under active research.

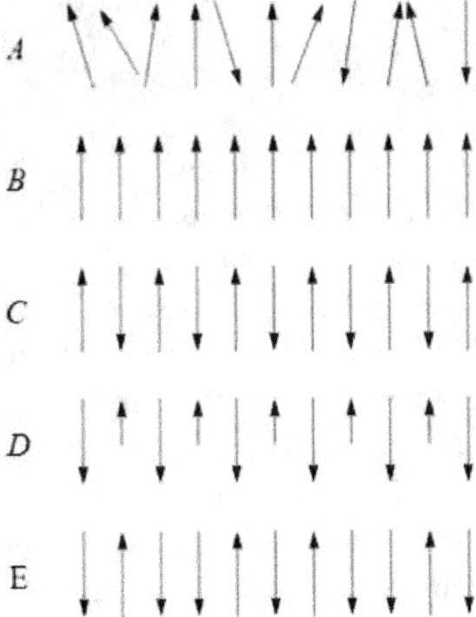

Fig. 2.42: Various magnetic patterns. From top to bottom: spin glass, ferromagnet, simple antiferromagnet, ferrimagnet, and a structure combing a complex antiferromagnetic order with a ferromagnetic one.

The magnetic unit cell. The ferromagnetic order in line B of Fig. 2.42 has the same periodicity as the underlying lattice, since the moments are identical. Hence the unit cell contains a single atom, as in the original lattice. Though the unit cell is not changed, the symmetry group does, because the ferromagnetic lattice has no rotational symmetry, as opposed to the underlying structural lattice. The symmetry with respect to reflections through mirrors perpendicular to the lattice still persists. On the other hand, the periodicity as well as the symmetry groups of the structures in rows C, D, and E, differ from those of the original lattice, and the unit cell contains more atoms (or ions). The structure of the simple antiferromagnet (line C), and also that in line D (which shows a **ferrimagnet**: a magnetic pattern with a ferromagnetic character – the total magnetic moment does not vanish – combined with an antiferromagnetic feature) is similar to the one that appears in the second panel of Fig. 2.1, or to the structure of table salt displayed in Fig. 2.7 (or in the right panel of 2.17): the unit cell is doubled as compared to the unit cell of the simple lattice. The symmetry of the simple antiferromagnet C is higher than the one of the structure in D.

Problem 2.34.
Determine the magnetic unit cells of each of the structures C, D, and E, in Fig. 2.42. List the symmetry operations corresponding to each of them.

Problem 2.35.
The magnetic moments of the atoms in a material whose structural lattice is a simple cubic one, of a lattice constant a, are all directed along the same axis, and are given by the expression $\mu(\mathbf{R}) = \mu_0 \cos(\mathbf{Q} \cdot \mathbf{R})$. The vector $\mathbf{Q}$ is $\mathbf{Q} = (1,1,1)(\pi/a)$, and $\mathbf{R}$ is a vector to a lattice site. Determine the structure of the magnetic lattice and find the primitive unit cell. What would have been the unit cell were the vector $\mathbf{Q}$ given by $(1,1,1)(\tau/a)$, where τ is the golden ratio?

Frustration and three-partite patterns. An important comment is called for: the magnetic cell of the antiferromagnet shown in line C of Fig. 2.42 contains two moments, and the magnetic lattice can be viewed as a combination of two intercalated simple sublattices. The same is true for the example described in problem 2.35: the lattice there resembles the one of table salt, with the two ions corresponding to the two moments (say, Cl for the positive moment, and Na for the negative one). The magnetic lattice again consists of two sublattices, each containing identical moments. Such a lattice is called **bi-partite.** Though this configuration pertains to many other lattices, it is not possible for the triangular lattice; as seen in Fig. 2.43 (and also in Fig. 2.12), the triangular lattice is built of three identical sublattices (hence it is **tripartite**); each site is surrounded by three neighbors that belong to one of the other two sublattices. When a magnetic moment that can point only along one of two allowed directions (towards "up" or "down", as in Fig. 2.42) is assigned to each site, and when the magnetic interaction "prefers" nearest-neighbor moments to be antiparallel, the result is a frustrated situation. The moments simply cannot be arranged such that all magnetic bonds are "satisfied" (i.e., connect antiparallel moments). There are quite a number of materials in Nature that have triangular or hexagonal symmetry and they all "suffer" from this phenomenon. However, had the moments been allowed to point at any direction in the plane, then there is an optimal arrangement: the magnetic moments in each of the three sublattices point along directions that form an angle of 120° with each other (see the arrows in Fig. 2.43). This arrangement still leaves a certain amount of frustration, as the moments are not exactly antiparallel. This point is revisited in Chapter 4.

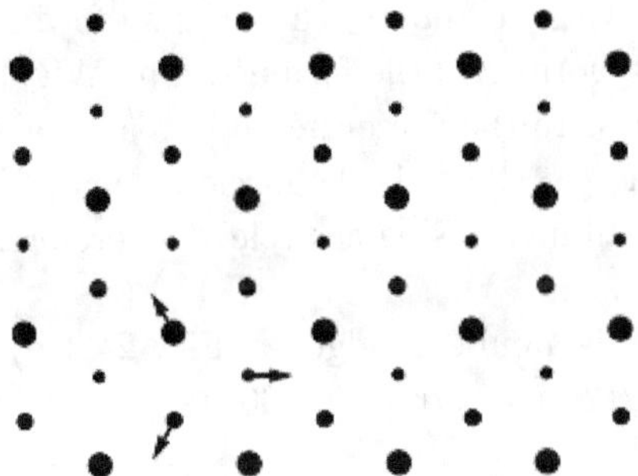

Fig. 2.43: Dividing the triangular lattice into three sublattices; the arrows represent the directions of the magnetic moments in each sublattice in the energetically-optimal state for interactions that prefer an antiparallel arrangement for each pair of nearest neighbors.

Problem 2.36.
Determine the primitive unit cell of the lattice in Fig. 2.43 and find its base.

2.11 Answers for the problems in the text

Answer 2.1.
a. The packing ratio is the total area of all circles divided by the area of the plane. Since the lattice is periodic, this ratio is also the area of a single circle divided by the area of the unit cell, for example, the hexagonal Wigner-Seitz cell depicted in Fig. 2.5. Each of the hexagons there consists of six equilateral triangles. The height of each triangle is the radius of the circle it encloses, r, and therefore the area of the triangle is $r^2/\sqrt{3}$. The hexagon contains a single circle, and hence the packing ratio is $\pi r^2/(6r^2/\sqrt{3}) = \pi/(2\sqrt{3}) = 0.9069$. Another way to obtain the same result: consider the equilateral triangle created by the two lattice vectors displayed in Fig.

2.5, and the line joining their tips. The edge of the triangle is $a = 2r$, and hence its area is $a^2\sqrt{3}/4 = r^2\sqrt{3}$. The triangle contains three sectors of a circle, spanned by $60°$. The area of each such sector is $\pi r^2/6$, and consequently the packing ratio is $3(\pi r^2/6)/(r^2\sqrt{3}) = \pi/(2\sqrt{3})$.

b. The corresponding Wigner-Seitz unit cell is a prism whose base is the hexagon mentioned in part (a), and whose height is $2r$. Hence the volume of the parallelogram containing a single sphere is $(2r)(6r^2/\sqrt{3}) = 12r^3/\sqrt{3}$. The volume packing ratio is therefore $(4\pi r^3/3)/(12r^3/\sqrt{3}) = \pi/(3\sqrt{3}) = 0.6046$.

Answer 2.2.

a. The planar Wigner-Seitz cell can be obtained as follows. One connects an arbitrary lattice site to all its neighbors; the lines so formed are bisected at their mid points by other lines, normal to the formers. The Wigner-Seitz cell is confined by the cutting normal lines. In the present case it is a square; whose edge length is the diameter of each circle, $a = 2r$ (r is the circle radius), Fig. 2.44(a).

b. Each such square contains a single circle. Therefore, the packing ratio is $\pi r^2/(2r)^2 = \pi/4 \approx 0.7854$.

c. The Wigner-Seitz cell is a cube of edge $2r$, Fig. 2.44(b), and consequently the packing ratio is $(4\pi r^3/3)/(2r)^3 = \pi/6 \approx 0.5236$.

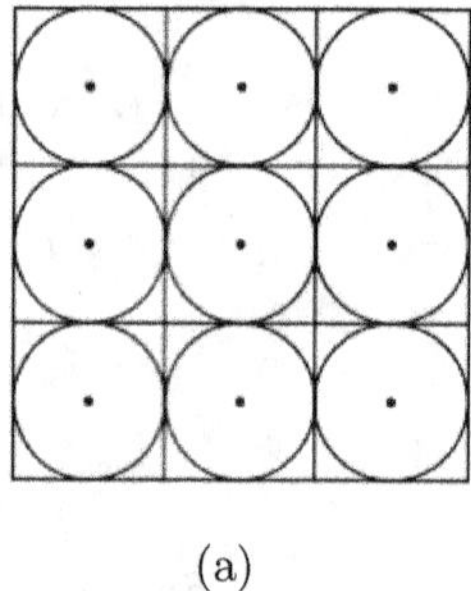

(a)

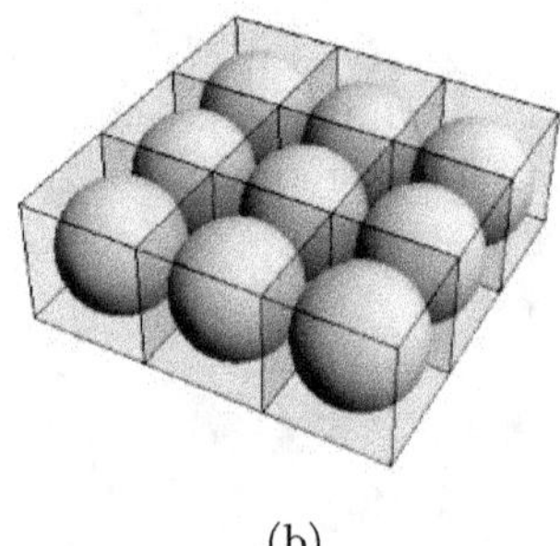

(b)

Fig. 2.44

Answer 2.3.

a. The length of the diagonal of the dashed unit cell is $2(r_< + r_>)$, and hence the length of the edge of the cell, which equals the lattice constant a, is $a = 2(r_> + r_<)/\sqrt{2} = \sqrt{2}(r_> + r_<)$. The structure portrayed in Fig. 2.7(a) exists as long as this edge is longer than the diameter of the large circle, $a = \sqrt{2}(r_> + r_<) > 2r_>$, i.e., as long as $x = r_</r_> > \sqrt{2} - 1$. In this range the packing ratio is $\rho = \pi(r_>^2 + r_<^2)/[\sqrt{2}(r_> + r_<)]^2 = \pi(x^2 + 1)/[2(x + 1)^2]$.

b. The large circles are tangential to each other when $a = \sqrt{2}(r_> + r_<) = 2r_>$, i.e., when $x = \sqrt{2} - 1$.

c. For $x < \sqrt{2} - 1$, the smaller circles are located in the spaces in-between the larger ones, and are not tangential to them. Hence, the lattice constant is still $a = 2r_>$, and the packing ratio is $\rho = \pi(r_>^2 + r_<^2)/[2r_>]^2 = \pi(x^2 + 1)/4$. Figure 2.45(a) displays the dependence of the packing ratio on x. In the $x = 0$ limit (and also in the $x = 1$ one) the system becomes a simple square lattice and therefore the result is $\rho = \pi/4 \approx 0.7854$, the same as the answer to problem 2.2. The maximum, $\rho \approx 0.92$, is attained at the mid point $x = \sqrt{2} - 1$.

d. The lattice constant is a, as calculated above.

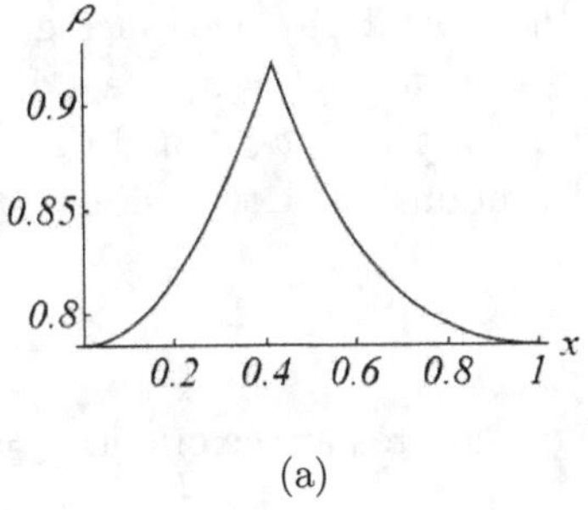

(a)

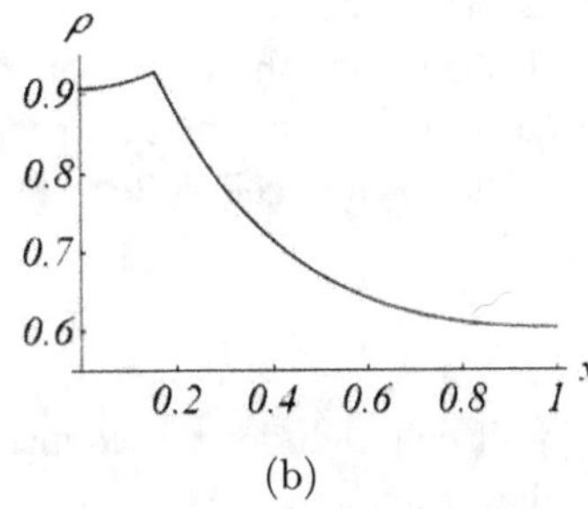

(b)

Fig. 2.45: (a) The packing ratio for answer 2.3. (b) The packing ratio for answer 2.4.

Answer 2.4.

Denote the lattice constant (the distance between the centers of two large circles) by a. The distance between the centers of two nearest-neighbors, a small circle and a large one, is 2/3 of the height of the triangle built of the centers of three large neighboring circles, and is therefore $a/\sqrt{3}$. The area of the rhombic unit cell is $a^2\sqrt{3}/2$, and it contains a large circle and a small one. The packing ratio is the sum of the areas of the two circles, divided by the area of the rhombus, $\rho = \pi(r_>^2 + r_<^2)/[a^2\sqrt{3}/2]$. The large circles do not touch each other when $2r_> < a$; it is assumed that in this case the large circle is tangential to the smaller one, and therefore the distance in-between them is such that $r_> + r_< = a/\sqrt{3}$. Then, $\rho = 2\pi(x^2 + 1)/[3\sqrt{3}(x + 1)^2]$. The condition $2r_> < a$ implies that $x > 2\sqrt{3}/3 - 1 \approx 0.1547$. On the other hand, when $x < 2\sqrt{3}/3 - 1$, then the large circles touch each other, and consequently $a = 2r_>$. Then $\rho = \pi(x^2 + 1)/[2\sqrt{3}]$. The result resembles qualitatively the one obtained in problem 2.3: the packing ratio is maximal, with the value 0.9286 for $x = 2\sqrt{3}/3 - 1$, and it attains the value of the ordinary triangular lattice, $\rho = \pi/[2\sqrt{3}] \approx 0.9069$ when $x = 0$. For graphene $x = 1$ and then $\rho = \pi/[3\sqrt{3}] \approx 0.6046$. This number is exactly two thirds of the packing ratio of the ordinary triangular lattice, as can indeed be expected: the hexagonal lattice is obtained from the triangular one when one third of its sites are obliterated. The packing ratio as a function of x is depicted in Fig. 2.45(b).

Answer 2.5.

a. The length of the rhombus edge in Fig. 2.4 is the lattice constant, a, and its height is $a\sqrt{3}/2$. Hence the area of the rhombus is $a^2\sqrt{3}/2$. The hexagonal unit cell in Fig. 2.5 consists of six equilateral triangles; the height of each is $a/2$ and hence the edge length is $x = a/\sqrt{3}$. The area of the hexagon is thus $6ax/4 = a^2\sqrt{3}/2$. Indeed, the two answers are identical.

b. In the case of the square lattice, the two unit cells are squares of edge a, and therefore their areas are both a^2.

c. Denote the lattice vectors of the simple square lattice (e.g., in Fig. 2.6 or in Fig. 2.14) by $\mathbf{a}_1$ and $\mathbf{a}_2$. The lattice vectors of the crystal displayed in Fig. 2.7 (the dashed lines there) are then given by $\mathbf{a}'_1 = \mathbf{a}_1 + \mathbf{a}_2$ and $\mathbf{a}''_2 = \mathbf{a}_1 - \mathbf{a}_2$. If the original lattice constant is a then the "new" lattice constant is $a\sqrt{2}$, and the area of the "new" unit cell is $2a^2$, as it should, since it contains two lattice sites or two "atoms".

Answer 2.6.

This unit cell includes two identical lattice points, one located at the origin, $\mathbf{r}_1 = 0$, and the other at $\mathbf{r}_2 = (\mathbf{a}_1 + \mathbf{a}_2)/2$. The area of the unit cell is $2a^2$, where a is the lattice constant of the original lattice. This indeed is the specific area that represents two lattice sites.

Answer 2.7.

Denote the lattice constant of the triangular lattice by a. In cartesian coordinates, the lattice vectors are $\mathbf{a}_1 = (a, 0)$ and $\mathbf{a}_2 = (a, a\sqrt{3})/2$. The distance between nearest neighbors is $R = a/\sqrt{3}$ and thus $a = R\sqrt{3}$. In the cell marked by the solid lines in Fig. 2.8, one base point is at the origin, $\mathbf{r}_1 = 0$, and the other is at $\mathbf{r}_2 = (\mathbf{a}_1 + \mathbf{a}_2)/3$. For the unit cell shown in the mid panel of Fig. 2.9, one point is located at the origin, but the other is at $\mathbf{r}_2 = 2(\mathbf{a}_1 + \mathbf{a}_2)/3$.

Answer 2.8.

a. The volume is a^3. The volume packing-ratio is $\pi/6 \approx 0.524$ (as in problem 2.2). The coordination number is 6: two nearest-neighbors along each of the three axes.

b. The Wigner-Seitz cell is a cube, similar to the previous unit cell but shifted such that the center of the cube is at the center of the sphere contained in it and tangential to all its faces (as in problem 2.2).

c. Use the cartesian lattice vectors shown in Fig. 2.15(b); then another set of lattice vectors is $\mathbf{a}_1$, $\mathbf{a}_2$, and $(\mathbf{a}_1 + \mathbf{a}_3)$. These build a parallelogram whose volume is again a^3, and it also represents a single lattice site. Another option, of the same volume, is the set of vectors $\mathbf{a}'_1 = \mathbf{a}_1$, $\mathbf{a}'_2 = (\mathbf{a}_1 + \mathbf{a}_2)$, $\mathbf{a}'_3 = (\mathbf{a}_1 + \mathbf{a}_2 + \mathbf{a}_3)$. In this example an arbitrary lattice site is given by $\mathbf{R} = n'_1\mathbf{a}_1 + n'_2(\mathbf{a}_1 + \mathbf{a}_2) + n'_3(\mathbf{a}_1 + \mathbf{a}_2 + \mathbf{a}_3)$, n'_1, n'_2, and n'_3 are arbitrary integers. This ensemble of points is identical to the one obtained from Eq. (2.1), with different integral coefficients: $n_1 = n_1' + n_2' + n_3'$, $n_2 = n_2' + n_3'$, $n_3 = n_3'$.

Answer 2.9.
The cube in the CsCl lattice contains eight 1/8's of the big spheres, at the corners, and a smaller sphere at its center. Hence, the total volume of the spheres within the cube is $4\pi(r_>^3 + r_<^3)/3$, while the volume of the cube is a^3. The packing ratio is therefore $\rho = 4\pi(r_>^3 + r_<^3)/(3a^3)$. The discussion around Fig. 2.16 is confined to the case in which the sphere at the center is small, $x < (\sqrt{3} - 1)$. Then the lattice constant is $a = 2r_>$, yielding $\rho = \pi(x^3+1)/6$. When $x > (\sqrt{3}-1)$, the larger spheres are not tangential to each other, and therefore $a > 2r_>$. However, in this range the sphere at the center of the cube is tangential to its big neighbors, and therefore the length of the cube diagonal is $a\sqrt{3} = 2(r_> + r_<)$. Hence, $\rho = \pi\sqrt{3}(x^3+1)/[2(x+1)^3)]$. Figure 2.46 shows the packing ratio as a function of the radii ratio x. For $x = 0$, the packing ratio is that of the **SC** lattice, $\rho = \pi/6 \approx 0.5236$, and for $x = 1$ the packing ratio is that of the **BCC** lattice, $\rho = \pi\sqrt{3}/8 \approx 0.68$. The maximal packing ratio, 0.729, is reached when all spheres are tangential to each other, $x = (\sqrt{3}-1) \approx 0.732$.

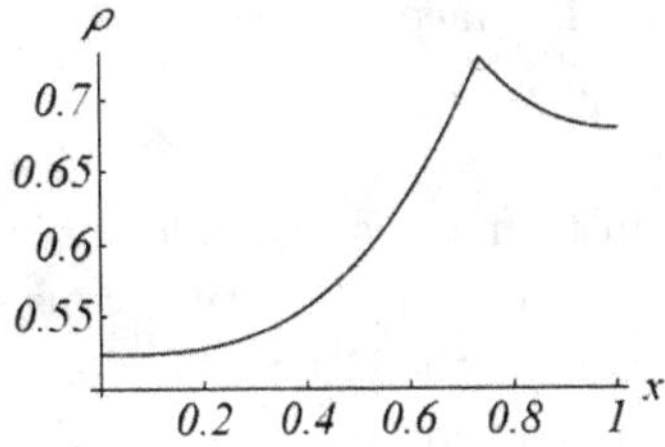

Fig. 2.46

Answer 2.10.
a. The volume of this unit cell is given by Eq. (2.2),

$$V = |\mathbf{a}_1 \cdot [\mathbf{a}_2 \times \mathbf{a}_3]| = \begin{vmatrix} -a/2 & a/2 & a/2 \\ a/2 & -a/2 & a/2 \\ a/2 & a/2 & -a/2 \end{vmatrix} = a^3/2 \,. \tag{2.10}$$

This answer can be also derived by realizing that the cubic unit cell includes two lattice sites, and therefore its volume is twice that of the primitive unit cell.
b. This cell includes a single lattice site, and therefore its volume is $a^3/2$ and it is a primitive unit cell. One may get this result directly: this cell is a parallelogram whose base is a square of edge a and its height equals $a/2$. The lattice vectors are not primitive, as they are not the shortest ones connecting a lattice site to its neighbors.

Answer 2.11.

a. $a^3/4$. This answer can be reached by two routes, either by inserting Eq. (2.6) into Eq. (2.2), or by noting that the volume of the cubic unit cell of the **FCC** lattice, which includes four lattice sites, is a^3.

b. The cubic unit cell contains four spheres of radius $r = a\sqrt{2}/4$ (obtained from the diagonal of the cube face), and hence the volume packing ratio is $4 \times (4\pi r^3/3)/a^3 = \pi\sqrt{2}/6 \approx 0.740$.

Answer 2.12.

One may obtain the result by examining the dashed square in Fig. 2.19(a). The mid points of the edges of this square are each located on the mid orthogonals of the lines connecting the vertices of this square with the origin at the center of the cube. The square joining these mid points is built of four diagonals of squares which are the faces of the Wigner-Seitz cell. As the edge of the dashed square is a, each diagonal is $a/\sqrt{2}$, and consequently the length of the edge of the face of the Wigner-Seitz cell is $a/2$. Another method is based on the rectangle marked by the thick dashed lines. The length of the smaller edge of this rectangle is $a/\sqrt{2}$, and it also equals the length of the diagonal of the face of the Wigner-Seitz cell. As a result, the length of the edge of that face is $a/2$.

Answer 2.13.

The Bravais lattice of strontium titanate is a simple cubic, with a lattice constant equal to the distance between neighboring strontium ions, a. The surroundings of each of the strontium ions which form the lattice are identical. The surroundings of each of the titanium ions, which also create a simple-cubic lattice, are identical as well. On the other hand, the oxygen ions do not all share the same environment. It is only when an oxygen ion is displaced by a lattice vector of the afore-mentioned cubic lattice that it comes to a location with identical surrounding. Hence, the primitive unit cell is the cube shown in Fig. 2.21(a). Upon slightly shifting the cube along one of its diagonals (e.g., along the negative direction of all axes), it is found that in the displaced cube there remain a single strontium ion (at the origin), one titanium ion (at the center of the cube, much like in cesium-chloride lattice), and three oxygen ions, at the points $(a/2, a/2, 0)$, $(a/2, 0, a/2)$, and $(0, a/2, a/2)$. Therefore, the base of the crystal contains exactly the chemical formula of the material, $SrTiO_3$.

Answer 2.14.

By counting, the tetragonal cell in Fig. 2.21(b) contains twice the chemical formula of the material: two copper ions, four lanthanums, and eight oxygens. The advantage of the tetragonal cell is the right angles between its lattice vectors. Since the cell contains a copper ion also at its center, the lattice is terms **"body-centered tetragonal"**, for obvious reasons. Nonetheless, the surrounding of the copper ion at the center of the cell is identical to that of the copper at the low left corner; it is therefore possible to build a primitive unit cell that contains half of the ions appearing in the figure, i.e., the chemical formula. The primitives unit cell consists of the two vectors $(a, 0, 0)$ and $(0, a, 0)$, which create the square of copper ions in the basis of the cell, and of a third vector, $(a/2, a/2, c/2)$, which connects the copper ion at the center with the origin. The base of the crystal contains a copper ion at the origin, two lanthanum ions at $(a/2, a/2, c/6)$ and $(0, 0, c/3)$, and four oxygen ions at $(a/2, 0, 0)$, $(0, a/2, 0)$, $(0, 0, c/6)$, and $(a/2, a/2, c/3)$. This base generalizes the one discussed in problem 2.10(b). The lattice vectors building the primitive cell are the shortest possible ones, provided that $(a^2 + c^2)/4 > a^2$, i.e., $c > a\sqrt{3}$. When $c < a\sqrt{3}$, the shortest lattice vectors are similar to those in the left panel of Fig. 2.16 and in Eq. (2.5): $\mathbf{a}_1 = (a/2)(-\hat{\mathbf{x}} + \hat{\mathbf{y}}) + (c/2)\hat{\mathbf{z}}$, $\mathbf{a}_2 = (a/2)(\hat{\mathbf{x}} - \hat{\mathbf{y}}) + (c/2)\hat{\mathbf{z}}$, $\mathbf{a}_3 = (a/2)(\hat{\mathbf{x}} + \hat{\mathbf{y}}) - (c/2)\hat{\mathbf{z}}$.

Answer 2.15.

a. Consider the unit cell of the **HCP** lattice, indicated by the thick lines in Fig. 2.24(b). Recall that the lattice constant of the planar triangular lattice is $a = 2r$, where r is the radius of the spheres in the dense planar packing. The volume of the unit cell is then $ca^2\sqrt{3}/2$, where c is the height of the cell. The cell contains two lattice sites; the total volume of the spheres centered on these two sites is thus $8\pi r^3/3$. (This can also be found by "dividing" each sphere at the corner of the unit cell among all neighboring cells; in the lower plane and in the upper one there are altogether four spheres "contributing" $1/6$ of their volume and four others "contributing" $1/12$ of their volume. Their total contribution is then $4 \times (1/6) + 4 \times (1/12) = 1$ of the sphere volume, to which the volume of the "internal" sphere is added.) Therefore, the packing ratio is $\rho = (8\pi r^3/3)/(ca^2\sqrt{3}/2) = 2\pi a/(3\sqrt{3}c)$. In the densest lattice $c = 2a\sqrt{2/3} \approx 1.633a$, and then $\rho = \pi/(3\sqrt{2}) \approx 0.74$.

b. The packing ratio of the **FCC** is $\rho = (16\pi r^3/3)/a^3 = \pi/(3\sqrt{2}) \approx 0.74$, identical to the result above. This is not surprising, since in both lattices each layer lies in the "valleys" of its predecessor, and the density is not affected by shifting the layer among these "valleys".

Answer 2.16.

Since the arrangement marked by A repeats itself only after four layers, the unit cell is a prism whose basis is the rhombic unit cell of the planar triangular lattice, but its height is twice that of the simple **HCP** structure, i.e., $c = 4a\sqrt{2/3} \approx 3.266a$. The distance between neighboring layers is determined by the condition that the spheres are tangential to each other; it thus remains the same. Indeed, the height of the prism is doubled, but so is also the number of spheres (four in place of two), and the packing ratio is the same as before, $\rho \approx 0.74$.

Answer 2.17.

The unit cell is identical to that of the **HCP** lattice, but the base contains two zinc ions, at the points $(a/2, a\sqrt{3}/6, c/8)$ and $(0, 0, 5c/8)$ and two sulphur ions, at $(0, 0, 0)$ and $(a/2, a\sqrt{3}/6, c/2)$.

Answer 2.18.

This lattice is identical to the square lattice, rotated relative to the former by $45°$. If the lattice constant of the original square lattice is a, then the lattice constant of the square-centered lattice is $a/\sqrt{2}$ (half the diagonal of a square of edge a).

Answer 2.19.

The base-centered tetragonal lattice consists of layers of square-centered lattices placed one on top of the other. As seen in problem 2.18, the square-centered lattice is identical to a square lattice rotated by $45°$ whose edge length is smaller by a factor of $\sqrt{2}$. Hence, the based-centered tetragonal lattice is identical to the tetragonal lattice. Quite similarly, the face-centered tetragonal lattice is identical to a body-centered tetragonal lattice also rotated by $45°$. The lattice sites at the centers of the faces become the sites at the centers of the body-centered tetragonal lattice.

Answer 2.20.

In both cases $a_1 = a_2 = a_3$ and $\alpha = \beta = \gamma$, and consequently the two primitive unit cells construct **trigonal** lattices. The angle in-between the two lattice vectors can be derived by using the relation $\mathbf{a}_1 \cdot \mathbf{a}_2 = a_1 a_2 \cos\gamma$. Equations (2.5) and (2.6) thus yield $\alpha = \beta = \gamma = 109.47°$ for the **BCC** lattice, and $\alpha = \beta = \gamma = 60°$ for the **FCC** one. This latter result can be directly observed in Fig. 2.18: each face of the primitive unit cell is a rhombus comprising two equilateral triangles; the length of the shorter diagonal equals the length of the triangles' edge (and the distance between nearest-neighbor lattice sites).

Answer 2.21.

Equation (2.1) which defines a Bravais lattice, gives the location of a lattice site as $\mathbf{R} = n_1\mathbf{a}_1 + n_2\mathbf{a}_2 + n_3\mathbf{a}_3$, where the coefficients are arbitrary integers. Inversion through the origin transforms this point into $-\mathbf{R} = -n_1\mathbf{a}_1 - n_2\mathbf{a}_2 - n_3\mathbf{a}_3$. The coefficients here are integers as well, and hence this point is also a lattice point.

Answer 2.22.

a. Denote a rotation axis of order n by A_n, a mirror plane by m, and an inversion center by i. The tetragonal lattice has one rotation axis of order 4, normal to the square face of its unit cell and threading it at the center, and four axes of order 2, perpendicular to the rectangular faces of the unit cell, again passing through their centers. In addition, this lattice possesses three mirror planes, parallel to the faces and passing through the mid points between them, and two mirror planes connecting the diagonals of opposite faces. Finally, this lattice has an inversion center at the center of the unit cell. Thus, the symmetry group of the tetragonal lattice includes the elements i, $1A_4$, $4A_2$, $5m$.

b. The square basis of the unit cell is transformed into a rectangular one whose edges are not all equal. Therefore the rotation of order 4 and the two mirror planes passing through the diagonals of this basis are eliminated from the group.

c. In this case all faces are squares, and consequently the mirror planes through the diagonals of those faces have to be added to the group, and also all rotations around axes passing through the mid points of the faces are of order 4. In addition, there are now the rotations A_3 around each of the cube diagonals. Altogether, the symmetry group of the cubic lattice contains three axes of order 4, six axes of order 2, four axes of order 3, nine mirror planes, and one inversion center.

Answer 2.23.

a. The general oblique lattice is displayed in Fig. 2.47(a), for the case where $|\mathbf{a}_2| = a_2$ is much longer than $|\mathbf{a}_1| = a_1$. The lattice then comprises separate parallel rows of circles, tangential to each other within each row, but not touching the circles in the neighboring rows. The area of the unit cell is $a_1 a_2 \sin \gamma$, and the area of each circle is $\pi r^2 = \pi(a_1/2)^2$ ($r = a_1/2$ is the radius of each circle). The packing ratio is thus $\rho = [\pi a_1^2/4]/[a_1 a_2 \sin \gamma] = \pi a_1/(4 a_2 \sin \gamma)$. When $a_2 < a_1$, one needs just to interchange a_1 and a_2.

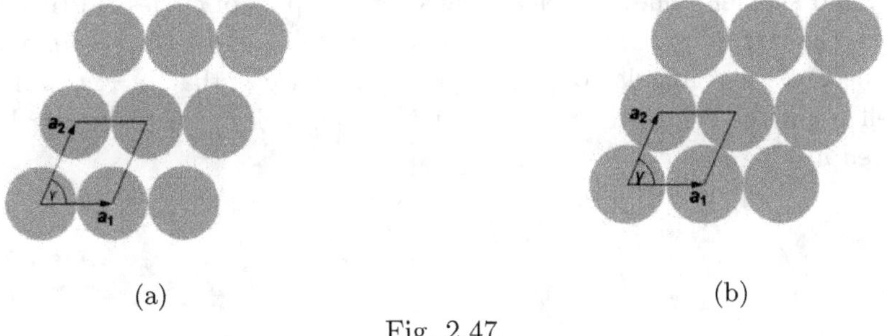

Fig. 2.47

The figure shows that the packing ratio can be increased by pushing the rows closer to each other. When $60° \leq \gamma \leq 120°$, one may decrease a_2 until the circles in different rows touch each other, like in Fig. 2.47(b). In this case, $a_1 = a_2 = 2r$, and the packing ratio of the oblique oblique lattice for angles in this range is $\rho = \pi/(4 \sin \gamma)$. In particular, this result reproduces the packing ratio of the square lattice, $\rho = \pi/4$, for $\gamma = \pi/2$, and the packing ratio of the triangular lattice, $\rho = \pi/(4 \sin 60°) = \pi/[2\sqrt{3}]$ when $\gamma = \pi/3$. One may increase the packing ratio by decreasing $\sin \gamma$. However, the figure shows that γ cannot be reduced below $\gamma = \pi/3 = 60°$, because in that case the circles in two neighboring rows overlap. For the same reason, γ cannot be increased beyond $\gamma = 2\pi/3 = 120°$. It follows that the maximal packing ratio of the oblique lattice with angles in the range $60° \leq \gamma \leq 120°$ is achieved in these two limits, $\gamma = \pi/3 = 60°$ or $\gamma = 2\pi/3 = 120°$. In both cases there emerges the triangular lattice displayed in Fig. 2.48(c), with the packing ratio $\rho = \pi/(4\sin 60°) = \pi/[2\sqrt{3}]$.

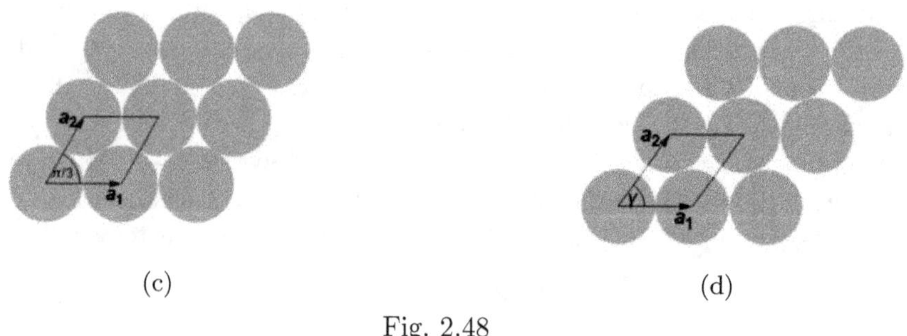

(c)

(d)

Fig. 2.48

When the angle γ is in the range $\gamma < 60°$, a_2 cannot be made equal to a_1 because then again the circles in neighboring rows overlap. The optimal configuration of this case is shown in Fig. 2.48(d). The vector $\mathbf{a}_2$ is now the basis of an isosceles triangle (with one of the equal-length sides being the vector $\mathbf{a}_1$), and hence $a_2 = 2a_1 \cos\gamma$, leading to $\rho = \pi/(4\sin 2\gamma)$. This result pertains only to the range $30° \leq \gamma < 60°$, because when γ reaches the value $30°$ the lattice becomes the one shown in Fig. 2.47(e), and smaller values of γ again cause overlapping of circles. In this range of γ values, the largest packing ratio is achieved at the ends of the range, where the lattice becomes a triangular one. The only difference is that in Fig. 2.49(e) the unit cell is not the primitive one: the primitive unit cell of the triangular lattice is portrayed in Fig. 2.48(c).

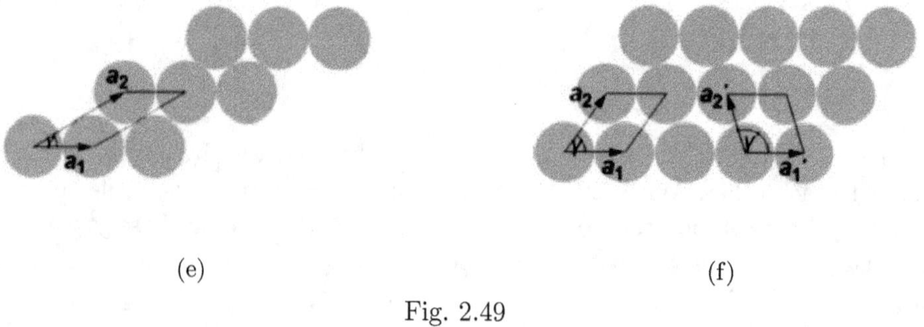

(e)

(f)

Fig. 2.49

The lattice in Fig. 2.48(d) can be described by a primitive unit cell like the one shown in Fig. 2.49(f) (right side). This unit cell is a reflection image of the cell in Fig. 2.47(b); the results obtained from that figure apply to this case as well.

Answer 2.24.

a. A rotation axis of order 2 normal to the lattice and passing through any lattice site and through any mid point between nearest neighbors, a reflection through a mirror plane placed normal to the lattice at these points, and an inversion with respect to any of those points.

b. A rotation axis of order 2 normal to the lattice and passing through any lattice point, a reflection through a mirror perpendicular to the lattice at any lattice point, and an inversion with respect to any lattice point.

c. No symmetry operations except for translations.

d. A rotation axis of order 2 threading the lattice perpendicularly through a mid point between two sites, a reflection through a mirror normal to the lattice at the mid point between two neighboring sites, and an inversion with respect to any of the mid points between two lattice sites.

e. A reflection through a mirror placed at any lattice site normal to the lattice.

f. A rotation of order 2 around an axis normal to the lattice plane and passing through the points A or C, a reflection by a mirror normal to the lattice placed at the points B or D, and an inversion with respect to the points A or C.

Answer 2.25.

Most of the symmetry operations do not exist. The one that remains in both the triangular and the square lattices is a reflection through a mirror perpendicular to the lattice plane passing through the vector $(\mathbf{a}_1 + \mathbf{a}_2)$.

Answer 2.26.

When the two ions in the unit cell portrayed in the first panel Fig. 2.2 have opposite signs, their displacement relative to the structure shown in the first panel of Fig. 2.1 creates a dipole moment, directed along the direction of the lattice. This dipole moment is shown by the arrow in Fig. 2.50(a). A similar situation occurs in the crystal displayed in the second panel of Fig. 2.2, where an electric dipole moment appears along the direction of the relative motion of the ions, see Fig. 2.50(b).

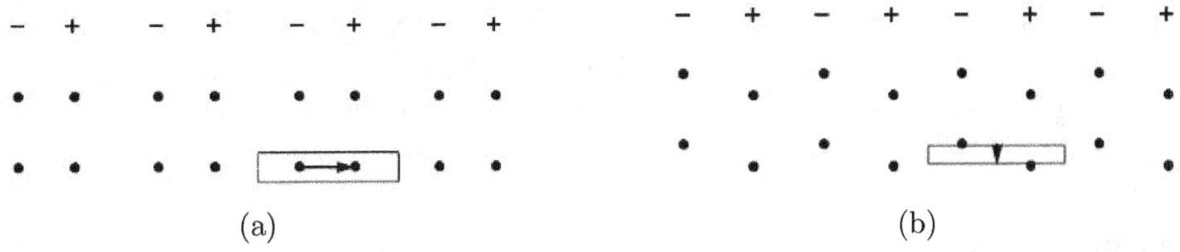

(a)   (b)

Fig. 2.50: (a) The electric dipole moment for the lattice in the first panel of Fig. 2.2; (b) The electric dipole moment for the lattice in the second panel there.

Answer 2.27.

a. Divide Eq. (2.8) by F_{n-1}, to obtain $t_n = 1 + 1/t_{n-1}$. Assuming that the ratio t_n does approach a constant limit, $t_n \to \tau$, the equation becomes $\tau = 1 + 1/\tau$, and the solution is indeed the golden ratio. Subtracting the two equations from one another, the result is $|t_n - \tau| = |t_{n-1} - \tau|/(t_{n-1}\tau)$. As the denominator on the right hand-side is larger than 1, the difference $|t_n - \tau|$ decreases at each generation and eventually approaches zero.

b. Denote the number of long and short segments in the nth generation by $N_L(n)$ and $N_S(n)$, respectively. The Fibonacci rule, mapping S onto L and L onto LS, yields the relations $N_S(n+1) = N_L(n)$ and $N_L(n+1) = N_L(n) + N_S(n)$. Dividing these two equations by one another gives $x(n+1) = 1 + 1/x(n)$, where $x(n) = N_L(n)/N_S(n)$. In the $n \to \infty$ limit, corresponding to the infinite lattice, both $x(n)$ and $x(n+1)$ equal $x(\infty)$, and consequently the equation of part (a) is reproduced, with the solution $x(\infty) = \tau$.

Answer 2.28.

Assuming that the thick line passes through the lattice point (0,0), as in Fig. 2.51, its equation is $y = x/\tau$, and the equations of the dotted lines are $y = (x/\tau) \pm y_0$, with $2y_0 = w/\cos\alpha$ being the projected distance between these two lines on the $\hat{\mathbf{y}}$−axis. For the choice $w = a(\cos\alpha + \sin\alpha)$, $y_0 = a(1+\tau)/(2\tau) = a\tau/2$. Denoting an arbitrary lattice site located in-between the two dotted lines by $\mathbf{r}_n = m_1(n)\mathbf{a}_1' + m_2(n)\mathbf{a}_2'$, the $n+1$ successive lattice site along the $\hat{\mathbf{y}}$−axis has $m_2(n+1) = m_2(n)$ and $m_1(n+1) = m_1(n)+1$, and the one along the $\hat{\mathbf{x}}$−axis has $m_2(n+1) = m_2(n)+1$ and $m_1(n+1) = m_1(n)$. In both options, $m_1(n+1)+m_2(n+1) = m_1(n)+m_2(n)+1$, so that one may write $m_1(n) + m_2(n) = n$. The lattice point $\mathbf{r}_n$ is indeed between the two dotted lines provided that $m_1 a/\tau - y_0 < m_2 a < m_1 a/\tau + y_0$ and $(m_2 a - y_0)\tau < m_1 a < (m_2 a + y_0)\tau$. These two inequalities are identical; since $\tau = 1 + 1/\tau$, they yield $m_1 - 1/2 < n/\tau < m_1 + 1/2$, leading to $m_1 = ||n/\tau||$. Hence, $x_n = m_1(n)\mathrm{L}+m_2(n)\mathrm{S}= n\mathrm{S}+m_1(n)(\mathrm{L\text{-}S})=\mathrm{S}[n + ||n/\tau||/\tau]$. Alternatively, $m_1 < n/\tau + 1/2 < m_1 + 1$, and consequently $m_1 = [n/\tau + 1/2]$, where $[y]$ is the integral part of the number y. In order to prove that this is indeed a Fibonacci sequence, one exploits an inflation transformation. Figure 2.51(a) shows Fig. 2.33 with an additional band, in-between the solid (red) lines, whose width is w/τ. The new band does not include any of the vertical steps of the original line, whose projections gave the segments of length S. The lattice sites within this new band give rise solely to projections of length L'=L+S or S'=L and the new sequence (starting at the left edge) is L'S'L'S'L'L'S'L'L'S', which is also a Fibonacci lattice, with a length scale larger by τ.

Answer 2.29.

a. With the choice of the origin as given in the problem, the equation of the thick line is $y = x\tan\alpha$. All lattice sites are either $x = n_1 a$, or $y = n_2 a$. It follows that

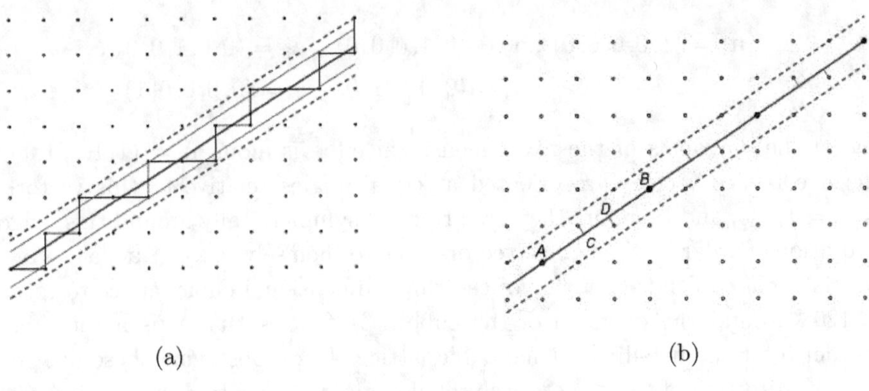

Fig. 2.51

the ratio $x/y = n_1/n_2$ is rational; it never becomes $\tan(\alpha) = 1/\tau$, and therefore there are no more lattice sites on this line.

b. When the slope is a rational number, the thick line can pass through a lattice site; this line then passes also through all lattice sites obeying $n_1/n_2 = p/q$. Figure 2.51(b) shows the result for $\tan(\alpha) = 2/3$. As seen, the square-lattice sites that lie on the thick line are projected onto a periodic lattice, with a lattice constant $a' = a\sqrt{p^2 + q^2}$. For the band between the two dashed lines, each such unit cell contains also the projections of two additional points, at $x_1 = a[2 \pm \sin(\alpha)\cos(\alpha)]/[2\sin(\alpha)]$; the distances on the thick line are $AB = 2a/\sin(\alpha)$, $CD = a\cos(\alpha)$. This lattice can be described by the periodic sequence LSLLSLLSLLS. More complicated bases are possible for other widths of the band.

c. Since 1.6=8/5, 1.61=161/100, and 1.618=809/500, the lattice constants of the corresponding periodic lattices are $\sqrt{8^2 + 5^2} \approx 9.43$, $\sqrt{161^2 + 100^2} \approx 189.5$, and $\sqrt{809^2 + 500^2} \approx 951$, all in units of the square-lattice constant. Each unit cell represents a base comprising an increasing number of lattice sites, with irrational coordinates within the cell. In the infinite limit, the unit cell is identical to the whole lattice.

Answer 2.30.

a. An example for such a basis appears in Fig. 2.52(a), which shows five equal-length vectors, with a $72°$−angle in-between them. Each of the edges in Fig. 2.35 is parallel to one of these five vectors: when the origin is placed at the locus of the five central rhombi in Fig. 2.52(b), their vertices are reached by these five vectors. Any edge of the bright rhombus is parallel to an edge of a dark rhombus, and therefore it is also equal to one of the proposed lattice vectors.

b. The d–dimensional cubic lattice can be described by a cartesian basis, in which the nth lattice vector is a vector of d components, all zero except for the nth one, which is 1. In five dimensions these vectors are

$$\mathbf{a}_1 = (1,0,0,0,0) \ , \ \mathbf{a}_2 = (0,1,0,0,0) \ , \ \mathbf{a}_3 = (0,0,1,0,0) \ ,$$
$$\mathbf{a}_4 = (0,0,0,1,0) \ , \mathbf{a}_5 = (0,0,0,0,1) \ .$$

One of the diagonals of the five-dimensional cube is along $\hat{\mathbf{u}} = (1,1,1,1,1)/\sqrt{5}$. The five lattice vectors are oriented at equal angles relative to this vector, i.e., $\mathbf{a}_n \cdot \hat{\mathbf{u}} = 1/\sqrt{5}$, and therefore they are arranged symmetrically around this diagonal. A rotation of order 5 brings these vectors back to themselves, $\mathbf{a}_1 \to \mathbf{a}_2$, $\mathbf{a}_2 \to \mathbf{a}_3$, *etc.* (this is similar to the symmetry of the three-dimensional cubic lattice to rotations by 120° around the diagonal of the cube). It follows that $\hat{\mathbf{u}}$ is a rotation axis of order 5 for the five-dimensional cubic lattice. Upon observing these five vectors from the direction of $\hat{\mathbf{u}}$, one notes the equal angles between two neighboring vectors. Therefore their projections on a plane normal to $\hat{\mathbf{u}}$ appears as in Fig. 2.52(a).

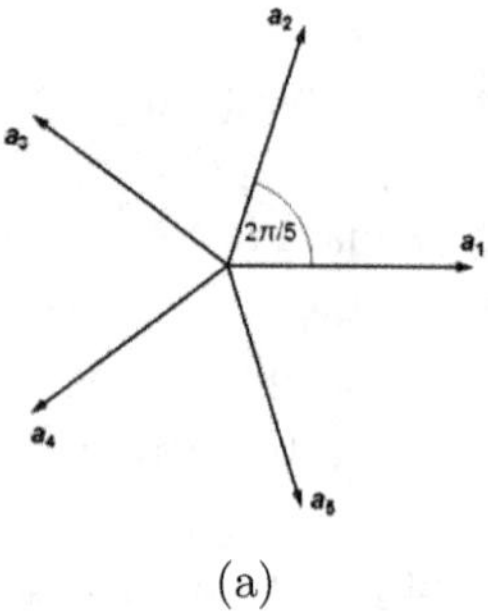

(a)

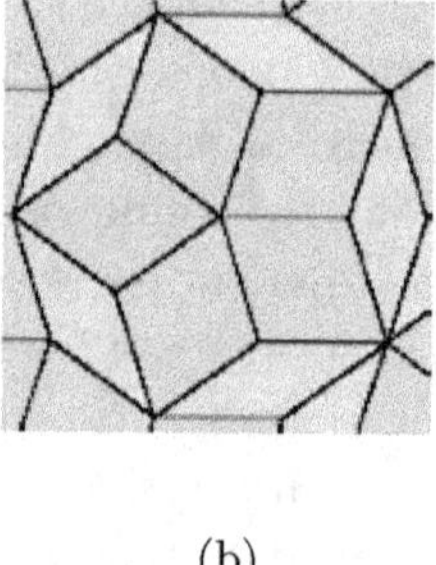

(b)

Fig. 2.52

Answer 2.31.
Assume that the joint unit cell contains m_1 circles on the substrate, and m_2 ones in the layer above. The length of the unit cell is hence $2m_1r_1 = 2m_2r_2$ ($2r_1$ and $2r_2$ are the respective diameters). It follows that $m_1/m_2 = r_2/r_1 = p/q$, and the smallest values obeying this relation are $m_1 = p$ and $m_2 = q$. The length of the unit cell is hence $2pr_1 = 2qr_2$. The base contains p circles on the substrate and q ones in the layer above.

Answer 2.32.
The adsorbed atoms create a triangular lattice, with a rhombic unit cell whose edge is of length $3a$, where a is the distance between nearest neighbors in the graphite substrate, see Fig. 2.53. Shifting slightly this rhombus down to the left shows that it includes a single krypton atom (at the lower left corner) and also six carbon

atoms, encompassing the inner hexagon. Thus the base comprises a krypton atom and six carbon ones.

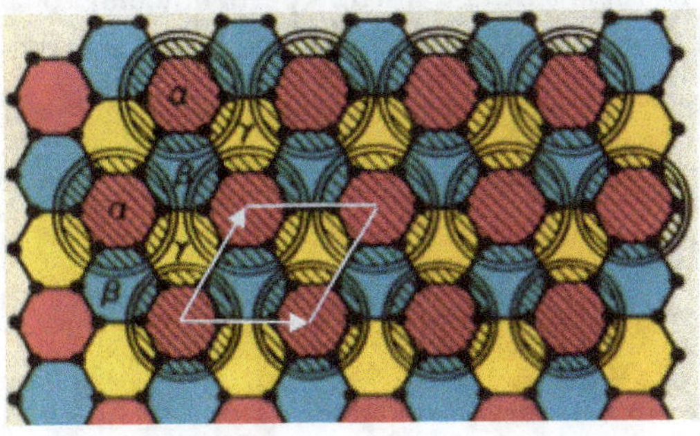

Fig. 2.53

Answer 2.33.
The tetragonal unit cell of the structure in Fig. 2.39 contains two cubic-perovskite cells [Fig. 2.21(a)], one above the other. Along the direction normal to the layers, this array includes a SrO plane, a MnO_2 plane, a LaO plane, and another MnO_2 plane, altogether a single strontium ion, a lanthanum ion, two manganese and six oxygen ions (see Fig. 2.54). When $x = 1/3$, there should be two layers of LaO for each layer of SrO, and hence the tetragonal unit cell contains a Sr ion, two La ones, three Mn ions and nine oxygen ions.

Fig. 2.54

Answer 2.34.
The unit cells of the structures C and D contain two lattice sites, while there are five lattice sites in the unit cell of E. The symmetry group of C includes rotations of order 2 around axes normal to the lattice and passing through the mid points between nearest neighbors, and inversions through these points. It also includes reflections

through mirrors normal to the lattice and passing through the lattice points. The symmetry group of D includes only the latter reflections. The symmetry group of E includes solely reflections through mirrors perpendicular to the lattice at the mid points between two downwards-oriented moments, as displayed in Fig. 2.55.

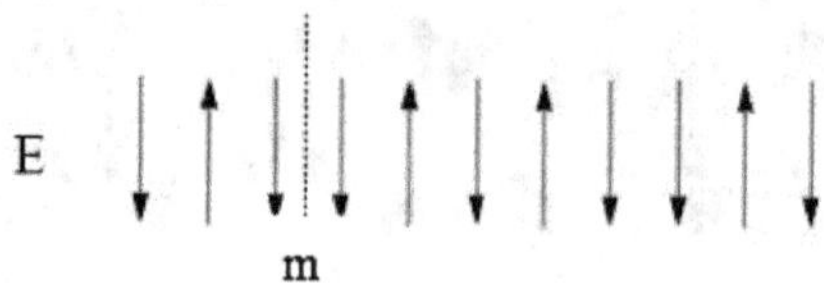

Fig. 2.55

Answer 2.35.

The magnetic moment at the origin $\mathbf{R} = 0$ is $+\mu_0$; its sign alternates upon propagating along each axis. Thus, the magnetic lattice resembles that of table salt: the "chloride" sites represent the moment $+\mu_0$, and the "sodium" ones the moment $-\mu_0$. It follows that the base of the magnetic lattice contains two sites, one with the moment $+\mu_0$, the other with $-\mu_0$ (like the pair in the ellipse of Fig. 2.17), and the lattice is an **FCC** one. As opposed to the lattice of table salt, where the two sublattices consist of different ions and therefore no symmetry operations can inter relate them, in the antiferromagnet case one may map one sublattice on the other by a combination of adequate symmetry operations. For instance, a reflection by a mirror passing through the mid points of the cube's faces. When $\mathbf{Q} = (1,1,1)(\tau/a)$ and $\mathbf{R} = (n_1, n_2, n_3)a$, then $\mu(\mathbf{R}) = \mu_0 \cos[\tau(n_1 + n_2 + n_3)]$. The moment at the origin is still $+\mu_0$, but there is no other lattice point with this moment, as no integer product of τ can be an (even) integer product of π; the ratio τ/π is irrational. Thus the magnetic order in this case is not commensurate with the structural lattice, and the magnetic unit cell is infinitely large.

Answer 2.36.

The unit cell is the same rhombus that has been the base for the triangular lattice, Fig. 2.4, or for the hexagonal lattice, Fig. 2.8. However, in the present case it contains a base of three sites, each of them belonging to a different sublattice– see Fig. 2.12. Exploiting this unit cell, the base includes the sites at $\mathbf{r}_1 = 0$, $\mathbf{r}_2 = (\mathbf{a}_1 + \mathbf{a}_2)/3$, $\mathbf{r}_3 = 2(\mathbf{a}_1 + \mathbf{a}_2)/3$.

2.12   Problems for self-evaluation

s.2.1.
Draw the Wigner-Seitz cells belonging to each of the lattices in Fig. 2.27.

s.2.2.
Figure 2.56 displays two lattices called Kagomé (right panel) and Cairo (left panel). The first is built of equilateral triangles; the second consists of pentagons, each comprising four identical edges and a shorter one, with three $120°-$angles and two right ones. Identify the corresponding primitive unit cells and the lattice vectors. Write down the radius vectors of the points that belong to the base when the unit cell includes more than a single lattice site (a junction in the figures).

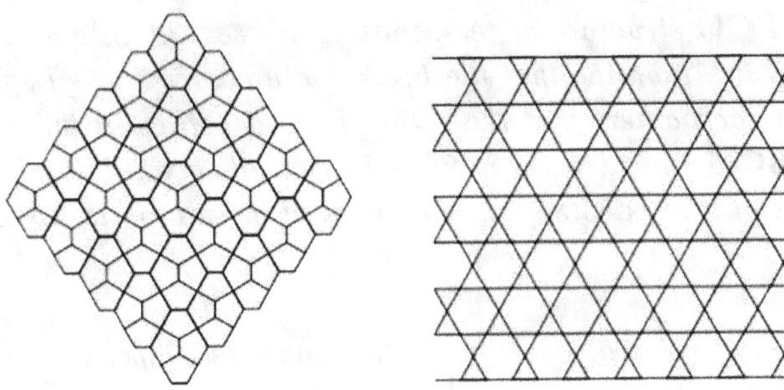

Fig. 2.56

s.2.3.
Determine which of the structures in the following list are Bravais lattices. Indicate the lattice vectors of those which are Bravais lattices, and describe the lattice as a Bravais one with a base for those that are not; write down the lattice vectors and the coordinates of the points in the base within the primitive unit cell.
a. *A square lattice with an additional lattice site at the mid point of each edge.*
b. *A simple cubic lattice with additional lattice points at the centers of the bases (the horizontal faces).*
c. *A simple cubic lattice with additional lattice points at the centers of the four vertical faces.*
d. *A simple cubic lattice with additional points at the mid point of each edge.*

s.2.4.
Which Bravais lattice is created of the points (n_1, n_2, n_3) in cartesian coordinates

when:

a. n_i *are all even, or are all odd;*

b. $(n_1 + n_2 + n_3)$ *is always even;*

c. $(n_1 + n_2 + n_3)$ *is always odd.*

s.2.5.

Figure 2.19 portrays the Wigner-Seitz cells of the **BCC** *and* **FCC** *lattices. Prove that the four-edge polygons on the faces of the Wigner-Seitz cell of the* **BCC** *are all squares, that the hexagons are all equilateral, and that the length of each edge is $a\sqrt{2}/4$, where a is the edge length of the cubic cell.*

s.2.6.

a. *Sodium crosses from the* **BCC** *structure to a perfect* **HCP** *one at $23°K$. Assuming that the density does not change, find the new lattice constant [in units of the* **BCC** *one, $a(\text{BCC}) = 4.23Å$].*

b. *Iron has the* **FCC** *structure at temperatures above $900°C$, and a* **BCC** *one at lower temperatures. Assuming that the two structures are built of spheres of equal radii adjacent to each other, find the ratio of the densities of the two structures. Given that the* **BCC** *iron density is $7900\ Kg/m^3$, determine the lattice constant (the atomic mass of iron is 55.85, and the mass of the proton is $1.6726\times10^{-27}Kg$.)*

s.2.7.

a. *Find the packing ratio of a crystal of the table-salt type, in which the larger "chloride" ions are adjacent to each other and the radii ratio between the "sodium" and "chloride" ions is x. Determine the relevant range of the x values.*

b. *Find the packing ratio when the ions of the two types touch each other, but the "chloride" ions do not.*

c. *Provide a drawing of the dependence of the packing ratio on the radii ratio.*

s.2.8.

a. *Find the packing ratio of a zinc-blende lattice, in which the bigger "sulphur" ions are contiguous, and the ratio of the radii of the "zinc" and "sulphur" ions is x. What is the relevant range of x values?*

b. *Find the packing ratio when the ions of the two types touch each other, but the "sulphur" ions do not.*

c. *Provide a drawing of the dependence of the packing ratio on the radii ratio. Where is the diamond in this figure? Compare the results with those obtained for structures of the table-salt type and the CsCl one. Which packing is denser for various values of x?*

d. *The distance between the center of the sulphur ion and the center of the zinc one in zinc-blende is 0.234nm. Find the density of this material (the atomic weights of zinc and sulphur are 65.38 and 32.064, respectively, and the mass of the proton is $1.6726\times10^{-27}\ Kg$).*

s.2.9.

Figure 2.57 displays the double-perovskite structure, $A_2BB'O_6$, in which half of the B ions in the original lattice [Fig. 2.21(a)] are alternatively replaced by other ions B' (the figure shows Sr_2MoFeO_6]. The left part shows only the ions B and B' (filled and empty circles, respectively), and the right depicts the octahedra of the oxygens encapsulating each one of them. Determine the lattice corresponding to this structure. Find the primitive unit cell and the base.

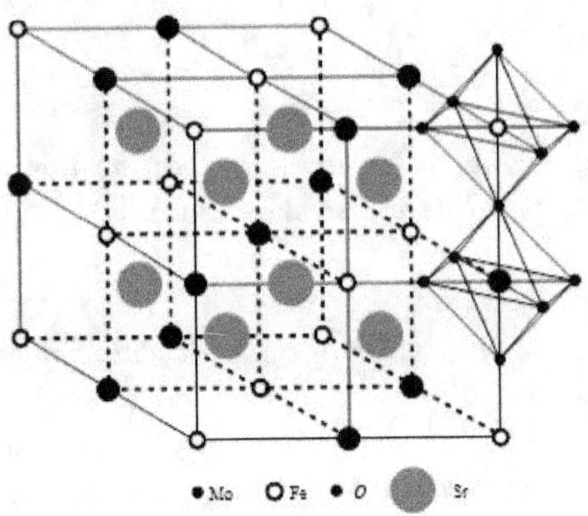

Fig. 2.57

s.2.10.

a. *Determine the primitive unit cell of the graphite lattice displayed in Fig. 2.9; find the lattice vectors of this cell and its base. Determine the coordinates of the atoms that belong to it (in terms of the lattice vectors).*

b. *The planes created in Fig. 2.12 once the atoms of type A, B, or C are eliminated are denoted by the letters A, B, and C, respectively. The discussion following Fig. 2.9 shows that the graphite lattice is described by the arrangement BCBCBC..., where C is the configuration shown in Fig. 2.8(b) and B is the one in the mid panel of Fig. 2.9. Determine the same quantities as in part (a) for the configuration ABCABCABC... of the graphene layers.*

s.2.11.

Referring to Fig. 2.28, express the volume of the primitive unit cell of the following lattices, in terms of the three lattice constants and the angles in-between them.

a. *face-centered orthorhombic;*

b. *triclinic;*

c. *monoclinic;*

d. *trigonal.*

s.2.12.

The parameters of a three-dimensional crystal are $a_1 = a_2 = a$, $a_3 = c = a[1 + \Delta(\sqrt{8/3} - 1)]$, $\alpha = \beta = 90°$, $\gamma = 30°(3 - \Delta)$. The base contains two spheres, at the origin and at $\mathbf{r}_2 = [(1 - \Delta/3)(\mathbf{a}_1 + \mathbf{a}_2) + \mathbf{a}_3]/2$. The two spheres touch one another. Draw the packing ratio of this crystal as a function of Δ for $0 \le \Delta \le 1$, and find the maximal packing ratio. Which lattice structures correspond to the two extreme values, $\Delta = 0$ and $\Delta = 1$?

s.2.13.

Determine the structure that is created by repeating the projection procedure described in Fig. 2.32, with a line whose slope is $\tan(\alpha) = 1/\sqrt{2}$.

s.2.14.

*Figure 2.58 illustrates the adsorption of spherical "atoms" on two planar cuts of the **FCC** lattice of iron, so that the adsorbed atoms are located in the dips formed among the atoms of the substrate. The small dashed circles represent the iron atoms, and the full large ones the adsorbed atoms. Figures A and B show the adsorption on a plane passing through the face of the cubic lattice shown in the mid panel of Fig. 2.17 (e.g., the upper face), while figures C and D refer to the adsorption on a plane passing through the diagonals of two adjacent faces, originated at the same lattice site [e.g., the green points in the lower part of Fig. 2.24(a)]. The plane is normal to the principal diagonal of the cubic unit cell.*

a. *Determine the two-dimensional lattice of each of these planes (before the adsorption takes place), and the corresponding primitive lattice constant (in terms of the original lattice constant of the **FCC**, a).*

b. *Determine the primitive unit cells of each of the lattices displayed in Fig. 2.58, comprising the substrate and the adsorbed planes, the lattice constants and the base.*

c. *Find the maximal radius of the adsorbed atoms in each of the figures, which is compatible with the structure shown there.*

d. *For which values of the radius of the adsorbed atoms are all the dips in the substrate occupied? In that case, what are the answers to part (b)?*

s.2.15.

Graphene can be described as a hexagonal lattice (Fig. 1.6) of adjacent spheres of radius r_0. Assuming that spheres of radius r_1 are adsorbed on that plane, that are attracted stronger to the substrate atoms than to each other, find the densest packing possible for the adsorbed atoms (within a single planar layer), when the ratio $x = r_1/r_0$ is between 1 and 3.

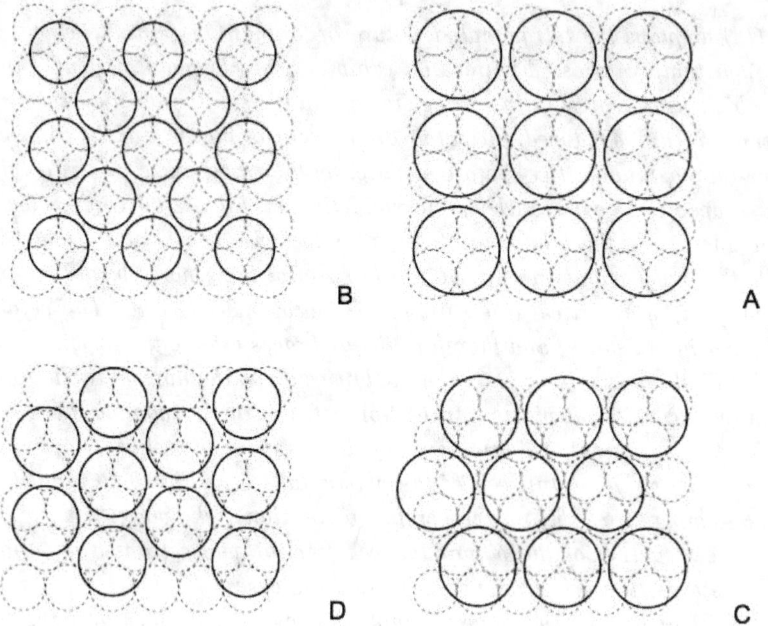

Fig. 2.58

s.2.16.

Figure 2.59 displays an antiferromagnetic arrangement of magnetic moments on a square lattice. Determine the magnetic primitive unit cell of this lattice, and its base.

Fig. 2.59

s.2.17.

Figure 2.21(b) displays the tetragonal structure of lanthanum cuprate, which exists at high enough temperatures. Figure 2.60 exhibits a single plane of CuO_2 in this lattice. The big points represent the copper ions, and the small ones the oxygen ions. When the temperature is reduced, the octahedra encompassing each of the copper ions (in the three-dimensional lattice) rotate slightly, by the small angle φ: the octahedra around the cooper ions connected by the solid line in Fig. 2.60 rotate clockwise around that line, while the others rotate counterclockwise around the dashed lines. As a result, the oxygen ions marked by the horizontal lines move below the plane, and those marked by a vertical line shift to new locations above it. The octahedra preserve their original shape, and therefore the distances among the oxygens are not changed. The final result is a squeezing of the lattice along the normal to the rotation axes, like in an accordion; it yields a transition between the tetragonal phase (which exists at high temperatures) and another, new lower-symmetry phase.

a. Determine the primitive unit cell of the planar lattice in Fig. 2.60, and its base at the higher temperatures, before the octahedra rotations take place.

b. What is the structure of the planar lattice after the phase transition, what are the lattice constants, the primitive unit cell, and the base?

c. At an even lower temperature there is another phase transition, where the copper ions arrange themselves in an antiferromagnetic order. All ions connected by solid lines in Fig. 2.60 attain a positive magnetic moment $+1$, while those connected by dashed lines attain a negative magnetic moment -1 (in the suitable units). Determine the planar magnetic unit cell.

d. Each of the copper oxide planes in Fig. 2.21(b) undergoes the above-mentioned phase transitions, with the rotation axes and the magnetic moments displaced in parallel between the planes. The copper ions in the middle plane of the tetragonal cell in Fig. 2.21(b) remain above the center of the rhombus created in the plane below due to the squeezing. Ignoring possible deformations of the lattice along the normal to the basis, what is the three-dimensional lattice structure, the unit cell, and its base at low temperatures?

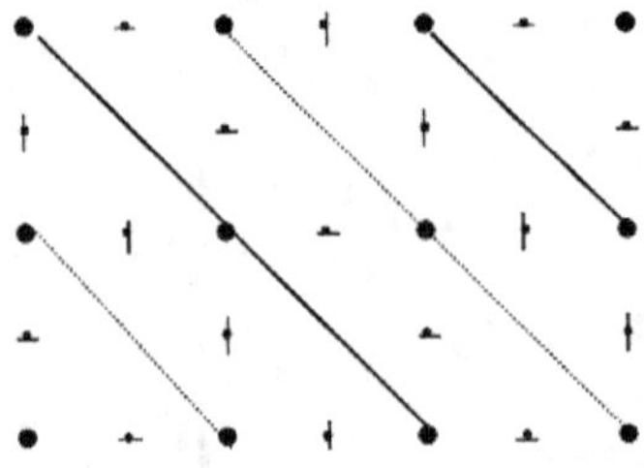

Fig. 2.60

2.13   Answers for the self-evaluation problems

Answer s.2.1.
The Wigner-Seitz cell of the triangular lattice is displayed in Fig. 2.5. The Wigner-Seitz cells of the square lattice and the rectangular lattice are those composed of their lattice vectors, with the origin displaced into the center of the original cell, see problem 2.2. The Wigner-Seitz cell of the oblique lattice is presented in Fig. 2.61, in thick lines. Each such line is normal to the dashed line connecting the site at the center with one of its neighbors, bisecting it at the middle.

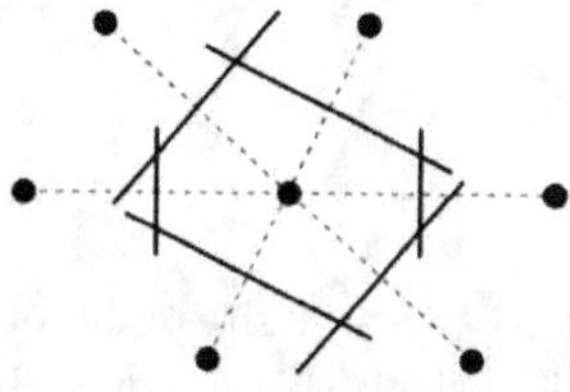

Fig. 2.61

Answer s.2.2.
The Kagomé (the Japanese term for artisan paper cutting) lattice is triangular, with a base that contains two triangles, see the rhombus marked by thick lines in the right panel of Fig. 2.62. The environments of the corners of the unit cell are identical; hence it is the primitive unit cell. Shifting slightly the rhombus downwards to the left leaves three lattice sites within the cell–these form the base of this lattice. These points are the origin, $\mathbf{a}_2/2$, and $(\mathbf{a}_1 + \mathbf{a}_2)/2$, where the lattice vectors are the two edges of the unit cell.

The Cairo lattice appears in several versions in which the angle between the longer edge and the horizontal axis is α or $(90° - \alpha)$. The one illustrated in the figure has $\alpha = 30°$. Assuming that the length of the longer edge is 1, the cartesian coordinates of all lattice points can be expressed in terms of α. As seen in the left panel of Fig. 2.62, the unit cell of the Cairo lattice is a square (marked by thick lines), independent of α: the surroundings of all corners of the square are identical. The base includes the six lattice points as marked in the figure (and can be seen by slightly displacing the unit cell towards the left). The lattice vectors are $\mathbf{a}_1 = 2\cos(\alpha)(\hat{\mathbf{x}} + \hat{\mathbf{y}})$ and $\mathbf{a}_2 = 2\cos(\alpha)(\hat{\mathbf{x}} - \hat{\mathbf{y}})$, and the base points are the origin, and the points $x_1^n \mathbf{a}_1 + x_2^n \mathbf{a}_2$, with

$$(x_1^n, x_2^n) = \frac{(1+w, 1-w)}{4} \;,\; \frac{(1+w, 1-w)}{2} \;,\; \frac{(w, 1)}{2} \;,\; \frac{(1, 2-w)}{2} \;,\; \frac{3+w, 3-w}{4} \;,$$

where $w = \tan(\alpha)$.

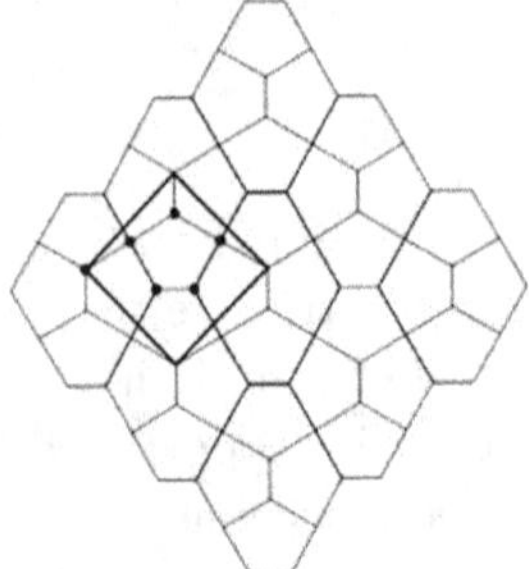 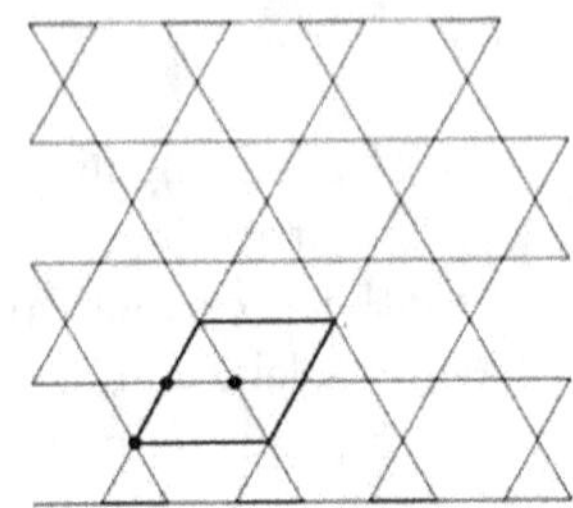

Fig. 2.62

Answer s.2.3.

a. The environments of the two types of sites (the original ones and those at the centers of the edges) are not identical; this is not a Bravais lattice. The primitive unit cell is identical to that of the original square lattice, and the base consists of three points, for instance, those encircled in Fig. 2.63(a). The lattice vectors are $\mathbf{a}_1 = (a, 0)$ and $\mathbf{a}_2 = (0, a)$, where a is the lattice constant of the square lattice. The base comprises the points at $(0, 0)$, $(a/2, 0)$, and $(0, a/2)$.

b. This is a bravais lattice. Figure 2.63(b) shows the unit cell of the tetragonal lattice which corresponds to this case. Denoting the cubic lattice constant by a, the lattice vectors are $\mathbf{a}_1 = (a/2, -a/2, 0)$, $\mathbf{a}_2 = (a/2, a/2, 0)$, and $\mathbf{a}_3 = (0, 0, a)$.

c. This is not a Bravais lattice. The surroundings of the points at the centers of the vertical faces are not the same as those of the corners of the cube. The primitive unit cell is the unit cell of the original cubic lattice, and the base includes three points, e.g., $(0, 0, 0)$, $(a/2, 0, a/2)$, and $(0, a/2, a/2)$.

d. This one is also not a Bravais lattice. The primitive cell is the original cubic one, and the base includes four points, $(0, 0, 0)$, $(a/2, 0, 0)$, $(0, a/2, 0)$, and $(0, 0, a/2)$.

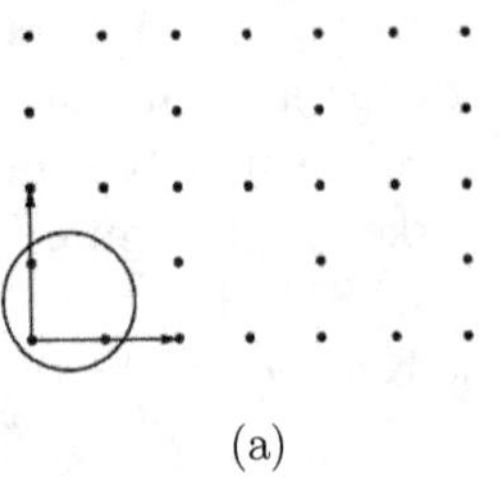

(a)

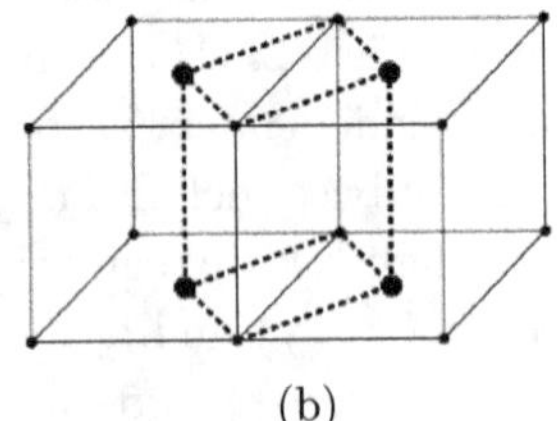

(b)

Fig. 2.63

Answer s.2.4.

a. In the first case one uses $n_i = 2n_i'$ and obtains $\mathbf{R}_i = 2(n_1', n_2', n_3')$; this is a simple cubic lattice with a lattice constant of length $2a$, where a is the lattice constant of the original lattice. In the second case one denotes $n_i = 2n_i' + 1$ and obtains $\mathbf{R}_i = (1,1,1) + 2(n_1', n_2', n_3')$. This again is a simple cubic lattice, with the same lattice constant $2a$, whose origin is at the point (1,1,1).

b. $n_1 + n_2 + n_3$ is even when all three n_i are even, or when two of them are odd and the third one is even. The first option corresponds to the first part of (a) (**SC** whose lattice constant is $2a$). The second yields three lattices characterized by the points $\mathbf{R}_i^m = \mathbf{R}^m + 2(n_1', n_2', n_3')$, where $\mathbf{R}^1 = (1,1,0)$, $\mathbf{R}^2 = (1,0,1)$, and $\mathbf{R}^3 = (0,1,1)$. These points are the centers of the faces of the unit cell. When all types of points are connected together, the result is an **FCC** lattice.

c. The requirement that $n_1 + n_2 + n_3$ is odd is equivalent to the requirement that this sum is even, provided that the vector $(1,1,1)$ is added to $\mathbf{R}^m$. This is again an **FCC** lattice, shifted relative to the one in (b).

Answer s.2.5.

The left panel in Fig. 2.64 shows the rectangle created by connecting together two opposite edges in the cube. It is also displayed by the thick lines on the cubic unit cell of the reciprocal lattice in the right panel. Its edges are AG=a, and AF=$a\sqrt{2}$. The diagonal of this rectangle is the diagonal of the cube, and hence AO=$a\sqrt{3}/2$. The line BC is normal to AO (passing through its mid point), which connects the two lattice points A and O, and therefore it lies on the face of the Wigner-Seitz cell. This face, in turn, touches the face of the cube at point C. From the similarity between the triangles ABC and AOD it follows that AB/AC=AD/AO=$\sqrt{2/3}$. Hence, AC=$3a\sqrt{2}/8$, and CD=AD-AC=$a\sqrt{2}/8$. The latter is equal to half of the edge of the square face of the cell, which lies on the face. Repeating the above considerations for all planes joining two opposite edges of the cube shows that all faces are indeed squared, and completes the proof. This also ensures that all hexagons are equilateral.

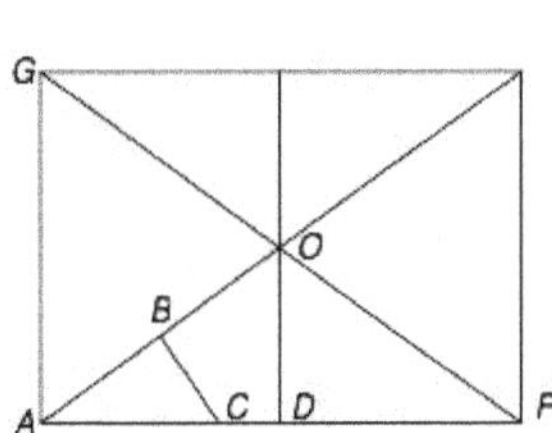
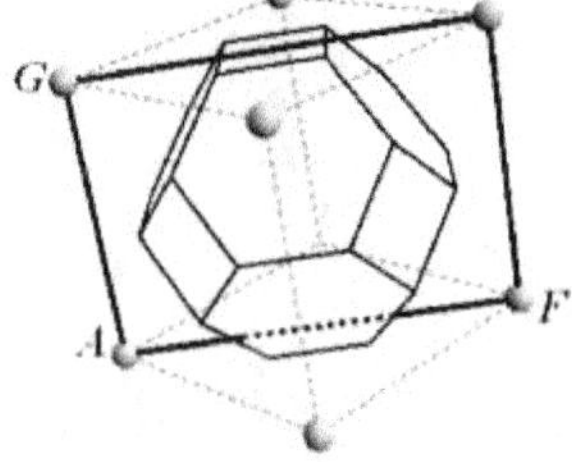

Fig. 2.64

Answer s.2.6.

a. Each cell in the **BCC** contains two atoms, and consequently the atomic density (number of atoms per unit volume) is $\rho(BCC) = 2/[a(BCC)]^3$. The solution to problem 2.15 states that the volume of the unit cell of the **HCP**, which is denser, is $V(HCP) = a^2 c\sqrt{3}/2 = \sqrt{2}[a(HCP)]^3$. This cell contains two atoms and hence $\rho(HCP) = 2/V(HCP) = \sqrt{2}/[a(HCP)]^3$. Comparing the two densities yields $a(HCP) = a(BCC) \times 2^{-1/6} \approx 3.77\text{Å}$.

b. The discussion following Eq. (2.5) implies that if the atom radius is r, then the edge of the cubic cell of the **BCC** is $a(BCC) = 4r/\sqrt{3}$. As each cell contains two atoms, the atomic density is $\rho(BCC) = 2/[a(BCC)]^3 = 3\sqrt{3}/(32r^3)$. The solution to problem 2.11 gives $a(FCC) = 2\sqrt{2}r$, and thus the atomic density for the **FCC** is $\rho(FCC) = 4/[a(FCC)]^3 = 1/(4\sqrt{2}r^3)$. Consequently, $\rho(FCC)/\rho(BCC) = 8/(3\sqrt{6}) \approx 1.08866$. The mass of the iron atom is the proton's mass times the atomic mass of iron, i.e., $M(Fe) = 55.85 \times 1.6726 \times 10^{-27}$ Kg$= 9.34 \times 10^{-26}$ Kg. It follows that $\rho(BCC) = 7900(\text{Kg/m}^3)/M(Fe) = 8.46 \times 10^{28}/\text{m}^3$; the relation $\rho(BCC) = 2/[a(BCC)]^3$ implies $a(BCC) = [2/\rho(BCC)]^{1/3} = 0.287$ nm$=2.87\text{Å}$.

Answer s.2.7.

a. Consider first the $x = 0$ limit. For the dense **FCC** lattice, $a = 2\sqrt{2}r_>$, as in the solution to problem 2.11. This relation persists as long as the small ion enters in-between the bigger ones on the cube edge, i.e., $a \geq 2(r_> + r_<)$, or $x \leq \sqrt{2} - 1$. In this range the packing ratio is

$$\rho = 4 \times (4\pi/3)(r_>^3 + r_<^3)/a^3 = \pi(x^3 + 1)/(3\sqrt{2}) \ .$$

b. When the two ions touch one another then $a = 2(r_< + r_>)$, while the diagonal on the face is such that $a\sqrt{2} \geq 4r_>$, i.e., $x \geq \sqrt{2} - 1$. In this range

$$\rho = 4 \times (4\pi/3)(r_>^3 + r_<^3)/a^3 = 2\pi(x^3 + 1)/[3(x + 1)^3] \ .$$

c. Figure 2.65 shows the result. The packing ratio varies between the one of the **FCC**, $\rho \approx 0.74$, for $x = 0$, and that of **SC**, $\rho \approx 0.52$ for $x = 1$. The packing ratio is maximal, $\rho \approx 0.793$, when all ions are contiguous, i.e., when the ratio is $x = \sqrt{2} - 1 \approx 0.414$.

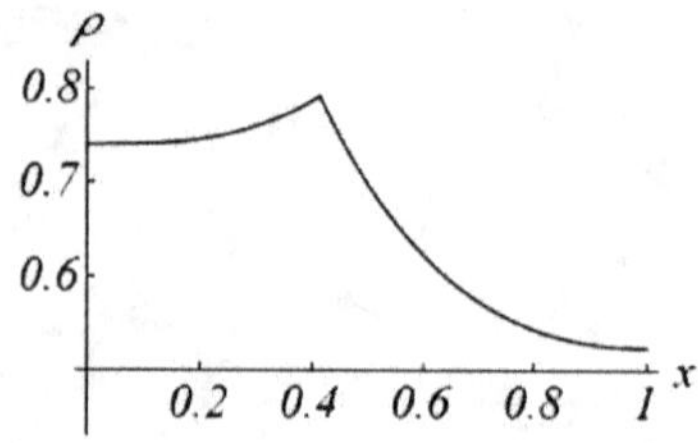

Fig. 2.65

Answer s.2.8.

a. The value $x = 0$ corresponds to the dense **FCC** lattice, for which $a = 2\sqrt{2}r_>$ as in the solution to problem 2.11. This relation persists as long as the small ion fits into the tetrahedron shown in Fig. 2.20. As the distance between the centers of two neighboring ions is a quarter of the diagonal of the cube, the bigger ions touch each other as long as $a\sqrt{3}/4 \geq (r_> + r_<)$, or $x \leq \sqrt{6}/2 - 1$. In this range the packing ratio is

$$\rho = 4 \times (4\pi/3)(r_<^3 + r_>^3)/a^3 = \pi(x^3 + 1)/(3\sqrt{2}) \ .$$

b. When the ions touch one another, then $a\sqrt{3}/4 = (r_< + r_>)$, while the diagonal on the face is such that $a\sqrt{2} \geq 4r_>$, i.e., $x \geq \sqrt{6}/2 - 1$. In this range

$$\rho = 4 \times (4\pi/3)(r_<^3 + r_>^3)/a^3 = \pi\sqrt{3}(x^3 + 1)/[4(x+1)^3] \ .$$

c. Figure 2.66(a) displays the packing ratio. Diamond corresponds to $x = 1$ and has $\rho = \pi\sqrt{3}/16 \approx 0.34$. The maximal packing ratio, $\rho \approx 0.749$, is reached when all ions are contiguous, $x = \sqrt{6}/2 - 1 \approx 0.225$. Figure 2.66(b) displays the packing ratios of the three types of crystals. When $x = 1$ the cesium chloride crystal is the densest. As x is reduced, the table-salt type is the densest in the range $\sqrt{6}/2 - 1 \leq x \leq 4^{1/3} - 1$. When $x \leq \sqrt{6}/2 - 1$, table salt and zinc-blende have the same packing ratios.

d. From the results of part (b) for the distance between neighboring ions: $a\sqrt{3}/4 = (r_< + r_>) = 0.234$ nm. Hence, $a = 0.540$ nm, and the density is $4[m(\text{Zn}) + m(\text{S})]/a^3 = 4(65.38 + 32.064) \times 1.6726 \times 10^{-27}$ Kg$/a^3 = 4140$ Kg/m^3.

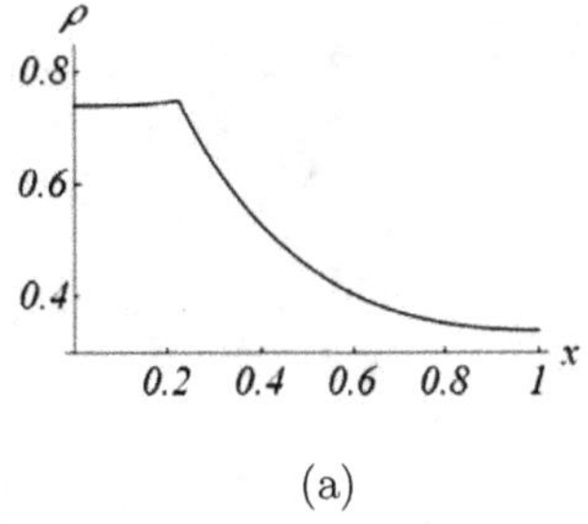

(a)

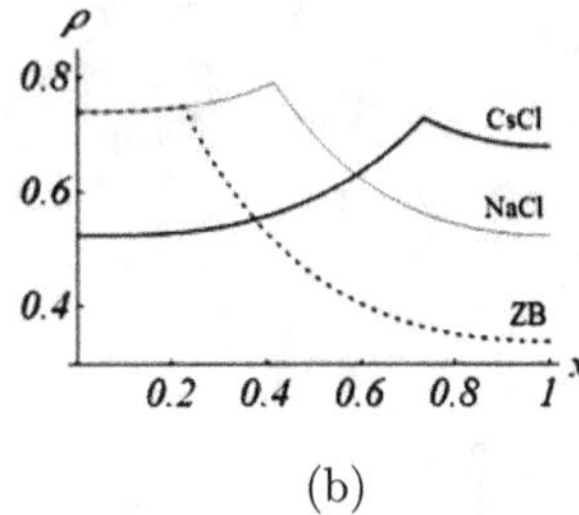

(b)

Fig. 2.66

Answer s.2.9.

The ions B and B' by themselves have the structure of table salt (a cubic lattice on which the locations of the two ions alternate). They therefore form an **FCC** lattice, with a base comprising a B ion and a B' ion. It remains to add all other ions. Denoting the lattice constant of this **FCC** by a, the ions B and B' can be located, e.g., in the points $(a/4, a/4, a/4)$ and $(a/4, a/4, 5a/4)$, that is, at the centers of two simple perovskite cells placed one on the top of the other. In these notations, the unit cell contains also two A ions, at $(0,0,0)$ and $(0,0,a/2)$, and six oxygen ions, at $(a/4,a/4,0)$, $(a/4,0,a/4)$, $(0,a/4,a/4)$, $(a/4,a/4,a/2)$, $(a/4,0,5a/4)$, and $(0,a/4,5a/4)$. All in all, the unit cell contains the chemical formula.

Answer s.2.10.
a. The unit cell is displayed in Fig. 2.9. The lattice vectors in the lower and upper planes are those shown in Fig. 2.8, i.e., $\mathbf{a}_1 = (a,0,0)$ and $\mathbf{a}_2 = (a/2,a\sqrt{3}/2,0)$. The third vector is normal to the planes, and its length is the distance in-between them c, hence $\mathbf{a}_3 = (0,0,c)$. The base of the crystal contains two carbon ions in the lower plane, $\mathbf{r}_1 = (0,0,0)$ and $\mathbf{r}_2 = (a/2,a\sqrt{3}/6,0)$, and two carbon ions in the mid plane, $\mathbf{r}_3 = (0,0,c/2)$ and $\mathbf{r}_4 = (a,2a\sqrt{3}/3,c/2)$. The two latter appear in the planar unit cell shown in the mid panel of Fig. 2.9.
b. Figure 2.67 displays the rhombic unit cells of the three types of planes. In a structure of the form ...ABCABC... the unit cells are placed one on top of the other. The lattice repeats itself in the fourth layer, whose distance from the first one is c; the distance between two successive planes is thus $c/3$. In these new notations, the lattice vectors are $\mathbf{a}_1 = (a,0,0)$, $\mathbf{a}_2 = (a/2,a\sqrt{3}/2,0)$ and $\mathbf{a}_3 = (0,0,c)$. Each unit cell contains a base of six atoms, at $\mathbf{r}_1 = (a/2,a\sqrt{3}/6,0)$, $\mathbf{r}_2 = (a,2a\sqrt{3}/3,0)$ from the first plane (A), $\mathbf{r}_3 = (0,0,c/3)$, $\mathbf{r}_4 = (a,2a\sqrt{3}/3,c/3)$ from the second plane (B), and $\mathbf{r}_5 = (0,0,2c/3)$, $\mathbf{r}_6 = (a/2,a\sqrt{3}/6,2c/3)$ from the third plane (C).

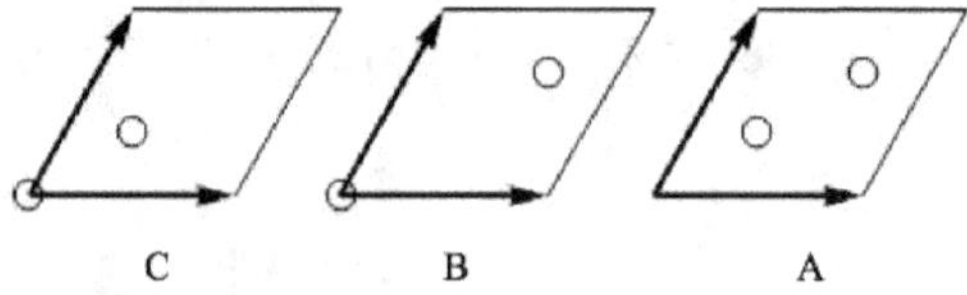

Fig. 2.67

Answer s.2.11.
In all answers below, the angles are defined according to Fig. 2.13.
a. The volume of the orthorhombic cell is abc. It contains four lattice sites, and hence the volume of the primitive unit cell (which contains a single site) is $abc/4$.
b. Choose the triclinic lattice vectors along the $\hat{\mathbf{x}}$-axis, in the XY plane, and along an arbitrary third direction. Then, $\mathbf{a}_1 = a(1,0,0)$, $\mathbf{a}_2 = b(\cos\gamma,\sin\gamma,0)$, $\mathbf{a}_3 = c(\cos\beta,(\cos\alpha - \cos\beta\cos\gamma)/\sin\gamma,z)$, where $z^2 = 1 - \cos^2\beta - (\cos\alpha - \cos\beta\cos\gamma)^2/(\sin\gamma)^2$ [this is obtained from the equation $\mathbf{a}_3 \cdot \mathbf{a}_3 = c^2$]. The y-component of $\mathbf{a}_3$ is derived from $\mathbf{a}_2 \cdot \mathbf{a}_3 = bc\cos\alpha$. Exploiting Eq. (2.2) yields

$$V = abcz\sin\gamma = abc\sqrt{1 - \cos^2\alpha - \cos^2\beta - \cos^2\gamma + 2\cos\alpha\cos\beta\cos\gamma}\ .$$

c. Using $\alpha = \gamma = 90°$ yields that the volume of the monoclinic unit cell is $abc\sin\beta$.

d. The trigonal lattice is a special case of the triclinic one, with $a = b = c$ and $\alpha = \beta = \gamma$, and hence, according to (b), $V = a^3\sqrt{1 - 3\cos^2\alpha + 2\cos^3\alpha}$.

Answer s.2.12.

The volume of the unit cell is $V = a^2 c \sin(\gamma)$. The diameter of each sphere is the distance between the centers of the two spheres in the base, $(2r)^2 = \mathbf{r}_2^2 = [2(1 - \Delta/3)^2(1 + \cos(\gamma)) + (c/a)^2]a^2/4$. As in each unit cell there are two spheres, the packing ratio is $\rho = 8\pi r^3/(3V)$. Figure 2.68 displays this packing ratio. One finds that $\Delta = 0$ corresponds to a **BCC** lattice, while $\Delta = 1$ yields the quintessential **HCP** one. Indeed, the maximal ratio, $\rho = 0.74$, is obtained for the latter.

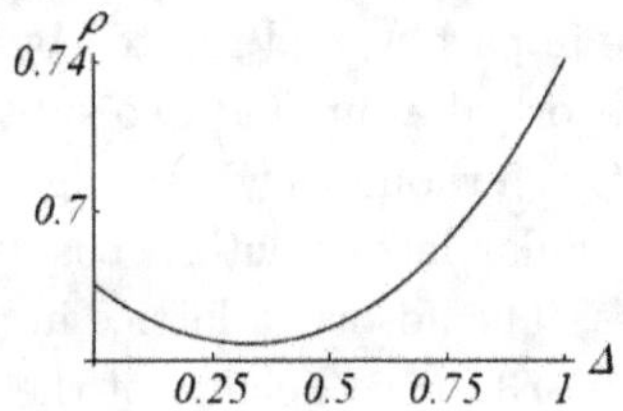

Fig. 2.68

Answer s.2.13.

Figure 2.69 displays a projection of the square lattice on a solid line of a slope $\tan(\alpha) = 1/\sqrt{2}$. One observes that there are only two projections, of lengths $S = a\sin(\alpha)$ and $L = a\cos(\alpha)$. Examining the segments on the solid line, it is seen that their ordering (beginning with the left edge) is $LSLSLLSLSLSLLSL\ldots$. This is neither a Fibonacci sequence, nor a periodic one. It describes a quasicrystals different than the Fibonacci one.

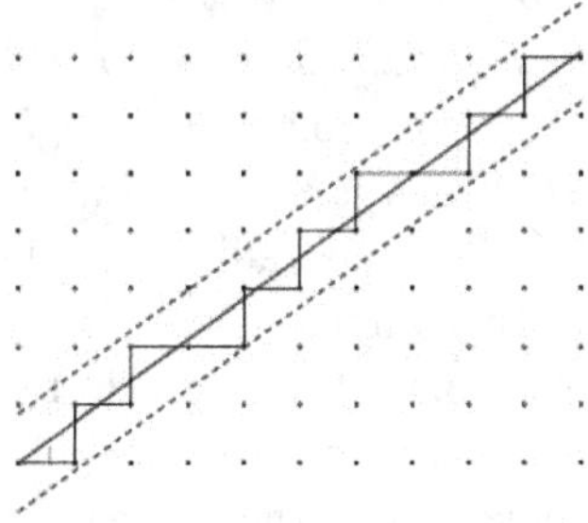

Fig. 2.69

Answer s.2.14.

a. The substrate in parts A and B of Fig. 2.58 is a square lattice with a lattice constant $a/\sqrt{2}$ (half of the diagonal on the face of the cube). The unit cell is represented by the full line in the right panel of the upper row in Fig. 2.70. The substrate in parts C and D is a triangular lattice with the same lattice constant (the diagonal of each face in the **FCC** lattice contains two diameters of the substrate atoms). The unit cell is the rhombus in the right panel in the lower row of Fig. 2.70.

b. The lattice adsorbed on the substrate in part A is a square one, with a lattice constant $a\sqrt{2}$ (the square in part A of Fig. 2.70). Each unit cell contains a base comprising one adsorbed atom and four substrate atoms. The adsorbed lattice in part B is a square one as well, but with a lattice constant a (that is, half of the diagonal of the square lattice in part A, and rotated by 45° relative to the former). The unit cell contains one adsorbed atom and two substrate ones. The unit cell of the adsorbed lattice in part C is a rhombus whose edge length is $\sqrt{2}a$, and its angle is 60°. It is therefore a triangular lattice with a base that contains one adsorbed atom and four substrate ones. The adsorbed lattice in part D is again a triangular one, whose lattice constant is $\sqrt{3}a$ (two heights of the triangles in the substrate). The unit cell is rotated by 30° relative to the formers, and is represented by the rhombus in the corresponding panel of Fig. 2.70.

c. In each of the cases, the maximal radius of the spheres that can occupy the displayed structures is half of the lattice constant.

d. When the radius of the adsorbed atoms is smaller than that of the substrate atoms, they can occupy all the dips between the substrate atoms. Then the adsorbed lattice is identical to the substrate lattice. The two lattices are shifted relative to one another by the distance between the center of an atom in the substrate and the center of the nearest dip. The unit cell is identical to that of the substrate as well, but includes also an adsorbed atom in addition to a substrate atom.

Answer s.2.15.

The spheres that represent the atoms in the substrate are marked by dashed curves in Fig. 2.71. The distance between nearest neighbors on this hexagonal lattice is $2r_0$. The centers of the hexagons, (the sites for the adsorbed atoms), create a new triangular lattice, whose lattice constant is $2\sqrt{3}r_0$. Figure 2.71(b) shows a sphere (in solid curve) whose radius is $r_1 = \sqrt{3}r_0$ around each such center. For this choice of the radius, the spheres are adjacent. Hence, when the radius of the adsorbed atom is larger than this value, i.e., $x = r/r_1 \geq \sqrt{3}$, the adsorbed atoms cannot assume the structure displayed in Fig. 2.71(b). On the the other hand, $x \leq \sqrt{3}$ allows an adsorbed atom in each hexagonal dip of the substrate; when $x \leq 1$ the adsorbed sphere hovers over the center of each hexagon, and is in the same plane as the substrate atoms. For larger values of x, the adsorbed spheres touch the six adsorbing atoms around each hexagon; the centers of the adsorbed

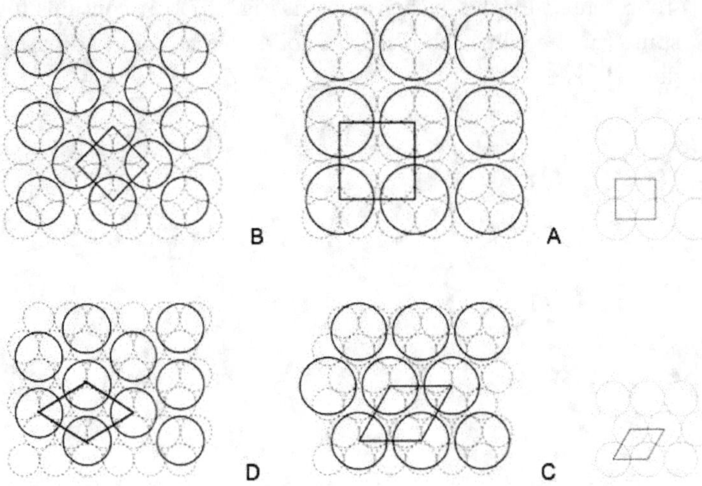

Fig. 2.70

ones move up, until $x = \sqrt{3}$. For $\sqrt{3} \leq x \leq 3$ the adsorbed spheres can occupy next nearest-neighbor hexagons on the triangular lattice of the hexagons' centers. They then create a hexagonal lattice of a lattice constant $6r_0$. Figure 2.71(a) displays the adsorbed lattice for the limiting case $x = 3$. For $x \geq 3$ this lattice cannot be realized, and the adsorbed spheres arrange themselves on a lattice of an even larger lattice constant. For the two limiting cases, $x = \sqrt{3}$ and $x = 3$, the adsorbed atoms form a triangular lattice, and in both cases the packing ratio is the same, the maximal one.

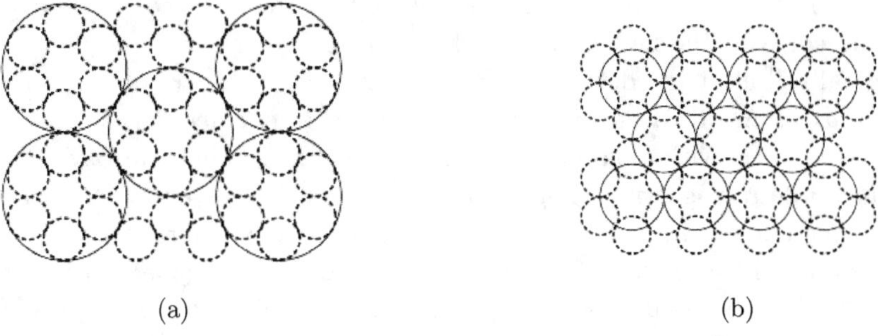

(a)   (b)

Fig. 2.71

Answer s.2.16.

The unit cell is represented by the dashed lines in Fig. 2.72. It contains four sites, two with "up" spins (in the ellipse shown by a solid curve), and two with "down" ones (in the dashed ellipse).

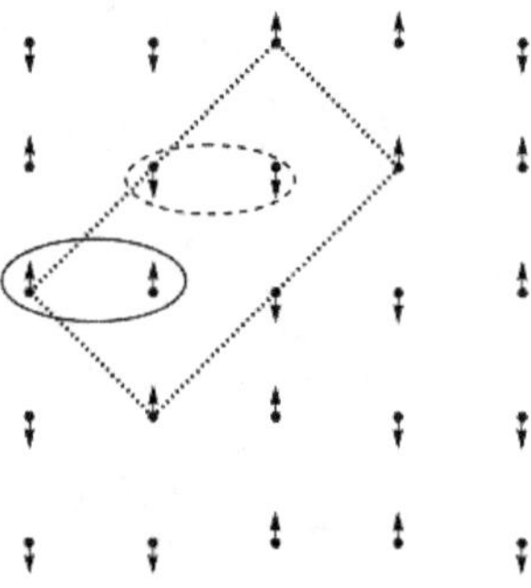

Fig. 2.72

Answer s.2.17.

a. The structure of the high-temperature phase, before the phase transition takes place, is similar to that discussed in part (a) of problem s.2.3: the lattice is a square one, with a base comprising a copper ion and two oxygen ones.

b. Because of the twisting of the plane, each edge of the original square is squeezed and its length becomes $a \cos \varphi$; the diagonals are perpendicular to one another, but the edges are not. The two diagonals of the square are not equal to one another anymore; the square turns into a rhombus. The diagonal along the direction of the rotation axis preserves its original length $\sqrt{2}a$, but the second one is shortened and becomes $2\sqrt{(a\cos\varphi)2 - (a\sqrt{2}/2)^2} = a\sqrt{2\cos(2\varphi)}$ (Pythagoras theorem for a triangle built of two half diagonals and a squeezed edge of the original square). The surroundings of the copper ions marked by stars in the upper panel of Fig. 2.73 are identical but are different from those of the ions that are not marked in this way. The edges of the new unit cell, shown in that figure, are the diagonals of the original cell after the deformation; hence their lengths are $a_1 = a\sqrt{2\cos(2\varphi)}$ and $a_2 = a\sqrt{2}$. As the diagonals are perpendicular to one another, the result is a rectangular lattice, with a base containing two copper ions and four oxygen ones.

c. In the antiferromagnetic arrangement all copper ions marked by stars have the same magnetic moment, and therefore the unit cell is identical to that in (b).

d. The lower panel of Fig. 2.73 exhibits a projection of the copper ions' lattice on the the plane of the basis at low temperatures. The unit cell of that plane is the dashed rectangle, which contains two copper ions as in part (b). Along the normal to this plane, at the height of the original unit cell, there is another plane identical to the basis; the new unit cell is thus an orthorhombic one, with a rectangular basis and of the same height as that of the original tetragonal unit cell (ignoring distortions along this direction). The copper ions in the plane located at half height (which were before at the centers of the tetragonal unit cells) are now located at the centers of the faces of the orthorhombic unit cell. The new lattice is hence a face-centered orthorhombic one, and its base comprises four copper ions (at the origin and at the centers of the three faces close to it). As each copper ion is surrounded by the chemical formula of the material, the new base includes that chemical formula four times: four copper ions, eight lanthanum ones, and sixteen oxygen ions.

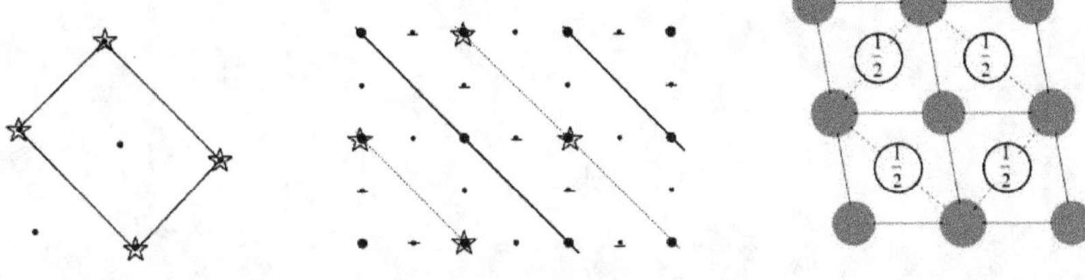

Fig. 2.73

Appendix A

The Platonic bodies

Plato's polyhedra. Symmetries have enchanted people since antiquity and symmetric bodies feature in many ancient art objects. Their namesake, Plato (*circa* 400 BC), had classified **polyhedra**: three-dimensional bodies with two-dimensional faces, one-dimensional edges, and point-like corners (vertices). A **perfect platonic body** is a concave polyhedron built of identical equilateral polygons with identical edges and identical vertices; it thus has **perfect symmetries**. Neighboring faces of a perfect polyhedron touch each other along one common edge; identical numbers of polygons share each corner. As seen in problem (A.1), there exist **solely five bodies** that comply with these rules. These are displayed in Fig. A.1; one can verify that they all obey the **Euler theorem**, $V - E + F = 2$, where V, E, and F are the numbers of corners, edges, and faces, respectively. There are many proofs of this important theorem, which pertains in fact to any polyhedron.

Problem A.1.
Each face of a perfect polyhedron is bounded by S edges, and P faces join together at each of its corners. Explain why the sum of the P angles on the faces that meet at a certain corner must be smaller than $360°$ and why does this fact entail that $P(1 - 2/S) < 2$. Exploit the result to prove that there exist only five platonic bodies.

Symmetries. Inspection of Fig. A.1 shows that a platonic body is invariant under many symmetry operations which form its symmetry group. The symmetry group of the octahedron coincides with that of the cube: the octahedron in Fig. A.1 is not altered by any of the symmetry operations of the cube. In particular, the octahedron consists of three squares and eight triangles which are invariant under rotations of 90° and 120°, respectively. Each of the platonic polyhedra possesses a dual polyhedron, obtained upon joining the centers of neighboring faces. The two dual bodies share the same symmetries. The octahedron is dual to the cube, the dodecahedron is dual to the icosahedron, and the tetrahedron is dual to itself.

The symmetry group of the platonic bodies can be derived by exploiting the fact that each of the symmetry operations transforms the corners of the polyhedron

Fig. A.1: The platonic bodies, from left to right: the tetrahedron (4 equilateral triangles, 6 edges, and 4 corners), the octahedron (8 equilateral triangles, 12 edges, and 6 corners), the cube (6 squares, 12 edges, and 8 corners), the icosahedron (20 equilateral triangles, 30 edges, and 12 corners), and the dodecahedron (12 equilateral pentagons, 30 edges, and 20 corners).

to themselves. For example, the tetrahedron has four vortices, and therefore there are $4! = 24$ permutations of them, which produce 24 elements in the tetrahedron's symmetry group. This argument is valid only for the tetrahedron, where each corner is a neighbor of all others. For the bigger bodies that are not dual to themselves, not all permutations are allowed: in some of them neighboring corners do not remain as such after the permutation. In general, each of the symmetry operations consists of a combination of rotations and reflections of the polyhedron. For instance, if each of the faces of a polyhedron is bounded by S edges, and at each of its corners meet P faces, then the line joining the center of the polyhedron with the center of each face is a rotation axis of order S, and the line joining that center with any corner is likewise a rotation axis of order P.

Problem A.2.
Describe the symmetry operations of the tetrahedron as rotations and reflections.

The symmetry group of the icosahedron. A glance at Fig. A.1 shows that each triangular face is opposite to another triangular one, rotated by 180°. Similarly, opposite to each pentagon face of the dodecahedron there is a parallel one whose pentagon is rotated by 180°. In addition, opposite to each of the edges of these polyhedra there is another, parallel, edge. The argument given above implies that the symmetry group of the icosahedron contains four rotations of order five (rotations by $2\pi\ell/5$, with $\ell = 1$, 2, 3, and 4) around the axes connecting six pairs of opposite corners, and two rotations of order three around each of the axes that connect ten pairs of oppositely-located faces. There are also the second-order rotations around each of the axes connecting the 15 midpoints of the oppositely-located edges. Together with the unity, all these operations amount to a symmetry group comprising 60 elements. The **icosahedron and the dodecahedron** do not appear in periodic lattices, but do play a decisive role in quasicrystals (see Sec. 2.8). The reason is that the point symmetry-group of the icosahedron contains rotations of order five, as indeed is observed in experiments carried out on quasicrystals.

Figure 2.19 and problem 3.11 mention the **rhombic dodecahedron**. Though this body consists of identical square faces, its corners do not share the same environment: there are corners with three faces, and others that combine four faces. Hence, the rhombic dodecahedron is not a prefect platonic body. Nonetheless, one may convince oneself that the symmetry of the rhombic dodecahedron is the same as that of the cubic lattice.

A.1   Answers

Answer A.1.

Joining one of the vertices of an equilateral polygon with the $S-3$ vertices that are not its neighbors partitions the polygon into $S-2$ triangles. Therefore, the sum over the angles of the polygon is $(S-2)180°$. As the polygon possesses S corners, the angle at each corner is $(S-2)180°/S$. The sum of the angles around each corner is thus $P(S-2)180°/S$ and since the polygon is concave, this sum must be smaller than $360°$. It follows that $P(1-2/S) < 2$. When $P(S-2)180°/S = 360°$ the faces lie on a plane, and when $P(S-2)180°/S > 360°$, the body becomes convex. The number of the edges S must exceed 3 (lest it will not be a polygon) and P must exceed 3 in order to build a corner. For $S = 3$ the inequality $P(1-2/S) < 2$ is obeyed solely for $P = 3$, $P = 4$, or $P = 5$, values that correspond to the tetrahedron, the cube, and the icosahedron. For $S = 4$ and $S = 5$ the inequality is obeyed solely for $P = 3$, and these values agree with the cube and the dodecahedron. There are no other values of S and P that obey $P(1-2/S) < 2$; there are thus only five platonic bodies.

Answer A.2.

The line joining each vertex of the tetrahedron with the center of the opposite face is the height of the tetrahedron. Around each of these heights, there are two rotations of order 3, for instance by the angles $\pm 2\pi/3$, altogether eight rotations. On each of the four faces there are three heights, altogether 12 heights. Joining each of the heights with the vertex opposite to the face results in a planar mirror; a reflection through which leaves the tetrahedron unchanged. The eight rotations correspond to permutations of the three vertices, and the 12 reflections correspond to permutations of pairs of vertices. In addition there are three symmetry operations transforming all four vertices simultaneously, e.g., ABCD→DABC. Each of these operations can be accomplished in two steps; e.g., ABCD→ADBC→DABC. The first step is a rotation of the triangle BCD, and the second is a reflection that interchanges between AD and DA. Together with the identity element there are 24 symmetry operations.

Chapter 3

Radiation scattering off crystals

This chapter focuses on experimental methods to identify the crystalline structures of various materials. The paramount one to be described below exploits scattering of certain types of radiation off crystals, notably X-rays, but also scattering of neutron or electron beams. The radiation scattered by the periodic lattice creates a diffraction pattern on a screen. The laws that determine the scattering angles which give rise to constructive interference, named after Bragg and von Laue, are examined, as well as ubiquitous tools and techniques, as monochromatic scattering off powders or from a single crystal upon rotating it, or multi-chromatic radiation. The intensity of the scattered radiation comprises distinct peaks ("Bragg peaks") that appear at particular scattering angles, for which the difference between the impinging wave vector and the scattered one equals one of the vectors of the reciprocal lattice. Accordingly, the properties of reciprocal lattices are studied in detail. These particular scattering angles correspond to diffraction from families of planes in the lattice. The intensity of the scattered radiation is related to the structure factor, i.e., the Fourier transform of the scatterers' density in the material. This is the density of the electrons for X-ray scattering, of the nuclei or the magnetic moments for neutron scattering. Additional topics surveyed in this chapter include scattering off quasicrystals, scattering of electrons from planar samples, and scattering of neutrons off magnetic systems. The appendix contains a terse summary of Fourier series on periodic lattices.

3.1 Introduction: identifying crystalline structures

Scanning tunneling microscope (STM). There are several **experimental methods to identify crystalline structures**. A relatively modern one, in particular suitable for investigations of **two-dimensional surfaces**, is based on the **scanning tunneling microscope, STM**. In this method, a very thin metallic tip that hovers above a solid surface, is moved around. Upon applying a voltage drop between the tip and the surface [generated by contacting the surface and the tip to the terminals of a voltage source, see Fig. 3.1(a)] the current that flows between

113

them increases as the tip approaches closer the surface: the space in-between the tip and the surface forms a quantum potential barrier, which the charge carriers (i.e., the electrons) overcome by **quantum tunneling**. This potential barrier is reduced as the tip approaches closer to the surface, leading to an increase in the current. Measuring the current as a function of the location of the tip enables the mapping of the surface. Alternatively, the current is kept constant but the tip is connected to a spring that allows its vertical motion to follow the varying height of the surface. In that case one measures the voltage that monitors the height. Figure 3.1(b) displays such a mapping of graphene (the graphene structure is described in Sec. 2.3, see Figs. 2.8 and 2.10, and also Fig. 1.6).

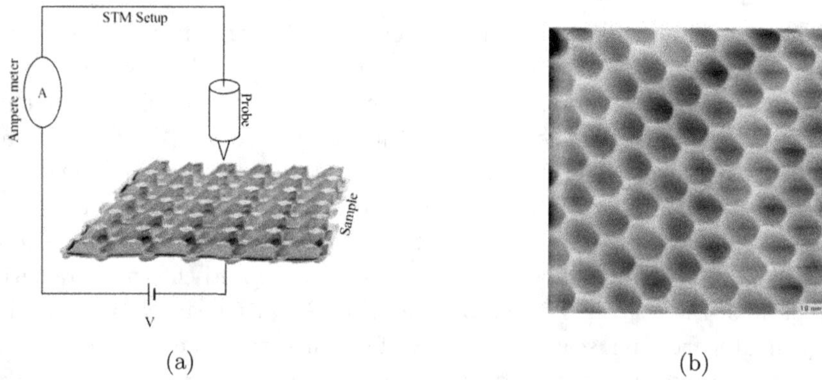

Fig. 3.1: (a) Illustration of the scanning tunneling microscope, STM. (b) Mapping of graphene by STM (the brighter color corresponds to a larger current, and hence to higher regions of the surface). The dark regions represent valleys at the centers of the hexagons of the graphene lattice. Courtesy of E. Andrei, Rutgers Univ.

The atomic force microscope (AFM). A complementary version of the STM is the **atomic force microscope**, AFM, which is based on a cantilever with a very sharp needle-like edge (thickness of the order of nanometers), see Fig. 3.2. The cantilever hovers over the surface; its tip is sensitive to the force exerted on it by the surface, stemming, e.g., from the mechanical contact, or from the electric or magnetic potentials of the surface. This force increases as the cantilever comes closer to the surface, causing the latter to move up and down. These vibrations are detected by a laser beam reflected from the upper side of the cantilever; the reflection angle is in particular sensitive to the distance of the cantilever from the surface.

Scattering of radiation off a crystal. The methods described above are useful for the study of **thin films** or **surfaces** of three-dimensional crystals; they do not convey much information about the internal structure of the bulk itself. Such knowledge is gathered by **scattering radiation off crystals**. Each of the atoms in the crystal scatters the impinging waves; the outgoing radiation comprises in-

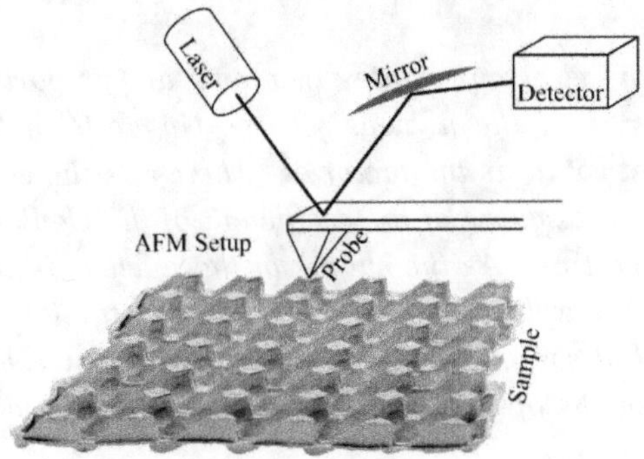

Fig. 3.2: Atomic force microscope.

terference of all scattered waves, from all atoms. Conspicuous diffraction patterns are obtained when the wave length of the radiation is of the order of the distances between the scattering objects, i.e., of the order of several angstroms for solid crystals. Electromagnetic waves of frequency ν and wave length λ consist of photons of energy $\varepsilon = h\nu = hc/\lambda$, where h is the Planck constant and c is the speed of light. Hence,

$$\lambda(\mathring{A}) = 12.4/\varepsilon(\text{KeV}) \,, \tag{3.1}$$

where the wave length is measured in angstroms ($1\mathring{A} = 10^{-8}$ cm), and the energy of the photon is measured in kilo electron-volt, KeV (1 eV$=1.602\times10^{-19}$ Joule). Energies of the order of several KeV's yield wave lengths of the order of the distances in the lattice; such energies correspond to **X-rays**. These rays are created, for instance, when an electron in a lower atomic level is "kicked" out, and another electron from an upper level moves "down" to occupy it, emitting in the process a photon of energy equal to the difference between the two levels. The emitted radiation contains waves of well-defined, discrete wavelengths. The quantum states of electrons in the atom are characterized by the quantum numbers n, ℓ, and j, which represent the energy, the orbital angular momentum, and the total angular momentum (orbital and spin), respectively. The ground state of a single electron is given by $n = 1$, $\ell = 0$, and $j = 1/2$. This state is termed K, or $1s_{1/2}$ (the notation s represents $\ell = 0$). When this state is empty, it can be filled by an electron coming from a higher energy state, upon emitting a photon of the X-ray radiation. This transition can take place from excited levels with $\ell = 1$, which are denoted by np_j (the notation p represents $\ell = 1$). In particular, the transitions from the levels $2p_{3/2}$, $2p_{1/2}$, and $3p_{3/2}$ are called $K_{\alpha 1}$, $K_{\alpha 2}$, and $K_{\beta 1}$, respectively.

Problem 3.1.

a. *The ground-state energy of an electron in copper is -8979 eV, and the energies of the excited states $2p_{3/2}$, $2p_{1/2}$, and $3p_{3/2}$ are -952 eV, -933 eV, and -76 eV,* *respectively. Determine the wave lengths corresponding to the three transitions $K_{\alpha 1}$, $K_{\alpha 2}$, and $K_{\beta 1}$.*

b. *The energies of the electronic states in atoms are proportional to z^2, where z is the atomic number, i.e., the nucleus' charge (in units of the electron charge). [The Rydberg constant contains the factor e^4, where e is the electron's charge. This factor originates from the square of the coefficient of the Coulomb potential between the nucleus and the electron. As the charge of the nucleus is ze, it follows that the Coulomb potential is multiplied by z and consequently the levels' energies are multiplied by z^2.] Potential sources for X-rays are chromium, cobalt, or molybdenum. What are the wave lengths of the three transitions for these sources?*

Synchrotron radiation. "Standard" sources of X-rays are based on the acceleration of electrons in vacuum tubes. Electrons emitted from a hot cathode are accelerated in a high voltage drop, to impinge on the anode, where they release electrons from low-energy levels. In this way the X-ray radiation described in problem 3.1 is created. The intensity of this radiation is not particularly strong. Higher-intensity radiation, involving photons of energies of the order of several kilo electron-volts, can be obtained e.g., from **electrons colliding with solids**; the deceleration of these electrons results in the emission of photons. Another option is **synchrotron radiation**: the radiation emitted from charged particles moving on a circle. Such particles emit photons because of their acceleration. In both cases the emitted spectra comprise **a continuum of energies**, but discrete wave lengths can be achieved upon using suitable filters. Synchrotron radiation can be of intensity much higher than the one emitted from vacuum tubes.

Scattering of electrons or neutrons. Diffraction patterns can also be achieved by scattering **massive particles** off the crystal, e.g., **neutrons or electrons** (found in nuclear reactors or in accelerators), exploiting the duality of massive particles and waves in the quantum realm. The de Broglie wave vector and wave length of a particle of mass m and momentum $\mathbf{p}$ are $\mathbf{k} = \mathbf{p}/\hbar$ and $\lambda = h/|\mathbf{p}|$, respectively; the energy of each particle is $\varepsilon = h^2/(2m\lambda^2)$. The wave length of electrons is thus given by

$$\lambda(\text{Å}) = 12/[\varepsilon \ (\text{eV})]^{1/2} \, , \qquad (3.2)$$

and that of the neutrons is

$$\lambda(\text{Å}) = 0.28/[\varepsilon \ (\text{eV})]^{1/2} \qquad (3.3)$$

(check!). The corresponding energies are of the order of several electron-volts for electrons, and parts of electron-volts for neutrons. Note though that the atomic scattering cross-section for electrons is quite large, and therefore the electrons are mostly scattered by the atoms close to the surface of the specimen, and do not penetrate deep into the crystal. For this reason electrons' scattering is used mainly in explorations of surfaces (see Sec. 3.12). Neutrons, on the other hand, have two unique properties that distinguish them from photons: (i) they are scattered mostly by the nuclei and not by the electrons in the material, and thus are suitable for extracting information on the nuclei's locations; (ii) they possess a magnetic moment and are electrically neutral, and hence are useful for investigations aimed to study the orientations of the magnetic moments in the material (see Sec. 3.13).

3.2 Bragg's law

Elastic scattering from a crystal. In Chapter 4 it is explained that the sum of the forces between the atoms (or ions, or molecules) in the crystal vanishes at equilibrium, when the atoms are located on the sites of a perfectly periodic lattice. In reality the atoms oscillate around these lattice sites (see Chapter 5), in the same way as a quantum harmonic oscillator vibrates around the minimum of its potential energy, even in the ground state. These oscillations are ignored in most of this Chapter: it is assumed that the atoms stay in their average locations, i.e., at the lattice sites. Figure 3.3 displays the two main configurations exploited in experiments that monitor the scattering of radiation off crystals. As the atoms rest in the lattice sites, there is no energy exchange between them and the radiation, and hence the **scattering** is **elastic**, much like a classical ball that hits a wall is reflected from it with no energy exchange. The kinetic energy of the scattered particles (photons, neutrons, or electrons) is conserved, and the wave length of the scattered wave is the same as that of the incoming one. Below are presented several methods to derive the condition for constructive interference of elastically-scattered radiation, all equivalent to each other.

Scattering off a single crystal. A scattering experiment can be carried out on powder, which contains many tiny crystals (grains) of the material, or on a single crystal. Figure 3.3(a) exhibits scattering off a **single crystal**. As found below, when the scattered radiation contains constructive interference, it can be associated with scattering from a **family of parallel planes** comprising atoms of the crystal. For simplicity, assume that the surface of the sample is planar, and contains one of those planes. The crystal is oriented such that the angle between the impinging beam and that plane, denoted by θ, and the angle between the scattered beam and the plane, are equal [see Fig. 3.3(a)]. The angle between the normal to the plane and the impinging wave, $90° - \theta$, is called in optics the **impinging angle**. The angle between the impinging beam (incoming from the left) and the outgoing one [hitting the monitor in Fig. 3.3(a) or the counter in Fig. 3.3(b)] is 2θ, and is called the **scattering angle**. In the geometry adopted here, the surface plane of the sample is perpendicular to bisector of the angle in-between the incoming and the outgoing waves. Usually, the sample is much smaller than the distances between it and the radiation source and the monitor, and the wave packet is very narrow (though wide compared to the distances between atoms within the crystal). Then all the rays parallel to each other that hit the counter are focused there into a single point, and hence the amplitude of the scattered wave, called the **scattering**

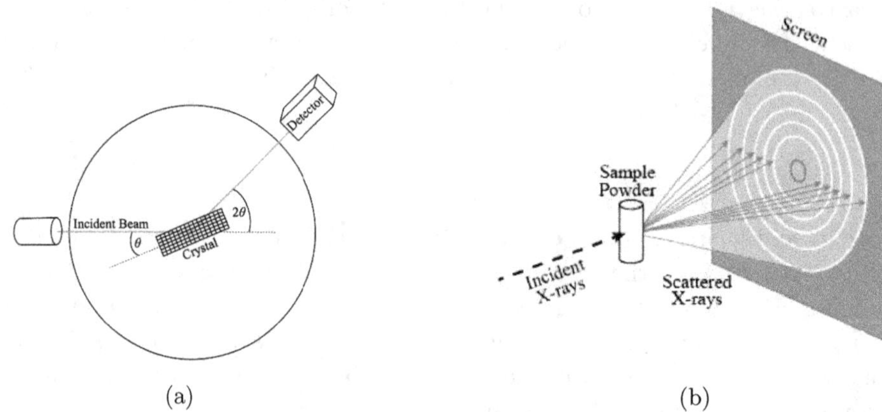

Fig. 3.3: Illustration of radiation scattering. (a) The beam incident from left is scattered from a single crystal, and then impinges on the radiation counter; the latter can be moved on the surface of a sphere enclosing the crystal (represented by the circle). The parallel lines inside the crystal represent the planes scattering the radiation. (b) The beam scattered from a powder of the investigated material impinges on a screen, where each circle corresponds to a particular scattering angle; the various points on each circle result from different grains in the powder.

amplitude, is the sum over all their amplitudes. The **intensity** of the scattered radiation is proportional to the absolute value squared of this amplitude. The scattered waves interfere; a constructive interference is realized only for discrete values of the wave length (when the scattering angle is fixed) or for discrete values of the scattering angle (when the wave length is fixed and the data are taken at various orientations). Each such constructive interference is related to scattering off a certain family of planes in the crystal, and therefore the detection of those discrete values allows for the identification of the structure of the scattering crystal (see below). One may use **monochromatic radiation**, which has a single wave length, moving the counter over the sphere shown in Fig. 3.3(a), and so measure the intensity of the scattered radiation for various scattering angles. Alternatively, one may locate the counter at a certain point and use **multi-chromatic radiation** which comprises a continuum of wave lengths. Constructive interference is again achieved only for certain wave lengths. The counter may be replaced by a screen on which there appear discrete points whenever the interference is constructive. These points create the **diffraction pattern**.

Scattering off powders. Figure 3.3(b) displays scattering off a powder: the sample is chopped into tiny grains (but still much larger than the distances between atoms in the crystal). The lattice planes in each grain are oriented differently relative to the impinging beam, resulting in circular rings of radiation on the screen, at specific scattering angles obeying Bragg's law. Figure 3.4 exhibits such rings with the corresponding intensities of the scattered radiation.

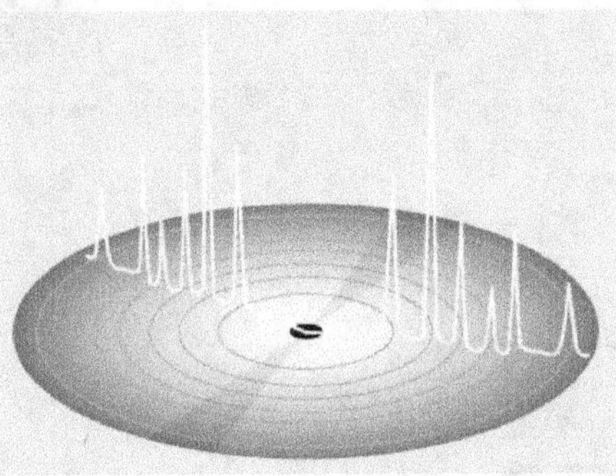

Fig. 3.4: Diffraction pattern obtained from a powder. The circular plane is the screen on which the scattered radiation is impinging; the bright peaks give the average intensity of each ring. Each peak corresponds to a specific scattering angle for which there is a constructive interference.

Bragg's theory of scattering off planes. The diffraction patterns generated by scattering off crystals were discovered by **von Laue**. More than a hundred years after this discovery the science of **crystallography**, which explores the structures of various materials by scattering off radiation, is still quite active. Figure 3.5 summarizes the hallmarks of the progress in this field up to 2014, a year which was declared the "international year of crystallography".

The analysis of the diffraction patterns generated by scattering off crystals was accomplished by **William Henry Bragg and by his son, William Lawrence Bragg**. They assumed that the material consists of parallel planes, as illustrated in Fig. 3.3(a), and that the diffraction pattern is created by scattering off these planes. This is not quite so, as each crystal is built of discrete elements (atoms, molecules, or more complicate bases). The analysis carried out by the two Braggs pertains to the planes on which those elements are situated. Nonetheless, calculations of the scattering amplitude off discrete points in the lattice yield the same results (see below).

Planes in the lattice. As mentioned in Sec. 2.9, planar cross-sections of a three-dimensional lattice can assume various structures. In fact, a three-dimensional lattice can be described as an ensemble of parallel planes, where each plane is a two-dimensional lattice. For example, the graphite lattice in Fig. 2.9 is a collection of planar hexagonal layers, (where each plane is the two-dimensional graphene lattice). The cubic lattice in Fig. 2.15 is built of square-lattice planes, parallel to each other, and the **FCC** and **HCP** lattices in Fig. 2.24 comprise planar layers, each being a triangular lattice. The **FCC** lattice can also be described as built of parallel square-

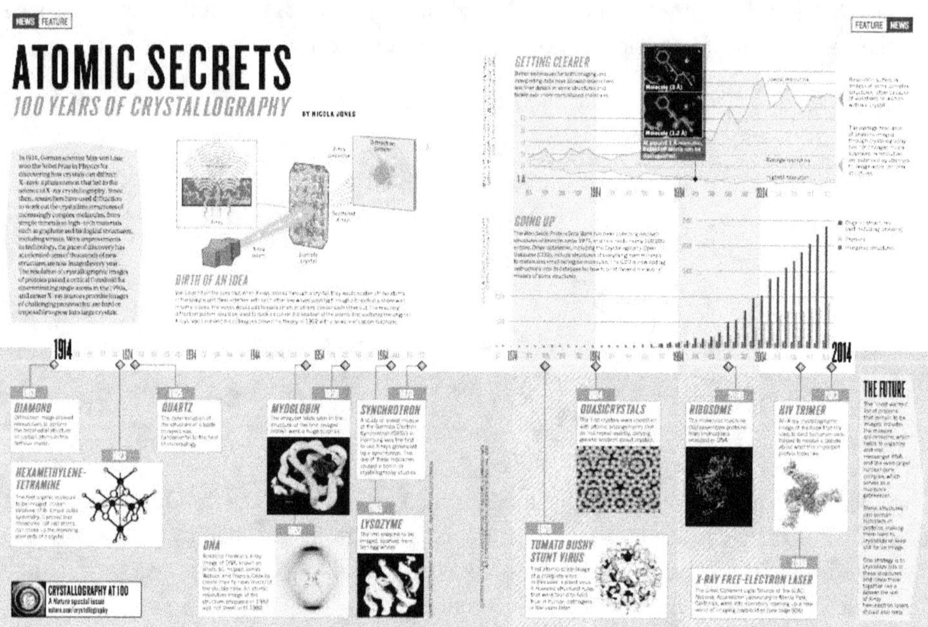

Fig. 3.5: Hundred years of crystallography. [N. Jones, Nature **505**, 602 (2014).]

lattice planes [e.g., the plane of the cube's face in the central panel of Fig. 2.17]. A point in the three-dimensional lattice can hence be denoted by the serial number of the plane on which it is located, and by its location within that plane. Figure 3.6 shows that a lattice can be described by **different families of planes**. Figure 3.6(a) displays five such families for the rectangular lattice (where each "plane" is in fact a line); Fig. 3.6(b) portrays three families of planes for the cubic lattices. It is seen that each site in the lattice belongs to a certain plane, in each of the families. The families differ from each other by the distances among the planes, and by the density of points on each plane.

A lattice may be partitioned into families of planes as follows. One chooses at random three lattice points (that are not lying on a straight line). These three points define a plane in space. As each pair of them is connected by a vector comprising lattice vectors, two (out of three) of the lattice vectors allow one to reach an infinite number of other lattice points, all located on the same plane. Choosing a point on the plane and identifying the lattice point closest to it which is not on the same plane, and then using the very same planar lattice vectors to reach other lattice points results in the creation of a new plane, parallel to its predecessor. Repeating this procedure eventually assigns all lattice points to such planes.

As an example, consider the planes of the **FCC** lattice in Fig. 3.6 marked by (200). Assign to this lattice the cubic lattice vectors, $\mathbf{a}_1 = a\hat{\mathbf{x}}$, $\mathbf{a}_2 = a\hat{\mathbf{y}}$, and $\mathbf{a}_3 = a\hat{\mathbf{z}}$, and choose the three points $(0,0,0)$, $(0,a,0)$, and $(0,0,a)$. These points

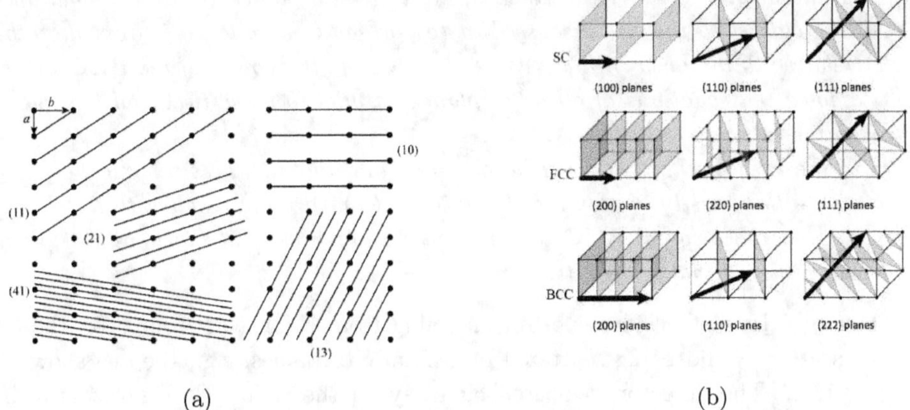

(a) (b)

Fig. 3.6: Partition of lattices into families of planes. (a) The rectangular lattice; (b) the three cubic lattices, **SC**, **FCC**, and **BCC** in descending order. The thick arrows show vectors normal to each family of planes (see text).

are located on the YZ plane (the left plane in the figure). The lattice points on this plane form a square lattice, rotated by 45° relative to the cartesian coordinates, with the lattice vectors $\mathbf{a}_2' = (\mathbf{a}_2 + \mathbf{a}_3)/2$ and $\mathbf{a}_3' = (\mathbf{a}_2 - \mathbf{a}_3)/2$. Any other point in this plane can be reached by a linear combination of these lattice vectors, with integer coefficients. Noting that the distance of the point at $\mathbf{R}_2 = (a/2, a/2, 0)$ (located at the center of the cube near the origin) from the YZ plane is $a/2$, which is the smallest possible distance of a lattice point from this plane, identifies the nearest plane. All points on the plane parallel to first one, which are on the same plane as $\mathbf{R}_2$, are located at $\mathbf{R}_2 + n_2 \mathbf{a}_2' + n_3 \mathbf{a}_3'$, with integer coefficients. Similarly, the points on the next plane are located at $\mathbf{R}_3 + n_2 \mathbf{a}_2' + n_3 \mathbf{a}_3'$, where $\mathbf{R}_3 = (a, 0, 0)$. The distance of the point at $\mathbf{R}_3$ to the second plane is $a/2$, and so forth. It is easy to be convinced that the ensemble of points on all the planes is identical to that of the points on the original lattice. A different choice of the initial three points constructs another family of planes. The various families, as well as the notations in Fig. 3.6, are sorted out below.

Problem 3.2.

a. *Find the equation of the plane that contains the three points at the ends of the edges of the lattice vectors in Eq. (2.5) and in the left panel Fig. 2.16,*

$$\mathbf{R}_1 = \frac{a}{2}(-\hat{\mathbf{x}} + \hat{\mathbf{y}} + \hat{\mathbf{z}}) , \quad \mathbf{R}_2 = \frac{a}{2}(\hat{\mathbf{x}} - \hat{\mathbf{y}} + \hat{\mathbf{z}}) , \quad \mathbf{R}_3 = \frac{a}{2}(\hat{\mathbf{x}} + \hat{\mathbf{y}} - \hat{\mathbf{z}}) .$$

*Determine the lattice vectors that describe the points of the **BCC** lattice located on this plane. Find the structure of the planar lattice formed by these points, and the area of the unit cell. At which points does this plane cut the axes of the cubic lattice? Does this plane appear in Fig. 3.6?*

b. *Find the equation of the plane parallel to the one discussed in (a) that passes through the origin. Find the distance between the two planes. Find the equation of the plane closest to the one discussed in (a), which is parallel to it from the other side (relative to the origin), that passes also through the points of the* **BCC** *lattice. Write down the equations of all other planes parallel to these two, and that passes through the lattice points.*

c. *Use the area of the unit cell of each plane and the distances among the planes to compute the density of points of the lattice. Compare the result with the density of points on the original* **BCC** *lattice, and demonstrate that the family of planes indeed contains all lattice points.*

Bragg's law. Consider a beam incoming from a far-away source on the left of the scatterer, and being scattered by a family of planes, e.g., the one shown in Fig. 3.7(a). The detector is located far away on the right. Each horizontal line in this figure represents a plane normal to the page, and connects lattice sites on this plane. The grey areas (in the page plane) represent the impinging beam and the scattered one, which are both oriented at an angle θ relative to each of the planes. It is the same angle for the scattered and the impinging beam since – like for particles colliding with a planar surface–the momentum of the radiation particles along the tangent to the surface is not changed; momentum is exchanged only along the normal to the surface. The two rays marked by thick arrows are scattered by points located on neighboring planes. Denoting the distance between the planes by d, the difference between the lengths of these two rays is $2d\sin\theta$. [This is the sum of the two bold segments in the figure, each being the side opposite to the angle θ in a right triangle (the dashed lines), whose hypotenuse is d.] Wave mechanics implies that for the two rays to interfere constructively the difference of their lengths should be an integer number of wave lengths, λ, i.e.,

$$2d\sin\theta = n\lambda \,, \tag{3.4}$$

where n is an integer. Equation (3.4) represents **Bragg's law**. Monochromatic rays (i.e., of a single wave length λ) that are scattered from a certain family of planes (i.e., with a given value of d) interfere constructively only for discrete angles, for which $\sin\theta = n\lambda/(2d)$. One has to rotate the crystal or the counter in Fig. 3.3(a) until constructive interference is achieved. Alternatively, for a given family of planes and a specific scattering angle, the interference is constructive only for certain wave lengths $\lambda = 2d\sin\theta/n$. The Bragg condition, although derived for continuous planes, is valid also for scattering off discrete points (see below).

Problem 3.3.

Show that condition (3.4) is obtained also when the two rays are scattered from **any** *pair of points in two neighboring planes, not only from the particular one displayed in Fig. 3.7(a).*

von Laue law. An alternative derivation of the condition for constructive interference is carried out by introducing a vector **d** of length d that is normal to

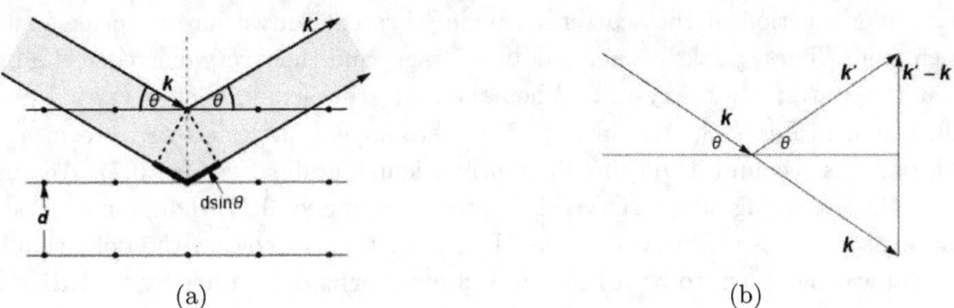

(a)   (b)

Fig. 3.7: (a) A beam scattered from planes in the lattice. (b) The wave vectors of the impinging and the scattered waves, and their vectorial difference.

the planes in Fig. 3.7(a), and assigning to the incoming and to the scattered beams the wave vectors $\mathbf{k}$ and $\mathbf{k'}$, respectively. The latter point along the propagation directions of the waves, and their lengths ("wave vectors") are $|\mathbf{k}| = |\mathbf{k}| = k = 2\pi/\lambda$. The wave lengths of the rays are identical since the scattering is elastic, i.e., it conserves energy. The wave vectors are related to the momentum of the photons (or in general, of the particles scattered off the crystal), e.g., $\mathbf{p} = \hbar\mathbf{k}$. The momentum of the scattered particles is not conserved; the difference between the momentum of the impinging wave and that of the scattered one is absorbed by the crystal. Examining Fig. 3.7(a), one notes that $\mathbf{k'} \cdot \mathbf{d} = -\mathbf{k} \cdot \mathbf{d} = kd\sin\theta$. The product of the wave number k and the length of the traversed path is the **optical path**; these paths are marked by the thick segments in Fig. 3.7(a). The difference between the optical paths of the two rays is the phase difference between the two waves. The two waves interfere constructively when the phase difference is an integer product of 2π. It follows that the condition for constructive interference, i.e., the Bragg law, can take the form

$$(\mathbf{k'} - \mathbf{k}) \cdot \mathbf{d} = 2\pi n \ . \tag{3.5}$$

This form is known as the **von Laue law**.

As seen in Fig. 3.7(b), the vector $\mathbf{k'} - \mathbf{k}$ is normal to the planes, i.e., it is parallel to the vector $\mathbf{d}$. Hence, the scalar product on the left hand-side of Eq. (3.5) is the product of the lengths of the two vectors, $|\mathbf{k'} - \mathbf{k}|d = 2\pi n$. The smallest momentum difference is obtained for $n = 1$,

$$\min|\mathbf{k'} - \mathbf{k}| = 2\pi/d \ . \tag{3.6}$$

Bragg peaks. Note that the scattered beam is in the plane that contains the impinging beam, which is normal to the crystalline planes. Therefore, a screen located perpendicular to this scattering plane "captures" the scattered beam in discrete points, at angles which obey the Bragg (or the von Laue) law, with various values of the integer n. This resembles a well-known phenomenon of optics: each peak in the diffraction pattern corresponds to a difference of the optical paths comprising an integer number of wave lengths. A plot of the intensity of the scattered radiation as a function of the location of the corresponding point on the screen (i.e., as a function of the scattering angle) thus contains a narrow peak for each such point. These peaks are named after Bragg, and their very existence is a proof that the scatterer is a crystal. When the scattering experiment is carried out as illustrated in Fig. 3.3(b) or in Fig. 3.4, there appear circles on the screen, as the scattering is accomplished from differently-oriented grains (see Sec. 3.7). An analysis of the scattering off quasicrystals is presented in Sec. 3.11; though not periodic, those also give rise to narrow peaks. Following their discovery, the definition of a crystal was modified, to be **a material characterized by narrow and discrete Bragg peaks**, whether it is periodic or else. The discussion in the following is focused on periodic crystals.

3.3 Scattering off a point-like Bravais lattice

Scattering off points. The discussion so far dealt with the condition of constructive interference; the **radiation intensity** has not been calculated. The amplitude of the scattered wave is the sum of the amplitudes of the waves scattered off various elements in the crystal. Consider first the simplest case, in which each point $\mathbf{R}$ on a Bravais lattice accommodates a single, **point-like scatterer**; more general configurations are discussed below. The scattered waves emerge only from these points. Assuming that the amplitude of the wave outgoing from the (arbitrarily chosen) origin is $a = a_0 \exp[i\psi_0]$ (where a_0 is a real number), the amplitude of the wave emerging from a lattice point at $\mathbf{R}$ is $a(\mathbf{R}) = a_0 \exp[i\psi(\mathbf{R})]$. The phase difference between the two waves is the difference between the optical paths. As seen in Fig. 3.8, this difference is the sum of the optical path from the origin to point A (which is $\mathbf{k}' \cdot \mathbf{R}$), and the optical path from B to the origin (which is $-\mathbf{k} \cdot \mathbf{R}$). Hence, $\psi(\mathbf{R}) = \psi_0 + (\mathbf{k}' - \mathbf{k}) \cdot \mathbf{R}$, and the total amplitude of the scattered beam, called the **scattering amplitude**, is

$$A = a_0 e^{i\psi_0} \sum_{\mathbf{R}} e^{i(\mathbf{k}'-\mathbf{k})\cdot\mathbf{R}} \equiv a_0 e^{i\psi_0} \mathcal{Z}(\mathbf{q}) \, , \tag{3.7}$$

where $\mathbf{q} = \mathbf{k}' - \mathbf{k}$ is, up to the multiplicative factor $\hbar$, the momentum difference between the impinging and scattered particles (e.g., photons). In Eq. (3.7),

$$\mathcal{Z}(\mathbf{q}) = \sum_{\mathbf{R}} e^{i\mathbf{q}\cdot\mathbf{R}} \, . \tag{3.8}$$

Vectors in the reciprocal lattice. The intensity of the scattered radiation is proportional to the absolute value squared of the total amplitude, i.e., to $|A|^2 = a_0^2 |\mathcal{Z}(\mathbf{q})|^2$. This intensity is maximal when all terms in the sum (3.8) have the same phase, e.g., 0 (for any other constant phase, say ψ, the factor $\exp[i\psi]$ can be taken out of the sum, and then it does not change the absolute value). When this is the case, the sum equals the number of lattice sites, $\mathcal{Z}(\mathbf{q}) = N$, and $|A|^2 = a_0^2 N^2$. This

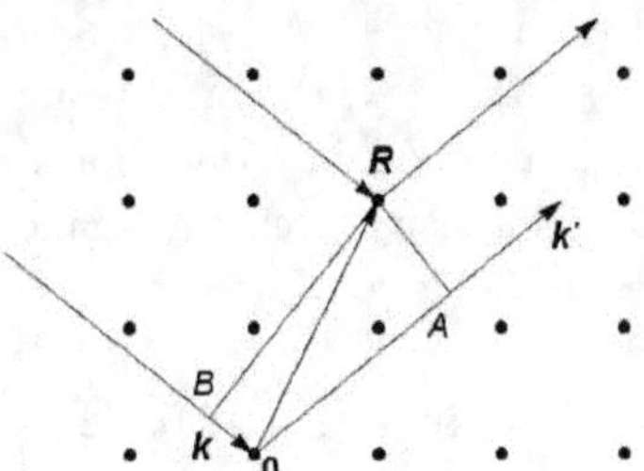

Fig. 3.8: Interference of two waves scattered from the points $\mathbf{0}$ and $\mathbf{R}$ in the lattice.

maximal value is reached when

$$e^{i\mathbf{q}\cdot\mathbf{R}} = 1 \,, \tag{3.9}$$

for **any** lattice point $\mathbf{R}$. Such an equality is realized only for a specific sequence of vectors $\mathbf{q}$. Indeed, a collection of vectors obeying Eq. (3.9) for **any** $\mathbf{R}$ can be found for any Bravais lattice, as proven in Sec. 3.4. Denote the solutions of Eq. (3.9) by $\{\mathbf{q} = \mathbf{G}\}$, where with each solution $\mathbf{q} = \mathbf{G}$ there appears also the solution $\mathbf{q} = -\mathbf{G}$. This collection of vectors builds a periodic Bravais lattice in "momentum" space, which is called the **reciprocal lattice**. It follows that maximal scattering intensity is reached when $\mathbf{q}$ equals one of the vectors of the reciprocal lattice. For a given direction of the wave vector $\mathbf{k}$ of the impinging radiation, the peaks in the intensity appear for the vectors $\mathbf{k}'$ such that $\mathbf{q} = \mathbf{k}' - \mathbf{k} = \mathbf{G}$ is one of the vectors in reciprocal lattice. These are the aforementioned **Bragg peaks**. When this is not the case, i.e., $\mathbf{q}$ is not one of the vectors of the reciprocal lattice, the sum $\mathcal{Z}(\mathbf{q})$ contains terms with various phases, and therefore possesses real and imaginary terms that may be of opposite signs and so are partially cancelled. As a result, the intensity of the scattered radiation is much smaller.

Scattering off large but finite lattices. When the sample in the scattering experiment is very large, the Bragg peaks are quite narrow; when the size approaches infinity their width tends to zero. This property is proven as follows. Inserting into the sum (3.8) the expression for $\mathbf{R}$ in terms of the lattice vectors, $\mathbf{R} = n_1\mathbf{a}_1 + n_2\mathbf{a}_2 + n_3\mathbf{a}_3$ [with integer coefficients, Eqs. (2.1)], yields

$$\mathcal{Z}(\mathbf{q}) = \sum_{\mathbf{R}} e^{i\mathbf{q}\cdot\mathbf{R}} = \sum_{n_1} e^{iu_1 n_1} \sum_{n_2} e^{iu_2 n_2} \sum_{n_3} e^{iu_3 n_3} \,, \qquad u_i = \mathbf{q}\cdot\mathbf{a}_i \,, \tag{3.10}$$

where the sums run over the entire lattice. Assume that there are N_i atoms in the direction of $\mathbf{a}_i$ and exploit periodic boundary conditions at the surfaces of the specimen [i.e., the lattice repeats itself ad infinitum at each direction (assumptions on the edges in a macroscopically-large sample have almost no effect)]. This type of boundary conditions is named after **Born and von Karman**. The periodic boundary conditions allow one to place the origin at an arbitrary lattice point.

Hence, when N_i is odd, it can be taken at the middle of the sample, and then the sum encompasses values in the range $-(N_i - 1)/2 \le n_i \le (N_i - 1)/2$. (For even N_i the range is $-N_i/2 \le n_i \le N_i/2 - 1$; the final result is identical for both cases provided that the sample is large enough. Any choice of the origin produces the same result for the scattering intensity – check.) Since each of the sums in Eq. (3.10) is a geometrical series, it yields

$$\mathcal{Z}_i(u_i) = \sum_{-(N_i-1)/2}^{(N_i-1)/2} \exp[iu_i n_i] = \exp[-i(N_i - 1)u_i/2]\frac{1 - e^{iN_i u_i}}{1 - e^{iu_i}} = \frac{\sin(N_i u_i/2)}{\sin(u_i/2)} \ .$$

$$(3.11)$$

The function $S_i = |\mathcal{Z}_i|^2$ is depicted in Fig. 3.9 for two values of N_i. It has narrow peaks, $S_i = N_i^2$, whenever the denominator in Eq. (3.11) vanishes, i.e., for $\mathbf{q} \cdot \mathbf{a}_i = u_i \to 2\pi m_i$ with an integer m_i. This condition is identical to the one in Eq. (3.9), for which the scattering intensity is maximal (check!). At the vicinity of each such peak, the ratio on the right hand-side of Eq. (3.11) equals $\sin(N_i w_i/2)/\sin(w_i/2)$, with $w_i = u_i - 2\pi m_i$. This ratio equals N_i when $w_i = 0$, and is approaching zero when $\sin(N_i w_i/2) = 0$ and $w_i \ne 0$. The first zero appears for $w_i = \pm 2\pi/N_i$. The ratio is sharply peaked around $w_i = 0$, with height N_i and width (at its base) $4\pi/N_i$. Hence, the height increases with N_i while the width decreases; the area under the peak, however, does not change and is close to 2π (like the area of a triangle of base $4\pi/N_i$ and height N_i). The function remains relatively small for other values of u_i in-between the peaks, where the numerator in Eq. (3.11) oscillates wildly between 1 and -1, and the denominator remains finite. As the samples in the experiment are always of a finite size, any experimentally-detected Bragg peak has a finite width. Nonetheless, the width is much narrower as compared to the intervals between the reciprocal-lattice vectors.

In three dimensions, Eq. (3.10) yields $\mathcal{Z}(\mathbf{q}) = \mathcal{Z}_1(u_1)\mathcal{Z}_2(u_2)\mathcal{Z}_3(u_3)$, and therefore Bragg peaks appear only when all three equations are satisfied, $\mathbf{q} \cdot \mathbf{a}_i = u_i = 2\pi m_i$, $i = 1, 2, 3$, namely when $\mathbf{q} \cdot \mathbf{R} = \mathbf{q} \cdot (n_1\mathbf{a}_1 + n_2\mathbf{a}_2 + n_3\mathbf{a}_3) = 2\pi(n_1 m_1 + n_2 m_2 + n_3 m_3)$, or $\exp[i\mathbf{q} \cdot \mathbf{R}] = 1$. The discussion following Eq. (3.9) implies that $\mathbf{q}$ must be one of the reciprocal-lattice vectors.

Scattering off an infinite lattice. As in most cases the measured lattices are rather large, one may well describe them as having an infinite size. To consider this limit, recall first the properties of the **Dirac delta-function**, $\delta(\mathbf{r})$. Mathematically speaking, the delta function is a distribution and not a function, but the difference is of no importance for the present purposes. In one dimension, the delta function is defined by the its property,

$$\int_{-\epsilon}^{\epsilon} dx f(x)\delta(x) = f(0) \ , \ \text{for any continuous function } f(x) \text{ and for any } \epsilon > 0 \ .$$

$$(3.12)$$

When $f(x) = 1$ Eq. (3.12) gives $\int_{-\epsilon}^{\epsilon} dx\delta(x) = 1$ and therefore the integral over the delta function must be 1 over **any** segment that includes the origin. As this equality

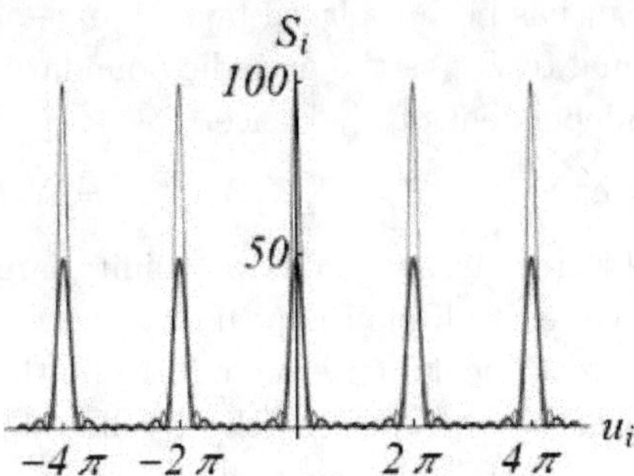

Fig. 3.9: The function $S_i = |\mathcal{Z}_i|^2$ for $N_i = 7$ (thick curve) and $N_i = 10$ (thin curve).

must hold for **any** ϵ, including $\epsilon \to 0$, the delta function diverges at the origin and vanishes elsewhere (such that its integral over a segment that does not include the origin vanishes). Similarly, the three-dimensional delta function is defined by

$$\int d^3 r \, f(\mathbf{r}) \delta(\mathbf{r}) = f(0) \, , \tag{3.13}$$

for any continuous function $f(\mathbf{r})$, when the integration is over a volume which contains the point $\mathbf{r} = 0$.

Equation (B.27) in Appendix B demonstrates that for an infinite periodic lattice,

$$\mathcal{Z}(\mathbf{q}) = \sum_{\mathbf{R}} e^{i\mathbf{q}\cdot\mathbf{R}} \to \frac{(2\pi)^3}{V} \sum_{\mathbf{G}} \delta(\mathbf{q} - \mathbf{G}) \, , \tag{3.14}$$

where the sum over $\mathbf{R}$ contains all lattice points, and the sum over $\mathbf{G}$ includes all points in the reciprocal lattice. This sum, the "three-dimensional Dirac's comb", contains peaks of infinite heights and zero widths that occur at points where the vector $\mathbf{q} = \mathbf{k}' - \mathbf{k}$ equals one of the reciprocal-lattice vectors. In other words, Bragg peaks, e.g., those shown in Fig. 3.9, become delta functions.

In summary, **the scattering amplitude of point-like scatterers is sharply peaked when the vector $\mathbf{q} = \mathbf{k}' - \mathbf{k}$ coincides with one of the reciprocal-lattice vectors.** As each Bravais lattice has its own reciprocal lattice (see below), monitoring the locations in momentum space where the Bragg peaks appear enables the unequivocal identification of the lattice structure.

The scattering intensity. The scattering intensity is given by the absolute value squared of the amplitude, Eq. (3.7), $|A|^2 = a_0^2 |\mathcal{Z}|^2$. This intensity (in three spatial dimensions) is proportional to the product of the three functions $S_i = |\mathcal{Z}_i|^2$, Eq. (3.10), where each of them is

$$S_i = |\mathcal{Z}_i(u_i)|^2 = \left[\sum_{n_i} e^{in_i u_i}\right]\left[\sum_{n_i'} e^{-in_i' u_i}\right] = \sum_{n_i, n_i'} e^{i(n_i - n_i')u_i} = \sum_{n_i', n_i''} e^{in_i'' u_i} = N_i \mathcal{Z}_i \, .$$

$$\tag{3.15}$$

(The summation index n_i has been replaced by $n_i'' = n_i - n_i'$ in the last two steps.) The resulting sum is equal to $\mathcal{Z}_i$ (as the periodic boundary conditions allow a shift of the origin), and is independent of n_i'. Hence,

$$|A|^2 = a_0^2 N \mathcal{Z} \ , \quad \mathcal{Z} = \mathcal{Z}_1 \mathcal{Z}_2 \mathcal{Z}_3 \ , \quad N = N_1 N_2 N_3 \ , \tag{3.16}$$

(check!) Equation (3.14) (or its analogue for a finite large sample) implies that the intensity of the scattered radiation is peaked when the difference of the wave vectors equals one of the reciprocal-lattice vectors. Note, though, that by Eq. (3.14) the intensity of all peaks is the same (as the coefficients of the delta functions are identical). As shown in the following, this result is valid only for point-like scatterers (one per unit cell). When the unit cell contains more than a single point-like scatterer, the Bragg peaks do appear at the same locations as for the point-like scatterers, but the intensity of each of them is different; it thus conveys information about the scatterers within the unit cell.

3.4   The reciprocal lattice

The von Laue equations. As found in Sec. 3.3, the radiation intensity is peaked when the momentum difference between the impinging particles and the scattered ones equals a vector $\mathbf{G}$, which obeys the equation

$$\exp[i\mathbf{G} \cdot \mathbf{R}] = 1 \ , \tag{3.17}$$

where $\mathbf{R}$ is the radius vector to **any** of the lattice points. The vectors $\mathbf{G}$ form a periodic lattice, called the **"reciprocal lattice"**, whose definition is Eq. (3.17). For the vector $\mathbf{G}$ to obey the condition (3.17), it has to satisfy

$$\mathbf{G} \cdot \mathbf{R} = 2\pi m \ , \tag{3.18}$$

where m is an arbitrary integer and $\mathbf{R}$ is an arbitrary lattice site. Thus, Eq. (3.18) is satisfied for any $\mathbf{R}$ if it is obeyed for each of the lattice vectors,

$$\mathbf{G} \cdot \mathbf{a}_i = 2\pi m_i \ , \tag{3.19}$$

where (in three dimensions) $i = 1, 2, 3$ and m_i is an integer. [Substituting $\mathbf{R} = n_1 \mathbf{a}_1 + n_2 \mathbf{a}_2 + n_3 \mathbf{a}_3$, Eq. (2.1), in Eq. (3.18) yields $\mathbf{G} \cdot \mathbf{R} = \sum_i n_i \mathbf{G} \cdot \mathbf{a}_i = 2\pi \sum_i n_i m_i$.] Equations (3.19) are identical to those obtained in the discussion following Fig. 3.9. Combining these equations with the requirement $\mathbf{k}' - \mathbf{k} = \mathbf{q} = \mathbf{G}$ that follows from Eq. (3.9) results in $(\mathbf{k}' - \mathbf{k}) \cdot \mathbf{a}_i = 2\pi m_i$; these are the **von Laue equations**.

Equation (3.19) comprises three equations, each imposing the tip of the vector $\mathbf{G} = \mathbf{k}' - \mathbf{k}$ to lie on a plane normal to the lattice vector $\mathbf{a}_i$, such that the projection of $\mathbf{G}$ on that lattice vector is $2\pi m_i / a_i$. Constructive interference is achieved only when all three equations are obeyed together, i.e, all three planes cut each other. Such a joint cut exists only at a discrete number of points. In other words, three equations for three unknowns (which are the components of the vector $G = \mathbf{k}' - \mathbf{k}$) have at most a single solution. For this reason there appear only discrete points on the monitor, e.g., in the geometry of Fig. 3.3(a). It is worthwhile to note that for a one-dimensional sample there is only a single plane, while for a two-dimensional specimen there are two planes. In these cases the diffraction pattern can include lines in place of points. This issue is revisited in Sec. 3.12, where scattering off planes is analyzed.

The vectors of the reciprocal lattice. When two vectors, $\mathbf{G}_1$ and $\mathbf{G}_2$, obey Eq. (3.19), their sum, $\mathbf{G}_1 + \mathbf{G}_2$, fulfills it as well (prove!). It thus suffices to find three independent vectors obeying Eq. (3.19); then any linear combination of the three vectors, with integer coefficients, also fulfills that equation. To determine **all** these vectors, one needs to find the **shortest** independent vectors. (When along a certain direction the vector is not the shortest one, then there exists another vector pointing along this direction which also satisfies the equation but cannot be written as a linear combination of the three vectors with integer coefficients!) Denote the required vectors by $\mathbf{b}_1$, $\mathbf{b}_2$, and $\mathbf{b}_3$. As it is possible to choose any three integer coefficients, clearly the linear combinations of these vectors create an infinite lattice, which is the **reciprocal lattice.**

The shortest vectors that obey the aforementioned equations are found by choosing the smallest values of the integers in Eqs. (3.19), $m_i = 1, 0$. A sufficient condition is then

$$\mathbf{b}_i \cdot \mathbf{a}_i = 2\pi\delta_{ij} \ , \quad i,j = 1,2,3 \ , \tag{3.20}$$

where

$$\delta_{ij} = \begin{cases} 1 \ , \ i = j \\ 0 \ , \ i \neq j \end{cases}$$

is the **Kronecker delta.** As the vectors $\mathbf{b}_i$ obey Eqs. (3.19), they are all **vectors in reciprocal lattice. Any other vector in that lattice is a linear combination of them, i.e.,**

$$\mathbf{G} = h\mathbf{b}_1 + k\mathbf{b}_2 + \ell\mathbf{b}_3 \ , \tag{3.21}$$

where h, k, and ℓ are arbitrary integers. Indeed, inserting Eq. (3.21) in Eq. (3.19) yields, e.g., $\mathbf{G} \cdot \mathbf{a}_i = h\mathbf{b}_1 \cdot \mathbf{a}_1 + k\mathbf{b}_2 \cdot \mathbf{a}_1 + \ell\mathbf{b}_3 \cdot \mathbf{a}_1 = 2\pi h$, as required by Eq. (3.19). Moreover, Eq. (3.21) [upon comparing it with Eq. (2.1)] shows that the collection of vectors $\mathbf{G}$ forms a Bravais lattice, with lattice vectors $\mathbf{b}_1$, $\mathbf{b}_2$, and $\mathbf{b}_3$. Since constructive interference is achieved only for $\mathbf{q} = \mathbf{k}' - \mathbf{k} = \mathbf{G}$, and as the lattice vectors are discrete, then only a discrete collection of scattering angles or wave lengths gives rise to such interference.

Simple examples. In one-dimension there is only a single equation in (3.20), $ba = 2\pi$, and therefore the length of the reciprocal-lattice vector is $b = 2\pi/a$. Note that the dimensions of the vectors in reciprocal lattice are length^{-1} (and when multiplied by $\hbar$, after the de Broglie's rule, the units are those of momentum; this is the reason for referring to the reciprocal lattice as momentum space). A similar result is obtained for the **two-dimensional rectangular lattice** for which the reciprocal-lattice vectors are $\mathbf{b}_1 = (2\pi/a_1)\hat{\mathbf{x}}$ and $\mathbf{b}_2 = (2\pi/a_2)\hat{\mathbf{y}}$, or for the **three-dimensional orthorhombic lattice.** In the latter case, denoting the original lattice vectors (of the "direct" lattice in real space) by $\mathbf{a}_1 = a_1\hat{\mathbf{x}}$, $\mathbf{a}_2 = a_2\hat{\mathbf{y}}$, and $\mathbf{a}_3 = a_3\hat{\mathbf{z}}$, the solution of Eqs. (3.20) is

$$\mathbf{b}_1 = (2\pi/a_1)\hat{\mathbf{x}} \ , \quad \mathbf{b}_2 = (2\pi/a_2)\hat{\mathbf{y}} \ , \quad \mathbf{b}_3 = (2\pi/a_3)\hat{\mathbf{z}} \ . \tag{3.22}$$

Figure 3.10 displays the rectangular lattice and its reciprocal.

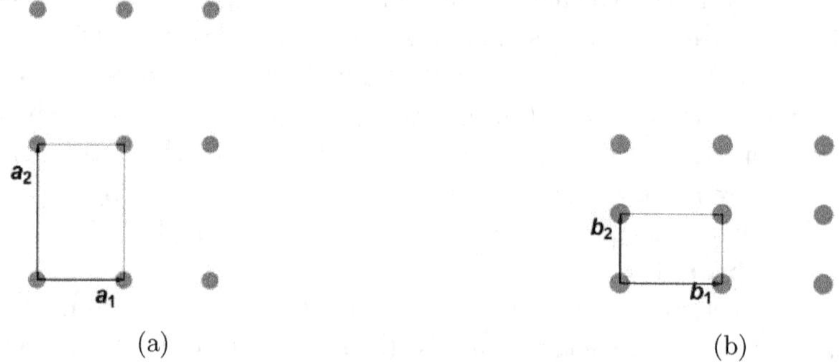

Fig. 3.10: (a) The rectangular lattice, with lattice constants $a_1 < a_2$. (b) The reciprocal lattice of the rectangular lattice, with lattice constants $b_1 = 2\pi/a_1 > 2\pi/a_2 = b_2$.

Three-dimensional lattices. The solution of Eqs. (3.20) for the general three-dimensional lattice is more complex. The vector $\mathbf{b}_1$ is perpendicular to both $\mathbf{a}_2$ and $\mathbf{a}_3$; one may conclude that it is parallel to the vectorial product $\mathbf{a}_2 \times \mathbf{a}_3$. Similar reasons determine the orientations of the other two lattice vectors. The lengths of all three is dictated by the additional equation $\mathbf{b}_i \cdot \mathbf{a}_i = 2\pi$. Assuming that $\mathbf{b}_1 = \beta\mathbf{a}_2 \times \mathbf{a}_3$, then $\mathbf{b}_1 \cdot \mathbf{a}_1 = \beta[\mathbf{a}_2 \times \mathbf{a}_3] \cdot \mathbf{a}_1$, and Eq. (3.20) yields $\beta = 2\pi/(\mathbf{a}_1 \cdot [\mathbf{a}_2 \times \mathbf{a}_3])$. The volume of the unit cell in the original lattice, from Eq. (2.2), is $V = |\mathbf{a}_1 \cdot [\mathbf{a}_2 \times \mathbf{a}_3]|$. When the sign of the triple product is negative, one has just to invert the signs of the lattice vectors. It is therefore customary to choose the reciprocal lattice-vectors as

$$\mathbf{b}_1 = 2\pi\mathbf{a}_2 \times \mathbf{a}_3/V \ , \quad \mathbf{b}_2 = 2\pi\mathbf{a}_3 \times \mathbf{a}_1/V \ , \quad \mathbf{b}_3 = 2\pi\mathbf{a}_1 \times \mathbf{a}_2/V \ . \tag{3.23}$$

The reciprocal-lattice vectors are at times defined without the 2π factor, i.e., by the condition $\mathbf{b}_i \cdot \mathbf{a}_j = \delta_{ij}$. Indeed, the reciprocal-lattice vectors of the orthorhombic lattice, Eqs. (3.22), obey the more general relations (3.23) (check!).

The connection with the distance between planes. Interestingly enough, the length of the vector $\mathbf{b}_3$ equals the product of 2π and the ratio of the area of the "base" of the unit cell, $S = |\mathbf{a}_1 \times \mathbf{a}_2|$, and the volume of that cell. Hence, $|\mathbf{b}_3| = 2\pi/d(001)$, where $d(001)$ is the height of the unit cell, i.e., the distance between the base plane and the plane parallel to it that contains the opposite face of the unit cell (which passes through the tip of $\mathbf{a}_3$, Fig. 2.13). Problem 3.2 gives an example for such a distance. Indeed, the length of each vector in reciprocal lattice is related to a certain distance among planes in the lattice; this explains the notation (001) and similar notations that appear in Fig. 3.6.

Two-dimensional lattices. The vectors of the reciprocal lattice of a **two-dimensional lattice**, whose base includes the vectors $\mathbf{a}_1$ and $\mathbf{a}_2$, are found by introducing a unit vector normal to the lattice plane, $\hat{\mathbf{z}}$. The reciprocal-lattice vectors are then given by $\mathbf{b}_1 = 2\pi\mathbf{a}_2\times\hat{\mathbf{z}}/S$ and $\mathbf{b}_2 = 2\pi\hat{\mathbf{z}}\times\mathbf{a}_1/S$, where $S = |\mathbf{a}_1\times\mathbf{a}_2|$ is the area of the unit cell [Eq. (2.3)].

The triangular lattice. The reciprocal lattice of the **triangular lattice** is found as follows. As in Fig. 2.4, the (direct) lattice vectors are chosen to be $\mathbf{a}_1 = a\hat{\mathbf{x}}$ and $\mathbf{a}_2 = (\hat{\mathbf{x}} + \sqrt{3}\hat{\mathbf{y}})a/2$. [One may choose also $\mathbf{a}_2 = (-\hat{\mathbf{x}} + \sqrt{3}\hat{\mathbf{y}})a/2$, but the physical results are unaltered.] Hence, $S = |\mathbf{a}_1 \times \mathbf{a}_2| = \sqrt{3}a^2/2$, and consequently $\mathbf{b}_1 = 2\pi\mathbf{a}_2 \times \hat{\mathbf{z}}/S = 2\pi(\hat{\mathbf{x}} - \hat{\mathbf{y}}/\sqrt{3})/a$ and $\mathbf{b}_2 = 4\pi\hat{\mathbf{y}}/(a\sqrt{3})$. The triangular lattice and its reciprocal are displayed in Figs. 3.11. The reciprocal lattice is triangular too, with a lattice constant $b = 4\pi/(a\sqrt{3})$ (check!). From the scalar product of the two lattice vectors one finds that the angle between them is $120°$ (as opposed to the lattice vectors of the original lattice, where the corresponding angle is $60°$). As can be expected, the reciprocal-lattice vectors are perpendicular to those of the original lattice (or alternatively, rotated relative to them by $\pm30°$).

(a)   (b)

Fig. 3.11: (a) The triangular lattice, with lattice vectors $\mathbf{a}_1 = a\hat{\mathbf{x}}$ and $\mathbf{a}_2 = (\hat{\mathbf{x}} + \sqrt{3}\hat{\mathbf{y}})a/2$. (b) The reciprocal lattice of the triangular lattice, with lattice vectors $\mathbf{b}_1 = 2\pi(\hat{\mathbf{x}} - \hat{\mathbf{y}}/\sqrt{3})/a$ and $\mathbf{b}_2 = 4\pi\hat{\mathbf{y}}/(a\sqrt{3})$. It is a triangular lattice whose lattice constant is $b = 4\pi/(a\sqrt{3})$.

Problem 3.4.
a. *Prove that the reciprocal-lattice vectors of an arbitrary two-dimensional lattice*

are given by

$$\mathbf{b}_1 = 2\pi[|\mathbf{a}_2|^2\mathbf{a}_1 - (\mathbf{a}_1 \cdot \mathbf{a}_2)\mathbf{a}_2]/S^2 \ ,$$

$$\mathbf{b}_2 = 2\pi[|\mathbf{a}_1|^2\mathbf{a}_2 - (\mathbf{a}_1 \cdot \mathbf{a}_2)\mathbf{a}_1]/S^2 \ ,$$

where S is the area of the original unit cell.

b. *What is the area of the unit cell of the planar reciprocal lattice?*

Problem 3.5.

Find the reciprocal of the simple hexagonal lattice (which is built of triangular lattices placed on top of each other).

Non-primitive cubic lattices. A more complex example is the face-centered cubic lattice, **FCC**, whose lattice vectors are given in Eq. (2.6) and in Fig. 2.18. For this lattice $\mathbf{a}_1 \cdot [\mathbf{a}_2 \times \mathbf{a}_3] = -a^3/4 = -V$. Inserting the vectors (2.6) into Eq. (3.23) yields, e.g., $\mathbf{b}_1 = 2\pi(a/2)^2[(\hat{\mathbf{x}} + \hat{\mathbf{z}}) \times (\hat{\mathbf{y}} + \hat{\mathbf{z}})]/(-V) = (2\pi/a)(\hat{\mathbf{x}} + \hat{\mathbf{y}} - \hat{\mathbf{z}})$. Hence,

$$\mathbf{b}_1 = \frac{2\pi}{a}(\hat{\mathbf{x}} + \hat{\mathbf{y}} - \hat{\mathbf{z}}), \ \mathbf{b}_2 = \frac{2\pi}{a}(-\hat{\mathbf{x}} + \hat{\mathbf{y}} + \hat{\mathbf{z}}), \ \mathbf{b}_3 = \frac{\pi}{a}(\hat{\mathbf{x}} - \hat{\mathbf{y}} + \hat{\mathbf{z}}) \ . \tag{3.24}$$

A comparison of these expressions with the ones in Eq. (2.5), shows that they are the lattice vectors of a **BCC** lattice, whose cube edge is $4\pi/a$. Hence, **the reciprocal of the FCC lattice is a BCC lattice.**

Problem 3.6.

Prove that in three dimensions, the product of the volumes of the two unit cells, the one of the original lattice and that of the reciprocal one, equals $(2\pi)^3$. What are the analogous relations in one and two dimensions?

Problem 3.7.

What is the reciprocal of the reciprocal lattice?

Problem 3.8.

*What is the reciprocal of the **BCC** lattice? Can one deduce this result without carrying out the detailed calculation?*

As found above, the reciprocal of the orthorhombic lattice is orthorhombic as well, and so is the case also for the triangular and the hexagonal lattices. The reciprocal of the cubic **FCC** and **BCC** lattices are also cubic (**BCC** and **FCC**, respectively). In all these examples the symmetries of the original lattices are similar to those of the reciprocal ones. This feature is common to all Bravais lattices as can be proven by computing the reciprocal lattices of all 14 Bravais lattices (see also problem s.3.2 at the end of the chapter). Since scattering experiments probe directly the reciprocal lattice, this property enables the unique identification of the corresponding Bravais lattice.

Ewald sphere. A geometric picture which clarifies the relations among the vectors of the impinging and scattered radiations and those of the reciprocal lattice, is displayed in Figs. 3.12. Figure 3.12(a) shows a planar cross section of the reciprocal lattice. Assume that the impinging wave vector **k** lies on the horizontal axis in space, and locate its tip on a reciprocal-lattice site, O. A sphere of radius equal the length of **k**, i.e., $2\pi/\lambda$, is drawn around the other tip of **k** (placed at point B and marked by ×). This is the **Ewald sphere**. As the outgoing wave has the same wave length, its wave vector **k′** connects the point B with another point on the same sphere (marked by C). The discussions following Eqs. (3.9) and (3.19) imply that the difference **k′** − **k** is identical to a reciprocal-lattice vector **G**. The figure demonstrates that this difference connects the point O and C. Hence, the scattered wave that propagates along **k′** produces a point on the screen only when it passes through a reciprocal-lattice site on the sphere (C in the present example). In other words, one plots the Ewald sphere for a certain impinging wave **k**, and identifies the directions of the scattered radiation that can produce points on the screen, like in Fig. 3.12(b). The points where these directions cut the sphere are sites of the reciprocal lattice. Other wave lengths give rise to spheres of different radii, which contain other sites of the reciprocal lattice. Hence, all points on the screen are related to reciprocal-lattice sites, and their lot should reflect the symmetry of that lattice. In particular, when the impinging ray is directed along one of the rotation axes of the lattice (for instance, of order 4), then the points on the screen possess this symmetry as well (in the present example, the diffraction pattern does not change upon rotating the sphere by 90° along that axis).

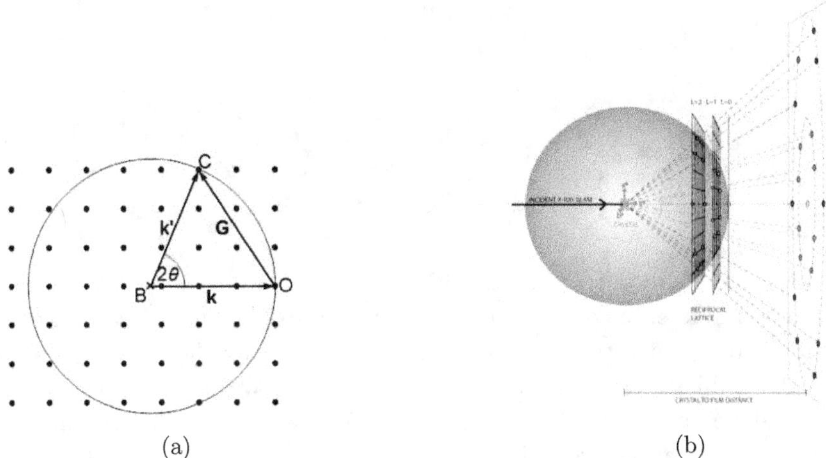

(a)   (b)

Fig. 3.12: The Ewald sphere. (a) The vectors of the impinging and scattered waves, and the reciprocal lattice vector joining them, **G** = **k′** − **k**. (b) The diffraction pattern on the screen. Whenever the scattered wave passes through a reciprocal-lattice site on the sphere there appears a Bragg peak on the screen (Courtesy of M. Sawaya).

3.5 Brillouin zones

As discussed, constructive interference is achieved only when the vectors of the impinging and scattered waves obey Eq. (3.9), i.e.,

$$\mathbf{k}' = \mathbf{k} + \mathbf{G} \ , \tag{3.25}$$

where $\mathbf{G}$ is an arbitrary vector of the reciprocal lattice. Squaring this equation and using the relations $k^2 = k'^2 = (2\pi/\lambda)^2$ give $2\mathbf{k} \cdot \mathbf{G} = -G^2$, i.e.,

$$\mathbf{k} \cdot \hat{\mathbf{G}} = -G/2 \ , \tag{3.26}$$

where $\hat{\mathbf{G}}$ is a unit vector along $\mathbf{G}$. The left hand-side is the projection of $\mathbf{k}$ on the direction of $\mathbf{G}$. As this projection is equal to half of the length of $\mathbf{G}$ (the sign is irrelevant), it follows that when the vectors $\mathbf{k}$ and $\mathbf{k}'$ emerge from the same point O in reciprocal space and end at the opposite tips of $\mathbf{G}$ (see Fig. 3.13), then O lies on a plane normal to $\mathbf{G}$ and bisecting it (only vectors emerging from such a point and ending at the tip of $\mathbf{G}$ have projections on it whose length is $G/2$, Fig. 3.13). Hence, **scattering of nonzero intensity is achieved only when the vector of the impinging wave (and hence also the vector of the scattered radiation) emerges from a point on the plane normal to one of the reciprocal-lattice vectors at its mid point and ends at the tip of that vector.**

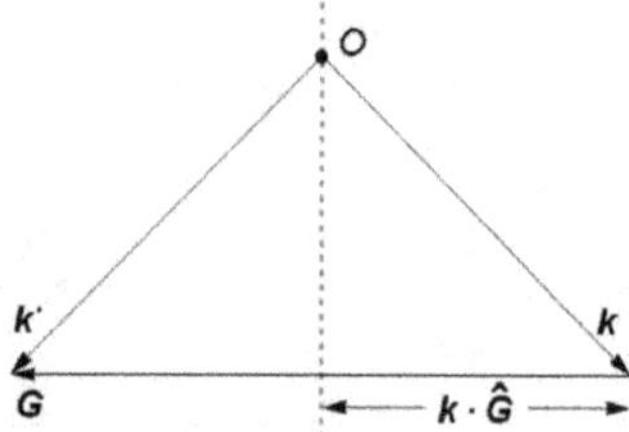

Fig. 3.13: The wave vectors of the impinging and scattered radiation in the constructive-interference configuration. $\mathbf{G}$ is a vector of the reciprocal lattice, and the point O from which the two wave vectors emerge is on the normal to it (that cut it at the middle). Then the projection of each of the wave vectors on $\mathbf{G}$ is equal to $G/2$, as dictated by Eq. (3.26). The point O may be located anywhere on the plane normal to $\mathbf{G}$ at its mid point.

Once the coordinate origin is assigned to one of the reciprocal-lattice sites, then reciprocal-lattice vectors emerge from it to all other sites in the reciprocal lattice. The planes normal to the mid points of the vectors connecting the origin with its nearest neighbors encompass the volume of the reciprocal lattice that contains all points closer to the origin than to any other site there. This volume is analogous to the Wigner-Seitz unit cell, introduced in Chapter 2; it is the unit cell of the reciprocal lattice, that exists in momentum space. This is the **first Brillouin zone**. One may repeat this construction for the vectors which join the origin with its next-nearest neighbors, and thus create the second Brillouin zone, and so forth. Figure 3.14 displays several Brillouin zones of the two-dimensional square lattice. Figure 3.14(a) portrays the sites of the reciprocal lattice, and Fig. 3.14(b) shows the boundaries among the various Brillouin zones. In two dimensions these are lines, each being perpendicular to one of the lattice vectors. As shown in the solution of problem 2.2, the Wigner-Seitz cell of the square lattice is a square encompassing a single lattice point, whose edge length is the lattice constant. Therefore, the first Brillouin zone of the square lattice is a square whose edge's length is $2\pi/a$, enclosing a single reciprocal-lattice site. It is the small square enclosing the origin in Fig. 3.14(b), indicated by "1", whose boundaries are lines perpendicular to the reciprocal-lattice vectors connecting the origin with its four nearest neighbors, bisecting them at the middle ($\mathbf{G} = \pm\mathbf{b}_1$ or $\mathbf{G} = \pm\mathbf{b}_2$). The outer borders of the second Brillouin zone are the lines perpendicular to the reciprocal-lattice vectors connecting the origin with the four next-nearest neighbors ($\mathbf{G} = \pm\mathbf{b}_1 \pm \mathbf{b}_2$). As seen, the latter form a square rotated relative to the axes by 45°, whose area is twice that of the first Brillouin zone. Note though, that the second Brillouin zone contains the lattice sites that are within the large square but **are not** in the first zone. Hence, the actual area of the second Brillouin zone, which consists of the four triangles marked in the figure by "2", is identical to that of the first Brillouin zone. Similarly, the lines normal to the vectors $\mathbf{G} = \pm 2\mathbf{b}_1$ or $\mathbf{G} = \pm 2\mathbf{b}_2$, which connect the origin with the third nearest-neighbors, are parallel to the axes, and together with the lines defined previously contain eight triangles which form the third Brillouin zone. Its area is equal to that of its two predecessors.

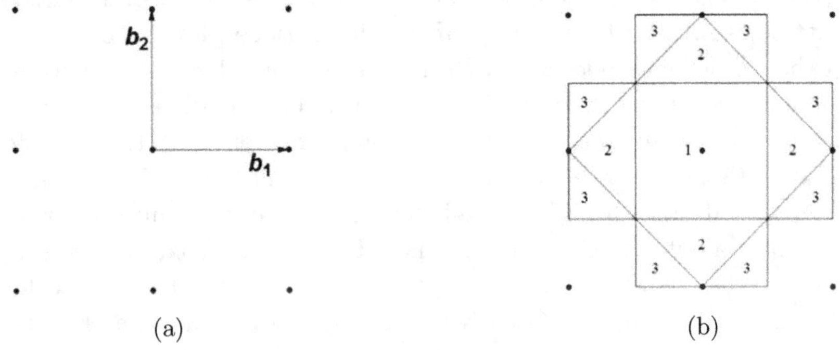

(a)   (b)

Fig. 3.14: (a) The reciprocal of the square lattice. (b) Brillouin zones of the square lattice. The numbers within each area indicate the serial number of the Brillouin zone to which this area belongs. The straight lines are normal to the lattice vectors connecting the origin with its neighbors, for instance $\mathbf{G} = \pm 2\mathbf{b}_1$, $\mathbf{G} = \pm 2\mathbf{b}_2$, $\mathbf{G} = \pm\mathbf{b}_1$, $\mathbf{G} = \pm\mathbf{b}_2$, or $\mathbf{G} = \pm\mathbf{b}_1 \pm \mathbf{b}_2$.

Problem 3.9.
Draw the first three Brillouin zones of the triangular lattice.

The construction of the Brillouin zones of the simple cubic lattice resembles the one carried out for the square lattice. This task becomes more laborious for the more complicated Bravais lattices. For instance, as shown in Sec. 3.4, the reciprocal of the **BCC** lattice is an **FCC** one, and *vice versa*. Hence, the first Brillouin zones of these two lattices are those that appear in Figs. 2.19: the first Brillouin zone of the **BCC** in Fig. 2.19(a), and the one of the **FCC** in Fig. 2.19(b). In both cases the length of the edge of the cube is $4\pi/a$.

In summary, **constructive interference is achieved only when the scattered wave vector emerges from a point on one of the planes that separates different Brillouin zones, and ends at the tip of a reciprocal-lattice vector to which this plane is perpendicular (cutting it at its middle).** The following chapters emphasize other properties of the first Brillouin zone: the frequencies of the vibrations of the atoms around their equilibrium locations (Chapter 5) or the electronic motion (Chapter 6) are characterized by wave numbers; it suffices to explore those only for wave vectors in the first Brillouin zone (their values in the other zones are identical to the ones in the first zone).

3.6 The reciprocal lattice and lattice planes

Planes perpendicular to reciprocal-lattice vectors. As mentioned, there are several ways to specify the directions along which the scattered waves interfere constructively. The relation between these directions and families of lattice planes is discussed in Sec. 3.2. Section 3.3 connects those directions with the reciprocal-lattice vectors. These two descriptions are equivalent: **each vector in the reciprocal lattice is related to a family of parallel lattice planes, and** *vice versa*. Though the various considerations yield the same result (luckily enough), each is based on a different point of view. The experimental observations are conveniently described in terms of scattering angles, but equally convenient is the identification of each angle that gives rise to constructive interference with the corresponding lattice vector, and, in addition, the family of lattice planes to which it is related.

The reciprocal-lattice vectors are given in Eq. (3.18). That equation may be understood in the following manner: given the integer m and the reciprocal-lattice vector $\mathbf{G}$, then the tips of all vectors $\mathbf{R}$ that obey Eq. (3.18) lie on a plane perpendicular to $\mathbf{G}$; hence Eq. (3.18) is an equation of a plane. Indeed, two vectors, $\mathbf{R}$ and $\mathbf{R}'$, that obey Eq. (3.18) also satisfy $\mathbf{G} \cdot (\mathbf{R} - \mathbf{R}') = 0$ and therefore the difference $\mathbf{R} - \mathbf{R}'$ is normal to $\mathbf{G}$; it hence lies in a plane normal to $\mathbf{G}$. This is true for all vectors obeying Eq. (3.18), with the same values of m and of $\mathbf{G}$. Other values of m represent other planes, all normal to $\mathbf{G}$, and consequently parallel to each other. This collection of planes is **the family of planes normal to G**. Suppose that the origin is located on one of these planes, say the plane for which $m = 0$. The vector $\mathbf{R}$ in any other plane can be written in the from $\mathbf{R} = \mathbf{d} + \mathbf{R}_\perp$, where $\mathbf{d} \| \mathbf{G}$ and $\mathbf{R}_\perp \perp \mathbf{G}$. Equation (3.18) becomes $\mathbf{G} \cdot \mathbf{d} = 2\pi m$. Equation (3.25), $\mathbf{k}' = \mathbf{k} + \mathbf{G}$, is identical to Eq. (3.5), making the reciprocal-lattice vectors identical to the vectors of momentum difference as given by the laws of von Laue and Bragg (see Sec. 3.2).

The distance between neighboring planes. There are many reciprocal-lattice vectors that are normal to a certain family of planes. Each of the thick arrows in Fig. 3.6 represents the direction of a vector normal to each of the families of planes displayed there. There are numerous reciprocal-lattice vectors pointing along this direction, since $n\mathbf{G}$ (where n is an integer) is also a vector in the reciprocal lattice. It is customary to characterize the family of planes by the shortest reciprocal-lattice vector of this type, denoted $\mathbf{G}_0$. Since Eq. (3.18) can be written in the form $\mathbf{G}_0 \cdot \mathbf{R} = 2\pi m$, it is clear that successive values of m represent successive planes in that family. The distance between the planes corresponds to $m = 1$, $G_0 d = \mathbf{G}_0 \cdot \mathbf{R} = 2\pi$ (recall that $\mathbf{d} \| \mathbf{G}_0$), and therefore the distance between neighboring planes is

$$d = 2\pi/G_0 \ . \tag{3.27}$$

This result is equivalent to Eq. (3.6). Bragg's law, Eq. (3.4), can hence take the form

$$\sin\theta = nG_0\lambda/(4\pi) = G\lambda/(4\pi) \ . \tag{3.28}$$

Measurements of the angles for which the scattered waves interfere constructively yield the "allowed" values of the length of vectors in the reciprocal lattice (that give rise to constructive interference). These lengths are equivalent to the "allowed" distances between planes in the families of planes that exist in the crystal. Calculations of angles corresponding to Bragg peaks for various lattices are presented below, e.g., at the beginning of Sec. 3.7.

Miller indices. Each of the vectors $\mathbf{G}$ in the reciprocal lattice is specified by three integers, h, k, and ℓ, Eq. (3.21). The corresponding family of planes is characterized by the shortest vector, $\mathbf{G}_0$, which is also given by such three integers. Consider now two neighboring planes in the family, of which one passes through the origin and the other is a nearest neighbor parallel to it. The equation of this second plane is $\mathbf{G}_0 \cdot \mathbf{r} = (h\mathbf{b}_1 + k\mathbf{b}_2 + \ell\mathbf{b}_3) \cdot \mathbf{r} = 2\pi$ [the discrete points $\mathbf{R}$ are replaced by all (continuous) points on the plane, denoted by $\mathbf{r}$]. This plane cuts the lattice vectors of the original lattice. Denote the coordinates of these cuts by the vectors $x_1\mathbf{a}_1$, $x_2\mathbf{a}_2$, and $x_3\mathbf{a}_3$. Inserting each of these into the equation of the second plane, and using Eq. (3.20), yields

$$x_1 h = x_2 k = x_3 \ell = 1 \ . \tag{3.29}$$

The three numbers h, k, and ℓ (that belong to $\mathbf{G}_0$) therefore characterize the **cuts of the planes with the major directions in the lattice (i.e., along the lattice vectors).** In particular, the lattice vector $\mathbf{a}_1$ is cut by h planes that belong to the family (for which $x_1 = 1/h$, $2/h$, $\ldots, 1$), *etc.* Consequently, the family of planes is identified by $(hk\ell)$. The three numbers h, k, and ℓ are called **the indices of Miller**. Examples for lattice planes are displayed in Fig. 3.6. In the upper part of Fig. 3.6(a) appear lines (in a two-dimensional planar lattice these replace the planes of the three-dimensional lattice). For instance, the lines in the upper left part of Fig. 3.6(a) cut the axes at $\mathbf{a}_1$ and $\mathbf{a}_2$, and therefore are marked by (11). The lines in tFig. 3.6(a) are parallel to the lattice vector $\mathbf{a}_2$, and therefore cut this axis only at infinity; hence $k = 1/x_2 = 0$. These lines cut the $\mathbf{a}_1-$axis at $x_1 = 1$, and hence are marked as (10). Similar considerations apply to the other lines. When one of the numbers, e.g., x_2, is negative, one adds a minus sign above the respective index, for example, $(4, -1) \Leftrightarrow (4, \bar{1})$ (as in the lower left part of the rectangular lattice in the figure). Figure 3.6(b) also displays a family of planes for the three cubic lattices. In Sec. 3.2, and in particular in problem 3.2, several such planes are identified. The reader is welcome to examine the Miller indices of those.

Problem 3.10.

The equations which describe a certain family of planes in the **BCC** *lattice are discussed in problem 3.2, where the distances between the planes are also calculated. Identify the reciprocal-lattice vector that corresponds to this family, and verify that the results obtained in the solution of that problem are identical to the ones expressed in terms of this vector.*

The density of sites in the plane. The specific volume of a primitive Bravais lattice (the volume per each lattice point) is that of the unit cell, V. Since the distance between neighboring planes is $d = 2\pi/G_0$, the planar density of points (per unit area) in each plane is $d/V = 2\pi/(VG_0)$. Hence, as the reciprocal-lattice vector becomes longer, the distance between nearest-neighbor planes is shortened, and the density of points in each plane decreases. Similar conclusions pertain to the densest lines in a planar lattice. Several such densities are computed in problem 3.2. The reader is asked to confirm (or else) those results by exploiting the expressions given here.

Multiplicity of plane. Different planes may be related to each other by symmetry operations, i.e., one is obtained from the other by applying one of the symmetry operations of the lattice. For instance, the plane $(hk\ell)$ in the cubic lattice is equivalent to the planes for which the indices are interchanged, e.g., $(h\ell k)$ or $(kh\ell)$. Likewise, each of the eight planes $(\pm h, \pm k, \pm \ell)$ is also similar to that plane. The family of all those equivalent planes is denoted $\{hk\ell\}$. Symmetry considerations imply that all planes in such a family yield the same scattering angles. The number of triplets $(hk\ell)$ which gives rise to the same scattering angles is called the **multiplicity** of the $\{hk\ell\}$ family of planes.

Problem 3.11.

a. *Find the multiplicity of the following families of planes in a cubic lattice: $\{00k\}$, $\{hk\ell\}$, $\{hh\ell\}$, $\{hhh\}$, $\{k\ell 0\}$, $\{kk0\}$, where h, k, and ℓ are all different and nonzero.*

b. *Plot all planes close to the origin in a cubic lattice, that belong to the $\{111\}$ family. What shape do they form?*

c. *Describe all planes close to the origin in a cubic lattice, that belong to the $\{110\}$ family. Determine the shape created by them.*

d. *Find the maximal multiplicity of different planes that belong to the same family.*

Problem 3.12.

An alternative way to obtain the Miller indices is the following. Define a plane by the points where it cuts the three lattice vectors, $\mathbf{a}_1/h$, $\mathbf{a}_2/k$, and $\mathbf{a}_3/\ell$, where h, k, and ℓ are three integers. Prove that the vector $\mathbf{G} = h\mathbf{b}_1 + k\mathbf{b}_2 + \ell\mathbf{b}_3$ is normal to this plane, and that the distance of the plane from the origin is $2\pi/G$.

Problem 3.13.

*A family of planes in the **BCC** lattice is related to the reciprocal-lattice vectors $\mathbf{G} = h\mathbf{b}_1 + k\mathbf{b}_2 + \ell\mathbf{b}_3$, where $\mathbf{b}_{1,2,3}$ are the lattice vectors found in problem 3.8. Show that these vectors can be represented in terms of the reciprocal-lattice vectors of the **SC**, e.g., those given in Eq. (3.22), $\mathbf{b}_1' = (2\pi/a)\hat{\mathbf{x}}$, $\mathbf{b}_2' = (2\pi/a)\hat{\mathbf{y}}$, and $\mathbf{b}_3' = (2\pi/a)\hat{\mathbf{z}}$, $\mathbf{G} = h'\mathbf{b}_1' + k'\mathbf{b}_2' + \ell'\mathbf{b}_3'$, provided that the new indices $\{h'k'\ell'\}$ cannot assume all integral values. What are the limitations on these new indices?*

The hexagonal lattice. The planes of the hexagonal lattice can be described in terms of the three Miller indices from Eq. (3.21), with the reciprocal-lattice vectors given in Eq. (3.18). The literature at times describes these planes with **four** indices, $(ijk\ell)$. The fourth index represents the cuts of these planes with the $\hat{\mathbf{z}}$−axis, like the third index in the previous description. The first three indices represent the cuts of the planes with three axes in the plane, with angles of $120°$ in-between them (see Fig. 3.15). These three indices are not independent, e.g., $k = -i - j$. Nonetheless, the characterization with four indices is based on three symmetric axes, and thus has certain advantages.

3.7   Experimental methods

The scattering angles and lattice planes. Inserting Eq. (3.27), $d = 2\pi/G$, into the Bragg law, Eq. (3.4), and making use of Eq. (3.21) yield

$$\sin\theta = \lambda|h\mathbf{b}_1 + k\mathbf{b}_2 + \ell\mathbf{b}_3|/(4\pi) \ . \tag{3.30}$$

The angle θ is between the wave vector of the impinging (or of the scattered) radiation and the scattering planes. Recall that the scattering angle is defined as 2θ (see Figs. 3.3 and 3.7). As the lattice vectors of the reciprocals of the 14 Bravais lattice are known (see Sec. 3.4 and problem s.3.2), one may use them in Eq. (3.30), and obtain the scattering angles for various planes $(hk\ell)$, for which the interference is constructive. Alternatively, when the scattering angle is fixed but the wave vector of the radiation is varied, one may identify the discrete wave lengths

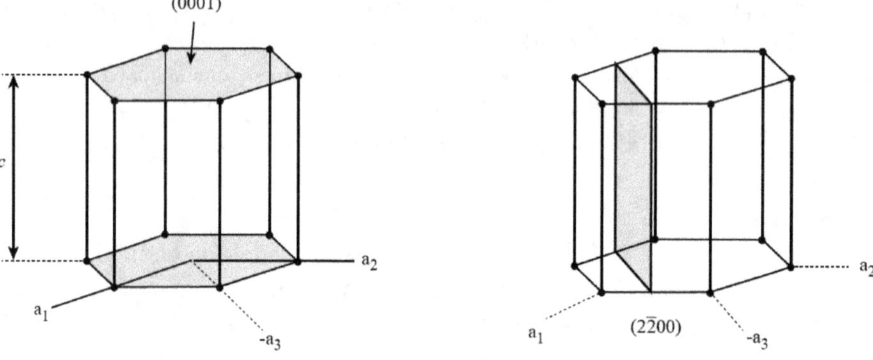

Fig. 3.15: (a) The four directions in the simple hexagonal lattice that define the four Miller indices for this lattice. (b) The (2̄200) plane, that is parallel to the $\hat{\mathbf{z}}$–axis and to the $\mathbf{a}_3$–axis, and cut the $\mathbf{a}_1$ and $\mathbf{a}_2$ axes at the middle of the corresponding lattice vectors.

that give rise to constructive interference. A comparison of the experimentally measured values (of the angles or of the wave lengths) with those calculated for each of the lattices (whose number is finite) enables the unique determination of the crystalline structure and its lattice constants. Examples for the calculation of the scattering angles, or alternatively, the determination of the crystal structure from its diffraction pattern, are discussed below.

Scattering off a simple-cubic lattice. The reciprocal lattice vectors of the **SC** lattice are $\mathbf{G} = (2\pi/a)(h\hat{\mathbf{x}} + k\hat{\mathbf{y}} + \ell\hat{\mathbf{z}})$ and therefore

$$\sin^2\theta = \frac{\lambda^2}{4a^2}(h^2 + k^2 + \ell^2) \,. \tag{3.31}$$

For a monochromatic wave, of a single wave length λ, and for a given lattice (whose lattice constant is a), constructive interference is achieved only for discrete values of θ, that correspond to certain triplets h, k, and ℓ. For instance, consider the range of values $h^2 + k^2 + \ell^2 \leq 24$. (This range is chosen arbitrarily to demonstrate the disparity among the cubic crystals; it is relevant for wave lengths such that $\lambda < 2a/5$, i.e., $\sin^2\theta \leq 1$.) In this range, the combination $(h^2 + k^2 + \ell^2)$ can attain all integers between 1 and 24, except for 7, 15, and 23 [see problem 3.14]. In practice, one prepares a list of the measured scattering angles. For a **SC** lattice, the smallest angle is given by $h^2 + k^2 + \ell^2 = 1$, i.e., $\sin\theta_{\min} = \lambda/(2a)$, and the other angles are then given by

$$\sin^2\theta/\sin^2\theta_{\min} = h^2 + k^2 + \ell^2 \,, \tag{3.32}$$

where the right hand-side assumes only the aforementioned allowed values. When the measurement yields the allowed values (and numbers higher than 24) for the ratio on the left hand-side, then one may conclude that the measured crystal is a **SC**. The value of the minimal θ yields the lattice constant a. For the treatment of other lattices, see problem s.3.2. As shown in Sec. 3.9, in other lattices of a cubic symmetry (e.g., **BCC**, **FCC**, or diamond) the scattering intensity from certain families of planes may vanish because of destructive interference (among the rays scattered from different base points); this in fact enables the identification of the internal structure of the unit cell.

Scattering off powders. The experimental technique to monitor **all** scattering angles in a single measurement is based on scattering off powders. In this method, illustrated in Fig. 3.3(b), a **monochromatic radiation**, of a **single wave length**, is scattered off a powder that consists of many grains directed at random. Each grain is large enough to provide a good diffraction picture, but is also sufficiently small so that there are many grains in the sample. For this reason, it is safe to assume that there are enough grains to scatter the impinging beam along all allowed directions; this means that the points that appear on the screen in Fig. 3.3(b) are due to all allowed angles according to Eq. (3.30). Since the lattice vectors in each grain are oriented at random relative to the direction of the beam (and to the directions of the other grains), the points on the screen are at different locations on circles determined by those angles. As a result, the pattern displayed on the screen is that of continuous circles, as in Fig. 3.3(b) or in Fig. 3.4. In other words, given the direction of the wave vector of the impinging radiation, $\mathbf{k}$, the wave vector of the scattered radiation, $\mathbf{k}'$, lies on a cone, whose axis is along $\mathbf{k}$ and its aperture angle is one of the allowed values of the scattering angle 2θ. The circles in Fig. 3.4 are created by the cut of the cone with the screen. In the configuration of Fig. 3.3(b), where only the forward-scattered radiation is examined, the circles appear only for $2\theta < \pi/2$. This condition limits the values of G which are available for the analysis. A less stringent condition pertains to the geometry of Fig. 3.3(a): there $2\theta < \pi$.

Multiplicity and intensity. Equation (3.31) implies that for a cubic lattice, the scattering angle depends solely on the combination $h^2+k^2+\ell^2$. That is, the same scattering angle corresponds to all planes belonging to the family $\{hk\ell\}$, as defined in the previous section. When the scattering is off a powder, all possible planes contribute to it; as a result, the intensity for each scattering angle is proportional to the number of planes belonging to the family (assuming point-like scatterers). For instance, the $\{100\}$ family contains six planes, the $\{110\}$ one contains 12, and the $\{211\}$ family comprises 24 equivalent planes. It is therefore expected that the scattering intensities at angles corresponding to these indices have the ratios 1:2:4. When the combination $h^2 + k^2 + \ell^2$ increases there might appear triplets of different Miller indices that give the same scattering angles; for example, the triplets (221) and (300). All reciprocal-lattice vectors whose tips lie on a sphere whose equation is $|\mathbf{G}|^2 = (2\pi/a)^2(h^2 + k^2 + \ell^2)$ and its center is at the origin, give rise to scattering at the same angle.

Problem 3.14.

Determine the scattering angles and their multiplicities for a **SC** *lattice whose lattice constant is trice the wave length of the radiation. [Consider only forward scattering,* *like in Fig. 3.3(b)].*

Other factors that affect the scattering intensities. Beside the multiplicity, the intensity of the scattered radiation at each of the Bragg peaks is sensitive to other factors.

- As mentioned at the end of Sec. 3.3, **the internal structure of the unit cell** (rather than scattering off point-like scatterers) affects the intensity. This issue is discussed in Sec. 3.9.
- In a finite specimen the **number of grains is finite** and therefore it is not obvious that the full multiplicity is reached for each angle. At times, the intensity is multiplied by a factor that takes into account the possible preference of certain grains' directions (this may happen for grains of specific geometrical shapes, a fact which restricts their random arrangement in the powder).
- Each of the grains has **a finite size**, and therefore the Bragg peaks attain a finite width (as explained in Sec. 3.3, e.g., Fig. 3.9). The intensities to be compared with each other are determined by their integrals under each peak (and not only their maximal heights).
- There are two possible **polarization directions** for the electromagnetic fields of the X-rays, in the plane perpendicular to $\mathbf{k}$. Assuming that the radiation is unpolarized, then the intensity has to be averaged over the two directions.
- As the radiation penetrates the sample, its intensity gradually decays due to **inelastic** scattering of the photons (or the scattered electrons) by the material electrons. (The photons collide with the latter, transfer them to higher energy levels and lose part of their energy in the process; therefore their wave length is modified.) This effect adds a multiplicative factor to the intensity of the elastically-scattered radiation that decays exponentially with the path each photon traverses in the material (which also depends upon the angle θ).

It follows that the extraction of the "pure" intensity to be compared with the calculated one is not a simple task and requires special experimental dexterity. This section focuses on the analysis of the scattering angles; these allow for the determination of the primitive lattice of the measured material. The factors that determine the intensity of the scattering, which carry the information on the internal structure (i.e., the base of the unit cell), are considered in the following.

Problem 3.15.
a. *Derive equations that replace Eq. (3.30) for the orthorhombic lattice and for the tetragonal one.*
b. *X-rays, obtained from the transitions $K_{\alpha 1}$ and $K_{\beta 1}$ of copper (problem 3.1), are scattered off a powder of a tetragonal lattice whose lattice constants are $a_1 = a_2 =$*

$2\mathring{A}$, $a_3 = 3\mathring{A}$. *Find the smallest eight scattering angles for which the interference is constructive.*

Problem 3.16.

a. *The smallest scattering angle measured in a scattering experiment off a powder of a* **SC** *lattice is $40°$. How many scattering circles are detected? At which angles? Assume that each unit cell contains a single point-like scatterer, and that the scattering intensity of all of them is the same [Eq. (3.14)]. Find the scattering intensity in each ring (accounting only for the multiplicities, and not for the other factors that affect the intensity).*

b. *As the temperature decreases, the lattice described in part (a) becomes tetragonal, with the ratio $a_3/a_2 = 1.1$ (the lattice constants $a_1 = a_2 = a$ remain the same). How are modified the scattering angles and the intensities?*

c. *Describe qualitatively the scattering angles and the intensities when there is an additional transition, to the orthorhombic structure.*

Measurement by rotating a single crystal. Another experimental method exploits **monochromatic radiation**, of a single wave length; the scattering angle is varied by **rotating the crystal**. Constructive interference is achieved only for angles for which there exist suitable families of planes. The setup is displayed in Fig. 3.16: the beam impinges normal to the axis around which the crystal is rotated, and the scattered beam is collected on a cylindrical screen encompassing the rotation axis. Denoting that axis by $\hat{\mathbf{x}}$, and the direction of the incoming beam by $\hat{\mathbf{z}}$, then the vector $\mathbf{G}$ representing a family of planes parallel to the rotation axis is perpendicular to that axis. Thus the vectors $\mathbf{k}$ and $\mathbf{G}$ are in the YZ plane. It then follows that the vector $\mathbf{k}' = \mathbf{k} + \mathbf{G}$ lies in the same plane, and therefore there appears a point on the screen whenever the scattering angle fulfils the Bragg law, i.e., $\sin\theta = \lambda G/(4\pi)$. When the planes are inclined by an angle γ relative to the rotation axis, then the vector $\mathbf{G}$ forms an angle $90° - \gamma$ with that axis; hence its $x-$component is fixed, $G_x = |\mathbf{G}| \sin\gamma$, independent of the scattering angle. The $x-$component of $\mathbf{k}' = \mathbf{k} + \mathbf{G}$ is fixed as well, and all radiation points resulting from these planes are on a plane parallel to the YZ planes. The distances between the planes can be deduced from the scattering angles.

The von Laue method. A third experimental method, named after **von Laue**, uses a single crystal irradiated by a beam comprising a **continuum** of wave lengths, spanned over several angstroms. The scattered beam hits a screen, like in Fig. 3.3(b). For an arbitrary impinging direction it may well happen that a monochromatic beam will not obey the Bragg law. However, when the radiation consists of many wave lengths, there is a finite probability that one (or several) of them will give rise to constructive interference. Therefore, as the crystal is rotated radiation points appear on the screen, due to various vectors in the reciprocal lattice. These reflect the symmetry of the reciprocal lattice, and, in turn, that of the original lattice. Consequently, one may ensure that the crystal is placed along a

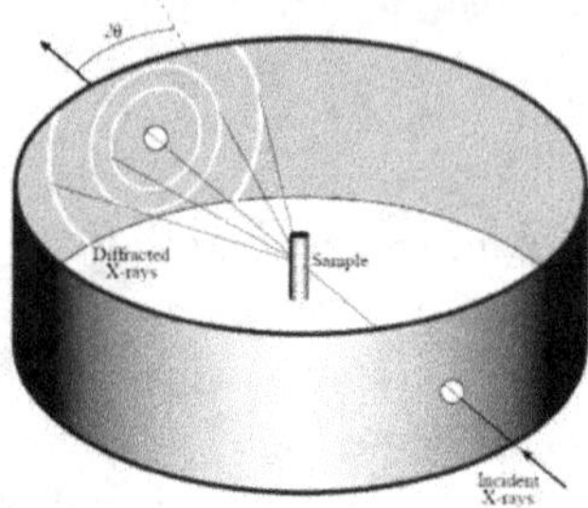

Fig. 3.16: Diffraction pattern obtained by rotating a crystal placed on the axis of a cylinder. The radiation impinges normal to the axis, and is scattered onto the cylindrical screen.

direction that corresponds to a certain symmetry operation. This method is indeed mainly exploited to identify the symmetries of the crystal and is not usually used for precision measurements of the structure.

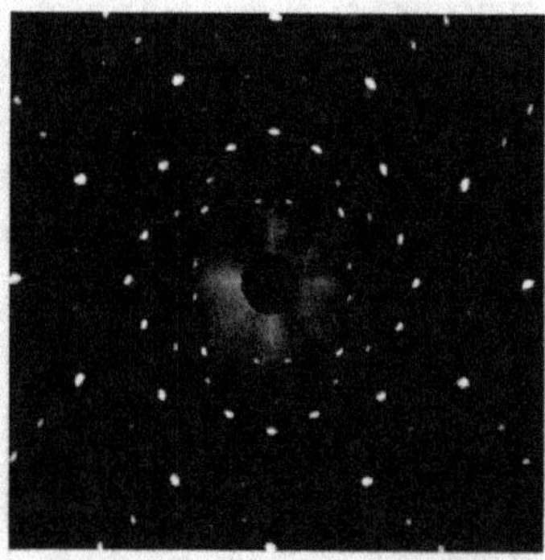

Fig. 3.17: Diffraction pattern of table salt. [D. Halliday, R. Resnick, and K. S. Krane, *Physics*, vol. 2, 5th edition, John Wiley and Sons, 2002.]

Figure 3.17 displays a typical diffraction picture, obtained by scattering X-rays off a table-salt crystal (which has an **FCC** structure) using the von Laue method. The impinging ray is parallel to a rotation axis of 90°, and the diffraction pattern reflects this symmetry. A detailed analysis of the radiation points for this structure is presented in Sec. 3.9.

3.8 Fourier series and reciprocal lattice

Periodic functions on the crystal. Leaving the physical discussion aside for a little while, this section considers the properties of **functions that have the periodicity of the lattice.** Appendix B surveys some formal aspects of the **Fourier theorem**; the implications of this theorem are exploited here. As mentioned in Chapter 2, any physical property $f(\mathbf{r})$ of the periodic crystal is periodic as well. That is, it has identical values at points whose distance from each other is one of the lattice vectors,

$$f(\mathbf{r} + \mathbf{R}) = f(\mathbf{r}) , \tag{3.33}$$

for **any** lattice vector $\mathbf{R} = n_1\mathbf{a}_1 + n_2\mathbf{a}_2 + n_3\mathbf{a}_3$. To apply this condition to a finite crystal, it is assumed that the crystal obeys periodic boundary conditions (those of Born and von Karman), and hence properties on a certain face of the specimen are identical to those on the opposite face; this assumption ensures that Eq. (3.33) is satisfied also near the surfaces [see also the discussion preceding Eq. (3.11)]. The results obtained under these circumstances are valid for very large crystals, but do not pertain to nanometric ones (for which one may not ignore the effect of the surface). The periodic boundary conditions imply that the specimen repeats itself time and again and therefore they are equivalent to its periodic continuation over the entire space.

Fourier series. Consider the function $\Phi_\mathbf{k}(\mathbf{r}) = \exp[i\mathbf{k} \cdot \mathbf{r}]$, which represents a **plane wave in space**, with wave vector $\mathbf{k}$. For this function to satisfy Eq. (3.33), it should be such that $\Phi_\mathbf{k}(\mathbf{r}) = \Phi_\mathbf{k}(\mathbf{r} + \mathbf{R}) = \exp[i\mathbf{k} \cdot (\mathbf{r} + \mathbf{R})] = \Phi_\mathbf{k}(\mathbf{r}) \exp[i\mathbf{k} \cdot \mathbf{R}]$. This function has the periodicity of the lattice provided that $\exp[i\mathbf{k} \cdot \mathbf{R}] = 1$ for any lattice vector $\mathbf{R}$. As defined in the context of Eq. (3.9), this condition requires that the wave vector $\mathbf{k}$ coincides with one of the reciprocal-lattice vectors $\mathbf{G}$ [see also Eq. (3.17)]. The **generalized Fourier theorem** states that the collection of functions $\{\Phi_\mathbf{G}(\mathbf{r}) = \exp[i\mathbf{G} \cdot \mathbf{r}]\}$ **spans the space that comprises all functions which have the periodicity of the lattice (3.33)**; i.e., **each such function can be written as a series,**

$$f(\mathbf{r}) = \sum_\mathbf{G} \widetilde{f}(\mathbf{G})e^{i\mathbf{G} \cdot \mathbf{r}} , \tag{3.34}$$

with coefficients $\widetilde{f}(\mathbf{G})$ determined by the function, and the sum runs over **all** reciprocal-lattice vectors. This series is named after **Fourier**. The equation is identical to Eq. (B.24) in Appendix B, where it is obtained in a different way. The coefficients are given by Eq. (B.25); another possibility to derive them is by multiplying both sides of Eq. (3.34) by $\exp[-i\mathbf{G}' \cdot \mathbf{r}]$ and then integrating over a unit cell, of volume V,

$$\int_V d^3r f(\mathbf{r})e^{-i\mathbf{G}' \cdot \mathbf{r}} = \sum_\mathbf{G} \widetilde{f}(\mathbf{G}) \int_V d^3r e^{i(\mathbf{G} - \mathbf{G}') \cdot \mathbf{r}} . \tag{3.35}$$

When $\mathbf{G} = \mathbf{G}'$ the integrand on the right hand-side is unity, and the integral yields V. In contrast, when $\mathbf{G} \neq \mathbf{G}'$, the integral vanishes. This can be shown as follows. Since the integrand is periodic and satisfies Eq. (3.33), the integral is unchanged when the unit cell is shifted by an arbitrary vector $\mathbf{d}$. This shift is equivalent to a shift in the integration variable by $\mathbf{d}$, without changing the integration range,

$$\int_V d^3 r\, e^{i(\mathbf{G}-\mathbf{G}')\cdot\mathbf{r}} = \int_V d^3 r\, e^{i(\mathbf{G}-\mathbf{G}')\cdot(\mathbf{r}+\mathbf{d})} = e^{i(\mathbf{G}-\mathbf{G}')\cdot\mathbf{d}} \int_V d^3 r\, e^{i(\mathbf{G}-\mathbf{G}')\cdot\mathbf{r}} \ .$$

As $\mathbf{d}$ is arbitrary and $\mathbf{G} \neq \mathbf{G}'$, the factor $\exp[i(\mathbf{G} - \mathbf{G}') \cdot \mathbf{d}]$ is not equal to 1, and the integral must vanish. Hence,

$$\int_V d^3 r\, e^{i(\mathbf{G}-\mathbf{G}')\cdot\mathbf{r}} = V\delta_{\mathbf{G},\mathbf{G}'} \ , \tag{3.36}$$

where $\delta_{\mathbf{G},\mathbf{G}'}$ is the Kronecker delta function [mentioned after Eq. (3.20)]. It follows that the sum on the right hand-side of Eq. (3.35) includes just the term $\mathbf{G} = \mathbf{G}'$, i.e.,

$$\tilde{f}(\mathbf{G}) = \frac{1}{V} \int d^3 r\, f(\mathbf{r}) e^{-i\mathbf{G}\cdot\mathbf{r}} \ , \tag{3.37}$$

just like Eq. (B.25). Thus, **in order to obtain the coefficients of the Fourier series of a function periodic on the crystal it suffices to compute the integral (3.37) on a single unit cell.**

Problem 3.17.
A function $f(\mathbf{r})$ is periodic on a planar square lattice, whose lattice constant is a. The function equals $1/b^2$ within the square $0 \leq x \leq b$ and $0 \leq y \leq b$ and zero otherwise, where $b \leq a$. Find the Fourier series of this function. What is the answer in the limit $b \to 0$?

Examples. The length of the reciprocal-lattice vector in **one dimension** is $b = 2\pi/a$. It follows that the coefficients in the Fourier series of a one-dimensional periodic function, $\tilde{f}(q)$, are nonzero only for $q = G = 2\pi\ell/a$, where ℓ is an arbitrary integer, and Eq. (3.37) becomes

$$\tilde{f}(\ell) = (1/a) \int_0^a dx\, f(x) e^{-2\pi i\ell x/a} \ . \tag{3.38}$$

Thus, the **Fourier series of the periodic function** $f(x)$ is

$$f(x) = \sum_{\ell=-\infty}^{\infty} \tilde{f}(\ell) e^{2\pi i\ell x/a} \ . \tag{3.39}$$

These are precisely the expressions of the Fourier theorem in one dimension, Eqs. (B.2) and (B.4) in Appendix B. A similar result is obtained for the rectangular lattice in **two dimensions,** or for the orthorhombic lattice in **three dimensions.**

3.9   The structure factor

Several scatterers in the unit cell. So far only scattering off point-like scatterers is considered, for which all Bragg peaks have the same intensity, e.g., Eq. (3.14). In reality, X-rays are scattered mostly by the electrons within the atoms while neutrons are more sensitive to the nuclei density and/or to the magnetic moment of each atom. Even when the unit cell contains a single atom, its electrons are not confined to a single point and therefore the assumption of a point-like scatterer is not entirely correct. Likewise, the nuclei occupy a finite volume and are not precisely confined to the center of the atom. It follows that Eq. (3.7) has to be modified. When the scatterers are all identical (e.g., they are all electrons) then their scattering amplitudes are also the same; denote that amplitude by $a_0 \exp[i\psi_0]$. In addition to the summation over all unit cells, as carried out in Eq. (3.7), it is also required to sum over all contributions to the amplitude A of the scattered wave, coming from each of the scatterers **within** each unit cell,

$$A = a_0 e^{i\psi_0} \sum_{\mathbf{R}} \sum_{i} e^{i(\mathbf{k}'-\mathbf{k})\cdot(\mathbf{R}+\mathbf{r}_i)} , \tag{3.40}$$

where $\mathbf{r}_i$ is the location of the i−th scatterer in the unit cell, relative to the lattice point $\mathbf{R}$ that determines the origin there.

   The structure factor. Equation (3.40) can be written as

$$A = a_0 e^{i\psi_0} \Big[\sum_{\mathbf{R}} e^{i\mathbf{q}\cdot\mathbf{R}} \Big] \Big[\sum_{i} e^{i\mathbf{q}\cdot\mathbf{r}_i} \Big] = a_0 e^{i\psi_0} \mathcal{Z}(\mathbf{q}) F(\mathbf{q}) , \tag{3.41}$$

where $\mathbf{q} = \mathbf{k}' - \mathbf{k}$, and $\mathcal{Z}(\mathbf{q}) = \sum_{\mathbf{R}} \exp[i\mathbf{q} \cdot \mathbf{R}]$ is defined in Eq. (3.8). The new function,

$$F(\mathbf{q}) = \sum_{i} e^{i\mathbf{q}\cdot\mathbf{r}_i} , \tag{3.42}$$

is a sum over the locations of the scatterers in the unit cell, and therefore carries the information on the structure of the base contained in it. This function is called the **structure factor**. In Sec. 3.3 it is shown that the function $\mathcal{Z}(\mathbf{q})$ is sharply peaked when $\mathbf{q}$ approaches one of the reciprocal-lattice sites, i.e., when $\mathbf{q} \to \mathbf{G}$. These peaks turn into delta functions for the infinite lattice [Eq. (3.14)]. Hence, **when the unit cell possesses an internal structure that is not point-like, the diffraction pattern is still peaked at the reciprocal-lattice points.** The structure factor, though, adds a dependency of the scattering intensity on the reciprocal-lattice vector, $|F(\mathbf{G})|^2$, so that each Bragg peak has a different intensity. As demonstrated below, there are even cases where the contributions of the various constituents within the unit cell to the structure factor (3.42) for a certain reciprocal-lattice vector cancel each other, and the intensity approaches zero. In such cases the waves scattered from the various elements in the unit cell interfere destructively. Monitoring the scattering intensities at many -reciprocal-lattice vectors provides information on $|F(\mathbf{G})|^2$, that enables one to deduce the scatterers' locations $\{\mathbf{r}_i\}$ and thus the internal structure of the base. As the number of those scatterers increases more data points must be collected from a larger number of reciprocal-lattice vectors. The spatial distribution of the electrons within each atom induces a decay of $|F(\mathbf{G})|^2$ as $|\mathbf{G}|$ becomes larger (see below). It follows that measurements at large $\mathbf{G}'$s necessitate high intensities of the incoming radiation (to ensure sufficient intensities for large reciprocal-lattice vectors).

Integral representation. A more general representation of the scattering amplitude, that allows for a **continuous volume-density of scatterers**, is derived by transforming the sums in Eq. (3.40) into a continuous integral over the entire volume of the crystal. Suppose that an infinitesimal volume-element d^3r around an arbitrary point $\mathbf{r}$ contains $n(\mathbf{r})d^3r$ scatterers (e.g., electrons or nucleons). Equation (3.40) then becomes

$$A = a_0 e^{i\psi_0} \int d^3r\, n(\mathbf{r}) e^{i(\mathbf{k}'-\mathbf{k})\cdot\mathbf{r}} \ , \qquad (3.43)$$

where the integral is carried out over the crystal volume NV, and where $\exp[i(\mathbf{k}' - \mathbf{k}) \cdot \mathbf{r}]$ represents the phase difference between the wave scattered from the point $\mathbf{r}$ and the one scattered from the origin, see Fig. 3.18. The function $n(\mathbf{r})$ is the **scatterers' density** at the point $\mathbf{r}$. When the unit cell contains a single atom that contains z electrons, and when it is assumed that all electrons reside at the center of the atom, then the density function is a sum over delta functions, [or a "comb" of delta, Eq. (B.26)],

$$n(\mathbf{r}) = \sum_{\mathbf{R}} z\delta(\mathbf{r} - \mathbf{R}) \ . \qquad (3.44)$$

Substituting this in Eq. (3.43), and making use of the main property of the delta function [Eq. (3.13)], yield $A = za_0 \exp[i\psi_0] \sum_{\mathbf{R}} \exp[i(\mathbf{k}' - \mathbf{k}) \cdot \mathbf{R}]$; this is the scattering amplitude of a point-like scatterer, Eq. (3.7). Similarly, substituting $n(\mathbf{r}) = \sum_{\mathbf{R},j} z\delta(\mathbf{r} - \mathbf{R} - \mathbf{r}_j)$ reproduces Eq. (3.40).

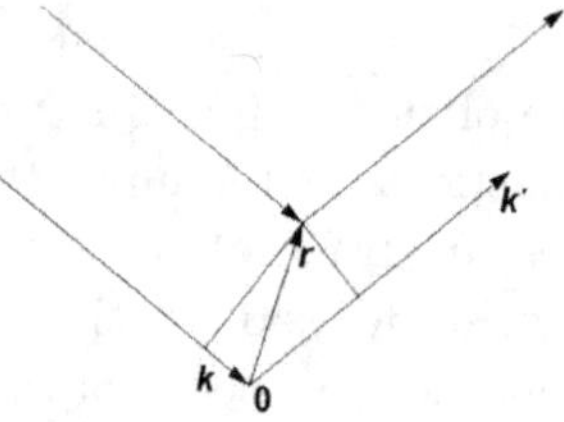

Fig. 3.18: Scattering off an arbitrary point in a continuous specimen. The phase difference between the wave scattered from the origin, 0, and the one scattered from the scatterers located within a volume d^3r around the point $\mathbf{r}$ (both within the specimen) is $(\mathbf{k}' - \mathbf{k}) \cdot \mathbf{r}$ and the total amplitude of these scatterers is $a_0 \exp[i\psi_0] n(\mathbf{r}) d^3r$.

The scattering amplitude and Fourier series. In a periodic lattice the density function $n(\mathbf{r})$ follows the periodicity (of the lattice), and therefore it satisfies

Eq. (3.33),

$$n(\mathbf{r} + \mathbf{R}) = n(\mathbf{r}) \ . \tag{3.45}$$

Divide the integration volume in Eq. (3.43) into N unit cells of the sample, and write down the integral as the sum of integrals over each of them, which contains the lattice point $\mathbf{R}$. The result is

$$\int_{NV} d^3 r\, n(\mathbf{r}) e^{i(\mathbf{k'}-\mathbf{k})\cdot\mathbf{r}} = \sum_{\mathbf{R}} \left[\int_{V(\mathbf{R})} d^3 r\, n(\mathbf{r}-\mathbf{R}) e^{i(\mathbf{k'}-\mathbf{k})\cdot(\mathbf{r}-\mathbf{R})} \right] e^{i(\mathbf{k'}-\mathbf{k})\cdot\mathbf{R}}$$

$$= \mathcal{Z}(\mathbf{k'}-\mathbf{k}) \int_{V(0)} d^3 r\, n(\mathbf{r}) e^{i(\mathbf{k'}-\mathbf{k})\cdot\mathbf{r}} \ ,$$

where the first step uses Eq. (3.45), and in the second each member is multiplied and divided by $\exp[i(\mathbf{k'}-\mathbf{k})\cdot\mathbf{R}]$. By translating the origin in each of the integrals within the square brackets, it is found that they are all equal to each other; this results in the last step of the calculation. Substituting in Eq. (3.43) reproduces then Eq. (3.41), with the structure factor given by

$$F(\mathbf{q}) = \int_V d^3 r\, n(\mathbf{r}) e^{i\mathbf{q}\cdot\mathbf{r}} \ . \tag{3.46}$$

As in Eq. (3.41), the factor $\mathcal{Z}(\mathbf{q})$ is nonzero only when $\mathbf{q} = \mathbf{k'} - \mathbf{k}$ approaches a reciprocal-lattice vector, $\mathbf{q} \to \mathbf{G}$. It therefore suffices to find the structure factor at reciprocal-lattice sites, $F(\mathbf{G})$. Equation (3.45) implies that $n(\mathbf{r})$ obeys Eq. (3.33) and hence it satisfies Eq. (3.34) as well, $n(\mathbf{r}) = \sum_{\mathbf{G}} \tilde{n}(\mathbf{G}) \exp[i\mathbf{G}\cdot\mathbf{r}]$, with coefficients given by Eq. (3.37),

$$\tilde{n}(\mathbf{G}) = \frac{1}{V} \int_V d^3 r\, n(\mathbf{r}) e^{-i\mathbf{G}\cdot\mathbf{r}} \equiv \frac{F(-\mathbf{G})}{V} \ . \tag{3.47}$$

Therefore, **the structure factor is proportional to the Fourier coefficient of the scatterers' density.** When the scattering is off discrete points in the unit cell, the integral in Eq. (3.47) becomes a discrete sum, precisely as in Eq. (3.42).

   The correlation function. The intensity of the scattered radiation of the Bragg peaks at $\mathbf{k'} - \mathbf{k} = \mathbf{G}$ is given by the absolute value squared of the amplitude, Eq. (3.43). Indeed, for $\mathbf{k'} - \mathbf{k} = \mathbf{q} = \mathbf{G}$,

$$|A|^2 = a_0^2 \int_{NV} d^3 r\, n(\mathbf{r}) e^{i\mathbf{G}\cdot\mathbf{r}} \int_{NV} d^3 r'\, n(\mathbf{r'}) e^{-i\mathbf{G}\cdot\mathbf{r'}}$$

$$= a_0^2 \int_{NV} d^3 r \int_{NV} d^3 r''\, n(\mathbf{r}) n(\mathbf{r}+\mathbf{r''}) e^{-i\mathbf{G}\cdot\mathbf{r''}} \ .$$

The periodic boundary conditions of Born and von Karman are used in the second step to interchange the integration variable, $\mathbf{r'} \to \mathbf{r''} = \mathbf{r'} - \mathbf{r}$. This expression can be written as

$$|A|^2 = a_0^2 (NV)^2 \tilde{\Gamma}(\mathbf{G}) \ , \tag{3.48}$$

where

$$\widetilde{\Gamma}(\mathbf{G}) = [1/(NV)] \int_{NV} d^3r \Gamma(\mathbf{r}) \exp[-i\mathbf{G} \cdot \mathbf{r}]$$

is the coefficient in the Fourier series of the **average correlation function**, defined by

$$\Gamma(\mathbf{r}'') = \frac{1}{NV} \int_{NV} d^3r\, n(\mathbf{r}) n(\mathbf{r} + \mathbf{r}'') = \langle n(0) n(\mathbf{r}'') \rangle . \tag{3.49}$$

(Here the integrations are carried out over the entire sample; this enables the consideration of samples with no inner periodicity, like liquids or gases.) **The average correlation function** gives the average probability per unit volume to find a scatterer at $\mathbf{r}''$ when it is known that there is a scatterer at $\mathbf{r} = 0$. The angular brackets represent an average over the origin location [the integral in Eq. (3.49) implies a summation over all possible origins, and a division by their total number, and therefore it represents an average over them]. The correlation function is discussed in Chapter 1, and Fig. 1.2 illustrates the disparity among the correlation functions of a gas, a liquid, and a solid. Measuring radiation scattered off various materials yields the Fourier transform of the correlation function, and thus provides information on the specific phase of the material and its structure. In particular, a periodic solid (and also a quasi-periodic one) gives discrete Bragg peaks, as opposed to gases, liquids, and glasses, which yield a continuous diffraction pattern with broad peaks.

The Fourier coefficient of the scatterers' density is

$$\widetilde{n}(\mathbf{q}) = \frac{1}{NV} \int_{NV} d^3r\, n(\mathbf{r}) e^{-i\mathbf{q} \cdot \mathbf{r}} , \tag{3.50}$$

and therefore $|A|^2 = a_0^2 (NV)^2 |\widetilde{n}(\mathbf{q})|^2$. This result also follows from the relation $\widetilde{\Gamma}(\mathbf{q}) = |\widetilde{n}(\mathbf{q})|^2$. The Fourier coefficients of the correlation function are the absolute values squared of the coefficients of the scatterers' density function. This can be seen directly from Eq. (3.49), which states that the **average correlation function is a convolution of the density function with itself**, $\Gamma(\mathbf{r}) = n(\mathbf{r}) \otimes n(\mathbf{r})$, see Eq. (B.12).

Scattering off a single atom in the unit cell. Suppose that **each unit cell contains a single atom**. This atom is located at the center of the Wigner-Seitz unit cell; the density function $n(\mathbf{r})$ is thus expected to decay away from that center. For instance, in the ground state of the hydrogen atom, the quantum probability density to find the electron at a distance $\mathbf{r}$ from the nucleus is $n(\mathbf{r}) = |\psi(\mathbf{r})|^2 = \exp[-2r/a_B]/(\pi a_B^3)$, where $a_B = \hbar^2/(me^2) = 0.529\text{Å}$ is the Bohr radius. Such an exponential decay is obtained also for other atoms, with an atomic radius of the same order of magnitude. Because of this strong decay outside the Wigner-Seitz cell, the integral in Eq. (3.47) can be approximated by an integral carried out over the entire space, and not only within the cell. (In more precise computations one begins from the electrons' distribution within the unit cell, takes into account the potentials of the other atoms in the lattice and the boundary conditions on the faces of the cell—

and then calculates the integrals only within the unit cell.) When the atomic radius is significantly smaller than the size of the unit cell, then $r \simeq a_B \ll a \sim 2\pi/G$ (using the fact that the length of the reciprocal-lattice vector is of the order of $2\pi/a$, where a is one of the lattice constants). Under these circumstances, the oscillating term $\exp[-i\mathbf{G}\cdot\mathbf{r}]$ is roughly 1, and consequently the integral in Eq. (3.47) gives the total number of electrons in the atom, z. The structure factor becomes independent of the reciprocal-lattice vector $\mathbf{G}$, and the intensities at the points of constructive interference are identical (up to the multiplicity). This is the case of scattering off point-like scatterers, analyzed in Sec. 3.3. Nonetheless, as $|\mathbf{G}|$ increases, the factor $\exp[-i\mathbf{G}\cdot\mathbf{r}]$ oscillates more wildly within the atomic radius, and the integral decreases (inverse signs of this factor cancel each other's contribution). These oscillations diminish the contribution to the integral from far-away regions, and validate the integration over the entire space. As the scattering angle increases [and with it the length of the vector $\mathbf{G}$, see Fig. 3.12(a)], the intensity of the radiation points collected on the screen decreases. Hence, when many points with not-too-small intensities are needed for extracting information about the density function, the intensity of the impinging radiation has to be increased. For instance, good scattering data off complicate molecules, e.g., ribosome, located at lattice points, requires the use of radiation emitted from a synchrotron.

Problem 3.18.

a. *Prove that the structure factor of a charge distribution symmetric under reflection, $n(\mathbf{r}) = n(-\mathbf{r})$, is real.*

b. *Prove that for a spherically-symmetric electrons' distribution in an atom (or molecule), and when the integration is over the entire space, the structure factor depends solely on the length of the vector $\mathbf{G}$, and is given by $F(\mathbf{G}) = (4\pi/G)\int_0^\infty dr\ r\ n(r)\sin(Gr)$.*

c. *Find the structure factor for the ground state of the hydrogen atom. Determine the way the answer is modified for the helium atom (ignoring the interaction between the two electrons).*

Scattering off a crystal with a base. When each unit cell contains a base with n_B atoms (or ions) [Eq. (2.4)], the electrons' density in the unit cell is a sum, $n(\mathbf{r}) = \sum_{i=1}^{n_B} n_i(\mathbf{r} - \mathbf{r}_i)$, where the function $n_i(\mathbf{r} - \mathbf{r}_i)$ is the density of electrons around the i-th atom, located at $\mathbf{r}_i$. Inserting this sum into Eq. (3.47), and displacing the origin in each of the summands to the center of the i-th atom (the integration variable in each of the summands is changed from $\mathbf{r}$ to $\mathbf{r} - \mathbf{r}_i$) yields

$$F(\mathbf{G}) = \int_V d^3r \sum_{i=1}^{n_B} n_i(\mathbf{r} - \mathbf{r}_i)e^{i\mathbf{G}\cdot\mathbf{r}} = \sum_{i=1}^{n_B} F_i(\mathbf{G})e^{i\mathbf{G}\cdot\mathbf{r}_i}\,, \qquad (3.51)$$

where F_i is the structure factor of the i-th atom,

$$F_i(\mathbf{G}) = \int_V d^3r\, n_i(\mathbf{r})e^{i\mathbf{G}\cdot\mathbf{r}}\,, \qquad (3.52)$$

and the integration bounds were shifted to the Wigner-Seitz cell encompassing the i–th atom. As in the preceding discussion, it is plausible that the density $n_i(\mathbf{r})$ decays with the distance form the center of the i–th atom, and therefore the integration region can be extended to cover the entire space (though this approximation is not really necessary here). Equation (3.51) contains the structure factors of all atoms in the base of the crystal and their locations within the unit cell. As the scattering intensity is related to the reciprocal-lattice vector $\mathbf{G}$ (which in turn determines the structure factor) then when enough data points are collected, it is possible in principle to identify the atoms in the base and their locations.

Scattering off a rectangular-centered lattice. A simple two-dimensional example is that of the **rectangular-centered lattice**, described by the lattice vectors $\mathbf{a}_1 = a_1\hat{\mathbf{x}}$ and $\mathbf{a}_2 = a_2\hat{\mathbf{y}}$, with a base comprising two identical atoms at the origin and at the point $\mathbf{r}_2 = (\mathbf{a}_1 + \mathbf{a}_2)/2$ (these are not the Bravais lattice vectors displayed in Fig. 2.27). The reciprocal of the rectangular lattice is rectangular as well (Fig. 3.10), with $\mathbf{G} = 2\pi(h\hat{\mathbf{x}}/a_1 + k\hat{\mathbf{y}}/a_2)$. In this case, Eq. (3.52) yields

$$F(\mathbf{G}) = F_0(\mathbf{G})(1 + \exp[i\mathbf{G} \cdot \mathbf{r}_2]) = F_0(\mathbf{G})(1 + \exp[i\pi(h + k)]) \ ,$$

where $F_0(\mathbf{G})$ is the structure factor of a single atom. Hence, $F(\mathbf{G}) = 2F_0(\mathbf{G})$ when $h + k$ is even, and $F(\mathbf{G}) = 0$ when $h + k$ is odd. The Bragg peaks appear only for $h + k$ even, e.g., $(hk) = (\pm 1 \ \pm 1)$, $(\pm 2 \ 0)$, $(0 \ \pm 2)$, $(\pm 1 \ \pm 3)$, …. The collection of these points forms a rectangular-centered lattice, with edges of lengths $4\pi/a_1$ and $4\pi/a_2$ (check!).

This result can be derived in a different way. The lattice vectors of the primitive unit cell are $\mathbf{a}'_1 = a_1\hat{\mathbf{x}}$ and $\mathbf{a}'_2 = (a_1\hat{\mathbf{x}} + a_2\hat{\mathbf{y}})/2$ (Fig. 2.27), and its area is $a_1a_2/2$. The reciprocal-lattice vectors are $\mathbf{b}'_1 = 2\pi(\hat{\mathbf{x}}/a_1 - \hat{\mathbf{y}}/a_2)$ and $\mathbf{b}'_2 = 4\pi\hat{\mathbf{y}}/a_2$. Hence $\mathbf{G} = h'\mathbf{b}'_1 + k'\mathbf{b}'_2 = 2\pi[h'\hat{\mathbf{x}}/a_1 + (2k' - h')\hat{\mathbf{y}}/a_2]$, leading to $h = h'$ and $k = 2k' - h'$. Indeed, $h + k$ is even, as found previously. This identity is quite general: the measured structure factor is obviously independent of the choice of the unit cell. Sometimes it is more convenient to compute it in the primitive unit cell, while there are situations in which it is more expedient to find it in a bigger unit cell (which is more symmetric). Other such examples appear in all cubic lattices (e.g., **FCC**, **BCC**, diamond).

Scattering off hexagonal lattices. A somewhat more complex case is that of the **hexagonal lattice**, which describes the planar crystals of **graphene**. As seen in Fig. 2.8, this is a triangular lattice with two atoms in the unit cell, at the origin and at $\mathbf{r}_2 = (\mathbf{a}_1 + \mathbf{a}_2)/3$. Figure 3.11 shows that the reciprocal of the triangular lattice is also triangular, with the lattice vectors $\mathbf{b}_1 = 2\pi(\hat{\mathbf{x}} - \hat{\mathbf{y}}/\sqrt{3})/a$ and $\mathbf{b}_2 = 4\pi\hat{\mathbf{y}}/(a\sqrt{3})$, and hence an arbitrary reciprocal-lattice vector is $\mathbf{G} = h\mathbf{b}_1 + k\mathbf{b}_2$. It follows that

$$F(\mathbf{G}) = F_0(\mathbf{G})(1 + \exp[i\mathbf{G} \cdot \mathbf{r}_2]) = F_0(\mathbf{G})(1 + \exp[2i\pi(h + k)/3])$$

(check!). The expression in the brackets has the values 2 or $(1 \pm i\sqrt{3})/2$, when $h + k$ is an integral multiple of 3 or else, respectively. Figure 3.19 displays the reciprocal of the triangular lattice. The big points there are those for which $h + k$ is an integral multiple of 3. For instance, the two arrows in the figure are for $(hk) = (12)$, (21). These are the lattice vectors of a triangular lattice containing all big points, whose lattice constant is $b' = b\sqrt{3} = 4\pi/a$ [where $b = 4\pi/(a\sqrt{3})$ is the reciprocal-lattice constant in Fig. 3.11]. The scattering intensity at these points is four times that of the other sites in the reciprocal lattice (marked by small points in the figure). As shown in Fig. 2.12, the triangular lattice is naturally divided into three sublattices. The diffraction pattern from the hexagonal lattice breaks the symmetry among those sublattices, by yielding higher scattering intensities on one of them. Notice that the other two sublattices form a hexagonal lattice, such that the hexagonal symmetry is conserved in the diffraction pattern. The points of the higher intensity are at the centers of the hexagons. For a simple triangular lattice, the intensity is the same at all points.

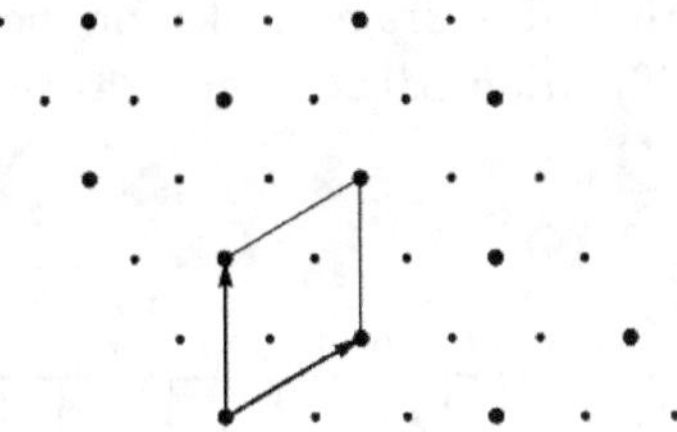

Fig. 3.19: Lattice points in the reciprocal of the hexagonal lattice. The scattering intensity at the big points is larger. The intensities at all big (or all small) points are identical. The arrows confine the enlarged unit cell of this lattice, which contains one big point and two small ones.

Problem 3.19.

Assume that the two points in the unit cell of the hexagonal lattice represent two different point-like atoms, whose structure factors are F_A and F_B. Determine the structure factor of this crystal. Can the intensity at part of the reciprocal-lattice sites vanish?

Problem 3.20.

Determine the structure factor of graphite, shown in Fig. 2.9 (see also problem s.2.10). Are there any sites in the reciprocal lattice where the structure factor is zero?

Scattering off cubic lattices with a base. An example for a **three-dimensional lattice with a base** is the **BCC** lattice, the left panel in Fig. 2.16. Its base comprises two sites, at the origin and at the center of the cube, $a(\hat{\mathbf{x}} + \hat{\mathbf{y}} + \hat{\mathbf{z}})/2$. The reciprocal-lattice vector is $\mathbf{G}_0 = (2\pi/a)(h\hat{\mathbf{x}} + k\hat{\mathbf{y}} + \ell\hat{\mathbf{z}})$. The structure factor (3.51) is thus proportional to $\{1 + \exp[i\pi(h + k + \ell)]\}$ and is nonzero when $h + k + \ell$ is even, i.e., when all three indices are even, or when one of them is even and the other two are odd. In both options the allowed values of $h^2 + k^2 + \ell^2$ are all even, and the structure of the crystal may be identified by measuring the ratios $\sin^2\theta / \sin^2\theta_{\min} = (h^2 + k^2 + \ell^2)$, Eq. (3.32). Compared with the simple cubic lattice, the diffraction pattern of the **BCC** one contains far less Bragg peaks.

The **FCC** lattice, Fig. 2.18, has four atoms in the unit cell, at the origin and in the points $(\hat{\mathbf{y}}+\hat{\mathbf{z}})a/2$, $(\hat{\mathbf{x}}+\hat{\mathbf{y}})a/2$, and $(\hat{\mathbf{z}}+\hat{\mathbf{x}})a/2$. A similar calculation shows that the structure factor is proportional to $\{1+\exp[i\pi(h+k)]+\exp[i\pi(h+\ell)]+\exp[i\pi(k+\ell)]\}$. Hence, constructive interference is realized only when the three integers h, k, and ℓ are all even or are all odd. From the discussion around Eq. (3.31), the only values of the sum $h^2 + k^2 + \ell^2$ that yield constructive interference off the **FCC** lattice, up to 24, are 3, 4, 8, 11, 12, 16, 19, 20, and 24 (check!) In both cases, the list of angles that yield constructive interference is much shorter than that of the **SC** lattice, and one may deduce the type of the lattice from this list. Figure 3.20 summarizes the values of $h^2 + k^2 + \ell^2$ that give Bragg peaks for the ubiquitous cubic lattices. The disparity among the diffraction patterns enables the experimental distinction between them.

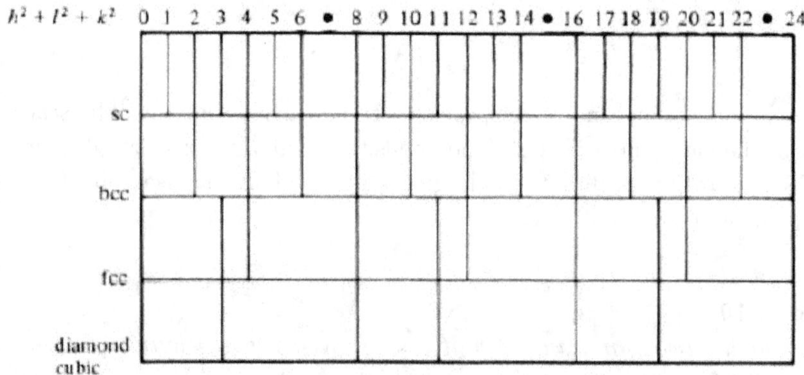

Fig. 3.20: The values of $h^2 + k^2 + \ell^2$ that yield Bragg peaks for the **SC**, the **BCC**, the **FCC** lattices, and the diamond lattice, mentioned in problem 3.25.

Problem 3.21.
*Derive the structure factor of the **FCC** lattice directly, using its primitive lattice vectors, Fig. 2.18.*

Problem 3.22.
a. *Determine the reciprocal-lattice vectors that belong to the planes of the highest site densities which lead to constructive interference off the **BCC** lattice.*
b. *Find the planar lattice that describes these planes.*
c. *What is the densest line on each plane?*

d. *What are the answers for the* **FCC** *lattice?*

Scattering off a crystal with several types of atoms. Consider the **table salt** crystal, Fig. 2.17. As discussed in Sec. 2.5, table salt is described by the **FCC** lattice, with a base comprising a chlorine ion at the origin and a sodium one at the mid point of the cube's edge, e.g., along the $\hat{\mathbf{y}}$−axis. Instead of calculating the structure factor for the primitive unit cell, it is more convenient to derive it for the cubic unit cell, that contains four chlorine ions [at the origin, and at $(\hat{\mathbf{x}} + \hat{\mathbf{y}})a/2$, $(\hat{\mathbf{y}}+\hat{\mathbf{z}})a/2$, and $(\hat{\mathbf{z}}+\hat{\mathbf{x}})a/2$] and four sodium ions (displaced away from each of these four points by the vector $a\hat{\mathbf{y}}/2$). For the cubic lattice whose lattice constant is a, the reciprocal-lattice vectors are given by $\mathbf{G}_0 = (2\pi/a)(h\hat{\mathbf{x}} + k\hat{\mathbf{y}} + \ell\hat{\mathbf{z}})$. It follows that Eq. (3.51) for table salt takes the form

$$F(hk\ell) = [1 + e^{i\pi(h+k)} + e^{i\pi(h+\ell)} + e^{i\pi(k+\ell)}](F_{\text{Na}} + F_{\text{Cl}}e^{i\pi k}) . \qquad (3.53)$$

The expression in the square brackets is the one of the **FCC** lattice with a single element at each lattice site. The other factor in Eq. (3.53) is the structure factor of the crystal, which contains contributions from the two ions forming the base. This term yields $(F_{\text{Na}} + F_{\text{Cl}})$ for even k's (and then all indices are even) and $(F_{\text{Na}} - F_{\text{Cl}})$ for the odd k's (and then all indices are odd). These two types of points give rise to disparate intensities. The chlorine and sodium ions are quite different and therefore both types produce nonzero intensities. As opposed, potassium chloride, KCl, yields a different result: each of the ions K^+ and Cl^-, possesses the same number (18) of electrons, like the noble gas argon (see the periodic table in Fig. 2.22). Therefore, although the charges of their nuclei are not the same (and, as a result, the radii are slightly different) their structure factors are quite close to one another, and the radiation intensity at the odd points is rather small. A similar conclusion pertains to CeSb: an inspection of the periodic table, together with the assumption that three of the cerium electrons have been transferred to the antimony, reveals that the electrons' numbers on the two ions are 55 and 54; they both are similar to the xenon atom, and their structure factors are also close to one another. Notice that the similarity of the structure factors of potassium and chlorine (as well as that of the Ce and Sb ions) pertains only for X-ray scattering. For comparison, neutrons are scattered mainly from the nuclei; the scattering cross sections of neutrons from Cl and K are significantly dissimilar. Thus, much can be learnt from comparisons between the diffraction patterns of X-rays and neutrons, taken from the same crystal.

Figure 3.21 displays experimental values of the intensity of the peaks obtained by scattering X rays, with wavelength $\lambda \approx 1.545 \mathring{A}$ that come from the $K_{\alpha 1}$ line of copper, off table salt (see problem 3.1). The data are deduced from an analysis of the intensities of rings on a screen, as illustrated in Fig. 3.4. The digits on the vertical axis, which denote the intensity (i.e., the heights of the Bragg peaks) are determined up to a multiplicative factor, since the measured intensity is proportional to, e.g., the duration of the irradiation, and also to other experimental details. As the interpretation of the data follows from **ratios** of intensities, this multiplicative factor is insignificant. Each peak is marked by the Miller indices that represent the relevant reciprocal-lattice vector. The intensities indeed decay when the indices become larger [see the discussion following Eq. (3.50)]. Nonetheless, the dependence on the scattering angle is not always monotonic, indicating additional dependencies of the structure factors on the reciprocal-lattice vectors; the intensities are also affected by the multiplicities of each of the family of planes, and other factors. The peaks have a finite width also due to the finite size of each grain in the powder (as discussed in Sec. 3.3).

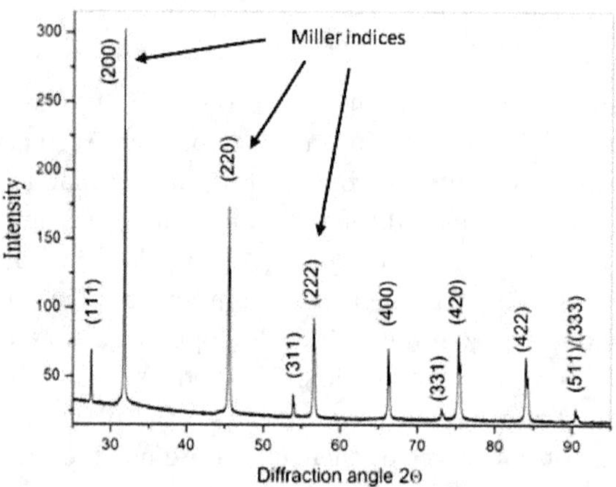

Fig. 3.21: Intensity of the scattering off a powder of table salt as a function of the scattering angle. The scattering is of X-rays emitted from copper, with wave length $\lambda \approx 1.545 \text{Å}$. [courtesy of Y. Rosenberg, Tel Aviv University.]

Problem 3.23.

a. *Show that the scattering angles in Fig. 3.21 indeed correspond to an* **FCC** *lattice, and deduce from them the lattice constant of the table-salt crystal.*

b. *Exploit the first two peaks in Fig. 3.21 to estimate the ratio $F_{\text{Na}}/F_{\text{Cl}}$, assuming that it is real and independent of h, k, and ℓ. What are the conditions for this assumption to be true?*

c. *What can be done when this ratio is complex?*

d. *Assume that the electrons' density in the i–th ion is approximately given by $n_i(r) = z_i \exp[-2r/a_i]/(\pi a_i^3)$, where z_i is the number of electrons on this ion and a_i is its "radius". Find the dependence of the structure factor on h, k, and ℓ, and obtain the ratio between the two first peaks. Given that $a_{\text{Na}} = 0.95 \text{Å}$ and $a_{\text{Cl}} = 1.81 \text{Å}$, calculate the ratio $z_{\text{Na}}/z_{\text{Cl}}$.*

Problem 3.24.

Show, by a calculation, that were the structure factors of potassium and chlorine precisely identical in the KCl crystal (which is similar to the table-salt one) then its diffraction pattern would have been identical to that of a simple cubic lattice with a lattice constant a/2. Explain this result (without carrying out the calculation).

Problem 3.25.

Determine the structure factors of CsCl, of diamond, and of zinc-blende.

Problem 3.26.

a. *Determine the structure factor of the* **HCP** *lattice.*

b. *Assuming that all scatterers are point-like, what are the values of the intensities in the various Bragg peaks (ignoring the multiplicities)?*

c. *Which Miller indices give the smallest scattering angles for the ideal* **HCP** *lattice, with $a_3/a_1 = \sqrt{8/3}$? (Express the result in terms of the ratio λ/a_1.)*

d. *Determine the structure factor of Wurtzite.*

Problem 3.27.

Determine the structure factors of the body-centered orthorhombic lattice, of the face-centered one, and of the base-centered one.

Problem 3.28.

Which of the angles derived in problem 3.15 correspond to constructive interference for the body-centered tetragonal lattice?

Scattering off large organic and biological molecules. The examples studied so far are of crystals whose unit cells contain only a few ions, and therefore their structures can be deduced from a relatively small number of data points of the intensities $|F(\mathbf{G})|^2$. When the unit cell contains many scatterers, the deciphering of the experimental results becomes more complex and requires many Bragg peaks. The problems at the end of the chapter present several examples of more complicated crystals. The treatment of such crystals is carried out by various methods. One possibility is to conjecture the structure of the unit cell and then to calculate structure factors for each reciprocal-lattice vector. The electron density in each cell is computed by numerically solving the Schrödinger equation, including the electronic interactions. These numerically-computed structure factors are compared with the experimental data; the initial conjecture is then improved and the procedure repeats itself till a satisfactory agreement with the data is achieved.

Another method is based on an initial identification of the lattice symmetry (for instance, by scattering multi-chromatic radiation which contains many wave lengths off a single crystal, i.e., by using the von Laue technique); then the scattering angles are deduced, for example by scattering monochromatic radiation off powders. The results are then compared with the symmetries found in the first stage. In the next step, the scattering intensities at each of the Bragg peaks are analyzed. As mentioned in Sec. 3.7, the intensities contain many multiplicative factors that have to be included in the analysis. At the end of this procedure one obtains the values of $\widetilde{\Gamma}(\mathbf{G})$ [Eq. (3.48)], for a finite numbers of vectors $\mathbf{G}$. The inverse Fourier transform of these values produces the correlation function of the crystal,

$$\langle n(\mathbf{0})\mathbf{n}(\mathbf{r})\rangle = \mathbf{\Gamma}(\mathbf{r}) = \sum_{\mathbf{G}} \widetilde{\mathbf{\Gamma}}(\mathbf{G})e^{i\mathbf{G}\cdot\mathbf{r}} . \tag{3.54}$$

This function provides valuable information on the relative distribution of the scatterers among pairs of sites in the crystal. Since this information is collected from lattice vectors smaller than a certain one, $\mathbf{G}_{\mathrm{max}}$, determined by the experimental setup, and consequently is collected from wave lengths longer than $2\pi/|\mathbf{G}_{\mathrm{max}}|$, it is reliable only for distances such that $r > 2\pi/G_{\mathrm{max}}$. The precision is enhanced by increasing G_{max}, for instance by using more intensive radiation.

Notice that the intensity at the Bragg peaks gives the absolute values of the structure factors of the entire unit cell (each of them is a sum over atomic structure factors with phases) and hence does not convey information about their phases. The structure factors of atoms which have reflection symmetry are real (see problem 3.18). Hence, for such crystals it is possible to obtain $\widetilde{n}(\mathbf{G}) = \sqrt{\widetilde{\Gamma}(\mathbf{G})}$, and then derive the distribution of the scatterers in the crystal, $n(\mathbf{r}) = \sum_{\mathbf{G}} \widetilde{n}(\mathbf{G})\exp[i\mathbf{G}\cdot\mathbf{r}]$. This procedure is usually not applicable for **organic chemistry** or for **molecular biology**, i.e., when the crystals contain big molecules lacking reflection symmetry. In that case the function $\widetilde{F}(\mathbf{G}) = \sqrt{\widetilde{n}(-\mathbf{G})}$ [Eq. (3.47)] is complex, and measurements of the intensity do not provide information about its phase. This is called the **phase problem**. These phases can be identified by various methods (for instance, by replacing chemically part of the atoms in the unit cells by others, and exploiting information obtained from the correlation functions). Once those phases are known, the inverse Fourier transform of the structure factor yields the distribution of the scatterers within the unit cell, $n(\mathbf{r})$. The latter enables the identification of the atoms or ions in the base, and consequently the structure of the molecules in each unit cell.

This procedure is an important research tool in **molecular biology**: experimental methods to explore the structure of large biological molecules are based on scattering off crystals made of such molecules. The analysis of the diffraction patterns yields the chemical and the geometrical structures of the molecules. The structures of DNA and penicillin were discovered in this way. Another conspicuous example concerns the ribosomes' molecules that contain several tens of proteins. For instance, the ribosome 50S contains two molecules of RNA, one comprising ≈ 2900 atoms and the other ≈ 120. Figure 3.22 displays a diffraction pattern measured on a single crystal of the ribosome; the crystal was rotated while being irradiated by synchrotron radiation, of wave length $\lambda = 1.51\text{Å}$. The analysis of this pattern reveals a triclinic lattice with relatively large lattice constants, $a = (182 \pm 5)\text{Å}$, $b = (584 \pm 10)\text{Å}$, $c = (186 \pm 5)\text{Å}$, $\beta = 109°$. The base comprises two molecules, of molecular mass 1.6×10^6 (in units of the proton mass).

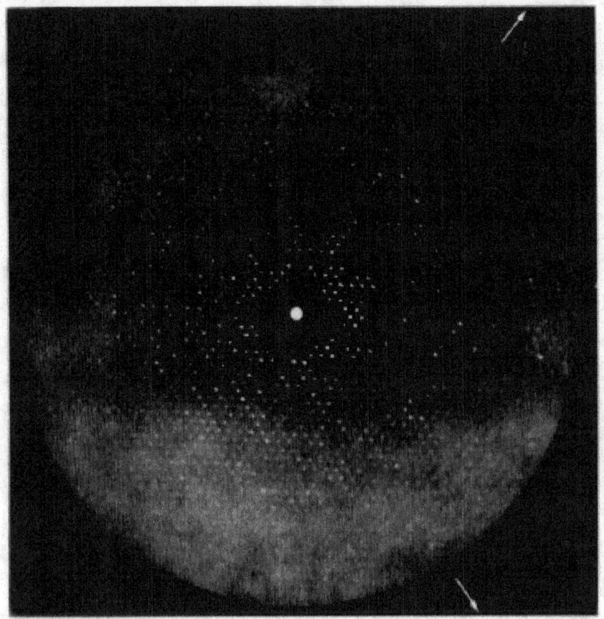

Fig. 3.22: Diffraction pattern of a ribosome molecule, taken from halobacterium marismortui. [I. Makowski, F. Frolow, M. A. Saper, M. Shoham, H. G. Wittman, and A. Yonath, *Single Crystals of Large Ribosomal Particles from Halobacterium marismortui Diffract to 6 Å*, J. Mol. Biol. **193**, 819 (1987).]

3.10 The temperature dependence; the Debye-Waller factor

Atomic vibrations in the crystal. The discussions in the previous chapters are based on the tacit assumption that the centers of mass of the atoms (or the ions) rest precisely on the lattice sites (or at the sites of the base). In fact, both the electrons and the nuclei are constantly in motion around the centers of the mass. In order to compare with the experimental data, the previous expressions have to be averaged over these motions; the average over the electrons' motion around the nuclei is carried out in the previous section, e.g., in part (c) of problem 3.18. The average over the nuclei's motion is discussed in this section. This average is accomplished in two steps: first it is done over the quantum states of each degree of freedom. Recall that quantum mechanics yields the probability distribution for a degree of freedom to attain a certain value. Each of these values appears with its proper weight – the absolute value squared of the quantum wave functions. Second, the various quantum states are averaged upon with their Boltzmann weights. The Boltzmann weight of a state of energy E (measured relative to the ground-state energy), at temperature T, is proportional to $\exp[-E/k_BT]$, where k_B is the Boltzmann constant. When the energies of all states exceed by far the thermal energy k_BT, then these states can be ignored and it suffices to quantum-mechanically average over the degrees of freedom in the ground state. Usually these energies are indeed high for the excited states of the electrons in the atom. This is the reason for the assumption that the "density" of the electrons is equal to the quantum probability to find them in the ground state at each point in space (i.e., equal to the absolute value squared of the wave function of their ground state).

The centers of mass of the atoms (i.e., their nuclei) reside at minima of the potential energy, see Chapter 4. But quantum mechanics implies that microscopic particles are never exactly at the sites of their minimal potential energy, and therefore their locations have to be averaged upon. Even in the ground state of the harmonic oscillator the wave function differs from zero for many a value of the particle location, indicating that there is a finite probability to find the particle away from the minimum of the potential energy. Consider the vibrations of the centers of mass of the atoms in the particular case of a single atom in each unit cell. As opposed to electrons, the energies of these vibrations are not high compared to k_BT, and thus at finite temperatures the excited states have also to be included in the average, each with its (temperature dependent) Boltzmann weight. The instantaneous location of the center of mass of the atom can be written in the form $\mathbf{R}' = \mathbf{R} + \mathbf{u_R}$, where $\mathbf{R}$ is the location on the periodic lattice and $\mathbf{u_R}$ is the instantaneus distance away from that site. As a result, the scatterers' density is different in each cell, $n(\mathbf{r}) = \sum_{\mathbf{R}} n_0(\mathbf{r} - \mathbf{R} - \mathbf{u_R})$, where $n_0(\mathbf{r})$ is the density of the scatterers around a single atom (placed at the origin). It follows that Eq. (3.50) for $\tilde{n}(\mathbf{G})$ has to be revised,

$$\tilde{n}(\mathbf{G}) = \frac{1}{NV} \sum_{\mathbf{R}} \int_{V(\mathbf{R})} d^3r\, n_0(\mathbf{r} - \mathbf{R} - \mathbf{u_R}) e^{-i\mathbf{G}\cdot(\mathbf{r}-\mathbf{R}-\mathbf{u_R})} e^{-i\mathbf{G}\cdot\mathbf{u_R}}$$

$$= \frac{F(-\mathbf{G})}{NV} \sum_{\mathbf{R}} e^{-i\mathbf{G}\cdot\mathbf{u_R}} \ . \tag{3.55}$$

In the first step the integral is separated into a sum of integrals on each of the unit cells, displacing the origin in each term to the instantaneous location of the atom center in the relevant unit cell (and exploiting the identity $\exp[-i\mathbf{G}\cdot\mathbf{R}] = 1$). As the duration of the experiment is much longer than the period of the vibrations, the measurement is sensitive only to the average value of these vibrations. It is therefore plausible to replace each summand by its average, $\langle \exp[-i\mathbf{G}\cdot\mathbf{u_R}]\rangle$, where the angular brackets denote the double average, the quantum average and thermal one. Usually the probabilities for the displacements $\mathbf{u_R}$ and $-\mathbf{u_R}$ are identical, and consequently $\langle \exp[-i\mathbf{G}\cdot\mathbf{u_R}]\rangle = \langle \cos(\mathbf{G}\cdot\mathbf{u_R})\rangle$ is real. As $|\mathbf{G}|$ increases, the expression within the average oscillates faster, and reduces the average. Since it is smaller than 1, it is customary to denote $\langle \exp[-i\mathbf{G} \cdot \mathbf{u_R}]\rangle = \exp[-W]$. This quantity, which suppresses the structure factor, is called the **Debye-Waller factor**. Assuming that the averages over all atoms are the same, one retrieves Eq. (3.47), but with

$$F(\mathbf{G}) \Rightarrow F(\mathbf{G})\langle \exp[-i\mathbf{G} \cdot \mathbf{u_R}]\rangle \equiv F(\mathbf{G})\exp[-W] \ . \tag{3.56}$$

Calculating the Debye-Waller factor. The Debye-Waller factor can be estimated by assuming that the vibrations are small; the exponential can then be expanded in a Taylor series: $\exp[-i\mathbf{G} \cdot \mathbf{u_R}] \approx 1 - i\mathbf{G} \cdot \mathbf{u_R} - (\mathbf{G} \cdot \mathbf{u_R})^2/2 + \dots$. Assuming in addition that the vibrations of each component of $\mathbf{u_R}$ are independent, and that their average vanishes, the first term in the expansion disappears. The second yields $\langle (\mathbf{G} \cdot \mathbf{u_R})^2 \rangle = G^2(\langle \mathbf{u_R}\rangle^2 \cos^2\phi) = G^2\langle (\mathbf{u_R})^2\rangle/3$, where ϕ is the angle between $\mathbf{G}$ and $\mathbf{u_R}$; the average of $\cos^2\phi$ is $1/3$ when all directions are identical (see problem 3.29 with $\mathbf{G}$ directed along one of the axes). Using the approximation $\exp[-W] \approx 1 - W$ gives

$$W \approx G^2(\langle \mathbf{u_R}\rangle^2\rangle)/6 \ . \tag{3.57}$$

The quantity W determines the **Debye-Waller factor**. Within the present approximation, this factor decays exponentially with the square of the length of the reciprocal-lattice length, G^2. This decay hinders the collection of data from many Bragg peaks: the peaks of the large G's practically disappear, unless high-radiation intensity is used.

Problem 3.29.

Prove that the average over all directions of an arbitrary vector in a d-dimensional space, pointing along an arbitrary direction that makes an angle ϕ with one of the axes, is $\langle \cos^2\phi \rangle = 1/d$.

It remains to find the average of the amplitude squared, i.e., $\langle (\mathbf{u_R})^2 \rangle$. This is carried out in detail in Chapter 5, which focuses on the theory of lattice vibrations. A qualitative answer can be obtained by assuming that the vibration of the atom at $\mathbf{R}$ is independent of those of the other atoms, and that there exists solely a single vibrational frequency, ω (as in the model of Einstein, discussed in Chapter 5). Assuming that (i) the vibration's amplitude is small, and (ii) the symmetry around the minimum (of the potential energy) is spherical, the potential that induces the vibrations, in the harmonic approximation, is $V = M\omega^2|\mathbf{u_R}|^2/2$, where M is the mass of the atom and ω is the angular frequency of the harmonic oscillation around the minimum ($K = M\omega^2$ is the second derivative of the potential at the minimum, or the "spring constant"). Decomposing the vector $\mathbf{u_R}$ into its cartesian components, x, y, and z, then $V = M\omega^2(x^2 + y^2 + z^2)/2$, i.e., the sum of three potentials of one-dimensional, independent harmonic oscillators. It remains to find the standard deviation of such an oscillator, $\langle x^2 \rangle = \langle y^2 \rangle = \langle z^2 \rangle$.

As mentioned, the averages are carried out in each quantum state of the oscillator, with weights given by the absolute value squared of the wave function of the quantum state, and over the ensemble of quantum states with the appropriate Boltzmann weights. There are two limits where the result is simple. At zero temperature the oscillator is in its ground state, and then its average potential energy is the quantum average of the standard deviation there, $3M\omega^2\langle x^2\rangle/2 = 3\hbar\omega/4$. Indeed, as the wave function in the ground state is $\psi_0 = C_0\exp[-(x/x_0)^2/2]$, where $x_0 = \sqrt{\hbar/(M\omega)}$, it follows that $\langle x^2\rangle = \int dx x^2|\psi_0|^2/\int dx|\psi_0|^2 = x_0^2/2$, and the standard deviation of $\mathbf{u_R}$ represents the **zero-point motion** of the oscillator, which is larger as the mass and/or the frequency are smaller. [As $\langle x\rangle = 0$ in the oscillator states, the square of the standard deviation is $\langle(x-\langle x\rangle)^2\rangle = \langle x^2\rangle$.] In this limit, $2W = \hbar G^2/(2M\omega)$.

At high temperatures, when $\hbar\omega \ll k_\mathrm{B}T$, the thermodynamic equipartition theorem can be used; it implies that the average energy of each degree of freedom is $k_\mathrm{B}T/2$. Thus, in this limit $M\omega^2\langle x^2\rangle/2 = k_\mathrm{B}T/2$, which yields $2W = G^2k_\mathrm{B}T/(M\omega^2)$. The Debye-Waller factor is smaller than 1 even at zero temperature, and becomes even smaller as the temperature increases.

Intermediate temperatures require a full calculation of both the quantum and the thermal averages. The one-dimensional harmonic oscillator is described by the potential energy $V(x) = M\omega^2x^2/2$. It is known that the quantum energy levels of this oscillator are given by $E_n = \hbar\omega(n+1/2)$, where n is a non negative integer. The kinetic energy and the potential energy of a classical oscillator are equal to one another. This equality persists also for the average quantum values of these energies in the n−th energy level of the oscillator, i.e., $\langle E_\mathrm{K}\rangle_n = \langle V\rangle_n = E_n/2 = \hbar\omega(n+1/2)/2$, hence the quantum average in the n−th state is $\langle x^2\rangle = \hbar(n+1/2)/(M\omega)$. This quantum average has to be further thermally-averaged. The thermal probability to find a system at temperature T with energy E_n is the **Boltzmann factor**, $p_n = \exp[-\beta E_n]/Z$, where $\beta \equiv 1/(k_\mathrm{B}T)$, and where $Z = \sum_n \exp[-\beta E_n]$ is the normalization, ensuring that the sum of all probabilities is 1 (Z is called the **partition function**). At zero temperature the probability to find the system at its ground states is 1, while at infinite temperature there is an equal probability to find the system at any of its quantum states.

By symmetry, $\langle x\rangle_n = 0$; therefore the average standard deviation of x is obtained by averaging $\langle(x-\langle x\rangle)^2\rangle_n = \langle x^2\rangle_n$ with the weights $p_n = \exp[-\beta E_n]/Z$,

$$\langle x^2\rangle = \sum_n p_n\langle x^2\rangle_n = \frac{\hbar}{M\omega}\frac{1}{Z}\sum_n(n+\frac{1}{2})e^{-\beta\hbar\omega(n+\frac{1}{2})} = -\frac{\hbar}{M\omega}\frac{\partial}{\partial(\beta\hbar\omega)}\ln Z \ . \quad (3.58)$$

The last step is a mathematical trick that facilitates the calculation. One finds Z as a sum of an infinite geometrical series,

$$Z = \sum_n \exp[-\beta\hbar\omega(n+1/2)] = 1/[2\sinh(\beta\hbar\omega/2)] \ .$$

The differentiation yields $\langle x^2 \rangle = [\hbar/(2M\omega)]\coth(\hbar\omega/[2k_\mathrm{B}T])$, hence

$$2W = \frac{\hbar G^2}{2M\omega}\coth\left(\frac{\hbar\omega}{2k_\mathrm{B}T}\right). \tag{3.59}$$

One observes that W indeed increases monotonously between its value at zero temperature and its linear temperature dependence at high temperatures [part (a) of problem 3.30]; thus, the intensity of the Bragg peaks decays as the temperature is raised.

The Lindemann criterion for melting. When the average vibration amplitude of the atom in the lattice is large (larger than the lattice constant) the solid becomes unstable and eventually crosses over to the liquid phase. In other words, the material melts. The melting temperature can be related to the size of the average vibration's amplitude for which melting takes place. Examining various melting temperatures and lattice vibrations led Lindemann in 1910 to conjecture that cubic lattices melt once the amplitude of the vibration exceeds a certain percentage of the lattice constant a, i.e., when $\langle (\mathbf{u_R})^2 \rangle^{1/2} > c_L a$. It is customary to use $c_L = 0.1$ for simple lattices; the experimental data is scattered around this value but at times this estimate yields correctly just the order of magnitude. In more complicated lattices the lattice constant is replaced by the distance between nearest neighbors. When there are several lattice constants a is replaced by a typical length scale, for example, the cubic root of the unit cell volume. As mentioned, at very high temperatures $\langle (\mathbf{u_R})^2 \rangle = 3k_\mathrm{B}T/(M\omega^2)$, and melting is expected to occur at the melting temperature T_M, where $k_\mathrm{B}T_\mathrm{M} \approx c_L^2 M\omega^2 a^2/3$. This **criterion** is called after **Lindemann**. Note that the melting temperature increases with the spring constant $(M\omega^2)$, that characterizes the potential V, i.e., when the material becomes more rigid (as the spring constant is enhanced, it is harder to stretch it, i.e., it is harder to deform the crystal).

Crystal with a base. For a crystal with a base, Eq. (3.56) is generalized to be

$$F(\mathbf{G}) = \sum_i F_i(\mathbf{G})e^{-i\mathbf{G}\cdot\mathbf{r}_i}e^{-W_i(\mathbf{G})}. \tag{3.60}$$

Thus, the relative weights of the contributions to the structure factor from the various atoms in the unit cell may vary among the Bragg peaks, and that variation is temperature-dependent.

Problem 3.30.
a. *Prove that Eq. (3.59) reproduces the limits of W, for zero temperature and for very high temperatures.*
b. *Show that $W = B\sin^2\theta/\lambda^2$, where 2θ is the scattering angle and λ is the wave length, and find the coefficient B. Which properties of the crystal determine it?*
c. *Use the Lindemann criterion to show that at high temperatures, $k_\mathrm{B}T \gg \hbar\omega$, the intensity of the Bragg peaks is a function of T/T_M alone, and that the Debye-Waller factor includes just the ratio of the lattice constants and not their specific values.*

Problem 3.31.
a. *How is the Debye-Waller factor modified when all atoms of a certain type are replaced by another isotope of them?*
b. *How should the Debye-Waller factor be modified when the surrounding of each atom is not spherically-symmetric?*

3.11   Scattering off quasicrystals

As mentioned, there exist diffraction patterns that show rotational symmetry of order 10, which is forbidden for periodic crystals. Such a diffraction pattern is displayed in Fig. 3.23; it was obtained by Shechtman in 1982 by scattering X-rays off Mg-Al alloys. The figure shows diffraction patterns taken at various directions of the crystal relative to the impinging beam. The different angles represent scattering from different planes; the figures reflect the symmetries that pertain to each family of planes. In all cases, the Bragg peaks are narrow, but along certain directions, for instance, at 63.43° (upper figure) the pattern possesses a rotation axis of order 10, i.e, rotations by 36°. **This symmetry is forbidden for a periodic Bravais lattice.** Already in 1982 there were theoretical calculations of the diffraction patterns of a three-dimensional arrangement that consists of two types of three-dimensional structures which generalize the Penrose tiling; the result resembles very much the patterns in Fig. 3.23. This comparison confirms the conjecture that three-dimensional quasicrystals can be described by combinations of more than three lattice vectors, with irrational relations in-between them.

Scattering off one-dimensional quasicrystals. The Fibonacci lattice can serve to exemplify the diffraction pattern off a quasi-periodic one-dimensional crystal. The scattering amplitude of point-like scatterers arranged on a one-dimensional lattice is given in Eqs. (3.7) and (3.8), $A \propto \mathcal{Z} = \sum_n \exp[iqx_n]$, where q is the component of $\mathbf{q} = \mathbf{k}' - \mathbf{k}$ along the lattice, and $\{x_n\}$ denotes the locations of the lattice sites. In a Fibonacci lattice the locations are given by $x_n = S[n+||n/\tau||/\tau]$; τ is the golden ratio and the integer $m = ||n/\tau||$ is in the range $n/\tau - 1/2 < m < n/\tau + 1/2$ [see Eq. (2.8); S is the length of the shorter segment in the Fibonacci sequence]. It follows that

$$A \propto \mathcal{Z} = \sum_n e^{iqx_n} = \sum_{n,m} e^{iqS(n+m/\tau)} \Theta(m - n/\tau + 1/2)\Theta(n/\tau + 1/2 - m) \,,$$

$$(3.61)$$

where $\Theta(x)$ is the step function: it is equal to 1 for $x \geq 0$ and to 0 otherwise.

The result of the summations is given in Eq. (3.64). It is derived by converting the sums into integrals

$$\mathcal{Z}(q_x, q_y) = \int dx \int dy\, e^{i(q_x x + q_y y)} f_1(x,y) f_2(x,y) \,, \qquad (3.62)$$

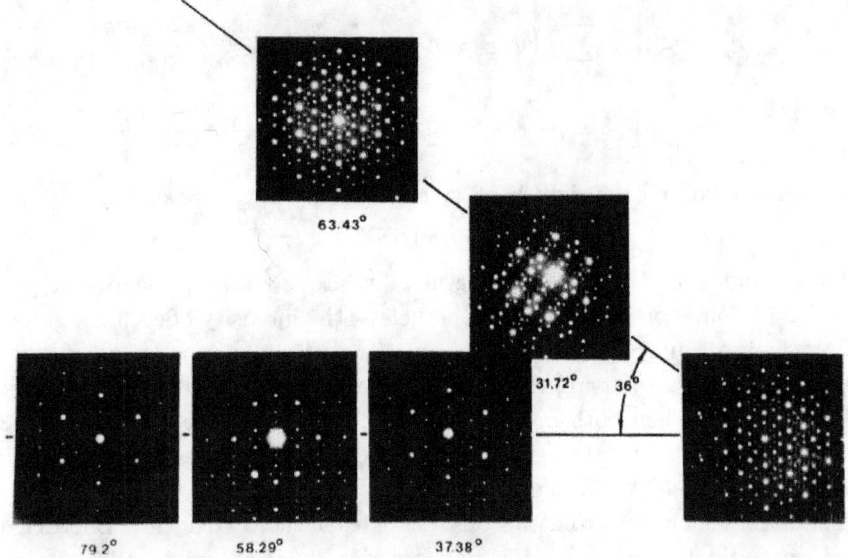

Fig. 3.23: Diffraction patterns of a quasicrystal. [D. Shechtman, I. Blech, D. Gratis, and J. W. Cahn, *Metallic Phase with Long-Range Orientational Order and No Translational Symmetry*, Phys. Rev. Lett. **53**, 1951 (1984).]

with

$$f_1(x,y) = \sum_{n,m} \delta(x-n)\delta(y-m) \ , \quad f_2(x,y) = \Theta(y - x/\tau + 1/2)\Theta(x/\tau + 1/2 - y) \ ,$$

and $q_x = qS$, $q_y = qS/\tau$. The integral in Eq. (3.62) is the Fourier transform of a product of two functions; the theory of Fourier transforms [see in particular Eq. (B.11)] gives for this transform a convolution, i.e.,

$$\mathcal{Z}(q_x, q_y) = \int dq'_x \int dq'_y \widetilde{f}_1(q'_x, q'_y)\widetilde{f}_2(q_x - q'_x, q_y - q'_y) \ . \tag{3.63}$$

The Fourier transform of f_1 is the two-dimensional version of Eq. (3.14),

$$\widetilde{f}_1(q_x, q_y) = \sum_{n,m} e^{i(q_x n + q_y m)} = \sum_{n,m} \delta(q_x - 2\pi n)\delta(q_y - 2\pi m) \ .$$

The Fourier transform of f_2 is a simple integral,

$$\widetilde{f}_2(q_x, q_y) = \int_{-\infty}^{\infty} dx \int_{y_-}^{y_+} dy\, e^{i(q_x x + q_y y)} = \int_{-\infty}^{\infty} dx\, e^{i(q_x + q_y/\tau)x} 2\sin(q_y/2)/q_y \ ,$$

$$y_\pm = x/\tau \pm 1/2 \ .$$

Inserting the two transforms into Eq. (3.63), and using the identity

$$\int_{-\infty}^{\infty} dx \exp[ikx] = 2\pi\delta(x) \ ,$$

results in

$$\mathcal{Z}(q_x, q_y) = 2(2\pi)^3 \sum_{n,m} \delta\left(q_x - 2\pi n + \frac{q_y - 2\pi m}{\tau}\right) \frac{\sin(q_y/2 - \pi m)}{q_y - 2\pi m}$$

$$= 2(2\pi)^3 \sum_{n,m} \delta\left(qS[1 + \frac{1}{\tau^2} - 2\pi[n + m/\tau]\right) \frac{\sin(qS/(2\tau) - \pi m)}{qS/\tau - 2\pi m} \ . \tag{3.64}$$

The delta function on the right hand-side indicates a Bragg peak when

$$q = 2\pi(n + m/\tau)/[S(1 + 1/\tau^2)] \ . \tag{3.65}$$

As both n and m can attain any integral value, Eq. (3.65) represents a very dense collection of points on the $q-$axis. Nonetheless, the intensity of each Bragg peak is proportional to $[\sin(qS/(2\tau) - m\pi)/(qS/\tau - 2\pi m)]^2$, and as such is significant only when $q \approx 2\pi m\tau/S$. Inserting this into Eq. (3.65) implies that $n/m \approx \tau$. This can be the case only when both integers are quite large. An inspection of Eq. (3.65) reveals that the q values that give rise to significant Bragg peaks are themselves a Fibonacci sequence! (see problem 3.32).

The main conclusion from this analysis is that the **diffraction pattern of a quasi-periodic lattice contains narrow peaks, very much like the usual Bragg peaks**, but these peaks do not correspond to a simple reciprocal lattice. They have the symmetries of quasicrystals in momentum space, that are decisively not those of a periodic lattice. Similar considerations for two- and three-dimensional quasicrystals reach the same conclusion. The diffraction pattern of a quasicrystal reflects the symmetry of the original crystal, much like the diffraction patterns of periodic crystals, which display their symmetries.

Problem 3.32.

a. *Show that the condition $n/m \approx \tau$ is equivalent to the result of problem 2.27 part (b), that gives the ratio of the numbers of the long and short segments in a Fibonacci sequence. What conclusions can be deduced about the sequence of the high peaks in the diffraction pattern?*

b. *Compose a code that prepares a table of pairs of values, one for $q = \pi(n + m/\tau)/(\tau + 1/\tau)$ and the other for $A = [\sin(q - \pi m)/(q - \pi m)]^2$, for a range of integral values of n and m (e.g., $n, m = 1, 2, \ldots, 100$) and plot these values in the $q - A$ plane. Insert into the code the condition that the allowed A values are only those within the highest percentage of all values obtained for these points, and mark down the corresponding values of q and the intervals in-between them. What are the relations between these intervals? What is the conclusion drawn about the sequence of the q values?*

3.12 Scattering of electrons off surfaces

Diffraction of low-energy electrons. As mentioned, for instance in Sec. 2.9, the surface of a crystal may be considered as one of the planes forming that crystal.

Therefore the surface can be denoted by the Miller indices of the family of planes that contains the surface plane. At times one wishes to explore the surface separately from the three-dimensional bulk below. In the example discussed in Sec. 2.9 the focus is on the thin layer adsorbed on the surface. Usually X-rays are not useful for this purpose since they penetrate deep into the material (and thus convey information on its three-dimensional structure). A possible tool to investigate the structure of surfaces, already mentioned in Sec. 3.1, is the scanning tunneling microscope. Another option is to exploit **diffraction of electrons at low energies; this method is termed low-energy electron diffraction, or LEED**. The typical energies involved are 10-200 eV, i.e., wave lengths of the order of $\lambda \sim 0.8 - 3.8\text{Å}$ (see Sec. 3.1). Due to the large cross section for inelastic scattering of such electrons off the crystalline electrons, they do not penetrate far beneath the surface, and are mainly scattered from the outer atomic layers.

Constructive interference of waves scattered from planes. Several examples for scattering off two-dimensional crystals are presented in Sec. 3.9. However, the expressions derived there are based on the assertion that the wave vectors, the impinging and the scattered ones, are also in the plane. In order to include the possibility that the wave vectors are in space, one may use Eq. (3.9), under the condition that the lattice vectors $\mathbf{R}$ are restricted to atoms belonging to the outer layer of the crystal. When the surface is represented by the reciprocal-lattice vector $\mathbf{G}_0$, then the surface itself obeys the equation $\mathbf{G}_0 \cdot \mathbf{R} = 0$ (see Sec. 3.6). Decomposing the vector $\mathbf{q} = \mathbf{k}' - \mathbf{k}$, that gives the momentum difference between the incoming and the outgoing waves, into the component $\mathbf{q}_\perp$, perpendicular to the surface (i.e., parallel to $\mathbf{G}_0$) and the component $\mathbf{q}_\parallel$, that is parallel to the surface, $\mathbf{q} = \mathbf{q}_\perp + \mathbf{q}_\parallel$, and using $\mathbf{q}_\perp \cdot \mathbf{R} = 0$, Eq. (3.9) implies

$$\mathbf{q}_\parallel \cdot \mathbf{R} = \mathbf{q} \cdot \mathbf{R} = 2\pi m \ . \tag{3.66}$$

It follows that the vector $\mathbf{q}_\parallel$ is given by a linear combination with integer coefficients of the reciprocal-lattice vectors of the plane, $\mathbf{q}_\parallel = \mathbf{G}_{\text{surface}} = h\mathbf{b}_1 + k\mathbf{b}_2$ (which contains the information on the structure of the surface plane), but there is no restriction on the component normal to the surface, $\mathbf{q}_\perp$, that thus may attain continuously any value. Radiation with wave vector $\mathbf{k}$ gives rise to constructive interference on **lines in momentum space**, normal to the surface plane at discrete points that belong to the reciprocal lattice of the plane,

$$\mathbf{k}' - \mathbf{k} = \mathbf{q} = \mathbf{G}_{\text{surface}} + \mathbf{q}_\perp \ . \tag{3.67}$$

Assuming that the impinging beam is normal to the planar sample, constructive interference is expected whenever the scattered beam hits one of the lines perpendicular to the sample at one of its reciprocal-lattice sites. The points on the right hand-side in Fig. 3.24 represent these sites, and the circle there (whose radius equals the length of the impinging wave vector and its center is at the origin of that vector) is a manifestation of energy conservation, $|\mathbf{k}'| = |\mathbf{k}|$. This circle is part of the three-dimensional Ewald sphere. As opposed to Fig. 3.12, where constructive interference is realized only for the point C on the sphere, in the present case such interference emerges whenever one of the parallel lines normal to the plane of the sample at its reciprocal- lattice sites cuts the sphere. Thus, there are many more points on that sphere, as compared to the three-dimensional case depicted in Fig. 3.12, for which the interference is constructive. When the point B is connected with each of these points (marked by $*$), and the lines are continued to a spherical screen around the point B, the scattering points formed on that screen reflect the structure of the reciprocal lattice.

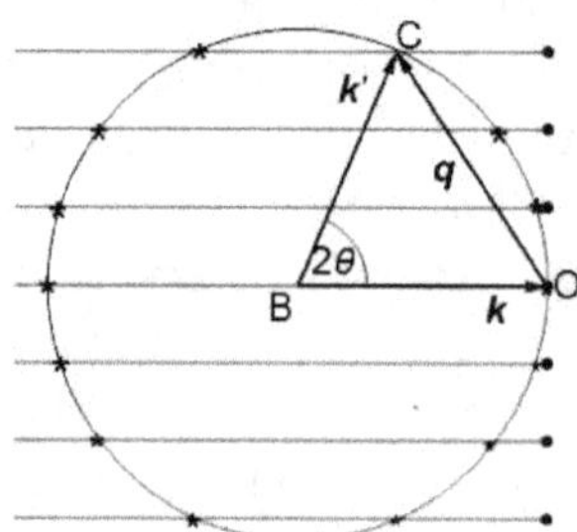

Fig. 3.24: The Ewald sphere for scattering off a two-dimensional lattice. The points on the right represent the plane of the measured sample.

Problem 3.33.
Referring to the geometry of Fig. 3.24, express the direction of the vector $\mathbf{k}'$ in terms of the indices h and k that characterize the planar reciprocal-lattice vector which scatters the radiation, the angle θ, and the wave length of the impinging wave, and use these data to determine the locations of the radiation points on a planar screen parallel to the measured sample.

The diffraction picture reflects the sites of the planar reciprocal-lattice of the measured sample. When it is a planar layer adsorbed on a planar cut of another crystal, the intensities of the various Bragg peaks indicate the composite structure of both planes. An example is the structure of krypton placed on the (001) plane of graphite, displayed in the upper part of Fig. 2.38. As explained in problem 2.32, the krypton lattice is a triangular one, whose lattice constant is $a' = 3a$, where a is the distance between nearest-neighbor carbon atoms in the upmost layer of the graphite. The reciprocal lattice of this triangular lattice is illustrated in Fig. 3.11. Ignoring the height differences between the krypton atoms and the carbon ones in the adsorbing layer, the unit cell contains a krypton atom at the origin and six carbon ones at the sites $\pm\mathbf{a}_1/3$, $\pm\mathbf{a}_2/3$, $\pm(\mathbf{a}_2 - \mathbf{a}_1)/3$. The planar structure factor is then $F(hk) = F_{\mathrm{Kr}} + 2F_{\mathrm{C}}[\cos(2\pi h/3) + \cos(2\pi k/3) + \cos(2\pi(k - h)/3]$. It attains three values, $F(hk) = F_{\mathrm{Kr}} + 6F_{\mathrm{C}}$, F_{Kr}, $F_{\mathrm{Kr}} - 3F_{\mathrm{C}}$. The magnitudes of the points in Fig. 3.25 display these three values, according to the appropriate Miller indices. It is illuminating to compare this figure with Fig. 3.19, that portrays the intensities of the Bragg peaks in the diffraction pattern of graphene. In the latter, there appears a hexagonal lattice of identical intensities (the small points in Fig. 3.19); at the center of each hexagon there is a peak with a different intensity. The hexagonal lattice is still there in Fig. 3.25, but inside the hexagons there are now points of two different intensities, arranged alternately.

When the actual height, u, of the krypton atom above the substrate is accounted for, the structure factor F_{Kr} is multiplied by $\exp[iq_\perp u] = \exp[ik_0[1 - \cos(2\theta)]u]$ (see problem 3.33 and Fig. 3.24). It follows that the structure factor at each site is a complex number, and the intensity attains an additional dependency on the scattering angle.

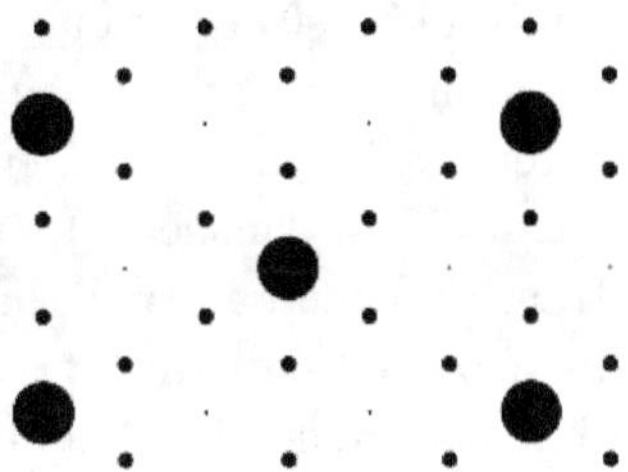

Fig. 3.25: The intensities of the Bragg peaks on the reciprocal lattice of the triangular lattice that represent the periodic arrangement of krypton on graphite.

Problem 3.34.
Describe the diffraction pattern of electrons scattered off a one-dimensional lattice.

3.13 Neutron scattering off magnetic crystals

Advantages of neutron scattering. Scattering by neutrons differs from that by X-rays since the latter are scattered by the electrons while the neutrons are scattered by the nuclei or by the atomic magnetic moments. Since the nuclei are much smaller than the atoms, they can be considered as point-like scatterers; for this reason the atomic structure factor for neutron scattering hardly decays as the reciprocal-lattice vector $\mathbf{G}$ increases [see the discussion of scattering off a single atom following Eq. (3.50)]. As opposed to scattering of X-rays, which gives rise to structure factors that increase with the atomic number of the scatterer (this parameter determines the number of the scattering electrons), the cross section for neutron scattering is not sensitive to the atomic number. One of the main advantages of neutrons is that the scattering intensities, even from light atoms like hydrogen (which is hardly accessed by X-rays), are large. Another important factor is the penetration depth: the mean free-path (the distance along which the intensity is reduced by the factor $1/e$, where

$e \approx 2.73$ is the Euler number) of X-rays is of the order of a few millimeters, while that of neutrons can exceed one centimeter.

Identification of magnetic structures. As mentioned in Sec. 2.10, magnetic materials undergo phase transitions: above the magnetic transition temperature the magnetic moment of each atom (or ion) is free to align along an arbitrary direction, such that the average moment of the entire sample vanishes. Below that temperature, however, these moments arrange themselves in various periodic patterns. Neutrons possess a magnetic moment; for this reason they are sensitive to the **magnetic structure** of the measured crystal; the scattering amplitude depends on the angle in-between the magnetic moment of the neutron and that of the atom (or ion). The magnetic moments of atoms are on the nuclei and on the electrons, and only the neutrons are sensitive to both types of moments; neutron scattering is thus the paramount tool for studying magnetic structures. Since above the transition temperature the atoms (on the average) do not possess a magnetic moment, the neutrons are scattered solely from the periodic structure of the nuclei (this structure is termed "chemical" or "nuclear"), very much like X-rays. On the other hand, below the transition temperature the magnetic moments of the nuclei and/or the electrons are arranged periodically, and then there is a magnetic component to the scattering intensity, reflecting the magnetic structure.

Sources of neutrons. Neutrons are largely obtained from nuclear reactors. Inside these reactors there is a multitude of thermal neutrons, with kinetic energy of the order of $k_B T$. At room temperature, i.e., $T \approx 300 - 400$ degrees K [that is, $k_B T = (0.025 - 0.033)\text{eV}$] these neutrons have wave lengths of the order of 1-2$\mathring{A}$ [Eq. (3.3)]. The disadvantage of this source is the low intensity, that necessitates the use of relatively large samples (whose typical size is a few centimeters). Larger-intensity neutrons are obtained from **pulsed spallation sources**: accelerated protons collide with heavy metals and release in the process quite a number of neutrons. The neutron flux from such sources is about a hundred times more than that coming from reactors.

The discovery of Shull: antiferromagnetism in manganese oxide. A neutron scattering experiment carried out in 1949 (with wave length $\lambda = 1.057\mathring{A}$) on a powder of manganese oxide, MnO, had been the first to prove the existence of the **antiferromagnetic structure**. The Bragg peaks had been monitored at two temperatures [displayed in Fig. 3.26(a)]. The mere fact that the low-temperature data contains peaks that are not there at the higher temperature indicates that this material has two different periodic structures. The experimenters, Shull and collaborators, concluded that at the higher temperature the magnetic moments are not yet periodically-arranged, and therefore the Bragg peaks represent only the scattering off the "chemical" periodic structure of the atoms in the lattice. These peaks imply an **FCC** lattice, similar to that of table salt, in which the manganese and the oxygen ions replace the sodium and the chlorine ones, respectively (Fig. 2.17), with high intensities for odd indices, i.e., $(hk\ell) = (111)$, (311), and weak ones for even indices, $(hk\ell) = (200)$, (220), and a lattice constant $a = 4.426\mathring{A}$ [Eq. (3.53)]. The figure differs from Fig. 3.21 (portraying the data of X-ray scattering off table salt, which is also an **FCC** lattice with a base comprising two ions) because, as opposed to the electronic structure factors of the X-ray data, the nuclear structure factors F_O and F_{Mn} are opposite in sign. This explains the disparity between the intensities of the odd-indices peaks and the even-indices ones: for X-ray the intensity is higher at the even indices, while in the case of neutron scattering it is higher at the odd ones. The high-temperature data of the neutron scattering is also in accord with X-ray scattering off the same powder (allowing for the different intensities).

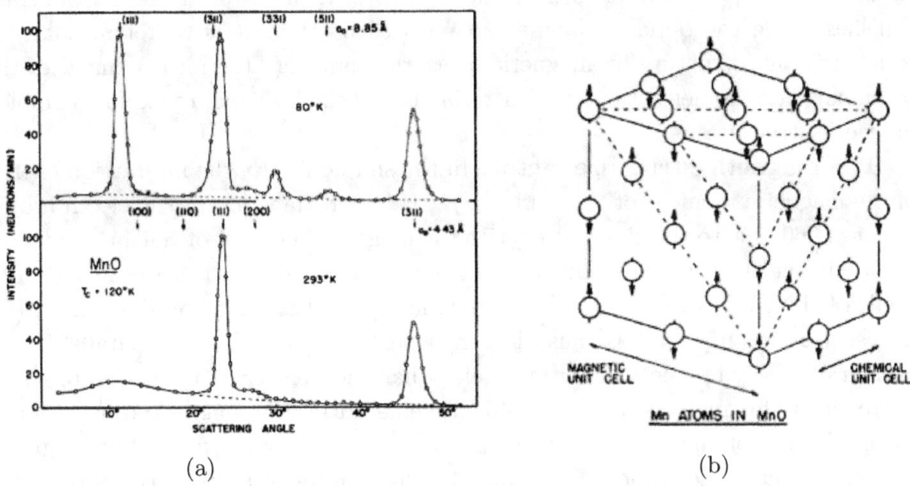

(a)   (b)

Fig. 3.26: (a) Bragg peaks of neutron scattering off a powder of manganese oxide at two different temperatures. (b) The magnetic structure of the manganese ions at low temperatures, as deduced by Shull and coworkers. [C. G. Shull, W. A. Strauser, and E. O. Wollan, *Neutron Diffraction by Paramagnetic and Antiferromagnetic Substances*, Phys. Rev. **83**, 333 (1951).]

The "chemical" peaks remain in their places in the upper plot of Fig. 3.26(a), taken at 80°K. However, this plot contains more peaks, which can be related to an **FCC** lattice with a doubly-larger lattice constant, $a' = 8.85\mathring{A}$. These peaks belong to indices which are all even or all odd (in the new basis of the larger **FCC** lattice), $(hk\ell) = (111)$, $(3\overline{1}\overline{1})$, $(33\overline{1})$, $(511) + (333)$. These were interpreted as due to the appearance, at a low temperature, of a periodic pattern of the magnetic moments which is not there at the higher temperature. As the magnetic moment of the neutron interacts differently with "upward" directed moments and "downward"-directed ones, the two moments behave as two different scattering agents on the lattice, and increase the size of the unit cell (see Sec. 2.10): the scattering angles identify the double unit cell. It remains to find out the magnetic structure within each unit cell. Testing several options, Shull and coworkers have concluded that the intensities of the peaks agree with the structure drawn in Fig. 3.26(b). In this structure, the magnetic moments of the manganese ions are arranged antiferromagnetically: half of them points "upwards" and the other half "downwards". The figure contains dashed triangles, corresponding to planes of the (111) type in the original cubic lattice (compare with Fig. 2.24). Within each plane the moments are parallel to each other, but are antiparallel to those in the neighboring plane. Along any axis of the cube (as well as upon moving from one plane to the other) the neighboring moments are anti parallel to each other, as in part C of Fig. 2.42. This picture is re-visited at the end of the section, once simpler structures are explained. It is worthwhile to note that the total moment of an antiferromagnet vanishes, since the opposite moments cancel each other; it is therefore not easy to identify this structure by magnetic measurements, e.g., the measurement of the total magnetic moment. Neutron scattering is still the best tool to explore complex magnetic structures.

The magnetic structure factor. In the simplest case, the interaction energy of the magnetic moment of the neutron, $\boldsymbol{\mu}_n$ (smaller than that of the electron by a factor of 960, due to the mass ratio) with the magnetic moment of the ion located at point $\mathbf{r}$ in the crystal (that is due to the magnetic moments of electrons in the outer shells of this ion) is proportional (approximately) to the scalar product $\boldsymbol{\mu}_n \cdot \boldsymbol{\mu}(\mathbf{r})$. The scattering amplitude is thus different when these moments are parallel or are antiparallel. A detailed calculation yields that the density, $n(\mathbf{r})$, in the previous expressions, [for instance in Eq. (3.46)] is replaced by the magnetization (defined as the density of the magnetic moment) at the same point, $\boldsymbol{\mu}(\mathbf{r})$. The magnetic scattering amplitude is then determined by the **magnetic structure factor,**

$$\mathbf{F}_{\mathrm{M}}(\mathbf{G}) = \int d^3 r \boldsymbol{\mu}(\mathbf{r}) e^{i\mathbf{G}\cdot\mathbf{r}} \, , \tag{3.68}$$

which replaces Eq. (3.46). Note that, as opposed to the scalar nuclear structure factor, the magnetic structure factor is a vector. When the unit cell comprises several discrete moments $\boldsymbol{\mu}_i$, whose locations are $\mathbf{r}_i$, the magnetic structure factor is given by a generalization of Eq. (3.51),

$$\mathbf{F}_{\mathrm{M}}(\mathbf{G}) = \sum_{i=1}^{n_B} \boldsymbol{\mu}_i e^{i\mathbf{G}\cdot\mathbf{r}_i} \, . \tag{3.69}$$

Simple examples. Consider first the one-dimensional arrangements depicted in Fig. 2.42. All the moments in the ferromagnetic structure B are identical to each other; the magnetic unit cell thus coincides with the "chemical" or the "nuclear" one. Neutron scattering then produces the same diffraction pattern as in the absence of a magnetic order (e.g., when the moments are randomly arranged as in panel A: there, each moment attains various random directions during the measurement time and is averaged to zero). On the other hand, the antiferromagnetic order in panel C is characterized by a double-sized unit cell, with a lattice constant $a' = 2a$, and with a base comprising two magnetic ions of opposite moments (see Fig. 2.1, second line). The reciprocal lattice of this lattice is twice as small as its predecessor, and its lattice constant is $b' = 2\pi/a' = \pi/a$. When the "positive" moment, μ_0, resides at the origin and the "negative" one, $-\mu_0$, is located at $\mathbf{r}_1 = a\hat{\mathbf{x}}$, the magnetic structure factor is

$$F_{\mathrm{M}}(G) = \mu_0(1 - e^{-iGa}) = \mu_0(1 - e^{-i(\pi h/a)a}) = \mu_0[1 - (-1)^h] \ . \qquad (3.70)$$

(It is a scalar since the moments are parallel or antiparallel to one another, and have a single nonzero component.) The diffraction pattern of the magnetic structure contains only odd values of h, i.e., $G = (\pi/a)[1, 3, 5, \ldots]$. On the other hand, the scattering off the nuclei gives rise to the same structure factor identical for all the ions, whose contribution is $F_{\mathrm{n}}(G) = n_0(1 + \exp[-iGa]) = n_0[1 + (-1)^h]$. This interference is constructive for even values of h, $G = (\pi/a)[2, 4, 6, \ldots]$; these are exactly the reciprocal-lattice vectors of the original nuclear lattice whose lattice constant is $b = 2\pi/a$ and therefore $G = (2\pi/a)[1, 2, 3, \ldots]$. The magnetic order produces additional Bragg peaks at the odd multiples of π/a. The appearance of these additional peaks in the diffraction picture implies that the unit cell is doubled: the smallest reciprocal-lattice vector is twice as small. An analysis of the scattering intensities indicates that the neighboring moments are of identical magnitudes but opposite in signs, and consequently this is the magnetic order of panel C in Fig. 2.42. As opposed, the scattering off the structure D in Fig. 2.42, with two different moments μ_1 and μ_2, yields $F_{\mathrm{M}}(G) = \mu_1 - \mu_2 \exp[-iGa] = \mu_1 - \mu_2(-1)^h$, with the intensities $(\mu_1 - \mu_2)^2$ or $(\mu_1 + \mu_2)^2$, for an even or odd h, respectively. In addition to the appearance of new peaks, the scattering intensities at the nuclear peaks is modified as well (due to the magnetic intensity added to them).

Likewise, the lattice constant of the pattern marked by E in Fig. 2.42 is $a' = 3a$, and its base comprises three magnetic moments, for instance, positive, negative, negative (*cf.* the lower panel in Fig. 2.1). Thus,

$$F_{\mathrm{M}}(G) = \mu_0(1 - e^{2\pi ih/3} - e^{4\pi ih/3}) = \mu_0[1 - 2\cos(2\pi h/3)] \ .$$

The possible values are $F_{\mathrm{M}}(h) = -\mu_0$ when h is an integer product of 3, or $F_{\mathrm{M}}(h) = 2\mu_0$ for any other value. As opposed, the nuclear structure factor is $F_{\mathrm{n}} = n_0[1 + 2\cos(2\pi h/3)]$, vanishing unless h is an integer multiple of 3. This result agrees with the reciprocal-lattice vectors of the original lattice. The magnetic order gives rise to new Bragg peaks, which indicate that the unit cell is tripled. The intensities enable one to identify the magnetic order within this new unit cell.

Figure 3.27 displays three types of simple antiferromagnetic structures on the simple cubic lattice. In the arrangement of the G−type, all the neighbors of a given moment are opposite to it. In the A−type one, the moments within each horizontal plane are parallel to each other, but are opposite to those in the nearest-neighbor planes. In the C−type order, the moments along each axis in the vertical direction are parallel to each other and opposite to those on the neighboring axes. In all three types, half of the moments point along one direction, and the other half point along the opposite one, so that the total moment of the sample vanishes. Nonetheless, the primitive unit cells, their bases and their diffraction patterns are different from the "chemical" ones.

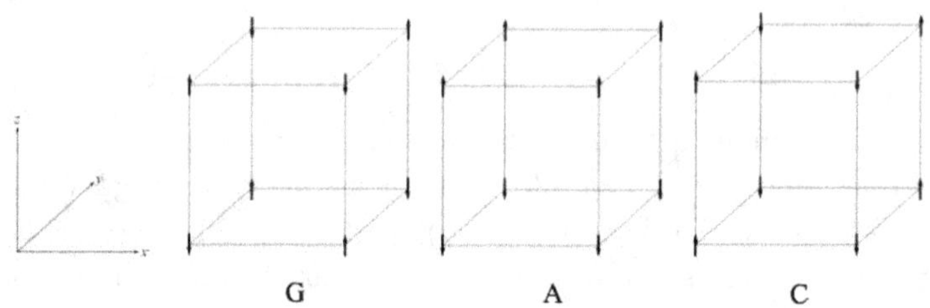

Fig. 3.27: Antiferromagnetic arrangements of the types C, A, and G.

The magnetic moment at lattice point $\mathbf{R}$, in each of the structures in Fig. 3.27, can be written as

$$\boldsymbol{\mu}(\mathbf{R}) = \boldsymbol{\mu}_0 e^{i\mathbf{K}\cdot\mathbf{R}} . \tag{3.71}$$

The vectors $\mathbf{K}$ in each of the patterns are

$$\mathbf{K}_A = (\pi/a)\hat{\mathbf{z}} , \quad \mathbf{K}_C = (\pi/a)(\hat{\mathbf{x}} + \hat{\mathbf{y}}) , \quad \mathbf{K}_G = (\pi/a)(\hat{\mathbf{x}} + \hat{\mathbf{y}} + \hat{\mathbf{z}}) . \tag{3.72}$$

For example, for $\mathbf{R} = (n_1\hat{\mathbf{x}} + n_2\hat{\mathbf{y}} + n_3\hat{\mathbf{z}})a$, the factor $\exp[i\mathbf{K}\cdot\mathbf{R}]$ equals $(-1)^{n_3}$ for the type-A order, $(-1)^{n_1+n_2}$ for the type C, and $(-1)^{n_1+n_2+n_3}$ for the type G. In the one-dimensional pattern of Fig. 2.42, the ordering C is given by Eq. (3.71) with $\mathbf{K} \cdot \mathbf{R} = (\pi/a)(na) = \pi n$.

None of the vectors in Eq. (3.72) is a reciprocal-lattice vector of the **SC** lattice, which represents the "nuclear" crystal [these vectors obey Eq. (3.17), $\exp[i\mathbf{G} \cdot \mathbf{R}] = 1$, and no sign change can occur]. This crystal is built of the vectors $\mathbf{G} = (2\pi/a)(h\hat{\mathbf{x}} + k\hat{\mathbf{y}} + \ell\hat{\mathbf{z}})$. Nevertheless, the vectors in Eq. (3.72) are parallel to the reciprocal-lattice vectors, and therefore each of them is normal to a family of planes in the "nuclear" lattice. The factor $\exp[i\mathbf{K}\cdot\mathbf{R}]$ in Eq. (3.71) has a certain sign on each plane in the family, and the opposite sign on the neighboring planes. For instance, inspecting the structure G (the left panel in Fig. 3.27) one observes that $\mathbf{K}_G$ is normal to all (111) planes of the original cubic lattice, but the signs of the magnetic moments alternate from one plane to the next. In a similar fashion, the signs change alternately as one passes among the (001) planes in structure A, and among the (110) planes in structure C. Since the moments repeat themselves at each second plane, the magnetic unit cell is twice that of the nuclear one, along the corresponding vector $\mathbf{K}$.

This conclusion is based on the notion that in the lattices discussed here, the magnetic crystal is built of two sublattices, one comprising the positive moments, the other the negative ones. Therefore it can be viewed as a lattice with a base of two opposite moments. For example, the structure G resembles the lattice of table salt shown in Fig. 2.17, and hence it is an **FCC** lattice where each lattice site represents a base of two opposite moments. In the cubic representation of the **FCC** lattice, the lattice constant is twice that of the original cubic lattice. Similar to the situation with the table salt crystal [Eq. (3.52)], the magnetic structure factor is

$$F_{\mathrm{M}}(hk\ell) = \mu_0[1 + e^{i\pi(h+k)} + e^{i\pi(h+\ell)} + e^{i\pi(k+\ell)}](1 - e^{i\pi\ell}) \ . \tag{3.73}$$

It yields Bragg peaks only when all Miller indices are odd; the "nuclear" peaks necessitate even Miller indices.

Problem 3.35.

Determine the magnetic structure factors of the antiferromagnetic A and C arrangements in Fig. 3.27. What are the conditions for the appearance of the nuclear and magnetic Bragg peaks?

Manganese oxide. Returning to the cubic unit cell of manganese oxide, Fig. 3.26(b), one recalls that the lattice constant of this cell, denoted a', is twice that of the original lattice. It follows that this cell contains eight unit cells of the nuclear **FCC** original lattice. Since that cell comprises four manganese ions, the magnetic unit cell contains 32 ions of Mn. The coordinates of these ions are given by $\mathbf{r}_i = x_1^i\mathbf{a}_1' + x_2^i\mathbf{a}_2' + x_3^i\mathbf{a}_3' = a'(x_1^i\hat{\mathbf{x}} + x_2^i\hat{\mathbf{y}} + x_3^i\hat{\mathbf{z}})$, where $4(x_1^i + x_2^i + x_3^i)$ is an even number, and where $4x_m^i = 0$, 1, 2, 3, with $m = 1$, 2, 3 (check!). As noted, the planes containing parallel moments are parallel to the (111) plane. It is then found that the antiferromagnetic order proposed by Shull *et al.* obeys Eq. (3.71), with $\mathbf{K} = (2\pi/a')(\hat{\mathbf{x}} + \hat{\mathbf{y}} + \hat{\mathbf{z}})$. Therefore $\mu(\mathbf{r}_i) = \mu_0(-1)^{2(x_1^i+x_2^i+x_3^i)}$, as can indeed be seen in the figure. Hence

$$F_{\mathrm{M}}(hk\ell) = \mu_0 \sum_{\mathbf{r}_i} e^{i(\mathbf{K}-\mathbf{G})\cdot\mathbf{r}_i}$$

$$= \mu_0[1 - (-1)^h][1 - (-1)^k][1 - (-1)^\ell][1 - (-1)^{(h+k)/2} - (-1)^{(k+\ell)/2} - (-1)^{(h+\ell)/2}]$$

$$= \mu_0[1 + (-1)^{h+k} + (-1)^{h+\ell} + (-1)^{k+\ell}]$$

$$\times [1 - (-1)^{h+k+\ell}][1 - (-1)^{(h+k)/2} - (-1)^{(k+\ell)/2} - (-1)^{(h+\ell)/2}] \ .$$

All 32 terms are listed in the second step. The first square brackets in the third step represent the **FCC** lattice, as e.g., in Eq. (3.53), and the factors multiplying them represent the eight manganese ions of the base of this lattice. Hence all Miller indices must be odd, but in addition, also $(h+\ell)/2$, $(k+\ell)/2$, and $(h+k)/2$ have to be odd as well. The result is $F_{\mathrm{M}}(hk\ell) = 32\mu_0$, while for any other combination of indices the result is zero. Indeed, the indices identified by Shull and his collaborators comply with these conditions. Note also that the magnetic symmetry implies different results for (311) and for $(3\bar{1}\bar{1})$: the structure factor is zero for the first, while the one for the second yields the maximal amplitude.

Problem 3.36.

Show that the scattering angles in Fig. 3.26 indeed correspond to the lattice constants and the "nuclear" and magnetic structures of manganese oxide, as deduced by Shull and coworkers. Recall that $\lambda = 1.057\mathring{A}$.

The work of Shull and his collaborators opened a novel era in the research of magnetism. Nonetheless, later studies that exploited stronger neutron sources have revealed that some of the lines are split, for example, (115) and (333), and therefore the actual structure of manganese oxide is rhombohedral, with $\alpha = 90.62°$. It has also been concluded, in contrast with the previous understanding, that the magnetic moment vectors lie within the (111) planes.

More complex magnetic structures. The antiferromagnetic structures discussed above are all based on **two sublattices** built of opposite-sign moments, but there exist other configurations. The base of the structure E in Fig. 2.42 contains three magnetic ions; hence a crystal of this type contains **three sublattices**. The structure portrayed in Fig. 2.43 and discussed in problem 2.36 also contains more than two sublattices. When the different sites in the triangular lattice indeed contain magnetic moments with an angle 120° in-between each pair of them, then the magnetic form factor is a vector,

$$\mathbf{F}_{\mathrm{M}}(hk) = \boldsymbol{\mu}_1 + \boldsymbol{\mu}_2 e^{2\pi i(h+k)/3} + \boldsymbol{\mu}_3 e^{4\pi i(h+k)/3} \ .$$

Due to the $120°-$angle, $\boldsymbol{\mu}_i \cdot \boldsymbol{\mu}_j = -\mu_0^2/2$ for $i \neq j$, which implies that $|\mathbf{F}_{\mathrm{M}}|^2 = 3\mu_0^2\{1 - \cos[2\pi(h+k)/3]\}$. Thus, there are no Bragg peaks when $h+k$ is an integer multiple of 3.

Another, rather extreme, situation is described in problem 2.35. The nuclear lattice is a **SC** crystal with a lattice constant a, but the magnetic moments are given by $\mu(\mathbf{R}) = \mu_0 \cos(\mathbf{Q} \cdot \mathbf{R})$, where $\mathbf{Q} = (\hat{\mathbf{x}} + \hat{\mathbf{y}} + \hat{\mathbf{z}})\tau/a$. Since τ is irrational, the unit cell is infinite. The magnetic scattering amplitude is thence $\sum_{\mathbf{R}} \mu(\mathbf{R}) \exp[-i\mathbf{q}\cdot\mathbf{R}] = \mu_0 \sum_{\mathbf{R}} (\exp[i(\mathbf{Q}-\mathbf{q})\cdot\mathbf{R}] + \exp[-i(\mathbf{Q}+\mathbf{q})\cdot\mathbf{R}])/2$. Each of these two sums resembles those appearing in Eq. (3.14), and thus

$$\sum_{\mathbf{R}} \mu(\mathbf{R})e^{-i\mathbf{q}\cdot\mathbf{R}} = \mu_0 \frac{(2\pi)^3}{V} \sum_{\mathbf{G}} [\delta(\mathbf{q} - \mathbf{Q} - \mathbf{G}) + \delta(\mathbf{q} + \mathbf{Q} - \mathbf{G})]/2 \ . \tag{3.74}$$

The **Bragg peaks are "incommensurate"**: they appear in pairs around each point in the reciprocal lattice, $\mathbf{q} = \mathbf{G} \pm \mathbf{Q}$. The Bragg peaks that belong to the "nuclear" lattice still appear at the reciprocal-lattice sites.

3.14 Answers for the problems in the text

Answer 3.1.
a. Equation (3.1) yields $1.545\mathring{A}$, $1.541\mathring{A}$, and $1.39\mathring{A}$, respectively.
b. The atomic numbers of copper, chromium, cobalt, and molybdenum are 29, 24, 27, and 42, respectively. It follows that the three wave lengths given in part (a) have to be multiplied by $(29/z)^2$ to obtain the wave lengths of the atom with the atomic number z. For chromium, one finds $2.256\mathring{A}$, $2.250\mathring{A}$, and $2.03\mathring{A}$, respectively; for cobalt $1.782\mathring{A}$, $1.778\mathring{A}$, and $1.60\mathring{A}$, respectively; for molybdenum $0.737\mathring{A}$, $0.735\mathring{A}$, and $0.66\mathring{A}$, respectively.

Answer 3.2.

a. As the three points are on the plane, the vectors connecting each two of them are also on that plane. Therefore the two vectors, $\mathbf{a}_1' = \mathbf{R}_2 - \mathbf{R}_1 = a(\hat{\mathbf{x}} - \hat{\mathbf{y}})$ and $\mathbf{a}_2' = \mathbf{R}_3 - \mathbf{R}_2 = a(\hat{\mathbf{y}} - \hat{\mathbf{z}})$ can be chosen as the lattice vectors on the plane. The length of each such vector is $|\mathbf{a}_1'| = |\mathbf{a}_2'| = a\sqrt{2}$, and the angle in-between them is given by $\cos\gamma = \mathbf{a}_1' \cdot \mathbf{a}_2'/(|\mathbf{a}_1'||\mathbf{a}_2'|) = -a^2/(2a^2) = -1/2$, i.e., $\gamma = 120°$. It follows that this is a triangular lattice. Its unit cell is, e.g., a rhombus like in Fig. 2.4, whose area is $S = |\mathbf{a}_1' \times \mathbf{a}_2'| = |\mathbf{a}_1'||\mathbf{a}_2'|\sin\gamma = a^2\sqrt{3}$. It is possible to reach an arbitrary point on the plane upon beginning, for instance, at the point $\mathbf{R}_1$ and adding to it arbitrary vectors oriented along the planar lattice vectors,

$$\mathbf{r} = \mathbf{R}_1 + y_1\mathbf{a}_1' + y_2\mathbf{a}_2' = \frac{a}{2}[(2y_1 - 1)\hat{\mathbf{x}} + (2y_2 - 2y_1 + 1)\hat{\mathbf{y}} + (1 - 2y_2)\hat{\mathbf{z}}]\,,$$

where y_1 and y_2 are numbers (not necessarily integers). This is the equation of the plane. This plane cuts the $\hat{\mathbf{x}}$–axis at the point whose coefficients of $\hat{\mathbf{y}}$ and $\hat{\mathbf{z}}$ vanish, i.e., for $y_1 = 1$ and $y_2 = 1/2$, and hence at $\mathbf{r} = (a/2)\hat{\mathbf{x}}$. Clearly this is not a lattice point, but it is located at the mid point of the cubic-lattice vector along this direction. Similarly, the plane cuts the two other axes at $\mathbf{r} = (a/2)\hat{\mathbf{y}}$ and $\mathbf{r} = (a/2)\hat{\mathbf{z}}$. A plane that cuts the axes at these points is indeed shown in Fig. 3.6, as one of the planes marked there by (222), in the **BCC** panel. As explained, the notation (222) indicates that the plane closest to the origin cuts each of the cubic axes at the mid points of the cubic lattice vectors.

b. As the plane in question should pass through the origin, its equation is $\mathbf{r} = y_1\mathbf{a}_1' + y_2\mathbf{a}_2'$. Its distance from the previous plane equals the height of the pyramid whose base is the triangle connecting the three points chosen at the initial stage in (a), and its apex is at the origin. The length of each edge on the base of the pyramid is $|\mathbf{a}_1'| = |\mathbf{a}_2'| = a\sqrt{2}$, and the length of each of the faces' edge is $|\mathbf{R}_1| = |\mathbf{R}_2| = |\mathbf{R}_3| = a\sqrt{3}/2$. Each such edge forms a right triangle, together with the height of the pyramid and 2/3 of the height of the equilateral triangle in the base, whose length is $(2/3)(a\sqrt{2})\sqrt{3}/2 = a\sqrt{2/3}$. By the Pythagorean theorem, the distance between the two planes is $d = \sqrt{3a^2/4 - 2a^2/3} = a/(2\sqrt{3})$. In order to find the equation of the next plane, one has to locate a lattice point whose distance from it is the smallest. From Fig. 3.6 it is seen that such a point is $\mathbf{R}_4 = a\hat{\mathbf{x}}$, and therefore the equation of the plane is $\mathbf{r} = \mathbf{R}_4 + y_1\mathbf{a}_1' + y_2\mathbf{a}_2'$. As the first plane cuts the $\hat{\mathbf{x}}$–axis at $(a/2)\hat{\mathbf{x}}$, and the second plane at $a\hat{\mathbf{x}}$, it follows that the equation of the m–th plane is $\mathbf{r} = m(a/2)\hat{\mathbf{x}} + y_1\mathbf{a}_1' + y_2\mathbf{a}_2'$. Several of these planes are depicted in Fig. 3.6 where they are denoted by (222).

c. The planar density of lattice points is $1/S = 1/(a^2\sqrt{3})$. The distance between two neighboring planes is $d = a/(2\sqrt{3})$, and therefore the density of lattice points in space is $1/(Sd) = 2/a^3$. This is precisely the density of the **BCC** lattice, whose cubic unit cell (of volume a^3) contains two points. Hence, the family of planes indeed includes all lattice points.

Answer 3.3.

The rays, as depicted in Fig. 3.28, hit the points A and B, whose distance from one another is L. The angle between AB and the distance between the planes, $d =$ AC, is α. The segments AF and AD are normal to the incoming ray and to the scattered one, respectively. The angles between AD and AC and between AF and AC are both θ. As seen in the figure, the difference between the two paths that the rays traverse is

$$\mathrm{FB} + \mathrm{BD} = L\sin(\theta - \alpha) + L\sin(\theta + \alpha) = 2L\sin\theta\cos\alpha = 2d\sin\theta \ .$$

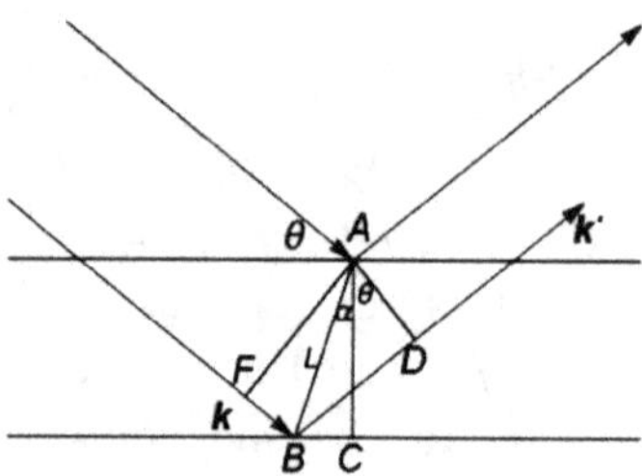

Fig. 3.28

Answer 3.4.

a. Since the unit vector $\hat{z}$ is normal to the plane of the lattice, it can be written as $\hat{z} = \mathbf{a}_1 \times \mathbf{a}_2/S$. Exploiting the identity

$$\mathbf{u} \times [\mathbf{v} \times \mathbf{w}] = (\mathbf{u} \cdot \mathbf{w})\mathbf{v} - (\mathbf{u} \cdot \mathbf{v})\mathbf{w} \ , \tag{3.75}$$

(prove!) yields

$$\mathbf{b}_1 = 2\pi \mathbf{a}_2 \times \hat{z}/S = 2\pi \mathbf{a}_2 \times [\mathbf{a}_1 \times \mathbf{a}_2]/S^2 = 2\pi[|\mathbf{a}_2|^2 \mathbf{a}_1 - (\mathbf{a}_1 \cdot \mathbf{a}_2)\mathbf{a}_2]/S^2 \ .$$

The second equality is proven in the same way.

b. Using part (a) gives

$$\mathbf{b}_1 \times \mathbf{b}_2 = (2\pi)^2[|\mathbf{a}_1|^2|\mathbf{a}_2|^2 - (\mathbf{a}_1 \cdot \mathbf{a}_2)^2]\mathbf{a}_1 \times \mathbf{a}_2/S^4 \ .$$

Exploiting in addition the identity $S^2 = |\mathbf{a}_1 \times \mathbf{a}_2|^2 = |\mathbf{a}_1|^2|\mathbf{a}_2|^2 - (\mathbf{a}_1 \cdot \mathbf{a}_2)^2$ yields the area of the unit cell of the reciprocal lattice, $S_{\mathrm{rec}} = |\mathbf{b}_1 \times \mathbf{b}_2| = (2\pi)^2/S$.

Answer 3.5.

In addition to the vectors displayed in Fig. 3.11, the lattice vectors of the simple hexagonal lattice include $\mathbf{a}_3 = c\hat{z}$, with $a_3 = c$. Using Eq. (3.23) yields the same results for the two vectors in the plane, and in addition $\mathbf{b}_3 = 2\pi\hat{z}/c$.

Answer 3.6.

The volume of the unit cell of the reciprocal lattice is (up to a sign) $V_{\rm rec} = [\mathbf{b}_1 \times \mathbf{b}_2] \cdot \mathbf{b}_3$. The vectorial product is, from Eq. (3.75),

$$\mathbf{b}_1 \times \mathbf{b}_2 = (2\pi/V)^2 [\mathbf{a}_2 \times \mathbf{a}_3] \times [\mathbf{a}_3 \times \mathbf{a}_1] = [(2\pi)^2/V]\mathbf{a}_3 \ .$$

Consequently, since $\mathbf{a}_3 \cdot \mathbf{b}_3 = 2\pi$, one finds $V_{\rm rec} = (2\pi)^3/V$.

Another method to solve this problem is the following. The volume $V = \mathbf{a}_1 \cdot [\mathbf{a}_2 \times \mathbf{a}_3]$ is equivalent to the determinant of a 3×3 matrix, whose rows contain the cartesian components of the lattice vectors $\mathbf{a}_1$, $\mathbf{a}_2$, and $\mathbf{a}_3$. Similarly, the volume $V_{\rm rec} = [\mathbf{b}_1 \times \mathbf{b}_2] \cdot \mathbf{b}_3$ is the determinant of a matrix whose columns contain the vectors $\mathbf{b}_1$, $\mathbf{b}_2$, and $\mathbf{b}_3$. (Note that interchanging the rows with the columns in a square matrix leaves the determinant unaltered.) The product of these two determinants is the determinant of the product of the matrices. Equation (3.20) states that the product matrix is diagonal, with the value 2π for each diagonal entry. Therefore the determinant of the product is $(2\pi)^3$, *qed*. In one dimension $ab = 2\pi$. One may repeat the proof presented above for two dimensions, to obtain $SS_{\rm rec} = (2\pi)^2$, as has been found in problem 3.4. Hence, in d dimensions, the result is $VV_{\rm rec} = (2\pi)^d$.

Answer 3.7.

A straightforward calculation of the lattice vectors of the reciprocal to the reciprocal lattice is to use Eq. (3.23). One of these vectors is $(2\pi/V_{\rm rec})\mathbf{b}_1 \times \mathbf{b}_2$, where $V_{\rm rec}$ is the volume of the unit cell in the reciprocal lattice. The solution of problem 3.6 gives $\mathbf{b}_1 \times \mathbf{b}_2 = [(2\pi)^2/V]\mathbf{a}_3$. Hence, $(2\pi/V_{\rm rec})\mathbf{b}_1 \times \mathbf{b}_2 = (2\pi/V_{\rm rec})[(2\pi)^3/V]\mathbf{a}_3 = \mathbf{a}_3$. The last step exploits the relation $VV_{\rm rec} = (2\pi)^3$, see problem 3.6. The other two lattice vectors, $\mathbf{a}_1$ and $\mathbf{a}_2$, are obtained in a similar way. It follows that **the reciprocal of the reciprocal lattice is the original lattice.** An alternative solution is based on the symmetry of Eqs. (3.20) for the three lattice vectors $\mathbf{a}_1$, $\mathbf{a}_2$, and $\mathbf{a}_3$, and the three vectors $\mathbf{b}_1$, $\mathbf{b}_2$, and $\mathbf{b}_3$. This symmetry implies that the first triplet contains the vectors of the reciprocal to the lattice whose vectors form the second triplet.

Answer 3.8.

The lattice vectors of the **BCC** lattice are given in Eq. (2.5) and the volume of the unit cell is $V = a^3/2$. Exploiting Eq. (3.23) gives $\mathbf{b}_1 = (2\pi/a)(\hat{\mathbf{y}} + \hat{\mathbf{z}})$, $\mathbf{b}_2 = (2\pi/a)(\hat{\mathbf{z}} + \hat{\mathbf{x}})$, and $\mathbf{b}_3 = (2\pi/a)(\hat{\mathbf{x}} + \hat{\mathbf{y}})$. A comparison with Eq. (2.6) shows that this is an **FCC** lattice, where the length of the cube's edge is $4\pi/a$. Alternative method: the reciprocal of the **BCC** lattice is the **FCC** one. According to the solution of problem 3.7, the reciprocal of the **BCC** lattice is the **FCC** one. One has still to determine the lattice constant. Denoting the lattice constant of the reciprocal lattice of the **BCC** lattice (which is an **FCC** lattice) by b, then $V_{\rm rec} = b^3/4$. On the other hand, $V = a^3/2$, and so, according to the solution of problem 3.6, $b^3/4 = V_{\rm rec} = (2\pi)^3/V = 16\pi^3/a^3$, which yields $b = 4\pi/a$.

Answer 3.9.

As seen in Fig. 3.11, the reciprocal of the triangular lattice is a triangular lattice as well. Figure 2.5 demonstrates that the Wigner-Seitz cell of the triangular lattice is a hexagon. It therefore follows that the first Brillouin zone of the triangular reciprocal lattice is a hexagon. The first three Brillouin zones of this lattice are depicted in Fig. 3.29. Each of the lines is perpendicular to a reciprocal-lattice vector that connects the origin with one of the lattice sites (the dark circles), cutting it at its middle. The nth Brillouin zone is reached after crossing $n-1$ border lines.

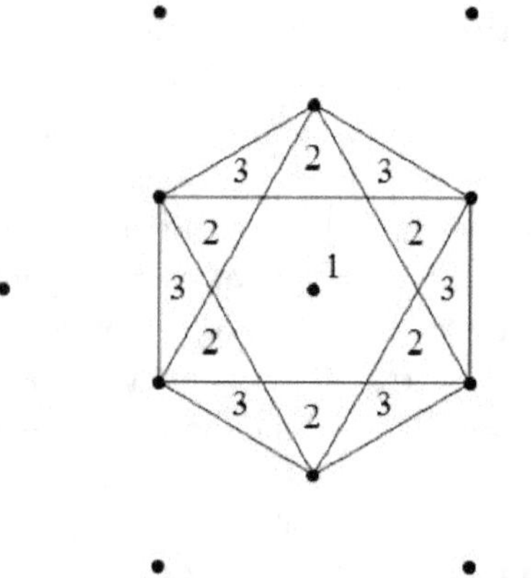

Fig. 3.29

Answer 3.10.

As discussed in problem 3.2, an arbitrary plane in this family of planes is given by the equation $\mathbf{r} = m(a/2)\hat{\mathbf{x}} + x_1\mathbf{a}'_1 + x_2\mathbf{a}'_2$, where $\mathbf{a}'_1 = a(\hat{\mathbf{x}} - \hat{\mathbf{y}})$ and $\mathbf{a}'_2 = a(\hat{\mathbf{x}} - \hat{\mathbf{z}})$. The plane closest to the origin cuts the lattice vectors at $\mathbf{a}_1/2 = a\hat{\mathbf{x}}/2$, $\mathbf{a}_2/2 = a\hat{\mathbf{y}}/2$, and $\mathbf{a}_3/2 = a\hat{\mathbf{z}}/2$. It follows from Eq. (3.29) that $x_1 = x_2 = x_3 = 1/2$, and the Miller indices are $h = k = \ell = 2$. Indeed, these planes are marked in Fig. 3.6 by (222), in the line belonging to the **BCC** lattice. The reciprocal-lattice vector that corresponds to these indices is $\mathbf{G}_0 = h\mathbf{b}_1 + k\mathbf{b}_2 + \ell\mathbf{b}_3 = (4\pi/a)(\hat{\mathbf{x}} + \hat{\mathbf{y}} + \hat{\mathbf{z}})$. A scalar product of this vector with the equation of the mth plane, $\mathbf{r} = m(a/2)\hat{\mathbf{x}} + x_1 a(\hat{\mathbf{x}} - \hat{\mathbf{y}}) + x_2 a(\hat{\mathbf{x}} - \hat{\mathbf{z}})$, indeed yields $\mathbf{G}_0 \cdot \mathbf{r} = 2\pi m$, like in Eq. (3.18). It is found in problem 3.2 that the distance between neighboring planes is $d = a/(2\sqrt{3})$, as indeed obtained from Eq. (3.27), $d = 2\pi/G_0$, with $G_0 = 4\pi\sqrt{3}/a$.

Answer 3.11.

a. Inverting the sign of any nonzero index produces a factor of 2 in the multiplicity. Another factor enumerates all possible permutations of the triplet. Hence, the answers are 6, 48, 24, 8, 24, and 12, respectively.

b. There are eight such planes. For example, the (111) plane joins the tips of the lattice vectors along the positive directions of the cartesian axes (as in the lower right part of Fig. 3.6, for the **SC** lattice). All other planes in the family join such tips along all other possible directions. The structure thus created is the octahedron shown on the left in Fig. 3.30. The origin is located at the center of the octahedron, and the vertices are at the points $\pm a\hat{\mathbf{x}}$, $\pm a\hat{\mathbf{y}}$, and $\pm a\hat{\mathbf{z}}$.

c. The (110) plane is parallel to the $\hat{\mathbf{z}}-$axis, and passes through two opposite edges of the cubic unit cell (as in the lower central part of Fig. 3.6, for the **SC** lattice). There are four such planes that are parallel to each axis, and hence altogether 12 planes that encompass the origin. They form a symmetric body with 12 square faces, as displayed in the right panel of Fig. 3.30. This body is the **rhombic dodecahedron**; it is identical to the Wigner-Seitz cell of the **FCC** lattice, Fig. 2.19(a). See also Appendix A.

d. When all indices h, k, and ℓ differ from each other, there are six permutations interchanging them. Each permutation with nonzero indices possesses eight planes obtained by inverting the signs of each of the indices. Hence the maximal number of planes in the family is 48.

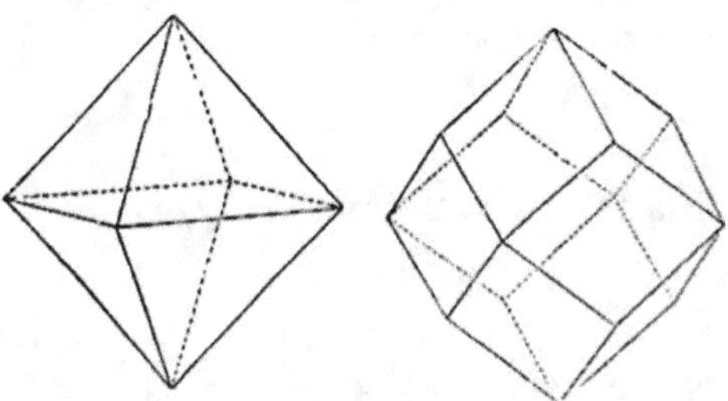

Fig. 3.30

Answer 3.12.
Since the plane passes through the points $\mathbf{a}_1/h$, $\mathbf{a}_2/k$, and $\mathbf{a}_3/\ell$, it follows that the independent vectors $(\mathbf{a}_1/h - \mathbf{a}_2/k)$ and $(\mathbf{a}_1/h - \mathbf{a}_3/\ell)$, which connect pairs of such points, are located on the plane. Then the scalar product of each of these two vectors with $\mathbf{G}$ vanishes, and therefore $\mathbf{G}$ is normal to the plane. The distance between the origin and the plane is thus equal to the projection of each of the original vectors, e.g., $\mathbf{a}_1/h$, on the direction of $\mathbf{G}$. This projection is $(\mathbf{a}_1/h) \cdot \mathbf{G}/G = 2\pi/G$, using Eqs. (3.20) and (3.21).

Answer 3.13.
It is found in problem 3.8 that $\mathbf{b}_1 = 2\pi(0,1,1)/a$, $\mathbf{b}_2 = 2\pi(1,0,1)/a$, and $\mathbf{b}_3 = 2\pi(1,1,0)/a$, and hence $\mathbf{G} = (2\pi/a)(k+\ell, h+\ell, h+k)$. It follows that $h' = k+\ell$, $k' = h+\ell$, and $\ell' = h+k$. Adding together these three relations results in

$h' + k' + \ell' = 2(h + k + \ell)$. Thus, the sum of the "new" indices is even: either they are all even or two of them are odd and the third is even.

Answer 3.14.

The table below lists all possible triplets $\{hk\ell\}$ with $h \leq k \leq \ell$ (so that each triplet appears only once), the values of $S = h^2 + k^2 + \ell^2$, the multiplicity p, and the angle θ [the latter is derived from Eq. (3.30), $\sin\theta = [\lambda/(2a)]\sqrt{S} = \sqrt{S}/6$]. The multiplicity for $\{00\ell\}$ is 6 since there are three positions for the nonzero index ℓ, and it can have two signs, ± 1; $p(0kk) = 12$ since there are three places to put the zero, and each of the k's can have two signs; $p(kkk) = 8$, as each of the indices can have two signs. For $k \neq \ell$ $p(0k\ell)=24$, as there are 6 permutations of the three indices, and the two nonzero ones can have two signs; $p(kk\ell) = 24$ for $k \neq \ell$ because there are three places for the index ℓ, and each of the three indices can have two signs. When all three indices are nonzero and differ from each other the multiplicity is 48 (six permutations and two signs for each of the indices). For each of the values $S = 9, 17, 18$ there are two sets of indices that yield the same scattering angle. The values $S = 7, 15$ do not appear since there are no three integers that give them. Since the scattering angle is 2θ, it cannot exceed $90°$ in forward scattering and therefore there are no scattering points for $S > 18$.

Answer 3.15.

a. In the orthorhombic lattice $\mathbf{G}_0 = 2\pi(h\hat{\mathbf{x}}/a_1 + k\hat{\mathbf{y}}/a_2 + \ell\hat{\mathbf{z}}/a_3)$ [Eq. (3.22)]. Hence,

$$\sin^2\theta = \frac{\lambda^2}{4}\left(\frac{h^2}{a_1^2} + \frac{k^2}{a_2^2} + \frac{\ell^2}{a_3^2}\right).$$

In the tetragonal lattice $a_1 = a_2$, and thus

$$\sin^2\theta = \frac{\lambda^2}{4}\left(\frac{h^2}{a_1^2} + \frac{k^2}{a_1^2} + \frac{\ell^2}{a_3^2}\right).$$

b. For a given wave length, the smallest angles are due to the smallest values of the expression in the brackets. In the present example there are two wave lengths, 1.39Å and 1.54Å, and consequently the scattering angles come in pairs. The smallest ones result from the smallest values of $h^2 + k^2 + \ell^2(a_1/a_3)^2 = h^2 + k^2 + 4\ell^2/9$, i.e., $(hk\ell) = (001)$, (100), (101), and (002), for which $h^2 + k^2 + 4\ell^2/9 = 4/9$, 1, $15/9$, and $16/9$. Hence, the smallest eight angles obey $\sin^2\theta = 0.54$, .066, .121, .148, .174, .214, .215, and .242. Note that the fourth angle of the longer wave length is smaller than the third one of the shorter wave length. The angles themselves are $\theta = 13.4°$, $14.9°$, $20.3°$, $22.6°$, $24.7°$, $27.56°$, $27.6°$, and $29.4°$.

Answer 3.16.

a. In the geometry of Fig. 3.3(b), the scattered beam does not hit the screen when the scattering angle is greater than $\theta_{\max} = 45°$. Using $\theta_{\min} = 20°$ yields $\sin^2\theta_{\max}/\sin^2\theta_{\min} = 4.27$, and hence $h^2+k^2+\ell^2 \leq 4$. For the **SC** lattice, this condition allows solely the families $\{100\}$, $\{110\}$, $\{111\}$, and $\{200\}$. The scattering angles

h	k	ℓ	S	p	θ
0	0	1	1	6	9.6
0	1	1	2	12	13.6
1	1	1	3	8	16.8
0	0	2	4	6	19.5
0	1	2	5	24	21.9
1	1	2	6	24	24.1
0	2	2	8	12	28.1
0	0	3	9	6	30
1	2	2	9	24	30
0	1	3	10	24	31.8
1	1	3	11	24	33.6
2	2	2	12	8	35.3
0	2	3	13	24	36.9
1	2	3	14	48	38.6
0	0	4	16	6	41.8
0	1	4	17	24	43.4
2	2	3	17	24	43.4
0	3	3	18	12	45
1	1	4	18	24	45
1	3	3	19	24	-
0	2	4	20	24	-
1	2	4	21	48	-

are then $2\theta = 40°$, $58°$, $73°$, and $86°$, respectively. Taking into account the multiplicities (see problem 3.14) yields for the intensities $I = 6I_0$, $12I_0$, $8I_0$, and $6I_0$, where I_0 is a multiplicative factor that expresses the scattering intensity off each lattice point.

b. As found in problem 3.15, $\sin^2\theta/\sin^2\theta(100) = h^2 + k^2 + \ell^2(a_1/a_3)^2$. Therefore, the scattering angles for $(hk\ell) =(100)$, (010), (110), (200), and (020) are not changed; there appear scattering rings for three of the previous angles, $2\theta = 40°$, $58°$, and $86°$. However, the scattering intensity of each is $4I_0$, since the multiplicities are lower. In addition, the angles for which

$$\sin^2\theta/\sin^2\theta(100) = (a_1/a_3)^2,\ 1 + (a_1/a_3)^2,\ 2 + (a_1/a_3)^2,\ \text{and } 4(a_1/a_3)^2 ,$$

gives rise to constructive interference. These angles are $2\theta = 36°$, $55°$, $70°$, and $77°$, with the scattering intensities $I = 2I_0$, $8I_0$, $8I_0$, and $2I_0$. Altogether, there are seven scattering circles in place of four.

c. In this case $\sin^2\theta/\sin^2\theta(100) = h^2 + k^2(a_1/a_2)^2 + \ell^2(a_1/a_3)^2$. Some of the previous angles attain new values, while others are left unchanged. In addition the peaks resulting originally from $(hk\ell) = (010)$, (011), and (020) are split, and altogether there are ten circles. Assuming that $a_1 < a_2 < a_3$, the new rings are due to the indices $(hk\ell) = (001)$, (010), (100), (011), (101), (110), (111), (002), (020), and (200). The only remaining multiplicity is that of sign changes of each of the indices, and therefore the intensities are $I = 2I_0$, $2I_0$, $2I_0$, $4I_0$, $4I_0$, $4I_0$, $8I_0$, $2I_0$, $2I_0$, and $2I_0$, respectively.

Answer 3.17.

From Eq. (3.37),

$$\widetilde{f}(\mathbf{G}) = \frac{1}{S}\int_S d^2r f(\mathbf{r})e^{-i\mathbf{G}\cdot\mathbf{r}} = \frac{1}{a^2}\int_0^b dx \frac{1}{b^2}\int_0^b dy e^{-i(G_x x + G_y y)} \; ,$$

where $S = a^2$ is the area of the unit cell. The integral is a product of two integrals, each of the form

$$\int_0^b dx e^{-iG_x x} = (e^{-iG_x b} - 1)/(-iG_x) = 2e^{-iG_x b/2}\sin(G_x b/2)/G_x \; ,$$

and therefore

$$\widetilde{f}(\mathbf{G}) = 4e^{-i(G_x + G_y)b/2}\frac{\sin(G_x b/2)}{abG_x}\frac{\sin(G_y b/2)}{abG_y} \; .$$

In the $b \to 0$ limit one finds $\sin(G_x b/2)/(G_x b) \to 1/2$ and therefore $\widetilde{f}(\mathbf{G}) \to 1/a^2$. In this limit Eq. (3.34) yields $f(\mathbf{r}) = (1/a^2)\sum_{\mathbf{G}}\exp[i\mathbf{G}\cdot\mathbf{r}]$, i.e., the "intensity" is the same for all plane waves whose wave-vector differences are reciprocal-lattice vectors. Indeed, in this limit the function $f(\mathbf{r})$ in each cell approaches a delta function (it diverges to infinity at the origin and vanishes elsewhere, and its integral equals unity). Adding together the contributions from all cells gives $f(\mathbf{r}) \to \sum_{\mathbf{R}}\delta(\mathbf{r}-\mathbf{R})$, and therefore the function is a delta comb, as encountered in Sec. 3.3: the result obtained here is similar to Eq. (3.14).

Answer 3.18.

a. Consider first the one-dimensional integral $I = \int_{-X}^{X} dx f(x)$. Changing the integration variable from x to $-x$ yields $I = \int_{X}^{-X} d(-x)f(-x) = \int_{-X}^{X} dx f(-x)$. The signs of integration bounds are inverted in the first step, and in the second they are interchanged, which results in the opposite sign of the integral. Inverting the sign of the integration variable in the expression for the structure factor gives

$$F(\mathbf{G}) = \int_V d^3r n(\mathbf{r})e^{i\mathbf{G}\cdot\mathbf{r}}$$

$$= \int_V d^3r n(-\mathbf{r})e^{-i\mathbf{G}\cdot\mathbf{r}} = \int_V d^3r n(\mathbf{r})e^{-i\mathbf{G}\cdot\mathbf{r}} = F(-\mathbf{G}) = F^*(\mathbf{G}) \; .$$

The last step uses the fact that the scatterers' density, $n(\mathbf{r})$, is symmetric, $n(\mathbf{r}) = n(-\mathbf{r})$. The integral in the third step is over the unit cell that is the reflection of the original one, but the two cells are identical.

b. Using spherical coordinates in Eq. (3.46), with the $\hat{\mathbf{z}}$−axis chosen to be along $\mathbf{G}$ and with the notation $\mu = \cos\theta$, yields

$$F(\mathbf{G}) = 2\pi\int_0^\infty dr r^2 n(r)\int_{-1}^{1} d\mu \exp[iGr\mu] \; .$$

The result depends solely on $G = |\mathbf{G}|$. The angular integration is

$$\int_{-1}^{1} d\mu \exp[iGr\mu] = 2\sin(Gr)/(Gr) ,$$

as required.

c. In this case $n(\mathbf{r}) = |\psi_{100}(\mathbf{r})|^2 = \exp[-2r/a_B]/(\pi a_B^3)$ is spherically symmetric. Exploiting the result of the previous part yields

$$F = \frac{4}{Ga_B^3} \int_0^\infty dr\, r e^{-2r/a_B} \sin(Gr) = \frac{1}{[1 + (a_B G/2)^2]^2} .$$

Indeed, the structure factor decays for long reciprocal-lattice vectors, but the dependence on $G = |\mathbf{G}|$ is weak as G is smaller than $1/a_B$. The helium atom contains two electrons in the ground state, but the "radius" of their orbits is $a_B/2$ (because of the atomic number $z = 2$). Therefore the result is $F = 2/[1 + (a_B G/4)^2]^2$. The electrons are closer to the nucleus, and therefore the decay of the structure factor with the length of the reciprocal-lattice vector is slower.

Answer 3.19.

The structure factor is

$$F(\mathbf{G}) = F_A + F_B e^{i\mathbf{G}\cdot\mathbf{r}_2} = F_A + F_B e^{2i\pi(h+k)/3} ,$$

and therefore the scattering intensity is

$$|F(\mathbf{G})|^2 = F_A^2 + F_B^2 + 2F_A F_B \cos[2\pi(h+k)/3] .$$

There are two possible values for this intensity, $|F(\mathbf{G})|^2 = (F_A + F_B)^2$ when $h+k$ is an integer multiple of 3, and $|F(\mathbf{G})|^2 = (F_A + F_B)^2 - 3F_A F_B$ when it is not. This expression reproduces the result for graphene when $F_A = F_B$, and yields a vanishing intensity at the points where $h + k$ is an integer multiple of 3 and $F_A = -F_B$. In the latter case, the reciprocal lattice is the hexagonal one.

Answer 3.20.

Figure 2.9 displays the hexagonal lattice built of a hexagonal lattice in the XY plane and an additional lattice vector, $\mathbf{a}_3 = c\hat{\mathbf{z}}$. The reciprocal of the simple hexagonal lattice is given in problem 3.5. The present case is a lattice with a base: the unit cell contains four carbon atoms, at the origin, and at the points $\mathbf{r}_2 = (\mathbf{a}_1 + \mathbf{a}_2)/3$, $\mathbf{r}_3 = \mathbf{a}_3/2$, and $\mathbf{r}_4 = \mathbf{r}_3 + 2\mathbf{r}_2$. The structure factor is hence

$$F(\mathbf{G}) = F_0\left(1 + e^{2i\pi(h+k)/3} + e^{i\pi\ell}(1 + e^{4i\pi(h+k)/3})\right) .$$

It follows that $F(\mathbf{G}) = 2F_0(1 + \cos[2\pi(h+k)/3])$ when ℓ is even. On the other hand, when ℓ is odd, then $F(\mathbf{G}) = 2iF_0 \sin[2\pi(h+k)/3]$, and therefore the intensity vanishes when $h + k$ is an integer multiple of 3, similar to the situation described in the previous problem.

Answer 3.21.

The lattice vectors of the reciprocal to the **FCC** lattice are given in Eq. (3.24). Using them, with the coefficients h, k, and ℓ, yields

$$\mathbf{G} = h\mathbf{b}_1 + k\mathbf{b}_2 + \ell\mathbf{b}_3 = (2\pi/a)(h'\hat{\mathbf{x}} + k'\hat{\mathbf{y}} + \ell'\hat{\mathbf{z}}) \,,$$

where the right-side equality is written in terms of the reciprocal-lattice vectors of the simple cubic lattice, with $h' = h - k + \ell$, $k' = h + k - \ell$, $\ell' = -h + k + \ell$. Substituting the latter into the right hand-side of the equation for $\mathbf{G}$ shows that $h' + k' = 2h$, $h' + \ell' = 2\ell$, and $k' + \ell' = 2k$. The sum of two integers is even only when they both are even or they are both odd. Hence, constructive interference necessitates the cubic coefficients h', k', and ℓ' to be all even or all odd, as obtained also from the calculation for the cubic unit cell.

Answer 3.22.

a. The inter-planar distance in a cubic lattice, for the family given by the indices h, k, and ℓ is $d(hk\ell) = 2\pi/G = a/\sqrt{h^2 + k^2 + \ell^2}$. As mentioned in Sec. 3.6, the site density in each plane is d/V, and therefore the largest density is attained at the smallest value of $h^2 + k^2 + \ell^2$. In the **BCC** lattice, the sum $h + k + \ell$ should be even and hence the smallest value of $h^2 + k^2 + \ell^2$ is realized when two of the indices are 1 and the third is 0, e.g., (110). Then, the inter-plane distance is $d(110) = a/\sqrt{2}$, and the site density in each plane is $\sqrt{2}/a^2$ (recall that $V = a^3/2$).

b. The plane (110) in the **BCC** lattice is parallel to the $\hat{\mathbf{z}}-$axis, and passes through two opposite edges of the cube (see Fig. 3.6). This cut creates a rectangle, whose edges are a and $b = a\sqrt{2}$; it contains an additional lattice site at its center. Therefore, it is a rectangular-centered lattice. The site density on it is $\sqrt{2}/a^2$, and the inter-planar distance is $a/\sqrt{2}$.

c. Within the plane, the lines can be marked by the reciprocal-lattice vectors normal to them. As this is a rectangular-centered lattice, these vectors are $\mathbf{G} = 2\pi(h\hat{\mathbf{x}}/a + k\hat{\mathbf{y}}/b)$, with $b = a\sqrt{2}$. The inter-line distance is $d = 2\pi/G = a/\sqrt{h^2 + k^2/2}$, with the maximum distance attained for $(hk) = (01)$, $d_{\max} = a\sqrt{2}$. The vector $\mathbf{G}$ is then parallel to the $\hat{\mathbf{y}}-$axis; consequently the densest line is parallel to the $\hat{\mathbf{x}}-$axis, i.e., to the $\hat{\mathbf{z}}-$axis of the original cubic lattice. The inter-site distance on the densest line is a.

d. In the **FCC** lattice, the indices must be either all even or all odd. The lowest value of $h^2 + k^2 + \ell^2$ is obtained for $h = k = \ell = 1$. Then $d(111) = a/\sqrt{3}$ and consequently the planar density is $4/(a^2\sqrt{3})$ (the volume of the unit cell is $V = a^3/4$). The (111) planes in the **FCC** lattice are those containing the red or the green spheres in the lowest row of Fig. 2.24(a). As explained there, these planes comprise triangular lattices, which indeed are the densest planar ones, with the density $4/(a^2\sqrt{3})$. The reciprocal-lattice vectors of this lattice are $\mathbf{G} = 2\pi[h\hat{\mathbf{x}}' + (2k - h)\hat{\mathbf{y}}'/\sqrt{3}]/a$, where $\hat{\mathbf{x}}'$ and $\hat{\mathbf{y}}'$ define the coordinate system in that plane (see Fig. 3.11). It follows that $d = 2\pi/G = a\sqrt{3}/[2\sqrt{h^2 + k^2 - hk}]$, and $d_{\max} = a\sqrt{3}/2$. This result is obtained for $(hk) = (0\pm1)$, $(hk) = (\pm10)$, and for $(hk) = \pm(11)$. The corresponding vectors, $\mathbf{G}(10) = \pm\mathbf{b}_1$, $\mathbf{G}(01) = \pm\mathbf{b}_2$, and $\mathbf{G}(\pm 1 \pm 1) = \pm(\mathbf{b}_1 + \mathbf{b}_2)$, are normal to the vectors $\mathbf{a}_2$, $\mathbf{a}_1$, and $(\mathbf{a}_1 - \mathbf{a}_2)$. It follows that the densest lines on the triangular plane are along the directions connecting nearest-neighbors on this lattice, as indeed can be expected without carrying out the calculation.

Answer 3.23.

a. Figure 3.21 shows that the interference is constructive for $2\theta \approx 27°, 31°, 46°, 54°$, and $56°$, which correspond to $h^2 + k^2 + \ell^2 = 3$, 4, 8, 11, and 12. This sequence indeed fits the **FCC** lattice (Fig. 3.20); the values confirm the Miller indices that appear in the figure. They yield approximately $\sin^2\theta/(h^2 + k^2 + \ell^2) \approx 0.018$. Using this result in Eq. (3.31) yields $[\lambda/(2a)]^2 \approx 0.018$; hence $a \approx 1.545\text{Å}/2/0.134 \approx 5.76\text{Å}$. This is only an approximation, because it is hard to read the precise numbers in the figure. The value cited in the literature is 5.65Å.

b. The intensities decay with $|\mathbf{G}|^2$, which in turn is proportional to $h^2 + k^2 + \ell^2$. Ignoring this decay, the first scattering angle corresponds to the family of planes $\{111\}$, and thus is obtained from 8 planes. The second angle comes from the family $\{200\}$, comprising 6 planes. Hence, the intensities ratio is $I(111)/I(200) = 8|F(111)|^2/(6|F(200)|^2)$. On the other hand, $F(111)/F(200) = (F_{\text{Na}} - F_{\text{Cl}})/(F_{\text{Na}} + F_{\text{Cl}})$. Subtracting the background (about 30 in the arbitrary relative units of the figure), one finds $I(111) \approx 40$ and $I(200) \approx 270$. Assuming the two ions to be point-like scatterers, their structure factors are real and independent of $\mathbf{G}$, i.e., $x = F_{\text{Na}}/F_{\text{Cl}}$ is real. Then the equation $(8/6)(x - 1)^2/(x + 1)^2 \approx 40/270$ yields two solutions, $F_{\text{Na}}/F_{\text{Cl}} \approx 2$ and $F_{\text{Na}}/F_{\text{Cl}} \approx 0.5$. As chlorine has more electrons, it seems that the lower value should be chosen. Sodium has 10 electrons while chlorine possesses 18. Therefore a rough estimate would have yielded $F_{\text{Na}}/F_{\text{Cl}} \approx 10/18 = 0.55$. Thus, the estimate found here is quite plausible.

c. The assumption that $F_{\text{Na}}/F_{\text{Cl}}$ is real is plausible, as the electrons' distribution around each ion is spherically-symmetric (see problem 3.18). Assuming nonetheless that this ratio is complex, one obtains $|F(111)/F(200)|^2 = (|x|^2 + 1 - 2|x|\cos\phi)/(|x|^2 + 1 + 2|x|\cos\phi)$, where $F_{\text{Na}}/F_{\text{Cl}} = |x|\exp[i\phi]$. There are now two variables to be determined, and more information is required; this issue is known in the literature as **"the phase problem"**.

d. From the given densities it follows that $F(hk\ell) = z_i/[1 + \alpha_i(h^2 + k^2 + \ell^2)]^2$, where $\alpha_i = (\pi a_i/a)^2$ (see problem 3.18). Exploiting the value $a = 5.65\text{Å}$ (from the literature), one obtains $F_{\text{Na}}(hk\ell) = z_{\text{Na}}/[1 + .279(h^2 + k^2 + \ell^2)]^2$ and $F_{\text{Cl}}(hk\ell) = z_{\text{Cl}}/[1 + 1.01(h^2 + k^2 + \ell^2)]^2$. Adding to this the ratio $I(111)/I(200) = 8[F_{\text{Na}}(111) - F_{\text{Cl}}(111)]^2/\{6[F_{\text{Na}}(200) + F_{\text{Cl}}(200)]^2\}$ yields $z_{\text{Na}}/z_{\text{Cl}} \approx .1$, and .47. The solution .47 is closer to the value .55 mentioned above, but does not improve the previous estimate. As mentioned in the main text, there are other factors which affect the measured intensities, that are not included in this analysis.

Answer 3.24.

In this case, the expression in last brackets of Eq. (3.53) is proportional to $(1 + \exp[i\pi k])$. This expression is nonzero for even k, and then h and ℓ are even as well. Substituting $\ell = 2\ell'$, $k = 2k'$, and $h = 2h'$ in the expression for $\mathbf{G}$ yields $\mathbf{G} = (4\pi/a)(h'\hat{\mathbf{x}} + k'\hat{\mathbf{y}} + \ell'\hat{\mathbf{z}})$, i.e., the reciprocal-lattice vectors of a simple cubic lattice with a lattice constant $a/2$. Indeed, replacing the two types of atoms in the right panel of Fig. 2.17 by identical atoms produces such a **SC** lattice.

Answer 3.25.

In CsCl, the atoms at the origin and at the center of the cube of the cubic lattice have different structure factors, and therefore the structure factor is $(F_{Cs}+F_{Cl}\exp[i\pi(h+k+\ell)])$. In contrast to the **BCC** lattice, here the interference is constructive for all values of h, k, and ℓ, but the scattering amplitude is smaller for odd values of $h+k+\ell$ (assuming that the ionic structure factors are real and have the same sign). In diamond, the structure factor is proportional to $(1+\exp[i\pi(h+k+\ell)/2])$, times the structure factor of the **FCC** lattice (explain!), which attains the value 2 when $h + k + \ell$ is an integer product of 4, the value $1 \pm i$ when this sum is odd, and the value 0 otherwise. Keeping in mind that the three indices have to be all even or all odd (since this is an **FCC** lattice), the allowed values of $h^2 + k^2 + \ell^2$ (up to 24) are 3, 8, 11, 16, 19 , and 24. Ignoring the dependence of the structure factors on the multiplicities and on the Miller indices, the ratio of the intensities in the three types of sites are 4, 2, and 0. In zinc-blende, the ions at the origin and at the point $a(1,1,1)/4$ are different, and consequently the structure factor is proportional to $(F_{Zn} + F_S \exp[i\pi(h+k+\ell)/2])$. The result is thus $F_{Zn} + F_S$, $F_{Zn} - F_S$, or $F_{Zn} \pm iF_S$ when $h+k+\ell$ is an integer multiple of 4, an odd multiple of 2, or is odd, respectively.

Answer 3.26.

a. Denote $a_1 = a$ and $a_3 = c$. The lattice vectors of the **HCP** lattice are $\mathbf{a}_1 = (a,0,0)$, $\mathbf{a}_2 = (a/2, a\sqrt{3}/2, 0)$ and $\mathbf{a}_3 = (0,0,c)$, see the discussion following Fig. 2.24. The unit cell contains a base comprising two atoms, at the origin and at $\mathbf{r}_2 = (a/2, a\sqrt{3}/6, c/2)$. Using Eq. (3.23) yields $\mathbf{b}_1 = 2\pi(\hat{\mathbf{x}} - \hat{\mathbf{y}}/\sqrt{3})/a$, $\mathbf{b}_2 = 4\pi\hat{\mathbf{y}}/(a\sqrt{3})$, and $\mathbf{b}_3 = 2\pi\hat{\mathbf{z}}/c$, as also found in problem 3.5. Hence, $\mathbf{r}_2 \cdot \mathbf{G} = \pi\ell + 2\pi(h + k)/3$, and the structure factor is $F(\mathbf{G}) = 1 + \exp[i\mathbf{r}_2 \cdot \mathbf{G}]$. It vanishes when $\ell + 2(h+k)/3$ is odd, e.g., $h+k = 3m$, where m is an arbitrary integer, and ℓ is odd. Interestingly enough, the same condition for a null structure factor is derived in problem 3.20 for graphite.

b. $\exp[i\pi\{\ell + 2(h + k)/3\}]$ can be 1, $\exp[\pm i\pi/3]$, $\exp[\pm 2i\pi/3]$, and -1, for which $|F(\mathbf{G})|^2 = 4$, 3, 1, and 0.

c. Since $\mathbf{G}(hk\ell) = 2\pi(h, (2k - h)/\sqrt{3}, \ell a/c)/a$, Eq. (3.30) gives

$$\sin^2 \theta = \lambda^2 |\mathbf{G}/(4\pi)|^2 = \lambda^2[(h^2 - hk + k^2)/(3a^2) + \ell^2/(4c^2)] \ .$$

For the ideal **HCP** lattice this expression is proportional to $(h^2 - hk + k^2 + 9\ell^2/32)$. The smallest values are for $(hk\ell) = (001)$, (100), (002), (101), (102). The values

(100) and (110) yield the same angle. Similarly, the pairs (101) and (111), and also (102) and (112), give rise to identical angles. According to part (a), the structure factor vanishes for the triplet (001). Hence the smallest angle corresponds to $(hk\ell) =$ (100) [or (110)], i.e., $\sin^2\theta_{\min} = \lambda^2/(3a^2)$. The next three angles, corresponding to $(hk\ell) =$ (002), (101), (102), are given by $\sin^2\theta/\sin^2\theta_{\min} = 9/8$, $41/32$, $17/8$.

d. The hexagonal unit cell of wurtzite (see Fig. 2.25 and problem 2.17) contains, in addition to the atoms at the origin and at $\mathbf{r}_2$ which represent, e.g., the sulphur ions, two additional zinc ions at $\mathbf{r}_3 = (0, 0, 5c/8) = (5/8)\mathbf{a}_3$ and at $\mathbf{r}_4 = (a/2, a\sqrt{3}/6, c/8) = (\mathbf{a}_1 + \mathbf{a}_2)/3 + \mathbf{a}_3/8$. The structure factor is thus

$$F(\mathbf{G}) = F_{\mathrm{S}}(1 + \exp[i\mathbf{r}_2 \cdot \mathbf{G}]) + F_{\mathrm{Zn}}(\exp[i\mathbf{r}_3 \cdot \mathbf{G}] + \exp[i\mathbf{r}_4 \cdot \mathbf{G}]) ,$$

where $\mathbf{r}_3 \cdot \mathbf{G} = 5\pi\ell/4$, and $\mathbf{r}_4 \cdot \mathbf{G} = \pi\ell/4 + 2\pi(h+k)/3$.

Answer 3.27.

The structure factor of the body-centered orthorhombic crystal is proportional to $(1 + \exp[i\pi(h+k+\ell)])$, precisely as that of the **BCC** lattice. The structure factor of the face-centered orthorhombic lattice is proportional to $(1 + \exp[i\pi(h+k)] + \exp[i\pi(h+\ell)] + \exp[i\pi(k+\ell)])$, the same as that of the **FCC** lattice. As opposed, the structure factor of the base-centered orthorhombic orthorhombic lattice is proportional to $(1 + \exp[i\pi(h+k)])$, and yields constructive interference only when $(h+k)$ is even.

Answer 3.28.

The conditions for constructive interference in the tetragonal lattices are identical to those of the rhombohedral ones, and hence are also identical to those of the cubic lattices. The body-centered tetragonal lattice possesses constructive interference only when $(h+k+\ell)$ is even. In problem 3.15 this condition is fulfilled for the four largest angles of the eight ones.

Answer 3.29.

Denote the vector by $\mathbf{u} = \sum_{i=1}^{d} u_i \hat{\mathbf{x}}_i$, where $\hat{\mathbf{x}}_i$ is a unit vector along the i-th axis. The corresponding coefficient is $u_i = \mathbf{u} \cdot \hat{\mathbf{x}}_i = u\cos\phi_i$, and the angle between the vector and this axis is ϕ_i. Since $u^2 = \mathbf{u} \cdot \mathbf{u} = \sum_{i=1}^{d} u_i^2 = u^2 \sum_{i=1}^{d} \cos^2\phi_i$, it follows that $\sum_{i=1}^{d} \cos^2\phi_i = 1$, leading to $\langle \sum_{i=1}^{d} \cos^2\phi^i \rangle = 1$. As the probabilities of all directions are identical, all terms in the sum are identical as well, and hence each of them equals $1/d$. One may also calculate this average explicitly, exploiting spherical coordinates. For instance, in three dimensions

$$\langle \cos^2\theta \rangle = \int_0^\pi d\theta \sin\theta \cos^2\theta \Big/ \int_0^\pi d\theta \sin\theta = \int_{-1}^1 d\mu\,\mu^2 \Big/ \int_{-1}^1 d\mu = 1/3 .$$

Answer 3.30.

a. At zero temperature, the argument of the coth diverges, $\coth(\infty) = 1$, and consequently $2W = [\hbar/(2M\omega)]G^2$. At high temperatures, $k_{\mathrm B}T \gg \hbar\omega$, one may use $\coth(x) \approx 1/x$, to obtain $2W = [k_{\mathrm B}T/(M\omega^2)]G^2$.

b. As $\sin^2\theta = \lambda^2 G^2/(4\pi)^2$, [see Eq. (3.30)] one finds the relation given in the problem, with $B = 8\pi^2\langle(\mathbf{u_R})^2\rangle/3$. Equation (3.59) implies that this coefficient is independent of the lattice constants; it depends on the temperature, the atoms' mass, and the frequency.

c. At high temperatures $2W = [k_{\mathrm B}T/(M\omega^2)]G^2$. Substituting the Lindemann criterion yields $2W = [c_L^2 T/(3T_{\mathrm M})](aG)^2$, and the temperature appears solely as T/T_M. According to Eq. (3.21), $\mathbf{G} = h\mathbf{b}_1 + k\mathbf{b}_2 + \ell\mathbf{b}_3$. For a cubic lattice, $\mathbf{b}_i = (2\pi/a)\hat{\mathbf{x}}_i$ and consequently $a\mathbf{b}_i = 2\pi\hat{\mathbf{x}}_i$, as well as $(aG)^2 = 4\pi^2(h^2 + k^2 + \ell^2)$, are independent of the lattice constant a. In the general case, the reciprocal-lattice vectors are given by Eq. (3.23), and hence $a\mathbf{b}_i$ is a dimensionless vector, determined only by the ratio between the lattice constants.

Answer 3.31.

a. The decay factor W is inversely proportional to the mass of the atom in the crystal, see Eq. (3.59). It follows that upon replacing the atom by a heavier isotope (e.g., replacing hydrogen by deuterium) the decay caused by the lattice vibrations is reduced.

b. Generally,

$$2W = \langle(\mathbf{G} \cdot \mathbf{u_R})^2\rangle = \sum_{\alpha,\beta} G_\alpha G_\beta \langle u_{\mathbf{R}\alpha} u_{\mathbf{R}\beta}\rangle = \sum_{\alpha,\beta} B_{\alpha\beta} G_\alpha G_\beta \ ,$$

where the indices denote the vectors' components in a certain basis. For instance, for orthorhombic symmetry, the only nonzero coefficients are the diagonal ones, and $2W = (2\pi)^2(B_{11}h^2/a_1^2 + B_{22}k^2/a_2^2 + B_{33}\ell^2/a_3^2)$. Thus, the decay is different for different lattice directions.

Answer 3.32.

a. The ratio between the numbers of the short and long segments in a Fibonacci sequence approaches the golden ratio, $x(n) = N_L(n)/N_S(n) \to \tau$, see problem 2.27. In the large n limit the distance from the origin to the n−th site contains $N_S(n)$ segments of length 1 and $N_L(n)$ segments of length τ. In a similar fashion, the distance from the origin to a point q in momentum space is proportional to $m + n\tau$, and hence the ratio between the number of long segments and the number of the short ones tends to τ. It follows that the points in momentum space which give rise to high peaks are arranged as in a Fibonacci sequence.

b. The MATHEMATICA code is shown below. The vector *vec* in the code, whose values are depicted in Figs. 3.32, keeps track of the pairs (q, A) for which $A \geq A1$. The right panel displays the case $A1 = 0$, and it contains all the points in the diffraction picture. As seen, peaks of various heights are indeed quite dense. The

```
m=.
τ=N[(1+Sqrt[5])/2]
NN=100
nn=0
A1=.99
vec=Table[{0,0},{nn,1,NN^2}];
dq=Table[0,{mm,NN^2}];
Do[Do[qq=π (n+m/τ)/(τ+1/τ);
   AA=If[qq==π m,1,(Sin[qq-π m]/(qq-π m))^2];
   If[AA>A1,nn=nn+1;vec[[nn]]={qq,AA};If[nn>1,
   dq[[nn-1]]=vec[[nn,1]]-vec[[nn-1,1]]],
   {n,1,NN}],{m,1,NN}]
ListPlot[vec,PlotRange→All,Ticks→{{50,100,150},{.5,1}},
LabelStyle→{Italic,FontSize→25},AxesLabel→{q,A}]
DQ=Table[dq[[n]]/dq[[1]],{n,1,nn-1}]
```

Fig. 3.31

left panel portrays only the points for which $A \geq A1 = 0.99$, i.e., very close to the maximal value $A = 1$. For these points the code computes also the vector DQ, that contains the intervals between successive q's, divided by the first difference. The result is

$$DQ = \{1., \; 1.61803, \; 1.00000, \; 1.61803, \; 1.61803, \; 0.99999, \; 1.61803,$$
$$1.61803, \; 1.00000, \; 1.61803, \; 1.00000, \; 1.61803, \; 1.61803, \}$$

These intervals attain solely the values 1 and τ, and their sequence fits the Fibonacci sequence $SLSLLSLLSLSLL$. Thus, the high peaks in the diffraction picture have the same symmetry as the original lattice.

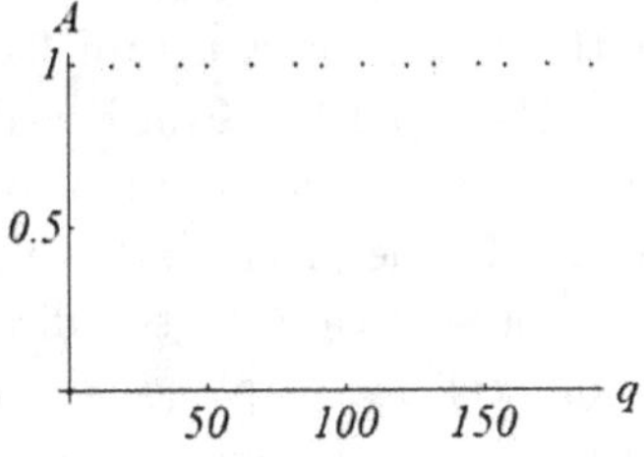 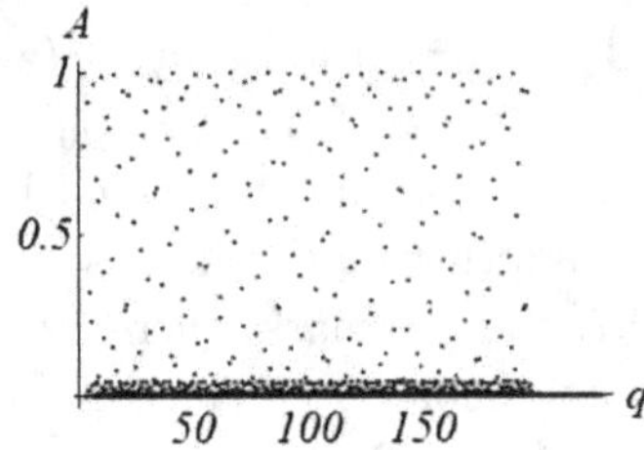

Fig. 3.32

Answer 3.33.

Place the measured sample (the collection of points on the right side of the figure) on the XY plane, and the impinging beam (BO) along the $\hat{\mathbf{z}}$−direction. The figure shows that the projection of the vector $\mathbf{k}'$ on the XY plane equals the projection of

$\mathbf{q}$ on the same plane, i.e., it is equal to $\mathbf{G}_{\text{surface}} = h\mathbf{b}_1 + k\mathbf{b}_2$, while its $z-$component (its projection on BO) equals $k_z' = k_0 \cos(2\theta)$, with $k_0 = |\mathbf{k}'| = 2\pi/\lambda$. On the other hand, $k_0^2 = (h\mathbf{b}_1 + k\mathbf{b}_2)^2 + (k_z')^2$, and therefore the angles at which the scattering is constructive are given by

$$(h\mathbf{b}_1 + k\mathbf{b}_2)^2 = k_0^2 - (k_z')^2 = (2\pi/\lambda)^2 \sin^2(2\theta) \ ,$$

where h and k are integers. The direction of the scattered beam is parallel to the direction of the unit vector $\hat{\mathbf{k}}' = \mathbf{k}'/k_0 = [\lambda/(2\pi)](h\mathbf{b}_1 + k\mathbf{b}_2) + \cos(2\theta)\hat{\mathbf{z}}$. When the planar screen is parallel to the XY plane and is located at a distance D from that plane, then the triangle comprising the beam hitting the screen and its projections on the plane of the screen and on the $\hat{\mathbf{z}}-$axis is similar to the triangle built of the vector $\hat{\mathbf{k}}'$ and its projections along the same directions. This similarity implies that the ratio between the projection of the scattered beam on the screen, $\mathbf{V}$, and the vector $[\lambda/(2\pi)](h\mathbf{b}_2 + k\mathbf{b}_2)$ equals the ratio between D and $\cos(2\theta)$. It follows that the beam along $\hat{\mathbf{k}}'$ hits the screen at a point whose coordinates on the screen plane are given by the vector $\mathbf{V} = D\lambda(h\mathbf{b}_1 + k\mathbf{b}_2)/[2\pi \cos(2\theta)]$.

Answer 3.34.

There is only one component of $\mathbf{q}$ which is parallel to the crystal in this case: $\mathbf{G}_{\text{surface}} = h\mathbf{b}$ ($\mathbf{b}$ is the reciprocal-lattice vector of the one-dimensional crystal). Concomitantly, the vector $\mathbf{q}_{\perp} = \mathbf{q} - h\mathbf{b}$ represents an entire plane normal to the one-dimensional reciprocal lattice. It follows that constructive interference is realized whenever one of these planes cuts the Ewald's sphere. Each such cut is a circle; hence there are full lines in the diffraction pattern.

Answer 3.35.

In structure A the unit cell is doubled along the $\hat{\mathbf{z}}-$direction, and therefore the magnetic unit cell is tetragonal, with edges a, a, and $2a$. The reciprocal-lattice vector is $\mathbf{G} = (2\pi/a)(h\hat{\mathbf{x}} + k\hat{\mathbf{y}} + \ell\hat{\mathbf{z}}/2)$, and the magnetic structure factor is $F_{\text{M}}(hk\ell) = \mu_0(1 - \exp[iG_z a]) = \mu_0[1 - (-1)^\ell]$. The magnetic Bragg peaks appear only for odd ℓ's, while for the nuclear Bragg peaks ℓ is even, like in the original **SC** lattice. In structure C, the base plane is similar to the planar lattice of table salt, Fig. 2.7: the two marks there represent in the present case the two directions of the magnetic moments. Thus, the magnetic unit cell in the plane is a square, with an edge $a' = a\sqrt{2}$. As the planes parallel to the base are identical, the lattice constant along the $\hat{\mathbf{z}}-$axis remains as in the original lattice. The result is again a tetragonal lattice, with the edges $a\sqrt{2}$, $a\sqrt{2}$, a (note that the volume remains as for structure A. Why is this?) The vectors of this lattice in the plane are rotated by $45°$ relative to the original ones, and the reciprocal-lattice vectors are

$$\mathbf{G} = \frac{2\pi}{a}\left(h\frac{\hat{\mathbf{x}} - \hat{\mathbf{y}}}{2} + k\frac{\hat{\mathbf{x}} + \hat{\mathbf{y}}}{2} + \ell\hat{\mathbf{z}}\right) \ .$$

It follows that $F_{\text{M}}(hk\ell) = \mu_0(1 - \exp[iG_x a]) = \mu_0[1 - (-1)^{h+k}]$. The magnetic Bragg peaks appear solely for odd $(h+k)$'s, while for the nuclear ones this sum has to be even. This condition reproduces the original cubic peaks, $\mathbf{G} = (2\pi/a)(h'\hat{\mathbf{x}} + k'\hat{\mathbf{y}} + \ell\hat{\mathbf{z}})$, with $h' = (h+k)/2$ and $k' = (k-h)/2$.

Answer 3.36.

At high temperatures there appear two angles, $2\theta \approx 23.8°$ and $46.5°$. Assuming that the structure is an **FCC** lattice, and hence the indices are all odd, these angles correspond to $(hk\ell) = (111)$ and (311), as indeed marked in the original figure. The peak at $(hk\ell) = (111)$ yields $a = \lambda\sqrt{3}/(2\sin\theta_{\min}) = 1.057\sqrt{3}/(2\sin 11.9°)\text{Å} = 4.44\text{Å}$, and the second peak gives rise to $\sin^2\theta(311)/\sin^2\theta(111) \approx 11/3$, as expected from an **FCC** lattice. At low temperatures there are additional peaks at $2\theta \approx 11.9°$, $22.7°$, $30°$, $36°$, which imply $(hk\ell) = (111)$, $(3\bar{1}\bar{1})$, $(33\bar{1})$, $(5\bar{1}\bar{1})$. The first peak yields the lattice constant, $a = \lambda\sqrt{3}/(2\sin\theta_{\min}) = 1.057\sqrt{3}/(2\sin 5.9°)\text{Å} = 8.85\text{Å}$, and the other peaks yield $\sin^2(hk\ell)/\sin^2(111) \approx 11/3$, $19/3$, $27/3$, in accordance with the **FCC** structure. In other words, the calculation verifies the identification of Shull and coworkers.

3.15 Problems for self-evaluation

s.3.1.
Prove that the reciprocal of the trigonal lattice is also trigonal, with a lattice constant $b = 2\pi/[a\sin\alpha\sin\alpha']$, where α' is given by the relation $\cos(\alpha'/2) = 1/[2\cos(\alpha/2)]$.

s.3.2.
Present explicit expressions for the distance between neighboring planes described by the Miller indices h, k, ℓ for the following lattices: triclinic, orthorhombic, tetragonal, hexagonal, monoclinic and trigonal. Determine the multiplicities of each family of planes.

s.3.3.
a. *Find the structure factor of a sphere of radius r_0 carrying a constant volume charge density, $n_0 = z/(4\pi r_0^3/3)$. What is the limit of $F(\mathbf{G})$ when $|\mathbf{G}| \to 0$?*
b. *Find the structure factor which corresponds to the first five Bragg peaks of scattering off densely-packed spheres on a square lattice (Fig. 2.6) and off densely-packed spheres on a simple-cubic lattice (Fig. 2.15). Assume a constant charge density within each sphere.*
c. *The structure factor of the molecule C_{60} [the buckminsterfullerene, or buckyball, see Fig. 1.9(b)] can be approximated by a constant charge density, with a total charge z, spread on the surface of a hollow sphere of radius $r_0 = 3.5\text{Å}$. Find the structure factor of each molecule.*
d. *These molecules create an **FCC** lattice, of lattice constant $a = 14.1\text{Å}$. Find the*

structure factor for scattering off the planes (111) and (200).

s.3.4.

Consider the orthorhombic lattice, whose edges are $5a$, $4a$, and $3a$. Atoms with a structure factor F_1 are located at the vertices; at the center of the smallest face there is another atom, with a structure factor F_2. The wave length of the X-rays is $\lambda = a$.

a. *Find the structure factor of the crystal.*
b. *Find the first six scattering angles for this lattice.*
c. *Which of these angles give constructive interference also when $F_1 = F_2$?*
d. *Which of those angles give constructive interference also when $F_1 = -F_2$?*

s.3.5.

Consider a two-dimensional material of a square-lattice structure, with a lattice constant a. X-rays of wave length $\lambda = 1.5\mathring{A}$ are scattered in two dimensions off a two-dimensional powder of this material. The smallest scattering angle is $2\theta_{min} = 60°$.

a. *Find the lattice constant a.*
b. *Find the second scattering angle.*
c. *Assuming point-like atoms, find the ratio between the intensity at the second angle and the intensity in the first one.*
d. *It is now assumed that each atom is described by a square of edge L with a uniform charge distribution,*

$$n(x, y) = \begin{cases} -ez/L^2 \, , & |x| \leq L/2 \text{ and } |y| \leq L/2 \, , \\ 0 \, , & \text{otherwise} \, . \end{cases}$$

Find an expression for the intensities' ratio between the scattering at the second angle and at the first angle, as a function of L/a.

s.3.6.

Figure 3.33 shows a diffraction pattern obtained by a theoretical calculation of X-rays scattering off a Bravais lattice with a single atom in the unit cell. The radiation source is copper (wave length $1.54\mathring{A}$).

a. *Determine the lattice structure and the lattice constant.*
b. *Find the scattering angle and the height of the next peak in the sequence shown in the figure (assuming that the experiment is carried out on a uniform powder, at zero temperature.)*

s.3.7.

The molecule carbon tetrachloride, CCl_4, is built of a tetrahedron with the chlorine ions at its vertices and the carbon one at its center. The distance between each Cl and the carbon is $1.78\mathring{A}$. It is assumed that $F_{Cl}/F_C = 3$.

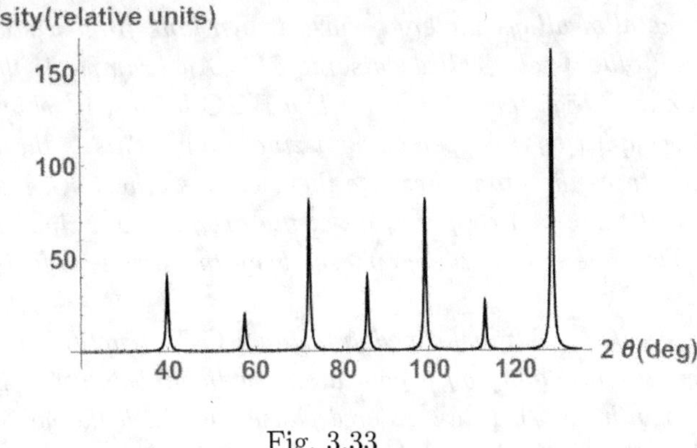

Fig. 3.33

a. *Find the structure factor of this molecule, assuming that the ions are point-like, and that the molecule is placed on a cubic lattice of lattice constant a, such that the carbons are located at the lattice sites and the chlorines on the opposite diagonals of the cube.*

b. *Upon cooling, this material becomes a solid of an* **FCC** *structure, with a lattice constant $a = 8.34\text{Å}$ (at an even lower temperature it crosses over into a monoclinic structure with 32 molecules in each unit cell). For X-rays of wave length $\lambda = 1.5\text{Å}$ scattered off the* **FCC** *phase, what are the two smallest angles for which the interference is constructive and what is the ratio of the intensities?*

s.3.8.

Here is a description of an experiment that measured the intensity of X-rays scattered off four crystalline materials. First it was found that when the materials were in the form of single crystals, the symmetry of three of them was cubic, and the one of the fourth crystal was hexagonal. When they were in the form of powders, the values of the scattering angles (2θ) of the four first diffraction rings were:

$$A:\ 42.2°,\ 49.2°,\ 72.0°,\ 87.3°,\qquad B:\ 30.0°,\ 31.9°,\ 34.1°,\ 44.3°,$$
$$C:\ 42.8°,\ 73.2°,\ 89.0°,\ 115.0°,\qquad D:\ 28.8°,\ 41.0°,\ 50.8°,\ 59.6°.$$

a. *Identify the four crystals. Is the identification unique? and if it is not, what else is needed to make the identification unique?*

b. *Assuming that the wave length is $\lambda = 1.5\text{Å}$, what are the lattice constants of these crystals?*

s.3.9.

Problem s.2.6 mentions phase transitions among various structures of sodium and iron. Find the first four scattering angles obtained from scattering off powders of each of the structures, for $\lambda = 1.5\text{Å}$.

s.3.10.

a. *Certain metallic alloys undergo phase transitions from a state (phase) called "ordered" to another state, called "disordered". An example is the alloy of copper and zinc, CuZn: at high temperatures it is a* **BCC** *lattice, in which each lattice site is randomly occupied by a copper or by a zinc atom. This is the disordered phase. At low temperatures the atoms arrange themselves in the CsCl structure, such that in each unit cell there is a copper atom at the origin and a zinc atom at the center of the cube. This one is the ordered phase. Find the structure factors of each of the phases.*

b. *Similarly, AuCu$_3$ is transformed from an* **FCC** *structure in which the lattice sites are occupied randomly by a gold atom (with probability 1/4) or by a copper ion (with probability 3/4), to an ordered phase, in which the gold atoms are at the corners of the cubic unit cell, and the copper atoms reside at the centers of the faces. What are the structure factors of the two phases?*

s.3.11.

Figure 2.29 displays the ferroelectric phase transition of barium titanate. At this transition the cubic perovskite structure is transformed into a tetragonal one where the lattice-constants ratio is $c/a = 1.0045$ (recall that $a_1 = a_2 = a$, and $a_3 = c$). The titanium and oxygen ions are displaced along the $\hat{z}-$axis, by $u_{Ti}c$ and $u_{O1}a$ (for the oxygens in the base plane, which contains also the barium ions) and $u_{O2}a$ in the mid-plane (which contains the Ti ions). Compare the scattering angles that give rise to Bragg peaks and the intensities in the two phases.

s.3.12.

a. *Which values of the Miller indices give constructive interference in a face-centered tetragonal lattice? Write down the expression for $\sin^2 \theta$ of this lattice.*

b. *Which lattice is obtained in the limit $a_3 = a_1$? Write down the seven lowest values of the ratio $\sin^2 \theta / \sin^2 \theta_{\min}$ for this case.*

c. *Write down the seven lowest values of the same ratio when $a_1 = a_3\sqrt{2}$. Determine the lattice structure deduced from the values you obtain. What is its lattice constant? Explain the result by geometrical arguments.*

s.3.13.

a. *The distance between two neighboring planes of CuO_2 and LaO in Fig. 2.21(b) is xc. Experiment yields $a_1 = a_2 = a \approx 3.8\text{Å}$ and $a_3 = c \approx 13\text{Å}$. Calculate the scattering amplitudes from this lattice and find the first five angles that yield Bragg peaks for $\lambda = 1.5\text{Å}$.*

b. *As described in problem s.2.17, at low temperatures this lattice becomes a face-centered orthorhombic. For small values of the angle ϕ defined there, what modifications are caused in the scattering angles?*

s.3.14.

X-ray scattering off a powder of a single-atom **FCC** *lattice yielded that the intensity of the (800) Bragg peak decreases with the temperature as follows: at $T[\mathrm{K}] = 4,\ 200,\ 400,\ 600$ the respective intensities were $I(800) = 13.14,\ 8.06,\ 3.60,\ 1.57$ (in arbitrary units). It is assumed that the atomic vibrations are described by the Einstein model, that is, by a single frequency ω.*

a. *Exploit the Lindemann approximation to find the melting temperature of this material.*

b. *Assuming that 4K is low enough to use the zero-temperature expressions, what is the frequency of the atomic vibrations in the crystal? Is the assumption about the temperature valid?*

c. *What is the intensity of the Bragg peak resulting from the (440) planes at $T = 200\mathrm{K}$? Note: $\hbar = 6.581 \times 10^{-16}$ eV sec, and $k_{\mathrm{B}} = 8.617 \times 10^{-5}$ eV/K.*

3.16 Answers for the self-evaluation problems

Answer s.3.1.

In the trigonal lattice all edges and all angles are equal to each other. Hence $|\mathbf{b}_1| = (2\pi/V)|\mathbf{a}_2 \times \mathbf{a}_3| = (2\pi/V)a^2 \sin\alpha$. The same answer is obtained for the other two reciprocal-lattice vectors, so their lengths are all equal. As shown in problem 3.7, the reciprocal of the reciprocal lattice is the original one, and therefore $a = |\mathbf{a}_1| = (2\pi/V_{\mathrm{rec}})|\mathbf{b}_2 \times \mathbf{b}_3| = (2\pi/V_{\mathrm{rec}})b^2 \sin\alpha'$. Multiplying these expressions by one another, and using the identity $VV_{\mathrm{rec}} = (2\pi)^3$ from problem 3.6, results in $b = 2\pi/[a \sin\alpha \sin\alpha']$. The angle α' is found from the scalar product of the lattice vectors

$$b^2 \cos\alpha' = \mathbf{b}_3 \cdot \mathbf{b}_1 = (2\pi/V)^2[\mathbf{a}_1 \times \mathbf{a}_2] \cdot [\mathbf{a}_2 \times \mathbf{a}_3] = (2\pi/V)^2 a^4(\cos^2\alpha - \cos\alpha) \ ,$$

[the right-hand side is derived from the identity

$$[\mathbf{u} \times \mathbf{v}] \cdot [\mathbf{w} \times \mathbf{y}] = (\mathbf{u} \cdot \mathbf{w})(\mathbf{v} \cdot \mathbf{y}) - (\mathbf{u} \cdot \mathbf{y})(\mathbf{v} \cdot \mathbf{w}) \tag{3.76}$$

check!] Upon inserting $b = |\mathbf{b}_1|$ one finds $\cos\alpha' = \cos\alpha(\cos\alpha - 1)/(1 - \cos^2\alpha) = -\cos\alpha/(1 + \cos\alpha)$. Substituting $\cos\alpha = 2\cos^2(\alpha/2) - 1$ in both sides yields $\cos(\alpha'/2) = 1/[2\cos(\alpha/2)]$ (assuming that both angles are acute).

Answer s.3.2.

Consider first the triclinic lattice; all other lattices are special limits of this case. One begins with the result of Eq. (3.27),

$$(2\pi/d)^2 = G^2 = h^2 b_1^2 + k^2 b_2^2 + \ell^2 b_3^2 + 2hk\,\mathbf{b}_1 \cdot \mathbf{b}_2 + 2h\ell\,\mathbf{b}_1 \cdot \mathbf{b}_3 + 2k\ell\,\mathbf{b}_2 \cdot \mathbf{b}_3 \ .$$

In an arbitrary triclinic lattice (e.g., Fig. 2.13 or the last row in Fig. 2.28) $|\mathbf{a}_1 \times \mathbf{a}_2|^2 = a_1^2 a_2^2 \sin^2\gamma$ and $\mathbf{a}_1 \cdot \mathbf{a}_2 = a_1 a_2 \cos\gamma$. Using the identity Eq. (3.76) one finds

$$[\mathbf{a}_1 \times \mathbf{a}_2] \cdot [\mathbf{a}_2 \times \mathbf{a}_3] = a_1 a_2^2 a_3(\cos\gamma \cos\alpha - \cos\beta) \ ,$$

and similar expressions for the other permutations of the vectors. Problem s.2.11 yields that the volume of the unit cell is

$$V = a_1 a_2 a_3 \sqrt{1 - \cos^2 \alpha - \cos^2 \beta - \cos^2 \gamma + 2 \cos \alpha \cos \beta \cos \gamma} \ .$$

Substituting all these expressions into the expression for G^2 gives

$$\frac{1}{d^2} = \frac{1}{V^2} \Big(h^2 a_2^2 a_3^2 \sin^2 \alpha + k^2 a_1^2 a_3^2 \sin^2 \beta + \ell^2 a_1^2 a_2^2 \sin^2 \gamma$$
$$+ 2hk a_1 a_2 a_3^2 (\cos \alpha \cos \beta - \cos \gamma) + 2k\ell a_1^2 a_2 a_3 (\cos \beta \cos \gamma - \cos \alpha)$$
$$+ 2h\ell a_1 a_2^2 a_3 (\cos \alpha \cos \gamma - \cos \beta) \Big) \ .$$

For the triclinic lattice, this expression is unchanged only when the signs of all three indices are inverted together, $(h \ k \ \ell) \leftrightarrow (\overline{h} \ \overline{k} \ \overline{\ell})$. The multiplicity is thus $p\{hk\ell\} = 2$. In the orthorhombic lattice, $\alpha = \beta = \gamma = 90°$; hence $1/d^2 = (h^2/a_1^2) + (k^2/a_2^2) + (\ell^2/a_3^2)$, with $p(\{hk\ell\}) = 8$ when all indices are nonzero (the sign of each one of them can be inverted separately), $p(\{0k\ell\}) = p(\{h0\ell\}) = p(\{hk0\}) = 4$, $p(\{h00\}) = p(\{0k0\}) = p(\{00\ell\}) = 2$. In the tetragonal lattice $a_1 = a_2$, and as a result there are more permutations among h and k, $p(\{hk\ell\}) = 16$, $p(\{hh\ell\}) = p(\{0k\ell\}) = p(\{hk0\}) = 8$, $p(\{hh0\}) = p(\{0k0\}) = 4$, $p(\{00\ell\}) = 2$. In the hexagonal lattice $a_1 = a_2 \neq a_3$, $\alpha = \beta = 90°$ and $\gamma = 60°$. It follows that $(1/d^2) = (4/3)[(h^2 - hk + k^2)/a_1^2] + (\ell^2/a_3^2)$ (sometimes γ is chosen to be $120°$, and then the sign of the hk term is reversed). This lattice is invariant with respect to rotations in the plane by $60°$, and then $p(\{00\ell\}) = 2$ (sign inversion), $p(\{hh0\}) = p(\{0h0\}) = 6$ [the cases $(\pm h, 0, 0)$, $(0, \pm h, 0)$, and $\pm(h, h, 0)$ are degenerate], $p(\{hh\ell\}) = p(\{h0\ell\}) = 12$ (including the sign inversion of ℓ), $p(\{hk0\}) = 4$, $p(\{hk\ell\}) = 8$. For the monoclinic lattice $\alpha = \gamma = 90°$ and then,

$$\frac{1}{d^2} = \frac{1}{\sin^2 \beta} \left(\frac{h^2}{a_1^2} + \frac{k^2 \sin^2 \beta}{a_2^2} + \frac{\ell^2}{a_3^2} - \frac{2h\ell \cos \beta}{a_1 a_3} \right) \ .$$

Here $p(\{hk\ell\}) = 4$, $p(\{h0\ell\}) = 2$, $p(\{0k0\}) = 2$. In the trigonal (or rhombohedral) lattice, $\alpha = \beta = \gamma \neq 90°$ and $a_1 = a_2 = a_3 = a$; then

$$\frac{1}{d^2} = \frac{(h^2 + k^2 + \ell^2) \sin^2 \alpha + 2(hk + h\ell + kl) \cos \alpha (1 - \cos \alpha)}{a^2 (1 - 3 \cos^2 \alpha + 2 \cos^3 \alpha)} \ .$$

Here, $p(\{hk\ell\}) = 12$ (6 permutations and sign inversion of all of them together) $p(\{hh0\}) = p(\{h00\}) = p(\{hh\ell\}) = p(\{0k\ell\}) = p(\{hk0\}) = 6$, $p(\{00\ell\}) = 2$. Note that in certain cases, for example, for the tetragonal lattice, the triplets (34ℓ) and (50ℓ) are degenerate.

Answer s.3.3.

a. Introduce into the result of problem 3.18 the constant value of the density, $n(r) = n_0$, and integrate over $0 \le r \le r_0$. The result is

$$F(\mathbf{G}) = \frac{4\pi n_0}{G} \int_0^{r_0} dr\, r \sin(Gr) = \frac{3z}{(Gr_0)^3} \Big(\sin(Gr_0) - Gr_0 \cos(Gr_0) \Big) \ .$$

In the $G \to 0$ limit, a Taylor expansion gives $\sin(Gr_0) - Gr_0 \cos(Gr_0) \approx (Gr_0)^3/3$, and hence $F(\mathbf{G}) \to z$, as could have been obtained also by substituting $G = 0$ in the original integral.

b. Denote $r_0 = a/2$. In a cubic lattice $(Gr_0)^2 = \pi^2(h^2 + k^2 + \ell^2)$, and for the first five Bragg peaks the sum $h^2 + k^2 + \ell^2$ attains integer values between 1 and 5. The respective values of $F(\mathbf{G})$ are $F(\mathbf{G})/z = 0.304,\ 0.0075,\ -0.081,\ -0.076$, and -0.039, respectively. In the square lattice $\ell = 0$ and therefore the lowest values of the sum are 1, 2, 4, 5, and 8, and hence $F(\mathbf{G})/z = 0.304,\ 0.0075,\ -0.076,\ -0.039$, and 0.0348. Interestingly enough, the intensities are not monotonic: the intensity of the second peak is in particular low, because of the interference of the waves within the sphere.

c. As the charge is spread over the surface, one uses $n(r) = \delta(r - r_0)$. The integration as obtained in problem 3.18 is only over the angles, which yields $F(\mathbf{G}) = z \sin(Gr_0)/(Gr_0)$, where z is the total number of electrons on the sphere. As each carbon atom contains 6 electrons, the result is $z = 60 \times 6 = 360$.

d. In the cubic lattice $G(hk\ell) = (2\pi/a)\sqrt{h^2 + k^2 + \ell^2}$. Using $r_0 = 3.5\text{\AA}$ and $a = 14.1\text{\AA}$, results in $G(111)r_0 = 2\pi\sqrt{3}r_0/a \approx 0.86\pi$, and $G(200)r_0 = 4\pi r_0/a \approx 0.99\pi$. Hence, $F(200) \approx 0.01z$ and $F(111) \approx 0.16z$. The low intensity for the (200) planes results from the proximity of $G(200)r_0$ to π; it has no special relation to any special symmetry within the unit cell.

Answer s.3.4.

a. The unit cell contains an atom at the origin and another one at $\mathbf{r}_2 = (\mathbf{a}_1 + \mathbf{a}_2)/2$; the structure factor is thus $(F_1 + F_2 \exp[i\pi(h + k)])$.

b. In the solution of problem 3.15 it is found that $\sin^2\theta = (\lambda^2/4)[(h^2/a_1^2) + (k^2/a_2^2) + (\ell^2/a_3^2)]$; it follows that $\sin^2\theta = (1/4)[h^2/9 + k^2/16 + \ell^2/25]$. The smallest angles are obtained for $(hk\ell) = (001)$, (010), (011), (100), (101), and (002), and their values are $\theta = 5.7°$, $7.2°$, $9.2°$, $9.6°$, $11.2°$, and $11.5°$.

c. In this case $F(\mathbf{G}) = F_1(1 + \exp[i\pi(h + k)])$, and therefore $(h + k)$ must be even (see also problem 3.25). This condition leaves just the first and the last angles.

d. In this case $F(\mathbf{G}) = F_1(1 - \exp[i\pi(h + k)])$ and therefore $(h + k)$ must be odd. Consequently, the scattering angles are the second, third, fourth, and fifth.

Answer s.3.5.

The scattering angles are given by
$$\sin^2\theta = [\lambda^2/(4a^2)](h^2 + k^2) \, .$$

a. The smallest scattering angle corresponds to $(hk) = (01)$, (10), and hence $\sin\theta_{min} = \lambda/(2a)$. As $\theta_{min} = 30°$, it follows that $a = \lambda$.

b. The second scattering angle corresponds to $(hk) = (11)$, and hence $\sin\theta_2 = \lambda/(\sqrt{2}a) = 1/\sqrt{2}$, yielding $\theta = 45°$.

c. The scattering intensity is independent of $\mathbf{G}$ for point-like scatterers, and therefore the intensities are all identical.

d. With the given density, the structure factor is

$$F(\mathbf{G}) = -\frac{ez}{L^2} \int_{-L/2}^{L/2} dx\, e^{iG_x x} \int_{-L/2}^{L/2} dy\, e^{iG_y y} = -ez\frac{sin u_x}{u_x}\frac{\sin u_y}{u_y}\ ,$$

with $u_x = G_x L/2 = \pi h L/a$ and similarly $u_y = G_y L/2 = \pi k L/a$. The first-peak intensity is thus $I_1 = |F(10)|^2 = (ez)^2[\sin(\pi L/a)/(\pi L/a)]^2$, and the second-peak one is $I_2 = |F(11)|^2 = (ez)^2[\sin(\pi L/a)/(\pi L/a)]^4$. Thus, $I_2/I_1 = [\sin(\pi L/a)/(\pi L/a)]^2$, tending to 1 when $L \to 0$, and to zero when the material covers the entire plane, $L \to a$.

Answer s.3.6.

The values of the scattering angles, as can be read from Fig. 3.33, are $2\theta \simeq 40°$, $57°$, $72°$, $86°$, $98°$, $113°$, $128°$, ... To identify the type of the lattice one prepares a list of the values $\sin^2\theta_n/\sin^2\theta_{\min} = 1$, 1.95, 2.95, 3.98, 4.87, 5.94, 6.91, ... In a cubic lattice, these values are of the form $(h^2+k^2+\ell^2)_n/(h^2+k^2+\ell^2)_{\min}$. The ratios obtained from the measurements fit those of the **BCC** sequence, for which it is known that the sum $h+k+\ell$ is even, and $(h^2+k^2+\ell^2)_n/(h^2+k^2+\ell^2)_{\min} = 1$, 2, 3, 4, 5, 6, 7, 8, ... Note that the difference between the **BCC** and the **SC** lattices is the seventh peak in the sequence, which does not exist for the **SC** lattice. To find the lattice constant, one exploits Eq. (3.31), $\sin^2\theta_{\rm in} = [\lambda^2/(4a^2)] \times 2$; the lattice constant is then $a \approx 3.2\,\mathring{A}$.

b. The next peak corresponds to the value 8 of the ratio $\sin^2\theta_n/\sin^2\theta_{\min}$, and therefore $2\theta_8 = 151°$. To estimate the relative height of this peak it is useful to study the multiplicities of each peak. As they correspond to $(hk\ell) = (110)$, (200), (211), (220), (310), (222), (321), (400), the respective multiplicities are 12, 6, 24, 12, 24, 8, 48, 6. The figure shows that the height of the first peak is approximately 120, and thus, according to the multiplicities ratio, the eighth one should be twice shorter, i.e., 60.

Answer s.3.7.

a. Locating the carbon ion at the origin, the four unit vectors

$$(1,1,1)/\sqrt{3}\ ,\ (1,-1,-1)/\sqrt{3}\ ,(-1,1,-1)/\sqrt{3}\ ,\ (-1,-1,1)/\sqrt{3}\ ,$$

create a tetrahedron around the origin. Denoting $\mathbf{G}(hk\ell) = (2\pi/a)(h,k,\ell)$ and $u = 2\pi r_0/(a\sqrt{3})$ (where r_0 is the distance between a chloride and the carbon), one finds

$$F(\mathbf{G}) = F_{\rm C} + F_{\rm Cl}(e^{iu(h+k+\ell)} + e^{iu(h-k-\ell)} + e^{iu(-h+k-\ell)} + e^{iu(-h-k+\ell)})\ .$$

b. The four vectors listed above point along the cube's diagonals, as assumed in the problem. For an **FCC** lattice, the indices h, k, and ℓ are all even or are all odd, and hence the smallest scattering angles are obtained from $(hk\ell) = (111)$, (200), i.e., $\sin^2\theta/[\lambda^2/(4a^2)] = 3$, 4. The angles are $2\theta \approx 18°$, $21°$. The scattering intensity in the first one is $8|F(111)|^2 = 367|F_{\rm C}|^2$, in the second is $6|F(200)|^2 = 9.65|F_{\rm Cl}|^2$;

the ratio is 38. (Note that the measured intensities include an arbitrary multiplicative factor and hence an individual intensity cannot be computed; the ratios of intensities, however, are independent of that factor.)

Answer s.3.8.

a. For the cubic crystals, as seen from Eq. (3.31), $\sin^2\theta = [\lambda^2/(4a^2)](h^2+k^2+\ell^2)$, leading to $\sin^2\theta/\sin^2\theta_{\min} = (h^2+k^2+\ell^2)/(h^2+k^2+\ell^2)_{\min}$. This ratio is 1, 1.34, 2.67, 3.67 for crystal A, 1, 1.13, 1.28. 2.12 for crystal B, 1, 2.67, 3.69, 5.34 for crystal C, and 1, 1.98, 2.97, 3.99 for crystal D. The values that belong to A, C, and D are very close to rational fractions, as required for cubic crystals. The values of crystal A seem to obey $\sin^2\theta/\sin^2\theta_{\min} = (h^2+k^2+\ell^2)/(h^2+k^2+\ell^2)_{\min} = 1, 4/3, 8/3, 11/3$, fitting the sequence $(h^2+k^2+\ell^2) = 3$, 4, 8, 11. From the discussion following problem 3.20 it follows that this is an **FCC** lattice (Fig. 3.20). Similarly, the values for crystal C are close to those following from $(h^2+k^2+\ell^2) = 3$, 8, 11, 16, which, from the solution of problem 3.25, correspond to diamond. For crystal D, the values fit $(h^2+k^2+\ell^2) = 1$, 2, 3, 4, that agree with a **SC** lattice. However, they also fit $(h^2+k^2+\ell^2) = 2$, 4, 6, 8, that agree with a **BCC** lattice (see the discussion after problem 3.20 and Fig. 3.20). The data points are not sufficient to distinguish between these two possibilities. Nonetheless, for the **SC** lattice the value $(h^2+k^2+\ell^2) = 7$ is forbidden, while the **BCC** lattice allows for 14, e.g., for $(hk\ell) = (123)$. One could have then identify the crystal unambiguously had one continued to the seventh scattering angle. As three of the crystals are cubic, one may deduce that crystal B is hexagonal (according to the conditions in the problem). It remains then to examine whether it is a simple hexagonal lattice, or may be it has a base. The solution to problem 3.26 shows that for an arbitrary hexagonal lattice $\sin^2\theta = [4\lambda^2/(3a_1^2)][h^2 - hk + k^2 + 3\ell^2 a_1^2/(4a_3^2)]$; consequently

$$\frac{\sin^2\theta}{\sin^2\theta_{\min}} = \frac{h^2 - hk + k^2 + 3\ell^2 a_1^2/(4a_3^2)}{[h^2 - hk + k^2 + 3\ell^2 a_1^2/(4a_3^2)]_{\min}}.$$

A simple hexagonal lattice allows for all values of $(hk\ell)$. Hence, the smallest scattering angle corresponds to $(hk\ell) = (001)$, where $h^2 - hk + k^2 + 3\ell^2 a_1^2/(4a_3^2) = 3a_1^2/(4a_3^2)$. The next angle is for $(hk\ell) = (100)$, yielding $\sin^2\theta/\sin^2\theta_{\min} = 4a_3^2/(3a_1^2)$. A comparison of this expression with the experimental value 1.13 implies that $a_3/a_1 = 0.92$. However, the next two experimental values cannot be fitted to any of the triplets $(hk\ell)$. On the other hand, the **HCP** structure forbids the triplet $(hk\ell) = (001)$, and hence $[h^2 - hk + k^2 + 3\ell^2 a_1^2/(4a_3^2)]_{\min} = 1$. The next angles all agree with those expected for the ideal **HCP** lattice, with $a_3/a_1 = \sqrt{8/3}$, as detailed in problem 3.26, $\sin^2\theta/\sin^2\theta_{\min} = 9/8$, $41/32$ $17/8$. This result is further confirmed by the difference between the second and the fourth ratios, which is about 1.

b. For the cubic lattices $\sin^2\theta_{\min} = [\lambda^2/(4a^2)](h^2+k^2+\ell)_{\min}$, and the expression in the brackets equals 3 for the A and C crystals, and 1 or 2 for the D one, when it is considered as a **SC** lattice or as a **BCC** lattice, respectively. In both cases

$a = \lambda\sqrt{3}/(2\sin\theta_{\min}) \approx 3.6\text{Å}$. For crystal D, $a(\mathbf{SC}) \approx 3\text{Å}$ or $a(\mathbf{BCC}) \approx 4.3\text{Å}$. For crystal B $\sin^2\theta_{\min} = \lambda^2/(3a_1^2)$; hence $a_1 = 3.35\text{Å}$ and $a_3 = a_1\sqrt{8/3} \approx 5.46\text{Å}$.

Answer s.3.9.

The lattice constants of sodium are $a(\mathbf{BCC})=4.23\text{Å}$ and $a(\mathbf{HCP}) \approx 3.77\text{Å}$. The **BCC** lattice obeys

$$\sin^2\theta = [\lambda(2a^2)]\{2,\ 4,\ 6,\ 8\}\ ,$$

and hence $2\theta = 30°,\ 43°,\ 54°,\ 62°$. Problem 3.26 cites

$$\sin^2\theta = [4\lambda^2/(3a^2)]\{1,\ 9/8,\ 41/32,\ 17/8\}\ ,$$

for the dense **HCP** lattice and therefore $2\theta = 27°,\ 29°,\ 31°,\ 40°$. The lattice constant of iron is $a(\mathbf{BCC})=2.87\text{Å}$ and therefore $a(\mathbf{FCC})=(2\sqrt{2}r/(4r/\sqrt{3}) = 3.51\text{Å}$. For the **BCC** lattice

$$\sin^2\theta = [\lambda^2/(4a^2)]\{2,\ 4,\ 6,\ 8\}\ ,$$

$2\theta = 45°,\ 65°,\ 82°,\ 99°$. For the **FCC** lattice,

$$\sin^2\theta = [\lambda/(4a^2)]\{3,\ 4,\ 8,\ 11\}\ ,$$

and hence $2\theta = 45°,\ 52°,\ 72°,\ 93°$. In both cases the fourth angle does not produce a ring on the screen as it is larger than $90°$.

Answer s.3.10.

a. In the disordered phase, the structure factor at each site of the lattice is an average of the atomic structure factors, i.e., $F_{\mathrm{av}} = (F_{\mathrm{Cu}} + F_{\mathrm{Zn}})/2$, as follows directly from Eq. (3.51) for the scattering amplitude at a reciprocal-lattice site. The structure factor measured in experiment is then identical to that of a usual **BCC** lattice, $F(hk\ell) = F_{\mathrm{av}}[1+(-1)^{h+k+\ell}]$. There are no Bragg peaks for odd values of $h+k+\ell$. On the other hand, the ordered structure is represented by a simple cubic lattice with a base, like CsCl, and the structure factor is $F(hk\ell) = F_{\mathrm{Cu}} + (-1)^{h+k+\ell}F_{\mathrm{Zn}}$. The scattering amplitude for even values of $h+k+\ell$ is identical to that obtained for the disordered phase; the one of odd values is nonzero as well, $F(hk\ell) = F_{\mathrm{Cu}} - F_{\mathrm{Zn}}$.
b. In the disordered phase, $F(hk\ell) = F_{\mathrm{av}}[1 + (-1)^{h+k} + (-1)^{h+\ell} + (-1)^{k+\ell}]$, with $F_{\mathrm{av}} = (F_{\mathrm{Au}} + 3F_{\mathrm{Cu}})/4$. Bragg peaks appear only when the three indices are all odd or are all even. In contrast, in the ordered phase the structure factor is $F(hk\ell) = F_{\mathrm{Au}} + [(-1)^{h+k} + (-1)^{h+\ell} + (-1)^{k+\ell}]F_{\mathrm{Cu}}$. Again, there are more peaks: indices that are not all odd or all even contribute as well, as in the case of table salt.

Answer s.3.11.

The first part of the problem is similar to problem 3.15: the scattering angles for a tetragonal lattice obey $\sin^2\theta/\sin^2\theta(100) = h^2 + k^2 + \ell^2(a_1/a_3)^2$, and therefore all Bragg peaks of the cubic lattice that contain a contribution from the index ℓ are split or displaced relative to their locations in the cubic lattice. For instance, the scattering angles for $(hk\ell) = (100)$, (010), (110), (200), (020), are not changed. Constructive interference also appears for the angles obeying $\sin^2\theta/\sin^2\theta(100) = (a_1/a_3)^2$, $1 + (a_1/a_3)^2$, $2 + (a_1/a_3)^2$, $4(a_1/a_3)^2$, due to the splitting of the Bragg peaks corresponding to $(hk\ell) = (001)$, (101), (111), (002). The scattering amplitude in the ferroelectric phases is given by

$$F(hk\ell) = F_{\mathrm{Ba}} + (-1)^{h+k+\ell} e^{2i\pi u_{\mathrm{Ti}}\ell} F_{\mathrm{Ti}}$$
$$+ [(-1)^{h+k} e^{2i\pi u_{\mathrm{O1}}\ell} + \{(-1)^h + (-1)^k\}(-1)^\ell e^{2i\pi u_{\mathrm{O2}}\ell}] F_{\mathrm{O}} .$$

It follows that the intensities at the points with $\ell = 0$ are not changed, while the other amplitudes acquire a small imaginary part. For instance, the peak at (100) is divided: $F(100) = F(010) = F_{\mathrm{Ba}} - F_{\mathrm{Ti}} - F_{\mathrm{O}}$, but $F(001) = F_{\mathrm{Ba}} - e^{2i\pi u_{\mathrm{Ti}}} F_{\mathrm{Ti}} + [e^{2\pi i u_{\mathrm{O1}}} - 2e^{2\pi i u_{\mathrm{O2}}}] F_{\mathrm{O}}$. The changes in the scattering angles and in the intensities become more pronounced as the ferroelectric deformation is enhanced. They also increase with ℓ.

Answer s.3.12.

a. The reciprocal-lattice vectors are $\mathbf{G} = (2\pi/a_1)(h\hat{\mathbf{x}} + \ell\hat{\mathbf{y}}) + (2\pi/a_3)k\hat{\mathbf{z}}$, and the atoms in the base are at the origin and at the points $(\mathbf{a}_1 + \mathbf{a}_2)/2$, $(\mathbf{a}_1 + \mathbf{a}_3)/2$, and $(\mathbf{a}_3 + \mathbf{a}_2)/2$. The structure factor is thence like in the **FCC** lattice, $\{1 + \exp[i\pi(h + k)] + \exp[i\pi(h + \ell)] + \exp[i\pi(k + \ell)]\}$, and the Miller indices should be all even or all odd. From the solution of problem 3.15, $\sin^2\theta = (\lambda/[4a_1^2])[h^2 + k^2 + (\ell a_1/a_3)^2]$.
b. For $a_1 = a_3$ the lattice becomes an **FCC** lattice, and the smallest angles are obtained from $h^2 + k^2 + \ell^2 = 3$, 4, 8, 11, 12, 16, 19. It follows that $\sin^2\theta_{\min} = 3\lambda^2/(4a_1)^2$, and the smallest angles are given by $\sin^2\theta/\sin^2\theta_{\min} = 1$, $4/3$, $8/3$, $11/3$, 4, $16/3$, $19/3$.
c. In this case $\sin^2\theta = [\lambda/(4a_1^2)][h^2 + k^2 + 2\ell^2]$, leading to $h^2 + k^2 + 2\ell^2 = 4$, 8, 12, 16, 20, 24, 28. Substituting $\sin^2\theta_{\min} = \lambda^2/a_1^2$ yields $\sin^2\theta/\sin^2\theta_{\min} = 1$, 2, 3, 4, 5, 6, 7. Figure 3.20 shows that this is a **BCC** lattice (the ratio 7 does not appear in a **SC** lattice). In a **BCC** lattice with a lattice constant a the result is $\sin^2\theta_{\min} = \lambda^2/(2a^2)$ (check!). It thus follows that $a_1 = a\sqrt{2}$. Figure 3.34 displays the projection on the base plane of the four unit cells of a **BCC** lattice. The lattice constant is a, and the atoms at the center of each cube are at $a/2$ above the center of the square. Consider the square marked by dashed lines: the edge length is $a_1 = a\sqrt{2}$. Using this square as a base for a prism of height a shows that the prism is the unit cell of the face-centered tetragonal lattice. The atoms that previously resided at the centers of the cells of the **BCC** lattice are now at the centers of the side faces of the tetragonal lattice.

Answer s.3.13.

a. The lattice is a body-centered tetragonal. As seen in the solution of problem 2.14, the unit cell contains a base with two atoms represented, e.g., by the locations of the copper ions: the one at the origin and the other at the center of the cell,

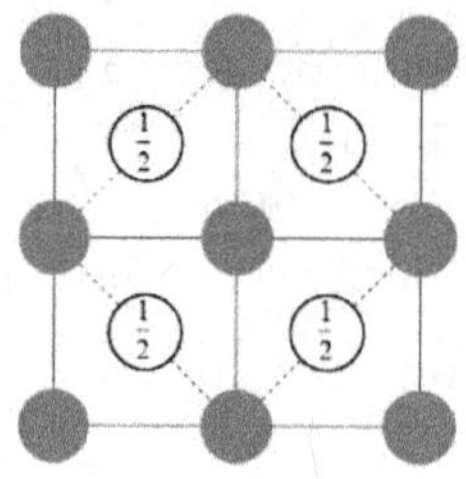

Fig. 3.34

$\mathbf{r}_2 = (\hat{\mathbf{x}} + \hat{\mathbf{y}})a/2 + \hat{\mathbf{z}}c/2$. Since the reciprocal-lattice vectors are given by $\mathbf{G} = (2\pi/a)(h\hat{\mathbf{x}}+k\hat{\mathbf{y}})+(2\pi/c)\ell\hat{\mathbf{z}}$, the structure factor is $F(hk\ell) = F_0(hk\ell)[1+(-1)^{h+k+\ell}]$, where $F_0(hk\ell)$ contains the information on the other ions in the base. Bragg peaks appear when $(h+k+\ell)$ is even, like in a **BCC** lattice. As described in the solution of problem 2.14, the base connected with each copper ion comprises a copper ion at the origin, two lanthanum ones at $(0,0,(1/2 - x)c)$ and $(a/2, a/2, xc)$, and four oxygens at $(a/2, a/2, (1/2 - x)c)$, $(0,0,xc)$, $(0, a/2, 0)$ and $a/2, 0, 0)$. It follows that $F_0(hk\ell) = F_{\mathrm{Cu}}$

$$+ F_{\mathrm{La}}[e^{i\pi(1-2x)\ell} + e^{i\pi(h+k+2\ell x)}] + F_{\mathrm{O}}[e^{i\pi h} + e^{i\pi k} + e^{2i\pi\ell} + e^{i\pi[h+k+(1-2\ell x)]}]$$

$$= F_{\mathrm{Cu}} + 2F_{\mathrm{La}}\cos[\pi(1 - 2x)\ell] + F_{\mathrm{O}}[(-1)^h + (-1)^k + 2\cos(2\pi x\ell)] \ ,$$

where the second step uses the fact that $h + k + \ell$ is even. As seen in the solution of problem 3.15, the Bragg peaks appear for the scattering angles

$$\sin^2\theta = \frac{\lambda^2}{4}\left(\frac{h^2}{a_1^2} + \frac{k^2}{a_2^2} + \frac{\ell^2}{a_3^2}\right) .$$

As $h + k + \ell$ is even, and given that $(c/a)^2 = (13/3.8)^2 \approx 11.7$, the lowest values of the indices are (002). The minimal angle is such that $\sin\theta_{\min} = \lambda/c = .115$, i.e., $2\theta_{\min} = 13.2°$. The next angles are for $(hk\ell) = (101)$, (004), (103), (110); thus, $2\theta = 13.2°$, $23.7°$, $26.6°$, $30.3°$, $32.3°$.

b. At low temperatures the material assumes the face-centered orthorhombic structure; the unit cell is thus larger, with lattice constants $a_1' = a\sqrt{2\cos(2\phi)}$, $a_2' = a\sqrt{2}$, $a_3' = c$, see problem s.2.17. This unit cell contains four copper ions, at the origin and at $\mathbf{r}_2 = (a_1'/2)\hat{\mathbf{x}}' + (a_3'/2)\hat{\mathbf{z}}'$, $\mathbf{r}_3 = (a_1'/2)\hat{\mathbf{x}}' + (a_2'/2)\hat{\mathbf{y}}'$, and $\mathbf{r}_4 = (a_2'/2)\hat{\mathbf{y}}' + (a_3'/2)\hat{\mathbf{z}}'$. The reciprocal-lattice vectors are

$$\mathbf{G} = \frac{2\pi}{a_1'}h'\hat{\mathbf{x}}' + \frac{2\pi}{a_2'}k'\hat{\mathbf{y}}' + \frac{2\pi}{a_3'}\ell'\hat{\mathbf{z}}' \ ,$$

where $\hat{\mathbf{x}}'$ and $\hat{\mathbf{y}}'$ are unit vectors, oriented along the dashed lines in the base of the cell in Fig. 2.60. It follows that $F(h'k'\ell') = F_0(h'k'\ell')[1 + (-1)^{h'+k'} + (-1)^{h'+\ell'} + (-1)^{k'+\ell'}]$ (as in an **FCC** lattice); Bragg peaks appear when the indices are all odd or they are all even. The scattering angles are given by

$$\sin^2\theta = \frac{\lambda^2}{4}\left(\frac{h'^2}{a_1'^2} + \frac{k'^2}{a_2'^2} + \frac{\ell'^2}{a_3'^2}\right) ,$$

and the smallest one is for $(h'k'\ell') = (002)$, i.e., $\sin\theta_{\min} = \lambda/c = .115$. By assumption, $a_3' = a_3 = c$, and therefore the minimal angle is identical to that obtained at high temperatures. Since

$$\frac{\sin^2\theta}{\sin^2\theta_{\min}} = \frac{1}{8}\left[\left(h'^2 + \frac{k'^2}{\cos(2\phi)}\right)\frac{c^2}{a^2} + 2\ell'^2\right],$$

the next angles correspond to $(h'k'\ell') = (111)$, (004), (113), (200), (020). Peaks for which $k' = 0$ correspond to the same angles as at high temperatures, while all the others are displaced or even split. Figure 3.35 shows the first six angles as functions of ϕ. In particular, the last three triplets, which give rise to a peak at the same angle in the tetragonal structure, yield two peaks at low temperatures, at two different angles. The second and fourth angles increase with the orthorhombic deformation, and the fifth angle is split into two angles.

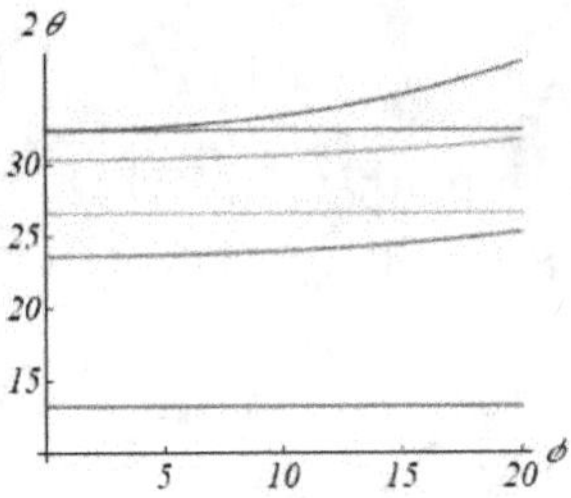

Fig. 3.35

The figure shows that the angles approach their values in the tetragonal phase when ϕ tends to zero. In this limit $\hat{\mathbf{x}}' = (\hat{\mathbf{x}} - \hat{\mathbf{y}})/\sqrt{2}$ and $\hat{\mathbf{y}}' = (\hat{\mathbf{x}} + \hat{\mathbf{y}})/\sqrt{2}$, and thus $\mathbf{G} = (\pi/a)[(h' + k')\hat{\mathbf{x}} + (k' - h')\hat{\mathbf{y}}] + (2\pi/c)\ell'\hat{\mathbf{z}}$. These vectors coincide with those of the tetragonal phase, once $h = (h' + k')/2$, $k = (k' - h')/2$ and $\ell = \ell'$ are used. Note that then $h + k + \ell = \ell' + k'$. As the new indices h', k', and ℓ' are all even or are all odd, this sum is always even, as it should be in the tetragonal phase. Substituting this in the expression for the scattering angles shows that indeed the two answers derived above are identical. In general, the structure factor of the base, $F_0(hk\ell)$, is modified due to the various displacements of the ions in the unit cell.

Answer s.3.14.
a. The intensity of the Bragg peaks is given by $I(hk\ell) = I_0 \exp[-2W]$. According to Eq. (3.59),

$$2W = \frac{\hbar}{2M\omega}\coth\left(\frac{\hbar\omega}{2k_{\mathrm{B}}T}\right)G^2 .$$

At high enough temperatures $W = (k_{\mathrm{B}}T/[M\omega^2])G^2$. In the present case $G^2 = (2\pi/a)^2(h^2 + k^2 + \ell^2) = 256\pi^2/a^2$, and therefore the high-temperature value is

$2W = 256\pi^2 k_{\rm B} T/(M\omega^2 a^2) = 64\pi^2 T/(75 T_{\rm M})$, where the Lindemann criterion with $c_{\rm L} = 0.1$, and $k_{\rm B} T_{\rm M} = c_{\rm L}^2 M\omega^2 a^2/3 = M\omega^2 a^2/300$ have been used. (For an **FCC** lattice, the distance between nearest neighbors is $a/\sqrt{2}$, and therefore $c_{\rm L} \to c_{\rm L}/\sqrt{2}$, which halves the melting temperature, but does not affect the results below.) A plot of $\ln(I/I_0)$ *vs.* the temperature T (with the given numbers) indeed reveals a linear dependence for the last three points. The slope can be deduced from the last two of them, $s = 64\pi^2/(75 T_{\rm M}) = \ln(3.60/1.57)/200[{\rm K}]$, and hence $T_{\rm M} \approx 2030{\rm K}$. From the relation $\ln I = \ln I_0 - sT$, and from the intensity of the last point follows $\ln I_0 = \ln 1.57 + s600 \approx 2.94$, yielding $I_0 \approx 18.9$.

b. At sufficiently low temperatures

$$2W = \frac{\hbar}{2M\omega}G^2 = \frac{\hbar\omega}{2M\omega^2 a^2}(aG)^2 = \frac{\hbar\omega}{600 k_{\rm B} T_{\rm M}}256\pi^2 \ .$$

Using the value of the intensity at 4 K gives

$$2W = \frac{\hbar\omega}{600 k_{\rm B} T_{\rm M}}256\pi^2 = \ln(I_0/I) = \ln(18.9/13.14) \ ,$$

and hence $\hbar\omega \approx 0.086 k_{\rm B} T_{\rm M} \approx 0.015{\rm eV}$. It follows that $\omega \approx 2.28 \times 10^{13}\ {\rm sec}^{-1}$. Therefore 4 K $k_{\rm B} T/(\hbar\omega) = T/(0.086 T_{\rm M}) = 0.023 \ll 1$, justifying the use of the low-temperature expression.

c. Inserting the information listed above into Eq. (3.59) gives

$$2W = \frac{\hbar\omega}{600 k_{\rm B} T_{\rm M}} \coth\left(\frac{\hbar\omega}{2k_{\rm B} T_{\rm M}}\frac{T_{\rm M}}{T}\right)G^2 \approx 0.00566 \coth\left(0.043\frac{T_{\rm M}}{T}\right)(h^2 + k^2 + \ell^2) \ .$$

For (440) and 200 K we find that $2W \approx 0.44$, and therefore $I(440) \approx 18.9\exp[-0.44] \approx 12.2$.

Appendix B

Fourier series

Periodic functions in one dimension. This Appendix presents **periodic functions as sums of trigonometric functions.** In one dimension, a periodic function $f(x)$ is defined by the relation $f(x) = f(x+a)$. Such a periodicity pertains, e.g., to the density of the material within each unit cell of a periodic lattice whose lattice constant is a, or to the potential exerted on electrons moving in a periodic lattice of ions. Examples for these functions are the trigonometric ones, $\sin(k_n x)$ or $\cos(k_n x)$, where $k_n = 2\pi n/a$ and $\{n\}$ are arbitrary integers, which describe instantaneous shapes of waves, with **wave numbers** k_n. **The Fourier theorem** states that the infinite ensemble of the trigonometric functions forms a **basis** for the description of **any** periodic function of this type. In other words, **any function** $f(x)$ **that obeys** $f(x+a) = f(x)$ **can be expressed by the infinite series**

$$f(x) = \frac{a_0}{2} + \sum_{n=1}^{\infty} [a_n \cos(k_n x) + b_n \sin(k_n x)] \;, \tag{B.1}$$

with constant coefficients that do not depend on x. This is a **Fourier series.**

It is expedient to re-write the Fourier series in terms of exponential functions, by using $\cos(k_n x) = (\exp[ik_n x] + \exp[-ik_n x])/2$ and $\sin(k_n x) = (\exp[ik_n x] - \exp[-ik_n x])/(2i)$, and then

$$f(x) = \sum_{n=-\infty}^{\infty} \tilde{f}_n e^{ik_n x} \;, \tag{B.2}$$

with $\tilde{f}_0 = a_0/2$, $\tilde{f}_n = (a_n - ib_n)/2$, $\tilde{f}_{-|n|} = (\tilde{f}_{|n|})^*$. To calculate the coefficients, one exploits the orthogonality of the exponential functions,

$$\int_{-a/2}^{a/2} dx (e^{ik_n x})^* e^{ik_m x} = \frac{e^{i(k_m - k_n)a/2} - e^{-i(k_m - k_n)a/2}}{i(k_m - k_n)} \equiv a\delta_{mn} \;, \tag{B.3}$$

where δ_{mn} is the Kronecker delta function. For $n = m$ the integrand in Eq. (B.3) equals 1, leading to the result a. (The limit in the second equality should be derived carefully.) For $n \neq m$ the denominator is finite while the numerator vanishes [since $(k_m - k_n)a = 2\pi(m - n)$]. Multiplying both sides of Eq. (B.2) by $\exp[-ik_m x]$, then

207

integrating from $-a/2$ to $a/2$, and using Eq. (B.3), one obtains the coefficients of the series

$$\widetilde{f}_m = \frac{1}{a} \int_{-a/2}^{a/2} dx\, f(x) e^{-ik_m x} \ . \tag{B.4}$$

Since the functions $\Phi_n(x) = \exp[ik_n x]/\sqrt{a}$ can serve as **basis vectors in a linear space** (in analogy with the unit vectors along the axes in a cartesian coordinate space), Eq. (B.2) implies that any other vector in this space can be described by a linear combination of the basis vectors, with complex coefficients. To complete the description of this space, one needs to define a **scalar product**, i.e., $\langle f|g \rangle \equiv \int_{-a/2}^{a/2} dx\, f^*(x) g(x)$. The **orthonormality of the basis vectors**, Eq. (B.3), is expressed by the relation $\langle \Phi_n|\Phi_m \rangle = \delta_{nm}$.

Problem B.1

a. *Prove that if $f(x) = f^*(x)$ is a real function, then $(\widetilde{f}_n)^* = \widetilde{f}_{-n}$, and if $f(x) = f(-x)$ is an even function then $\widetilde{f}_n = \widetilde{f}_{-n}$.*

b. *Use Eq. (B.4) to compute the coefficients in Eq. (B.1). Derive these coefficients in the cases mentioned in part* **a**.

c. *Derive the coefficients for $f(x) = A \sin^3(2\pi x/a)$.*

d. *Find the coefficients in Eq. (B.2) for the function*

$$f(x) = \begin{cases} -1, & na < x < (n+1/2)a \ , \\ 1, & (n-1/2)a < x < na \ , \end{cases}$$

where n is an arbitrary integer. Use the result to sum over the first five terms of the Fourier series, and compare the outcome with the original function.

Dirac delta function. Inserting the expression for the Fourier coefficients, Eq. (B.4), into the original expansion in Eq. (B.2) yields

$$f(x) = \sum_{n=-\infty}^{\infty} \left[\frac{1}{a} \int_{-a/2}^{a/2} dy\, e^{-ik_n y} f(y) \right] e^{ik_n x} \ , \tag{B.5}$$

where the expression in the square brackets is $\widetilde{f}_n$ (the integration variable is denoted y to avoid confusion). Interchanging the order of the summation and the integration, Eq. (B.5) becomes

$$f(x) = \int_{-a/2}^{a/2} dy \left[\frac{1}{a} \sum_{n=-\infty}^{\infty} e^{ik_n(x-y)} \right] f(y) \ . \tag{B.6}$$

Comparison with the definition of the Dirac delta function, Eq. (3.12), shows that the expression in the square brackets is that function itself,

$$\delta(x-y) = \frac{1}{a} \sum_{n=-\infty}^{\infty} e^{ik_n(x-y)} \ . \tag{B.7}$$

This identification is valid only for x within the integration range, i.e., $-a/2 < x < a/2$. As $f(x) = f(x+a)$, the same result is obtained upon integrating over any other segment shifted by na, where n is an arbitrary integer. Hence,

$$\sum_{n=-\infty}^{\infty} e^{ik_n(x-y)} = a \sum_{m=-\infty}^{\infty} \delta(x-y+ma) \ . \tag{B.8}$$

The sum in the right-hand side comprises narrow peaks at each lattice site; it is called at times the **"Dirac comb"**.

Fourier transform. The formalism outlined above can be extended to treat arbitrary functions that are not necessarily periodic. In other words, it is valid also when the periodicity (in the present context, the periodicity of the lattice) is infinitely large, $a \to \infty$. In that case the difference between successive wave numbers approaches zero, $\Delta k = k_n - k_{n-1} = 2\pi/a \to 0$. The Fourier transform of an arbitrary function $f(x)$ is then defined by

$$\widetilde{f}(k) = \lim_{a\to\infty} [a\widetilde{f}_m] = \int_{-\infty}^{\infty} f(x)e^{-ikx} \ . \tag{B.9}$$

In this limit Eq. (B.4) takes the form

$$f(x) = \frac{1}{2\pi}\left[\frac{2\pi}{a}\sum_{n=-\infty}^{\infty} a\widetilde{f}_n e^{ik_n x}\right] = \frac{1}{2\pi}\left[\sum_{n=-\infty}^{\infty} \Delta k(a\widetilde{f}_n)e^{ik_n x}\right] \to \frac{1}{2\pi}\int_{-\infty}^{\infty} dk\,\widetilde{f}(k)e^{ikx} \ . \tag{B.10}$$

In the last step the sum is converted into an integral, since when $a \to \infty$, the integration increment approaches zero. Equation (B.10) is the **inverse Fourier transform**.

Problem B.2

a. *Show that the area under the function $f(x)$ is $\widetilde{f}(0)$. What is the area under the function $\widetilde{f}(k)$?*

b. *Use $\widetilde{f}(k)$ to compute the integral $I_m = \int_{-\infty}^{\infty} dx\,x^m f(x)$.*

c. *Find the Fourier transform of $f(x) = \exp[-|x|/w]/(2w)$.*

d. *Express the Fourier transforms of the functions $h_1(x) = f(x-x_0)$, $h_2(x) = f(ax)$ ($a > 0$), and $h_3(x) = \exp[iqx]f(x)$, in terms of the Fourier transform $\widetilde{f}(k)$.*

e. *Prove that the Dirac delta function is the inverse Fourier transform of the constant $\widetilde{\delta}(x) \equiv 1$.*

f. *The generalization of the scalar product for functions defined on the infinite axis is $\langle f|g\rangle \equiv \int_{-\infty}^{\infty} dx\,f^*(x)g(x)$. Show that $\langle f|g\rangle = \frac{1}{2\pi}\int_{-\infty}^{\infty} dk\,\widetilde{f}^*(k)\widetilde{g}(k)$.*

Convolution. In certain cases, the Fourier transform of a function is a product of two transforms, i.e., $\widetilde{\widetilde{F}}(k) = \widetilde{f}(k)\widetilde{g}(k)$. Then

$$F(x) = \frac{1}{2\pi}\int_{-\infty}^{\infty} dk\,\widetilde{\widetilde{F}}(k)e^{ikx} = \frac{1}{2\pi}\int_{-\infty}^{\infty} dk\,\widetilde{f}(k)\widetilde{g}(k)e^{ikx}$$

$$= \frac{1}{2\pi}\int_{-\infty}^{\infty} dk\left[\int_{-\infty}^{\infty} dy\,f(y)e^{-iky}\right]\widetilde{g}(k)e^{ikx}$$

$$= \int_{-\infty}^{\infty} dy\,f(y)\left[\frac{1}{2\pi}\int_{-\infty}^{\infty} dk\,\widetilde{g}(k)e^{ik(x-y)}\right] = \int_{-\infty}^{\infty} dy\,f(y)g(x-y)\ . \qquad \text{(B.11)}$$

The values of the function $F(x)$ are therefore determined by those of $f(x)$ and $g(x)$ at all points on the axis. The expression on the right hand-side of Eq. (B.11) is called the **convolution** of these two functions and is usually denoted by

$$f(x)\otimes g(x) = \int_{-\infty}^{\infty} dy\,f(y)g(x-y)\ . \qquad \text{(B.12)}$$

Periodic functions in three dimensions. Consider the three-dimensional vector $\mathbf{r} = (x,y,z)$, expressed in terms of its cartesian coordinates; a function periodic along all three axes obeys

$$f(x+a_1,y,z) = f(x,y+a_2,z) = f(x,y,z+a_3) = f(x,y,z)\ . \qquad \text{(B.13)}$$

The Fourier theorem, Eqs. (B.2) and (B.4), for the $x-$dependence gives

$$f(x,y,z) = \sum_{n_x=-\infty}^{\infty} \widetilde{f}_{n_x}(y,z)e^{ik^x_{n_x}x}\ , \qquad \text{(B.14)}$$

where $k^x_{n_x} = 2\pi n_x/a_1$ and $\widetilde{f}_{n_x}(y,z) = (1/a_1)\int_{-a_1/2}^{a_1/2} dx\,f(x,y,z)\exp[-ik^x_{n_x}x]$. Using again the theorem for the $y-$dependence of $\widetilde{f}_{n_x}(y,z)$, and then again for the $z-$dependence, results in

$$f(x,y,z) = \sum_{n_x=-\infty}^{\infty}\sum_{n_y=-\infty}^{\infty}\sum_{n_z=-\infty}^{\infty} \widetilde{f}(n_x,n_y,n_z)e^{i\mathbf{k}(n_x,n_y,n_z)\cdot\mathbf{r}}\ , \qquad \text{(B.15)}$$

where

$$\mathbf{k}(n_x,n_y,n_z) = 2\pi(n_x/a_1,n_y/a_2,n_z/a_3)\ , \qquad \text{(B.16)}$$

and

$$\widetilde{f}(n_x,n_y,n_z) = \frac{1}{a_1a_2a_3}\int_{-a_1/2}^{a_1/2} dx\int_{-a_2/2}^{a_2/2} dy\int_{-a_3/2}^{a_3/2} dz\,f(x,y,z)e^{-i\mathbf{k}(n_x,n_y,n_z)\cdot\mathbf{r}}\ . \qquad \text{(B.17)}$$

Note that the vectors $\mathbf{k}(n_x,n_y,n_z)$ in Eq. (B.16) are vectors in the reciprocal space of the orthorhombic lattice whose lattice constants are a_1, a_2, and a_3.

The three-dimensional Fourier transforms are obtained upon taking the limit $a_m \to \infty$,

$$f(\mathbf{r}) = \int d^3k\,\widetilde{f}(\mathbf{k})e^{i\mathbf{k}\cdot\mathbf{r}}\ , \qquad \text{(B.18)}$$

with

$$\widetilde{\widetilde{f}}(\mathbf{k}) = \frac{1}{(2\pi)^3} \int d^3 r f(\mathbf{r}) e^{-i\mathbf{k}\cdot\mathbf{r}} \,, \tag{B.19}$$

where in both expressions the integrals are carried out over the entire space (the reciprocal and the direct one, respectively).

Periodic functions on an arbitrary Bravais lattice. The Bravais lattice can be defined by the property $f(\mathbf{r}+\mathbf{a}_1) = f(\mathbf{r}+\mathbf{a}_2) = f(\mathbf{r}+\mathbf{a}_3) = f(\mathbf{r})$. Presenting the vector $\mathbf{r}$ in terms of the unit vectors $\hat{\mathbf{a}}_m \equiv \mathbf{a}_m/|\mathbf{a}_m|$, as $\mathbf{r} = y_1\hat{\mathbf{a}}_1 + y_2\hat{\mathbf{a}}_2 + y_3\hat{\mathbf{a}}_3$, one obtains

$$f(y_1 + a_1, y_2, y_3) = f(y_1, y_2 + a_2, y_3) = f(y_1, y_2, y_3 + a_3) = f(y_1, y_2, y_3) \,. \tag{B.20}$$

This equation resembles Eq. (B.13), and therefore one may exploit the one-dimensional Fourier theorem for the three coordinates y_1, y_2, and y_3. For instance,

$$f(y_1, y_2, y_3) = \sum_{n_1=-\infty}^{\infty} \widetilde{f}_{n_1}(y_2, y_3) e^{i(2\pi/a_1)n_1 y_1} \,. \tag{B.21}$$

The final result is

$$f(y_1, y_2, y_3) = \sum_{n_1=-\infty}^{\infty} \sum_{n_2=-\infty}^{\infty} \sum_{n_3=-\infty}^{\infty} \widetilde{f}(n_1, n_2, n_3) \exp[i2\pi(\frac{n_1 y_1}{a_1} + \frac{n_2 y_2}{a_2} + \frac{n_3 y_3}{a_3})] \,, \tag{B.22}$$

with

$$\widetilde{f}(n_1, n_2, n_3) = \int_{-\frac{a_1}{2}}^{\frac{a_1}{2}} \frac{dy_1}{a_1} \int_{-\frac{a_2}{2}}^{\frac{a_2}{2}} \frac{dy_2}{a_2} \int_{-\frac{a_3}{2}}^{\frac{a_3}{2}} \frac{dy_3}{a_3} f(y_1, y_2, y_3) e^{-i2\pi(\frac{n_1 y_1}{a_1} + \frac{n_2 y_2}{a_2} + \frac{n_3 y_3}{a_3})} \,. \tag{B.23}$$

The integration in Eq. (B.23) is carried out over a single unit cell. Using Eqs. (3.20) yields $\mathbf{b}_m\cdot\mathbf{r} = 2\pi y_m/a_m$, where $\{\mathbf{b}_m\}$ are the vectors of the reciprocal lattice. Hence, the expression in the exponent is $2\pi(n_1 y_1/a_1 + n_2 y_2/a_2 + n_3 y_3/a_3) = \mathbf{G}\cdot\mathbf{r}$, where $\mathbf{G} = n_1\mathbf{b}_1 + n_2\mathbf{b}_2 + n_3\mathbf{b}_3$ is an arbitrary vector in reciprocal lattice. The function $\Phi_{\mathbf{G}}(\mathbf{r}) = \exp[i\mathbf{G}\cdot\mathbf{r}]$ is indeed periodic over the lattice, since from Eq. (3.17) $\Phi_{\mathbf{G}}(\mathbf{r} + \mathbf{R}) = \exp[i\mathbf{G} \cdot (\mathbf{r} + \mathbf{R})] = \exp[i\mathbf{G} \cdot \mathbf{r}] = \Phi_{\mathbf{G}}(\mathbf{r})$. Thus, Eq. (B.22) implies that **any periodic function on the lattice can be presented as a series of these functions, with all reciprocal-lattice vectors,**

$$f(\mathbf{r}) = \sum_{n_1, n_2, n_3} \widetilde{f}(n_1, n_2, n_3) e^{i\mathbf{G}\cdot\mathbf{r}} \,. \tag{B.24}$$

It remains to consider the coefficients, Eq. (B.23). In order to carry out the integration over the cartesian coordinates x, y, and z (in place of the coordinates defined by $\mathbf{r} = y_1\hat{\mathbf{a}}_1 + y_2\hat{\mathbf{a}}_2 + y_3\hat{\mathbf{a}}_3$) one uses the identity $dy_1 dy_2 dy_3 = J dx dy dz$, where the Jacobian J is

$$J = \begin{vmatrix} \partial y_1/\partial x & \partial y_1/\partial y & \partial y_1/\partial z \\ \partial y_2/\partial x & \partial y_2/\partial y & \partial y_2/\partial z \\ \partial y_3/\partial x & \partial y_3/\partial y & \partial y_3/\partial z \end{vmatrix} = \frac{a_1 a_2 a_3}{(2\pi)^3} \begin{vmatrix} b_{1x} & b_{1y} & b_{1z} \\ b_{2x} & b_{2y} & b_{2z} \\ b_{3x} & b_{3y} & b_{3z} \end{vmatrix} = \frac{a_1 a_2 a_3}{(2\pi)^3} V_{\text{rec}} = \frac{a_1 a_2 a_3}{V} \,.$$

The first step uses $\mathbf{b}_m \cdot \mathbf{r} = 2\pi y_m/a_m$. The second step uses the expression for the volume of a unit cell in reciprocal lattice, $V_{\rm rec} = \mathbf{b}_1 \cdot [\mathbf{b}_2 \times \mathbf{b}_3]$, and the third makes use of problem 3.6, $V_{\rm rec}V = (2\pi)^3$. Hence,

$$\widetilde{f}(n_1, n_2, n_3) = \frac{1}{V} \int d^3 r f(\mathbf{r})e^{-i\mathbf{G}\cdot\mathbf{r}} , \tag{B.25}$$

where the integration is carried out over the unit cell.

The relation between Fourier series and the reciprocal lattice. As mentioned, the periodic boundary conditions on the sample's edges are identical to its extension over the entire space. The function $f(\mathbf{r})$ can then be written as in Eq. (B.18), with the Fourier transform given in Eq. (B.19). Exploiting the periodicity, Eq. (3.33), enables one to displace all points in space by a lattice vector, and thus Eq. (B.19) yields

$$\widetilde{f}(\mathbf{k}) = \frac{1}{(2\pi)^3} \int d^3 r f(\mathbf{r} + \mathbf{R})e^{-i\mathbf{k}\cdot(\mathbf{r}+\mathbf{R})} = \widetilde{f}(\mathbf{k})e^{-i\mathbf{k}\cdot\mathbf{R}} .$$

It follows that $\widetilde{f}(\mathbf{k}) \neq 0$ only when $\exp[i\mathbf{k} \cdot \mathbf{R}] = 1$. From Eq. (3.9) one finds that this condition is fulfilled only when $\mathbf{k}$ equals one of the reciprocal-lattice vectors, i.e., when Bragg's condition is obeyed. Hence, $f(\mathbf{r})$ is a linear combination of the periodic functions $\{\exp[i\mathbf{G}\cdot\mathbf{r}]\}$, that is, a Fourier series, [see Eqs. (B.24) or (3.34)]. Two alternative calculations of the coefficients are given in the paragraphs leading to Eqs. (3.37) or (B.25).

The delta function and Dirac's comb in three dimensions. As in Eq. (B.5), inserting Eq. (B.25) in Eq. (B.24) yields

$$f(\mathbf{r}) = \sum_{\mathbf{G}} \Big[\frac{1}{V} \int_V d^3 r' f(\mathbf{r}')e^{-i\mathbf{G}\cdot\mathbf{r}'} \Big] e^{i\mathbf{G}\cdot\mathbf{r}} = \int_V d^3 r' \Big[\frac{1}{V} \sum_{\mathbf{G}} e^{i\mathbf{G}\cdot(\mathbf{r}-\mathbf{r}')} \Big] f(\mathbf{r}') .$$

Hence $(1/V) \sum_{\mathbf{G}} \exp[i\mathbf{G} \cdot \mathbf{r}] = \delta(\mathbf{r})$, where $\mathbf{r}$ is within the unit cell that contains the origin. Since the function $f(\mathbf{r})$ is periodic, a similar integration over any other unit cell yields the function at the origin of that cell; consequently

$$\frac{1}{V} \sum_{\mathbf{G}} e^{i\mathbf{G}\cdot\mathbf{r}} = \sum_{\mathbf{R}} \delta(\mathbf{r} - \mathbf{R}) . \tag{B.26}$$

This equation generalizes Eq. (B.8), and the right hand-side extends the Dirac comb to three dimensions. There is a complete symmetry between the "usual" (direct) space and the reciprocal one; one therefore may interchange $\mathbf{G} \to \mathbf{R}$, $\mathbf{r} \to \mathbf{q}$, and $V \to V_{\rm rec} = (2\pi)^3/V$ (problem 3.6). It follows that

$$\sum_{\mathbf{R}} e^{i\mathbf{q}\cdot\mathbf{R}} = V_{\rm rec} \sum_{\mathbf{G}} \delta(\mathbf{q} - \mathbf{G}) = \frac{(2\pi)^3}{V} \sum_{\mathbf{G}} \delta(\mathbf{q} - \mathbf{G}) . \tag{B.27}$$

B.1   Answers

Answer B.1.

a. Using Eq. (B.4), one finds that when $f(x) = f^*(x)$ then

$$(\widetilde{f}_n)^* = \frac{1}{a}\int_{-a/2}^{a/2} dx\, f^*(x)\exp[-ik_n x] = \frac{1}{a}\int_{-a/2}^{a/2} dx\, f(x)\exp[ik_{-n}x] = \widetilde{f}_{-n} \ .$$

Similarly, when $f(x) = f(-x)$ then

$$\widetilde{f}_n = \frac{1}{a}\int_{-a/2}^{a/2} d(-x) f(-x)\exp[-ik_n x] = \frac{1}{a}\int_{-a/2}^{a/2} dx\, f(x)\exp[ik_{-n}x] = \widetilde{f}_{-n} \ ,$$

where in the first step the integration variable x is replaced by $-x$, and the integration bounds are interchanged in the second, using the fact that the function is even.

b. From the expressions listed after Eq. (B.2). it follows that $a_0 = 2\widetilde{f}_0 = 2/a$. For $n > 0$ one finds

$$a_n = \widetilde{f}_n + \widetilde{f}_{-n} = \frac{2}{a}\int_{-a/2}^{a/2} dx\, \cos(k_n x) f(x) \ ,$$

$$b_n = i(\widetilde{f}_n - \widetilde{f}_{-n}) = \frac{2}{a}\int_{-a/2}^{a/2} dx\, \sin(k_n x) f(x) \ .$$

For a real function one finds $a_n = 2\mathrm{Re}[\widetilde{f}_n]$ and $b_n = -2\mathrm{Im}[\widetilde{f}_n]$. When the function is even then $b_n = 0$, and there are only the even functions, i.e., the cosines, in the Fourier expansion.

c. The identity $\sin(3\alpha) = 3\sin(\alpha) - 4\sin^3(\alpha)$ implies that $f(x) = A[3\sin(2\pi x/a) - \sin(6\pi x/a)]4$. Thus, this function represents a special case: all coefficients are zero save two, $b_1 = 3A/4$ and $b_3 = -A/4$. Alternatively, the coefficients are given by the integral

$$b_n = \frac{2A}{a}\int_{-a/2}^{a/2} dx\, \sin^3(k_1 x)\sin(k_n x) = \frac{A}{2a}\int_{-a/2}^{a/2} dx[3\sin(k_1 x) - \sin(k_3 x)]\sin(k_n x) \ .$$

Since $\int_{-a/2}^{a/2} dx\, \sin(k_m x)\sin(k_n x) = (a/2)\delta_{nm}$ by Eq. (B.3), the integral is identical to the aforementioned result.

d. By Eq. (B.4),

$$\widetilde{f}_n = -\frac{1}{ik_n a}(2e^{-ik_n a/2} - 1 - e^{-ik_n a} + e^{-ik_n a/2}) = \frac{(1 - e^{-i\pi n})^2}{i\pi n} \ ,$$

and therefore $\widetilde{f}_{|n|} = -\widetilde{f}_{-|n|} = 2i/(\pi n)$ for an odd n, and 0 else. Substituting in Eq. (B.2) yields $f(x) = -4\sum_{n=1}^{\infty}\sin[\pi(2n-1)x]/[(2n-1)\pi]$. Figure B.1 compares the original function with the sum of the first five terms.

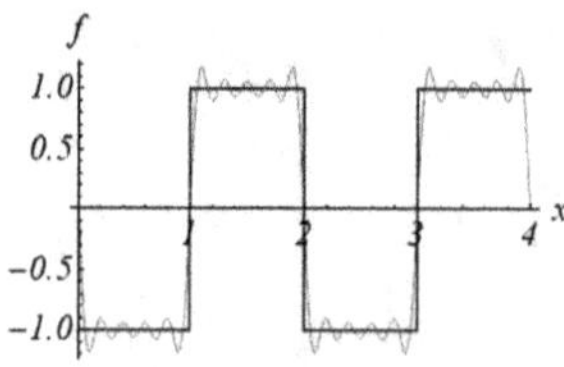

Fig. B.1

Answer B.2.

a. From Eq. (B.9), $\widetilde{f}(0) = \int_{-\infty}^{\infty} dx f(x)$, and the right-hand side is the desired area.
Similarly, Eq. (B.10) yields $2\pi f(0) = \int_{-\infty}^{\infty} dk \widetilde{f}(k)$.

b. Differentiating Eq. (B.9) yields

$$i^m \frac{\partial^m}{\partial k^m} \widetilde{f}(k) = \int_{-\infty}^{\infty} dx f(x) x^m e^{-ikx} \ ,$$

and hence $I_m = i^m \partial^m \widetilde{f}(k)/\partial k^m |_{k=0}$.

c. Equation (B.9) gives

$$\widetilde{f}(k) = \frac{1}{2w} \left[\int_{-\infty}^{0} dx e^{(x/w)-ikx} + \int_{0}^{\infty} e^{-(x/w)-ikx} \right]$$

$$= \frac{1}{2w} \left[\frac{1}{(1/w)-ik} + \frac{-1}{-(1/w)-ik} \right] = \frac{1}{1+w^2 k^2} \ .$$

The function on the right hand-side is called **"Lorentzian"**.

d. From Eq. (B.9),

$$\widetilde{h}_1(k) = \int_{-\infty}^{\infty} dx f(x-x_0)e^{-ikx} = \int_{-\infty}^{\infty} dy f(y)e^{-ik(y+x_0)} = e^{-ikx_0}\widetilde{f}(k) \ ,$$

$$\widetilde{h}_2(k) = \int_{-\infty}^{\infty} dx f(ax)e^{-ikx} = \frac{1}{a} \int_{-\infty}^{\infty} dy f(y)e^{-i(k/a)y} = \widetilde{f}(k/a)/a \ ,$$

$$\widetilde{h}_3(k) = \int_{-\infty}^{\infty} dx e^{iqx} f(x)e^{-ikx} = \int_{-\infty}^{\infty} dx f(x)e^{-k(k-q)x} = \widetilde{f}(k-q) \ .$$

e. One begins with Eq. (B.7). In the $a \to \infty$ limit it yields

$$\delta(x) = \frac{1}{2\pi} \left[\frac{2\pi}{a} \sum_{n=-\infty}^{\infty} e^{ik_n x} \right] \to \frac{1}{2\pi} \int_{-\infty}^{\infty} dk e^{ikx} \ ,$$

which is indeed the inverse Fourier transform of $\widetilde{\delta}(k) = 1$, as can be seen by comparing with Eq. (B.10).

f. One begins with the right hand-side and inserts there the expressions for the Fourier transforms of the two functions:

$$\int_{-\infty}^{\infty} dk \widetilde{f}^*(k)\widetilde{g}(k) = \int_{-\infty}^{\infty} dk \int_{-\infty}^{\infty} dx f^*(x)e^{ikx} \int_{-\infty}^{\infty} dy g(y)e^{-iky}$$

$$= \int_{-\infty}^{\infty} dx f^*(x) \int_{-\infty}^{\infty} dy g(y) \left[\int_{-\infty}^{\infty} dk e^{ik(x-y)} \right] = \int_{-\infty}^{\infty} dx f^*(x) \int_{-\infty}^{\infty} dy g(y) \left[2\pi\delta(x-y) \right]$$

$$= 2\pi \int_{-\infty}^{\infty} dx f^*(x)g(x) = 2\pi \langle f(x)|g(x) \rangle \ .$$

The intermediate step makes use of the result in part **e**.

Chapter 4

Crystal bonding

This chapter focuses on the various forces that cause the atoms or the molecules to prefer, at low enough temperatures, the periodic crystalline state. The binding energy of this state and the manner by which it determines the type of the preferred lattice are explained. The main considerations are based on electromagnetic forces that operate in-between the electrons and the nuclei, and on quantum mechanics, for instance on the uncertainty principle and on the Pauli exclusion principle. Analyses of the various bonds among the ions in the crystal (ionic, covalent, molecular, hydrogen, and metallic), that follow from these considerations are detailed; likewise, the energies which dominate the spatial arrangements of magnetic moments in magnetic solids are explained.

4.1   Preface

The binding energy of a crystal is the difference between two energies, the one of its constituents when they are far away from each other, and the energy of the arrangement in which they are close together in a periodic array. The crystal is formed when this difference is positive, i.e., when energy is "gained" upon the crystal creation. In other words, when the potential energy of the crystal is lower than the one of its separated constituents, then energy must be invested in order to disassemble it. The crystal is hence stable. The binding energy of a molecule is defined in a similar fashion: atoms join together to form a molecule when their total energy is reduced by so doing. In fact, the whole crystal may be viewed as a very big molecule. For this reason, the present chapter is based on quite a number of concepts borrowed from the realm of chemistry, that are needed for understanding the nature of the crystalline bond.

There exist several types of bonds among atoms, depending on the specific electronic structure of each of them. The potential energy of a pair of atoms (or ions) is qualitatively illustrated in Fig. 1.1: it corresponds to a repulsion at short distances, and to an attraction at long distances. Given this form, the total binding energy of each of the crystalline structures listed in Chapter 2 can be derived. Because

215

their number is finite, it is feasible to compare the various binding energies and to determine which of the structures has the highest binding energy. This is the array the material forms in its ground state. The results are often in agreement with the structures derived from scattering experiments (Chapter 3). Certain materials (as mentioned in Chapter 1) can exist in several different crystalline structures, called **polymorphs**. The discussion in this chapter is based only on energetic considerations, which are supposed to identify the structure at sufficiently low temperatures. Raising the temperature leads to a competition between the energy and the entropy, which in turn induces transitions to other crystalline patterns, as mentioned in Sec. 2.7. A qualitative survey of the various crystalline bonds is given below; the next sections offer quantitative analyses of each of them. Several of these analyses require some knowledge of quantum mechanics.

 The ionic bond. The simplest "crystal" is a **molecule** comprising two atoms. The nature of the bond between the atoms stems from their respective positions in the **periodic table.** The atoms at the far left side of the table (e.g., the alkali atoms in the first column of the table) have a small number of electrons in their outer shell. These electrons, termed **"valence electrons"**, are only weakly coupled to the atom, and are therefore easily detached from it. Once this happens, the atom becomes a positive ion with the same completely-full shell as the noble gas atom situated at the end of the previous row of the periodic table. As opposed, the outer shell of the atoms on the right side of the table (e.g., the halogens in the column second to the last one) is almost full, and they tend to absorb additional electrons in order to completely fill it and to attain the stable configuration of the noble gas situated after them in the periodic table. When two atoms from the two ends of the periodic table are brought close to one another, they "gain" energy when the atom from the left side (for instance, sodium) "donates" an electron to the other atom (for example, chlorine) such that two electrically-charged ions are created (Na^+ and Cl^-). The two ions electrically attract one another via the **Coulomb force**. Quantum mechanics prevents the ions from collapsing one of top of the other by the **Pauli principle** that forbids two electrons with identical quantum numbers to occupy the same spatial position. As each of the nuclei is surrounded by a "cloud" of electrons, this principle prevents the interpenetration of the "clouds". Thus, once each ion is assigned **an ionic radius** dictated by its electronic "cloud", the optimal structure is the one in which the corresponding spheres touch one another (as for example in Fig. 2.7 or in the first line of Fig. 4.3); the distance between their centers is the sum of the two ionic radii. At this distance, the gain in potential energy exceeds the energetic loss that results from the ionization and a molecule is formed (NaCl in the example above). This description can also be viewed as if there is an effective repulsion force, that prevents the ions to approach one another. The bond thus created between the two ions is called the **"ionic bond"**. This bond exists also in crystals built of such ions, for instance the table salt mentioned above, and other crystals comprising pairs of ions as described in the previous chapter: at each lattice site there is a point charge (positive or negative), and the competition between the overall Coulomb energy and the effective repulsion energy between neighboring ions (that results from the Pauli principle) determines the crystal structure with its lattice constants. Since the ions' clouds touch each other in the optimal configuration, the quintessential ionic lattice is described by close-packing of two types of spheres. In practice it is not always possible to achieve such packing, but the ratio between the radii of the two ions does determine the lattice structure. Indeed, explicit calculations along these lines do reproduce the structures described in Chapters 2 and 3.

The covalent bond. This bond is formed when the atoms that build the molecule reside in neighboring columns of the periodic table; in the extreme case the molecule consists of two identical atoms, e.g., oxygen, nitrogen, or hydrogen, which form diatomic molecules (O_2, N_2, or H_2). Each of the atoms "donates" its valence electrons (for instance, a single electron in the case of the hydrogen atom, or two electrons in the case of oxygen), and these electrons move around both ions. The quantum probability for the electrons to be between the two ions is larger than the probability to find them elsewhere (see below). This is because in the first case the electrons attract the two ions (the protons in the case of the hydrogen molecule) and cause the system to gain energy. The gain depends on the distance between the ions and consequently determines the molecule size. A bond of this type is termed **"covalent bond"**. For two identical ions the electrons' "cloud" is symmetric with respect to the center of the molecule, and then the covalent bond is unpolarized (the molecule lacks a dipole moment). When the two atoms are not identical (e.g., the CO or the H_2O molecule, see in the following) the electrons' "cloud" can be closer to one of the ions, and the molecule becomes electrically polarized. These two configurations, as well as the ionic molecule, are depicted in Fig. 4.1.

The covalent bond prevails also in numerous crystals. Then, the crystalline structure is dictated by the number of valence electrons of the atoms. Each of these electrons can form a covalent bond with an electron of another atom, and therefore the number of covalent bonds in which a certain atom participates, i.e., the coordination number of that specific atom, is just the number of its valence electrons. For example, in the diamond crystal (left panel in Fig. 2.20) each carbon "donates" four valence electrons such that each electron is bound to one of the four carbon neighbors; the coordination number is hence four. The four bonds around a carbon in diamond are also illustrated in the center of the upper line of Fig. 4.3. The diamond is in fact a huge molecule inside which any pair of carbons is coupled together by a covalent bond. In the graphene lattice (Fig. 2.8) each carbon has only three neighbors in the plane, with which it forms three covalent bonds. The fourth electron of each carbon is free to move around among the atoms in the crystal.

Hybridization and electronegativity. The ionic bond and the covalent one constitute two extremes of the **molecular bond.** In the first case, the atomic wave function of the electron is centered on the negative ion; in the second, the electron

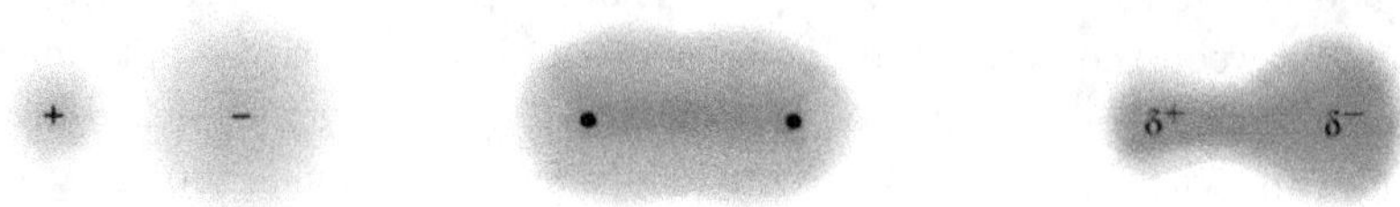

Fig. 4.1: The probability distribution "clouds" of the electrons' locations in a diatomic molecule: right panel–a polarized covalent bond; mid panel–an unpolarized
covalent bond; left panel–an ionic bond. In the right panel there is a positive
charge δ^+ and a negative charge δ^- on the two sides of the molecule, and therefore
it possesses a dipole moment.

is in a **quantum molecular state**, that belongs to both ions. In general, it may
well be that the quantum wave function of each electron is that of an intermediate
configuration, comprising an atomic wave function on one of the ions and a molecular function shared by both of them. This is the **"hybridization"** of the two types
of bonds. The chemists define the **"electronegativity"** of an atom, which is a
relative measure for its ability to attract valence electrons. The electronegativity
of an atom is a dimensionless number with values between 0.8 for the alkali metals
and about 3 for the halogens. When the difference between the electronegativity
values of two ions is small (for instance, smaller than 0.5), the bond between them
is mainly a covalent one. When that difference is large (e.g., larger than 1.8), the
bond is predominantly ionic. Intermediate values of electronegativity imply that
the two types of bonds are hybridized.

The metallic bond. Extending the physical picture of the covalent bond,
electrons can be found also in quantum states shared by more than two atoms. For
example, the benzene molecule, C_6H_6, has the form of a hexagonal ring built of six
carbon atoms (with a hydrogen atom attached to each of them), see Fig. 4.2. The
carbon atom possesses four valence electrons; three of them are used for covalent
bonds with its two carbon neighbors and one hydrogen atom (see the left panel in
the figure). There remain six "free" electrons (one from each carbon) that are shared
by all six carbons. The right panel exhibits the "clouds" of these electrons. In fact,
the wave function of such an electron is a linear combination of these atomic states,
with equal probabilities to find the electron at any of the carbons. These electrons
move freely around the ring. A further extension of this situation is **a metal:** each
ion in the metal "donates" its valence electrons, placing them in states where (see
the upper right panel of Fig. 4.3) the probability to find the electron around any
of the ions is the same. Under certain circumstances the electrons are thus free
to move around, hence the high electrical conductivity of metals. In a sense, the
metal is a huge molecule as well. The attraction between the electrons and the
nuclei reduces the potential energy, and consequently causes the binding energy to
increase. This is the basis of the **metallic bond.**

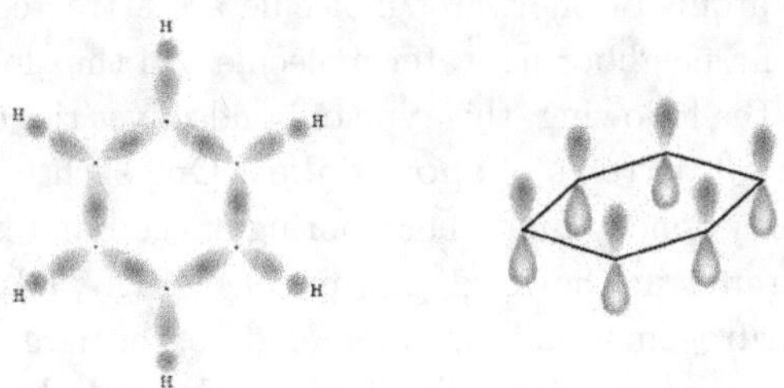

Fig. 4.2: Two projections of the benzene molecule. The carbon ions are located at the six vertices of the hexagon. The "clouds" represent the electrons participating in the covalent bonds (viewed from above in the left panel) and the free electrons common to all carbons (a side view). The different color tones indicate opposite signs of the wave functions.

The van der Waals bond. As opposed to the configurations surveyed above, where the atoms are relatively close to each other and so can share their electrons, the atoms of the **noble gases**, which have fully-occupied shells and lack valence electrons, cannot share electrons with other atoms. The same is true for **neutral molecules**, where the electrons remain within each molecule and are not transferred to any other. Nonetheless, the noble gases and numerous molecules do arrange themselves in a periodic array, since they do attract one another. The attraction is due to the **van der Waals force**, and the bond so created is termed the **"van der Waals bond"**. When two noble-gas atoms approach one another the electrons' cloud of each of them is shifted relative to the nucleus, creating a dipole moment. Figure 4.3 displays (in the lower panel on the right) two possibilities to form this dipole moment: in both cases the dipole moments attract one another. The van der Waals energy is an average over the energies of such configurations. This energy decays (see Sec. 4.4) with the distance R between the atoms as $-R^{-6}$. The Pauli principle again gives rise to an effective repulsion at short distances, which is customarily described by a steep potential increasing as R^{-12} (but any other steeply-rising potential will do). The sum of the two potential energies is the **Lennard-Jones potential**, illustrated in Fig. 1.1 [see also Eq. (4.32)]. When the potential energies of all pairs in the crystal are summed over, it turns out that the minimal total energy is negative (and consequently the bonding energy is positive) when the atoms are organized in the periodic crystalline structure. The noble gases arrange themselves on a **FCC** lattice, which has the highest bonding energy in their case. This is confirmed in experiment.

The hydrogen bond. Another way to bond atoms or molecules is by the **hydrogen bond**. The conspicuous example of this bond is the attraction among the molecules of **water**. These molecules possess a dipole electrical moment, since the two hydrogen atoms are situated on the same side of the oxygen atom (this follows from the nature of the covalent bonds within the molecule, see below).

Hence, the hydrogen atoms of one water molecule are attracted, via the Coulomb force, to the oxygen of a neighboring water molecule and thus form a bond between the two. As shown in the following, this type of bonds gives rise to various phases of **ice**, and to the beautiful patterns of **snow flakes**. One such pattern of crystalline ice, where the hydrogen bonds among neighboring oxygen and hydrogen atoms are perspicuous, is illustrated in the lower left panel of Fig. 4.3 (the small spheres that represent the hydrogen atoms are attracted to the large ones marking the oxygen atoms). The same bond can be formed whenever hydrogen ions in one molecule approach the negatively-charged face of another molecule (or even the same molecule, as happens in huge organic molecules). Since biological materials are rich in hydrogen, the hydrogen bonds play a very important role in biology. For instance, these bonds couple together the two strands of the double helix of DNA.

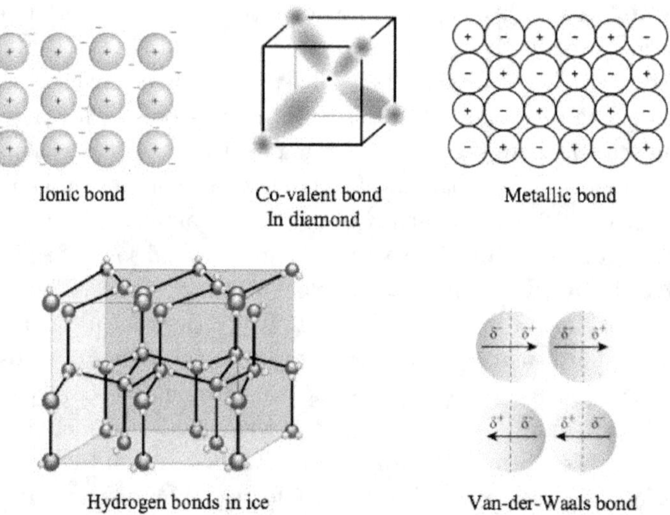

Ionic bond   Co-valent bond
In diamond   Metallic bond

Hydrogen bonds in ice   Van-der-Waals bond

Fig. 4.3: Summary of the various crystalline bonds. The upper panel shows the ionic bond (e.g., table salt), the covalent bond in diamond (the dark regions represent the electrons' "clouds" which create the bonds), and the metallic bond (in which the electrons move freely around). The lower panel exhibits the van der Waals bond (the median of the attraction energy of the dipoles in the two lines, represented by the opposite charges δ^+ and δ^-), and the hydrogen bond, in the cubic crystal of water.

Numerical values of the bonding energy and the melting temperature. Figure 4.3 summarizes the types of bonds mentioned so far. Quantitative values of the binding energy are derived in the next sections; these determine the specific crystalline structures. The calculations detailed in this chapter are often in agreement with most of the crystal types discussed in Chapter 2 and identified experimentally, as explained in Chapter 3. The binding energy per unit cell is of the order 1-15 eV for the ionic and the covalent bonds, 1-5 eV for the metallic one, 0.5 eV for the hydrogen bond, and 0.01-0.1 eV for the van der Waals bond. The strength of the crystalline bond, i.e., the actual value of the binding energy, is reflected in the **melting temperature** of the crystal. The larger is the bonding energy, the higher is the melting temperature; the crystal then exists at higher temperatures. Thus the melting temperature of diamond, copper, table salt, nitrogen, and neon are about $4000°K$, $1400°K$, $800°K$, $63°K$, and $25°K$, respectively. Indeed (see below) the atoms in the first three materials are coupled by covalent bonds, metallic bonds and ionic bonds, respectively, while those of the last two are van der Waals ones (between the nitrogen molecules or the noble gas atoms).

4.2   The ionic bond

The bonding energy of a salt comprising two ions. A simple example of an ionic crystal is a salt, which is built of an alkali metal from the first column of the periodic table (for instance, sodium or cesium) and an halogenic atom from the second to the last column (e.g., fluorine or chlorine). The sodium "donates" its external electron and remains in the configuration of the noble gas neon (the cesium ion resembles the xenon) with fully-occupied shells of electrons. The chlorine "absorbs" the electron and attains the configuration of argon, also with full shells of electrons. For both ions the electrons' probability density around the nucleus is approximately spherically-symmetric and the ion can be viewed as a point charge located at a lattice site. The calculation of the bonding energy therefore requires just the Coulomb energy of the charges representing the ions – it is the simplest calculation of a crystalline bond. Consider an infinite lattice that contains a pair of ions of opposite charges, $\pm q$, in each unit cell (e.g., table salt). The Coulomb energy of the lattice is given by an infinite double sum, $U_{\mathrm{C}} = \sum_{i \neq j} [\pm q^2/R_{ij}]/2$, where i and j denote two ions in the lattice, and R_{ij} is the distance in-between them. The factor $1/2$ arises since each pair of ions appears twice in the double sum over i and j. The sign of the ij term is determined by product of the signs of the charges at sites i and j. It is convenient to replace the distance between the ions by the dimensionless quantity $p_{ij} = R_{ij}/R_{01}$, where R_{01} is the distance between the two ions in the unit cell, i.e., the shortest distance between two ions of opposite charges. In the simple ionic lattice that comprises only two types of ions and has a single lattice constant, the ratio p_{ij} depends solely on the **lattice structure** and not on the **lattice constant** (convince yourselves that this is true). For instance, the distance between nearest neighbors sodium and chlorine ions in table salt is $R_{01} = a/2$, where a is the cubic lattice constant of the **FCC** lattice, Fig. 2.17. The distance between nearest-neighbor sodium ions (e.g., at the corner of the cube and at the center of a face) is $R_{02} = a/\sqrt{2}$ and consequently the ratio $p_{02} = R_{02}/R_{01} = \sqrt{2}$ is independent of the lattice constant and has the same value for all crystals of the table-salt type. One then finds $U_C = [q^2/(2R_{01})] \sum_{i \neq j} [\pm 1/p_{ij}]$ for all crystals of the table-salt type. When there are more than a single lattice constant, e.g., in the orthorhombic lattice, then p_{ij} is determined also by the ratio among the various lattice constants, but not by their specific values.

In an infinite lattice one may shift the origin in the sum over j and measure all distances in this sum relative to the origin at a point i (now marked by 0),

$$\sum_{i \neq j} [\pm 1/p_{ij}] = \sum_{i} \left(\sum_{j \neq 0} [\pm 1/p_{0j}] \right) .$$

For a crystal comprising two ions of opposite signs, the sum within the round brackets is independent of i (and attains the same value, irrespective whether the origin hosts a negative or a positive charge); therefore the result is just this sum multiplied by $2N$, where N is the number of unit cells in the lattice. The displacement of the origin alluded to above is justified only for the infinite crystal and hence $N \to \infty$. The result, divided by N, is **the specific bonding energy per unit cell**,

$$\frac{U_C}{N} = \frac{1}{N} \sum_{\langle ij \rangle} \left(\pm \frac{q^2}{R_{ij}} \right) = \frac{1}{2N} \frac{q^2}{R_{01}} \sum_{i} \left(\sum_{j \neq i} \left[\pm \frac{1}{p_{ij}} \right] \right)$$

$$= \frac{q^2}{R_{01}} \sum_{j \neq 0} \left(\pm \frac{1}{p_{0j}} \right) \equiv -\frac{q^2}{R_{01}} \alpha . \tag{4.1}$$

The last step introduces the **Madelung constant**,

$$\alpha = -\sum_{j \neq 0} \left(\pm \frac{1}{p_{0j}} \right) . \tag{4.2}$$

Dividing Eq. (4.1) by the volume of the unit cell yields the bonding energy per unit volume. As mentioned, in a crystal that contains only two types of ions and has a single lattice constant, the Madelung constant depends on the lattice structure alone. The numerical value of the Madelung constant is between 1 and 2, as demonstrated below.

A one-dimensional crystal. Figure 4.4 displays a one-dimensional ionic lattice, similar to the one in the second panel of Fig. 2.1: the unit cell contains two ions, and the lattice constant is twice the distance between nearest neighbors. At any distance $R_{0j} = jR_{01}$ there are precisely two ions (one on each side of the axis) and hence

$$\alpha = 2 \left(1 - \frac{1}{2} + \frac{1}{3} - \frac{1}{4} + \dots \right) = 2 \sum_{n=1}^{\infty} \frac{(-1)^{n-1}}{n} = 2 \ln 2 \approx 1.386 . \tag{4.3}$$

The last step makes use of the expansion $\ln(1 + x) = \sum_{n=0}^{\infty} (-x)^n/n$.

The series in Eq. (4.3) converges only slowly: the outcome of sums over a finite number of terms oscillates between the result of an even number and that of an

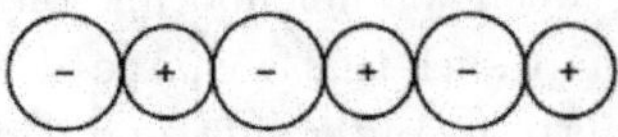

Fig. 4.4: One-dimensional ionic lattice built of negative ions (large circles) and positive ions (small ones).

odd number of summands; the convergence is slow because of the slow decay of the Coulomb interaction with the distance. At higher dimensions a sum like the one in Eq. (4.2) is even more problematic, since it depends on the surface up to which it is carried out. A surplus of positive or negative charges on the surface may well change the result of the summation. Such a situation is termed **"conditional convergence"**. The customary convention is to sum over finite samples for which the total charge on the surface vanishes. For instance, one may first sum over the contributions of the charges within the unit cell (the sum of these charges vanishes) and then add the contributions from all unit cells up to the surface. e.g., in the sum of Eq. (4.3), one first combines pairs of successive members, to obtain $\alpha = 2\sum_{n=1}^{\infty}\{1/[2n(2n-1)]\}$. In this sum the terms decay as $1/n^2$, which is faster than the $1/n$ decay in Eq. (4.3).

Arbitrary dimensions. For a lattice that comprises two ions of opposite charges, one first sums over the unit cells. Each cell contains two ions, e.g., a positive one at the origin and a negative one at $\mathbf{r}_1$ in the same unit cell. The distance in-between the two is R_{01}. Thus, a unit cell situated at $\mathbf{R}$ has a positive charge at $\mathbf{R}$, whose distance from the origin is $R = p(R)R_{01}$, and a negative one located at $\mathbf{R}+\mathbf{r}_1$, whose distance from the origin is $p(|\mathbf{R}+\mathbf{r}_1|)R_{01}$. (In the example given in Chapter 2, that of the table salt, $\mathbf{r}_1 = R_{01}\hat{\mathbf{y}}$.) It therefore follows that

$$\alpha = \frac{1}{p(r_1)} - \sum_{R\neq 0}\left[\frac{1}{p(R)} - \frac{1}{p(|\mathbf{R}+\mathbf{r}_1|)}\right], \tag{4.4}$$

where according to the definition above $p(\mathbf{r}_1) = 1$. There exist several ways to handle these sums. In the **Ewald method** the sum in the right hand-side is transformed into an integral in momentum space, which in turn becomes a sum over the reciprocal lattice; thus the method combines together a summation in space with a summation in momentum space. In other methods one sums over all ions within a sphere, adjusting the charges on the surface such that the total charge in the sphere vanishes. In the **Evjen method**, as applied to a cubic lattice, the summation is carried out on cubical shells encompassing a central ion, dividing the charges as to cancel the total charge within each shell. Figure 4.5 illustrates the Evjen method for the square lattice in the plane, which represents, e.g. the planar cross section of the table salt (Fig. 2.7). In cartesian coordinates, the lattice point (n, m) accommodates the charge $(-1)^{n+m}$ (in units of the electron charge, assuming that there is a single elementary charge in each lattice site). Consider the squares whose center is at the origin (the midpoint of the figure) and their edge length is

$2n$ (in units of the distance between two neighboring ions of opposite charges, R_{01}).
Divide the charges on each square such that a charge at the edge contributes half of
itself to the inner side, and half to its outer side, and a charge at the corner gives a
quarter of itself to the inner side and the rest, 3/4, to the outer side. The sum of all
charges within the square vanishes; the sum over the contributions to the Madelung
constant from square shells (between two successive squares) converges relatively
fast to the precise result (see problem 4.1).

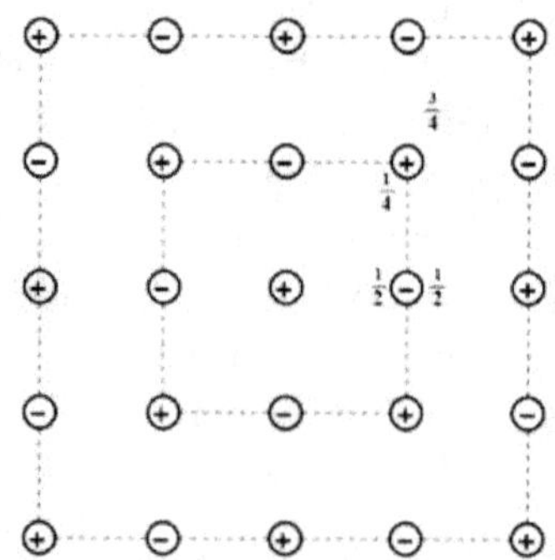

Fig. 4.5: The planar cross section of table salt. In the Evjen method one sums
over the contributions of square "shells" to the Madelung constant, for example,
the shell in-between the two dashed squares in the figure. The contributions of each
charge are split between the inner part and the outer part of the dashed line, as
marked in the figure.

Problem 4.1.

a. *Show that the total charge within each dashed square in Fig. 4.5 vanishes.*
b. *Compute the contribution of the charges in the inner side of the nth square to
the Madelung constant.*
c. *Derive an expression for the contribution of the charges located in-between the
nth and the $(n-1)$th squares to the Madelung sum, prove that all these are positive, and compute explicitly the contributions of the first six squares. How do these
contributions decay with n? [Suggestion: run your formula on your laptop, and find
how many terms are needed for a precision of three digits after the decimal point.]*

Three-dimensional crystals with two ions. In the simple cubic lattice each
site has six nearest neighbors (nn) at a distance 1 (in units of R_{01}), 12 neighbors in
the next spherical shell (next-nearest neighbors, nnn), at a distance $\sqrt{2}$, 8 neighbors
at a distance $\sqrt{3}$, 6 neighbors at a distance 2, and so on. In the structure of table
salt (right panel in Fig. 2.17) the shells contain successively positive and negative

charges, and hence

$$\alpha = \sum_{ijk}' \frac{(-1)^{i+j+k-1}}{\sqrt{i^2+j^2+k^2}} = 6/1 - 12/\sqrt{2} + 8/\sqrt{3} - 6/2 + 24/\sqrt{5} + \dots , \qquad (4.5)$$

where the sum runs over all (positive and negative) integers, and the notation $\sum'$ indicates that the term $i = j = k = 0$ does not appear. This sum converges only modestly, and its result does depend on the point where it is terminated. Nonetheless, careful computations for the ubiquitous structures of ionic crystals in three dimensions give

$$
\begin{aligned}
\text{NaCl}: &\qquad \alpha = 1.747565\dots \\
\text{CsCl}: &\qquad \alpha = 1.762675\dots \\
\text{ZnS (wurtzite)}: &\qquad \alpha = 1.64132\dots \\
\text{ZnS (zinc blende)}: &\qquad \alpha = 1.63806\dots . \qquad (4.6)
\end{aligned}
$$

The results for the two polymorphs of ZnS are quite close to one another. This is expected, since the corresponding structures are the Bravais lattices **HCP** and **FCC**, with a base comprising two ions. Recall (e.g., Fig. 2.24) that the coordination numbers of these lattices are identical (12), and the structural differences between them are minute. In fact, this is the reason for the natural appearance of this material in two forms; the specific structure it chooses is dictated by the growth conditions, *etc.* The same competition exists also in the (similar) materials ZnSe, ZnO, and ZnTe, for which the charges are $\pm q = \pm 2$ (in units of the electron charge).

Problem 4.2.
Write down the first five terms in the sum for the Madelung constant of the CsCl lattice.

Crystals with several ions in the unit cell. The salts considered above are of the AB type, with $\pm q$ charges on the A and B ions, respectively. In the more general case, the material is built of several ions which have different charges; then the Coulomb energies of the various ions in the unit cell are not equal to each other; the Coulomb potential energy is

$$U_{\text{C}} = \sum_{\langle ij \rangle} \left(\frac{q_i q_j}{R_{ij}} \right) = \frac{N}{2} \sum_{i \in \text{cell}} q_i \left(\sum_{j \neq i} \frac{q_j}{R_{ij}} \right) = \frac{N}{2} \sum_{i \in \text{cell}} q_i \phi_i , \qquad (4.7)$$

where the sums over i in the last two expressions run over the ions within a single unit cell; $\phi_i = \sum_{j \neq i}(q_j/R_{ij})$ is the electric potential which all other ions exert on the i-th ion. It is straightforward to verify that this expression produces the one in Eq. (4.1) when $q_1 = -q_2 = q$.

Problem 4.3.
A one-dimensional crystal is based on a unit cell comprising three ions, ABBAB-BABB ... (like in the third panel of Fig. 2.1, with identical B and C ions). The

charges of the ions are $q_A = 2e$ and $q_B = -e$, the lattice constant (the distance between two successive A ions) is a, and the distances within the unit cell between two B ions and between the ions A and B are $R_{BB} = a(1-2x)$ and $R_{AB} = R_{BA} = ax$. Find the Coulomb energy per unit cell (once all cells in the chain are summed upon). Hint: use the digamma function, defined by

$$\Psi(x) = -\gamma_E + \sum_{k=0}^{\infty}[1/(k+1) - 1/(k+x)] \ ,$$

where $\gamma_E \approx 0.577216$ is Euler's constant. The values of this function are documented in various softwares, and can also be obtained by carrying out the summation on a laptop.

The short-distance repulsion among the ions. The total negative Coulomb potential, Eq. (4.1) or Eq. (4.7), increases as the ions with opposite charges approach each other. Therefore, these ions "try" to to be close to each other as much as possible. The **Pauli principle**, as mentioned in Sec. 4.1, forbids the electron clouds of different ions to interpenetrate. This fact can be accounted for when each ion is assigned an **ionic radius**, such that outside this range the probability density to find electrons is low. The charged spheres of the positive and negative ions are then brought closer to one another in the unit cell (in order to reduce the distance in-between them). Measurements of the structure factors of ionic crystals indeed confirm this qualitative picture. (Recall that the structure factor is the Fourier transform of the electron density in the unit cell and its measurement conveys information about the volume around each ion where the electrons' cloud is concentrated.) Figure 4.6 displays typical values of the atomic radii and the different ionic radii in ionic crystals (the radii of the ions are also affected by their crystalline environment; the values listed in the table are approximate). As can be expected, the radii increase as one descends from row to row in the periodic table (and the number of electronic shells increases). Usually the radii of the elements become shorter along the row while the charge of the nucleus increases. The differences between the ionic radii (e.g., the cesium ion is much larger than the sodium one, and they are both smaller than the chlorine ion) explain the variety of crystalline structures (e.g., the difference between table salt and cesium chloride).

Comparison of various lattice structures. Assuming that the charged spheres that represent the positive and the negative ions touch one another enables the comparison of the resulting lattice structures. Figure 4.7 illustrates the charged spheres that represent table salt and cesium chloride. As seen, the size of the chlorine ions is almost twice that of the sodium ones; therefore the latter fit in the cavities of the **FCC** lattice of the chlorine ions. The cesium ions are bigger, and consequently the simple-cubic arrangement of the chlorine ions, with the cesium ions placed at the center of each cube, is preferred. This geometrical consideration can be given a quantitative basis. As mentioned, the Coulomb energy is largest when the positive and negative spheres are contiguous. Then the distance

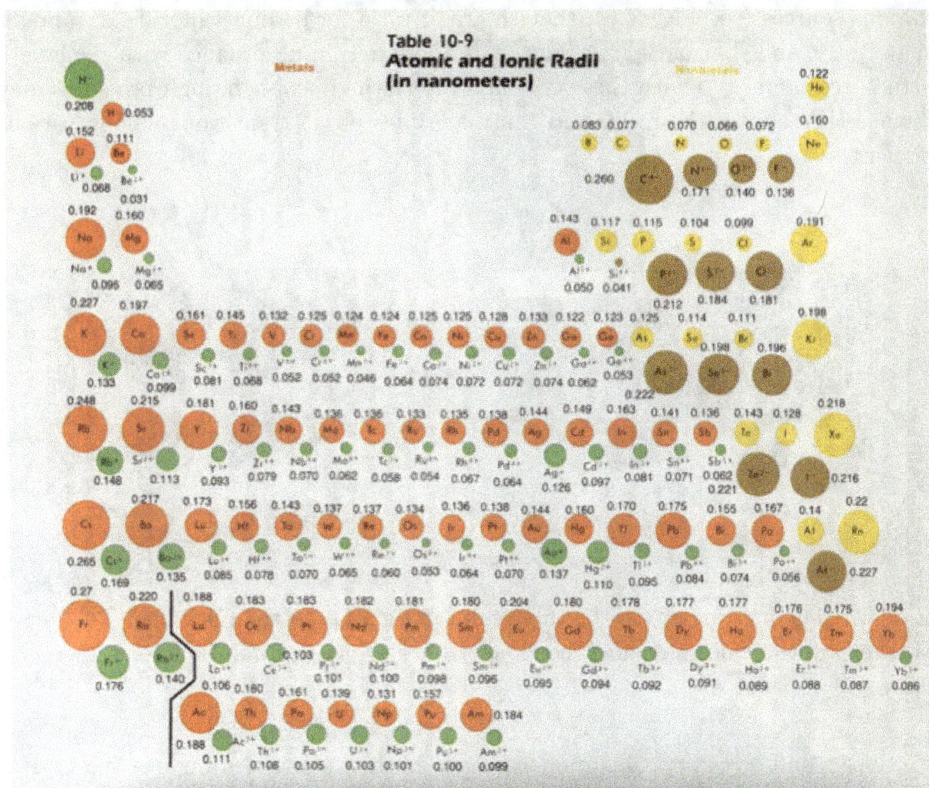

Fig. 4.6: Typical values (in nanometers) of atomic radii and ionic radii in ionic crystals. Taken from http://www.boomeria.org/chemlectures/textass2/table10-9.jpg.

between their centers is $R_{01} = r_< + r_>$, where $r_>$ is the larger ion radius (chlorine), and $r_<$ is the radius of the smaller one (sodium or cesium). It follows that the length of the diagonal of the cubic unit cell in the **cesium chloride** structure is $a\sqrt{3} = 2(r_< + r_>)$. This structure is feasible only when the lattice constant (which is the distance between two neighboring chlorine ions) is such that $a > 2r_>$, and therefore $1 < r_>/r_< < 1/(\sqrt{3} - 1) \approx 1.366$. On the other hand, in the structure of **table salt**, the sodium ion is located at the midpoint of the edge of the cubical unit cell, and therefore (since it is contiguous to the chlorine) the lattice constant must be $a = 2(r_< + r_>)$. The length of the diagonal of the face of the cubic cell is such that $a\sqrt{2} > 4r_>$; consequently $r_>/r_< < 1/(\sqrt{2} - 1) \approx 2.414$. Apparently, the **FCC** lattice fits better the table salt crystal for $1.366 < r_>/r_< < 2.414$, because for smaller ratios the cubic lattice with the higher Coulomb energy is preferred [see Eq. (4.6)], and this is the lattice of cesium chloride. A further increase of this ratio leads to the zinc-blende structure (ZnS), (Fig. 2.20, right panel), which may be the most suitable for $2.414 < r_>/r_< < 4.55$ (see problem 4.4). Indeed it is known, from various sources (see Fig. 4.6), that the radii ratio for sodium chloride is approximately 1.9, and that of cesium chloride is about 1.1, in accordance with the lattice structures associated with these crystals. Interestingly enough, high pressures may change the ratio of the radii, and cause at times phase transitions among various structures.

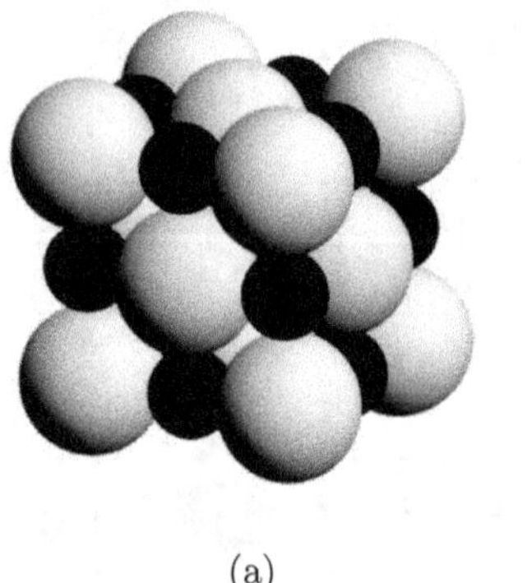
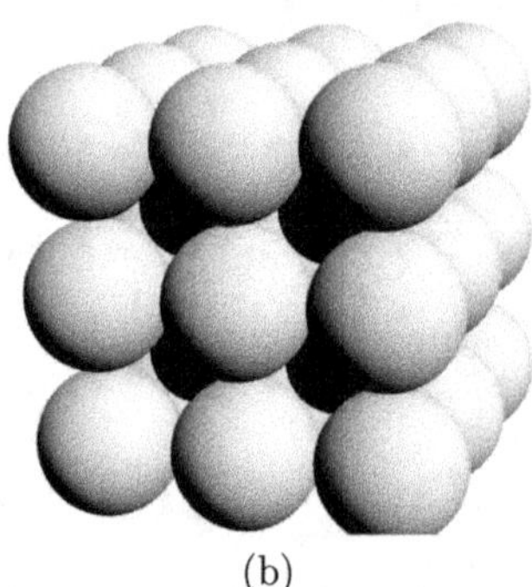

(a)   (b)

Fig. 4.7: The packing of (a) the table-salt lattice and (b) the CsCl one. The larger spheres are the chlorines, and the smaller ones are the sodium ions or the cesium ones.

So far only geometrical arguments have been exploited to deduce the crystal structure. One can also compare the Coulomb energies in various lattice structures. Consider a certain salt, in which the radii of the ions are $r_>$ and $r_<$. In the cesium-chloride structure of this salt, and when the ions of opposite charges in the unit cell are contiguous to one another, the distance between them is $R_{01} = r_< + r_>$. This is the preferred structure as long as $1 < r_>/r_< < 1/(\sqrt{3} - 1) \approx 1.366$. When $1.366 < r_>/r_<$ these ions cannot touch one another any more (since now the negative ions are contiguous). In this case, the distance between the neighboring positive and negative ions is half of the cube diagonal, $R_{01} = a\sqrt{3}/2 = \sqrt{3}r_>$. Therefore the Coulomb energy is $-N\alpha_{\mathrm{CsCl}}q^2/(r_> + r_<)$ in the first range of radii's length, and $-N\alpha_{\mathrm{CsCl}}q^2/(\sqrt{3}r_>)$ in the second. Had the same ions been arranged in the structure of the table salt, then $R_{01} = r_> + r_<$ for $r_>/r_< < 1/(\sqrt{2} - 1) \approx 2.414$. Since the Madelung constant of the cesium-chloride structure is slightly larger than that of the table salt, the preferred pattern is the one of the cesium-chloride in the first range. In the range $1.366 < r_>/r_< < 2.414$ the Coulomb energy of the table-salt structure is $-N\alpha_{\mathrm{NaCl}}q^2/(r_> + r_<)$, and this energy is lower than the one pertaining to the cesium-chloride structure only when

$$r_>/r_< > 1/[(\alpha_{\mathrm{NaCl}}/\alpha_{\mathrm{CsCl}})\sqrt{3} - 1] \approx 1.3943 , \quad \text{check!}$$

According to this argument, the cesium-chloride structure is still the "winner" in the range $1.366 < r_>/r_< < 1.3943$, though the ions with opposite charges are not contiguous there, and the table-salt structure exists only when $1.3943 < r_>/r_<$. However, when the energies of the two structures differ only slightly, other factors, like the repulsion between the ions at short ranges (though it has been assumed tacitly that ions of opposite charges are touching one another, the repulsion at short distances may take them apart), the van der Waals interaction (considered below), the quantum zero-point vibrational energy (discussed in the next chapter), and more, have to be accounted for. In most of the salts the bond is not purely ionic but is a combination (hybridization) of an ionic bond with a covalent one, and this also affects the result. Therefore, one cannot always predict the structure from simple geometrical considerations. Nonetheless, the outcome of these simple considerations is often confirmed in experiment.

Problem 4.4.

a. *Find the upper bound on $r_>/r_<$ up to which the ions can touch each other in the zinc-blende structure.*

b. *Determine the radii ratio at which the transition between the table-salt and the zinc-blende structures takes place, by comparing the Coulomb energies of the two structures.*

Problem 4.5.

The lattice constant of a zinc-blende crystal is 5.42Å.

a. *What is the distance R_{01} between the two ions in the unit cell?*

b. *Assuming a close-packed pattern, i.e., all nearest-neighbor ions are contiguous, find the radii of the two ions.*

c. *The same ions, with the same radii, create also the wurtzite lattice of the quintessential* **HCP** *structure. Assuming that the ions are still contiguous, determine the lattice constants of this lattice.*

d. *What is the specific volume of a single ZnS unit in each of the aforementioned cases?*

The bonding energy and the effective repulsive potential. A quantitative way to account for the effect of the **Pauli principle**, that prevents the electronic "clouds" of neighboring ions from "interpenetrating" each other, is to build an effective potential energy which represents a repulsive force between the neighboring ions in the unit cell, that keeps them apart. As mentioned, as long as this potential is steep enough, the results it yields are not sensitive to its details. It is customary to choose a convenient functional form for this repulsive potential, and use it to compute the distance between nearest neighbors that produces the minimal energy. A comparison of the outcome of this calculation with several measured quantities allows for the identification of the parameters that characterize the repulsive potential. Two ubiquitous forms for such a potential are

$$U_{P1} = C/R_{01}^m , \qquad U_{P2} = \lambda \exp[-R_{01}/\rho] . \tag{4.8}$$

The potential U_{P1} lacks any theoretical base, but is very convenient for computations. The potential U_{P2}, at times named after Born and Mayer, mimics the decay of the electronic wave functions around the atomic nucleus, and consequently also that of the probability density for an electron to be at a radial distance R_{01} away from the nucleus (for instance, in the ground state of the hydrogen atom). The decay length ρ is of the order of $r_< + r_>$. The total potential energy of the crystal takes the form

$$U_{\text{tot}}(R_{01}) = N[zU_P(R_{01}) - \alpha q^2/R_{01}] \,, \tag{4.9}$$

where z is the number of the negatively charged nearest-neighbors of each positive ion (and *vice versa*). The lattice structure is reflected in the coefficients α and z, and also in the parameters that characterize the effective potential U_P. The minimal value of U_{tot} as a function of R_{01} yields the optimal distance $\overline{R}$; inserting the latter into Eq. (4.9) gives the **binding energy per a unit cell of the lattice**, $u = -U_{\text{tot}}(\overline{R})/N$.

The minimal value of U_{tot} for the potential U_{P1} is at $\overline{R} = [mzC/(\alpha q^2)]^{1/(m-1)}$; the binding energy per unit cell is then $u = (1-1/m)\alpha q^2/\overline{R}$. Experiment shows that the binding energy of ionic lattices is approximately the Coulomb energy of nearest-neighbor ions, $u \approx \alpha q^2/\overline{R}$, and hence the parameter m is chosen to be much larger than 1. This verifies the assertion of a steep repulsive potential. Measurements of the binding energy and the distance $\overline{R}$ between nearest neighbors, allow for the identification of the parameters m and C when the crystal structure is known.

Assuming that all salts share the same value for m, the binding energy is determined by the distance $\overline{R}$ among nearest-neighbor ions in the unit cell, at equilibrium. As seen, this distance is mainly dictated by the radii of the two oppositely-charged ions. From the expression $\overline{R} = [mzC/(\alpha q^2)]^{1/(m-1)}$ follows that $zC \propto \alpha\overline{R}^{m-1}$. That is, the strength of the repulsive potential increases with the radii of the ions.

The relation $u \approx \alpha q^2/\overline{R}$ allows to estimate the magnitude of the binding energy in ionic crystals. It is expedient to measure $\overline{R}$ in units of the Bohr radius, $a_{\text{B}} = \hbar^2/(me^2) \approx 0.53\text{Å}$, and the binding energy in units of the Rydberg constant which determines the ground-state energy of the hydrogen atom, $R_{\text{y}} = m_0 e^4/(2\hbar^2) = e^2/(2a_{\text{B}}) \approx 13.6\text{eV}$ (m_0 is the electron mass, e is the electron charge). In these units $u/R_{\text{y}} = \alpha(q/e)^2(2a_{\text{B}}/\overline{R})$. Experiment shows that $\overline{R}$ is of the order of a few Bohr radii, so when $q = e$, u is of the order of the Rydberg constant, and is in the range 5-10 eV per unit cell. This energy increases with the amount of charge on the ion.

It is worth emphasizing that the binding energy is derived from a comparison of the energy of the ions when far apart from each other and the energy they acquire when arranged on a lattice. At times it is preferable to compare the crystal energy with the energy of the isolated atoms and not that of the ions. Then one has to account for the ionization energy. In the case of the table salt, for instance, the ionization of the sodium atom requires 5.1 eV, but 3.6 eV are gained since the chlorine atom is ionized (i.e., by "adding" an electron). The energy u is then about 7.9 eV. Therefore, the crystalline arrangement still leads to an energy gain, $(7.9 - 5.1 + 3.6)\text{eV}=6.4$ eV.

The bulk modulus. An independent identification of the parameters that characterize the effective repulsive potential between neighboring ions (i.e., C and m) is accomplished by a comparison of the measured and calculated values of the **bulk modulus**, defined as $B = -V(\partial P/\partial V)_T$, where P is the pressure and $V = Nv$ is the volume (v is the volume of a unit cell). The inverse of the bulk modulus, i.e., $1/B$, is the compressibility coefficient, that gives the relative change of the volume resulting from the applied pressure. As the pressure is given by $P = -dU_{\text{tot}}/dV$, with the total energy $U_{\text{tot}}(\overline{R}) = -Nu$, one finds

$$B = -v\frac{\partial}{\partial v}\left(\frac{\partial u}{\partial v}\right) = -v\frac{\partial^2 u}{\partial v^2} \ . \tag{4.10}$$

By a dimensional argument, the volume of the unit cell is $v = c\overline{R}^3$, where c is a dimensionless factor (of order one). For instance, in the table-salt crystal, the ions are arranged on an **FCC** lattice with a lattice constant $a = 2\overline{R}$ and a primitive unit cell of volume $v = a^3/4 = 2\overline{R}^3$, that is, $c = 2$. Upon changing the variables from v to $\overline{R}$,

$$\frac{\partial u}{\partial v} = \frac{1}{3c\overline{R}^2}\frac{\partial u}{\partial R_{01}}\bigg|_{R_{01}=\overline{R}} \ ,$$

and a further differentiation leads to

$$\frac{\partial^2 u}{\partial v^2} = \frac{1}{(3c\overline{R}^2)^2}\left(\frac{\partial^2 u}{\partial R_{01}^2}\bigg|_{R_{01}=\overline{R}} - \frac{2}{R_{01}}\frac{\partial u}{\partial R_{01}}\bigg|_{R_{01}=\overline{R}}\right) \ . \tag{4.11}$$

At equilibrium, $\overline{R}$ corresponds to the minimal energy, and therefore $(\partial u/\partial R_{01})_{R_{01}=\overline{R}} = 0$ and the second term in the brackets of Eq. (4.11) vanishes. Inserting into Eq. (4.9) the potential U_{P1} yields that for the table-salt structure

$$B = (m-1)\alpha q^2/(18\overline{R}^4) \ .$$

Measurements of the bulk modulus and the lattice constant, in conjunction with the identification of the lattice structure, allows to derive the decay factor m and to verify that the two parameters m and C of the potential U_{P1}, Eq. (4.8), indeed yield plausible values for $\overline{R}$, u, and B. For the alkali-halide crystals like table salt, this calculation gives for m values between 6 (for the smaller ions) and 10 (for the larger ones). Small errors in this parameter do not affect significantly the estimate of the binding energy, which attains values between 6 eV to 10 eV (per unit cell). As mentioned, the binding energy of ionic crystals is larger than the binding energies of the other types of bonds.

Problem 4.6.
Repeat the above calculation for the table-salt crystal, with the potential U_{P2}, Eq. (4.8): express the binding energy, the bulk modulus, and the parameter $z\lambda$, in terms of the distance in-between neighboring ions in the unit cell and the distance ρ.

4.3　Thecovalent bond

The Born-Oppenheimer approximation. A crystal comprises atoms (or ions) built of nuclei and electrons. As the nuclei are much heavier than the electrons, they move much slower. The **Born-Oppenheimer** approximation is based on this fact: one assumes that the nuclei are inert and calculates the quantum wave-functions of the electrons in the presence of the electric potential of these static nuclei. This approximation is used in computations of bond energies of molecules and solids. For a certain set of the nuclei locations, the electrons are in the state with the lowest energy (excitations to higher-energy levels have smaller Boltzmann probabilities, since the energy differences are large compared with the thermal energy). This ground-state energy, which depends on the aforementioned set, can be further decreased provided that those locations are chosen suitably: minimizing the energy determines the equilibrium locations in the molecule or in the solid. The next stage of the calculation (discussed in Chapter 5) accounts for the small vibrations around these equilibrium locations.

　　The hydrogen molecule. The simplest molecule is that of hydrogen, H_2. It contains two protons and two electrons. Instead of treating these two electrons together, it is advantageous to begin with a system of two protons and a single electron, as shown in Fig. 4.8. This figure displays the ion H_2^+. In the Born-Oppenheimer approximation the two nuclei, at distance R from one another, are placed on the $\hat{\mathbf{z}}-$ axis, (the points A and B in the figure) and the Schrödinger equation is solved for the wave function of the electron (marked in the figure by e). The ground-state energy resulting from this procedure is inevitably a function of R. An approximate calculation of this function is shown by the lower curve in Fig. 4.10: it has a single minimum as a function of R, and the value of R there, denoted $\overline{R}$, is the equilibrium distance between the two nuclei of the H_2^+ ion.

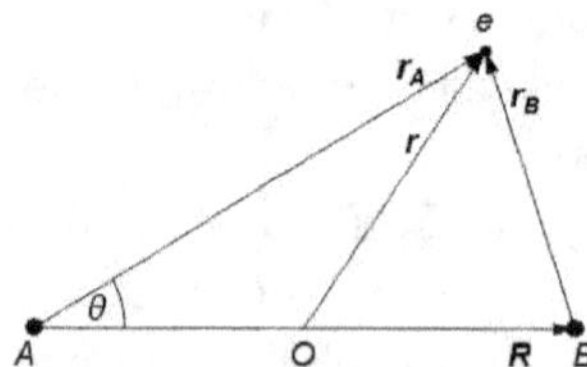

Fig. 4.8: The ion H_2^+. The two protons, each of a charge $+e$, are nailed at the points A and B. The third particle (marked by e) is the electron. The vector $\mathbf{R}$ connects nucleus A with nucleus B.

　　Placing the origin at the mid-point O between the two nuclei, and the electron at the point $\mathbf{r}$ relative to this origin (Fig. 4.8), the Hamiltonian of the electron in the H_2^+ ion is

$$\hat{\mathcal{H}} = -\frac{\hbar^2}{2m}\nabla^2 + U(\mathbf{r}) \, , \tag{4.12}$$

where the first term represents the kinetic energy (the gradient comprises derivatives with respect to the components of $\mathbf{r}$), and the electric potential (including also that between the nuclei) is

$$U(\mathbf{r}) = -\frac{e^2}{r_A} - \frac{e^2}{r_B} + \frac{e^2}{R} \, . \tag{4.13}$$

[Equation (4.13) exploits CGS units; to convert to the SI units, one has to divide the Coulomb energies by $4\pi\epsilon_0$, where ϵ_0 is the dielectric constant of the vacuum.] The dependence on $\mathbf{r}$ on the right-hand side of Eq. (4.13) is via $\mathbf{r}_A = (\mathbf{R}/2) + \mathbf{r}$, $\mathbf{r}_B = -(\mathbf{R}/2) + \mathbf{r}$.

The wave function for well-separated nuclei. When the distance in-between the nuclei exceeds considerably the Bohr radius, $R \gg a_B = 0.529\mathring{A}$, the electron can be in the ground state of the hydrogen atom near any of the nuclei. For instance, when the electron is near the A nucleus, then its wave function is $\psi_A = \psi_{100}(r_A) = \exp[-r_A/a_B]/\sqrt{\pi a_B^3}$, and the ground state energy is $E_1 = -R_y = -13.6\text{eV}$, where R_y is the Rydberg constant. The indices 100 are the quantum numbers, $n\ell m$, of this state (energy, angular momentum, and the $z-$component of the angular momentum). In such a configuration $r_B \approx R$, the last two terms in Eq. (4.13) nearly cancel one another and the Hamiltonian becomes that of the hydrogen atom around the nucleus A. Alternatively, the electron may be found near nucleus B, and then its wave function is $\psi_B = \psi_{100}(r_B)$, with the same energy. Each of these configurations describes an "ionic" state, in which the electron is "located" near one of the nuclei. These two states have the same energy and hence are degenerate. Therefore, any linear combination of them (with arbitrary complex coefficients), $\Psi(\mathbf{r}) = \alpha_A\psi_{100}(r_A) + \alpha_B\psi_{100}(r_B)$, is a solution of the same Schrödinger equation. The normalization of the wave function implies $|\alpha_A|^2 + |\alpha_B|^2 = 1$ (as is shown below, this expression for the normalization is valid only when the nuclei are far apart). $|\alpha_A|^2$ or $|\alpha_B|^2$ are the probabilities to find the electron around nucleus A or nucleus B, respectively. When the two nuclei are identical then $|\alpha_A|^2 = |\alpha_B|^2 = 1/2$ by symmetry. The state $\Psi(\mathbf{r})$, in which the electron is "shared" by the two nuclei, is called a **covalent state**. When the nuclei are not identical, the probabilities are different but still the energy of the covalent state can be lower than that of the **ionic states**, in which the electron is near one of the nuclei.

The symmetry of the wave functions. Denote the full solution of the molecular Schrödinger equation [including the full potential, Eq. (4.13)] by $\Psi(\mathbf{r}) = \Psi(\mathbf{r}_A, \mathbf{r}_B)$. Often, the symmetries of the wave functions can be deduced from the symmetries of the Hamiltonian, prior to undertaking the technical task of solving the corresponding differential equation. The Hamiltonian (4.12) possesses several symmetries: it is invariant under a rotation of the electron location around the molecular axis AB, upon interchanging the nuclei (i.e., replacing $\mathbf{R}$ by $-\mathbf{R}$), and also under a reflection through a mirror perpendicular to the molecular axis at point O (check!). The Hamiltonian is also unchanged when the sign of $\mathbf{r}$ is inverted (this latter symmetry is also obtained by a combination of the interchange of the nuclei locations and a rotation around the axis). This inversion is identical to the interchange $r_A \leftrightarrow r_B$. Hence, the Hamiltonian obeys $\hat{\mathcal{H}}(\mathbf{r}) = \hat{\mathcal{H}}(-\mathbf{r})$.

Each eigenfunction of such a Hamiltonian is either symmetric or antisymmetric with respect to this interchange. The Schrödinger equation, $\hat{\mathcal{H}}(\mathbf{r})\Psi(\mathbf{r}) = E\Psi(\mathbf{r})$, can be also written as $\hat{\mathcal{H}}(\mathbf{r})\Psi(-\mathbf{r}) = \hat{\mathcal{H}}(-\mathbf{r})\Psi(-\mathbf{r}) = E\Psi(-\mathbf{r})$, and therefore the two functions, $\Psi(\mathbf{r})$ and $\Psi(-\mathbf{r})$, solve the **same** Schrödinger equation, with the same energy E. It has been already found that in the limit of an infinite distance in-between the nuclei the ground state is doubly-degenerate. When the distance between the nuclei is finite, the corrections to the single-atom potential in Eq. (4.13) remove the degeneracy, and split the energy of the ground state into two different levels (see below). In the non-degenerate situation the two functions, $\Psi(\mathbf{r})$ and $\Psi(-\mathbf{r})$, are identical up to a multiplicative constant, $\Psi(-\mathbf{r}) = C\Psi(\mathbf{r})$ (since the Schrödinger equation may be multiplied by a constant). Therefore, $\Psi(\mathbf{r}) = C\Psi(-\mathbf{r}) = C^2\Psi(\mathbf{r})$; it follows that $C = \pm 1$, i.e., $\Psi(-\mathbf{r}) = \pm\Psi(\mathbf{r})$. **The function is even or is odd under the interchange** $r_A \leftrightarrow r_B$.

When the two nuclei are far apart, the even and the odd functions are given by

$$\Psi_{\pm}(\mathbf{r}) = A_{\pm}(\psi_A \pm \psi_B) = A_{\pm}[\psi_{100}(r_A) \pm \psi_{100}(r_B)] , \qquad (4.14)$$

where the plus (minus) sign denotes the limit of the symmetric (anti-symmetric) solution. The normalization of the wave function implies that $\alpha_B = \pm\alpha_A = \pm A_{\pm} = \pm 1/\sqrt{2}$, and therefore $|\alpha_A|^2 = |\alpha_B|^2 = 1/2$, as is expected from symmetry considerations; there is an equal probability to find the electron in the vicinity of each of the nuclei. These two functions and the corresponding probabilities are displayed in Fig. 4.9.

An approximate solution for the electron in the H_2^+ ion. Though the Schrödinger equation of this system can be solved exactly (by transforming it to elliptic coordinates), the solution presented here is obtained in the **variational approximation** which is further exploited below. This approximation is explained in Appendix C. Within the variational-method scheme one uses a trial wave function for the ground state, which is in fact a guess, and calculates the average of the quantum energy for this function. The trial wave functions used here, even when the nuclei are not located far away from one another, are those appearing in Eq. (4.14). The calculated energy is an upper bound on the exact energy of the desired ground state. The results of this calculation are presented in Fig. 4.10. As seen there, the energy E_+ is minimal for a certain distance between the nuclei R, and therefore the energy of the "true" ground state also possesses such a minimum. The conclusion is that energy can be "gained" by pushing the nuclei towards each other, thus allowing the electron to be simultaneously on both nuclei. This energetic

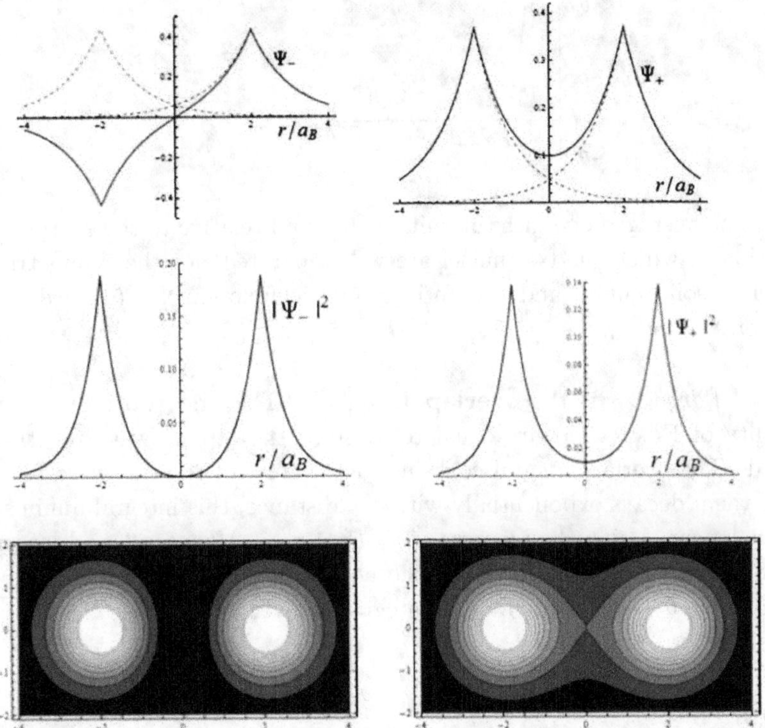

Fig. 4.9: The wave functions (top panel) and the corresponding probability density (mid panel), for $\theta = 0$ and $X = R/a_B = 4$. The dashed curves in the upper panel are the atomic wave functions on each of the nuclei. The lower panel displays equal-probability curves on a plane normal to the molecular axis (brighter regions correspond to higher probabilities).

"gain" (as compared with the configuration where the electron is around only one of the nuclei) is the binding energy of the molecule.

The variational approximation. As explained in Appendix C, the variational approximation is based on a calculation of the expectation value of the energy for a broad collection of trial wave-functions, and then choosing out of them the one that yields the lowest energy. Here it is applied to the two combinations of atomic wave functions [Eq. (4.14)], assuming an **arbitrary distance** in-between the nuclei (though these wave functions are derived for two nuclei which are far apart from one another). Then, the (real) atomic wave functions ψ_A and ψ_B are not orthogonal because they overlap, and consequently the normalization constants $A_\pm$ are determined by the requirement $\langle \Psi_\pm | \Psi_\pm \rangle = 1$, i.e.,

$$|A_\pm|^{-2} = \int d\mathbf{r}(|\psi_A|^2 + |\psi_B|^2 \pm 2\psi_A\psi_B) = 2(1 \pm S) , \qquad (4.15)$$

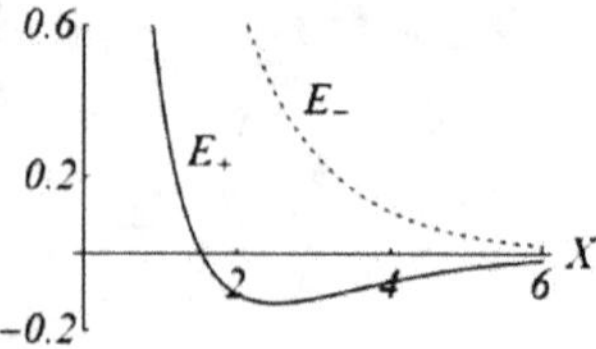

Fig. 4.10: The averaged energies (in units of R_y, and relative to the energies of the configuration in which the two nuclei are well separated) for the symmetric wave function Ψ_+ (solid curve) and the anti-symmetric function Ψ_- (dashed line), as functions of $X = R/a_{\rm B}$.

where $S = \int d\mathbf{r}\psi_A\psi_B$ is the **overlap integral**. The first two integrals in the mid equality of Eq. (4.15) are equal to 1, since the atomic wave functions are normalized. The overlap integral contains the product of the two wave functions; as each of them decays exponentially with the distance, this integral diminishes as R becomes longer. Using the identity $r_B^2 = R^2 + r_A^2 - 2Rr_A\cos\theta$, and integrating over $\mathbf{r}_A$ in spherical coordinates (with the notations $X = R/a_{\rm B}$, $x = r_A/a_{\rm B}$, and transforming the $\theta-$integration into an integration over $y = \sqrt{X^2 + x^2 - 2Xx\cos\theta}$, $dy = Xx\sin\theta d\theta/y$) gives

$$S = \frac{1}{\pi a_{\rm B}^3}\int d\mathbf{r}_A e^{-r_A/a_{\rm B}}e^{-r_B/a_{\rm B}}$$

$$= 2\int_0^\infty dx \int_0^\pi d\theta e^{-[x+\sqrt{X^2+x^2-2xX\cos\theta}]}x^2\sin\theta = e^{-X}(1 + X + X^2/3) \ . \quad (4.16)$$

Within the variational approximation, each of the functions, the symmetric Ψ_+ and the antisymmetric Ψ_-, can be chosen as a trial ground state. The energy of the first is then the upper bound on the lowest energy of a symmetric state, while the energy of the second is the upper bound on the lowest energy of an anti-symmetric state (the symmetric and the anti-symmetric states do not mix with one another as the eigen states have a definite symmetry). The upper panel in Fig. 4.9 displays the functions $\Psi_\pm = (e^{-r_A/a_{\rm B}} \pm e^{-r_B/a_{\rm B}})/\sqrt{2\pi a_{\rm B}^3(1 \pm S)}$ for the special case $\theta = 0$, with $r_{A,B} = |\mathbf{R}/2\pm\mathbf{r}|$. The dashed lines exhibit each of the summands. The mid panel of the figure shows the probability density to find the electron on the molecule axis for each of the wave functions, and the lower panel portrays lines of equal probability (of the electron location) in the plane normal to that axis. The results are not modified when this plane is rotated around the axis. These probability "clouds" are described below by such curves.

The variational method can be used to estimate the ground-state energy of the electron in the H_2^+ ion. Inserting the function Eq. (4.14) into Eq. (C.1) yields

$$\langle E_\pm \rangle = (2 \pm 2S)^{-1}[\langle\psi_A|\hat{\mathcal{H}}|\psi_A\rangle + \langle\psi_B|\hat{\mathcal{H}}|\psi_B\rangle \pm 2\langle\psi_A|\hat{\mathcal{H}}|\psi_B\rangle]$$

$$= (1 \pm S)^{-1}[\langle\psi_A|\hat{\mathcal{H}}|\psi_A\rangle \pm \langle\psi_A|\hat{\mathcal{H}}|\psi_B\rangle] \ . \quad (4.17)$$

The symmetry of the Hamiltonian under the interchange of the locations of the two nuclei is exploited in the second step. Using the Hamiltonian, Eqs. (4.12) and (4.13), gives

$$\langle \psi_A | \hat{\mathcal{H}} | \psi_A \rangle = E_1 + \frac{e^2}{R} - \langle \psi_A | \frac{e^2}{r_B} | \psi_A \rangle = E_1 + \frac{e^2}{R}\left(1 + \frac{R}{a_{\mathrm{B}}}\right) e^{-2R/a_{\mathrm{B}}} \ . \qquad (4.18)$$

The first term on the right hand-side of Eq. (4.18) arises since ψ_A is the ground-state wave function of the hydrogen atom near ion A,

$$\left(-\frac{\hbar^2}{2m}\nabla^2 - \frac{e^2}{r_A}\right)\psi_A = E_1\psi_A \ .$$

The second term results from a straightforward integration,

$$\langle \psi_A | \frac{e^2}{r_B} | \psi_A \rangle = e^2 \int d^3 r_A |\Psi_{100}(r_A)|^2 / r_B \ .$$

Similarly,

$$\langle \psi_A | \hat{\mathcal{H}} | \psi_B \rangle = \left(E_1 + \frac{e^2}{R}\right)S - \langle \psi_A | \frac{e^2}{r_B} | \psi_B \rangle = \left(E_1 + \frac{e^2}{R}\right)S - \frac{e^2}{a_{\mathrm{B}}}\left(1 + \frac{R}{a_{\mathrm{B}}}\right) e^{-R/a_{\mathrm{B}}} \ .$$
$$(4.19)$$

The ground-state energy of the electron when the nuclei are infinitely far away from one another is E_1, and the additional energy due to the bond between the two members of the molecular ion is $\langle E_\pm \rangle - E_1$. This difference, whose explicit expression is obtained upon inserting the results derived above into Eq. (4.17), is displayed in Fig. 4.10 for the two wave functions Ψ_+ and Ψ_-. As seen, the anti-symmetric wave function Ψ_- does not yield an energetic gain (at least, not in the approximation used here). For this reason, this function is called **"anti-bonding"**, and is not a suitable candidate for being the ground state of the molecule. On the other hand, the symmetric function Ψ_+ yields negative energies for a large range of $X = R/a_{\mathrm{B}}$ values; it is called a **"bonding"** function. The quantitative reason for this disparity can be seen from Fig. 4.9: in the symmetric state the electron has a larger probability to be in-between the two nuclei, it is then close to both of them, and "gains" energy from the Coulomb attraction of the two nuclei. As the energy calculated above is an upper bound on the true energy of the ground state, the latter is clearly more negative, and thus holds together the two parts of the molecule. This calculation yields an optimal distance of $1.32\text{\AA}$ and a binding energy of 1.76 eV, while the exact values are $1.06\text{\AA}$ and 2.79 eV. The approximation can be improved by exploiting better trial wave functions. For example, one may add atomic wave functions of higher levels to the functions ψ_A and ψ_B, as described in Appendix C, see Eq. (C.3). The main point is that the variational method proves that a bound state with a positive binding energy does exist. (Recall that the binding energy is the difference between the energy of the nuclei (with the electron) when far apart and the energy when they are in the bound state. Therefore a minimal negative energy in the bound state is identical to a positive binding energy.)

Problem 4.7.
Prove the results (4.16), (4.18), and (4.19).

Covalent bonds between different atoms. When the two atoms that participate in the covalent bond are not identical, the symmetry for interchanging them is lost and Eq. (4.14) is replaced by a linear combination of the two real atomic wave functions, $\Psi_{\pm} = A_{\pm}[\psi_A(\mathbf{r}_A) \pm \beta_{\pm}\psi_B(\mathbf{r}_B)]$. Assuming that $\beta_{\pm}$ are real and positive, the normalization yields

$$|A_{\pm}|^{-2} = \int d\mathbf{r}(|\psi_A|^2 + \beta_{\pm}^2|\psi_B|^2 \pm 2\beta_{\pm}\psi_A\psi_B) = 1 + \beta_{\pm}^2 \pm 2\beta_{\pm}S \ , \qquad (4.20)$$

and the condition for orthogonality gives

$$\int d\mathbf{r}[|\psi_A|^2 - \beta_+\beta_-|\psi_B|^2 + (\beta_+ - \beta_-)\psi_A\psi_B] = 1 - \beta_+\beta_- + (\beta_+ - \beta_-)S = 0 \ .$$
$$(4.21)$$

It follows that the choice $\beta_+ = \beta_- = 1$ made in Eq. (4.14) is still possible, though is not enforced by symmetry, and does not necessarily yield the lowest energy. Instead, one may consider the coefficient β_+ as a variational parameter, and choose it such as to minimize the energy (see problem 4.8). The other coefficient is determined by $\beta_- = (1 + S\beta_+)/(\beta_+ + S)$. The resulting energy turns out to be negative at large R, and hence corresponds to a bonding state. The ratio between the probability to find the electron near ion B to that of finding it near A in that bonding state is $|\beta_+|^2$.

Problem 4.8.
Extend Eq. (4.17) to allow for a bond between different states, and determine the coefficients β_+ and β_- which maximize the binding energy. Verify that the values obtained comply with the orthonormalization of the functions $\Psi_{\pm}$.

Molecular orbitals and the H_2 molecule. Since the bonding wave function of H_2^+ as described above (whether in the variational approximation or else) corresponds to symmetric states in which the electronic orbits are **shared by both atoms** (i.e., the electrons "move" around both nuclei simultaneously), it is called a **"molecular wave function"**, or a wave function of a **molecular orbital**. Consider the H_2 molecule, that has two nuclei and two electrons. This molecule is created when a second electron is added to the ion H_2^+. By the **Pauli principle**, each molecular state can be occupied by two electrons provided that their spin states are opposite to one another. This is similar to the ground state of the helium atom, in which there are two electrons in the lowest electronic orbital ψ_{100}. Ignoring the Coulomb repulsion between the two electrons, each of them is in a bonding ground state and the wave function of the two electrons is $\Psi_2(\mathbf{r}_1, \mathbf{r}_2) = \Psi_+(\mathbf{r}_1)\Psi_+(\mathbf{r}_2)$, where $\mathbf{r}_1$ and $\mathbf{r}_2$ are the coordinates of the two electrons. In this case the binding energy is roughly twice that of the ion H_2^+. The Coulomb repulsion can be accounted for by using $\Psi_2(\mathbf{r}_1, \mathbf{r}_2)$ as a variational wave function, and then the correction to the ground-state energy is the expectation value $\langle \Psi_2|e^2/r_{12}|\Psi_2\rangle$ of the repulsion energy. The same result is obtained to first order in perturbation theory.

This positive correction reduces the binding energy, but the total remains positive, higher than that of the H_2^+ ion.

The Heitler-London approximation for a covalent bond with two electrons. Inserting Eq. (4.14) in the expression $\Psi_2(\mathbf{r}_1, \mathbf{r}_2) = \Psi_+(\mathbf{r}_1)\Psi_+(\mathbf{r}_2)$ gives

$$\Psi_2(\mathbf{r}_1, \mathbf{r}_2) = \Psi_+(\mathbf{r}_1)\Psi_+(\mathbf{r}_2) = A_+^2 [\psi_A(\mathbf{r}_{1A})\psi_A(\mathbf{r}_{2A}) + \psi_B(\mathbf{r}_{1B})\psi_B(\mathbf{r}_{2B})$$
$$+ \psi_A(\mathbf{r}_{1A})\psi_B(\mathbf{r}_{2B}) + \psi_A(\mathbf{r}_{2A})\psi_B(\mathbf{r}_{1B})] , \tag{4.22}$$

where the radii $\mathbf{r}_{1A}$, $\mathbf{r}_{1B}$, $\mathbf{r}_{2A}$, and $\mathbf{r}_{2B}$ are the radius vectors of each of the electrons to each of the nuclei. The first two terms on the right hand-side of Eq. (4.22) represent states in which the two electrons are on the same nucleus, and therefore each of them is in an "ionic" state (one of the electrons moved from its "parent" atom to the other one). The last two terms describe states in which each electron resides on a different nucleus. In the next step of the calculation the energy is augmented by the average of the repulsion between the two electrons. As quite generally this one is higher when the two electrons are located on the same ion, it is customary to carry out the variational computation with different coefficients for the first two, "ionic" terms, and for the last two, "molecular" terms. The more similar the two atoms are, the higher is the weight of the molecular terms. In the **Heitler-London** approximation the two "ionic" functions are completely ignored, and the symmetric variational wave function is

$$\Psi_{2+}^{\mathrm{HL}}(\mathbf{r}_1, \mathbf{r}_2) = A_+ [\psi_A(\mathbf{r}_{1A})\psi_B(\mathbf{r}_{2B}) + \psi_A(\mathbf{r}_{2A})\psi_B(\mathbf{r}_{1B})] . \tag{4.23}$$

This function yields expressions like those in Eq. (4.17) (where each integral is carried out over the coordinates of both electrons), and energies whose qualitative dependence on R is similar to the ones shown in Fig. 4.10 (with different numerical values). The Heitler-London approximation gives an improved estimate of the ground-state energy as compared to the result obtained from the wave function based on the molecular orbitals, $\Psi_2(\mathbf{r}_1, \mathbf{r}_2) = \Psi_+(\mathbf{r}_1)\Psi_+(\mathbf{r}_2)$. Extending Eq. (4.23) to include the ionic terms (in the variational approximation),

$$\Psi_2^{\mathrm{ion}}(\mathbf{r}_1, \mathbf{r}_2) = B\Big(\psi_A(r_{1A})\psi_A(r_{2A}) + \psi_B(r_{1B})\psi_B(r_{2B})\Big) ,$$

with various coefficients, improves the approximation. Such a combination, already mentioned in Sec. 4.1, is termed **hybridization** of an ionic and a covalent bond.

Exchange energy, spin and exchange symmetry. The electronic Hamiltonian of the hydrogen molecule H_2 contains the single-electron Hamiltonian (for each of the two electrons), Eq. (4.12), and the interaction between the two electrons, e^2/r_{12}. This Hamiltonian is symmetric with respect to the interchange of the two electrons, and therefore it can be proved that the wave function of the two electrons is either symmetric or is antisymmetric under such an interchange. The proof resembles the one presented for the symmetry of interchanging the two nuclei in the H_2^+ molecule. Indeed, in the examples of two-electrons' Hamiltonian treated above, the wave function of the two electrons is symmetric with respect to their interchange, $\Psi_2(\mathbf{r}_1, \mathbf{r}_2) = \Psi_2(\mathbf{r}_2, \mathbf{r}_1)$. For instance, the variational energy obtained with the antisymmetric state, Eq. (4.23) with the negative sign between the two terms,

$$\Psi_{2-}^{\mathrm{HL}}(\mathbf{r}_1, \mathbf{r}_2) = A_-\left[\psi_A(\mathbf{r}_{1A})\psi_B(\mathbf{r}_{2B}) - \psi_A(\mathbf{r}_{2A})\psi_B(\mathbf{r}_{1B})\right] ,$$

is higher. This is since this wave function vanishes when $\mathbf{r}_1 = \mathbf{r}_2$, i.e., when the electrons are on the plane that bisects the line connecting the two nuclei. In the symmetric case the probability to find the two electrons in-between the two nuclei is higher; as a result there is an energy gain due to the Coulomb attraction between each of the nuclei and the electronic "cloud" located in-between them. As in Eq. (4.17), the difference between the energy of the symmetric state and that of the antisymmetric one contains the energy

$$\langle\psi_A(r_{1A})\psi_B(r_{2B})|\hat{\mathcal{H}}|\psi_A(r_{2A})\psi_B(r_{1B})\rangle ,$$

and the overlap integral, $\langle\psi_A(r_{1A})\psi_B(r_{2B})|\psi_A(r_{2A})\psi_B(r_{1B})\rangle$. In both expressions the electrons are interchanged between the wave functions on the left and on the right sides, and therefore this energy difference is termed **"exchange energy"**.

The considerations given so far pertain to the spatial part of the quantum wave function, which gives the probability density of the electrons in space. Electrons possess also the spin degree of freedom; the spin is described by the vector operator $\hbar\hat{\mathbf{S}}$. The square of this vector is $\hat{\mathbf{S}}^2 = s(s+1)$, with $s = 1/2$, and its $\hat{\mathbf{z}}-$ component attains two values, $\hat{S}_z = \pm s$. The Hilbert space of the spin states is thus spanned by two base vectors, denoted $|\pm s\rangle$. The Pauli principle dictates that when the spatial state of two electrons is symmetric with respect to their interchange (as in the examples discussed above), their spin state must be anti symmetric with respect to the spins' interchange, such that the total wave function is antisymmetric upon interchanging all degrees of freedom of the two electrons. The total spin state is then

$$\chi_S = \frac{1}{\sqrt{2}}\left(\left|\frac{1}{2}\right\rangle_1\left|-\frac{1}{2}\right\rangle_2 - \left|\frac{1}{2}\right\rangle_2\left|-\frac{1}{2}\right\rangle_1\right) .$$

The operator of the total spin of the two electrons is $\hat{\mathbf{S}}_{12} = \hat{\mathbf{S}}_1 + \hat{\mathbf{S}}_2$. Operating with $\hat{\mathbf{S}}_{12}^2$ on the state χ_S yields zero. The $\hat{\mathbf{z}}-$component of the total spin vanishes as well. There is only a single state with these properties, called **singlet**. Therefore, the ground state of the hydrogen molecule is a singlet.

The hydrogen molecule possesses also the antisymmetric spatial state, e.g., $\Psi_{2-}^{\mathrm{HL}}(\mathbf{r}_1, \mathbf{r}_2)$, whose energy is higher than that of the symmetric state. Since the spatial state is antisymmetric with respect to the exchange of the two electrons, then according to the Pauli principle the spin state is symmetric. There are three symmetric spin states of two electrons,

$$\chi_{T1} = \left|\frac{1}{2}\right\rangle_1\left|\frac{1}{2}\right\rangle_2 , \quad \chi_{T\bar{1}} = \left|-\frac{1}{2}\right\rangle_1\left|-\frac{1}{2}\right\rangle_2 ,$$

$$\chi_{T0} = \frac{1}{\sqrt{2}}\left(\left|\frac{1}{2}\right\rangle_1\left|-\frac{1}{2}\right\rangle_2 + \left|\frac{1}{2}\right\rangle_2\left|-\frac{1}{2}\right\rangle_1\right) .$$

These three states are termed **"triplet"**. They are eigenfunctions of $(\hat{\mathbf{S}}_{12})^2$, with the eigenvalue 2, and of $(\hat{S}_{12})_z$ with the eigenvalues 1, -1, and 0 (as the subscripts indicate). The triplet states correspond to states of unit (in the relevant units) spin angular-momentum. In certain situations, the ground state can be a triplet.

Heavier atoms. The calculation above shows that the covalent bond between two atoms is formed when two electrons "occupy" the same molecular orbital, which is built of a linear combination of atomic wave functions on each of the ions. This calculation is relatively simple since the atomic wave function of the hydrogen ground-state, ψ_{100}, is spherically symmetric, and because the orbital ground-state is not degenerate. It becomes more complex when the **valence electrons** participating in the covalent bond are in higher electronic shells, as such atoms can contribute to the bond more than one electron. The electronic states in the atom are usually identified by the quantum number ℓ of the angular momentum. The state with $\ell = 0$ is denoted by the letter s, those with $\ell = 1$ by p, with $\ell = 2$ by d, etc. The **electronic configurations** of hydrogen, helium, lithium, beryllium, boron, and carbon atoms are then $1s$, $1s^2$, $1s^2 2s$, $1s^2 2s^2$, $1s^2 2s^2 2p$, $1s^2 2s^2 2p^2$, respectively, where the "large" numbers on the left denote the quantum number n, and the small numbers (superscripts) indicate the number of electrons in the state with the corresponding angular momentum. For instance, hydrogen has a single electron in the shell with $n = 1$, $\ell = 0$, while carbon has two electrons in this first shell, and also four electrons in the next one ($n = 2$), two with $\ell = 0$ and two with $\ell = 1$. Each of the valence electrons in the $n = 2$ shell can form a covalent bond with an electron of another atom.

Examples for such configurations include, e.g., helium. In helium, the shell $n = 1$, $\ell = 0$ is occupied with two electrons which are then prohibited, by the Pauli principle, from forming covalent bonds. The next atom in the periodic table is lithium, with a single valence electron with $n = 2$, $\ell = 0$. Lithium often "contributes" this electron, forming an ionic bond. The atoms in the second row of the periodic table are considered in the following.

Covalent bonds of beryllium. In the lowest atomic state of beryllium, the two valence electrons (those in the outer shell, which are available for chemical bonding) are in the $2s$ state to which another electron cannot be added (because of the Pauli principle). Nonetheless, as the energy difference between this state and the $2p$ one is relatively small, it is sometimes "profitable" for one of the electrons to cross over to the $2p$ state, such that the electronic configuration becomes $1s^2 2s 2p$. Once this happens, the two valence electrons of the $n = 2$ shell can combine with electrons from other atoms and form **two** covalent bonds. The functions with $\ell = 0$ are spherically symmetric, e.g.,

$$\psi_{200} = 2[z/(2a_\mathrm{B})]^{3/2}[1 - zr/(2a_\mathrm{B})]\exp[-zr/(2a_\mathrm{B})]/\sqrt{4\pi}\;,$$

where z is the effective charge, due to the nucleus and the electrons in the first shell, as "seen" by the electrons in the second shell. On the other hand, the functions with $\ell = 1$ are not symmetric, as they depend on the spherical coordinates θ and φ, via the spherical harmonic functions $Y_{\ell m}(\theta, \varphi)$:

$$\psi_{21m}(\mathbf{r}) = R_{21}(r)Y_{1m}(\theta, \varphi) \ , \quad R_{21}(r) = \left(\frac{z}{2a_\mathrm{B}}\right)^{3/2}\left(\frac{zr}{a_\mathrm{B}}\right)\frac{\exp[-zr/(2a_\mathrm{B})]}{\sqrt{3}} \ ,$$

where $m = -1, 0, 1$ and R_{21} is the radial function. In place of the three functions Y_{1m} one may choose the real linear combinations,

$$\psi_z = \psi_{210} = R_{21}(r)\sqrt{\frac{3}{4\pi}}\cos\theta \ ,$$

$$\psi_x = \frac{1}{\sqrt{2}}(\psi_{21-1} - \psi_{211}) = R_{21}(r)\sqrt{\frac{3}{4\pi}}\sin\theta\cos\varphi \ ,$$

$$\psi_y = \frac{i}{\sqrt{2}}(\psi_{21-1} + \psi_{211}) = R_{21}(r)\sqrt{\frac{3}{4\pi}}\sin\theta\sin\varphi \ . \tag{4.24}$$

The angular parts of the three functions are identical to those that appear in the spherical coordinates, $x = r\sin\theta\cos\varphi$, $y = r\sin\theta\sin\varphi$, $z = r\cos\theta$. The regions of high probability of the electron's density are thus centered around each of the axes (for a given radial distance, the absolute value squared of each wave function is maximal along one of the axes). Figure 4.11 illustrates the probability "cloud" of the wave function ψ_x (denser points indicate higher probability to find the electron). The two valence electrons of beryllium can be in the state ψ_{200}, or in any of the three states (4.24).

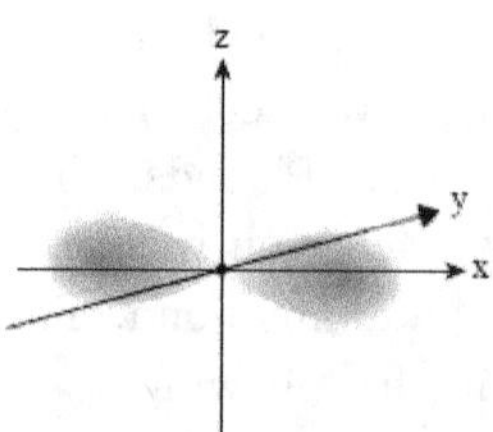

Fig. 4.11: The density of points reflects the probability density to find the electron in the ψ_x state. Only probabilities higher than a certain threshold are depicted.

Hybridization of the *sp* type. The arguments given above demonstrate that the larger is the probability to find the two electrons which participate in a covalent bond in-between the two nuclei, the higher is the binding energy. In the ψ_x state the probability to find the electron on the $\hat{\mathbf{x}}$–axis is the same on both sides of the nucleus; hence this is not the quintessential state. The probability density is concentrated on one side of the nucleus in the state $\Psi_\pm = (\psi_{200} \pm \psi_x)/\sqrt{2}$. Such functions are **"hybridized states of the *sp* type"** (i.e., linear combinations of $\ell = 0$ states with $\ell = 1$ ones). Since ψ_x has opposite signs on the two sides of the $\hat{\mathbf{x}}$–axis, Ψ_+ gives a larger probability on the positive side, while Ψ_- gives the larger probability on the negative side. The two states are described on the right side of Fig. 4.12. Their probability "clouds" are concentrated on the positive or on the negative side of the $\hat{\mathbf{x}}$−axis, as dictated by the relative sign. When each of the hybrid states $\Psi_\pm$ is occupied by a single electron, the average repulsion energy is smaller than in any of the original states, because in the hybridized state the electrons "reside" on the two sides of the nucleus, and hence are relatively far apart.

In the hybridized configuration, the beryllium atom can form covalent bonds with two other atoms. For instance, it can build the linear molecule BeH_2, in which the two hydrogens are on the $\hat{\mathbf{x}}$−axis on both sides of the beryllium. The electron of the right hydrogen hybridizes its hydrogen state with the state Ψ_+ of the beryllium, and the one of the left hydrogen hybridizes with the state Ψ_- (the coefficients of the wave functions of the beryllium and the hydrogen in each molecular state are not identical, see problem 4.9). Pictorially, this molecule is referred to as H−Be−H, where each hyphen represents a covalent bond comprising two electrons, one from each atom. The atoms in this molecule are located on a straight line (the angle between the two bonds is 180°), since bending the molecule brings closer the two electron clouds (as well as the two protons) on the two sides of the beryllium, and thus costs Coulomb energy.

In a similar fashion, a beryllium atom can form covalent bonds with two beryllium atoms on its two sides, thus creating the infinitely-long linear molecule ...Be−Be−Be−Be−Be− .... This is a simple realization of a **one-dimensional periodic lattice**, built of atoms connected by covalent bonds. The lattice constant and the bonding energy can be estimated from a calculation of the average variational energy for the (symmetric) molecular wave functions of the two electrons [based on $\Psi_+(r_A)$ in the left atom and $\Psi_-(r_B)$ in the right one]. The distance between the two atoms at which this energy is minimal corresponds to the lattice constant.

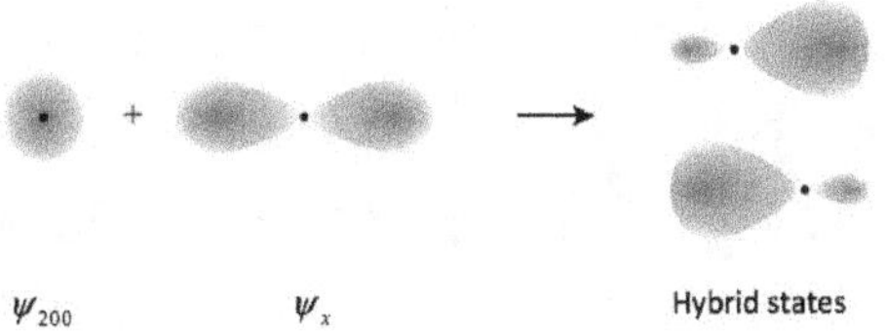

Fig. 4.12: From left to right: ψ_{200} state, ψ_x state, and the hybrid states $\Psi_\pm$.

Covalent bonds of boron, hybridization of the sp^2 type. The next atom in the periodic table is boron ($z = 5$). Again it is "better" for the three valence electrons to change the configuration $2s^2 2p$ into $2s 2p^2$, to enable all three to participate in covalent bonds. Assuming that the two $p-$electrons are in linear combinations of the states ψ_x and ψ_y, a preferable situation is that of hybrid states, which tend to separate the electrons' clouds and thus to decrease the average energy of the electric repulsion. These states, termed **hybridization states of the sp^2 type**, are

$$\Psi_1 = (\psi_{200} + \sqrt{2}\psi_x)/\sqrt{3} \ ,$$

$$\Psi_2 = (\psi_{200} - \sqrt{2}[\psi_x - \sqrt{3}\psi_y]/2)/\sqrt{3} \ ,$$

$$\Psi_3 = (\psi_{200} - \sqrt{2}[\psi_x + \sqrt{3}\psi_y]/2)/\sqrt{3} \ . \tag{4.25}$$

These three states are occupied by the three electrons of the boron. The choice of the coefficients is dictated by two considerations. First, the three functions should be orthogonal to each other, i.e., $\langle \Psi_i | \Psi_j \rangle \equiv \int d\mathbf{r} \Psi_i^* \Psi_j = \delta_{ij}$ as can be proven by the orthonormality of the atomic functions (check, see also problem 4.9); second, the probability clouds formed by them should be identical (up to rotations), as there is no reason for one electron to be different than the others. It is straightforward to show that the probability clouds of these three states, depicted in Fig. 4.13, attain their largest values along three directions, all in the $XY-$plane, and are oriented at an angle of 120° relative to each other. As the clouds are rather far apart, the repulsion energy, in a quantum state in which there is a single electron in each state, is relatively low.

Each such electron can be also bound covalently with an electron of a hydrogen atom, to form the molecule BH_3. This is a **planar** molecule, symmetric under rotations of order 3 around the nucleus of the boron (that is, the nuclei are all in-plane; the electrons' clouds can be also outside the plane). Each boron atom has three bonds of this type, and therefore boron can create an infinite hexagonal lattice, like that of graphene (Fig. 2.8); each atom in this lattice has three covalent bonds with three neighbors, making an angle of 120° in-between each pair. This type of covalent bonds characterizes numerous hexagonal lattices in Nature.

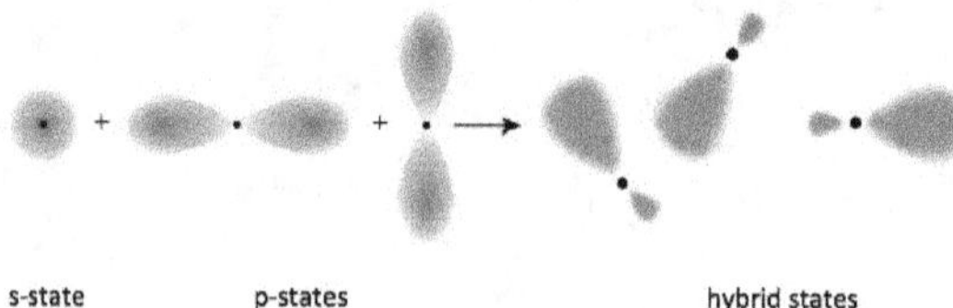

Fig. 4.13: From left to right: an $s-$state, $p-$states, and the hybrid states. Each "cloud" in the right panel illustrates one of the states listed in Eq. (4.25).

Covalent bonds of carbon, hybridization of the sp^3 type. The next atom in the periodic table is **carbon**; it participates in organic molecules, indispensable for many biological functions. In order for all four valence electrons of carbon to form covalent bonds, they must be in the configuration $2s2p^3$. Then hybridization among these states, where the probability clouds are distanced as far as possible from each other, can be established. These states are called **hybridization states of the sp^3 type,**

$$\Psi_{111} = (\psi_{200} + \psi_x + \psi_y + \psi_z)/2 \ ,$$
$$\Psi_{1\bar{1}\bar{1}} = (\psi_{200} + \psi_x - \psi_y - \psi_z)/2 \ ,$$
$$\Psi_{\bar{1}\bar{1}1} = (\psi_{200} - \psi_x - \psi_y + \psi_z)/2 \ ,$$
$$\Psi_{\bar{1}1\bar{1}} = (\psi_{200} - \psi_x + \psi_y - \psi_z)/2 \ . \tag{4.26}$$

The subscripts of each function represent the signs of the three p–states on the right hand-side, with $\bar{1}$ denoting -1. The four probability clouds are directed towards the four corners of a tetrahedron, shown by thick lines in Fig. 4.14 (see problem 4.9). The angle between the axes of each pair of clouds is $109.5°$ (check that this is indeed the angle in the tetrahedron). Indeed, in the simplest organic molecule, CH_4 (methane), Fig. 4.15(a), the hydrogen atoms are at the vertices of such a tetrahedron, with the electron of each one of them forming a covalent bond with one of the carbon electrons that occupies one of the states of Eqs. (4.26). The ion $(NH_4)^+$ has the same structure, as the nitrogen ion possesses four valence electrons like the carbon atom. As opposed, in the molecule C_2H_6 (ethane) [shown in Fig. 4.15(b)] each carbon is connected with three hydrogens and with another carbon, and thus the four bonds around each C are not identical. As a result, the angles between the bonds, as well as their lengths, are different. One may compute the angles by utilizing the variational approximation, with different coefficients for the terms in Eqs. (4.26) (see also problem 4.10).

Problem 4.9.

Two hybridizations of the $2s$ state with the $2p$ ones are $\Psi_{1,2} = A[\psi_{200} + \beta\hat{n}_{1,2} \cdot \mathbf{\Psi}]$, where $\hat{n}_{1,2}$ are two unit vectors in space, and $\mathbf{\Psi} = \{\psi_x, \psi_y, \psi_z\}$.
a. *Show that in the state Ψ_1, the probability to find the electron is centered around the direction of the vector $\hat{n}_1$.*
b. *Use the orthogonality of the original basis functions Eq. (4.24) to demonstrate that $\langle\psi_i|\psi_j\rangle = |A|^2(1 + |\beta|^2\hat{n}_i \cdot \hat{n}_j)$.*
c. *Utilize the result in part (b) to derive A and β when the hybrid states are orthonormal, and when the angle between the two unit vectors is α.*
d. *The hybridization of the sp^2 type corresponds to three states, with three unit vectors in the plane, $\hat{n}_1$, $\hat{n}_2$, and $\hat{n}_3$. Explain why the angles in-between these unit vectors have to be identical, and derive Eq. (4.25).*
e. *Derive in a similar way Eq. (4.26).*
f. *In a planar molecule like BH_2F the bonds are not entirely symmetric: the angle between the bonds of the boron with each of the hydrogens is γ, while the other two*

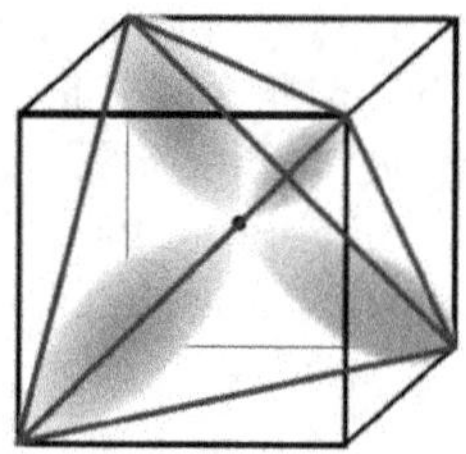

Fig. 4.14: The probability clouds of the four valence electrons of carbon (at the center of the cube), for the wave functions in Eq. (4.26). The probability clouds point at the corners of a tetrahedron (shown by thick lines). Alternatively, the probability clouds are directed at the opposite corners of the cube: two vertices are at the ends of one of the face diagonals, and the other two are at the ends of the diagonal on the opposite face, which is normal to the first one.

angles (in-between each BH bond and BF one) are $\alpha = \pi - \gamma/2$. Determine the hybrid functions that comply with this geometry.
g. *Given the angle $\alpha = 111.17°$ between the CC bond and the CH one in ethane [Fig. 4.15(b)], what is the angle γ between the two CH bonds? What can be deduced about the coefficients in the corresponding hybrid functions?*

Fig. 4.15: The methane molecule (a) and the ethane molecule (b). The bonds' lengths are measured in picometers ($10^{-2}\,\text{Å}$).

π**−bonds, double and triple bonds.** The ethene (C_2H_4) molecule is illustrated in Fig. 4.16(b). Around each carbon there are two hydrogens; this carbon is also coupled to the other carbon. These three bonds are described by the planar wave functions of Eq. (4.25), and are located on the same plane. Nonetheless, due to the difference between carbon and hydrogen, the three angles between the bonds are not precisely the same, and the coefficients in Eqs. (4.25) are slightly modified [see part (g) of problem 4.9]. Since carbon has four valence electrons, there remains in each carbon one electron, in the state ψ_z, whose probability cloud is normal to the plane. The hybridization of the functions ψ_z on the two carbons, $\Psi = A[\psi_{zA}(\mathbf{r}_A) + \psi_{zB}(\mathbf{r}_B)]$, creates an additional bond in-between them, as described in Fig. 4.16(a). This bond is weaker than the in-plane ones, as the overlap between the wave functions oriented normal to the plane is smaller than the one of two wave functions concentrated on the same axis. This bond is termed $\pi-$**bond**; the one in which the electron clouds are concentrated on the line connecting the nuclei is termed $\sigma-$**bond**. In the C_2H_4 molecule there is therefore a **double bond** between the carbons, comprising the two bonds. The double bond, marked by $=$, is stronger than a single bond, as it is due to four electrons, two from each atom. This is illustrated in Fig. 4.16(b). Note that without the $\pi-$bond it is possible to rotate one of the planes containing one carbon with the two hydrogens connected to it, such that the two planes will be perpendicular to one another. This will shift the hydrogens away from those of the second carbon; however, the overlap between the ψ_z functions will diminish, weakening the $\pi-$bond between the carbons. This is the reason for the hydrogens in the ethene molecule to be located on the same plane. Similarly, in acetylene (C_2H_2) there is a **triple bond** between the two carbons, comprising one $\sigma-$ bond (between the two linear functions $\Psi_\pm = (\psi_{200} \pm \psi_x)/\sqrt{2}$) and two $\pi-$bonds, consisting of the functions normal to this line (ψ_y and ψ_z). This molecule is denoted $H-C\equiv C-H$.

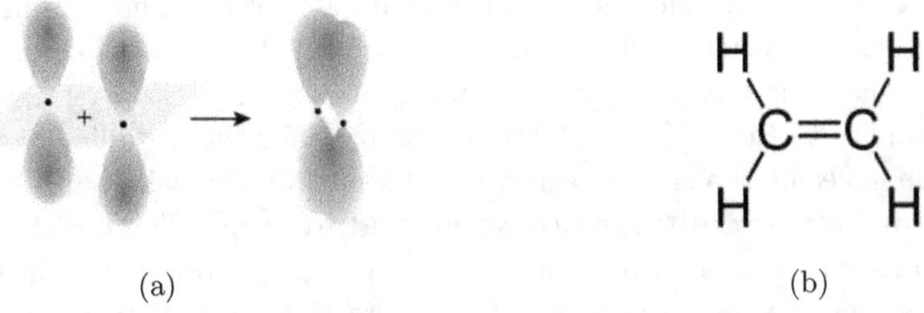

(a)   (b)

Fig. 4.16: (a) The $\pi-$bond between two p states for which the probability clouds are normal to the bond direction. The clouds on the right result from a combination of two atomic functions of the π type, shown separately on the left of that panel. (b) The planar molecule ethene, C_2H_4, with the double bond in-between the carbons.

Benzene, the metallic bond. In the benzene molecule, C_6H_6, mentioned in Sec. 4.1, each carbon is located on a corner of a hexagon, and is connected to a hydrogen atom (see Fig. 4.2). Three of the carbon electrons are in the plane of the hexagon, in wave functions similar to those in Eq. (4.25), with different angles for the different types of bonds: one electron participates in a covalent bond with the hydrogen, and the other two form $\sigma-$bonds with the neighboring carbons. The fourth electron is in the ψ_z state, with a probability cloud normal to the plane. By the Heisenberg uncertainty principle, the kinetic energy is decreased as the clouds' spatial extension is widened. (Increasing the uncertainty in the location reduces that of the momentum. When the average momentum vanishes, then the mean-square deviation equals the square of the momentum, and hence it is proportional to the kinetic energy. As the latter is reduced, the bonding energy is enhanced.) The uncertainty in the locations of the electrons is largest when their wave functions are spanned over the entire hexagon; extending Eq. (4.14),

$$\Psi = \sum_{\ell=1}^{6} a_\ell \psi_\ell(\mathbf{r}_\ell) , \qquad (4.27)$$

where the coefficients are chosen so that the wave function is normalized; $\mathbf{r}_\ell$ is the distance of the electron from the ℓ carbon, and ψ_ℓ is the ψ_z function of that carbon. The six ψ_z functions accommodate each a single electron and are orthonormal. As in the case of the hydrogen ion H_2^+, symmetry considerations imply that the probability to find the electron on any of the carbons is the same, and hence $|a_\ell|^2 \equiv A^2$ and is independent of ℓ. The complex coefficients differ from each other only in their phases (see Chapter 6). The covalent bonds between the perpendicular states, like in ethene (Fig. 4.16), are more effective when the probability clouds of these states are parallel, and therefore the other bonds around each carbon are in the same plane, as seen in Fig. 4.2. Since the probability to find the electron on any of the carbons is the same, these electrons are free to move around the hexagonal ring once subjected to external fields. For example, a uniform magnetic field causes the electrons to move around the ring, and to create a magnetic moment normal to the plane. The wave functions (4.27) contribute as well to the binding energy. This configuration resembles metals: in Chapter 6 it is found that in a metal each electron has the same probability to be near any of the ions in the crystal, and wave functions similar to those of Eq. (4.27) are the basis for the metallic bond. Thus, the benzene molecule is a simple realization of a metallic crystal.

Covalent bonds of nitrogen, oxygen, and fluorine. The $n = 2$ electronic shell contains four orbital states, one s and three p. Various hybridizations of these four states can be designed in order to create covalent bonds. The atoms in the second row of the periodic table possess 1-8 electrons. Up to carbon, each of the four states can be singly-occupied (or be empty), and each such electron can participate in a covalent bond. The next atom is **nitrogen**, which has five electrons in this shell. Consequently, two of these share the **same** orbital state, and therefore, by the Pauli principle that forbids the addition of another electron to this state, cannot participate in covalent bonds. Nitrogen has thus three free electrons, that may be in any of the hybridizations listed above. Similarly, **oxygen** has two doubly-occupied states and only two electrons that are free to form bonds. Fluorine possesses three states occupied by two electrons, and hence only a single electron that can be bonded. The configurations of the corresponding electronic states are shown in Fig. 4.17.

The water molecule. The **water** molecule is obviously of paramount importance. It is built of an oxygen atom and two hydrogens, H_2O. In the second shell

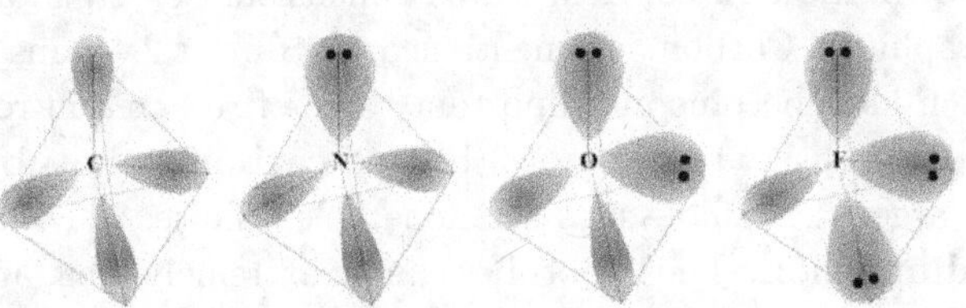

Fig. 4.17: From left to right: probability clouds to find the electrons in the second shell of carbon, nitrogen, oxygen, and fluorine. The bigger clouds contain two electrons each. The smaller ones accommodate a single electron, "free" to form a covalent bond.

of oxygen there are two orbital states which are doubly occupied, so only two of the electrons are available for forming bonds. In order to build the water molecule, it is expedient to begin with the four electronic states of Eq. (4.26), which create separate clouds (as far as possible from each other) as to reduce the Coulomb interaction among the electrons. Two of these states contain two electrons each (see Fig. 4.17), and the other two are coupled with the hydrogens as depicted in Fig. 4.18. The angle between the bonds with the hydrogens is 104.5°, slightly smaller than that of the tetrahedron of the symmetric case in Fig. 4.14 (see problem 4.10). This difference stems from the repulsion between the electrons taking part in the bonds with the hydrogens, and the other electrons of the oxygen, concentrated on the other side of this atom. Technically, the difference manifests itself in slight modifications of the coefficients of the base functions in Eq. (4.26). A similar difference is encountered in the ethane molecule, Fig. 4.15. As a result of this deviation from complete symmetry, the water molecule has an electric dipole moment.

Problem 4.10.
Describe the geometry of the electrons' clouds in the molecules N_2, O_2, F_2, NH_3, and FH. Which of those possesses an electric dipole moment?

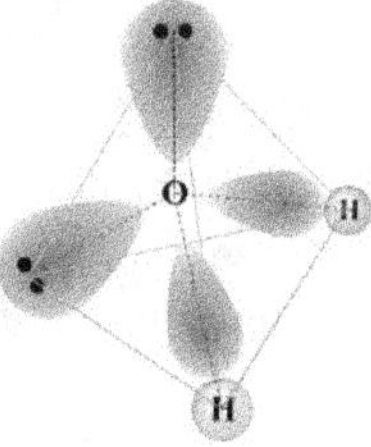

Fig. 4.18: The water molecule. The darker clouds represent the two doubly-occupied states of oxygen (indicated by the two black points). The angle between the OH bonds is 104.5°.

Giant carbon molecules. The three-dimensional crystals of carbon are diamond and graphite. Carbon, as mentioned in Sec. 1.2, forms also very large molecules, which have become an important topic of research in recent years. Examples are $\mathbf{C}_{60}$ [the Bucky ball, Fig. 1.9(b)] and **carbon nanotubes** [Fig. 1.9(a)]. Though quite large, these molecules are nonetheless finite; they are considered as being of **zero dimension**. The nanotubes can be extremely long, and then they are viewed as **one-dimensional lattices** (of finite thicknesses in the directions normal to the lattice). Each carbon in these molecules has three neighbors, and therefore the bonds are approximately of the sp^2−type. They are slightly distorted because of the deviation of the sphere or the tube from perfect planarity. Because of this deviation, the functions in Eq. (4.25) are augmented by a contribution from the wave function ψ_z, normal to the tube (or to the sphere), and then the electrons' clouds are not exactly in the same plane (see also problem 4.10). The wave functions of the surplus electrons may be spanned over the whole molecule, similar to Eq. (4.27).

Carbyne. The four valence electrons of carbon can be in any of the hybrid states described above, according to the geometric configuration of the covalent bonds at hand. Several allotropes of carbon, in various spatial dimensions, are mentioned in Chapter 1. Each such lattice requires different electronic states on the carbon. The simplest example of a **one-dimensional carbon lattice** is a **linear chain of carbons**, called carbyne. The chain may take one of two forms, cumulene, ...C=C=C=C=C..., with a double bond between two nearest neighbors, and polyyne, ...C−C≡C−C≡C..., in which each carbon has a single bond on one side and a triple one on the other. As explained in the context of beryllium, the σ−bonds between nearest-neighbor carbons cause the bonds to be on the same straight line.

Graphene. In two dimensions carbon creates the **hexagonal planar lattice** of graphene, Fig. 1.6 or Fig. 2.8. Similar to the bonds of boron, three of the carbon electrons are in the hybridized state of the type sp^2, Eq. (4.25). Each of these electrons forms a σ−bond with the neighboring carbon, and all these bonds are in the same plane. The remaining electron, as in benzene, is in a ψ_z state, normal to this plane. In order to reduce the kinetic energy, then according to the uncertainty principle of Heisenberg, these extra electrons occupy linear combinations of the ψ_z states spread on **all** carbons in the lattice, e.g., Eq. (4.27), with $|a_\ell|^2 = A^2$. These π−states contribute to the binding energy and are also responsible for the high electrical conductance and the metallic behavior of this material. The graphene lattice is planar; this is because the charge clouds of the π−states are parallel to each other. Such electronic states are discussed in Chapter 6.

Boron nitride, silicene and hexagonal germanium. The hexagonal lattice of graphene appears in other materials as well, e.g., boron nitride (BN). In this compound the boron and the nitrogen atoms occupy alternating sites of a hexagonal lattice, like the large and small points in Fig. 2.8. The electronic configurations of the borons and the nitrogens are $2s^2 2p^1$ and $2s^2 2p^3$, respectively; however, when in a bulk form, one electron is transferred from the nitrogen to the boron and the two ions attain the $2s^1 2p^3$ configuration, as in carbon. In this way bonds of the sp^2–type are formed, like in graphene, with the ψ_z–electrons common to all ions in the lattice. Because of the transferred electron, there are excess charges on the ions, and the resulting bonds comprise both covalent and ionic ones. To compute the binding energy of BN one utilizes hybridizations of the two types of wave functions, *cf.* the discussion following Eq. (4.23). Other examples of hexagonal lattices are **silicene** and **germanene**, Fig. 2.11. There, similar to BN, there are two sublattices which in this case are located on different planes. In the structure of Fig. 2.11 the electronic wave functions around each ion include also a component of ψ_z, in addition to the functions of Eq. (4.25). The deformation of the planar lattice stabilizes it mechanically (see also Chapter 5).

Graphite. Graphite is built of planar layers of graphene (see Fig. 2.11). The attraction between the layers is mostly due to the van der Waals interaction, which is related to the fluctuating dipole moments of the electrons in each carbon (see below). The inter-plane attraction is thus relatively weak and allows for the peeling of planar layers off the bulk graphite (this is the reason why graphite is used in pencils). Such a peeling (accomplished with a piece of celo-tape) enabled Geim and Novoselov to obtain single layers of **graphene**. The bond between atoms in each layer in graphite (and in graphene) is stronger even than that in diamond. This is the reason for graphite to be the preferred three-dimensional phase of carbon under ambient conditions. Diamonds are formed only under high pressures. Graphite is created upon cooling liquid carbon under atmospheric pressure below 4000C, while it is only when the liquid is cooled at pressures of about half a millions atms that diamond is created. The reason is related to the dependence of the binding energy on the distance between the carbons' nuclei in the crystal. Under pressure, the nuclei approach each other, and at a certain distance the bonding energy of diamond becomes smaller than that of graphite. Indeed, natural diamonds are found in places which were subjected to high pressures (due to certain geological phenomena); they are thus rather rare (and hence expensive).

The diamond crystal. The four directions of the probability clouds shown in Fig. 4.14 are identical to those between nearest-neighbor carbons in the **diamond lattice**. The cube in that figure is 1/8 of the unit cell of diamond (see Fig. 2.20). In this structure, each carbon contributes four electrons that form four covalent bonds of the sp^3–type with the four neighbors; each carbon is located at the center of a tetrahedron built of its four neighbors. This structure is very strong and not easily distorted, as any deformation brings together the electrons' clouds on the various bonds and requires considerable Coulomb energy.

Diamond and zinc-blende structures. Silicon and germanium, located at the same column of the periodic table as carbon, have the crystal structure of diamond. Interestingly enough, AlAs, GaAs, and InAs have the very similar zinc-blende structure (Fig. 2.20). The valence-electron configuration in aluminium, gallium, and indium is ns^2np^1 ($n = 3$, 4, and 5 for Al, Ga, and In, respectively) and that of arsenic is $4s^24p^3$. In the bulk one electron of As is transferred to the other atom of the compound, and then the two ions attain the ns^1np^3 configuration, which forms bonds of the sp^3-type. The resulting bond again comprises ionic and covalent bonds. GaAs is vastly exploited in electronic devices.

In summary, the simple geometrical picture of the various hybridizations of the p electrons explains the experimentally-determined structures (Chapters 2 and 3) of materials based on the elements in the fourth column in the periodic table, and of materials comprising elements from the third and fifth columns.

Covalent bonds with d electrons. Downwards from the third row of the periodic table, electrons can fill also the $3d$ shell, where their angular momentum is $\ell = 2$. The angular parts of the wave functions of these electrons are

$$\psi_{z^2} = \psi_{320} \propto 3z^2 - r^2 \ , \quad \psi_{xy} = \frac{\psi_{322} - \psi_{32-2}}{\sqrt{2}} \propto xy \ , \quad \psi_{yz} = \frac{\psi_{321} + \psi_{32-1}}{\sqrt{2}} \propto yz \ ,$$

$$\psi_{zx} = \frac{\psi_{321} - \psi_{32-1}}{\sqrt{2}} \propto zx \ , \quad \psi_{x^2-y^2} = \frac{\psi_{322} + \psi_{32-2}}{\sqrt{2}} \propto x^2 - y^2 \ . \tag{4.28}$$

The probability clouds of these states attain their maximal density at angles $90°$ or $180°$ with each other. [For instance, the probability density of the function proportional to xy is $(xy)^2 \propto (\sin^2\theta \cos\phi \sin\phi)^2$; it is thus maximal for $\theta = \pi/2$ and $\phi = \pm\pi/4$, $\pm3\pi/4$, i.e., when the vector $\mathbf{r} = r(\sin\theta\cos\phi, \sin\theta\sin\phi, \cos\theta)$ is along the diagonals in the XY−plane. The probability cloud proportional to $3z^2 - r^2$ is concentrated on the $\hat{\mathbf{z}}$−axis.] Hence, bonds based on these functions build up molecules with right angles between the bonds, or lattices with orthogonal base vectors.

Ions taking part in five covalent bonds often form **bonds of the sp^3d-type**, in which a single s−function combines with three p ones and a single d state. This combination leads to five functions whose peaks point at the corners of a trigonal bipyramid, along directions forming an angle of $120°$ in the plane (like in graphene) or along the normal to the plane, Fig. 4.19(a). In the PF$_5$ molecule, the configuration $3s^23p^3$ of phosphor changes into $3s^13p^33d^1$, such that each of the five states can contain two electrons. These five wave functions have trigonal symmetry, where the angles between their probability clouds are $120°$, $90°$, or $180°$. These clouds can therefore build e.g., hexagonal lattices.

Likewise, ions with six covalent bonds form them in the sp^3d^2-**type hybridization**, in which two d states are involved. The six functions so created point at the corners of an octahedron, as in Fig. 4.19(b). This type of bond exists, e.g., in the SF$_6$ molecule, where the $3s^23p^4$ configuration of the sulphur atom changes into $3s^13p^33d^2$, such that each of the states can accommodate two electrons and thus can participate in a covalent bond. These states have octahedral symmetry, and form lattices of this symmetry.

(a)   (b)

Fig. 4.19: (a) The five fluorines in the molecule PF_5, with bonds of the sp^3d–type; the phosphor is at the center of the planar triangle. (b) The six fluorines in SF_6, with bonds of the sp^3d^2–type; the sulphur is at the center of the square.

Problem 4.11.

a. *Propose two linear combinations of the functions ψ_z and ψ_{z^2} that can serve as the wave functions of electrons which take part in the covalent bonds normal to the plane, as those illustrated in Figs. 4.19(a) and (b).*

b. *Propose five functions, appropriate for the electrons in the sp^3d bonds as shown in Fig. 4.19(a).*

c. *In addition to the functions found in part (a), show that the wave functions of the four electrons participating in the planar bonds of Fig. 4.19(b) can be built from the linear combinations*

$$\Psi = A(\psi_{200} + \beta_x\psi_x + \beta_y\psi_y + \beta_{xy}\psi_{xy}) \ .$$

Propose numerical values for the coefficients of this linear combination.

Crystals with transition metals. There are many crystals in which the electrons that take part in the bonds are only in the d–shell, e.g., crystals which contain transition metals like chromium, manganese, iron, cobalt, nickel, copper, and zinc; the d–shell is partially occupied in these elements. For example, in the lanthanum-cuprate crystal (see Fig. 2.21) the copper ions have nine electrons in the $3d$–shell. One of those participates in the bonds with the nearest neighbors. Figure 4.20 illustrates the electrons' charge clouds in the Ti and O ions in the perovskite $LaTiO_3$. In this crystal, the Ti ion has a single valence electron in the $3d$–shell, and the figure displays the configuration in which the wave function of this electron is ψ_{xy} [Eq. (4.28)]. The electrons of the oxygens are in the states ψ_x, ψ_y, or ψ_z. In this configuration, nearest-neighbor Ti ions in the XY–plane and the oxygen located in-between them are hybridized. In this way the square structure of the plane can be stabilized. The bonds between the planes along the normal direction are formed by hybridization of the wave functions ψ_{xz} or ψ_{yz}.

Hybridization between ionic and covalent bonds. The ionic bond and the covalent one are the two extremal possibilities for the bonding of two ions in the Heitler-London approximation. The more ubiquitous bond is a combination of the two. For example, the zinc-blende and wurtzite crystals can be considered as ionic crystals. The sulphur ion "acquires" two electrons from the zinc; the occupations

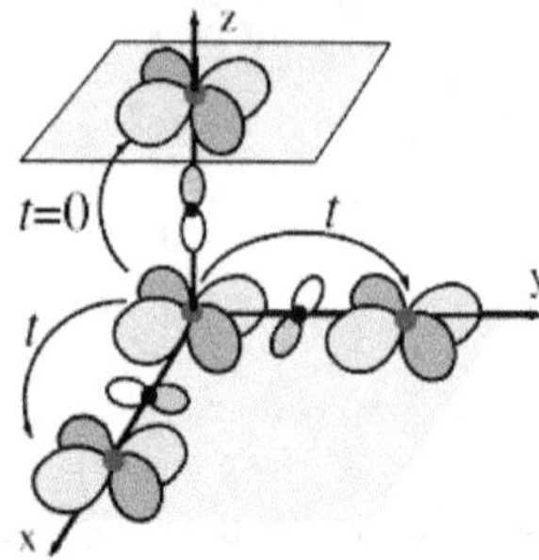

Fig. 4.20: The wave functions of the Ti ions (ψ_{xy}) and the oxygens in the XY-planes of the perovskite crystal LaTiO$_3$. The darker and lighter areas of the charge clouds indicate positive and negative functions, respectively. [A. B. Harris, T. Yildirim, A. Aharony, O. Entin-Wohlman, and I. Ya., Korenblit, *Unusual symmetries in the Kugel-Khomskii Hamiltonian*, Phys. Rev. Lett. **91**, 087206 (2003).]

of the lattice sites then alternate between charges of $+2$ or -2 each. On the other hand, since the electronic configuration of sulphur resembles that of carbon, the former can participate in four covalent bonds. Indeed, the coordination of the sulphur ion in both lattices, where each sulphur is surrounded by four zinc ions at the corners of tetrahedra, indicates that these bonds should be of the sp^3-type. As each zinc ion is surrounded by four sulphur ions, it is plausible that the zinc has also four covalent bonds of the same type. Glancing at the periodic table shows that the basic configuration of zinc includes $3d^{10}4s^2$. Exciting three electrons into the configuration $3d^8 4s^1 4p^3$ enables this ion to form sp^3 bonds. The result is a lattice similar to that of diamond, where all valence electrons are distributed over the covalent bonds between the zinc ions and the sulphur ones. This picture can be backed by, e.g., a variational calculation which includes both the ionic and the covalent descriptions, or by a numerical solution of the Schrödinger equation for the electrons. Such a mixture of ionic and covalent bonds appears also in more complex crystals, like the perovskite or the cuprate structures (Fig. 2.21). In many cases the bonds include also d-states, like in Eq. (4.28), which reinforces the square or tetragonal symmetry of these crystals.

4.4   Molecular crystals–van der Waals bond

Induced fluctuating dipoles. The bonds described so far have all their origin in the electric attraction between pairs of opposite charges. "Secondary" bonds exist in crystals built of neutral units, e.g., **noble gases** or **neutral molecules**. The first part of this section presents calculations that lead to the attraction force of van der Waals. As mentioned in Sec. 4.1, the coupling between two noble atoms results from their mutual polarization; the electrons' clouds in each atom are displaced relatively to the nucleus, such that each atom acquires a fluctuating electric dipole. When the moment on one atom is $\mathbf{p}_1 = Q\mathbf{r}_1$ (Q is the charge of the nucleus and $\mathbf{r}_1$ is the distance in-between the centers of the positive and negative charges) then at a distance R away there is an electric field, $\mathbf{E} \propto [(\mathbf{p}_1 \cdot \hat{\mathbf{R}})\hat{\mathbf{R}} - \mathbf{p}_1]/R^3$ (where $\hat{\mathbf{R}} \equiv \mathbf{R}/R$ is a unit vector along $\mathbf{R}$). This field exerts force on the electrons of the other atom, that causes them to move a little and create a dipole moment, $\mathbf{p}_2 = \alpha\mathbf{E}$, where α is the polarization coefficient. The magnitude of this moment is $p_2 \sim p_1/R^3$. Between the two dipoles operates a force, whose electrostatic energy is $\propto -p_2 p_1/R^3$; inserting here p_2 gives an energy proportional to $-p_1^2/R^6$. As the average of p_1^2 does not vanish (though the average of $\mathbf{p}_1$ does), there arises an energetic gain, proportional to R^{-6}. This gain is the origin of the van de Waals attraction between the two atoms.

Problem 4.12.

Classically (i.e., at temperatures higher than all energies of the problem), the van der Waals interaction can be derived from a thermodynamic average over the interaction between two classical dipoles, that may point along any direction in space. The energy of this dipole-dipole interaction is $U(\mathbf{R}) = (\mathbf{p}_1 \cdot \mathbf{p}_2)/R^3 - 3(\mathbf{p}_1 \cdot \mathbf{R})(\mathbf{p}_2 \cdot \mathbf{R})/R^5$, where R is the distance between the dipoles, and the average is carried out with the Boltzmann weight $\exp[-\beta U]$, where $\beta = 1/(k_B T)$. This means that $\langle U(\mathbf{R})\rangle = \int d\Omega_1 d\Omega_2 U(\mathbf{R}) \exp[-\beta U] / \int d\Omega_1 d\Omega_2 \exp[-\beta U]$, where the integrals are performed over all possible directions of the dipole moments. Show that at large distances $\langle U(\mathbf{R})\rangle \approx -4\beta p_1^2 p_2^2/(3R^6)$.

Quantum calculation of van der Waals force. The phenomenon of induced fluctuating dipole moments may be exemplified by considering two hydrogen atoms. Though hydrogen is not a noble gas, it has a single electron that can create a dipole moment. Figure 4.21 illustrates the coordinates of the two protons and the two electrons. When the nuclei are close enough to one another they create the molecule H_2, discussed in Sec. 4.3. For the present discussion it is assumed that the distance R between the nuclei is far longer than the Bohr radius (which is roughly the size of the atoms) so that there is almost no overlap of the electrons' clouds, and each electron is near its "own" nucleus. Similarly to the discussion in Sec. 4.3, the energy of the two atoms as a function of R is found within the Born-Oppenheimer approximation (one may then view the result as an effective potential between the two nuclei).

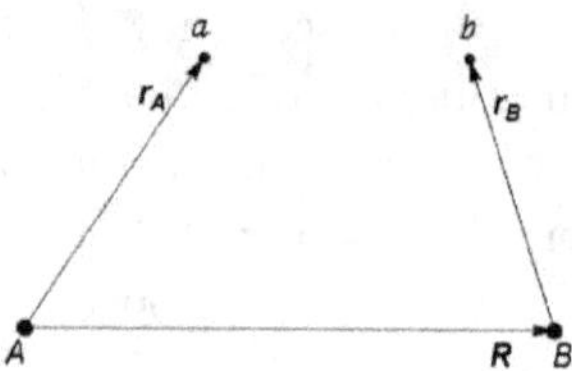

Fig. 4.21: Two hydrogen atoms whose nuclei are located at A and B, and their electrons at a and b.

Given the locations of the two nuclei, the Hamiltonian of the electrons is $\hat{\mathcal{H}} + \hat{\mathcal{H}}'$, where

$$\hat{\mathcal{H}} = -\frac{\hbar^2}{2m}\boldsymbol{\nabla}_A^2 - \frac{e^2}{r_A} - \frac{\hbar^2}{2m}\boldsymbol{\nabla}_B^2 - \frac{e^2}{r_B} ,$$

$$\hat{\mathcal{H}}' = e^2\Big(\frac{1}{R} + \frac{1}{|\mathbf{R}+\mathbf{r}_B-\mathbf{r}_A|} - \frac{1}{|\mathbf{R}+\mathbf{r}_B|} - \frac{1}{|\mathbf{R}-\mathbf{r}_A|}\Big) , \qquad (4.29)$$

(in CGS units; in the SI ones the e^2 terms in the Coulomb interaction are divided by $4\pi\epsilon_0$). The Hamiltonian $\hat{\mathcal{H}}$ describes two isolated hydrogens, whereas $\hat{\mathcal{H}}'$ pertains to the Coulomb interactions between the charges in these atoms. This interaction diminishes as the distance R becomes longer, and hence at large distances it is a small perturbation. When this perturbation is ignored, the electronic wave function in the ground state of $\hat{\mathcal{H}}$ is $\Psi_0(\mathbf{r}_A, \mathbf{r}_B) = \psi_{100}(\mathbf{r}_A)\psi_{100}(\mathbf{r}_B)$, with the energy $2E_1 = -2R_y = -27.2\text{eV}$, where $\psi_{n\ell m}(\mathbf{r})$ is the eigenfunction of the hydrogen atom with the quantum numbers n, ℓ, and m, and energy $E_n = -[13.6/n^2]\text{eV}$.

To account for the effect of $\hat{\mathcal{H}}'$, one takes $\hat{\mathbf{z}}$ to be the direction of $\mathbf{R}$, and expands the electrostatic interaction in $\mathbf{r}_A$ and $\mathbf{r}_B$. To leading order, the result is

$$\hat{\mathcal{H}}' = \frac{e^2}{R^3}\Big(\mathbf{r}_A \cdot \mathbf{r}_B - 3z_A z_B\Big) + \dots . \qquad (4.30)$$

The notation ... indicates higher-order terms, with higher powers of R in the denominator. As er_A and er_B are the dipole moments of the two atoms, Eq. (4.30) is the electric interaction between two dipoles. Within the variational approximation, the ground-state energy is augmented by the expectation value of $\hat{\mathcal{H}}'$ in the trial function. With the trial wave function of the ground state, i.e., $\Psi_0(\mathbf{r}_A, \mathbf{r}_B) = \psi_{100}(\mathbf{r}_A)\psi_{100}(\mathbf{r}_B)$, the second term in Eq. (4.30) contains $\langle\psi_{100}|z_A|\psi_{100}\rangle$. Since the wave function in the ground state of the hydrogen atom is symmetric with respect to reflection, this integral vanishes. For the same reason the first term does not contribute as well. To modify the ground-state energy (away from the one that pertains to the wave function Ψ_0) one adds to the trial wave function excited states of each atom, e.g., $\Psi(\mathbf{r}_A, \mathbf{r}_B) = A(\Psi_0 + c\Psi_1)$, where $\Psi_1 = \psi_{210}(\mathbf{r}_A)\psi_{210}(\mathbf{r}_B)$, with a small complex coefficient c. Expanding the expectation value of the energy in this state in c gives

$$\langle E \rangle = 2E_1 + c\langle\Psi_0|\hat{\mathcal{H}}'|\Psi_1\rangle + c^*\langle\Psi_1|\hat{\mathcal{H}}'|\Psi_0\rangle + 2c^*c(E_2 - E_1) + \dots , \qquad (4.31)$$

where the normalization condition, $\langle\Psi|\Psi\rangle = |A|^2(1+|c|^2)$, implies $|A|^2 \approx 1 - |c|^2$. The notation ... again indicates higher-order terms; the other terms vanish (see problem 4.14). Treating c and c^* as two independent variables, the expectation value is minimized with respect to both. Differentiating Eq. (4.31) then yields

$$2c^*(E_2 - E_1) + \langle\Psi_0|\hat{\mathcal{H}}'|\Psi_1\rangle = 0 \;, \quad 2c(E_2 - E_1) + \langle\Psi_1|\hat{\mathcal{H}}'|\Psi_0\rangle = 0 \;.$$

It follows that c is proportional to R^{-3}, and the correction to the energy is proportional to R^{-6}. The minimal variational energy is

$$\langle E\rangle \approx 2E_1 - \frac{|\langle\Psi_1|\hat{\mathcal{H}}'|\Psi_0\rangle|^2}{2(E_2 - E_1)} = 2E_1 - \frac{4e^4|\langle\psi_{100}(\mathbf{r})|z|\psi_{210}(\mathbf{r})\rangle|^4}{2R^6(E_2 - E_1)} \equiv 2E_1 - \zeta\frac{e^2}{a_{\rm B}}\left(\frac{a_{\rm B}}{R}\right)^6 \;,$$
$$(4.32)$$

where $\zeta = \mathcal{O}(1)$ is a dimensionless constant, determined by the wave functions and the energies of the hydrogen atom. This constant is calculated in problem 4.13, for both the trial wave function proposed above, and for more general trial functions. The correction in Eq. (4.32) is negative, showing that the trial wave function yields a lower energy. This correction is the van der Waals attractive potential energy between the two atoms.

Problem 4.13.
a. *Prove Eq. (4.31).*
b. *Calculate the coefficient ζ of Eq. (4.32).*
c. *Extend the calculation above by adding more excited states to the variational trial wave function of the ground state. Find the value of ζ in this case.*

Problem 4.14.
Prove the expansion in Eq. (4.30). Determine the next correction to the Hamiltonian when this expansion is extended to higher-order terms. What is the order of magnitude of this correction to the energy in Eq. (4.32)?

It is worthwhile to note that the discussion above (for both the leading order and its correction as analyzed in problem 4.13) is valid as long as the ground-state wave function is spherically symmetric. This is true for the noble gases, but not so for many molecules, see Sec. 4.3 and the end of this section.

Pauli's repulsion and the Lennard-Jones potential. The calculation carried out above is valid for R longer than the atomic radius (i.e., $a_{\rm B}$). When the distance R is shorter, the expansion in Eq. (4.30) is not valid anymore; a solution of the complete Hamiltonian, Eq. (4.29), is required. The resulting energy is positive at short distances; this can be interpreted as a repulsive potential. Such a repulsion follows from the Pauli principle, which prevents two atoms to approach one another too closely. Instead of carrying out this calculation, and similar to the discussion of the ionic bond in Sec. 4.2, the effective energy which represents the interaction between the two atoms is augmented by an effective repulsion term, that increases sharply as R decreases. Experiment shows that the detailed form of this term is not essential (see the examples in Sec. 4.2). It is customary to utilize the potential U_{P1} in Eq. (4.8), with $m = 12$. The combination of this repulsion with the van der Waals potential is the **effective Lennard-Jones potential**,

$$U(R) = 4\epsilon\left[\left(\frac{\sigma}{R}\right)^{12} - \left(\frac{\sigma}{R}\right)^{6}\right] . \tag{4.33}$$

This potential is depicted in Fig. 1.1. The parameters ϵ and σ [equivalent to the original coefficients in Eqs. (4.8) and (4.32)] set the energy and the length scales. They are chosen such that at $R = \sigma$ the effective potential vanishes. These parameters can be estimated from measurements of the thermodynamic properties of the gaseous and the liquid phases of the material (e.g., from the van der Waals equation of state). For the noble gases xenon, krypton, argon and neon, the values are $\epsilon = 0.0031, 0.0104, 0.0140, 0.0200$ eV and $\sigma = 2.74, 3.40, 3.65, 3.98$ Å, respectively. These numbers are slightly modified in the solid phase.

The bonding energy of the crystal. Similarly to Eq. (4.1), the potential of the crystal consists of

$$U_{\text{tot}} = \frac{1}{2}\sum_{i\neq j} U(R_{ij}) = 2N\epsilon\left[A_{12}\left(\frac{\sigma}{R_{01}}\right)^{12} - A_6\left(\frac{\sigma}{R_{01}}\right)^{6}\right] , \tag{4.34}$$

where R_{01} is the distance between nearest neighbors, and the coefficients A_n are given by the lattice sums

$$A_n = \sum_{j\neq 0}\frac{1}{p_{0j}^n} , \tag{4.35}$$

with $R_{0j} = p_{0j}R_{01}$. For the main cubic crystals

$$A_n^{\text{FCC}} = 12 + 6/2^{n/2} + 12/2^n + \dots ,$$
$$A_n^{\text{BCC}} = 8 + 6(3/4)^{n/2} + 12(3/8)^{n/2} + \dots ,$$
$$A_n^{\text{SC}} = 6 + 12/2^{n/2} + 8/3^{n/2} + 6/2^n + 24/5^{n/2} + \dots . \tag{4.36}$$

These sums converge relatively fast (check!). For large enough n they are almost entirely dominated by the first term (which results from the interactions of nearest neighbors, for which $p_{0j} = 1$), with a small correction due to the interactions with the next neighbors. In this limit $A_n \to z_{nn} + z_{nnn}/p_{02}^n$, where z_{nn} and z_{nnn} are the coordination numbers (for nearest and next-nearest neighbors), and where $p_{02}R_{01}$ is the distance to the next-nearest neighbors. Extended summations over numerous terms yield

$$A_{12}^{\text{FCC}} = 12.13188 , \qquad A_6^{\text{FCC}} = 14.45392 ,$$
$$A_{12}^{\text{HCP}} = 12.13229 , \qquad A_6^{\text{HCP}} = 14.45489 ,$$
$$A_{12}^{\text{BCC}} = 9.11418 , \qquad A_6^{\text{BCC}} = 12.25330 ,$$
$$A_{12}^{\text{SC}} = 6.20 , \qquad A_6^{\text{SC}} = 8.40 . \tag{4.37}$$

The minimal value of U_{tot} of Eq. (4.34) is reached for $\overline{R}/\sigma = (2A_{12}/A_6)^{1/6}$. For the four lattices mentioned above, $\overline{R}/\sigma \approx 1.090, 1,090, 1.068, 1.067$; the corresponding binding energies are $u/\epsilon = -U_{\text{tot}}(\overline{R})/(N\epsilon) = A_6^2/(2A_{12}) \approx 8.610, 8.611, 8.237, 5.69$.

The lattice structure of noble gases. Since the sums in Eq. (4.35) are dominated by the number of nearest neighbors, it is not surprising that the maximal binding energy is the one of the **FCC** and **HCP** lattices. These two lattices are the densest ones, with 12 nearest neighbors. The contributions from the third neighbor onwards differ in the two lattices, and give rise to a minute, 0.01%, energy gain, in favor of **HCP**. Nonetheless, the noble gases are usually arranged in the **FCC** structure. Apparently, the final structure is sensitive to small corrections in the binding energy, that are ignored in the considerations above. For instance, at finite temperatures the entropy has to be accounted for, in addition to the energy (i.e, contributions due to the number of degenerate states at the same energy, and to the Boltzmann weight-factors of excited states). It seems that the free energy (comprising the energy and the entropy) of the **FCC** structure is the lower one. Another energy ignored in the considerations above is the kinetic energy; this is due to the use of the Born-Oppenheimer approximation, which is based on the assumption that the nuclei are immobile at the lattice sites (where the the potential energy is minimal). The oscillations of the nuclei around these minima are discussed in Chapter 5. These oscillations exist even in the quantum ground state, similar to the zero-point motion of harmonic oscillators. The zero-point energy of the four noble gases alluded to above is relatively small; it is however significant enough to favor the **FCC** structure over the **HCP** one. (The full explanation necessitates an exact calculation.)

This argument breaks down in the case of helium: the small mass of the helium atom causes the energy of the zero-point motion to be relatively large, and leads to a quantum instability of the helium lattice. The **Lindemann criterion** (Sec. 3.10) examines the stability of the solid phase by comparing the average amplitude of the zero-point vibrations with the lattice constant. At low temperatures (and within the approximations used in Sec. 3.10) this criterion predicts that the crystal is unstable when $\langle \mathbf{u}_{\vec{R}}^2 \rangle = 3\hbar/(2M\omega) > (c_L a)^2$, with $c_L = 0.1$. Helium obeys this inequality. Under ambient pressure conditions, helium remains in the liquid phase down to zero temperature. Quantum effects, related to the fact that atoms of helium 4 are bosons (that is, they obey the Bose-Einstein statistics) turn this liquid at low temperatures into a **superfluid**, i.e., a liquid with no viscosity. High pressures (more than 25 atms) stabilize a solid phase of helium at temperatures of about $1°\mathrm{K}$, and then it is arranged in an **HCP** lattice. At much higher pressures (more than 1000 atms) there is a narrow temperature window (below the melting temperature) in which helium is an **FCC** lattice.

The disparity between the **HCP** structure, formed of ABABA... layers of triangular lattice, and the **FCC** one, where the layers take the ABCABC... structure (Fig. 2.24) is still being explored. The paper cited in the caption of Fig. 4.22 points out that the preference of **FCC** structure stems from the growth procedure of close-packed spheres. The authors followed the absorption dynamics of additional spheres on a dense planar layer of (other) spheres. As expected, the spheres sink into the dips in-between the ones in the previous layer. Figure 4.22 displays the pyramids formed by the spheres in the third and fourth layers, for each of the structures. The researchers found that the right pyramid is less stable than the left one, and the absorption of the upmost sphere frequently causes the runaway of spheres below (along the direction of the thick arrow in the figure). This does not happen in the left pyramid (there the arrow indicates the direction where runaway is the hardest). This leads to the preference of the **FCC** structure. It is somewhat surprising that after more than a hundred years of research, there are still open questions of this type. As mentioned, most of the noble gases are arranged on an **FCC** lattice, for which $\overline{R} = 1.09\sigma$ and $u = 8.6\epsilon$. These values result in a rather small binding energy, of the order of 0.03−0.17 eV per unit cell.

Problem 4.15.
*Find the bulk modulus of an **FCC** structure dominated by van der Waals interactions.*

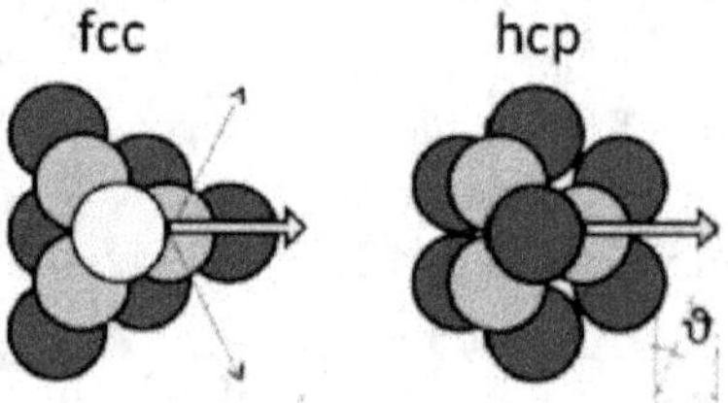

Fig. 4.22: Two pyramids formed during a dynamical growth of close-packed spheres. [S. Heitkam, W. Drenckhan, and J. Fröhlich, *Stability on Close-Packed Sphere Structures*, Phys. Rev. Lett. **108**, 148302 (2012).]

Van der Waals lattices of molecules. The discussion above pertains also for spherically-symmetric molecules. However, most of the molecules in Nature lack this symmetry. For example, the H_2 molecule, Sec. 4.3, is linear, with a charge distribution as displayed in Fig. 4.9 ($|\psi_+|^2$ there). Such linear molecules have a rotation symmetry around the molecule axis, and can be viewed as ellipsoids, as in the mid panel of Fig. 4.1. The ellipsoidal geometry introduces two modifications into the previous discussion. First, the close-packed arrangement of ellipsoids is not necessarily identical to that of spheres; recall that for a spherical symmetry, the van der Waals interaction prefers close-packing of spheres like in the **FCC** lattice. Second, each ellipsoidal molecule has a quadruple moment, proportional to the spatial average of $3(\mathbf{r}\cdot\hat{\mathbf{n}})^2 - r^2$, where $\hat{\mathbf{n}}$ is a unit vector along the molecule axis (see problem 4.16). Consequently, there exists a quadrupole-quadrupole interaction between any pair of molecules. As found in that problem, the energy of this interaction decays like R^{-5}, and depends on the angles between the two axes of the molecules and the line joining them.

Problem 4.16.
a. *Prove that the quadrupole-moment tensor of a molecule symmetric with respect to rotations around $\hat{\mathbf{z}}$ is diagonal, with the components $Q_{xx} = Q_{yy} = -Q_{zz}/2 = \sum_i \langle q_i(r_i^2 - 3z_i^2)/2 \rangle$, where the sum runs over all charges q_i located at $\mathbf{r}_i$ in the molecule, and where the angular brackets denote quantum averaging.*
b. *Show that the average electrical dipole moment of the hydrogen molecule is zero.*

The arrangements of linear molecules, viewed as very narrow ellipsoids, may be quite complicated. Figure 4.23 displays four structures of ellipsoids on the plane. In the first row the ellipsoids' axes are normal to the plane, and in the second they are in the plane or are inclined with respect to it. These structures are all realized in Nature; in three-dimensional patterns the successive planes can be identical and can also be different.

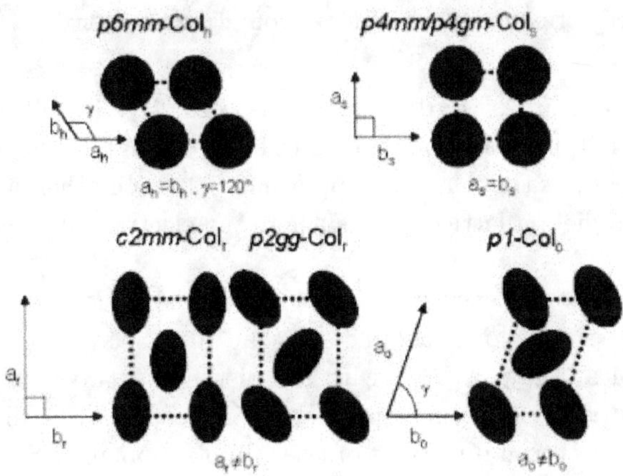

Fig. 4.23: Various planar lattices of ellipsoids. [B. Donio, S. Buathong, I. Bury, and D. Guillon, *Liquid crystalline dendrimers*, Chem. Soc. Rev. **36**, 1495 (2007).]

As for spherical molecules, the van der Waals interaction leads generally to an **FCC** lattice also for ellipsoidal molecules. These molecules, though, are not parallel to each other, and have different inclinations at different lattice points. Figure 4.24 illustrates one of the lattice structures of CO_2 molecules. Were these molecules just points, then this would have been an **FCC** lattice. The packing of the ellipsoids and the interaction between the quadruple moments cause the molecules in each XY−plane to arrange themselves parallel to each other, with their axes inclined relative to the plane. On the other hand, the molecules in the neighboring planes arrange themselves almost normal to each other; the emerging structure is then tetragonal, or even orthorhombic. The F_2 molecules have similar structures, and so do long organic molecules like those that form liquid crystals, or biological molecules, e.g., DNA.

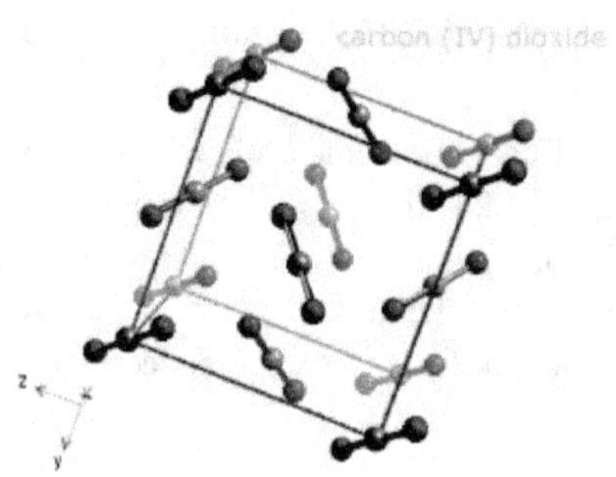

Fig. 4.24: One of the crystal structures of CO_2. Taken from `http://www.webelements.com/compounds/carbon/carbon_dioxide.html`

The crystal structure of **hydrogen molecules**, H_2, is still a topic of active research. Figure 4.25 portrays four possibilities to arrange H_2 in the plane. Under strong enough pressures the ions can approach closely each other, and the molecules then dismantle, to yield a lattice of ions where the electrons are shared by all ions, i.e., a metal.

The situation becomes even more subtle when the molecules possess a **dipole moment**, see the right panel in Fig. 4.1 (and also Sec. 4.3). The interaction between the dipoles decays as R^{-3} with the distance between them [Eq. (4.30)]. The arrangement of the molecules involves their locations, the directions of their axes, and also the vectorial directions of their dipole moments. Since dipoles sums, i.e.,

$$\sum_{<ij>} [\mathbf{p}_i \cdot \mathbf{p}_j - 3(\mathbf{p}_i \cdot \hat{\mathbf{R}}_{ij})(\mathbf{p}_j \cdot \hat{\mathbf{R}}_{ij})]/R_{ij}^3 \, , \tag{4.38}$$

are only conditionally convergent (much like the sums over interactions between point charge, Sec. 4.2), the result depends on the shape of the surfaces of the specimen. The dipolar energy varies with the dipoles' directions. When they are both parallel to the vector joining them, $\mathbf{R}_{ij}$, the energy is $-2p_ip_j/R_{ij}^3$, and hence energy is gained when the dipoles are parallel. As opposed, when they are both normal to $\mathbf{R}_{ij}$, the energy is p_ip_j/R_{ij}^3 and thus energy is gained when the dipoles are antiparallel. As an example, the results of a numerical minimization of the dipolar energy for an arbitrary planar lattice are shown in Fig. 4.26. The arrows there represent the projections of the dipole moments on the plane; the unit cell contains eight dipoles.

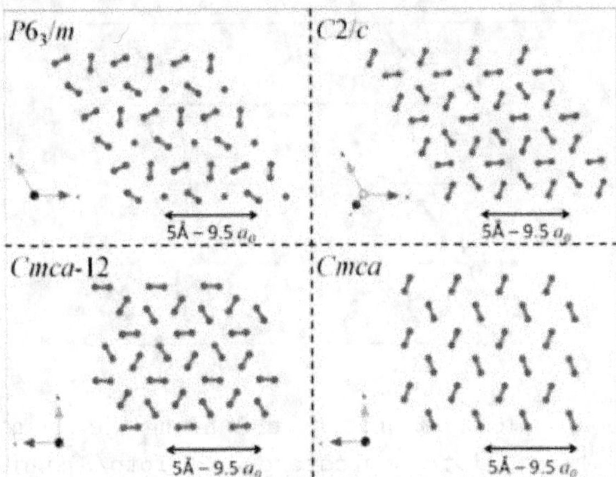

Fig. 4.25: Patterns of the hydrogen molecules in the plane. In the one labeled C2/c the triangular lattices in the figure arrange themselves in space as ABCDA. In all other cases, the spatial ordering is ABA. [V. Labet, P. Gonzales-Morelos, R. Hoffmann, and N. W. Ashcroft, *A fresh look at dense hydrogen under pressure. I. An introduction to the problem, and an index probing equalization of H-H distances*, J. Chem. Phys. **136**, 074501 (2012).]

4.5   The hydrogen bond

The hydrogen bond. Numerous molecules adopt a polar structure in which the positive and the negative charges are concentrated on the opposite sides of the molecule (see, e.g., the right panel in Fig. 4.1 and Sec. 4.3). Certain effects of the interactions among molecules that possess dipole moments are reviewed in Sec. 4.4. A more intricate situation arises when the molecules contain hydrogens. Since hydrogen has just a single valence electron, that is located on a covalent bond within the molecule, its nucleus (i.e., the proton) remains exposed on the outer side of the molecule. This proton can then be at the vicinity of the electrons' clouds on the negatively-charged sides of neighboring molecules, and then the system gains electrostatic energy. The binding energy involved is higher relative to that caused by the van der Waals interaction. This bond is termed **the "hydrogen bond"**; it is the dominant one in many natural materials, e.g., organic or biological molecules that contain many hydrogen atoms.

Hydrogen bonds among hydrogen-fluoride molecules. A simple polar molecule is that of hydrogen fluoride, HF. As seen in Fig. 4.17 and in problem 4.11, the electrons' clouds around the fluorine nucleus are directed towards the vertices of a slightly-distorted tetrahedron (a triangular pyramid built of isosceles triangles);

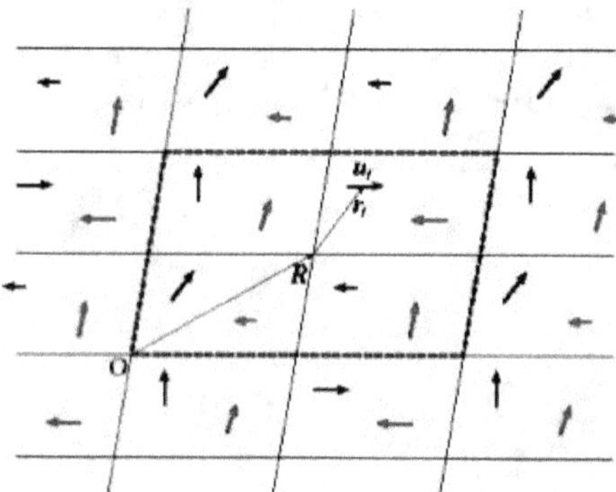

Fig. 4.26: The dipole pattern obtained by minimizing the energy on a planar lattice. Taken from `http://loto.sourceforge.net/loto/`, courtesy of Takeshi Nishimatsu.

three of the base vertices are occupied by identical doubly-occupied states of the fluorine, and the fourth accommodates the hydrogen ion, coupled to the fluorine by a covalent bond. The outer part of the positively-charged hydrogen nucleus is exposed and can then come closer to the negative charges on a neighboring molecule. When two such molecules approach one another, the hydrogen ion of one of them prefers to approach one of the doubly-occupied electrons' clouds of the other; a hydrogen bond between the molecules is then formed. The bond is along the axis that connects the two fluorine nuclei, such that on one side there is the hydrogen ion of one molecule, and the other side accommodates one of the doubly-occupied states, as shown in Fig. 4.27(a). Additional molecules can now approach the two molecules and form more hydrogen bonds with them. In the case of solid HF, the structure created is that of a zig-zag one-dimensional chain, as portrayed schematically in Fig. 4.27(b). On the line that joins the two fluorines, there is a single hydrogen ion. This hydrogen has a covalent bond with one of its nearest-neighbor fluorines, and a hydrogen bond with the negative charge clouds of the other. The angle between the covalent bond and the hydrogen one on the two sides of a fluorine ion in the chain is 116°. Each fluorine is surrounded by hybrid functions of three types, one that participates in the covalent bond, one, with a double occupation, is directed towards the hydrogen bond, and the other two are the remaining doubly-occupied states. A full calculation of the configuration requires an optimization of all angles among these bonds; however, the measured values are quite similar to the angles between other sp^3–bonds, e.g., those discussed in Sec. 4.3. Note that the chain has two possible states: the covalent and the hydrogen bonds between all pairs of fluorines can be interchanged while leaving the energy intact. In other cases the degeneracy is even larger.

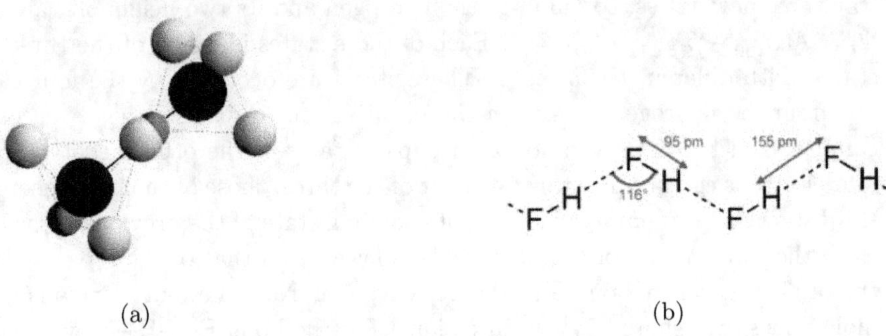

(a)   (b)

Fig. 4.27: (a) A pair of HF molecules; the larger spheres are the fluorine ions, the smaller are the hydrogen ions, and the intermediate-size ones represent the doubly-occupied electronic states of the fluorine. The straight thick line that connects the two fluorines is the hydrogen bond between the hydrogen ion which couples the right fluorine and one of the doubly-occupied states of the left fluorine. (b) A chain of such molecules. The short, solid lines describe covalent bonds, and the long dashed ones represent hydrogen bonds (the distances between the nuclei are given in picometers).

Lattice of hydrogen fluoride. At room temperature, hydrogen fluoride is a gas. When cooled below 190°K, it becomes a solid built of chains, see Fig. 4.27(b). The periodicity along the chain is of two molecules: each second fluorine has the same surrounding. In one of the three-dimensional structures of this material the chains are arranged parallel to each other along one of the axes of an orthorhombic lattice. Nevertheless, the dipolar interaction between the molecules, together with the van der Waals interaction, and the possibility to form more than a single hydrogen bond around each fluorine, give rise to a multitude of phases of this material, which are still being explored.

The hydrogen bond as a covalent bond. The hydrogen bond is particularly effective in binding hydrogen and ions of the upper right hand-side of the periodic table, e.g., nitrogen, oxygen, or fluorine. These ions contain only the first two electronic shells, and hence the hydrogen ion can be rather close to their nuclei. The hydrogen bond is presented above as resulting from the "classical" Coulomb attraction between the charge on the hydrogen and the charge on the negative ion in the neighboring molecule, very similar to the ionic bond. However, as in the case of HF, the hydrogen bond stems from the proximity of the hydrogen to a doubly-occupied state on the negative ion (e.g., F). Without this negative ion, the ψ_{100} state of H would have taken part in a molecular covalent bond with one of the states of the F1 ion, for example, ψ_{111}^{F1}; this joint state, termed "bonding state", is occupied by two electrons. The other state of the two ions (the "anti-bonding" one) is empty. Adding together the doubly-occupied state of the other negative ion, say ψ_{111}^{F2}, enables the formation of linear combinations of three states: the two that take part in the original covalent bond, and the state of the other negative ion. In this way three new states, common to the hydrogen and its two neighbors, are created: $\Psi_i = A(\psi_{100} + \beta_i \psi_{111}^{F1} + \gamma_i \psi_{11\bar{1}}^{F2})$. Each of these states is a sum of the three original states, with different coefficients. These states are occupied by the four electrons, one from the hydrogen, one from the negative ion to which it has been originally coupled, and two from the doubly-occupied states of the other negative ion. The four electrons can occupy two states out of the three, those of the lower energies; the third state remains empty, like the anti-bonding state of the previous case. In many cases the total energy of the new states is lower than that of the separate states in the original covalent bond and the negative ion. Now the four electrons are all in molecular states shared by the three ions. For the bond F−−H−F, where on both sides of H there is the same ion, these states are symmetric and the hydrogen is thus expected to be in the midpoint between the two F's. Figure 4.27, which is based on experiment, locates the hydrogen closer to one of the fluorines; this indicates that the actual bond joins together an ionic bond and a molecular one.

Ice. The paramount examples of hydrogen bonds are crystals of **water**, i.e., the various phases of **ice**. The angle between the two covalent bonds of the water molecule (between the oxygen and the two hydrogens) is 104.5° (Fig. 4.18). Each of the oxygens has in addition two pairs of electrons, which occupy hybrid states that point at the other two corners of the slightly-distorted tetrahedron. As for the molecule of HF, when two water molecules approach one another, the Coulomb attraction between a hydrogen ion (that possesses an extra positive charge, exposed towards the outer side of the molecule) of one of them and the doubly-occupied state of the oxygen of its neighboring molecule, leads to a hydrogen bond between those, as depicted in Fig. 4.28(a). Since each oxygen has two clouds of negative charges, it attracts a hydrogen ion from each of two other water molecules, thus surrounding itself by four other O's, in a tetrahedron-type structure. Two of the tetrahedral directions accommodate covalent bonds with the H's within the same molecule, and the other two support hydrogen bonds with neighboring molecules. Repeating these arrangements in all directions, it is not surprising that the oxygen ions create a diamond-like lattice. This is indeed the **cubic phase of ice**, shown in Fig. 4.28(b). This phase is rather rare, and appears at low enough temperatures. The ubiquitous phase is the **hexagonal phase of ice** (also built of tetrahedrons, like wurtzite) displayed in Fig. 4.28(c) which is obtained when water freezes under normal conditions. Viewed from above, it looks like graphite; however, each of the planes here is distorted, to accommodate the angles between the bonds in the water molecules.

Snowflakes. The hexagonal structure of ice leads to the beautiful patterns of snowflakes, like those displayed in Fig. 4.29. Typically, a snowflake grows out of a tiny ice crystal, that is frozen under ambient conditions. This tiny crystal has the shape of a hexagonal prism, with six faces normal to the base and oriented at 120° relative to each other (see Fig. 3.15). The little crystal is growing on each face, but water molecules that attempt to reach the space in-between the branches which

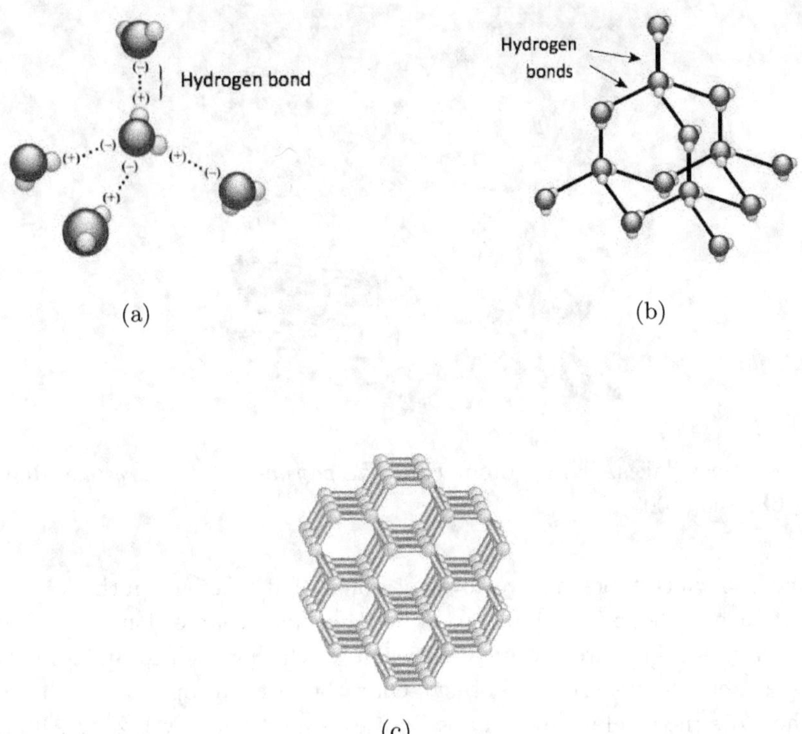

Fig. 4.28: (a) The bonds of a water molecule with its four neighbors. Each oxygen ion is surrounded by four hydrogen ones; it is coupled to two of them by covalent bonds and to the other two by hydrogen bonds. (b) The cubic (diamond-like) structure of ice. (c) The hexagonal structure of ice.

stem form each face collide with the latter, and are unable to enter inside. The result are bifurcations of the branches while the hexagonal symmetry is preserved.

Ice and entropy. The main reason for the numerous phases of ice (sixteen different phases have been identified so far) is related to the large **entropy** of the crystalline structures described above: each oxygen is surrounded by two covalent bonds and two hydrogen ones, and thus the same structure can be achieved by many different configurations in which these bonds are interchanged. The entropy of the ground state, as calculated by Pauling, is determined by the **"ice rules"**, according to which the ice crystallines are formed under two constraints. First, around each oxygen ion there are precisely two covalent and two hydrogen bonds. There are six possibilities to choose two bonds out of four, and for N water molecules there are all in all 6^N such possibilities. Second, on each bond between two oxygens there resides a single hydrogen ion. The lattice contains $2N$ bonds (around each oxygen there are four bonds but each of them is shared by two oxygens). The 6^N possibilities include

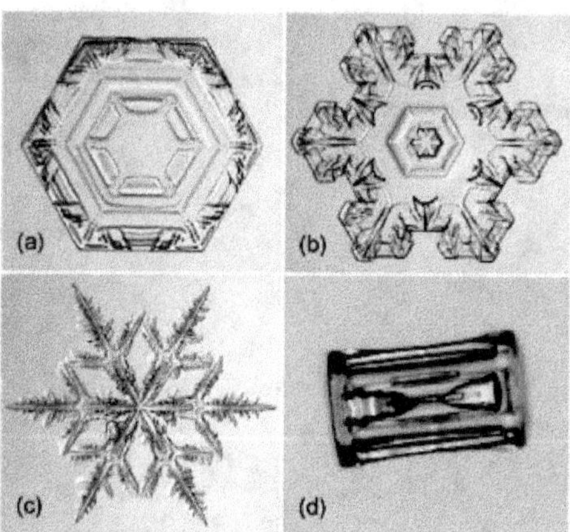

Fig. 4.29: Snowflakes. [K. G. Libbrecht, *The physics of snow crystals*, Rep. Prog. Phys. **68**, 855 (2005).]

also those in which there are two covalent bonds that take part in the coupling (one bond at each of the oxygens), or two doubly-occupied charge clouds (one near each O). In addition, there are two possibilities in which there is a covalent bond on one side and a charge cloud on the other. Only the last two options can be realized, and therefore the total number of possibilities is multiplied by $1/2^{2N}$. All in all the number of configurations which obey the ice rules is $W = 6^N/2^{2N} = (3/2)^N$. By Boltzmann's definition, the entropy is extensive, $S = k_{\mathrm{B}} \ln W = k_{\mathrm{B}} N \ln(3/2)$, i.e., it is proportional to the size of the specimen. Experiment indeed yields numbers close to this estimate.

Problem 4.17.
Consider the formaldehyde molecule, CH_2O.
a. *What is the shape of this molecule? Determine the type of the hybridization of the wave functions that form the covalent bonds.*
b. *Show that these molecules can construct a rectangular-centered lattice, in which each oxygen has two hydrogen bonds with neighboring molecules. Is this the only possible structure?*

4.6 Metallic crystals

Insulators and metals. In the crystals studied so far, the electrons are localized near their "parent" atoms, either in the atom itself, like in the noble gases, or in the ions (in ionic bonds) or on the covalent bonds shared by two atoms. (Exceptions are the π−electrons in graphene, whose wave functions are spread over the entire plane.) As the electrons are not free to move around, they can hardly respond to an external electric field; those materials are hence all **insulators**. As opposed (and as mentioned in Sec. 4.1) a **metal** can be considered as a huge molecule, in which each atom "contributes" its valence electrons; these electrons are described by wave functions spread on all ions in the lattice, as in Eq. (4.27) that corresponds to electrons shared by all carbons in the benzene molecule. The kinetic energy of an electron whose wave function is spread over the entire lattice is reduced considerably. This effect enables the electrons to move around under the influence of an electric field, and the material attains good **electrical conductivity**.

Localized and delocalized states. Figure 4.30 illustrates **localized states** of the electrons on the ions (on the left) and **delocalized states** where the wave function of each electron is spread over the entire lattice (on the right). When each atom contributes a single valence electron (e.g., sodium) then a delocalized state is a linear combination of the N atomic states, and thus there are N independent such states. Pauli principle dictates that each electronic state can accommodate at most two electrons. The energies of these states are gradually increasing, and at low temperatures the states are filled up to an energy called **"the Fermi energy"** (see chapter 6).

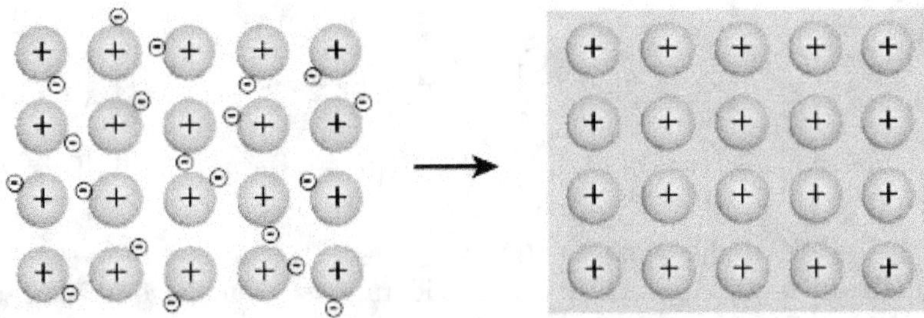

Fig. 4.30: The passage from localized electronic states (each electron is at the vicinity of its parent atom, on the left) to delocalized states (each electron is "spread" over the entire lattice, on the right).

Qualitative discussion: the bonding energy of metals. The detailed calculation of the energies of metals necessitates the knowledge of the quantum properties of electrons in the periodic potential of the ions, see Chapter 6. Here are mentioned general considerations that determine the crystal structure and the binding energy. In crystals of alkali-metals (e.g., sodium or lithium), each atom contributes a single electron. A useful length scale in this context is the radius of a sphere representing the average volume occupied by an electron in the lattice, in this case the volume of the unit cell, $V = 4\pi r_s^3/3$. Calculations yield that the momentum of an electron at the Fermi level (i.e., the momentum of an electron whose energy is the Fermi energy) is of the order of $\hbar/r_s$ and therefore the average kinetic energy of an electron in a metal is $C_1/r_s^2 = \overline{C}_1(\hbar/r_s)^2/(2m)$; m is the electron's mass and $\overline{C}_1$ is a dimensionless constant. (Apart from this coefficient, the general form is dictated by dimensional considerations: when r_s is the only length scale in the problem, the momentum must be proportional to $\hbar/r_s$.) The detailed calculation [Eq. (6.34)] leads to $\overline{C}_1 = (3/5)(9\pi/4)^{2/3}$. This positive energy is augmented by the Coulomb potential energy, of the electrons with the nuclei, and of the electrons with themselves. The calculation of the potential energy resembles that of the Madelung sum, Sec. 4.2, with continuous integrals over the electrons' locations. The answer can be written as $-\alpha e^2/(2r_s)$, where α is a constant that characterizes the lattice. With the assumption that the electrons are distributed uniformly in space, one finds $\alpha \approx 1.792$ for the three dense lattices **FCC**, **HCP**, and **BCC**, and smaller values for **SC** (1.760) and diamond (1.671). The higher values cause most of the metals to be arranged in a dense structure (see Fig. 2.22). The total energy includes also contributions due to the **exchange energy**, which is related to the overlap of the electronic wave function near a certain ion with those of the neighboring ones [see, e.g., Eq. (4.19)]. The exchange terms, which also stem from the Coulomb potentials, contribute the energy $-\alpha' e^2/(2r_s)$, where $\alpha' \approx \alpha/2$ (this can be expected, again from dimensional arguments: recall that the only length scale in the problem is r_s and the only charge scale is e). All in all, the energy is

$$E = \frac{C_1}{r_s^2} - (\alpha + \alpha')\frac{e^2}{2r_s} \ .$$

$$(4.39)$$

The minimal value is reached at $r_s/a_{\mathrm{B}} = 4C_1/[(\alpha + \alpha')e^2 a_{\mathrm{B}}] \approx 1.6$. This is smaller than the typical values of alkali metals, which are between 2 and 6. The error is due to the rather simplistic considerations.

Transition metals. The heuristic considerations given above account solely for the valence electrons in the outer shell of each atom. In reality, there are interactions with the electrons in the inner shells. For example, transition metals' ions like copper, iron, manganese, or titanium, have $d-$electrons available for bonding, and these can indeed create covalent bonds with the neighbors (similar to Fig. 4.19), in addition to the metallic bond between the $s-$electrons. The ions may also form van der Waals bonds with other ions in the lattice, and so further strengthen the total binding energy.

4.7   Magnetism

The atomic dipole moment. Magnetic systems deserve a separate treatise; nonetheless, it is illuminating to present several of the physical principles of magnetism. As mentioned in Sec. 2.10, many materials undergo magnetic phase transitions. This phase transition can occur when each atom or ion possesses an average magnetic dipole moment, that vanishes above the transition temperature and increases as the temperature is reduced towards zero below the transition temperature; the value of this moment at low temperatures is $\boldsymbol{\mu}_i$, where i labels the lattice site. Several examples of magnetic patterns formed by the moments are presented in Secs. 2.10 and 3.13; these arrangements usually reduce the symmetry (as compared to that of the disordered phase above the transition temperature) and increase the unit cell. Neutron-scattering experiments that identify the magnetic structures are described in Sec. 3.13. The magnetic moment of an electron is related to the angular momentum $\mathbf{L}$ and the spin $\mathbf{S}$ (measured in units of $\hbar$), $\boldsymbol{\mu} = -\mu_{\rm B}(\mathbf{L} + g_e\mathbf{S})$ where $g_e \approx 2.00232$ is the **gyromagnetic coefficient** of an electron and $\mu_{\rm B} = e\hbar/(2mc) = 5.7884 \times 10^{-9}$ eV/Gauss is the **Bohr magneton**.

The magnetostatic interaction between two magnetic dipole moments, $[\boldsymbol{\mu}_i \cdot \boldsymbol{\mu}_j - 3(\boldsymbol{\mu}_i \cdot \hat{\mathbf{R}}_i)(\boldsymbol{\mu}_j \cdot \hat{\mathbf{R}}_j)]/R_{ij}^3$ is of the order of $\mu_{\rm B}^2/R^3$, approximately 10^{-4} eV for $R = 2\mathring{A}$ (check!) Except in special situations, this energy is much lower than $k_{\rm B}T_c$, where T_c is the typical transition temperature of a magnetic phase transition; this interaction therefore cannot explain the existence of such transitions. Usually, the transition temperature of a phase transition is of the order of the energy of the interaction that leads to this transition (divided by the Boltzmann constant $k_{\rm B}$). The conclusion is that the main source of magnetism in solids is not the dipole-dipole interaction, but quantum effects.

Singlet and triplet energies. The spin state of two electrons is a singlet or a triplet, in correspondence with their symmetric or the antisymmetric orbital state (with respect to the interchange of the locations of the two electrons, see Sec. 4.3). The energy difference between these two states is the exchange energy J, $J = E_T - E_S$ ($J < 0$ for the hydrogen molecule). The square of the total spin of two electrons can attain only two values, 0 for the singlet and 2 for the triplet. The total energy of the two states can therefore be formally expressed as $E = E_S - J\mathbf{S}_{12}^2/2$, where $\mathbf{S}_{12}$ is the total spin of the two electrons. This energy is E_S for the singlet state and E_T for the triplet one. Recall that the energy difference stems from the Coulomb energies of the electrons in space; the expression in terms of the spins is just formal. Using

$$\mathbf{S}_{12}^2 = \mathbf{S}_1^2 + \mathbf{S}_2^2 + 2\mathbf{S}_1 \cdot \mathbf{S}_2 = \frac{3}{2} + 2\mathbf{S}_1 \cdot \mathbf{S}_2 \ ,$$

gives

$$E = E_0 - J\mathbf{S}_1 \cdot \mathbf{S}_2 \ , \tag{4.40}$$

where $E_0 = E_S - 3J/4 = (E_S + 3E_T)/4$ is the "center of mass" of the two energies, weighed by their ratio of degeneracies, 1:3.

When $J < 0$ (as for the hydrogen molecule) the singlet state is the preferred one. Then the scalar product of the two spins is negative,

$$\mathbf{S}_1 \cdot \mathbf{S}_2 = (\mathbf{S}_{12}^2 - \mathbf{S}_1^2 - \mathbf{S}_2^2)/2 = -3/4 \ ,$$

and they are antiparallel to each other in the ground state. On the other hand, sometimes the energy of the triplet state is the lower one (this is determined by the various integrals on the wave functions and the Coulomb energies, that are more complicated than in the hydrogen molecule). When $J > 0$, the scalar product of the two spins is positive,

$$\mathbf{S}_1 \cdot \mathbf{S}_2 = (\mathbf{S}_{12}^2 - \mathbf{S}_1^2 - \mathbf{S}_2^2)/2 = 1/4 \;,$$

and the two spins are parallel to one another in the ground state. (In fact, the angle between the two spins, though being acute, is not strictly zero due to quantum effects, and the spins are not precisely parallel. Inspection shows that in two of the three triplet states the two electrons' spin projections on the $\hat{\mathbf{z}}$−axis are identical but in the third one the projections have opposite signs.)

Super-exchange. Another process that can contribute to the effective energy between two neighboring spins, of the form (4.40), was proposed by Kramers and significantly developed later by Anderson. Consider two neighboring ions in the lattice, each having a singly-occupied orbital state. When the electron is confined to the vicinity of its parent ion, the kinetic energy is relatively large (due to the uncertainty principle, the momentum is of the order of $\hbar/R$, where R is the size of the confinement region). On the other hand, were this electron to hop over to the neighboring ion, and thereby reduce the kinetic energy, it would have created there an excited state, with a higher potential energy because of the Coulomb repulsion between that electron and the other one, already present there. The Pauli principle allows this configuration only when the spins of the two electrons are opposite to one another. Exploiting Eq. (C.3), with the excited state as one of the n−states in the sum, yields an energy similar to Eq. (4.40), with a negative J. The same result can be derived by the variation method: one begins with the state in which there is an electron on each ion, and adds an excited state in which the two electrons reside on the same ion. The latter, due to the Pauli principle, is a singlet. The correction it causes to the ground-state energy is negative, as for example in Eq. (4.32). It thus follows that it "pays" for the two electrons to be in the singlet state. Equation (4.40), with a negative J, is then valid.

This process is frequently encountered in transition-metal oxides, where the electron tunnels from a metal ion to the oxygen, and then to the neighboring metal ion. In fact, Kramers proposed this mechanism to explain antiferromagnetism in manganese oxide (see Sec. 3.13). This process also contributes to the magnetism of the lanthanum cuprate, as discussed in problem s.2.17. The relative geometry of the various ions' locations can cause the respective interaction constant J to be positive or negative.

Arbitrary angular momentum. When there are numerous electrons in the atom (or in the ion), their magnetic moments are summed over according to the addition rules of quantum angular momenta. In the context of magnetism in solids, the total angular momentum of atom i is frequently presented as a (dimensionless) vector $\mathbf{S}_i$ [the dipole moment is $\boldsymbol{\mu}_i = g\mu_{\mathrm{B}}\mathbf{S}_i$, where g is the (dimensionless) gyromagnetic coefficient (of order unity)]. The magnetic energy may include various combinations of these angular momenta. The treatment of the general case is beyond the scope of this book; note though, that the main features of the magnetic structures presented in Secs. 2.10 or 3.13 may be explained by exploiting the super-exchange energy. The remaining part of this section is devoted to certain simple models, and discusses solely the "classical" ground states of magnetic moments on atoms in the crystal, ignoring quantum effects.

The Heisenberg model. In the Heisenberg model, the energy of the spin degrees of freedom is [analogous to Eq. (4.40)]

$$\hat{\mathcal{H}}_M = -\sum_{<ij>} J_{ij}\mathbf{S}_i \cdot \mathbf{S}_j . \qquad (4.41)$$

The quantum number that corresponds to the vector $\mathbf{S}$ can be an integer or a half-integer, with $|\mathbf{S}|^2 = S(S+1)$. In the ground state the quantum averages of the vectors $\mathbf{S}_i$, and consequently of the magnetic moments $\boldsymbol{\mu}_i = g\mu_B\mathbf{S}_i$, attain definite values which determine the magnetic arrangement of the system.

A complete calculation of the energy spectrum of the Hamiltonian (4.41) is rather complicated. However, a qualitative estimate of the ground-state energy can be derived by treating the vectors $\mathbf{S}_i$ as classical vectors, i.e., as vectors of a definite length and a certain direction in space. The problem is then to find the configurations of these vectors that minimize the energy (4.41). If all coefficients J_{ij} are positive, the energy is minimal when all vectors $\mathbf{S}_i$ are parallel to each other, like in Fig. 2.42B; this arrangement constitutes the **ferromagnetic phase**: the total magnetic moment of the crystal is then the sum over the average magnetic moments on all atoms (ions).

The exchange energy originates from integrals over wave functions in which the locations of the two electrons are interchanged between the two atoms; it thus decays exponentially with the distance in-between the atoms. Hence, it suffices to include in the sum only pairs of close-by neighbors. For example, when only nearest neighbors are included, and when the coefficients J_{ij} are all negative, the energy is minimal when $\mathbf{S}_i$ and $\mathbf{S}_j$ on nearest-neighbor atoms are **antiparallel** to one another. All neighboring moments of a certain moment are antiparallel to it, e.g, as in Fig. 2.42C, or the configuration G in Fig. 3.27. This is the simplest **antiferromagnetic** state. Antiferromagnetic interactions between both nearest neighbors and next-nearest neighbors (i.e., the corresponding J's are negative) give rise to "frustration" (Sec. 2.10), which in turn implies complicated magnetic structures. This point is demonstrated on a simple-cubic lattice, see Fig. 3.27. Let the Hamiltonian be of the form (4.41), with an exchange energy J_1 between nearest neighbors (nn), and an exchange energy J_2 between next-nearest neighbors (nnn). The minimal energy of this classical Hamiltonian is derived as follows. According to Eq. (3.71), the three structures are described by $\mathbf{S}_i = \mathbf{S}_0 \exp[i\mathbf{K} \cdot \mathbf{R}_i]$. This is a plane wave of wave vector $\mathbf{K}$ (various magnetic structures can be described by such waves). The wave vectors of Eq. (3.72) are such that $\exp[2i\mathbf{K} \cdot \mathbf{R}_i] = 1$, and therefore the scalar product $\mathbf{S}_i \cdot \mathbf{S}_j = |\mathbf{S}_0|^2 \exp[i\mathbf{K} \cdot (\mathbf{R}_i - \mathbf{R}_j)]$ depends only on the distance in-between the pair of moments. It follows that all contributions of the neighbors to site i are identical to those of the neighbors of the origin; hence

$$E_M = -\frac{N}{2}\Big(J_1 \sum_{j=nn} \mathbf{S}_0 \cdot \mathbf{S}_j + J_2 \sum_{j=nnn} \mathbf{S}_0 \cdot \mathbf{S}_j\Big)\,, \tag{4.42}$$

where the first sum runs over the six nearest neighbors of the origin, $\mathbf{R}_j = \pm a\hat{\mathbf{x}},\ \pm a\hat{\mathbf{y}},\ \pm a\hat{\mathbf{z}}$, and the second over the twelve next-nearest neighbors, $\mathbf{R}_j = \pm a\hat{\mathbf{x}} \pm a\hat{\mathbf{y}},\ \pm a\hat{\mathbf{y}} \pm a\hat{\mathbf{z}}$, and $\pm a\hat{\mathbf{z}} \pm a\hat{\mathbf{x}}$. The factor of $1/2$ in front of the sums appears since the sum over pairs, $\sum_{<ij>}$ is transformed into two sums, over i and over j; thus each pair appears twice. Using the explicit expressions for the scalar product, and the normalization $|\mathbf{S}_0|^2 = 1$ of the spins' lengths, the energy per unit cell is

$$\begin{aligned}
e(\mathbf{K}) =&\frac{E(\mathbf{K})}{N} = -J_1[\cos(K_x a) + \cos(K_y a) + \cos(K_z a)] \\
&- J_2[\cos(K_x a + K_y a) + \cos(K_y a + K_z a) + \cos(K_z a + K_x a) \\
&+ \cos(K_x a - K_y a) + \cos(K_y a - K_z a) + \cos(K_z a - K_x a)] \\
&= -J_1[\cos(K_x a) + \cos(K_y a) + \cos(K_z a)] \\
&- 2J_2[\cos(K_x a)\cos(K_y a) + \cos(K_y a)\cos(K_z a) + \cos(K_z a)\cos(K_x a)]\,.
\end{aligned}$$
$$\tag{4.43}$$

At zero temperature the entropy contribution is absent, and stable state is reached when the energy is minimal. Indeed, the derivative of the energy with respect to any of the components of $\mathbf{K}$ yields

$$\sin(K_x a)\{J_1 + 2J_2[\cos(K_y a) + \cos(K_z a)]\} = 0\,, \tag{4.44}$$

and two more equations like (4.44) in which the roles of the axes are interchanged. The energy is extremal when $\sin(K_x a) = 0$ or when $\cos(K_y a) + \cos(K_z a) = -J_1/(2J_2)$, and two more similar conditions obtained by interchanging the axes. At the end of the day, Eqs. (4.44) give the following possibilities:

$$\mathbf{K}(\pi\pi\pi)a = (\pi,\pi,\pi),\ \mathbf{K}(0\pi\pi)a = (0,\pi,\pi),\ \mathbf{K}(00\pi)a = (0,0,\pi),\ \mathbf{K}(000)a = (0,0,0),$$
$$\mathbf{K}(\phi\phi\phi)a = (\phi,\phi,\phi),\ \mathbf{K}(\pi\theta'\theta')a = (\pi,\theta'\theta'),\ \mathbf{K}(0\theta\theta)a = (0,\theta,\theta)\,, \tag{4.45}$$

where $\cos\theta = -1 - J_1/(2J_2)$, $\cos\theta' = 1 - J_1/(2J_2)$, and $\cos\phi = -J_1/(4J_2)$. The last three solutions do not comply with $\exp[2i\mathbf{K}\cdot\mathbf{R}_i] = 1$, and therefore are not allowed in the present case. Upon inserting these three solutions into the expression for the energy, Eq. (4.43), they do not yield a minimum of the energy. As for the other options, one notes that together with each solution for which one of the axes differs from the others, there are also solutions with interchanged axes. The corresponding energies are

$$e(000) = -3J_1 - 6J_2,\ e(00\pi) = -J_1 + 2J_2,\ e(0\pi\pi) = J_1 + 2J_2,\ e(\pi\pi\pi) = 3J_1 - 6J_2,$$
$$e(\pi\theta'\theta') = -J_1 + 2J_2 + J_1^2/(2J_2),\ e(0,\theta,\theta) = J_1 + 2J_2 + J_1^2/(2J_2),$$
$$e(\phi\phi\phi) = -3J_1^2/(8J_2)\,. \tag{4.46}$$

Figure 4.31(a) shows these six energies (in units of J_2), each in the range where the corresponding state exists, as functions of J_1/J_2. When both J_1 and J_2 are positive, the minimal energy is that of the ferromagnetic state, with $\mathbf{K}(000) = 0$; the two types of bonds prefer the spins to be parallel. When $J_1 < 0$ but $J_2 > 0$, the minimal energy is that of a simple antiferromagnet (G in Fig. 3.27), with $\mathbf{K}(\pi\pi\pi)a = (\pi, \pi, \pi)$ (the spins on nearest neighbors are antiparallel, and the spins on next-nearest ones are parallel; there is no competition between the magnetic interaction of nearest neighbors and that of next nearest-neighbors). When $J_2 < 0$, the signs of the energies in Fig. 4.31 have to be reversed, and then the ferromagnetic state is realized for $J_1 > -4J_2 > 0$, and the simple antiferromagnetic one for $J_1 < 4J_2 < 0$ (in the two configurations the energy of the nearest-neighbor interaction overcomes that of the next nearest-neighbor ones). On the other hand, in the range $4J_2 < J_1 < 0$ the preferred state is the antiferromagnetic one of type C (right side of Fig. 3.27), with $\mathbf{K}(0\pi\pi)a = (0, \pi, \pi)$, and in the range $0 < J_1 < -4J_2$, the ground state is the antiferromagnetic state of the A type (mid panel in Fig. 3.27), with $\mathbf{K}(00\pi)a = (0, 0, \pi)$. (The axes can be interchanged in both cases.). Figure 4.31(b) depicts the four regions in the $J_1 - J_2$ plane, where each of the aforementioned structures can exist. Such a figure is called a "phase diagram", and each region in it represents a different phase of the magnetic material. Similar considerations enable the identification of more complex structures. for instance, that of the manganese oxide.

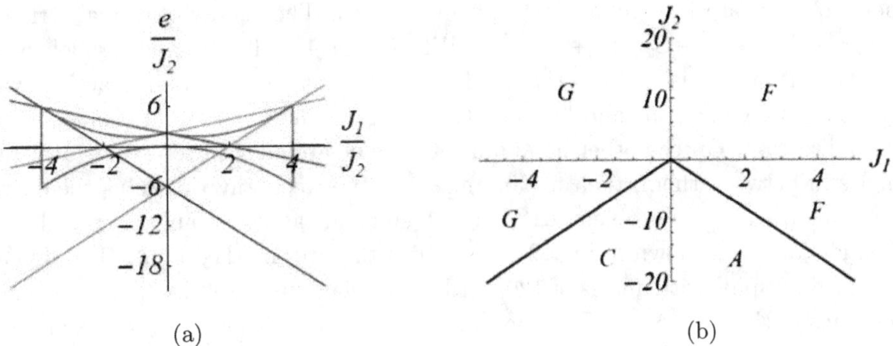

(a) (b)

Fig. 4.31: (a) The energies of Eq. (4.46), in units of J_2, as functions of the ratio J_1/J_2, when $J_2 < 0$. (b) The phase diagram of the minima of Eq. (4.43). The letters F, G, C, and A represent the four options for magnetic structures in the ground state of the Hamiltonian (4.42).

The Ising model. The Heisenberg model does not suffice to determine the **spatial directions of the spin** on each ion. Those are dictated by anisotropic terms in the energy, due e.g., to the coupling between the orbital angular momentum and the spin, that is, the **spin-orbit interaction**. This interaction relates the orientations of the magnetic moments to the spatial lattice vectors and determines special directions called **"easy axes"**, along which the spins point. The dipole-

dipole interaction is also capable of relating the directions of the magnetic moments and the lattice vectors. This interaction is given by Eq. (4.38), when the electric dipoles there are replaced by magnetic dipoles or spins, that is, $\mathbf{p}_i \to \boldsymbol{\mu}_i \to \mathbf{S}_i$. As described in Sec. 4.4, the spin components parallel to the spatial vector connecting them tend to be parallel to one another, while the normal components prefer to be antiparallel to one another). In the extreme situation, the anisotropic interaction that determines the easy axes is very strong, such that the magnetic moments are confined to point along a single axis, e.g., the $\hat{\mathbf{z}}$–axis. In that case, Eq. (4.41) becomes

$$\hat{\mathcal{H}}_M = -\sum_{<ij>} J_{ij} S_i^z S_j^z \ . \tag{4.47}$$

In the simplest situation, each spin may attain one of two values, $S_i^z = \pm 1$. This expression for the energy is called **the Ising model**. The spin patterns implied by this energy differ from those that emerge from the Heisenberg model. For instance, all solutions listed in the second line of Eq. (4.45) cannot be realized in the Ising model.

The triangular lattice and frustration. Another example is related to magnetic moments arranged on a triangular lattice. When Eq. (4.41) is applied to such a lattice, with identical exchange couplings, $J < 0$ for each pair of nearest neighbors (and zero otherwise), it is found that there is no configuration for which the moments of all nearest-neighbor pairs are antiparallel. The energy of a single triangle is $E = -J(\mathbf{S}_1 \cdot \mathbf{S}_2 + \mathbf{S}_2 \cdot \mathbf{S}_3 + \mathbf{S}_3 \cdot \mathbf{S}_1)$. When spins 1 and 2 are antiparallel to one another the energy is $-|J|$, as the third spin is **"frustrated"**: its interactions with the other two cancel one another, and the energy of the tiny system is independent of its direction. On the other hand, once the spins are allowed to assume arbitrary angles in-between them, then for an angle 120° between any two spins the energy is minimal, $-3|J|/2$ (problem 4.18). The triangular lattice is built of triangles; its ground state is as shown in Fig. 2.12, in which the sites marked by A, B, and C are occupied by spins with angles of 120° with each other (see also Fig. 2.43). Choosing the spins in the sites A and B to have an angle of 120° with respect to one another, the spin in the sites C should point out along the third direction. Each of the edges of the triangle ABC determines the spin at the point that closes the triangle. In this way all spins' directions in the ground state of the triangular lattice are determined.

Problem 4.18.
Prove that the energy of a triangle with antiferromagnetic couplings is optimal when the angle between each pair of spins is 120°.

Frustration and degeneracy. The solution of the triangle derived in problem 4.18 is valid for the Heisenberg model, but not for the Ising one. In the latter, all spins point out along the same direction, and therefore the frustration is not removed. The Ising model gives rise to many patterns with the same energy; consequently the ground state possesses a finite entropy (like ice). For instance, in

Fig. 2.43 all spins on one of the sublattices (e.g., the large points) can be chosen to point "upwards", and all those on the other (e.g., the intermediate points) to point "downwards". The couplings between these two sublattices are hence not frustrated, and each of them contribute an energy $-|J|$ per site. Each spin on the third sublattice is coupled antiferromagnetically to three spins that belong to each of the previous sublattices. These triplets are opposite to one another, and thus that spin is frustrated; it can be directed along any of the two possible directions without affecting the energy. This causes a degeneracy of at least $2^{N/3}$ of the ground state energy (where N is the total number of the spins in the system). In fact the degeneracy is much higher, as this energy is shared by many more arrangements. The degeneracy is even higher for other lattices, like the Kagomé one (problem s.2.2) in which each triangle can assume several magnetic patterns.

Problem 4.19.

A one-dimensional magnet is described by the Hamiltonian (4.41), with exchange energy J_1 between nearest neighbors (nn) and J_2 between next-nearest neighbors (nnn). These couplings may be positive or negative. The spins are normalized, i.e., $|\mathbf{S}_i| = 1$ (one may absorb $|\mathbf{S}_i|^2$ in the exchange energies). Determine the magnetic order in the classical ground state, as a function of J_1/J_2. How is this answer modified for the Hamiltonian (4.47)?

4.8 Answers for the problems in the text

Answer 4.1.

a. Within the first square, there is a $+1$ charge at the origin (the center of the square) four $+1$ charges at the four corners (at a distance $\sqrt{2}$ from the center; each one contributes $1/4$) and four -1 charges at the midpoints of the edges (at a distance 1 from the center, each one contributes $1/2$). The total sum is $1+4(1/4)-4(1/2) = 0$. The perimeter of the nth square includes four $+1$ charges at the corners (at a distance $n\sqrt{2}$ from the center, each contributes $1/4$ to the inner side of this square), four $(-1)^n$ charges at the midpoints of the edges, on the axes (at a distance n from the center, each gives $1/2$), and eight $(-1)^{n+m}$ charges at the eight points $(\pm n, \pm m)$, which are located on the edges at a distance $\sqrt{n^2 + m^2}$ from the center, with $1 \le |m| \le n - 1$ (each one contributes $1/2$). Therefore, the total contribution from the inner side of this square is

$$Q_{\mathrm{in}} = (1/4)4 + (1/2)(-1)^n + (1/2)8 \sum_{m=1}^{n-1} (-1)^{n+m} = -1 \ .$$

In a similar fashion, the outer part of the $(n-1)$th square contributes $Q_{\mathrm{out}} = Q_{\mathrm{in}} + 2 = 1$ (the additional factor of 2 is due to the different weights of the four charges at the corners, each giving $3/4$ to the outer side instead of $1/4$ to the inner side). Adding up the two contributions yields that the total charge in between the

$(n-1)$th and the nth squares vanishes. The sum of all charges within the nth square, as obtained from a sum over all the regions, vanishes.

b. The contribution of the inner part of the first square to the Madelung constant α is $S_1 = (1/2)4 - (1/4)4/\sqrt{2} = 1.29289$. The contribution to α from the inner side of the nth square $(n > 1)$ is

$$S_n = -[(1/4)4/\sqrt{2n^2} + (1/2)4(-1)^n/n + (1/2)8 \sum_{m=1}^{n-1}(-1)^{n+m}/\sqrt{n^2 + m^2}] \ .$$

c. The net contribution of the charges in between the two squares to the coefficient α is $S_n + S_{n-1} + 2/\sqrt{2(n-1)^2}$ (the last term is due to the corners of the inner square; the contribution of each of them to the outer side is $3/4$). Hence, $\alpha = 1.29289 + 0.313981 + 0.003648 + 0.002988 + 0.000981 + \ldots = 1.61544$. The sum of the first two terms yields 1.61, and the sum of the first six gives 1.615. The value 1.61554 is obtained from the summation over 26 terms. Plotting the terms as a function of n shows that they decay as n^{-4}.

Answer 4.2.

CsCl consists of two simple cubic sublattices. For a positive ion at the origin, the positive charges are located at distances $\sqrt{i^2 + j^2 + k^2}$ (there are 6 ions at a distance 1, 12 at a distance $\sqrt{2}$, 8 at a distance $\sqrt{3}$, 6 at a distance 2, *etc.*) and the negative ones are situated at distances $\sqrt{(i+.5)^2 + (j+.5)^2 + (k+.5)^2}$ (there are 8 at a distance $\sqrt{3}/2$, 24 at a distance $\sqrt{11}/2$, 24 at a distance $\sqrt{19}/2$, *etc.*) from the origin. The distance between the ions with opposite charges within the unit cell is $R_{01} = \sqrt{3}/2$. (All distances are in units of the lattice constant.) Hence,

$$\alpha_{\text{CsCl}} = -\frac{\sqrt{3}}{2}\left(\frac{6}{1} + \frac{12}{\sqrt{2}} + \frac{8}{\sqrt{3}} + \cdots - \frac{8}{\sqrt{3}/2} - \frac{24}{\sqrt{11}/2} - \cdots\right)$$

$$= 8 + 24\sqrt{3/11} + \ldots - 3\sqrt{3} - 6\sqrt{3/2} - 4 - \ldots \ .$$

Answer 4.3.

Inspection of the various distances in the unit cell yields

$$\phi_A = \frac{2e}{a}\sum_{k=1}^{\infty}\left(\frac{2}{k} - \frac{1}{k-x} - \frac{1}{k-1+x}\right) = \frac{2e}{a}[2\gamma_E + \Psi(x) + \Psi(1-x)] \ ,$$

(the factor of 2 multiplying the sum accounts for the charges on both sides of A); similarly

$$\phi_B = \frac{e}{a}\sum_{k=1}^{\infty}\left(\frac{2}{k-x} - \frac{1}{k} - \frac{1}{k-2x} + \frac{2}{k-1+x} - \frac{2}{k-1+2x} - \frac{1}{k}\right)$$

$$= \frac{e}{a}[-2\gamma_E - 2\Psi(1-x) - 2\Psi(x) + \Psi(2x) + \Psi(1-2x)] \ .$$

The Coulomb energy per unit cell is $(U_C/N) = (q_A\phi_A + 2q_B\phi_B)/2 = e(\phi_A - \phi_B)$. The binding energy, in units of e^2/a, $u = -(U_C/N)/(e^2/a)$ is given as a function of

x in Fig. 4.32. This energy diverges near $x = 0$, because of the attraction between the ions A and B; it changes sign around $x = 1/2$, where the two negative ions are close to one another. The crystal is stable for $u > 0$, i.e., for $x < 0.45$. In reality, the lattice constants a, $R_{AB} = R_{BA} = ax$, and $R_{BB} = a(1 - 2x)$ are determined by the repulsion among the ions at short distances.

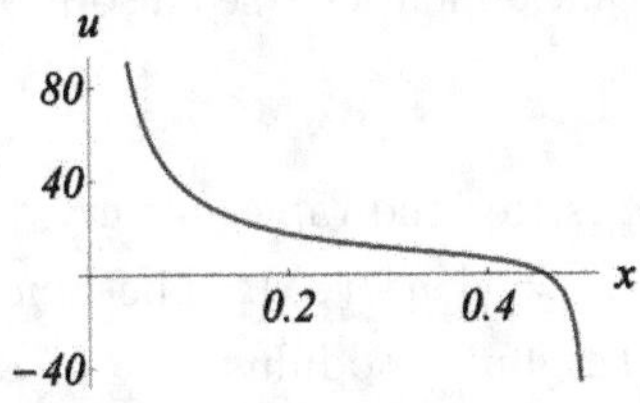

Fig. 4.32

Answer 4.4.

a. Placing the bigger ion, of radius $r_>$, at the origin, then one of the nearest neighbors (with the opposite charge and of radius $r_<$) is located at $a(1/4, 1/4, 1/4)$. The distance between the centers when the two spheres touch one another is a quarter of the cube diagonal, $R_{10} = r_< + r_> = a\sqrt{3}/4$. Each face of the unit cell contains one of the bigger ions, again at its center (this is an **FCC** lattice) and therefore the diagonal on the face is such that $a\sqrt{2} > 4r_>$. It follows that $r_>/r_< < 2 + \sqrt{6} \approx 4.45$.

b. When $2.414 < r_>/r_<$ and the structure is that of the table-salt crystal, the two ions of opposite charges cannot touch one another any more. Instead, the bigger negative ions are contiguous, and then the distance between them is half of the lattice constant, $R_{10} = a/2 = \sqrt{2}r_>$. Consequently, the Coulomb energy of this structure is $-N\alpha_{\mathrm{NaCl}}q^2/(\sqrt{2}r_>)$. The Coulomb energy of the zinc-blende structure is $-N\alpha_{\mathrm{ZB}}q^2/(r_> + r_<)$. This latter energy "wins" when

$$\frac{r_>}{r_<} > \frac{1}{\sqrt{2}(\alpha_{\mathrm{ZB}}/\alpha_{\mathrm{NaCl}}) - 1} \approx 3.047 \ .$$

The table-salt structure is realized in the range $2.414 < r_>/r_< < 3.047$, unless other considerations are involved. Indeed, the MgS crystal has the same structure as table salt, albeit the ionic radii there obey $r_>/r_< \approx 2.83 > 2.414$.

Answer 4.5.

a. The distance between the ions is a quarter of the diagonal, $R_{10} = a\sqrt{3}/4 \approx 2.35\text{Å}$.

b. In the close-packing arrangement, the diagonal of the face equals four radii of the larger ion, hence $r_> = a\sqrt{2}/4 \approx 1.92\text{Å}$. The sum of the two radii is the distance between nearest neighbors, hence $r_< = R_{10} - r_> \approx 0.43\text{Å}$. The ratio of the radii is 4.47, slightly higher than the limiting value $r_>/r_< = 2 + \sqrt{6} \approx 4.45$.

c. As in the zinc-blende structure, in the one of wurtzite the zinc ions occupy the cavity at the center of the tetrahedron formed by the sulphur ions. The edge of the tetrahedron is the lattice constant a of the **HCP** lattice; therefore, $a = R_{10}\sqrt{8/3} \approx 1.633 R_{10} \approx 3.84\text{Å}$. The lattice is a perfect **HCP** one; thus $c = a\sqrt{8/3} = 8R_{10}/3 \approx 6.27\text{Å}$.

d. The cubic unit cell contains four base units, and therefore the specific volume per unit cell is $a^2 c\sqrt{3}/4 \approx 40(\text{Å})^3$. Both structures are close packed; it is hence not surprising that their densities are identical (see also problem 2.15).

Answer 4.6.

The distance between the ions (for the table-salt crystal) is given by the solution of the equation $\overline{R}^2 \exp[-\overline{R}/\rho] = \rho\alpha q^2/(z\lambda)$. The binding energy per unit cell is $u = \alpha q^2(1 - \rho/\overline{R})/\overline{R}$, and the bulk modulus is $B = \alpha q^2/(18\overline{R}^4)[(\overline{R}/\rho) - 2]$. As mentioned, the measured binding energy is about $u \approx \alpha q^2/\overline{R}$, implying $\overline{R} \gg \rho$.

Answer 4.7.

The overlap integral is

$$S = \langle \psi_A | \psi_B \rangle = \frac{1}{\pi a_B^3} \int d\mathbf{r}_A e^{-r_A/a_B} e^{-r_B/a_B} = \frac{1}{\pi} \int d^3x\, e^{-x-y} \ .$$

In spherical coordinates, where the $\hat{\mathbf{z}}$–direction is along molecule axis, $d^3x = x^2 \sin\theta dx d\theta d\varphi$. The integrand is independent of the rotation angle φ around the axis, and hence its integration yields 2π. The substitution $y^2 = X^2 + x^2 - 2Xx\cos\theta$ yields $2ydy = 2Xx\sin\theta d\theta$, and consequently

$$S = \frac{2}{X} \int_0^\infty dx\, xe^{-x} \int_{|X-x|}^{X+x} dy\, ye^{-y}$$

$$= \frac{2}{X}\left(\int_0^X dx\, xe^{-x} \int_{X-x}^{X+x} dy\, ye^{-y} + \int_X^\infty dx\, xe^{-x} \int_{x-X}^{x+X} dy\, ye^{-y} \right) ,$$

where the limits on the y–integrals are derived from those on θ, $0 \le \theta \le \pi$. Making use of the (indefinite) integrals

$$\int dx\, xe^{-2x} = -\frac{1+2x}{4}e^{-2x} , \qquad \int dx\, x^2 e^{-2x} = -\frac{1+2x+2x^2}{4}e^{-2x} ,$$

yields Eq. (4.16). In a similar fashion,

$$\langle \psi_A | \frac{e^2}{r_B} | \psi_A \rangle = \frac{e^2}{\pi a_B^3 r_B} \int d\mathbf{r}_A e^{-2r_A/a_B} = \frac{2e^2}{X a_B} \int_0^\infty dx\, xe^{-2x} \int_{|x-X|}^{x+X} dy$$

$$= \frac{e^2}{\pi a_B^3 r_B} \int_0^\infty dx\, xe^{-2x}(x + X - |X - x|) = \frac{e^2}{R}[1 - (1 + X)e^{-2X}] ,$$

which yields Eq. (4.18). Finally,

$$\langle \psi_A | \frac{e^2}{r_B} | \psi_B \rangle = \frac{e^2}{\pi a_B^3 r_B} \int d\mathbf{r}_A e^{-r_A/a_B} e^{-r_B/a_B} = \frac{2e^2}{X a_B} \int_0^\infty dx\, xe^{-x} \int_{|x-X|}^{x+X} dy\, e^{-y} ,$$

and a similar algebra leads to Eq. (4.19).

Answer 4.8.

The variational average of the energy is

$$\langle E_\pm \rangle = (1 + \beta_\pm^2 \pm 2\beta_\pm S)^{-1}(\langle \psi_A | \hat{\mathcal{H}} | \psi_A \rangle + \beta_\pm^2 \langle \psi_B | \hat{\mathcal{H}} | \psi_B \rangle \pm 2\beta_\pm \langle \psi_A | \hat{\mathcal{H}} | \psi_B \rangle) \ .$$

The derivatives with respect to $\beta_\pm$ vanish when $a\beta_\pm^2 \pm b\beta_\pm + c = 0$, with

$$a = S\langle \psi_B | \hat{\mathcal{H}} | \psi_B \rangle - \langle \psi_A | \hat{\mathcal{H}} | \psi_B \rangle \ , \quad b = \langle \psi_B | \hat{\mathcal{H}} | \psi_B \rangle - \langle \psi_A | \hat{\mathcal{H}} | \psi_A \rangle \ ,$$

$$c = \langle \psi_A | \hat{\mathcal{H}} | \psi_B \rangle - S\langle \psi_A | \hat{\mathcal{H}} | \psi_A \rangle \ .$$

It is straightforward to verify that in the symmetric case, where $\langle \psi_B | \hat{\mathcal{H}} | \psi_B \rangle = \langle \psi_A | \hat{\mathcal{H}} | \psi_A \rangle$, one finds $\beta_+ = \beta_- = \pm 1$, reproducing the bonding and the anti bonding states discussed in the main text. In the general case one has to choose the solution $\beta_\pm = (\mp b + \sqrt{b^2 - 4ac})/(2a)$ (otherwise the roots are such that $\beta_+ = -\beta_-$). The chosen roots do fulfil the orthonormality condition, $1 - \beta_+\beta_- + S(\beta_+ - \beta_-) = 0$.

Answer 4.9.

a. Using Eq. (4.24) gives

$$\psi(\mathbf{r}) = \sqrt{\frac{3}{4\pi}} R_{21}(r)\frac{\mathbf{r}}{r} = f(r)\hat{\mathbf{r}} \ ,$$

where $\hat{\mathbf{r}}$ is a unit vector along the $\mathbf{r}$–direction. It follows that

$$\Psi_1(\mathbf{r}) = A[\psi_{200}(r) + \beta f(r)\hat{\mathbf{n}}_1 \cdot \hat{\mathbf{r}}] \ .$$

The second term is proportional to the cosine of the angle between $\mathbf{r}$ and $\hat{\mathbf{n}}_1$, and hence (assuming that β is positive) is maximal for $\mathbf{r} \parallel \hat{\mathbf{n}}_1$. The two terms are positive for $r < 2a_B/z$, and thus their sum is maximal along the vector $\hat{\mathbf{n}}_1$. When $r > 2a_B/z$, the two terms have opposite signs; in that case the square of the sum is maximal along $-\hat{\mathbf{n}}_1$. In both configurations the sum is maximal on the axis through which passes the vector $\hat{\mathbf{n}}_1$.

b. The quantum scalar product of the two states is

$$\langle \Psi_1 | \Psi_2 \rangle = |A|^2 \langle \psi_{200} + \beta\hat{\mathbf{n}}_1 \cdot \boldsymbol{\Psi} | \psi_{200} + \beta\hat{\mathbf{n}}_2 \cdot \boldsymbol{\Psi} \rangle = |A|^2 (1 + |\beta|^2 \sum_{ij} n_{1i} n_{2j} \langle \psi_i | \psi_j \rangle) \ ,$$

where $i, j = x, y, z$ indicate the cartesian components of the vectors, and $\langle \psi_{200} | \psi_i \rangle = 0$. Using $\langle \psi_i | \psi_j \rangle = \delta_{ij}$ yields the desired result.

c. For the hybrid states to be orthonormal, the scalar product should be $\langle \Psi_i | \Psi_j \rangle = \delta_{ij}$; consequently, for $i \neq j \ |\beta|^2 = -1/\hat{\mathbf{n}}_i \cdot \hat{\mathbf{n}}_j = -1/\cos\alpha$, and for $i = j \ |A|^2 = 1/(1 + |\beta|^2) = \cos\alpha/(\cos\alpha - 1)$.

d. As there is no reason to distinguish between the three bonds, symmetry implies that the angles between pairs of unit vectors are identical. This is also the configuration in which the clouds of negative charge are as far away from each other as possible. Since the sum of the three angles is $360°$, each of them is $120°$. Such a triplet of unit vectors is, for instance, $\hat{\mathbf{n}}_1 = \hat{\mathbf{x}}$, $\hat{\mathbf{n}}_{2,3} = -\hat{\mathbf{x}}/2 \pm \sqrt{3}\hat{\mathbf{y}}/2$. Inserting these into part (c) gives $|\beta|^2 = -1/\cos 120° = 2$ and $|A|^2 = \cos 120°/(\cos 120° - 1) = 1/3$, from which follows Eq. (4.25).

e. The sp^3 hybridization allows for four bonds; this necessitates four unit vectors, with an identical angle in-between each pair of them. For example,

$$\hat{\mathbf{n}}_{111} = \sqrt{\frac{1}{3}}(\hat{\mathbf{x}} + \hat{\mathbf{y}} + \hat{\mathbf{z}}) \,, \quad \hat{\mathbf{n}}_{1\bar{1}\bar{1}} = \sqrt{\frac{1}{3}}(\hat{\mathbf{x}} - \hat{\mathbf{y}} - \hat{\mathbf{z}}) \,,$$

$$\hat{\mathbf{n}}_{\bar{1}1\bar{1}} = \sqrt{\frac{1}{3}}(-\hat{\mathbf{x}} + \hat{\mathbf{y}} - \hat{\mathbf{z}}) \,, \quad \hat{\mathbf{n}}_{\bar{1}\bar{1}1} = \sqrt{\frac{1}{3}}(-\hat{\mathbf{x}} - \hat{\mathbf{y}} + \hat{\mathbf{z}}) \,.$$

The angle between any two unit vectors is given by $\cos\alpha = \hat{\mathbf{n}}_i \cdot \hat{\mathbf{n}}_j = -1/3$, leading to $|A|^2 = 1/4$ and $|\beta|^2 = 3$. The coefficients in Eq. (4.26) are chosen to be real; it follows that $A = 1/2$ and $\beta = \sqrt{3}$.

f. In the general case $\Psi_i = A_i[\psi_{200} + \beta_i\hat{\mathbf{n}}_i \cdot \boldsymbol{\psi}]$, with a unit vector $\hat{\mathbf{n}}_1$ along the bond with the fluorine, and unit vectors $\hat{\mathbf{n}}_2$ and $\hat{\mathbf{n}}_3$ along the bonds with the hydrogens. Hence $\langle\Psi_i|\Psi_j\rangle = A_i^* A_j(1 + \beta_i^*\beta_j\hat{\mathbf{n}}_i \cdot \hat{\mathbf{n}}_j)$. The bonds with the two hydrogens are identical; hence the angles between them and the bond with the fluorine are equal as well, $\beta_1^*\beta_2 = \beta_1^*\beta_3 = -1/\cos\alpha$, and $\beta_2^*\beta_3 = -1/\cos\gamma$. It follows (from the first relation) that $\beta_2 = \beta_3$. Choosing the coefficients to be real, then $\beta_2 = \beta_3 = \sqrt{-1/\cos\gamma}$ and $\beta_1 = -\sqrt{-\cos\gamma/\cos^2\alpha} = -\sqrt{-\cos\gamma/(\cos^2(\gamma/2))}$. The other coefficients are obtained from $|A_i|^2 = 1/(1 + |\beta_i|^2)$.

g. Figure 4.33 illustrates the triangular pyramid whose base is formed by the three hydrogens (H_1, H_2, H_3), and the carbon C_1 is at its apex. The height of the pyramid is the line YC_1, the continuation of the straight line connecting the two carbons, C_1C_2. The angle between YC_1 and C_1H_3 is $180° - \alpha$. Denoting the length of C_1H_3 by d, then the length of YH_3 is $d\sin\alpha$. Denoting the distance between two hydrogens by a, the height of the base triangle, XH_3, is $a\sqrt{3}/2$. The point Y is the locus of the heights of the isosceles triangle of the base, and therefore it is located at the third of the height. It follows that $d\sin\alpha = a\sqrt{3}/3$, i.e., $a = d\sqrt{3}\sin\alpha$. The angle can be derived from the triangle XC_1H_1: $\sin(\gamma/2) = a/(2d) = (\sqrt{3}/2)\sin\alpha$. Using the value given in the problem, $\gamma \approx 107.7°$. In general, the coefficients of the hybrid functions are such that $\beta_1^*\beta_2 = \beta_1^*\beta_3 = \beta_1^*\beta_4 = -1/\cos\alpha$, where the index 1 refers to the CC bond, and the other three indices refer to the CH bonds. In addition, $\beta_2^*\beta_3 = \beta_2^*\beta_4 = \beta_3^*\beta_4 = -1/\cos\gamma$. Hence, $|\beta_2|^2 = |\beta_3|^2 = |\beta_4|^2 = -1/\cos\gamma$, and $\beta_1 = \sqrt{-\cos\gamma/\cos^2\alpha}$. In the case of ethane, $\beta_1 \approx 1.53$ and $\beta_2 \approx 1.81$.

Answer 4.10.

Nitrogen has five electrons in the second shell, and therefore two of them have to be in the same orbital state. The other three are available for forming covalent bonds (Fig. 4.17). As the nitrogen molecule comprises two atoms, three electrons of each atom form a triple bond like in acetylene. One of those is a $\sigma-$bond, of the sp type, like in Fig. 4.12, and the other two are $\pi-$bonds, with the two $p-$states perpendicular to this bond (Fig. 4.16). The remaining two electrons in each nitrogen are in the hybrid state concentrated on the line joining the two nuclei, on the two far sides of the molecule. **Oxygen** has six electrons in the second shell,

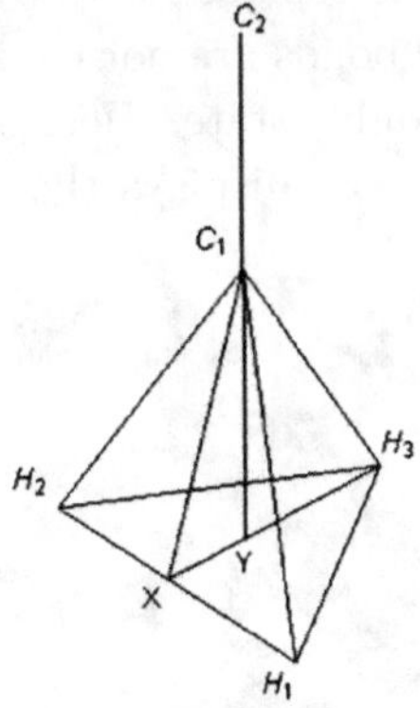

Fig. 4.33

four of which are in doubly-occupied states and the other two bond the two atoms (Fig. 4.17). Therefore the oxygen molecule is planar, with hybridization of the sp type, in each atom. The double connection between the two oxygens consists of a $\sigma-$bond and a $\pi-$bonds, like in ethene (Fig. 4.16), but each hydrogen there is replaced here by an oxygen state accommodating two electrons. Six of the **fluorine** electrons are in three orbital states, which have to be as far as possible from each other, and the seventh electron occupies the state forming the bond (Fig. 4.17). It follows that four different directions (three for the doubly-occupied states and one for the bond) are required. The hybridization on each fluorine atom must be of the sp^3-type, and the geometry of the F_2 molecule is similar to that of methane (Fig. 4.15), where each of the six hydrogens there is replaced here by a doubly-occupied state.

The diatomic molecules analyzed above are all symmetric under the interchange of the two nuclei, and therefore lack an electric dipole moment. The next two molecules do have such a moment.

In the **FH** molecule there are four directions around the fluorine, one for the covalent bond and three for the doubly-occupied states (the right panel of Fig. 4.17). As shown there, the three clouds of the doubly-occupied states terminate on an equilateral triangle which is a base of a triangular pyramid of equilateral edges, with the hydrogen at its apex. As these clouds are oriented towards the right in the figure, i.e., in the opposite direction to that of the bond with the hydrogen (which is oriented to the left) there is a considerable surplus of negative charges on the right side, which causes this molecule to have an electric dipole moment parallel to its axis (the line connecting the nuclei). In the ammonia molecule, **NH$_3$**, there are three different directions of the bonds with the hydrogens, and another direction is needed for the doubly-occupied state of the nitrogen. Therefore, the suitable hybridization is that of the sp^3-type, like in ethane (Fig. 4.15). The difference is that one of the hydrogens there is replaced here by a doubly-occupied state, and therefore the angles between the bonds are not equal to the angle between each of them and the direction of the double state. Here again there is a dipole moment along the molecule axis. Figure 4.34 displays the electrons' probability clouds of this molecule.

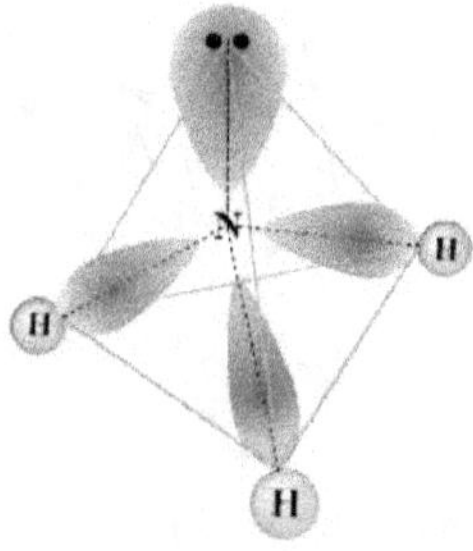

Fig. 4.34

Answer 4.11.

a. The probability clouds of the two functions $(\psi_z \pm \psi_{z^2})/\sqrt{2}$ point along the two directions of the $\hat{z}$−axis.

b. Equation (4.24) contains three wave functions that are maximal at the vertices of an equilateral triangle, as required in Fig. 4.19(a).

c. First, by symmetry considerations $|\beta_x| = |\beta_y|$. Second, the solution to problem 4.9 shows that the probability clouds of the functions $\psi_{200} \pm \psi_x \pm \psi_y$ point along the diagonals of the planar square, $[\pm 1, \pm 1]/\sqrt{2}$. These four functions, though, are not orthonormal to each other (in the quantum-mechanical sense). To make them orthonormal, one begins with a base of four functions, and adds to it the function ψ_{xy}, which yields maximal probability at the directions of all these diagonals. The four combinations

$$\Psi_{110} = (\psi_{200} + \psi_x + \psi_y + \psi_{xy})/2 \, ,$$
$$\Psi_{1\bar{1}0} = (\psi_{200} + \psi_x - \psi_y - \psi_{xy})/2 \, ,$$
$$\Psi_{\bar{1}\bar{1}0} = (\psi_{200} - \psi_x - \psi_y + \psi_{xy})/2 \, ,$$
$$\Psi_{\bar{1}10} = (\psi_{200} - \psi_x + \psi_y - \psi_{xy})/2 \, ,$$

are orthonormal to each other and yield maximal probabilities along the diagonals. They are also normal to the functions of part (a); thus these six functions together can form a base for bonds of the sp^3d^2−type.

Answer 4.12.

The average of U is given by a logarithmic derivative,

$$\langle U(\mathbf{R})\rangle = -\partial\{\ln \int d\Omega_1 d\Omega_2 \exp[-\beta U]\}/\partial\beta \ .$$

Taking $\mathbf{R}$ to lie along the $\hat{\mathbf{z}}$–axis leads to $U(\mathbf{R}) = (\mathbf{p}_1 \cdot \mathbf{p}_2 - 3p_{1z}p_{2z})/R^3$. In a Taylor expansion, $\exp[-\beta U] \approx 1 - \beta U + \beta^2 U^2/2 + \dots$. As $\int d\Omega_1 p_{1n} = 0$ for any cartesian component n (the positive values cancel the negative ones), $\int d\Omega_1 p_{1n}p_{1m} = \mathbf{p}_1^2 \delta_{nm}/3$ (since $p_{1x}^2 + p_{1y}^2 + p_{1z}^2 = \mathbf{p}^2$, and the averages of all terms are identical) one finds that

$$\int U = 0 \ ,$$

$$\int U^2 = \int [(\mathbf{p}_1 \cdot \mathbf{p}_2)^2 - 6(\mathbf{p}_1 \cdot \mathbf{p}_2)p_{1z}p_{2z} + 9p_{1z}^2 p_{2z}^2]/R^6 = 4p_1^2 p_2^2/(3R^6) \ ,$$

(the integrals are carried out over the spatial angles). It follows from the logarithmic derivative that $\langle U(\mathbf{R})\rangle = -4\beta p_1^2 p_2^2/(3R^6)$.

Answer 4.13.

a. A complete calculation of Eq. (4.31) yields

$$\langle E\rangle = \langle \Psi|\hat{\mathcal{H}} + \hat{\mathcal{H}}'|\Psi\rangle = |A|^2 \langle \Psi_0 + c\Psi_1|\hat{\mathcal{H}} + \hat{\mathcal{H}}'||\Psi_0 + c\Psi_1\rangle$$

$$= |A|^2 \Big(\langle \Psi_0|\hat{\mathcal{H}} + \hat{\mathcal{H}}'|\Psi_0\rangle + c^*\langle \Psi_1|\hat{\mathcal{H}} + \hat{\mathcal{H}}'|\Psi_0\rangle + c\langle \Psi_0|\hat{\mathcal{H}} + \hat{\mathcal{H}}'|\Psi_1\rangle + |c|^2 \langle \Psi_1|\hat{\mathcal{H}} + \hat{\mathcal{H}}'|\Psi_1\rangle\Big) \ .$$

Symmetry considerations imply that

$$\langle \Psi_0|\hat{\mathcal{H}}'|\Psi_0\rangle = \langle \Psi_1|\hat{\mathcal{H}}'|\Psi_1\rangle = \langle \Psi_0|\hat{\mathcal{H}}|\Psi_1\rangle = \langle \Psi_1|\hat{\mathcal{H}}|\Psi_0\rangle = 0 \ ,$$

since the integrands of all these expressions are antisymmetric with respect to reflection. In addition,

$$\langle \Psi_0|\hat{\mathcal{H}}|\Psi_0\rangle = 2E_1 \ , \quad \langle \Psi_1|\hat{\mathcal{H}}|\Psi_1\rangle = 2E_2 \ ,$$

since Ψ_0 and Ψ_1 are eigenfunctions of $\hat{\mathcal{H}}$. Taylor expanding the coefficient, $|A|^2 \approx 1 - |c|^2$, and discarding terms of order three or four in c leads to Eq. (4.31).

b. Equation 4.30) yields

$$\hat{\mathcal{H}}' = \frac{e^2}{R^3}(x_A x_B + y_A y_B - 2z_A z_B) \ .$$

It follows that

$$\langle \Psi_0|\hat{\mathcal{H}}'|\Psi_1\rangle = \frac{e^2}{R^3}\Big(\langle \psi_{100}|x|\psi_{210}\rangle^2 + \langle \psi_{100}|y|\psi_{210}\rangle^2 - 2\langle \psi_{100}|z|\psi_{210}\rangle^2\Big) \ .$$

The integrals are products of integrals; they can be carried out within each atom separately. For the chosen function, $\langle \psi_{100}|x|\psi_{210}\rangle = \langle \psi_{100}|y|\psi_{210}\rangle = 0$ (the wave functions are independent of the angle ϕ, which causes the integrals over $\sin\phi$ and $\cos\phi$ to vanish). It remains to consider the term

$$\langle \psi_{100}|z|\psi_{210}\rangle = \int_0^\infty r^2 dr \int_0^\pi \sin\theta d\theta \int_0^{2\pi} d\phi \frac{e^{-r/a_{\mathrm{B}}}}{a_{\mathrm{B}}^{3/2}\sqrt{\pi}} r\cos\theta \frac{e^{-r/(2a_{\mathrm{B}})}}{a_{\mathrm{B}}^{3/2}4\sqrt{2\pi}} \frac{r}{a_{\mathrm{B}}}\cos\theta$$

$$= \frac{a_{\mathrm{B}}}{3\sqrt{2}} \int_0^\infty du\, u^4 e^{-3u/2} = \frac{256}{243\sqrt{2}} a_{\mathrm{B}} \equiv \zeta a_{\mathrm{B}} \ ,$$

where the following relations are used

$$\int_0^{2\pi} d\phi = 2\pi \ , \quad \int_0^{\pi} d\theta \sin\theta \cos^2\theta = 2/3 \ , \quad \int_0^{\infty} du\,u^4 e^{-3u/2} = \frac{256}{81} \ ,$$

(with the variable $u = r/a_B$). The second term in Eq. (4.32) is

$$\frac{4e^4|\langle\psi_{100}(\mathbf{r})|z|\psi_{210}(\mathbf{r})\rangle|^4}{2R^6(E_2 - E_1)} = \frac{2e^4\xi^4 a_B^4}{R^6 R_y(1 - 1/4)} = \frac{16\xi^4 e^2 a_B^5}{3R^6} = \zeta\frac{e^2}{a_B}\left(\frac{a_B}{R}\right)^6 \ ,$$

with $\zeta = 16\xi^4/3 = 0.115$.

c. In the general case,

$$\Psi = A\left(\Psi_0 + \sum_{n,n'>1}\sum_{\ell,m}\sum_{\ell',m'} c_{n\ell m,n'\ell'm'}\psi_{n\ell m}(\mathbf{r}_A)\psi_{n'\ell'm'}(\mathbf{r}_B)\right) \ .$$

The coefficients are obtained by the variational method. The result is

$$\langle E\rangle = 2E_1 - \frac{e^2}{a_B}\left(\frac{a_B}{R}\right)^6\zeta \ ,$$

$$\zeta = \frac{e^2}{a_B^5}\sum_{n,\ell,m}\sum_{n'\ell'm'}\frac{|\langle\psi_{100}(\mathbf{r}_A)\psi_{100}(\mathbf{r}_B)|3z_A z_B - \mathbf{r}_A\cdot\mathbf{r}_B|\psi_{n\ell m}(\mathbf{r}_A)\psi_{n'\ell'm'}(\mathbf{r}_B)\rangle|^2}{E_n + E_{n'} - 2E_1} \ .$$

This result is clearly larger than the one obtained before, as all terms in the sum are positive. The sum can be estimated once $E_n + E_{n'}$ in the denominator is ignored. Then, $\zeta \approx 6$.

Answer 4.14.

The expansion is based on

$$\frac{1}{\sqrt{1+x}} = 1 - \frac{x}{2} + \frac{3x^2}{8} - \frac{5x^3}{16} + \frac{35x^4}{128} + \ldots \ .$$

Upon expanding each the terms in Eq. (4.29), e.g., $|\mathbf{R}+\mathbf{r}|^{-1} = R^{-1}[1 + 2z/R + (r/R)^2]^{-1/2}$ and collecting terms, one finds that the next term in the expansion (4.30) is

$$\frac{3e^2}{2R^4}\left([2\mathbf{r}_A\cdot\mathbf{r}_B - 5z_A z_B](z_A - z_B) + r_A^2 z_B - r_B^2 z_A\right) + \mathcal{O}(r_{AB}^4/R^6) \ .$$

Each of the terms here contains a product of the dipole moment on one of the atoms with a quadrupole moment on the other. These terms are antisymmetric with respect to reflection, and therefore their quantum averages vanish in the symmetric ground state. As in the calculation leading to Eq. (4.32), one has to include in the variational wave function excited states; the correction is then quadratic in the coefficient of the additional wave function. The contribution to Eq. (4.32) is of the order of $(a_B/R)^8$. As is always the case in variational calculations, the contribution to the ground-state energy is negative. It is a small correction to the leading order and therefore is frequently ignored. Note that higher orders in the multipole expansion yield corrections of the order of $(a_B/R)^{10}$, $(a_B/R)^{12}$, etc.

Answer 4.15.

For the **FCC** lattice, $a = \overline{R}/\sqrt{2}$; hence $v = a^3/4 = \overline{R}^3/\sqrt{2}$. Inserting these into Eqs. (4.10) and (4.11) yields

$$B = 4(\epsilon/\sigma^3)A_{12}(A_6/A_{12})^{5/2} = 75\epsilon/\sigma^3 \; .$$

Answer 4.16.

a. The potential created at a point $\mathbf{R}'$ due to charges q_i located at $\mathbf{r}_i$ can be expanded as

$$\sum_i \frac{q_i}{|\mathbf{r}_i - \mathbf{R}'|} = \frac{1}{R'}\sum_i \frac{q_i}{\sqrt{1 + (r_i/R')^2 - 2\mathbf{r}_i \cdot \mathbf{R}'/R'^2}}$$

$$= \frac{1}{R'}\sum_i q_i\left(1 + \frac{\mathbf{r}_i \cdot \mathbf{R}'}{R'^2} + \frac{3(\mathbf{r}_i \cdot \mathbf{R}')^2 - r_i^2 R'^2}{2R'^4} + \dots\right)$$

$$= \frac{Q}{R'} + \frac{\mathbf{P} \cdot \mathbf{R}'}{R'^3} + \frac{\sum_{\alpha\beta} Q_{\alpha\beta}(3R'_\alpha R'_\beta - \delta_{\alpha\beta}R'^2)}{6R'^5} + \dots \; .$$

In the first stage the potential is expanded in the charges' coordinates; the total charge, $Q = \sum_i q_i$, the dipole moment, $\mathbf{P} = \sum_i q_i \mathbf{r}_i$, and the components of the quadrupole-moment tensor, $Q_{\alpha\beta} = \sum_i q_i(3r_{i\alpha}r_{i\beta} - \delta_{\alpha\beta}r_i^2)$, are introduced in the second stage. Next, the quantum-average over the coordinates of the electrons in the molecule is carried out, with weights determined by the absolute-value squared of their wave functions. Due to the cylindrical symmetry assumed in the problem, this average yields $\langle x_i \rangle = \langle y_i \rangle = 0$, and $\langle x_i y_i \rangle = \langle z_i x_i \rangle = \langle y_i z_i \rangle = 0$. In addition, $\langle x_i^2 \rangle = \langle y_i^2 \rangle = \langle r_i^2 - z_i^2 \rangle/2$. Therefore, the dipole moment has at most a component along the molecule axis, and the quadrupole moment obeys the relations given in the problem.

b. The hydrogen molecule is also invariant to the inversion of $\mathbf{r}$ (the coordinate of the electron, see Fig. 4.8 and the discussion following it); therefore the probability to find the electron near any of the nuclei is the same, and the dipole moment vanishes.

Answer 4.17.

a. Oxygen has two valence electrons; both participate in a double bond with the carbon, built of a $\sigma-$bond and a $\pi-$one. Therefore both the oxygen and the carbon have two additional hybrid states per each, that are farther away from the bond and from one another, available for a planar hybridization of the sp$^2-$type, as in Fig. 4.13 or in Eq. (4.25). The remaining two states in the oxygen are occupied by two electrons each, while in the carbon they take part in covalent bonds with the hydrogens. The result resembles the planar ethene molecule, Fig. 4.16, with one of the carbons replaced by the oxygen, and the two adjacent hydrogens replaced by clouds containing two electrons each.

b. The answer is in Fig. 4.35. The solid lines represent covalent bonds, the double ones are double bonds, and the dashed lines are hydrogen bonds. The carbons build a rectangular-centered lattice. It is not the only possible structure: for instance, one may build a planar lattice in which the molecules in neighboring columns are normal to each other (and not parallel, as in Fig. 4.35).

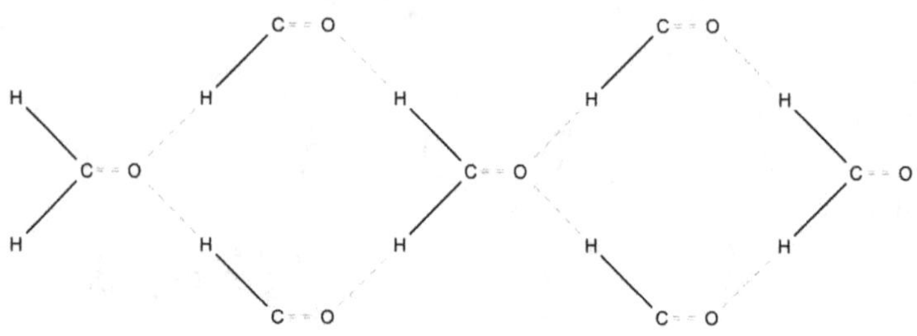

Fig. 4.35

Answer 4.18. The three spins are written in spherical coordinates,

$$\mathbf{S}_i = \{\sin\theta_i \cos\phi_i, \ \sin\theta_i \sin\phi_i, \ \cos\theta_i\}.$$

Without loss of generality, one of the spins, say 1, may be chosen to be $\mathbf{S}_1 = \{0,0,1\}$. The energy of the triangle is

$$E = |J|[\cos(\theta_2) + \cos(\theta_3) + \sin(\theta_2)\sin(\theta_3)\cos(\phi_2 - \phi_3) + \cos(\theta_2)\cos(\theta_3)] \ .$$

As $0 \le \theta_i \le \pi$, it follows that $\sin\theta_2 \sin\theta_3$ is positive, and thus the minimum is reached for $\cos(\phi_2 - \phi_3) = -1$, that is, $\phi_2 = \pi + \phi_3$. Hence, the spin components normal to the $\hat{\mathbf{z}}$−axis are antiparallel to each other. The energy is then

$$E = |J|[\cos(\theta_2) + \cos(\theta_3) + \cos(\theta_2 + \theta_3)] \ .$$

Differentiating with respect to θ_2 and to θ_3 gives

$$\sin(\theta_2) = -\sin(\theta_2 + \theta_3) = \sin(\theta_3) \ .$$

The equality $\sin(\theta_2) = \sin(\theta_3)$ has two solutions: $\theta_2 = \theta_3$ and $\theta_2 = \pi - \theta_3$. The energy corresponding to the second one is $E = -|J|$; the energy of the first is $E = |J|[2\cos(\theta_2) + 2\cos(2\theta_2)]$, which is extremal for

$$2\sin(\theta_2) + 2\sin(2\theta_2) = 2\sin(\theta_2)[1 + 2\cos(\theta_2)] = 0 \ ,$$

with the solutions $\theta_2 = \theta_3 = 0$ or π, or $\cos\theta_2 = -1/2$, implying $\theta_2 = \theta_3 = 2\pi/3$. The energy of the first two solutions is $E = 3|J|$ (a ferromagnetic state) or $E = -|J|$ (a frustrated state with two parallel spins and the third antiparallel to them). The energy of the last solution is $E = -3|J|/2$, the smallest possible energy of the problem. In this solution $\mathbf{S}_1 \cdot \mathbf{S}_3 = \mathbf{S}_1 \cdot \mathbf{S}_2 = -1/2$, and the corresponding angles are indeed $\theta_2 = \theta_3 = 2\pi/3$. Similarly, $\mathbf{S}_2 \cdot \mathbf{S}_3 = \cos(\theta_2 + \theta_3) = -1/2$ and hence the angle between these two spins is also 120°.

Answer 4.19.

The simplest cases are those for which $J_2 = 0$. The material is then a ferromagnet (all spins are parallel) for $J_1 > 0$, and a simple antiferromagnet where all neighboring spins are antiparallel, when $J_1 < 0$. When $J_2 > 0$, both arrangements are benefitted, as in both the next-nearest neighbors are parallel to each other; the total energy is $E = -N(|J_1| + J_2)$. The case $J_2 < 0$ is more subtle, since the interaction between next-nearest neighbors (that requires them to be antiparallel) competes with the nearest-neighbor interaction that prefers them to be parallel. In any event, it cannot be that the two types of bonds, between nearest neighbors and between next-nearest neighbors, will have their optimal energy. This is the situation called frustration in Sec. 2.10. To treat it, one may begin by presuming that the spins laying in the $XY-$plane, and that the angle between nearest neighbor spins is ϑ, i.e., $\mathbf{S}_n = (\cos[n\vartheta], \sin[n\vartheta])$. The total energy is $E = -N[J_1 \cos(\vartheta) + J \cos(2\vartheta)]$, where N is the number of spins in the system. Differentiating with respect to ϑ gives

$$J_1 \sin(\vartheta) + 2J_2 \sin(2\vartheta) = \sin(\vartheta)[J_1 + 4J_2 \cos(\vartheta)] = 0 \ .$$

Hence, either $\sin(\vartheta) = 0$ or $\cos(\vartheta) = -J_1/(4J_2)$. The second solution exists provided that $|J_1/(4J_2)| < 1$, and then the corresponding energy is $E_2 = NJ_2[1+(J_1/J_2)^2/8]$. The energy of the first solution is $E_0 = -N[J_1 + J_2]$ for a ferromagnetic structure ($\vartheta = 0$), or $E_\pi = N[J_1 - J_2]$ for an antiferromagnetic one ($\vartheta = \pi$). One easily convinces oneself that in its range of existence, the energy of the solution $\cos(\vartheta) = -J_1/(4J_2)$ is lower than that of the other two solutions: $E_2 = -N|J_2|[1 + (J_1/J_2)^2/8] < -N|J_2|(|J_1/J_2| - 1)$, when $|J_1/J_2| < 4$. The two energies are equal for $|J_1/J_2| = 4$; when $|J_1/J_2| > 4$, the second solution does not exist at all, and therefore either the ferromagnetic state (for $J_1 > 0$), or the antiferromagnetic one (for $J_1 < 0$) prevails. The dependence of ϑ on the ratio $j = J_1/|J_2|$ in the optimal situation (of the ground state) is depicted in Fig. 4.36. In the intermediate state, described by the second solution, the spins along the lattice rotate around the $\hat{\mathbf{z}}-$axis normal to the $XY-$ plane of the spins. When that plane is perpendicular to the one-dimensional lattice, then the edge-points of the spins create a screw that advances along the lattice. In most of the lattice points the ratio ϑ/π is not rational, and therefore this is an **incommensurate** pattern: the magnetic unit cell is infinite (see Sec. 2.9). For $j < -4$ the magnetic structure is antiferromagnetic and the magnetic unit cell coincides with the atomic one. At $j = \pm 4$ there occur continuous phase transitions, in which the angle ϑ loses its dependence on j.

For the Ising model, Eq. (4.47), the incommensurate solution is not valid. Again, the subtle configuration occurs for $J_2 < 0$. When $J_1 > 0$, the nearest neighbors "wish" to be parallel, and the next-nearest ones "wish" to be antiparallel. For strong enough J_1, the arrangement is ferromagnetic with the total energy $-NJ_1$. For large enough J_2, the next-nearest neighbors arrange themselves successively in opposite directions, creating the pattern $+ + - - + + - - + + - - + + \ldots$, in which the next-nearest neighbors are antiparallel (the first spin in the line can be

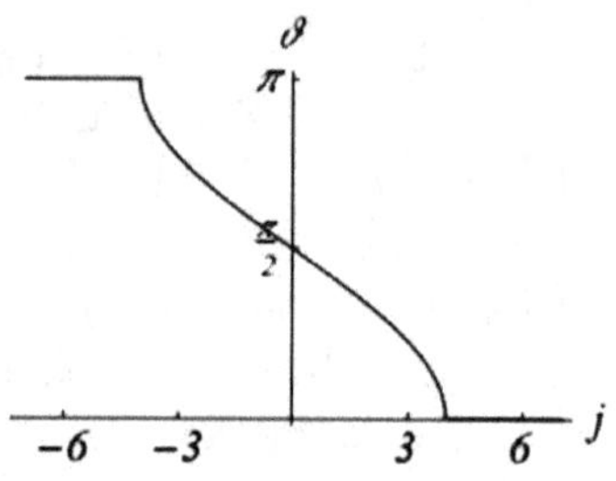

Fig. 4.36

"up" or "down" without affecting the energy). The total energy is $-N|J_2|$ as half of the nearest-neighbor pairs yields J_1, and the other half gives $-J_1$. The magnetic unit cell is four times larger than that of the original one. The transition between these two solutions occurs when $J_1 = |J_2|$. Alternatively, when $J_1 < 0$ then for $|J_2| < |J_1|$ the simple antiferromagnetic state is realized, $-+-+-+-+\ldots$, but for $|J_2| > |J_1|$ again appears the state $--++--++--++\ldots$. Therefore, for $J_2 < 0$ the simple antiferromagnetic state prevails over the entire range $j < -1$, and the ferromagnetic state is the ground state in the range $j > 1$, and in-between there is the new state $--++--++--++\ldots$.

4.9   Problems for self-evaluation

s.4.1.
The following values have been measured for the binding energy, the distance between nearest neighbors, and the bulk modulus of table salt: $u = 7.9\,eV$, $\overline{R} = a/2 = 2.82\,\text{Å}$, $B = 2.4 \times 10^{10}\,N/m^2$.
a. *Use these values to identify the parameters m and C in Eq. (4.8) and check whether they are not self-contradictory.*
b. *What is the pressure one should apply in order to squeeze table salt as to reduce its volume by 1% ?*

s.4.2.
a. *Assuming that the ions A, B, and O of the perovskite structure shown in Fig. 2.21 are spheres of radii r_A, r_B, and r_O, respectively, and denoting the lattice constant by a, obtain five inequalities that these radii have to obey.*
b. *With the assumption that the spheres of A and O are touching each other, and so are also the spheres of B and O (why is this advantageous?) find the range of allowed values for the ratios $x_A = r_A/r_O$, $x_B = r_B/r_O$.*
c. *What happens when the radii are not in that range? Is the perovskite structure still possible?*

d. *In the barium titanate compound, $a \approx 4\text{Å}$, and $r_A \approx r_O \approx 1.4\text{Å}$. Which of the conditions in part (a) becomes (approximately) an equality? What can be deduced about the radius of the titanium ion?*

e. *Explain how to derive the Coulomb energy of this crystal.*

s.4.3.

Assuming that the lattice constant of a certain alkali halide is determined by the radii of the ions, and that the binding energy is given by Eq. (4.9) with the repulsion U_{P2} of Eq. (4.8) with a constant ρ, find the value of λ. How does the repulsive potential vary as a function of the ions' radii?

s.4.4.

Sulphur appears in nature in several forms (allotropes); several of those are lattices (orthorhombic, rhombohedral, monoclinic, and more), with a base that comprises the molecule S_8 or the molecule S_{12}, depicted in Fig. 4.37.

a. *Determine the type of the bond between two neighboring atoms in these molecules. Find approximately the value of the angle between neighboring bonds in the molecule.*

b. *Do the molecules possess electrons that are free to move around them, like in benzene?*

c. *Occasionally (though rarely) sulphur appears in the molecular form S_2. What is the structure of this molecule?*

d. *There are six possible forms of the molecule S_4; in one of the them the atoms build a planar square. What is the electronic configuration of this molecule?*

e. *The left panel of Fig. 4.37 displays two versions of the structure of the polymer S_∞, which creates a spatial helix. What is the type of the bond along the molecule? Does the molecule conduct electricity? What is the lattice constant of the one-dimensional lattice it creates?*

f. *What are the types of the bonds between the molecules in crystals built of the molecules mentioned in the previous parts?*

g. *In rare cases it is possible to grow a lattice of sulphur atoms. Which types of bonds can be formed between the atoms in the lattice? Propose lattice structures based on these bonds.*

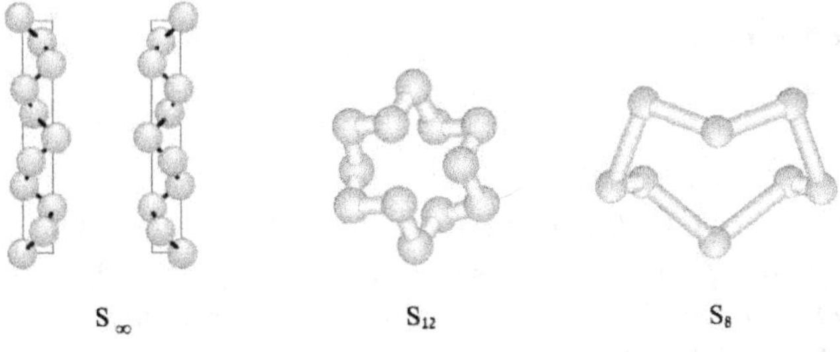

Fig. 4.37

s.4.5.

a. *Propose a general expression for the van der Waals binding energy of the rectangular lattice, and show that it depends solely on the ratio of the two lattice constants, $x = b/a$. For which value of x is it maximal? Is it possible to extend the expression to the family of simple three-dimensional orthorhombic lattices? What can be deduced about the symmetry of simple molecular crystals?*

b. *Find the expression for the binding energy due to van der Waals interactions in the oblique planar lattice, which has equal lattice vectors with an arbitrary angle between them.*

s.4.6.

The potential between pairs of molecules on a fictitious **BCC** *lattice is described by the Morse function*

$$U_{\text{tot}}(R) = NU_0(\exp[-2(R-\sigma)/r_0] - 2\exp[-(R-\sigma)/r_0]) \,,$$

where R is the distance between nearest neighbors. Find the distance between nearest neighbors at equilibrium, $\overline{R}$, and calculate the binding energy and the bulk modulus.

s.4.7.

The structure of the ammonia molecule, NH_3, is discussed in problem 4.11. Based on this structure, determine the type of the bonds among these molecules. Propose a crystal structure for a solid built of such molecules.

s.4.8.

The acetylene molecule, C_2H_2 is linear, with a triple bond between the carbons (see the discussion before Fig. 4.16). Can a hydrogen bond be formed between the hydrogen in one molecule and the $\pi-$states perpendicular to a neighboring molecule? Assuming that there are such bonds, describe the three-dimensional crystal that can be created. Discuss the hydrogen bonds based on a classical Coulomb interaction and on a quantum hybridization.

s.4.9.

Like in problem 2.35, the magnetic moments of spins on a square lattice are expressed in the form $\boldsymbol{\mu}(\mathbf{R}) = \boldsymbol{\mu}_0 \cos(\mathbf{Q} \cdot \mathbf{R})$. The magnetic Hamiltonian includes the exchange interaction J_1 between nearest neighbors (nn), and the exchange interaction J_2 between next-nearest neighbors (nnn). Determine the magnetic energy as a function of $\mathbf{Q}$, and find the magnetic order, i.e., the spin configuration, in the ground state, as a function of the exchange energies.

4.10 Answers for the self-evaluation problems

Answer s.4.1.

a. Changing the units, $B = 0.15\text{eV}/(\text{Å})^3$. Hence, $m = 1 + 18\overline{R}^4 B/(\alpha e^2) \approx 7.78$ [using the relations 1 Rydberg$=e^2/(2a_B) \approx 13.6\text{eV}$, $a_B \approx 0.53\text{Å}$, and $\alpha_{\text{NaCl}} = 1.747565$]. It follows that $u = (1-1/m)\alpha e^2/\overline{R} = 18\overline{R}^3/m \approx 7.8$ eV, in fair agreement with the measured binding energy. In addition, $zC = \overline{R}^{m-1}\alpha e^2/m \approx 3660\text{eV}(\text{Å})^m$, with $z = 6$.

b. By Eq. (4.10) and the discussion preceding it, the pressure is given by $p = -\int_{v_0}^v dv\, B(v)/v$, when the specific volume changes from v_0 to $v = xv_0$. Since $B = (m-1)\alpha e^2/(18\overline{R}^4) = B_0(v_0/v)^{4/3}$, one finds

$$p = -B_0 \int_1^{v/v_0} d(v/v_0)(v_0/v)^{7/3} = 3B_0(x^{-4/3} - 1)/4 \ .$$

Substituting $x = 0.99$ and the given bulk modulus yields $p = 2.4 \times 10^8 \text{N/m}^2$.

Answer s.4.2.

a. The distance between two centers of neighboring O ions is $a/\sqrt{2}$, and hence $2r_O < a/\sqrt{2}$. Similarly, $2r_A < a$, $2(r_A + r_O) < a\sqrt{2}$, $2(r_B + r_O) < a$, $2(r_B + r_A) < a\sqrt{3}$.

b. The Coulomb energy increases as the positive ions approach their negative neighbors. Thus, this energy is maximal when the positive ions touch the oxygen ions, and consequently two of the inequalities become equalities: $2(r_A + r_O) = a\sqrt{2}$ and $2(r_B + r_O) = a$. Substituting the expressions for a in the remaining inequalities yields six inequalities. The inequality $2r_O < a/\sqrt{2}$ gives $x_A > 1$ and $x_B > \sqrt{2} - 1$. Similarly, $2r_A < a$ gives $x_A < \sqrt{2} + 1$ and $x_A < x_B + 1$, and the third condition gives $x_A < (\sqrt{3} - 1)x_B + \sqrt{3}$ and $x_B < (\sqrt{6}/2 - 1)x_A + \sqrt{6}/2$. It follows that x_A is in the range $1 < x_A < \sqrt{2} + 1$, and x_B is in the range $\max\{\sqrt{2} - 1, x_A - 1\} < x_B < (\sqrt{6}/2 - 1)x_A + \sqrt{6}/2$.

c. In this case one of the conditions for the spheres to touch each other is abandoned, and a structure where the opposite-sign ions are as close as possible is to be sought for. The five conditions of part (a) leave room for many possibilities.

d. The numbers given in the problem show that (approximately) the barium ions are touching the oxygen ones, and the oxygen ions are touching each other. The remaining three conditions require that $r_{\text{Ti}} < a/2 - r_O \approx 0.6\text{Å}$. Figure 4.6 indeed shows that the titanium ion is smaller then the barium and the oxygen ions.

e. The Coulomb energy is $U_C = (N/2)\sum_{i\in\text{cell}} q_i\phi_i$, where $\phi_i = \sum_{j\neq i}(q_j/R_{ij})$ is the electric potential at i due to all other ions [see Eq. (4.7)]. Thus one has to find the three potentials for the ions A, B, and O, to obtain $U_C = (N/2)(q_A\phi_A + q_B\phi_B + 3q_O\phi_O)$. In the case of the barium titanate compound, the required potentials are for the barium, for which $q_A = +2$, for the titanium ($q_B = +4$) and for the oxygen

($q_O = -2$). Denote by $R = a/2$ the distance between the Ti and the O (this is the shortest distance between opposite-sign charges), and measure distances in units of R. Each barium ion is surrounded by 12 oxygens at a distance $\sqrt{2}/2$, 24 oxygens at a distance $\sqrt{3}/2$, 24 oxygens at a distance $\sqrt{10}/2$, *etc.* It is also surrounded by 6 barium ions at a distance 1, 12 bariums at a distance $\sqrt{2}$, eight at $\sqrt{3}$, *etc.*, and also by 8 Ti ions at a distance $\sqrt{3}/2$, 24 Ti's at $\sqrt{11}/2$, *etc.*. The electric potential on the barium is

$$\phi_{\text{Ba}} = \sum_j \frac{q_j}{R_{0j}} = -2\left(\frac{12}{(\sqrt{2}/2)} + \frac{24}{\sqrt{3}/2} + \frac{24}{(\sqrt{10}/2)} + \cdots\right)$$
$$+ 2\left(\frac{6}{1} + \frac{12}{\sqrt{2}} + \frac{8}{\sqrt{3}} + \cdots\right) + 4\left(\frac{8}{(\sqrt{3}/2)} + \frac{24}{(\sqrt{11}/2)} + \cdots\right).$$

Answer s.4.3.

The maximal binding energy is achieved when $z\lambda = (\alpha q^2/\rho) \exp[\overline{R}/\rho](\rho/\overline{R})^2$ (see the solution to problem 4.6). The function $\exp[x]/x^2$ has a minimum at $x = 2$. One generally expects that $\overline{R}/\rho > 2$, and in this range $z\lambda$ increases exponentially with $\overline{R}$. As mentioned, usually $\overline{R} = r_< + r_>$, and thus $z\lambda$ increases exponentially with the sum of the radii.

Answer s.4.4.

a. The ground configuration of the sulphur atom is similar to that of oxygen, $3s^2 3p^4$; as in oxygen, a possible hybrid state is $3s^1 3p^5$, which contains four states of the type given in Eq. (4.26) and displayed in Fig. 4.14. Two of these states are each doubly-occupied, and the other two have a single electron and can form covalent bonds like in Fig. 4.17 and similarly to the bonds of oxygen with the hydrogens in water – Fig. 4.18. In the symmetric case, Fig. 4.14, the angles between neighboring bonds is $109.5°$. In water, the presence of extra negative charges on the doubly-occupied states reduces the angle to $104.5°$. Since sulphur resembles oxygen, its atom too has four bonds of the type given in Eq. (4.26), two are doubly-occupied and two create covalent bonds, e.g., with the neighboring sulphur atoms in the molecules mentioned in the problem. One therefore expects the angle to be of the same order of magnitude as given above, about $105°$. In fact, this angle is $106°$. In order to close the loop in the S_8 molecule, the atoms have to move up and down successively, forming the zig-zag structure displayed in the figure, with four sulphur atoms in each plane. The same structure could have been realized for the S_{12} molecule. However, the Coulomb repulsion between the charges in the doubly-occupied states, which are located opposite to the covalent bonds, causes this molecule to adopt another structure in which there are six sulphur ions in the intermediate plane, and three atoms in the planes below and above that plane.

b. As the six electrons of each sulphur atom are in states localized near that atom (whether in doubly-occupied states or in a covalent bonds) there are no electrons free to move around the ring.

c. There are only two sulphur atoms in S_2 and thus its configuration is similar to that of the O_2 molecule. Each sulphur (or oxygen) has six electrons in the outer shell; all in all there are four orbital states (one s and three p). Two of the orbital states are doubly-occupied, and the other two (each accommodating a single electron) can form covalent bonds. In the previous example, similar to the water molecule, each atom has two covalent bonds along different directions, and therefore it is expedient to exploit the states given in Eq. (4.26). In the linear diatomic molecule both bonds have to be along the same direction, and therefore the bond is a double one. In this situation it is better to use the basis given in Eq. (4.25), depicted in Fig. 4.13. Two of the planar states are occupied by two electrons each and form an angle close to $120°$ with one another, and the third state forms a $\sigma-$bond with the other atom. The additional bond between the atoms is a π one, coupling the states perpendicular to the plane. The result resembles the ethene molecule, Fig. 4.16, with the hydrogens there replaced by doubly-occupied sulphur states.

d. As explained in part (d) of problem 4.10, one may assume arbitrary angles between the four hybrid states given by $\Psi_i = A_i[\psi_{200} + \beta_i\hat{\mathbf{n}}_i \cdot \boldsymbol{\psi}_i]$. If the sulphur atoms are arranged on a square, then one of the angles, (say between the states 1 and 2) is $90°$, and therefore $\beta_1 = \beta_2 = \infty$ and $|A_1|^2 = |A_2|^2 = 1/(1 + |\beta_1|^2) = 0$. These states are then ψ_x or ψ_y. From the scalar product of each of the other functions with the first two it follows that $\hat{\mathbf{n}}_3$ and $\hat{\mathbf{n}}_4$ are normal to $\hat{\mathbf{n}}_1$ and $\hat{\mathbf{n}}_2$, and hence $\Psi_{3,4} = [\psi_{200} \pm \psi_z]/\sqrt{2}$. The first two states are exploited for covalent bonds, and therefore the remaining two – which point along the positive and the negative directions of the $\hat{\mathbf{z}}-$axis – contain each two electrons. The binding energy of this configuration is not as high as in the previous cases, because of the larger proximity of the electrons' clouds. Alternatively, it is possible to excite the sulphur to the configuration sp^3d^2, and then the emerging bonds are with right angles, as displayed in Fig. 4.19. In this configuration there remain un-paired electrons, free to move around the square.

e. In the polymer molecule of sulphur each atom may retain the configuration sp^3, as described in part (a). Therefore, the angle between two neighboring bonds is approximately $106°$. Given two successive bonds like this, the third one has to form an identical angle with its predecessor; however, the atom at the edge can rotate in a plane perpendicular to the bond. Its final location is determined by the repulsion between the electrons' clouds in the two states that do not take part in the bond and the other electrons. This repulsion turns out to be minimal for the structures shown in Fig. 4.37, which differ from one another in their "screw" direction. According to various measurements and calculations, the third bond forms an angle of about $85.3°$ with the plane spanned by its two predecessors; in this way the screw structure that characterizes the helix is created. As seen in Fig. 4.37, the unit cell contains ten atoms. Like in part (a), the electrons are localized close to the sulphur atoms, and hence there is no electric conductivity along the molecule.

g. As all molecules are electrically neutral and have no electric dipole moment, the interactions among them are of the van der Waals type.

f. To build a lattice it is needed that each atom will be bound to at least three other atoms (like in graphene). For a three-dimensional bond even more neighboring atoms are needed. As the configurations $3s^13p^5$ or $3s^23p^4$ allow for no more than two bonds per atom, it is clear that a lattice structure requires excitations to higher hybridization states, e.g., $3s^13p^43d^1$ or $3s^13p^33d^2$. The latter configuration allows for six bonds normal to one another, see Fig. 4.19. Locating sulphur atoms at the end point of such a bond yields a simple tetragonal lattice. The first configuration requires that one of the five states [three in the plane, Eq. (4.25), and two perpendicular to the plane, e.g., $\psi_{z^2} \pm \psi_z$, see the solution to problem 4.11] will be doubly occupied, and therefore there remain only four states capable of forming bonds. When the double occupancy fills the states perpendicular to the plane, the resulting lattice is a hexagonal one, as in the case of graphene.

Answer s.4.5.

a. The binding energy is given by extending Eq. (4.33),

$$U_{\text{tot}} = 2N\epsilon[F_{12}(x)(\sigma/a)^{12} - F_6(x)(\sigma/a)^6] \,,$$

with $F_n = \sum_{n_1,n_2}(n_1^2 + n_2^2 x^2)^{-n/2}$. The minimum of U_{tot} yields $u = -U_{\text{tot}}(\text{min})/N = \epsilon F_6^2/(2F_{12})$, and indeed the result depends only on the ratio x. When the two directions in the lattice are interchanged, that is, the dummy indices in the sum of F_n are interchanged, then $U_{\text{tot}} = 2N\epsilon[F_{12}(1/x)(\sigma/b)^{12} - F_6(1/x)(\sigma/b)^6]$, and hence $u(x) = u(1/x)$. The derivative with respect to x gives $u'(x) = -u'(1/x)/x^2$, and thus $u'(1) = 0$: the function has an extremum at $x = 1$. Numerical test shows that this is indeed a maximum: the sum over $|n_1|, |n_2| \le 3$ gives

$$F_6(x) = \frac{47449}{23328} + \frac{47449}{23328x^6} + \frac{4}{(1+x^2)^3} + \frac{4}{(4+x^2)^3} + \frac{4}{(9+x^2)^3} + \frac{4}{(1+4x^2)^3}$$

$$+ \frac{4}{(4+4x^2)^3} + \frac{4}{(9+4x^2)^3} + \frac{4}{(1+9x^2)^3} + \frac{4}{(4+9x^2)^3} + \frac{4}{(9+9x^2)^3} + \cdots ,$$

$$F_{12}(x) = \frac{2177317873}{1088391168} + \frac{2177317873}{1088391168x^{12}}$$

$$+ \frac{4}{(1+x^2)^6} + \frac{4}{(4+x^2)^6} + \frac{4}{(9+x^2)^6} + \frac{4}{(1+4x^2)^6}$$

$$+ \frac{4}{(4+4x^2)^6} + \frac{4}{(9+4x^2)^6} + \frac{4}{(1+9x^2)^6} + \frac{4}{(4+9x^2)^6} + \frac{4}{(9+9x^2)^6} + \cdots .$$

Additional terms almost do not change the result. The right panel in Fig. 4.38 displays the dependence of u on x (in units of $\epsilon/2$). It is seen that the square lattice with $x = 1$ has the maximal binding energy. A heuristic way to understand this result is to note that the sums are dominated by their first term. The atom at the origin has two nearest neighbors at a distance a, and two at a distance b, and hence

$F_n \propto 2(1/a^n + 1/b^n)$. In this approximation $u = \epsilon F_6^2/F_{12} = \epsilon(1+y)^2/(1+y^2)$, where $y = 1/x^6$. This function is maximal for $y = 1$. For the orthorhombic lattice, denote $y = b/a$ and $z = c/a$, and then

$$u = -U_{\text{tot}}(\min)/N = \epsilon F_6^2(y,z)/[2F_{12}(y,z)] \ ,$$

where $F_n(y,z) = \sum_{n_1,n_2,n_3}(n_1^2 + n_2^2 y^2 + n_3^2 z^2)^{-n/2}$. Then $u(y,z) = u(1/y, z/y) = u(y/z, 1/z)$, and hence

$$u_y'(y,z) = -[u_y'(1/y, z/y) + zu_z'(1/y, z/y)]/y^2 \ ,$$
$$u_z'(y,z) = -[yu_y'(y/z, 1/z) + u_z'(y/z, 1/z)]/z^2 \ ,$$

yielding $u_y'(1,1) = -[u_y'(1,1) + u_z'(1,1)]$, and $u_z'(1,1) = -[u_y'(1,1) + u_z'(1,1)]$, i.e., $u_y'(1,1) = u_z'(1,1) = 0$. Again, a numerical check shows that $y = z = 1$ is a maximum. The number of nearest neighbors at the same distance is maximal for the cubic lattice (as compared with the orthorhombic one), and thus the binding energy is higher for this more symmetric lattice.

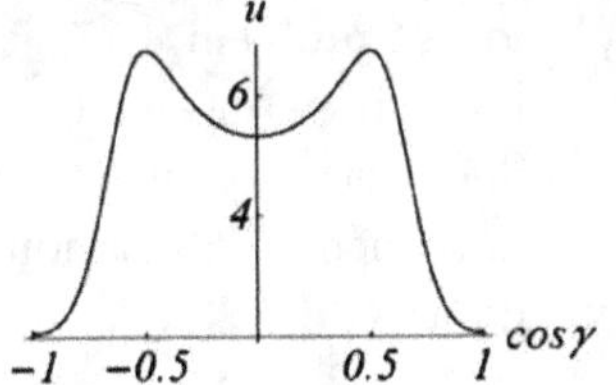

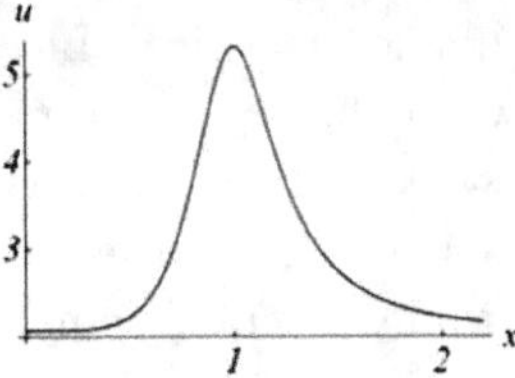

Fig. 4.38: Left panel: the binding energy of the oblique lattice; right panel: the binding energy of the rectangular lattice.

b. For the oblique lattice, the base vectors are $\mathbf{a}_1 = a\hat{\mathbf{x}}$, $\mathbf{a}_2 = a\cos\gamma\hat{\mathbf{x}} + a\sin\gamma\hat{\mathbf{y}}$, and therefore $F_n(\gamma) = \sum_{n_1,n_2}(n_1^2 + n_2^2 + 2n_1 n_2 \cos\gamma)^{-n/2}$. The left panel in Fig. 4.38 shows that the binding energy is maximal for $\cos\gamma = 1/2$, i.e., for the triangular lattice. Indeed, this is the densest planar lattice. In three dimensions, most materials with van der Waals interactions are arranged on an **FCC** or a **HCP** lattice, which are the densest lattices in three dimensions. Indeed, when only the neighbors next to the origin are taken into account, the atom at the origin has four neighbors at a distance a, and two at a distance $2a\sin(\gamma/2)$. In the crudest approximation, $F_n = 4/a^n + 2/[2a\sin(\gamma/2)]^n$, and hence $u \propto (4 + 2y)^2/(4 + 2y^2)$, where $y = 1/[2\sin(\gamma/2)]^6$. The maximal binding energy is now obtained for $y = 1$, i.e., $\gamma = 60°$.

Answer s.4.6.

The potential is extremal for

$$\frac{\partial U_{\text{tot}}(R)}{\partial R} = -2\frac{NU_0}{r_0}\left(e^{-2(R-\sigma)/r_0} - e^{-(R-\sigma)/r_0}\right) = 0 \ ,$$

and therefore $\overline{R} = \sigma$. At equilibrium, the binding energy is $u = -U_{\text{tot}}(\overline{R})/N = U_0$. In a **BCC** lattice, $\overline{R} = a\sqrt{3}/2$, and thus $v = c\overline{R}^3 = 8\overline{R}^3/(3\sqrt{3})$. Making use of Eqs. (4.10) and (4.11) yields

$$B = -v\frac{\partial^2 u}{\partial v^2}\bigg|_{\overline{R}=\sigma} = \frac{U_0}{4\sqrt{3}\sigma r_0} \ .$$

Answer s.4.7.

Each nitrogen ion is surrounded by three covalent bonds with the hydrogens, and a single cloud of a doubly-occupied state. The four states are described by hybridizations of the sp^3−type. Along each line that joins the nitrogen with a hydrogen it is possible to create a hydrogen bond with the doubly-occupied state near a neighboring nitrogen. In a similar fashion, along the line that accommodates the doubly-occupied state it is possible to form a hydrogen bond with a hydrogen from a neighboring molecule. Thus, each nitrogen ion creates four hydrogen bonds with four neighboring nitrogens, with the angles between the bonds that are similar to those of the diamond lattice. Like in the solution of problem 4.17, it can be assumed that when the crystal begins to grow from a single molecule, there is only one pattern which will be eventually formed. Such tetrahedral patterns exist in the diamond lattice and in the wurtzite one. Experimentally, it is found that the ammonia lattice has a cubic symmetry, like the diamond lattice.

Answer s.4.8.

Consider first a single molecule, H−C≡C−H. The triple bond comprises a σ−bond on the line that joins the carbons, e.g., along the $\hat{\mathbf{x}}$−axis, and two π−bonds perpendicular to the molecule, one along $\hat{\mathbf{y}}$ and another along $\hat{\mathbf{z}}$. The XZ plane then contains a cloud of electrons' probability around the $\hat{\mathbf{z}}$−axis, normal to the midpoint of the molecule (see Fig. 4.16). This negatively-charged cloud can attract the positive hydrogen ion located at the edge of another molecule, and therefore this other molecule lies along the line normal to the molecule in the XZ plane. Figure 4.39 displays this plane. Each thick line corresponds to a certain molecule; another molecule lies on the normal line, such that its hydrogen is close to the cloud of π−electrons along the same direction. A similar cloud is located on the other side of the molecule, and hence another molecule places itself there. The result is a planar lattice: the thin lines describe the square unit cell, which contains four molecules (at the lower left corner, at the mid-point of the edges close to it, and at the center of the square). The primitive unit cell is a square rotated by $45°$, with a base that contains two neighboring molecules, one horizontal and the other vertical. The crystal can be extended along the $\hat{\mathbf{y}}$−direction. Each of the molecules in the plane possesses another π−bond, normal to the plane at the mid-point of the molecule. Placing there another molecule yields a plane of molecules along the $\hat{\mathbf{y}}$−axis. Above this plane it is possible to locate the original plane, and so forth.

The result is a simple cubic lattice, with a base of eight molecules in the unit cell (four in the lower plane and four in the plane above).

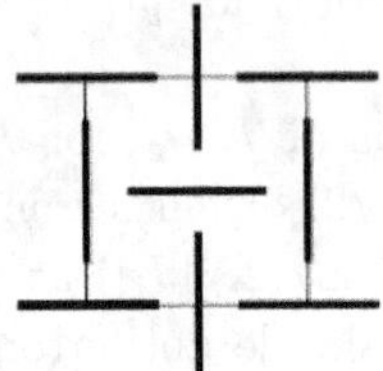

Fig. 4.39

So far the molecules are arranged such that energy is gained from the Coulomb attraction between the hydrogen ion and the negatively-charge electrons' clouds. However, recall that each π–bond in the plane is built of two functions of the ψ_z–type, from each of the carbons. Thus, in the molecule placed on each side of the $\hat{\mathbf{z}}$– axis there are two quantum states yielding an sp–bond between the hydrogen and the carbon. As a result there are six wave functions along this direction, two in each molecule. Six hybrid states can be formed out of those. Each molecule can also contribute two electrons: two from the sp–bond on the perpendicular molecules, and two from the π–bond. These six electrons can occupy the three new hybrid states of the lower energies, and thus form a covalent bond between the three molecules. The emerging bond is a joint one of an ionic bond in which the electrons remain within the molecule, and a covalent bond shared by all three molecules.

Answer s.4.9.
The nearest neighbors are located at $\mathbf{R} = \pm a\hat{\mathbf{x}} \pm a\hat{\mathbf{y}}$ and the next-nearest neighbors at $\mathbf{R} = \pm a(\hat{\mathbf{x}} + \hat{\mathbf{y}}),\ \pm a(\hat{\mathbf{x}} - \hat{\mathbf{y}})$. Hence the magnetic energy is

$$E = -N\{J_1\cos(Q_x a) + \cos(Q)_y a)] + J_2[\cos(Q_x a + Q_y a) + \cos(Q_x a - Q_y a)]\}$$
$$= -N\{J_1[\cos(Q_x a) + \cos(Q_y a)] + 2J_2\cos(Q_x a)\cos(Q_y a)\}\ ,$$

where N is the number of bonds (twice the number of sites) in the lattice. Differentiating with respect to the components of the wave vector yields

$$\sin(Q_x a)[J_1 + 2J_2\cos(Q_y a)] = 0\ ,\quad \sin(Q_y a)[J_1 + 2J_2\cos(Q_x a)] = 0\ .$$

These equations have five solutions:
a. a ferromagnet, $Q_x a = Q_y a = 0$, with energy $E(00) = -2N(J_1 + J_2)$;
b. an antiferromagnet along the $\hat{\mathbf{x}}$–direction, $Q_x a = \pi$, $Q_y a = 0$, with energy $E(\pi 0) = 2NJ_2$;
c. an antiferromagnet along the $\hat{\mathbf{y}}$–direction, $Q_y a = \pi$, $Q_x a = 0$, with energy $E(0\pi) = 2NJ_2$;

d. an antiferromagnet along the diagonal (11), $Q_x a = Q_y a = \pi$, with energy $E(\pi\pi) = 2N(J_1 - J_2)$;

e. an incommensurate antiferromagnet along the diagonal, $\cos(Q_x a) = \cos(Q_y a) = \cos\vartheta = -J_1/(2J_2)$, which can exist only for $|J_1/(2J_2)| \leq 1$, with energy $E(\vartheta\vartheta) = NJ_1^2/(2J_2)$.

A plot of $E/(2NJ_1)$ as a function of J_2/J_1, for $J_1 > 0$, reveals that the ground state is ferromagnetic for $J_2 > -J_1/2$, or antiferromagnetic along one of the axes when $J_2 < -J_1/2$. Similar plots of $E/(2N|J_1|)$ as a function of $J_2/|J_1|$ for $J_1 < 0$ show that the ground state is a simple antiferromagnet along the diagonal for $J_2 > -|J_1|/2$, or an antiferromagnet along one of the axes for $J_2 < -|J_1|/2$. Figure 4.40 displays the regions in parameter space, where in each region there appears the vector $(Q_x a, Q_y a)$, that indicates the magnetic order.

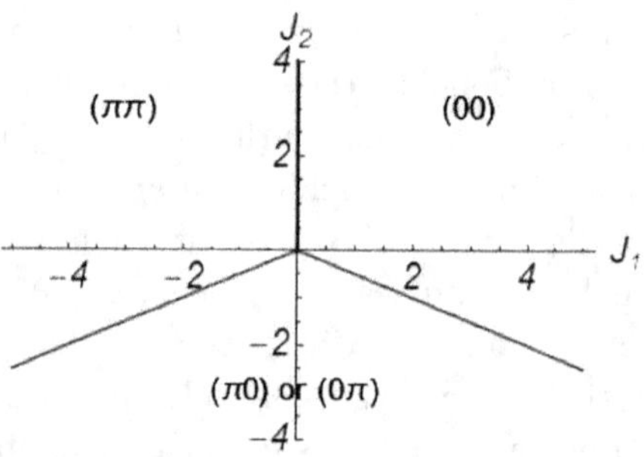

Fig. 4.40

Appendix C

Topics in quantum mechanics

Wave functions and Hilbert space. A quantum state is described by a complex wave function, Ψ, which is a function of the spatial coordinates of the particles in the system, and depends on time. The focus of this book is on stationary situations, and therefore the time dependence is omitted. Thus, the quantum state of a single particle in a three-dimensional space is described by the function $\Psi(\mathbf{r})$, where $\mathbf{r}$ is the particle's location in space; $|\Psi(\mathbf{r})|^2 d^3 r$ is the probability to find the particle in the volume element $d^3 r$ around $\mathbf{r}$. When there are many particles in the system, Ψ depends on the locations of all particles and $|\Psi|^2$ is the probability density to find the particles at these locations. Borrowing the terminology of linear algebra, all (allowed) wave functions form a **"linear space"**: if Ψ_1 and Ψ_2 are two wave functions that describe the system, then their linear combination, $\beta_1 \Psi_1 + \beta_2 \Psi_2$, with complex coefficients β_1 and β_2, is a possible wave function as well, and is also a member of that linear space, which is called the **Hilbert space**. This addition of the wave functions reflects the wavy nature of the quantum system: the linear combination represents the **interference** of the two waves described by the two wave functions. The **scalar product** of two wave functions is $\langle \Psi_1 | \Psi_2 \rangle \equiv \int \Psi_1^* \Psi_2$, where the integration is carried out over all coordinates on which the wave functions depend. When they refer to a single particle, the integration is over the entire three-dimensional space in which the particle exists. Since the sum of the probabilities over this space is 1, it follows that the wave functions are **normalized**, i.e., $\langle \Psi | \Psi \rangle \equiv \int |\Psi|^2 = 1$. (For a single particle, the integration is over the volume which contains the particle.)

The Schrödinger equation and the eigenenergies. A wave function is a solution of the Schrödinger equation. The time-independent functions solve $\hat{\mathcal{H}} \Psi_n = E_n \Psi_n$, where $\hat{\mathcal{H}}$ is the Hamiltonian operator of the system. (The "hat" symbol is used for a quantum operator.) The index n, i.e., the energy quantum number, denotes the different solutions. The Hamiltonian operator of a single particle, for instance, is given in Eq. (4.12). Each of the solutions Ψ_n is characterized by an **eigenvalue** of the equation, E_n. A measurement of the system energy yields one of these eigenvalues. The corresponding wave function Ψ_n is also termed **"eigenstate"** (or eigenfunction) of the energy. The set of wave functions is **or-**

thonormal, i.e., $\langle\Psi_n|\Psi_m\rangle = \delta_{nm}$, where δ_{nm} is the Kronecker delta-function. In many cases the energy levels are degenerate, i.e., a certain energy is an eigenenergy of several eigenstates. In such a situation, the degenerate eigenstates are distinguished by additional quantum numbers. The ensemble of eigenstates in a **basis for the Hilbert space**, which includes all possible wave functions of the system. As is the case of a cartesian coordinate system, the basis vectors are normalized (i.e., they are unit vectors) and are orthogonal to each other. This means that an arbitrary state of the system can be written as $\Psi = \sum_{n=0}^{\infty} a_n \Psi_n$, with complex coefficients a_n. As the wave function is normalized, $\langle\Psi|\Psi\rangle \equiv \int |\Psi|^2 = 1$, and the set of eigenfunctions is orthonormal, $\langle\Psi_n|\Psi_m\rangle = \delta_{nm}$, it follows that $\sum_{n=0}^{\infty} |a_n|^2 = 1$ (check!).

Free particles. The eigenstates of a free particle confined to a space of volume V are $\Psi_{\mathbf{k}} = \exp[i\mathbf{k}\cdot\mathbf{r}]/\sqrt{V}$, where V is the system volume and the wave functions obey periodic boundary conditions. The corresponding eigenenergies are $E(\mathbf{k}) = \hbar^2 k^2/(2m)$. By the Fourier theorem (see Appendix B), these functions are indeed a basis for all functions in space. Thus, the Fourier theorem proves the assertion stating that the solutions of the Schrödinger equation are a basis of the Hilbert space. In this case the energies are degenerate, because $E(\mathbf{k}) = E(-\mathbf{k})$. The various degenerate functions are distinguished from each other by the different values of $\mathbf{k}$.

Averages (expectation values). Applying the Hamiltonian operator to the function $\Psi = \sum_{n=0}^{\infty} a_n \Psi_n$, and carrying out the scalar product of the result with this very function, yields

$$\langle\Psi|\hat{\mathcal{H}}\Psi\rangle = \sum_{n,m} a_n^* a_m E_m \langle\Psi_n|\Psi_m\rangle = \sum_n |a_n|^2 E_n \ .$$

Since $\sum_n |a_n|^2 = 1$, the right-hand side can be interpreted as a **weighted average** of the values picked up in measurements of the energy, where $|a_n|^2$ is the probability to detect a certain eigenvalue E_n. It is therefore customary to denote $\langle E\rangle = \langle\Psi|\hat{\mathcal{H}}|\Psi\rangle$. (Here, the notation $\langle\Psi_1|\hat{\mathcal{H}}|\Psi_2\rangle \equiv \langle\Psi_1|\hat{\mathcal{H}}\Psi_2\rangle$ is introduced.) This average is termed the **expectation value** of the energy in the Ψ state.

The variational approximation. As all $|a_n|^2$ are non negative and their sum is 1, there exists the inequality

$$\langle E\rangle = \langle\Psi|\hat{\mathcal{H}}|\Psi\rangle = \int \Psi^*\hat{\mathcal{H}}\Psi = \sum_n |a_n|^2 E_n \geq E_0 \ . \qquad (\text{C.1})$$

(If all E_n in the sum are replaced by E_0 the result is reduced.) Therefore, **the average of the energy in an arbitrary quantum state is always larger or equal to the ground-state energy, E_0.** It follows that one may derive approximately the wave function of the ground state by calculating the average energy for a selection of wave functions, and then choosing the wave function with the lowest value of those averages. This value may well be higher than the true one; however, given an ample selection of wave functions, it might be a good approximation for the true ground-state energy. This is the **variational method** in quantum mechanics: **the expectation value of the energy in an arbitrary "guess" of**

the wave function is an upper bound on the ground-state energy of the system at hand. Improving the trial wave function, e.g., by trying several more functions, tends to diminish the computed energy and to push the result towards the true value of the ground-state energy.

Problem C.1.

a. *Given the Hamiltonian of a one-dimensional harmonic oscillator,*

$$\hat{\mathcal{H}} = -\frac{\hbar^2}{2m}\frac{d^2}{dx^2} + \frac{m\omega^2}{2}x^2 \, ,$$

and a trial function of the form $\psi(x) = (\alpha/\pi)^{1/4}\exp[-\alpha x^2/2]$, *find the expectation value of the energy, and its minimal value with respect to the parameter* α. *Compare the result with the exact ground-state energy of the one-dimensional harmonic oscillator.*

b. *Add to the Hamiltonian a "perturbation" term,* $\hat{\mathcal{H}}' = \lambda x^6$. *Find the expectation value of the energy with the same function as above. For which* α *is it minimal? Calculate the variational ground-state energy to leading order in* λ.

Perturbation theory for non degenerate states. A ubiquitous method to find approximately the eigenenergies and the eigenstates of a quantum system is termed **perturbation theory**. In many cases, the eigenstates and the eigenvalues of a Hamiltonian $\hat{\mathcal{H}}_0$, i.e., $\hat{\mathcal{H}}_0\Psi_n^{(0)} = E_n^{(0)}\Psi_n^{(0)}$, are known; but this Hamiltonian is augmented by a new term, $\lambda\hat{\mathcal{H}}'$, called 'perturbation', where λ is a small dimensionless real number. One wishes to find the eigenenergies and eigenstates of the new total Hamiltonian, $\hat{\mathcal{H}} = \hat{\mathcal{H}}_0 + \lambda\hat{\mathcal{H}}'$. Consider first the case in which the eigenvalues of the Hamiltonian $\hat{\mathcal{H}}_0$ are not degenerate, i.e., a single eigenstate, $\Psi_n^{(0)}$, corresponds to each eigenenergy, $E_n^{(0)}$. Solving the Schrödinger equation with the Hamiltonian $\hat{\mathcal{H}} = \hat{\mathcal{H}}_0 + \lambda\hat{\mathcal{H}}'$ might be a difficult task, and therefore one seeks approximate solutions. It is assumed that the dimensionless coefficient λ is smaller than 1; the perturbation theory yields the solutions as power series in λ. Denote the eigenfunctions of the new Hamiltonian by Ψ_n, and the corresponding eigenenergies by E_n, i.e., $\hat{\mathcal{H}}\Psi_n = E_n\Psi_n$. Expanding the wave function and the energy in the small parameter λ,

$$E_n = E_n^{(0)} + \lambda E_n^{(1)} + \lambda^2 E_n^{(2)} + \mathcal{O}(\lambda^3) \, , \quad \Psi_n = \Psi_n^{(0)} + \lambda\Psi_n^{(1)} + \mathcal{O}(\lambda^2) \, ,$$

and inserting the expansions into the Schrödinger equation $\hat{\mathcal{H}}\Psi_n = E_n\Psi_n$, one compares terms of the same order in λ. The ones to first order in λ are

$$\hat{\mathcal{H}}'\Psi_n^{(0)} + \hat{\mathcal{H}}_0\Psi_n^{(1)} = E_n^{(0)}\Psi_n^{(1)} + E_n^{(1)}\Psi_n^{(0)} \, .$$

Taking the scalar product with $\Psi_m^{(0)}$ yields

$$\langle\Psi_m^{(0)}|\hat{\mathcal{H}}'|\Psi_n^{(0)}\rangle = (E_n^{(0)} - E_m^{(0)})\langle\Psi_m^{(0)}|\Psi_n^{(1)}\rangle + E_n^{(1)}\delta_{nm} \, , \tag{C.2}$$

where the identity

$$\int[\Psi_m^{(0)}]^*\hat{\mathcal{H}}_0\Psi_n^{(1)} = \int[\hat{\mathcal{H}}_0\Psi_m^{(0)}]^*\Psi_n^{(1)} = E_m^{(0)}\langle\Psi_m^{(0)}|\Psi_n^{(1)}\rangle \, ,$$

which can be proven by integrating by parts, is used (the left equality expresses the fact that the Hamiltonian is a hermitian operator). The terms for which $n = m$ give

$$E_n^{(1)} = \langle \Psi_n^{(0)} | \hat{\mathcal{H}}' | \Psi_n^{(0)} \rangle \ .$$

For $n \neq m$ one obtains

$$\langle \Psi_m^{(0)} | \Psi_n^{(1)} \rangle = \langle \Psi_m^{(0)} | \hat{\mathcal{H}}' | \Psi_n^{(0)} \rangle / (E_n^{(0)} - E_m^{(0)}) \ ,$$

and hence

$$\Psi_n^{(1)} = \sum_m \langle \Psi_m^{(0)} | \Psi_n^{(1)} \rangle \Psi_m^{(0)} = \sum_{m \neq n} \frac{\langle \Psi_m^{(0)} | \hat{\mathcal{H}}' | \Psi_n^{(0)} \rangle}{E_n^{(0)} - E_m^{(0)}} \Psi_m^{(0)} \ .$$

The first equality can be checked by taking the scalar product with $\Psi_m^{(0)}$, and ignoring the term with $n = m$. Indeed, the normalization of the wave function yields

$$1 = \langle \Psi_n | \Psi_n \rangle = \langle \Psi_n^{(0)} + \lambda \Psi_n^{(1)} + \dots | \Psi_n^{(0)} + \lambda \Psi_n^{(1)} + \dots \rangle = 1 + 2\lambda \mathrm{Re}[\langle \Psi_n^{(0)} | \Psi_n^{(1)} \rangle] + \dots$$

(... represent higher orders in the expansion). It follows that $\langle \Psi_n^{(0)} | \Psi_n^{(1)} \rangle = iy$ is purely imaginary. Therefore in this order in λ one finds

$$\Psi_n = \Psi_n^{(0)} + i\lambda y \Psi_n^{(0)} + \cdots = \exp[i\lambda y] \Psi_n^{(0)} + \dots \ .$$

The phase factor $\exp[i\lambda y]$ is cancelled in any calculation of a physical quantity and therefore may be ignored.

In a similar fashion, the terms of order λ^2 yield

$$\hat{\mathcal{H}}' \Psi_n^{(1)} + \hat{\mathcal{H}}_0 \Psi_n^{(2)} = E_n^{(0)} \Psi_n^{(2)} + E_n^{(1)} \Psi_n^{(1)} + E_n^{(2)} \Psi_n^{(0)} \ .$$

Again, the scalar product with $\Psi_n^{(0)}$ and the use of the previous results yield

$$E_n = E_n^{(0)} + \lambda \langle \Psi_n^{(0)} | \hat{\mathcal{H}}' | \Psi_n^{(0)} \rangle - \lambda^2 \sum_{m \neq n} \frac{|\langle \Psi_m^{(0)} | \hat{\mathcal{H}}' | \Psi_n^{(0)} \rangle|^2}{E_m^{(0)} - E_n^{(0)}} + \dots \ . \qquad \text{(C.3)}$$

One may obtain the same results by the variational method. This can be demonstrated by considering the ground state of the full Hamiltonian. As the functions $\Psi_n^{(0)}$ are a basis for the space of all functions, the required solution can be expanded in terms of this basis, i.e., $\Psi_0 = \sum_{n'} a_{n'0} \Psi_{n'}^{(0)}$. The first coefficient, of $n' = 0$, is $a_{00} = 1 + \mathcal{O}(\lambda)$. All other coefficients are of order λ. Since the wave function is normalized, $\langle \Psi_0 | \Psi_0 \rangle = \sum_{n'=0}^{\infty} |a_{n'0}|^2 = 1$. The average energy in this ground state is

$$\langle E \rangle = \langle \Psi_0 | \hat{\mathcal{H}}_0 + \lambda \hat{\mathcal{H}}' | \Psi_0 \rangle = \sum_{n,m} a_{n0}^* a_{m0} \langle \Psi_n^{(0)} | \hat{\mathcal{H}}_0 + \lambda \hat{\mathcal{H}}' | \Psi_m^{(0)} \rangle$$

$$= E_0^{(0)} + \sum_{n \neq 0} |a_{n0}|^2 (E_n^{(0)} - E_0^{(0)}) + \lambda \sum_{n,m} a_{n0}^* a_{m0} \langle \Psi_n^{(0)} | \hat{\mathcal{H}}' | \Psi_m^{(0)} \rangle \ .$$

The last sum contains a single term of order λ, contributed by $n = m = 0$ (one may write $a_{00} = 1$, since the corrections to this coefficient are of higher orders in λ because of the normalization). The terms in the sum in which one of the two indices is zero yield contributions of order λ^2. The result is hence

$$\langle E \rangle = \langle \Psi_0^{(0)} | \hat{\mathcal{H}}_0 + \lambda \hat{\mathcal{H}}' | \Psi_0^{(0)} \rangle + \mathcal{O}(\lambda^2) \,.$$

Namely, to linear order the correction to the ground-state energy is the average of the additional Hamiltonian $\lambda \hat{\mathcal{H}}'$, calculated with the original ground-state function. This is a special case of Eq. (C.1), where the "trial" function is that of the unperturbed ground state. Equation (C.3) shows that this result is exact to linear order in λ.

The second-order (in λ) correction of the energy, within the variational method, is found from the values of a_{n0} which yield a minimal average energy. Since these are complex numbers, a_{n0} and a_{n0}^* are considered as two independent variables, and $\langle E \rangle$ is differentiated with respect to each of them. Up to linear order in λ, this procedure produces

$$\frac{\partial \langle E \rangle}{\partial a_{n0}} = a_{n0}^* (E_n^{(0)} - E_0^{(0)}) + \lambda \langle \Psi_0^{(0)} | \hat{\mathcal{H}}' | \Psi_n^{(0)} \rangle + \mathcal{O}(\lambda^2) = 0 \,,$$

where $a_{00} = 1$ is used in the second term. The derivative with respect to a_{n0}^* gives the complex conjugate of this equation. Using the solutions for a_{n0} and a_{n0}^* of these equations in the expression for $\langle E \rangle$, yields

$$\langle E \rangle = E_0 + \lambda \langle \Psi_0^{(0)} \hat{\mathcal{H}}' | \Psi_0^{(0)} \rangle - \lambda^2 \sum_{n \neq 0} \frac{|\langle \Psi_0^{(0)} | \hat{\mathcal{H}}' | \Psi_n^{(0)} \rangle|^2}{E_n^{(0)} - E_0^{(0)}} \,,$$

which is identical to Eq. (C.3) (for the ground state), derived by perturbation theory. The variational-method treatment of excited states necessitates also the requirement that the state with the minimal energy is orthogonal to the ground state found above. The results do reproduce Eq. (C.3) for the excited states.

Perturbation theory for degenerate states. The derivation above is restricted to the case where the eigenstates of $\hat{\mathcal{H}}_0$ are not degenerate. When they are, i.e., there are more than a single state with the same energy $E_n^{(0)}$, Eq. (C.3) cannot hold; the denominator in the last term there vanishes for pairs of degenerate states, and the sum diverges. That divergence is eliminated as follows. Consider Eq. (C.2) for the set of degenerate states alone. These are denoted $\Psi_{n\ell}^{(0)}$, where the index ℓ distinguishes among the degenerate states. Then

$$\langle \Psi_{n\ell}^{(0)} | \hat{\mathcal{H}}' | \Psi_{n\ell'}^{(0)} \rangle = E_{n\ell}^{(1)} \delta_{\ell\ell'} \,.$$

The index ℓ is added to the energy on the right-hand side, as the perturbation-theory correction need not be the same for all (original) degenerate states. This equation is valid only when the original states are eigenstates of the perturbation Hamiltonian, i.e., $\hat{\mathcal{H}}' \Psi_{n\ell}^{(0)} = E_{n\ell}^{(1)} \Psi_{n\ell}^{(0)}$, such that the matrix $\langle \Psi_{n\ell}^{(0)} | \hat{\mathcal{H}}' | \Psi_{n\ell'}^{(0)} \rangle$ is diagonal. Had not

$\Psi_{n\ell}^{(0)}$ possessed this property, one would have to construct linear combinations of them, $\Psi_{n\ell}^{(0)} = \sum_\alpha c_{\ell\alpha} \Psi_{n\alpha}^{(0)}$, with coefficients $c_{\ell\alpha}$ such that

$$\hat{\mathcal{H}}' \Psi_{n\ell}^{(0)} = E_{n\ell}^{(1)} \Psi_{n\ell}^{(0)} \, , \quad \text{or} \quad \sum_\alpha c_{\ell\alpha} \hat{\mathcal{H}}' \Psi_{n\alpha}^{(0)} = E_{n\ell}^{(1)} \sum_\alpha c_{\ell\alpha} \Psi_{n\alpha}^{(0)} \, .$$

Performing the scalar product (from left) with $\Psi_{n\alpha'}^{(0)}$ gives

$$\sum_\alpha \langle \Psi_{n\ell'}^{(0)} | \hat{\mathcal{H}}' | \Psi_{n\alpha}^{(0)} \rangle c_{\ell\alpha} = E_{n\ell}^{(1)} c_{n\alpha'} \, . \tag{C.4}$$

The coefficients' vector, $\{c_{\ell\alpha}\}$ has thus to be an eigenvector of the matrix $\langle \Psi_{n\alpha'}^{(0)} | \hat{\mathcal{H}}' | \Psi_{n\alpha}^{(0)} \rangle$, with an eigenvalue $E_{n\alpha}^{(1)}$, the first-order correction to the energy. In many cases these eigenvalues are not identical to each other, and thus the degeneracy of the Hamiltonian $\hat{\mathcal{H}}_0 + \lambda\hat{\mathcal{H}}'$ is smaller than that of the original one $\hat{\mathcal{H}}_0$.

Other operators. Any arbitrary physical quantity, A, is associated with an hermitian operator $\hat{A}$, with eigenfunctions Ψ_α and real eigenvalues α, $\hat{A}\Psi_\alpha = \alpha\Psi_\alpha$. The ensemble of functions $\{\Psi_\alpha\}$ is a possible basis for the Hilbert space of the system's states. For instance, the **momentum operator** of a single particle is $\hat{\mathbf{p}} = -i\hbar\boldsymbol{\nabla}$. In a box of volume V, on which periodic boundary conditions are applied, the eigenstates of the momentum are $\Psi_{\mathbf{k}} = \exp[i\mathbf{k} \cdot \mathbf{r}]/\sqrt{V}$, and the eigenvalues are $\hbar\mathbf{k}$ (with discrete values dictated by the boundary conditions). As mentioned, the state $\Psi_{\mathbf{k}}$ is also an eigenstate of the Hamiltonian $\hat{\mathcal{H}} = \hat{\mathbf{p}}^2/(2m)$, with the eigenvalue $E(\mathbf{k}) = (\hbar\mathbf{k})^2/(2m)$. This eigenstate of the energy is degenerate, since $E(\mathbf{k}) = E(-\mathbf{k})$. The corresponding eigenstates are distinguished from each other according to the eigenvalues of the momentum (that are not degenerate). Another example is that of a single particle in a central potential, e.g., in the hydrogen atom. The eigenstates of the Hamiltonian with energies E_n are characterized also by the quantum numbers ℓ and m, which represent the eigenstates of $\hat{\mathbf{L}}^2$ and $\hat{L}_z$, respectively, where $\hat{\mathbf{L}} = \hat{\mathbf{r}} \times \hat{\mathbf{p}}$ is the operator of the **angular momentum**.

Commutative operators. In the two examples mentioned above, the basis functions of the Hilbert space are simultaneously eigenfunctions of more than one operator; in the example of the free particle in a box these are the Hamiltonian and the momentum; and for the particle in the central potential these are the Hamiltonian, the square of the angular momentum, and its component along one of the axes. The reason for the appearance of such combinations stems from a theorem which states that **the basis of the Hilbert space can comprise states which are simultaneously eigenstates of several operators provided that those operators commute with each other.** Two operators, $\hat{A}$ and $\hat{B}$, are commutative when $\hat{A}\hat{B}f = \hat{B}\hat{A}f$ for any arbitrary function f in space.

proof: When all basis states $\Psi_{\alpha\beta}$ are eigenstates of the two operators, with eigenvalues α and β, $\hat{A}\Psi_{\alpha\beta} = \alpha\Psi_{\alpha\beta}$ and $\hat{B}\Psi_{\alpha\beta} = \beta\Psi_{\alpha\beta}$, then

$$\hat{A}\hat{B}\Psi_{\alpha\beta} = \alpha\beta\Psi_{\alpha\beta} = \hat{B}\hat{A}\Psi_{\alpha\beta} \, , \quad i.e., \quad \hat{A}\hat{B}\Psi_{\alpha\beta} = \hat{B}\hat{A}\Psi_{\alpha\beta} \, .$$

As this identity pertains to all basis states, it is valid for any other function in the Hilbert space.

This is a sufficient condition as well. Suppose all eigenfunctions of $\hat{A}$ are known, i.e., $\hat{A}\Psi_\alpha = \alpha\Psi_\alpha$. Operating on this equation with the operator $\hat{B}$ gives $\hat{B}\hat{A}\Psi_\alpha = \alpha\hat{B}\Psi_\alpha$. The left hand-side, due to the commutativity, can be also written as $\hat{B}\hat{A}\Psi_\alpha = \hat{A}\hat{B}\Psi_\alpha$, and therefore $\hat{A}[\hat{B}\Psi_\alpha] = \alpha[\hat{B}\Psi_\alpha]$. It follows that the function $\hat{B}\Psi_\alpha$ is an eigenstate of $\hat{A}$, with the same eigenvalue α. Provided that the eigenstates of $\hat{A}$ are non-degenerate, this implies that $\hat{B}\Psi_\alpha$ is proportional to Ψ_α, i.e., $\hat{B}\Psi_\alpha = \beta\Psi_\alpha$, where β is the proportionality coefficient. Hence, Ψ_α is an eigenstate of $\hat{B}$, with the eigenvalue β. This completes the proof for **non-degenerate states of $\hat{A}$.**

We now demonstrate the proof for the simplest degenerate case, when the degeneracy of the eigenvalue α is 2. In this case there exists a two-dimensional subspace of the Hilbert space, that includes all eigenstates of $\hat{A}$ with the eigenvalue α. It is always possible to choose a orthonormal basis for this subspace, such that the basis functions are also eigenstates of $\hat{B}$. Had one begun with a different basis, with basis functions $\Psi_{\alpha 1}$ and $\Psi_{\alpha 2}$, then any other linear combination of these two basis functions is another eigenstate, with the same eigenvalue α. In particular, the functions $\hat{B}\Psi_{\alpha 1}$ and $\hat{B}\Psi_{\alpha 2}$ are eigenfunctions of $\hat{A}$, with the eigenvalue α. That is, $\hat{B}\Psi_{\alpha 1} = \beta_{11}\Psi_{\alpha 1} + \beta_{12}\Psi_{\alpha 2}$ and $\hat{B}\Psi_{\alpha 2} = \beta_{21}\Psi_{\alpha 1} + \beta_{22}\Psi_{\alpha 2}$. An arbitrary state in the subspace, $\chi = c_1\Psi_{\alpha 1} + c_2\Psi_{\alpha 2}$, under the operation of $\hat{B}$, becomes

$$\hat{B}\chi = c_1\hat{B}\Psi_{\alpha 1} + c_2\hat{B}\Psi_{\alpha 2} = c_1(\beta_{11}\Psi_{\alpha 1} + \beta_{12}\Psi_{\alpha 2}) + c_2(\beta_{21}\Psi_{\alpha 1} + \beta_{22}\Psi_{\alpha 2}) .$$

χ is an eigenstate of $\hat{B}$, with the eigenvalue β, for $\hat{B}\chi = \beta\chi = \beta(c_1\Psi_{\alpha 1} + c_2\Psi_{\alpha 2})$; carrying out a scalar product of this equation with the functions $\Psi_{\alpha 1}$ and $\Psi_{\alpha 2}$ yields

$$\begin{bmatrix} \langle\Psi_{\alpha 1}|\hat{B}|\Psi_{\alpha 2}\rangle & \langle\Psi_{\alpha 1}|\hat{B}|\Psi_{\alpha 2}\rangle \\ \langle\Psi_{\alpha 2}|\hat{B}|\Psi_{\alpha 1}\rangle & \langle\Psi_{\alpha 2}|\hat{B}|\Psi_{\alpha 2}\rangle \end{bmatrix} \begin{bmatrix} c_1 \\ c_2 \end{bmatrix} = \beta \begin{bmatrix} c_1 \\ c_2 \end{bmatrix} . \tag{C.5}$$

This equation resembles Eq. (C.4): two linear, homogeneous equations for two unknowns. These have a solution provided that β is an eigenvalue of the matrix on the left hand-side. The matrix in fact possesses two eigenvalues, β_1 and β_2, (that need not be identical), and two eigenstates χ_1 and χ_2. These functions are simultaneously eigenfunctions of the two operators. The extension of this proof to degeneracies higher than two enlarges the matrix one has to diagonalize, but leads to the same result.

C.1 Answer

Answer C.1.

a. The average value is $\langle E \rangle = (\hbar^2\alpha + m^2\omega^2/\alpha)/(4m)$; the minimum is obtained for $\alpha = \alpha_0$,

$$\alpha_0 = m\omega/\hbar ,$$

and is $\hbar\omega/2$. It is equal to the exact result.

b. The expectation value of the perturbation Hamiltonian is

$$\langle\hat{\mathcal{H}}'\rangle = 2\lambda\sqrt{\frac{\alpha}{\pi}}\int_0^\infty dx\, x^6 e^{-\alpha x^2} = \frac{15\lambda}{8\alpha^3}\ .$$

It follows that

$$\langle E\rangle = \frac{\hbar^2}{4m}\left(\alpha + \frac{\alpha_0^2}{\alpha}\right) + \frac{15\lambda}{8\alpha^3} = \frac{\hbar\omega}{4}\left(u + \frac{1}{u} + \frac{\beta}{u^3}\right)\ ,\quad u = \frac{\alpha}{\alpha_0}\ ,\quad \beta = \frac{15\hbar\lambda}{2m^3\omega^4}\ . \quad \text{(C.6)}$$

The extrema are obtained from the equation $1 - 1/u^2 - 3\beta/u^4 = 0$, whose solutions are $u^2 = [1 \pm \sqrt{1 + 12\beta}]/2$. As the negative sign yields an imaginary u, the only solution is the one with the positive sign, $u = \sqrt{[1 + \sqrt{1 + 12\beta}]/2}$. Inserted into Eq. (C.6), it yields the variational approximation for the ground-state energy. Expanding in powers of β, one finds

$$u = 1 + \frac{3\beta}{2} + \dots\ ,$$

$$\langle E\rangle = \frac{\hbar\omega}{4}(2 + \beta + \dots) = \frac{\hbar\omega}{2} + \frac{15\lambda}{8\alpha_0^3} + \dots\ .$$

Hence, in this order, the correction to the energy is just the expectation value of the perturbation Hamiltonian, calculated with the ground-state function of the Hamiltonian without the perturbation.

Chapter 5

Lattice vibrations

This chapter discusses on the small oscillations of atoms (or ions) around their equilibrium locations. This picture is based on the harmonic approximation, in which the potential energy of each atom is replaced by a parabola centered around the minimum of that energy. The eigenmodes of the lattice are expressed in terms of Fourier transforms of the vibrations' amplitudes of individual atoms. The eigenmodes and the related frequencies are derived from both the classical Newtonian equations of motion, which give rise to waves similar to sound waves that propagate in the material, as well as by the quantization of the vibrational modes, i.e., the phonons (the particles dual to these waves). The lattice vibrations exist at finite temperatures, and even at zero temperature ("zero-point motion"). They contribute to the specific heat of the material, and affect its stability. Anharmonic effects, that give rise to the thermal expansion and to the heat conductance of crystals, are presented. The experimental methods by which eigenfrequencies of lattice vibrations are detected, e.g, using inelastic scattering of neutrons or of light, are described. Finally, spin waves, the excitations of magnetic materials that are analogous to lattice vibrations, are considered.

5.1 Classical equations of motion in one dimension

Preface. The **Born-Oppenheimer approximation**, based on the assumption that the nuclei are static, is used in Chapter 4 to calculate the quantum energies of atoms or molecules. Alternatively, Chapter 4 presents calculations of the Coulomb energy of the ions located right at the sites of the periodic lattice. In reality, the nuclei (or the centers of mass of the atoms, molecules, or ions of which the crystal is built) oscillate around the lattice sites; the latter are in fact just their **average locations**. The term "atoms" used below refers to the centers of mass. These vibrations contribute significantly to the specific heat, to the thermal expansion, to the heat conductance, to the sound propagation, and to the temperature dependences of many properties of the crystal (for instance, the electrical conductance, see Chapter 6). The lattice vibrations cause the decay of the intensity of radiation

309

scattered off the crystal as the temperature is raised (the Debye-Waller factor) and lead to the melting of the crystal (the Lindemann criterion, Sec. 3.10).

Longitudinal vibrations in one dimension. The simplest case is that of a one-dimensional lattice of lattice constant a, built of identical atoms. At equilibrium, the atoms are located on the periodic lattice, at $\mathbf{R}_n^0 = na\hat{\mathbf{x}}$, where n is an arbitrary integer. Consider small longitudinal oscillations around these equilibrium sites, such that the instantaneous position of the atom is $\mathbf{R}_n = \mathbf{R}_n^0 + \mathbf{u}_n = (na + u_n)\hat{\mathbf{x}}$ (see Fig. 5.1). The **longitudinal vibrations** are the displacements of the atoms that are parallel to the direction of the lattice. Transverse vibrations along the normal directions are discussed in the following. It is also assumed, for simplicity, that forces are active only between nearest neighbors, and that the force between two neighbors depends solely upon the distance in-between them. That is, the force is derived from a potential energy of the form $U(R_{n,n+1})$, where $R_{n,n+1} = R_{n+1} - R_n = a + u_{n+1} - u_n$. (The vector notations are superfluous when discussing a one-dimensional lattice.) The total potential energy is $U_{\text{tot}} = \sum_n U(R_{n,n+1})$, where the sum runs over the N atoms in the lattice. To avoid complications at the end points, it is assumed for the time being that the lattice is infinite. The boundary conditions are accounted for afterwards.

In Chapter 4 it is found that the equilibrium configuration corresponds to the pattern for which the potential energy is minimal. In the present context this is achieved when each of the summands is minimal, i.e., when the atoms reside in the periodic-lattice sites, $R_{n,n+1} = R_{n+1} - R_n = a$ or $u_n = 0$. At the minimum, the derivative vanishes, $U'(R)|_{R=a} = 0$. According to the Lindemann criterion, the deviation from the equilibrium location is smaller than the distance between nearest neighbors, lest the crystal melts into the liquid phase. Thus we assume that the deviation is much smaller than the lattice constant, $|u_n| \ll a$, and consequently the potential energy can be expanded as a power series in u_n,

$$U_{\text{tot}} = \sum_n [U(a) + U'(a)(u_{n+1} - u_n) + U''(a)(u_{n+1} - u_n)^2/2 + \ldots] ,$$

($\ldots$ indicate higher orders in the u_n's). In an infinite lattice the linear terms disappear, as u_n attains opposite signs in the $(n-1)$th term and in the nth one. (In the finite lattice, the boundary conditions ensure that these terms vanish.) Also, when the forces act only between nearest neighbors then $U'(a) = 0$ as no force is exerted at equilibrium. (Recall that $-\partial U_{\text{tot}}/\partial u_n$ is the force that acts on the nth atom, and this force vanishes at equilibrium when $u_n = 0$.) Keeping only the second-order terms,

$$U_{\text{tot}} = U_0 + \frac{D_L}{2}\sum_n (u_n - u_{n+1})^2 , \tag{5.1}$$

where $D_L = d^2U(x)/dx^2|_{x=0}$ (L indicates the **longitudinal vibrations**). Adding the kinetic energy of the vibrations, the Hamiltonian (that corresponds to the total energy) is

$$\mathcal{H} = \sum_n \frac{p_n^2}{2M} + U_{\text{tot}} , \qquad (5.2)$$

where p_n is the momentum of the nth atom, and M is the mass of each atom in the crystal.

The classical equations of motion. Newton's second law for the potential energy (5.1) yields

$$M\ddot{u}_n = -\frac{\partial U_{\text{tot}}}{\partial u_n} = -D_L(u_n - u_{n-1}) - D_L(u_n - u_{n+1}) . \qquad (5.3)$$

(A dot above the letter indicates a derivative with respect to time.) The right hand-side contains the forces exerted on the nth atom by its left and right neighbors. Had it been assumed that each pair of atoms is connected by a spring, with a spring constant D_L and an equilibrium length a, the result would have been the same: each of the forces is given by the spring constant times the elongation (or the shortening) of the spring as compared with its equilibrium length, as in **Hooke's law**. Hooke's law applies also to the oscillations of a **harmonic oscillator**; hence the term **"harmonic approximation"** for this description. The higher-order terms in the expansion of the energy are thus named **anharmonic terms** (they are discussed below). Figure 5.1 displays the equilibrium configuration of the lattice, as well as the instantaneous one.

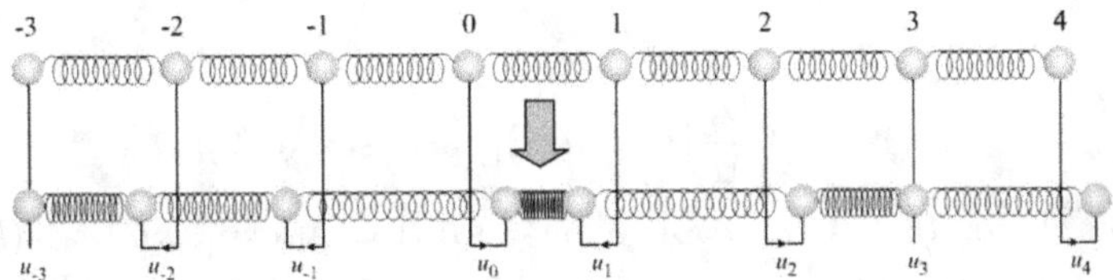

Fig. 5.1: Longitudinal oscillations of a one-dimensional, single-atom, lattice. Upper row: equilibrium configuration, the length of each spring (helical lines) is the lattice constant a. Lower row: an instantaneous configuration of the vibrating lattice, the nth atom is located at $x_n = na + u_n$. The deviations from the equilibrium sites are marked by the small arrows.

Dispersion laws. Techniques for solving equations of the form (5.3) are detailed in the following. Here it suffices to present the solution as a sum over "monochromatic" waves which oscillate in time with definite frequencies ω. Consider one of these monochromatic solutions, which can be written as $u_n(t) = A_n \exp[-i\omega t]$, where the complex amplitude A_n depends on the coordinate (i.e, on n), but not on time. It is convenient to represent the deviation $u_n(t)$ as a complex function; at the end of the calculation only the real part (or alternatively, the imaginary one) is to be taken into account. Both real and imaginary parts solve the same equation; the choice between them, or the proper linear combination, is made according to the boundary and initial conditions. With this form of the deviation, the equation of motion (5.3) becomes

$$M\omega^2 A_n = D_L(2A_n - A_{n-1} - A_{n+1}) \,. \tag{5.4}$$

In an infinite lattice, one may shift all lattice points by the lattice constant, to obtain the equation

$$M\omega^2 A_{n+1} = D_L(2A_{n+1} - A_n - A_{n+2}) \,.$$

Denoting $B_n = A_{n+1}$ shows that B_n solves precisely the same equation as A_n, i.e., Eq. (5.4): $M\omega^2 B_n = D_L(2B_n - B_{n-1} - B_{n+1})$. As a linear homogeneous equation has a single solution up to a multiplicative factor, one concludes that $B_n = CA_n$, where C is yet to be determined. Hence, the solution is such that $A_{n+1} = CA_n$. Repeating this procedure over and over again, assigning the origin to the point $n = 0$, yields $A_n = C^n A_0$. Had the absolute value of C been different from 1, then the solution $|A_n|$ would have attained very large values for positive (or negative) n's. This is not acceptable, since the deviation must be small. Hence C has to be a complex number, of unit absolute value, i.e., $C = \exp[ika]$, where k is real, leading to $A_n = \exp[ikna]A_0$. This argumentation is typical to **periodic systems**, and the result is a particular case of the **Bloch theorem** presented in Chapter 6. Inserting this result into Eq. (5.4) yields a relation between the vibration frequency ω and k,

$$M\omega^2 = D_L(2 - e^{-ika} - e^{ika}) = 2D_L[1 - \cos(ka)] = 4D_L \sin^2(ka/2) \,, \tag{5.5}$$

namely,

$$\omega(k) = 2\sqrt{\frac{D_L}{M}}\,|\sin(ka/2)| = \omega_0|\sin(ka/2)| \,. \tag{5.6}$$

The solution of Eq. (5.4) is $u_n = A_n \exp[-i\omega(k)t] = A_0 \exp[i(kna - \omega(k)t)]$, well-known from wave mechanics. It is a **wave**, with a **wave number** k and a **wave length** $\lambda = 2\pi/|k|$, propagating along the positive $\hat{x}$−axis when $k > 0$ with a **phase velocity** $v_{\text{ph}} = \omega/k$ [e.g., the maxima of the real part of u_n are located at the points where $kna - \omega t = 0$, and therefore they are moving to the right according to $na = (\omega/k)t = v_{\text{ph}}t$. The wave length is the distance in-between neighboring maxima.] Similarly, the function $u_n = A_0 \exp[i(kna - \omega(k)t)]$, with $k < 0$, that describes a wave moving along the opposite direction, also solves Eq. (5.4), with the same frequency. The function $\omega(k)$ is the **dispersion relation** relating the frequency to the wave number; it is displayed in Fig. 5.2. The frequency is symmetric with respect to the sign of the wave number, i.e., there are two solutions for each frequency.

Periodicity and Brillouin zones. An important property of the solution derived in the previous paragraph is the **periodicity** in reciprocal space: Eq. (5.6) obeys

$$\omega(k) = \omega(k + 2\pi h/a) = \omega(k + G) \,, \tag{5.7}$$

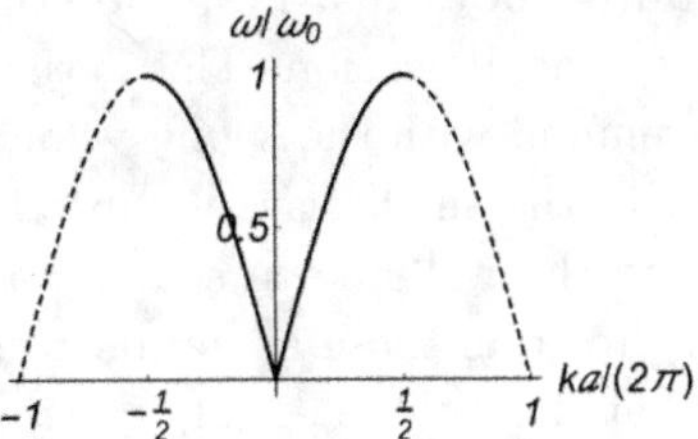

Fig. 5.2: Dispersion relation of a one-dimensional, single-atom lattice. The thick line marks the dispersion within the first Brillouin zone (see below). The frequency is measured in units of $\omega_0 = 2\sqrt{D_L/M}$, and the wave number in units of the basis vector of the reciprocal lattice, $G_0 = 2\pi/a$.

where $G = h(2\pi/a)$ is an arbitrary **reciprocal-lattice vector** of the one-dimensional lattice (h is an arbitrary integer). Recall (Sec. 3.5) that such a vector displaces a point from one Brillouin zone to another zone. Equation (5.7) expresses the fact that the frequency is a periodic function of the wave number, with a period $2\pi/a$ (as seen in Fig. 5.2). Another aspect of this periodicity refers to the wave functions u_n: the solution with the wave number k is also periodic between different Brillouin zones, i.e.,

$$u_n(k) = A_0 e^{i(kna - \omega(k)t)} = A_0 e^{i(kna + 2\pi hn - \omega(k)t)} = u_n(k + 2\pi h/a) = u_n(k + G) .$$

It follows that it suffices to obtain the solutions for wave numbers within the **first Brillouin zone** of the reciprocal lattice, $-\pi/a \le k \le \pi/a$. All other solutions are identical to those, and do not convey any additional information. As found below, this conclusion is quite general. In all cases **it suffices to calculate the dispersion laws and the wave functions within the first Brillouin zone.** This is the main reason for the introduction of Brillouin zones in Sec. 3.5

The continuum limit. The relation (5.6) has other remarkable properties. When the wave length is much longer than the lattice constant, $a \ll \lambda$ or $ka \ll 1$, the right hand-side may be expanded in a Taylor series in ka; this leads to a **linear dispersion relation**, $\omega = c|k|$, where $c = \omega_0 a/2 = a\sqrt{D_L/M}$ is the **velocity of the longitudinal wave**. This is the **sound velocity** in the material. (The sound velocity coincides with the phase velocity for long wave lengths.) Indeed, when the lattice constant is very small, the discrete nature of the material becomes approximately continuous along the $\hat{\mathbf{x}}$−axis, with a mass density per unit length $\rho = M/a$. The location of the atom can then be viewed as a continuous variable, $na \to x$; consequently, $2u_n - u_{n-1} - u_{n+1} \approx -u''(x)a^2$ (the double prime denotes the second derivative with respect to x). With these assumptions Eq. (5.3) becomes the continuum wave equation,

$$\rho \frac{\partial^2 u}{\partial t^2} = Y \frac{\partial^2 u}{\partial x^2} , \tag{5.8}$$

where $Y = D_L a$ is the **Young modulus, i.e., the compressibility** of the material, in units appropriate for one dimension. This is the equation for **longitudinal sound waves** in the continuum, with the sound velocity $c = a\sqrt{D_L/M} = \sqrt{Y/\rho}$ (see Appendix D). Hence, in the small wave-number limit, the wavy solutions of the equations of motion are the well-known sound waves in the continuum. For this reason it is customary to term those as **acoustic solutions**. As for acoustic waves $ka \ll 1$, it follows that the ratio $A_{n+1}/A_n = \exp[ika] \approx 1$, and therefore the displacements of nearest neighbors are quite close to one another, justifying the use of a continuous variable, $na \to x$. More details of the continuum limit are given in Appendix D.

"Forbidden" frequencies. As opposed to sound waves in the continuum, Eq. (5.6) yields solutions with a real wave number only for frequencies smaller than $\omega_0 = 2\sqrt{D_L/M} = 2c/a$. The sound velocity in solids is typically of the order of 10^5 cm/sec, the lattice constant is of the order of $1\mathring{A} = 10^{-8}$cm, and hence $\omega_0 \approx 2 \times 10^{13}$ sec^{-1}, like the frequencies of light waves in the infra red. When one tries to force upon a crystal a vibration of higher frequency, the solutions of Eq. (5.6) become complex, with a complex "wave number" $k = \pi/a + i\kappa$, where $\omega/\omega_0 = \cosh(\kappa a/2)$ (check!). However, inserting this expression into the wave function yields $|u_n| = |A_0| \exp[-n\kappa a]$, a solution which diverges to infinity when n is a very large positive or negative integer (depending on the sign of κ). This solution, being far away from the harmonic approximation (that assumes small displacements), is not acceptable in the infinite system. (Recall the Lindemann criterion which states that large displacements cause the lattice to disintegrate.) In a semi-infinite system these frequencies correspond to solutions that decay as the wave propagates into the lattice (the diverging solution is rejected on physical grounds). In other words, vibrations with such frequencies cannot propagate over large distances within the material. The maximal frequency that still allows for a wavy solution, ω_0, is reached at $ka = \pi$, i.e., when the wave vector is at the edge of the first Brillouin zone, or when $\lambda = 2a$. This is the shortest wave that can propagate in the system. (The physical origin is clear: shorter waves describe vibrations of the material in-between neighboring atoms, but in the lattice there are no atoms there; i.e., these vibrations are meaningless.) For this wave vector, $u_n(k = \pi/a) = A_0(-1)^n \exp[i\omega t]$; this is a **standing wave** in which neighboring atoms vibrate opposite to one another (contrary to the acoustic wave, for which neighboring atoms vibrate in unison along the same direction).

Wave packets. The appearance of a standing wave at the edge of the Brillouin zone can be understood by examining the **group velocity** of the waves. As mentioned, the solutions of the type $u_n = A_0 \exp[i(kna - \omega t)]$ describe propagating **monochromatic waves** (of a single frequency). As all these waves, with various frequencies, solve the original equations of motion (5.3), the general solution is a linear combination of them,

$$u_n(t) = \sum_k A(k) e^{i[ka - \omega(k)t]} , \qquad (5.9)$$

where $\omega(k)$ is given in Eq. (5.6), and the coefficients $\{A(k)\}$ are determined by the appropriate initial and boundary conditions. [For the finite-length system, the allowed values of k are discrete, and then Eq. (5.9) is a **Fourier series** of the vibration. A continuum of k values is allowed for the infinite system, and then the sum in Eq. (5.9) becomes an integral, which is the **Fourier transform** of the vibration, see Appendix B.] The solution (5.9) is a **wave packet**. The maximal intensity of this packet is reached at k values around $k = k_0$, where the amplitude $A(k)$ is maximal. As the amplitude is multiplied by the phase factor $\exp[ikna - i\omega(k)t]$, the intensity is larger when this phase factor is slowly varying around k_0. (For other values of the wave number the oscillating term changes its sign upon small variations of k, making the contribution to the sum much smaller.) This slow variation of the phase, and consequently this peak in the sum, are obtained for $(\partial/\partial k)[kna - \omega(k)t]_{k=k_0} = 0$, that is, for $na = v_{\mathrm{g}}t$, where the **group velocity** v_{g} is

$$v_{\mathrm{g}} = [\partial\omega/\partial k]|_{k=k_0} . \qquad (5.10)$$

The group velocity is the speed with which the peak of the wave packet evolves in time. In the first Brillouin zone, $(-\pi/a \leq k_0 \leq /\pi/a)$, this velocity is $v_{\mathrm{g}} = \pm c\cos(k_0 a/2)$, with the sound velocity $c = \omega_0 a/2 = a\sqrt{D_L/M}$, and with a sign given by that of k_0 [see Eq. (5.6)]. Thus, the group velocity is maximal (in its absolute value) and equals the sound velocity (of the acoustic limit) at the center of the Brillouin zone, and vanishes at the zone edges. Zero group velocity indeed implies standing waves, as found above.

Problem 5.1.
a. *Find the dispersion relation for the longitudinal vibrations of a single-atom, one-dimensional lattice with "spring constants" $D_{L,m}$ between neighbors located at a distance ma from one another. Find the group velocity, the condition for the dispersion law to be linear at small wave-numbers, and the sound velocity in that case.*
b. *Find the dispersion relations for the longitudinal vibrations of a single-atom, one-dimensional lattice with the Lenard-Jones interaction, Eq. (4.33), among the atoms. What form is taken by these relations at long wave-lengths?*
c. *Plot the dispersion relation for $D_{L,2} = D_{L,1}/2$, and $D_{L,m} = 0$ for $m \geq 2$. How many wavy solutions are there for each value of the frequency?*

Transverse vibrations. Suppose that the atoms can move out of the one-dimensional lattice, i.e., along the $\hat{\mathbf{z}}$ and $\hat{\mathbf{y}}$ directions. The location of the nth atom is then a three-dimensional vector, $\mathbf{R}_n = na\hat{\mathbf{x}} + \mathbf{u}_n$. Assuming **central forces** between pairs of atoms, the potential energy of the interaction between two atoms located at a distance $\mathbf{R}_{n,n+m} = ma\hat{\mathbf{x}} + \mathbf{u}_{n+m} - \mathbf{u}_n$ from one another depends solely on the length of this vector. For reasons to be clarified later on, the present analysis includes also pairs of atoms that are far apart. Taylor-expanding the length of the vector $R_{n,n+m} \equiv |\mathbf{R}_{n,n+m}|$ yields

$$R_{n,n+m} = ma + \delta u_x + (\delta u_\perp)^2/(2ma) + \dots \,,$$

with $\delta \mathbf{u} = \mathbf{u}_{n+m} - \mathbf{u}_n$, the square of the transverse displacement is $(\delta u_\perp)^2 = (\delta u_y)^2 + (\delta u_z)^2$, and ... represents higher-order powers of the deviations from equilibrium (check!). Inserting this expansion into the Taylor expansion of the corresponding potential yields

$$U(R_{n,n+m}) = U(ma) + U'(ma)\left(\delta u_x + \frac{(\delta u_\perp)^2}{2ma}\right) + U''(ma)\frac{(\delta u_x)^2}{2} + \dots \,. \quad (5.11)$$

The term linear in $\delta u_x = u_{n+m,x} - u_{n,x}$ vanishes as before, since at equilibrium the resultant force on each atom vanishes. The equation of motion for the transverse vibrations, e.g., along $\hat{\mathbf{y}}$, is $M\ddot{u}_{n,y} = -\partial U_{\text{tot}}/\partial u_{n,y}$. (The one for the longitudinal oscillations is identical to that derived in problem 5.1.) In the harmonic approximation,

$$M\ddot{u}_{n,y} = -\sum_m D_{T,m}(2u_{n,y} - u_{n-m,y} - u_{n+m,y}) \,, \quad \text{with} \quad D_{T,m} = U'(ma)/(ma) \,.$$

When the force is active only between nearest neighbors, $U'(a) = 0$, leading to $M\ddot{u}_{n,y} = 0$. That is, in this approximation any transverse displacement remains constant or increases with time until it becomes too long to be in the regime of small oscillations, and thus negates the harmonic approximation. This pitfall can be overcome by the inclusion of anharmonic terms, that give rise to nonlinear equations of motion. These are beyond the scope of this book (but do note Secs. 5.5 and 5.8). It can be also overcome by adding the effect of farther-away neighbors. Conjecturing that the solution is still wavy, of the form $u_{n,y} = B_0 \exp[i(kna - \omega_T t)]$, results in the dispersion relation

$$M\omega_T^2 = 4\sum_m D_{T,m}\sin^2(mka/2) = 4\sum_m U'(ma)\frac{\sin^2(mka/2)}{ma}$$

$$\Rightarrow \sum_m U'(ma)(ma)\left[k^2 - (ma)^2\frac{k^4}{12} + \dots\right] \,,$$

where the second expression pertains to the large wave-length limit. As the equilibrium energy per site, $u_{\text{tot}} = 2\sum_{m=1}^{\infty} U(ma)$, is minimal at the lattice constant, the derivative there vanishes, i.e., $\sum_{m>0} mU'(ma) = 0$. This implies that the first term on the right hand-side (of the large wave-length limit result) disappears, and there is no acoustic behavior (for which the frequency is proportional to the wave vector). The leading term yields a quadratic dependence of the frequency on the wave number, $\omega_T \propto k^2$, assuming that $\sum_m U'(ma)(ma)^3 < 0$. For instance, when there are only two terms in the sum, then $U'(2a) = -U'(a)/2$, and the sum equals $U'(a) + 8U'(2a) = -3U'(a)$. Generally, the sum is indeed negative. Had it been positive, ω_T would have been imaginary; the crystal is then unstable, and anharmonic terms must be added to stabilize it. In such cases there appears a new minimum in the energy, which describes a new crystalline structure.

Bending transverse vibrations. Another possibility to stabilize transverse vibrations is to add **non-central forces**, e.g., forces that depend on the **angles between neighboring bonds**. Consider for instance the covalent bond. Section 4.3 lists several hybridizations of wave functions on an atom in the lattice, e.g., of the sp type, or the sp^2 one, which are hybridized with wave functions on the neighboring atoms. In a linear chain like Be–Be–Be–Be–Be–... there are two electrons on each beryllium atom occupying the hybridized atomic states of the type sp, $\Psi_{\pm x} = [\psi_{200} \pm \psi_x]/\sqrt{2}$, and each of them creates a covalent bond with an electron on a neighboring atom. These bonds are located on the molecule axis, and the angle between nearest neighbors is $180°$. The energy of the bond depends upon the distance between the neighboring atoms, and its minimal value determines the lattice constant of this one-dimensional crystal. When the transverse displacements are ignored, an expansion of this potential, as done in Eq. (5.1) gives rise to longitudinal vibrations like the ones discussed at the beginning of this section. On the other hand, a traverse motion of an atom modifies the angle between the bonds that this atom has with its neighbors, say from π to γ; the hybridized states then assume the form $\Psi_i = A[\psi_{200} + \beta \hat{\mathbf{n}}_i \cdot \boldsymbol{\psi}]$ (see problem 4.9) where $\hat{\mathbf{n}}_{1,2}$ are unit vectors along the new directions of the two bonds. Part (f) of problem 4.9 details the dependence of the coefficients A and β on the angle γ. Because it is an angle between two bonds, the effective interaction under consideration depends on the coordinates of **three atoms**, extending the pair potential given in Eq. (5.1). The new terms in the interaction refer to **bending, flexural, or buckling deviations** of the lattice. As the angle γ decreases, the probability clouds of the two electrons in the atom approach one another, thus increasing the repulsive Coulomb energy. When the deviations are small one may express the energy as a power series of the small parameter $(1 + \cos\gamma)$; problem 5.2 demonstrates that the dispersion law obtained from this analysis is proportional to k^2 in the long-wave limit.

The transverse vibrations are in particular complex in a one-dimensional lattice, because there are no atoms surrounding the chain that can exert transverse forces and hinder the "runaway" of the atoms along the perpendicular directions. Similar difficulties arise for vibrations normal to the plane of a two-dimensional crystal, e.g., graphene. In that case again the lattice is stabilized by additional interactions, like those described above (see also Sec. 5.5). Coupling the lattice to the substrate upon which it lies may also stabilize the transverse vibrations. These difficulties disappear in three dimensions, where each atom is surrounded all around by other atoms. Below it is simply assumed that the spring constants of the transverse oscillations are finite.

Problem 5.2.

Write down an expression for the dependence of $(1+\cos\gamma_n)$ on the displacements of the atoms, where γ_n is the angle between two successive bonds in a one-dimensional *lattice, coupling the atoms at $n-1$, n, and $n+1$; find the leading terms of the result for small deviations (longitudinal and transverse). Write down the equations of motion for the transverse vibrations and obtain the dispersion relation.*

Crystal with a basis. Next consider a one-dimensional lattice with **two atoms (or ions) in the unit cell**, e.g., the one-dimensional version of table salt, Fig. 4.4. The two atoms have different masses, M and m, but all longitudinal springs are identical, with a spring constant D. The longitudinal displacements in the nth unit cell, u_n and v_n, are defined in Fig. 5.3. The equations of motion for those are

$$M\ddot{u}_n = -D(u_n - v_{n-1}) - D(u_n - v_n) \, ,$$
$$m\ddot{v}_n = -D(v_n - u_n) - D(v_n - u_{n+1}) \, . \tag{5.12}$$

These equations are invariant under the translation $n \to n+1$; one may expect then wavy solutions of the form $u_n = A\exp[i(kna - \omega t)]$ and $v_n = B\exp[i(kna - \omega t)]$. Inserting those into Eqs. (5.12) shows that these are indeed the solutions, provided that the coefficients obey the two equations

$$M\omega^2 A = D[2A - (e^{-ika} + 1)B] \, ,$$
$$m\omega^2 B = D[2B - (1 + e^{ika})A] \, . \tag{5.13}$$

These two homogeneous linear equations, for the two unknowns A and B, have a unique solution provided that the determinant of the coefficients vanishes, i.e.,

$$(M\omega^2 - 2D)(m\omega^2 - 2D) - 4D^2\cos^2(ka/2) = 0 \, . \tag{5.14}$$

This is a quadratic equation for ω^2, whose solutions are

$$\omega_\pm^2 = \frac{D}{\mu} \pm D\sqrt{\frac{1}{\mu^2} - \frac{4\sin^2(ka/2)}{Mm}} \, , \tag{5.15}$$

where $\mu = mM/(M+m)$ is the **reduced mass** of the atoms in the unit cell. The two solutions for the frequency are plotted in Fig. 5.4. They are both periodic with the periodicity given in Eq. (5.7), and hence it suffices to examine them in the first Brillouin zone, $-\pi/a \le k \le \pi/a$. As opposed to the example of a single-atom lattice, each wave vector of the two-atom chain corresponds to **two frequencies**. In the limit of small wave numbers the lower one is

$$\omega_-^2 \approx \frac{Da^2}{2(M+m)}k^2 = c^2 k^2 \, . \tag{5.16}$$

This is precisely the acoustic frequency of a sound wave in a continuum of mass density $\rho = (M+m)/a$ [see the discussion before Eq. (5.8)]. The higher frequency is approximately

$$\omega_+^2 \approx \frac{2D}{\mu} - \frac{Da^2}{2(M+m)}k^2 \, . \tag{5.17}$$

$\sqrt{2D/\mu}$ is a rather high frequency, of the order of the frequency of light in the infra red. Because of this, the lower branch of the dispersion relation is called the

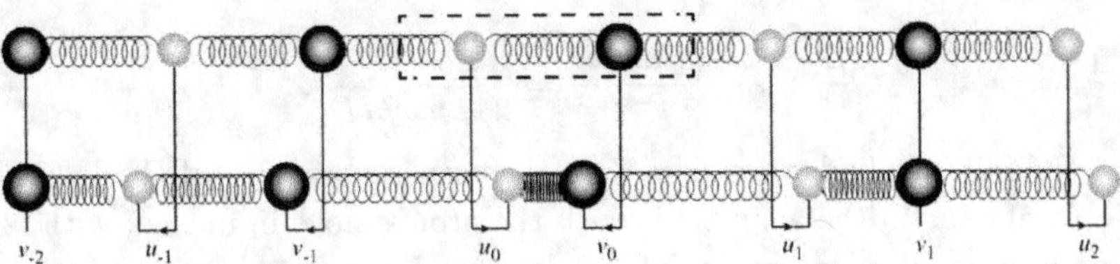

Fig. 5.3: Oscillations of a one-dimensional two-atom lattice. Top: the equilibrium configuration. The dashed frame encloses the unit cell at $n = 0$. The nth cell contains two atoms (or ions); the bright circle represents an atom of mass M and the dark one an atom with mass m. The equilibrium distance between the two atoms in the unit cell is b, and the lattice constant (the length of the dashed rectangle) is $a = 2b$. Bottom: an instantaneous glimpse at the vibrating lattice; the atoms in the nth cell are located at $na + u_n$ and $na + v_n + b$.

"acoustic branch" and the upper one is termed the **"optical branch"**. One has to keep in mind, though, that these are not electromagnetic waves.

As the wave number increases towards the edge of the Brillouin zone, the acoustic frequency increases while the optical one decreases. At $k = \pi/a$ the two frequencies are (assuming that $M > m$) $[\omega_-(\pi/a)]^2 = 2D/M$ and $[\omega_+(\pi/a)]^2 = 2D/m$. Thus, there is a **frequency gap** in-between the two: in the range $2D/M \leq \omega^2 \leq 2D/m$ there are no wavy solutions. All solutions in this range decay (or increase) exponentially with the distance. In addition, as in the single-atom one-dimensional lattice, solutions with frequencies above the maximum of the dispersion relation, i.e., above the optical branch $\omega^2 \geq \omega_0^2 = 2D/\mu$, do not propagate.

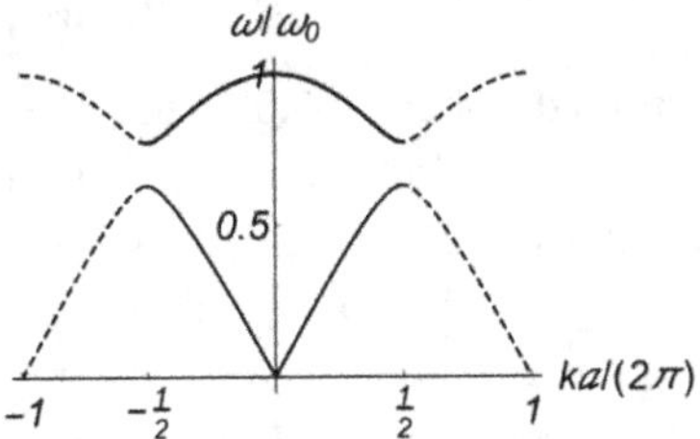

Fig. 5.4: Dispersion relation of a one-dimensional, two-atom lattice, with $M = 3m/2$. The solid line shows the dispersion in the first Brillouin zone of this crystal, whose unit cell contains two atoms. The frequency is measured in units of the maximal optical frequency, $\omega_0 = \sqrt{2D/\mu}$.

It is illuminating to consider the amplitudes themselves. For each of the frequencies $\omega = \omega_\pm$, Eqs. (5.13) yield only the ratio between the two amplitudes,

(explain!)

$$\frac{B}{A} = \frac{2 - M\omega^2/D}{1 + e^{-ika}} = \frac{1 + e^{ika}}{2 - m\omega^2/D} \ . \tag{5.18}$$

In the $k \approx 0$ limit, one finds that $A \approx B$ for the acoustic branch and $B \approx -(M/m)A$ for the optical one. In the acoustic branch the atoms move in unison, with almost equal amplitudes of displacements. This indeed can be expected for very long waves, for which the wave function changes only slightly over short distances. A similar situation occurs for the sound waves of the single-atom lattice. For this reason the sound velocity in Eq. (5.16) contains the total mass of the two atoms, $(M + m)$: the two atoms oscillate together as a single entity. On the other hand, in the optical branch the signs of the two amplitudes are opposite to one another. The two atoms in the unit cell vibrate one against the other; the instantaneous distance between them oscillates periodically while their center of mass remains immovable. [The displacement of the center of mass is proportional to $(MA + mB)$.] This is the reason for the optical frequency in Eq. (5.17) to depend on the reduced mass of the two atoms, $\mu = Mm/(M + m)$. A reduced mass always corresponds to a relative motion of two particles in the rest reference frame of the center of mass. The picture is significantly modified for wave numbers at the edge of the Brillouin zone, $k = \pi/a$. There $[\omega_-(\pi/a)]^2 = 2D/M$ and $[\omega_+(\pi/a)]^2 = 2D/m$, and therefore Eq. (5.18) yields $A = 0$ for the optical branch and $B = 0$ for the acoustic one. (One should be cautious with the order of limits, though, as both numerator and denominator vanish.) The other atom vibrates with an amplitude $(-1)^n$; this is again a standing wave, for which one of the atoms remains at rest and the other oscillates around its equilibrium location.

The limit of identical atoms. Equation (5.15) yields two frequencies for each wave number even for identical masses,

$$\omega_\pm^2 = 2D[1 \pm \cos(ka/2)]/M \ , \tag{5.19}$$

as shown on the right hand-side of Fig. 5.5. Since the masses are equal, it could have been expected that the solution coincides with that obtained for a single-atom chain, with a lattice constant $b = a/2$, and with a double-size Brillouin zone, $-2\pi/a = -\pi/b \le k \le \pi/b = 2\pi/a$, as portrayed in Fig. 5.2 and reproduced by the thick line in the left panel of Fig. 5.5. As implied from Eq. (5.5), $\omega^2 = 2D[1 - \cos(kb)]/M = 2D[1 - \cos(ka/2)]/M$ for the single-atom chain. This result coincides with the acoustic branch of the "new" frequency for $|k| \le \pi/a$, and with the optical one for $\pi/a \le |k| \le 2\pi/a$. Due to the periodicities of the two-atom chain solutions, the wave number may be shifted by a wave vector of the reciprocal lattice, $k \to k - 2\pi/a$, which transforms the left panel of Fig. 5.5 into the right one (or *vice versa*). When the two masses are the same, both descriptions are identical: the frequency is given either by the thick line in the left panel of Fig. 5.5 with the wider Brillouin zone of the single-atom chain, or by the two thick lines in the right panel there, with the narrower Brillouin zone of a lattice with a basis of two atoms. Recall that at times it is convenient to describe a crystal by a larger unit cell with several atoms rather than by the primitive one. In the present context, the result is a smaller Brillouin zone, which contains several branches. The conclusion is that **the number of longitudinal frequencies which pertain to each wave vector coincides with the number of atoms in the cell.** When the transverse frequencies are included, the total number of frequencies is the number of the atoms in the unit cell times the dimensions (i.e., times 3 in three dimensions). In three dimensions, three frequencies are acoustic (i.e., approach zero in the limit of long waves) and the rest are optical (i.e., tend to a finite, relatively large, constant frequency).

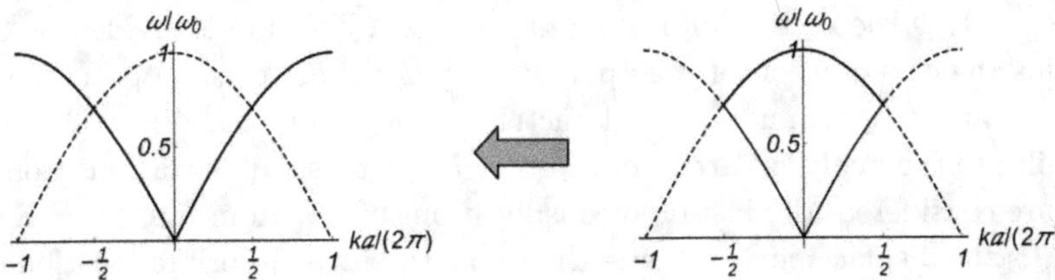

Fig. 5.5: Dispersion relations of two equal masses in the unit cell. The thick lines in the right panel display the optical and the acoustic branches within the first Brillouin zone of the lattice with the double-size unit cell, $-\pi/a \le k \le \pi/a$. The thick lines in the left panel are the dispersion relations of the single-atom lattice, whose Brillouin zone is $-2\pi/a = -\pi/b \le k \le \pi/b = 2\pi/a$.

Problem 5.3.
Show that Eq. (5.14) is the eigenvalue equation of a 2×2 matrix, and that Eq. (5.18) yields the eigenvectors. Determine the matrix.

Problem 5.4.
The two-atom crystal depicted in Fig. 5.3 may describe an ionic crystal, in which the ions of mass m are positively charged, with charge +e, while those of mass M are negative, with charge −e. An electromagnetic wave, whose electric field oscillates along the longitudinal direction with frequency ω, is applied on this system. The wave vector is $k = \omega/c_{\text{light}}$, where c_{light} is the speed of light. Thus, at a point $x = na$ the electric field is $\mathcal{E}(n,t) = \mathcal{E}_0 \exp[i(kna - \omega t)]$. Find the induced dipole moment. Find the frequency dependence of the dielectric function.

Problem 5.5.
Suppose that there are additional "springs", of spring constant D_2, between equal-mass atoms in neighboring cells of the two-atom one-dimensional lattice of Fig. 5.3. Find the dispersion relations of the longitudinal oscillations. Discuss the $D \to 0$ limit.

Problem 5.6.

As mentioned, the polyyne crystal contains two types of bonds ... $C - C \equiv C - C \equiv$
C .... [Fig. 1.7(b)] The spring constants of these two types differ from each other.
Write down the equations of motion for the longitudinal oscillations of this crystal.
Find the dispersion relations.

5.2   Classical equations of motion (arbitrary dimensions)

The harmonic approximation. Applying the harmonic approximation to dimensions higher than 1, for a potential energy that is a sum over the energies of pairs of atoms, is straightforward: the explicit expression is $U_{\text{tot}} = \sum_{i \neq j} U(\mathbf{R}_{ij})/2$, where $\mathbf{R}_{ij} = \mathbf{R}_i - \mathbf{R}_j$ is the vector connecting two atoms. (The sum is divided by 2 since it contains the contribution of each pair twice.) Qualitatively, the potential $U(\mathbf{R})$ is repulsive at short distances and attractive at longer ones, as illustrated in Fig. 1.1. Similar to the analysis carried out in Sec. 5.1, only small deviations from equilibrium are considered; the instantaneous locations of the atoms are $\mathbf{R}_i = \mathbf{R}_i^0 + \mathbf{u}_i$, where $\mathbf{u}_i$ is the displacement of the $i-$th atom away its equilibrium location on the lattice, $\mathbf{R}_i^0$ (where the potential is minimal). In the harmonic approximation the potential energy is expanded around its minimum in the small deviations $\mathbf{u}_i$'s (and therefore all first derivatives vanish),

$$U_{\text{tot}} = \frac{1}{2} \sum_{i \neq j} U(\mathbf{R}_{ij}^0 + \mathbf{u}_i - \mathbf{u}_j) = U_0 + \frac{1}{4} \sum_{i \neq j} \sum_{\mu,\nu} (u_{i\mu} - u_{j\mu}) D_{\mu\nu}(\mathbf{R}_{ij}^0)(u_{i\nu} - u_{j\nu}) + ..$$

$$= U_0 + \frac{1}{2} \sum_{i \neq j} \sum_{\mu,\nu} u_{i\mu} K_{\mu\nu}(\mathbf{R}_{ij}^0) u_{j\nu} + \dots . \qquad (5.20)$$

Here, μ and ν denote the cartesian components of the vectors, i.e., $\mu, \nu = 1, 2$ (or x and y) in two dimensions, $\mu, \nu = 1, 2, 3$ (or x ,y, and z) in three dimensions; $\mathbf{R}_{ij}^0 = \mathbf{R}_i^0 - \mathbf{R}_j^0$, and the definitions

$$K_{\mu\nu}(\mathbf{R}_{ij}^0) = \delta_{ij} \sum_{j' \neq i} D_{\mu\nu}(\mathbf{R}_{ij'}^0) - D_{\mu\nu}(\mathbf{R}_{ij}^0) ,$$

$$D_{\mu\nu}(\mathbf{R}_{ij}^0) = \partial^2 U/(\partial R_{ij,\mu} \partial R_{ij,\nu}) \Big|_{\mathbf{R}_{ij}=\mathbf{R}_{ij}^0} , \qquad (5.21)$$

have been introduced. The notation δ_{ij} stands for the Kronecker delta, $\delta_{ij} = 1$ when $i = j$ and 0 else. The constant term U_0 is the equilibrium-stateenergy (denoted in Chapter 4 as $U_0 = -Nu$, where u is the crystal binding energy per unit cell, and there are N unit cells. The boundaries are ignored in this discussion; they are accounted for in Sec. 5.3.) The ... in Eq. (5.20) indicate the **anharmonic terms**, of higher orders in the **u**'s; these are discarded in the **harmonic approximation**. This approximation allows one to obtain the temporal dependence of the small displacements $\mathbf{u}_i$ (in the classical approximation). In the particular situation in which the potential depends only on the distance between the two atoms, i.e., it is a central potential, the harmonic approximation amounts to joining each pair of atoms by a spring obeying Hooke's law.

The harmonic coefficients $D_{\mu\nu}(\mathbf{R}^0_{ij})$ are determined by the potentials, see Eqs. (5.21). In general, these coefficients are interrelated, and therefore only a small number of them is needed. For instance, Bravais lattices are invariant under inversion, $\mathbf{R}^0_{ij} \Leftrightarrow -\mathbf{R}^0_{ij}$ (Chapter 2). As a result, the terms in Eq. (5.20) appear in pairs, with the same coefficients for $\mathbf{R}^0_{ij}$ and $-\mathbf{R}^0_{ij}$. Another example is the simple cubic lattice, whose basis vectors are along the cartesian axes, $a\hat{\mathbf{x}}$, $a\hat{\mathbf{y}}$, and $a\hat{\mathbf{z}}$. The lattice is invariant under reflections through the planes (100), (010), or (001), that are normal to each of the basis vectors. The lattice is also invariant under rotations by 90° around each of the axes. It therefore follows that, for bonds between nearest neighbors, $D_{xx}(\pm a\hat{\mathbf{x}}) = D_{yy}(\pm a\hat{\mathbf{y}}) = D_{zz}(\pm a\hat{\mathbf{z}}) = D_L(\mathrm{nn})$ (nn denotes nearest neighbors, and L stands for a longitudinal displacement, parallel to the bond under consideration). Similarly, $D_{yy}(\pm a\hat{\mathbf{x}}) = D_{zz}(\pm a\hat{\mathbf{x}}) = D_{xx}(\pm a\hat{\mathbf{y}}) = D_{zz}(\pm a\hat{\mathbf{y}}) = D_{xx}(\pm a\hat{\mathbf{z}}) = D_{yy}(\pm a\hat{\mathbf{z}}) = D_T(\mathrm{nn})$, where T stands for transverse motion, along the direction normal to the bond. (The complexities related to transverse modes are ignored here, and arbitrary spring constants are assumed also along the transverse directions. Those, as shown, may vanish, or may be finite due to additional interactions.) For nearest neighbors in a **SC** lattice, all non-diagonal elements in the matrices $D_{\mu\nu}(\mathrm{nn})$ vanish, and the matrix comprises only two independent coefficients, $D_L(\mathrm{nn})$ and $D_T(\mathrm{nn})$. The bonds between next-nearest neighbors (denoted nnn) on this lattice, e.g., at $\pm a\hat{\mathbf{x}} \pm a\hat{\mathbf{y}}$, obey similar symmetries; the D matrix for them comprises three independent coefficients, one for the vibrations parallel to the bond, and two for those normal to it (check!).

Central potentials. The explicit values of the matrices $D_{\mu\nu}$ or $K_{\mu\nu}$ are found once the binding energy is known. As seen in Chapter 4, the potentials of pairs of ions (or atoms, or molecules) are often central, that is, they depend only on the length of the vector that joins the two members of the pair (this is not the case for covalent bonds). The harmonic coefficients $D_{\mu\nu}$ which result from a central potential can be derived from an extension of Eq. (5.11). Using the notation $\mathbf{u}_{ij} = \mathbf{u}_i - \mathbf{u}_j$,

$$R_{ij} = |\mathbf{R}^0_{ij} + \mathbf{u}_{ij}| = \sqrt{(\mathbf{R}^0_{ij})^2 + \mathbf{u}^2_{ij} + 2\mathbf{R}^0_{ij} \cdot \mathbf{u}_{ij}} \approx R^0_{ij} + \frac{\mathbf{u}^2_{ij}}{2R^0_{ij}}$$

$$+ \frac{\mathbf{R}^0_{ij} \cdot \mathbf{u}_{ij}}{R^0_{ij}} - \frac{(\mathbf{R}^0_{ij} \cdot \mathbf{u}_{ij})^2}{2(R^0_{ij})^3} + \cdots,$$

with $R^0_{ij} = |\mathbf{R}^0_{ij}|$. Adding to this the expansion

$$U(R_{ij}) \approx U(R^0_{ij}) + U'(R^0_{ij})(R_{ij} - R^0_{ij}) + \frac{1}{2}U''(R^0_{ij})(R_{ij} - R^0_{ij})^2 + \cdots,$$

leads, up to second order in the displacements, to

$$U(R_{ij}) \approx U(R^0_{ij}) + U'(R^0_{ij})\left[\hat{\mathbf{R}}^0_{ij} \cdot \mathbf{u}_{ij} + \frac{\mathbf{u}^2_{ij} - (\hat{\mathbf{R}}^0_{ij} \cdot \mathbf{u}_{ij})^2}{2R^0_{ij}}\right] + \frac{1}{2}U''(R^0_{ij})(\hat{\mathbf{R}}^0_{ij} \cdot \mathbf{u}_{ij})^2,$$

where $\hat{\mathbf{R}}_{ij}^0 = \mathbf{R}_{ij}^0/R_{ij}^0$ is a unit vector along the direction of the vector $\mathbf{R}_{ij}^0$.

The total potential energy is $U_{\text{tot}} = \sum_{i \neq j} U(\mathbf{R}_{ij}^0 + \mathbf{u}_{ij})/2$. In general, the sum runs over all lattice sites, but in most cases it is well converging, since the potential decays with the distance. The expansion of the total energy includes the linear term $\sum_{i \neq j} U'(R_{ij}^0)(\hat{\mathbf{R}}_{ij}^0 \cdot \mathbf{u}_{ij})/2$. All terms in this sum have to vanish, that is, the condition for equilibrium reads $\sum_{i \neq j} U'(R_{ij}^0)\hat{\mathbf{R}}_{ij}^0 = 0$. As a Bravais lattice is invariant with respect to inversion, $\mathbf{R}_{ij}^0 \Leftrightarrow -\mathbf{R}_{ij}^0$, this condition is always fulfilled by central potentials. The next term in the expansion yields

$$D_{\mu\nu}(\mathbf{R}_{ij}^0) = \delta_{\mu\nu}\frac{U'(R_{ij}^0)}{R_{ij}^0} + \left[U''(R_{ij}^0) - \frac{U'(R_{ij}^0)}{R_{ij}^0}\right]\hat{R}_{ij\mu}^0\hat{R}_{ij\nu}^0 \ . \qquad (5.22)$$

Problem 5.7.
a. *Show that in one dimension Eq. (5.22) reproduces Eq. (5.11). What are the longitudinal and transverse spring constants in this case?*
b. *Find the coefficients $D_{\mu\nu}(ma)$ for the Lennard-Jones potential, Eq. (4.33).*

The equations of motion and the dispersion relations. The classical equations of motion of a single-atom crystal are [see Eq. (5.20)]

$$M\ddot{u}_{i\mu} = -\sum_{j}\sum_{\nu} K_{\mu\nu}(\mathbf{R}_{ij}^0)\mathbf{u}_{j\nu} \ . \qquad (5.23)$$

Examination of the expression for the "spring constants", Eq. (5.21), reveals that $\sum_j K_{\mu\nu}(\mathbf{R}_{ij}^0) = 0$, and therefore all forces vanish when the entire crystal is displaced by a constant displacement, $\mathbf{u}_i = \mathbf{u}_0$ (check!). Indeed, forces appear only when the atoms move each relative to the others. One may now generalize the argument of Bloch: upon shifting all atoms by a basis vector of the lattice, e.g., $\mathbf{R}_i^0 \to \mathbf{R}_{i'}^0 = \mathbf{R}_i^0 + \mathbf{a}_i$, the difference $\mathbf{R}_{ij}^0 = \mathbf{R}_i^0 - \mathbf{R}_j^0$ remains as is, and the new variables $u_{i'\mu}$ obey the same equations. Barring exceptional cases, Eq. (5.23) has a unique solution up to a multiplicative factor, and the new variables are such that $u_{i'\mu} = Cu_{i\mu}$. For the above-mentioned displacement, the constant C is of the form $C = \exp[i\mathbf{k} \cdot \mathbf{a}_i]$ (see problem 5.8). Repeating the argument for each of the basis vectors results in $\mathbf{u}_i = \mathbf{A}\exp[i(\mathbf{k} \cdot \mathbf{R}_i^0 - \omega t)]$, which describes a planar wave in space (see problem 5.8). Inserting this expression for $\mathbf{u}_i$ into the equations of motion (5.23) shows that this indeed is a solution, provided that three equations (in three dimensions) are obeyed,

$$M\omega^2 A_\mu = \sum_{\nu}\left[\sum_{j} K_{\mu\nu}(\mathbf{R}_{ij}^0)e^{-i\mathbf{k}\cdot\mathbf{R}_{ij}^0}\right]A_\nu = \sum_{\nu}\widetilde{K}_{\mu\nu}(\mathbf{k})\,A_\nu \ , \qquad (5.24)$$

where

$$\widetilde{K}_{\mu\nu}(\mathbf{k}) = \sum_{j} K_{\mu\nu}(\mathbf{R}_{ij}^0)e^{-i\mathbf{k}\cdot\mathbf{R}_{ij}^0} = \sum_{j} D_{\mu\nu}(\mathbf{R}_{ij}^0)(1 - e^{-i\mathbf{k}\cdot\mathbf{R}_{ij}^0}) \ , \qquad (5.25)$$

is the Fourier transform of $K_{\mu\nu}(\mathbf{R}_{ij}^0)$ of Eq. (5.21) (check!). Equation (5.24) represents three linear equations for the three components of the vector $\mathbf{A}$; the latter are nonzero provided that the determinant of the coefficients, a polynomial of order three in ω^2, vanishes. As a result, there are three frequencies for each wave vector $\mathbf{k}$. In other words, the vectors $\mathbf{A}$ which solve these equations are eigenvectors of the matrix $\widetilde{K}_{\mu\nu}(\mathbf{k})$, with eigenvalues $M\omega^2$ (see also problem 5.3). One may convince oneself that Eq. (5.5) is a particular case of Eq. (5.25), for the one-dimensional configuration in which $\mathbf{A}$ is a one-dimensional vector, i.e., a number.

Problem 5.8.
Prove that the coefficient C in the equation $u_{i'\mu} = Cu_{i\mu}$ is given by $C = \exp[i\mathbf{k}\cdot\mathbf{a}_1]$; hence the wave function is of the form $\mathbf{u}_i = \mathbf{A}\exp[i(\mathbf{k}\cdot\mathbf{R}_i^0 - \omega t)]$.

When the unit cell contains several atoms, the wave function is described by several different amplitudes $\mathbf{A}^{(s)}$ for each atom in the unit cell [as in Eq. (5.14), which contains two such amplitudes]. For n_B atoms in the unit cell, there are $n_B d$ equations in d dimensions for the components of those vectors. It follows that there are $n_B d$ different frequencies for each wave vector $\mathbf{k}$. Out of all dispersion relations, d are acoustic (i.e., yield zero frequency in the limit of zero wave vector), and the remaining $(n_B - 1)d$ ones are optical.

Square lattice. A simple example is that of a single-atom square lattice (in the XY plane) with interactions just between nearest neighbors. The "spring constants" are then characterized by D_L for the vibrations along the bond in-between two neighbors, D_{T1} for the vibrations in the plane perpendicular to this bond, and D_{T2} for the vibrations along the normal to that plane. As mentioned, the transverse spring constants may vanish, unless the interactions are augmented by couplings with farther-away neighbors, or by interactions resulting from non-central potentials. There are four neighbors to each atom in the lattice, at $\pm a\hat{\mathbf{x}}$ and $\pm a\hat{\mathbf{y}}$. Exploiting the above-mentioned symmetries, one finds $D_{xx}(\pm a\hat{\mathbf{x}}) = D_{yy}(\pm a\hat{\mathbf{y}}) = D_L$, $D_{yy}(\pm a\hat{\mathbf{x}}) = D_{xx}(\pm a\hat{\mathbf{y}}) = D_{T1}$, and $D_{xx}(\pm\hat{\mathbf{z}}) = D_{yy}(\pm\hat{\mathbf{z}}) = D_{T2}$. Hence [see Eq. (5.21)] $K_{xx}(0) = K_{yy}(0) = 2(D_L + D_{T1})$, $K_{zz}(0) = 4D_{T2}$, and $K_{\mu\nu} = 0$ for $\mu \neq \nu$; the matrix in Eq. (5.25) is diagonal. Consequently one finds three equations,

$$\widetilde{K}_{xx}(\mathbf{k}) = 2D_L[1 - \cos(k_x a)] + 2D_{T1}[1 - \cos(k_y a)]$$

$$= 4D_L \sin^2(\frac{k_x a}{2}) + 4D_{T1}\sin^2(\frac{k_y a}{2}) ,$$

$$\widetilde{K}_{yy}(\mathbf{k}) = 2D_L[1 - \cos(k_y a)] + 2D_{T1}[1 - \cos(k_x a)]$$

$$= 4D_L \sin^2(\frac{k_y a}{2}) + 4D_{T1}\sin^2(\frac{k_x a}{2}) ,$$

$$\widetilde{K}_{zz}(\mathbf{k}) = 2D_{T2}[2 - \cos(k_x a) - \cos(k_y a)] = 4D_{T2}[\sin^2(\frac{k_x a}{2}) + \sin^2(\frac{k_y a}{2})] . \quad (5.26)$$

The μ−component of the amplitude $\mathbf{A}$ oscillates with a frequency given by $M\omega^2 = \widetilde{K}_{\mu\mu}(\mathbf{k})$. Figure 5.6 illustrates the dispersion relations as functions of the components of the wave vector, for transverse vibrations normal to the lattice plane ($\mu = z$) and for the ones along one of the plane axes ($\mu = x$). The results are displayed in the first Brillouin zone of the square lattice, as they are periodic among the zones. Methods for measuring the dispersion relations are discussed in the following. Instead of depicting the dispersion relations in a complex three-dimensional space as in Fig. 5.6, it is convenient to present them only along special routes in the Brillouin zone. Figure (5.7(b)) displays the dispersion for the three vibrational modes along the path $\Gamma \to M \to K \to \Gamma$, as defined in Fig. 5.7(a).

The dispersion relations of a simple cubic lattice in higher dimensions are derived in a similar fashion. Though the derivation is rather straightforward, the plot of the dispersion relations as functions of the three components of the wave vector requires a four-dimensional space. It is therefore customary to exhibit the dispersion as a function of the wave vector along special routes in the first Brillouin zone [an experimental plot of this kind is presented in Fig. 5.17(a) and discussed in problem (s.5.6)]

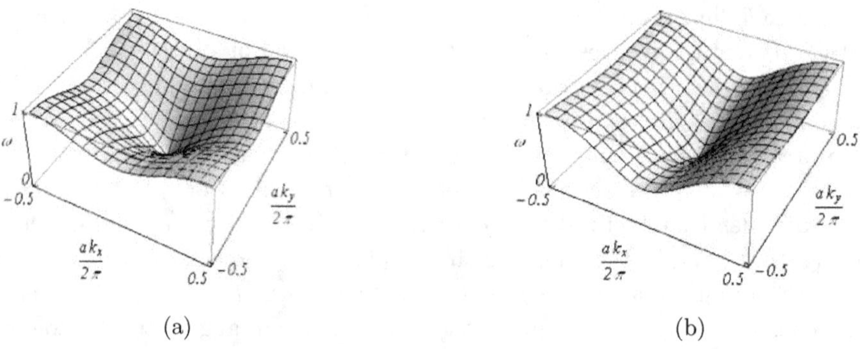

(a)   (b)

Fig. 5.6: Dispersion relations of the vibrations of a square lattice with nearest-neighbor interactions. The wave number is measured in units of $2\pi/a$. (a) Dispersion of oscillations normal to the lattice plane; the frequency is in units of $2\sqrt{2D_{T2}/M}$. (b) Dispersion of the vibrations along the $\hat{\mathbf{x}}$−axis, with $D_{T1} = D_L/4$; the frequency is in units of $2\sqrt{(D_L + D_{T1})/M}$ (note that while the vibrations are along $\hat{\mathbf{x}}$, $\mathbf{k}$ might have a component along $\hat{\mathbf{y}}$).

When the vibration is described by the wave function $\mathbf{u}_i = \mathbf{A}\exp[i(\mathbf{k}\cdot\mathbf{R}_i^0 - \omega t)]$, atoms residing on the plane normal to the wave vector $\mathbf{k}$ move in unison, without being displaced relative to each other within the plane. Therefore, such a wave corresponds to a motion of the planes one against the other, like in the one-dimensional examples given in Sec. 5.1. In particular, when the wave vector $\mathbf{k}$ lies along one of the special lines in Fig. 5.7(a) (e.g., $\Gamma \to M$), the wave describes a motion of planes normal to the reciprocal-lattice vector corresponding to that line [e.g., $\mathbf{G} = (2\pi/a)\hat{\mathbf{x}}$].

In one dimension, the longitudinal and the transverse vibrational modes are well separated, with distinct frequencies and sound velocities for each of them. In the present situation, however, such a separation pertains only to waves moving along one of the axes, e.g., when $\mathbf{k} = (k, 0)$. In that case the frequency of the longitudinal

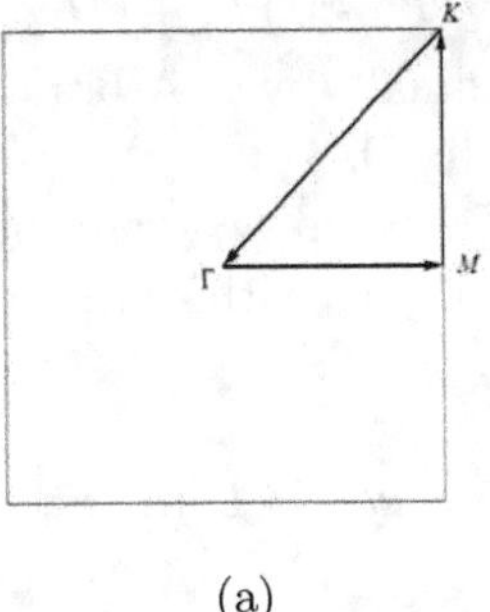

(a)

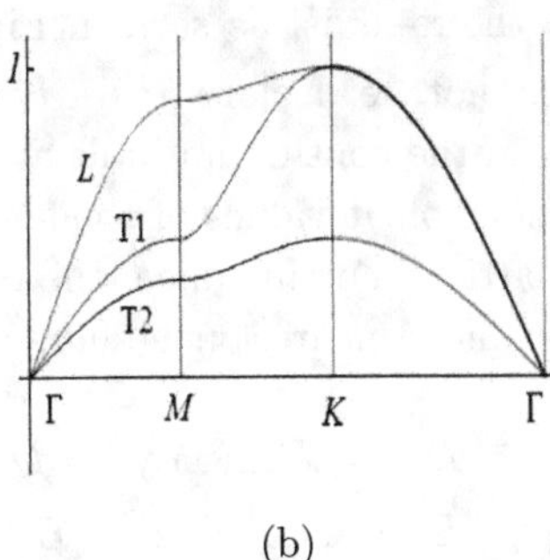

(b)

Fig. 5.7: (a) The first Brillouin zone of the square lattice. (b) Dispersion relations for the three vibrational modes along a path from the center of the zone, $\Gamma \equiv (0,0)$, to the center of the face, $M = (\pi/a, 0)$, then to the corner, $K = (\pi/a, \pi/a)$, and then back to the center. The spring constants are $D_{T2} = D_L/8$, $D_{T1} = D_L/4$.

wave with the amplitude A_x is given by $M\omega^2 = \widetilde{K}_{xx}(\mathbf{k})$ (on the path $\Gamma \to M$), and that of the transverse planar wave, of amplitude A_y, is given by $M\omega^2 = \widetilde{K}_{yy}(\mathbf{k})$. These results resemble the ones of the one-dimensional system, with one longitudinal wave and two transverse ones, and with three acoustic branches of the dispersion relation. This similarity is not accidental: in such a wave, all atoms residing on a lattice plane (a line in two dimensions) normal to the direction of the motion (i.e., to the $\hat{\mathbf{x}}$−axis) move in unison with the same displacement, and the vibrations become effectively one-dimensional. For a wave along $(1, 1)$, i.e., for a wave vector $\mathbf{k} = (k, k)/\sqrt{2}$ (the path $K \to \Gamma$ in Fig. 5.7), one can clearly identify a longitudinal and a transverse wave. In this case $\widetilde{K}_{xx} = \widetilde{K}_{yy} = 4(D_L + D_{T1})\sin^2(ka\sqrt{2}/4)$, and consequently the equations for A_x and A_y are identical. The longitudinal-wave amplitude is $(A_x + A_y)\sqrt{2}$ and the amplitude of the transverse one is $(A_x - A_y)\sqrt{2}$, both with the same frequency. However, the identification of three dispersion relations with one longitudinal wave and two transverse ones is generally meaningful only along special symmetry directions of the wave vector. In other directions there still exist three branches of dispersion relations (for the single-atom lattice), but there is no relation between a certain branch and the longitudinal wave (along the direction of the wave vector) or the transverse one (normal to the wave vector). It is also interesting to note that the degeneracy between the two transverse modes is lifted along the other paths in Fig. 5.7, where there are three different frequencies of vibrations along each of the lattice directions.

The dispersion relation is linear in the wave vector length at the vicinity of the center of the Brillouin zone, as expected for **sound waves**. The **sound velocity** of the vibrations normal to the plane is independent of the direction, $\omega \approx c|\mathbf{k}|$, with $c = a\sqrt{D_{T2}/M}$. On the other hand, the sound velocity of vibrations within the plane depends on the direction of the wave vector. The frequency of a wave oscillating along the $\hat{\mathbf{x}}$–axis, as seen in Fig. 5.6(b), is $\omega \approx a\sqrt{(D_L/M)[k_x^2 + (D_{T1}/D_L)k_y^2]}$, and cannot be declared as a transverse or a longitudinal wave. Arbitrary sound waves in the continuum limit are discussed in Appendix D.

The explicit values of the spring constants, D_L, D_{T1}, and D_{T2}, are specific for each lattice. For instance, the constants in a square lattice with central-potential forces among nearest neighbors are [see Eq. (5.22)]

$$D_{T1} = D_{xx}(\pm a\hat{\mathbf{y}}) = D_{yy}(\pm a\hat{\mathbf{x}}) = D_{T2} = D_{zz}(\pm a\hat{\mathbf{x}}) = U'(a)/a \ ,$$
$$D_L = D_{xx}(\pm a\hat{\mathbf{x}}) = D_{yy}(\pm a\hat{\mathbf{y}}) = U''(a) \ , \tag{5.27}$$

since the nearest neighbors lie along the axes. The equilibrium energy is $U_{\text{tot}} = 2NU(a)$; $U'(a) = 0$, and consequently $D_{T1} = D_{T2} = 0$: the frequencies of the transverse motions vanish for a central potential between nearest neighbors. Like in one dimension, it is customary to assume that the forces are not between nearest neighbors alone, and/or are not central, and to adjust the spring constants D_{T1} and D_{T2} so that the dispersion relations comply with the experimental values. As seen in problem 5.9 and also in the discussion of the one-dimensional configuration, additional interactions between farther-away neighbors give rise to transverse modes with non-vanishing sound velocities.

Problem 5.9.

a. *How are the solutions for the vibrations of the square lattice modified upon adding central-potential interactions in-between second neighbors, at distances $\mathbf{R}_{ij}^0 = a(\pm\hat{\mathbf{x}} \pm \hat{\mathbf{y}})$?*

b. *Find the elastic constants in the continuum limit of this system. Use the definitions of the elastic constants in Appendix D.*

Vibrations of a triangular lattice. The nearest neighbors of an atom on the triangular lattice are located at $\mathbf{R}_{nn}^0 = \pm\mathbf{a}_1, \pm\mathbf{a}_2, \pm(\mathbf{a}_2 - \mathbf{a}_1)$, where $\mathbf{a}_1 = a\hat{\mathbf{x}}$, and $\mathbf{a}_2 = (\hat{\mathbf{x}} + \sqrt{3}\hat{\mathbf{y}})a/2$ (see Fig. 3.11). It is assumed that the interactions are solely with the six nearest neighbors of each atom, and that the spring constants are given by

$$D_{\mu\nu}(\mathbf{R}_{nn}^0) = D_T\delta_{\mu\nu} + (D_T - D_L)\hat{\mathbf{R}}_{nn,\mu}^0\hat{\mathbf{R}}_{nn,\nu}^0 \ .$$

Treating D_L and D_T as given quantities, this specific form for the spring constants implies that the constant for the longitudinal vibrations is D_L, and that for the transverse ones is D_T. For instance, for $\mathbf{R}_{nn}^0 = \mathbf{a}_1 = a\hat{\mathbf{x}}$ one finds $D_{xx} = D_L$, and $D_{yy} = D_{zz} = D_T$. (The spring constants for the remaining neighbors are obtained by applying 60°–rotations on these coefficients.) The vibrations normal to the plane are described by the spring constant $D_{zz}(\mathbf{R}_{nn}^0) = D_T$ (the same for all neighbors). These vibrations are decoupled from those pertaining to the vibrations within the plane; their dispersion relation is

$$M\omega^2 = \widetilde{K}_{zz}(\mathbf{k}) = \sum_{nn} D_T(1 - e^{-i\mathbf{k}\cdot\mathbf{R}^0_{nn}})$$

$$= 2D_T[3 - \cos(\mathbf{k}\cdot\mathbf{a}_1) - \cos(\mathbf{k}\cdot\mathbf{a}_2) - \cos(\mathbf{k}\cdot(\mathbf{a}_2 - \mathbf{a}_1))]$$
$$= 4D_T[\sin^2(\mathbf{k}\cdot\mathbf{a}_1/2) + \sin^2(\mathbf{k}\cdot\mathbf{a}_2/2) + \sin^2(\mathbf{k}\cdot(\mathbf{a}_2 - \mathbf{a}_1)/2)] \ . \tag{5.28}$$

The frequency is periodic in the reciprocal lattice,

$$\omega(\mathbf{k}) = \omega(\mathbf{k} + h\mathbf{b}_1 + \ell\mathbf{b}_2) \ , \tag{5.29}$$

for arbitrary integers h and ℓ (the basis vectors in the reciprocal lattice are defined in Fig. 3.11). The dispersion relation obeys the symmetry of the reciprocal triangular lattice: a rotation of $\mathbf{k}$ by 60° leaves ω as is, since it just interchanges the vectors $\mathbf{a}_1$, $\mathbf{a}_2$, and $(\mathbf{a}_2 - \mathbf{a}_1)$ in Eq. (5.28). Figure 5.8(a) depicts the frequency as a function of the components of the wave vector $\mathbf{k}$; the periodicity and the symmetry are quite clear.

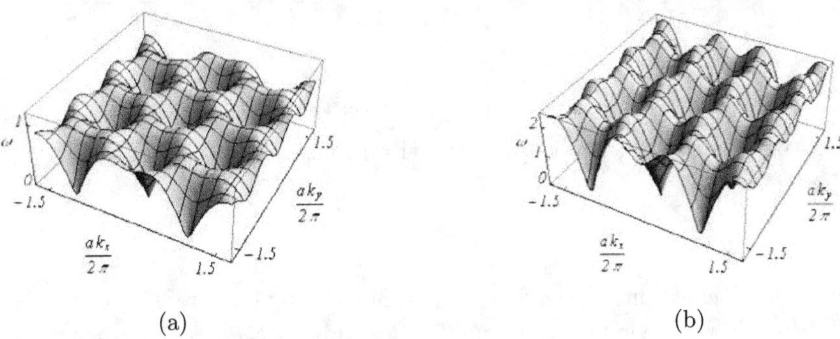

(a) (b)

Fig. 5.8: The dispersion law of vibrations in a triangular lattice with interactions between nearest neighbors, with $D_L = 4D_T$. (a) Vibrations normal to the plane, (b) vibrations of one of the modes within the plane. The frequency is measured in units of $2\sqrt{3D_T/M}$.

The vibrations within the plane are described by two coupled equations,

$$M\omega^2 A_x = \widetilde{K}_{xx}A_x + \widetilde{K}_{xy}A_y \ , \quad M\omega^2 A_y = \widetilde{K}_{xy}A_x + \widetilde{K}_{yy}A_y \ , \tag{5.30}$$

where

$$\widetilde{K}_{xx}(\mathbf{k}) = 4D_L \sin^2\left(\frac{\mathbf{k}\cdot\mathbf{a}_1}{2}\right) + (D_L + 3D_T)\left[\sin^2\left(\frac{\mathbf{k}\cdot\mathbf{a}_2}{2}\right) + \sin^2\left(\frac{\mathbf{k}\cdot(\mathbf{a}_2 - \mathbf{a}_1)}{2}\right)\right] \ ,$$

$$\widetilde{K}_{yy}(\mathbf{k}) = 4D_T \sin^2\left(\frac{\mathbf{k}\cdot\mathbf{a}_1}{2}\right) + (3D_L + D_T)\left[\sin^2\left(\frac{\mathbf{k}\cdot\mathbf{a}_2}{2}\right) + \sin^2\left(\frac{\mathbf{k}\cdot(\mathbf{a}_2 - \mathbf{a}_1)}{2}\right)\right] \ ,$$

$$\widetilde{K}_{xy}(\mathbf{k}) = \sqrt{3}(D_L - D_T)\sin\left(\frac{k_x a}{2}\right)\sin\left(\frac{\sqrt{3}k_y a}{2}\right) \tag{5.31}$$

(check!). The allowed frequencies result then from the equation $(M\omega^2 - \widetilde{K}_{xx})(M\omega^2 - \widetilde{K}_{yy}) = (\widetilde{K}_{xy})^2$. The solutions of the latter are qualitatively similar to those pertaining to the vibrations normal to the plane. One of them is portrayed in Fig. 5.8(b); it is periodic and symmetric as discussed above. Figure 5.9(b) displays the three dispersion relations on the path shown in Fig. 5.9(a).

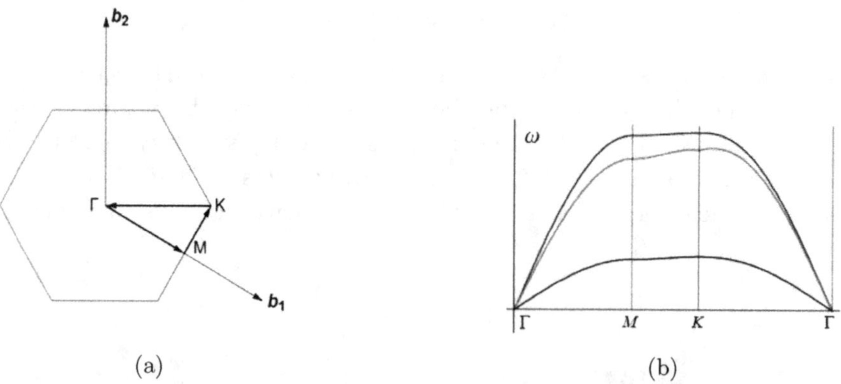

Fig. 5.9: (a) The path $\Gamma \to M \to K \to \Gamma$ in the first Brillouin zone of the triangular lattice. (b) The frequencies of the three vibrational modes along this path (from Fig. 5.8) .

The spring constants derived from a central potential are $D_L = U''(a)$ and $D_T = U'(a)/a$. The condition of equilibrium leads to $D_T = 0$, and hence to zero frequencies for the vibrations normal to the plane. Adding interactions with farther away neighbors gives rise to finite frequencies for those modes; the sound velocity, however, still vanishes (i.e., the frequency is not linear in the wave vector).

Vibrations of graphene. As illustrated in Fig. 2.8(a) and in the central panel of Fig. 2.9, graphene is represented by a triangular lattice with two atoms per unit cell. As a result, six dispersion laws ($n_B = 2$ times the three components of each displacement) are expected. The unit cells are indexed by (n, m), as defined by the lattice vectors $\mathbf{R}^0_{n,m} = n\mathbf{a}_1 + m\mathbf{a}_2$ [these lattice vectors are depicted in Fig. 2.8(a)]. Consider the vibrations normal to the plane. The normal-to-the-plane displacement of the point at the origin of the unit cell (i.e., at $\mathbf{R}^0_{n,m}$) is denoted $u_{n,m}$, and that of the second point in the cell, located at $\mathbf{R}^0_{n,m} + 2(\mathbf{a}_1 + \mathbf{a}_2)/3$, is labeled $v_{n,m}$, see Fig. 5.10. Ignoring all interactions except those between nearest neighbors (on the hexagonal lattice), then each point on one of the two sublattices is coupled by a spring to three neighboring points on the other sublattice. All spring constants for the vibrations normal to the plane are identical, and are denoted by D (at equilibrium, the length of this spring is the distance between nearest neighbors in the hexagonal lattice, $a/\sqrt{3}$). The equations of motion are

$$M\ddot{u}_{n,m} = -3Du_{n,m} + D(v_{n-1,m} + v_{n,m-1} + v_{n-1,m-1}) ,$$
$$M\ddot{v}_{n,m} = -3Dv_{n,m} + D(u_{n+1,m} + u_{n,m+1} + u_{n+1,m+1}) . \qquad (5.32)$$

Substituting the wavy solutions $u_{n,m} = A \exp[i(\mathbf{k} \cdot \mathbf{R}^0_{n,m} - \omega t)]$ and
$v_{n,m} = B \exp[i(\mathbf{k} \cdot \mathbf{R}^0_{n,m} - \omega t)]$, yields

$$M\omega^2 A = 3DA - D(e^{-i\mathbf{k}\cdot\mathbf{a}_1} + e^{-i\mathbf{k}\cdot\mathbf{a}_2} + e^{-i\mathbf{k}\cdot(\mathbf{a}_1+\mathbf{a}_2)})B ,$$
$$M\omega^2 B = 3DB - D(e^{i\mathbf{k}\cdot\mathbf{a}_1} + e^{i\mathbf{k}\cdot\mathbf{a}_2} + e^{i\mathbf{k}\cdot(\mathbf{a}_1+\mathbf{a}_2)})A . \qquad (5.33)$$

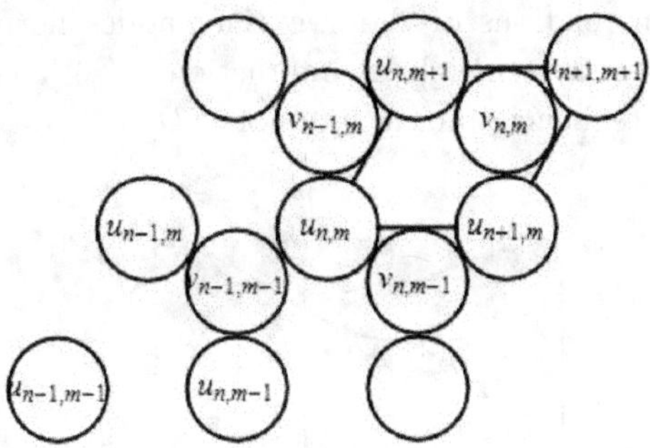

Fig. 5.10: The notations of the atoms' displacements in the hexagonal lattice. The indices (n, m) represent the rhombic unit cell marked by thick lines. Each such cell contains two atoms, whose displacements are u_{nm} and v_{nm}.

The dispersion relations are the solutions of the quadratic equation $(M\omega^2 - 3D)^2 = D^2|e^{i\mathbf{k}\cdot\mathbf{a}_1} + e^{i\mathbf{k}\cdot\mathbf{a}_2} + e^{i\mathbf{k}\cdot(\mathbf{a}_1+\mathbf{a}_2)}|^2$ (check!). These solutions are plotted in Fig. 5.11. The explicit values of the frequencies along the Brillouin-zone path defined in Fig. 5.9 are shown in Fig. 5.12. Note in particular the degeneracy of the two eigenfrequencies at the K point of the Brillouin zone. This degeneracy, as well as the linear in the wave-number dependence of the frequency around this point (see problem 5.10) are typical to graphene. Around the Γ point one may use the acoustic approximation to expand the dispersion relations in the wave vector. This yields $M\omega^2 \approx D(|\mathbf{k}|a)^2/4$, and hence the sound velocity is the same along all directions, $c = \omega/|\mathbf{k}| = a\sqrt{D/M}/2$. The thin lines in Fig. 5.12 display this approximation. The vibrational modes within the plane are given by four coupled equations (two atoms in the unit cell, and two directions of the displacement of each of them), and therefore the frequencies are derived from a quartic equation.

Problem 5.10.

a. *Prove that the two eigenfrequencies in graphene are degenerate at the K point.*
b. *Show that around this point the eigenfrequencies are linear in the distance from that point, $\omega(\mathbf{k}_K + \mathbf{q}) \approx \sqrt{(3D/M)}[1 \pm (\sqrt{3}/6)a|\mathbf{q}|]$.*

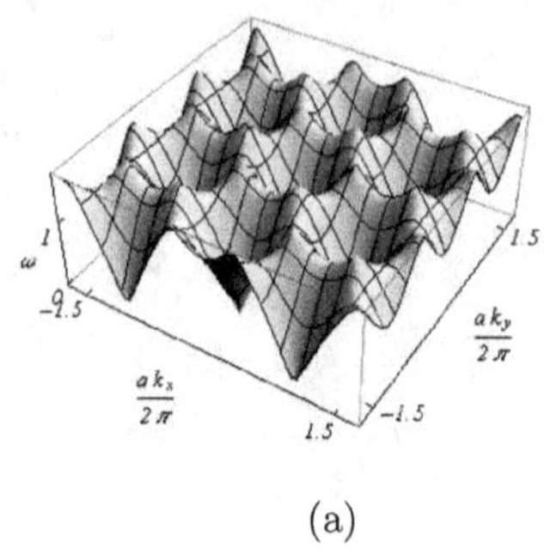

(a)

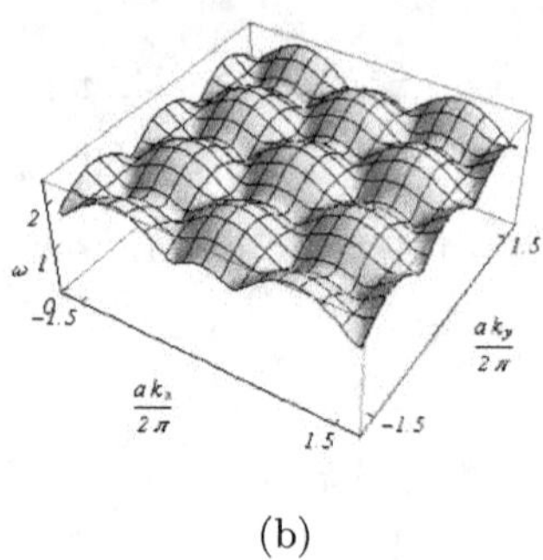

(b)

Fig. 5.11: Dispersion relations of the two eigenmodes normal to the plane of a graphene lattice with nearest-neighbor interactions. (a) The acoustic branch; (b) the optical branch. Frequencies are in units of $\sqrt{D/M}$.

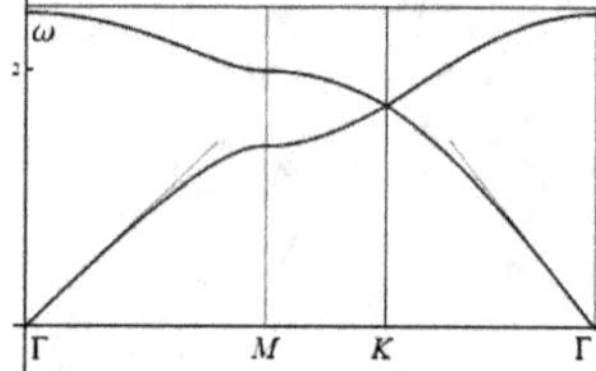

Fig. 5.12: The frequencies of the two eigenmodes normal to the plane in graphene, along the path displayed in Fig. 5.9. The thin lines pertain to the acoustic approximation.

Problem 5.11.

a. *Find the dispersion relations of the vibrations normal to the plane in graphene when there are also interactions with next-nearest neighbors, with spring constants D_2.*

b. *Prove that the dispersion relations of these vibrations are periodic according to Eq. (5.29), and are symmetric with respect to rotations of the wave vector by 60°.*

c. *Derive explicit expressions for the sound velocities around the center of the Brillouin zone.*

d. *Find the frequency of the optical branch at the center of the zone. What is the ratio of the amplitudes of the vibrations in the unit cell there?*

e. *Find the frequencies of the two branches at the M and K points. What are the ratios between the two amplitudes there?*

Problem 5.12.

Boron nitride, BN, has the same hexagonal structure as graphene: the boron and nitrogen ions occupy the two sites in the unit cell of the triangular lattice. Write down the equations of motion for the vibrations normal to the plane of this crystal, when solely nearest-neighbor interactions are accounted for. Find the dispersion relations. Point out the main difference between them and those of graphene.

5.3 The eigenmodes of a finite lattice

Boundary conditions in one dimension. The monochromatic-wave functions, $u_n = A_0 \exp[i(kna - \omega t)]$, and the wave packets in Eq. (5.9), are obtained by using the Bloch argument, which is based on certain shifts of the coordinates. While these shifts are harmless in an infinite system, they require special boundary conditions in a finite system. The ubiquitous choice are the **periodic boundary conditions of Born and von Karmann**. In one dimension they imply that any function u_n defined on the lattice sites is periodic

$$u_{n+N} = u_n \, , \tag{5.34}$$

where N is the number of unit cells in the finite system. The periodic boundary conditions validate Bloch's argument, and lead to wavy solutions even for a finite lattice. They correspond to atoms placed on a circle, that vibrate along the tangential to that circle (see problem 5.14).

Inserting $u_n = A_0 \exp[i(kna - \omega t)]$ into Eq. (5.34) gives rise to the condition $\exp[ikNa] = 1$, that is, the wave numbers are restricted to the N discrete values $k = k_\ell = 2\pi\ell/(Na)$, with $\ell = 0, 1, 2, \ldots, N-1$. Other integer values of ℓ, derived by displacing any of the former values $\ell \to \ell + N$, yield precisely the same wave functions. This numerical range covers the section $0 \le k_\ell < 2\pi/a$ at equal-length intervals, $\Delta k = 2\pi/(Na) = 2\pi/L$, where $L = Na$ is the length of the sample. The length of this range coincides with that of the first Brillouin zone. As mentioned, the wave functions are invariant under displacements by vectors in the reciprocal lattice, $\exp[i(k+G)na] = \exp[ikna]$, where $G = h(2\pi/a)$ is an arbitrary reciprocal-lattice vector. This implies that the range $-\pi/a \le k_\ell < \pi/a$ might be used as well, i.e., the wave vectors are confined to the first Brillouin zone. When N is even, then $\ell = -N/2, -N/2+1, -N/2+2, \ldots, N/2-1$, while for odd N the enumeration begins with $-(N-1)/2$ and spans a similar range. Below, N is chosen for simplicity to be even; the disparity between even and odd N's is negligible for large N's. The sums that appear below encompass this range of ℓ's.

The eigenmodes. According to the Fourier theorem (see Appendix B) the collection of functions $\{\exp[ik_\ell na], \ell = -N/2, \ldots, N/2-1\}$ is a basis for the Fourier series of functions defined on the lattice sites. One may therefore define the Fourier transform of the function u_n,

$$\widetilde{u}(k_\ell) = N^{-1/2} \sum_{n=1}^{N} u_n e^{-ik_\ell na} = N^{-1/2} \sum_{n=1}^{N} u_n e^{-2\pi i \ell n/N} \, , \tag{5.35}$$

and its inverse (i.e., the original function)

$$u_n = N^{-1/2} \sum_{\ell=-N/2}^{N/2-1} \widetilde{u}(k_\ell)e^{ik_\ell na} = N^{-1/2} \sum_{\ell=-N/2}^{N/2-1} \widetilde{u}(k_\ell)e^{2\pi i\ell n/N} \ . \tag{5.36}$$

Equation (5.36) is similar to Eq. (5.9); here, however, both u_n and $\widetilde{u}(k_\ell)$ are time-dependent. Inserting Eq. (5.36) into the equation of motion (5.3) yields the equation of motion of $\widetilde{u}(k_\ell)$,

$$M\ddot{\widetilde{u}}(k_\ell) = -D(2 - e^{-ik_\ell a} - e^{ik_\ell a})\widetilde{u}(k_\ell) = -M\omega^2(k_\ell)\widetilde{u}(k_\ell) \tag{5.37}$$

(check!). This is a separate equation for each of the variables $\{\widetilde{u}(k_\ell)\}$. Equation (5.37) describes the motion of a harmonic oscillator, i.e., Hooke's law, with the frequency $\omega(k_\ell)$ derived in Eq. (5.6). The transformation (5.35) thus maps the functions $\{u_n\}$ into $\{\widetilde{u}(k_\ell)\}$, which are the **vibrational eigenmodes** of the system.

An alternative way to obtain the vibrational eigenmodes is found by inserting into the Hamiltonian (5.2) the coordinates' transformation Eq. (5.36), and the parallel transformation for the momenta,

$$p_n = \frac{1}{N^{1/2}} \sum_\ell \widetilde{p}(k_\ell)e^{ik_\ell na} = \frac{1}{N^{1/2}} \sum_\ell \widetilde{p}(k_\ell)e^{2\pi i\ell n/N} \ . \tag{5.38}$$

(Recall that the sum over ℓ is confined to the first Brillouin zone.) The sum over the momenta in the kinetic-energy term becomes

$$\sum_{n=1}^{N} p_n^2 = \frac{1}{N^{1/2}} \sum_{n=1}^{N} \sum_\ell \sum_{\ell'} \widetilde{p}(k_\ell)\widetilde{p}(k_{\ell'})e^{i(k_\ell+k_{\ell'})na} = \sum_\ell |\widetilde{p}(k_\ell)|^2 \ . \tag{5.39}$$

[This manipulation uses the identity $\sum_{n=1}^{N} \exp[i(k_\ell + k_{\ell'})na] = N\delta_{k_\ell,-k_{\ell'}}$, which is derived from a summation of the geometrical series, with $\exp[i(k_\ell + k_{\ell'})Na] = 1$ (check!). In fact, this sum is N when $k_\ell + k_{\ell'} = G$ for any reciprocal-lattice vector G; therefore $\sum_{n=1}^{N} \exp[i(k_\ell + k_{\ell'})na] = N\sum_G \delta_{k_\ell,G-k_{\ell'}}$. However, because the analysis is confined to the first Brillouin zone, it suffices to use the identity with $G = 0$, as done above.] Note that the relation $\widetilde{p}(k) = [\widetilde{p}(-k)]^*$ [which follows directly from Eq. (5.38) and the fact that p_n is real] is used as well. Similarly, $u_{n+1} - u_n = N^{-1/2} \sum_\ell \widetilde{u}(k_\ell)(\exp[ik_\ell a] - 1) \exp[ik_\ell na]$, and therefore

$$\sum_{n=1}^{N} (u_{n+1} - u_n)^2 = \frac{1}{N^{1/2}} \sum_{n=1}^{N} \sum_\ell \sum_{\ell'} \widetilde{u}(k_\ell)\widetilde{u}(k_{\ell'})(e^{ik_\ell a} - 1)(e^{ik_{\ell'} a} - 1)e^{i(k_\ell+k_{\ell'})na}$$

$$= 4 \sum_\ell |\widetilde{u}(k_\ell)|^2 \sin^2(k_\ell a/2) \ . \tag{5.40}$$

Collecting these results, the Hamiltonian becomes a sum of N independent Hamiltonians. Each of those is the Hamiltonian of a harmonic oscillator whose coordinate is $\widetilde{u}(k_\ell)$, its conjugate momentum is $\widetilde{p}(k_\ell)$, and the frequency is $\omega(k_\ell)$. That is,

$$\mathcal{H} = \sum_{\ell=-N/2}^{N/2-1} \left[\frac{1}{2M}|\widetilde{p}(k_\ell)|^2 + \frac{M\omega^2(k_\ell)}{2}|\widetilde{u}(k_\ell)|^2 \right] \ . \tag{5.41}$$

Indeed, in the classical limit, each of these Hamiltonians gives rise to the equations of motion as in Eq. (5.37). Note that the N variables $\{\widetilde{u}(k_\ell)\}$ of Eq. (5.41) are each complex. However, with the notations $\widetilde{u}(k) = \widetilde{x}(k) + i\widetilde{y}(k)$, where $\widetilde{x}(k)$ and $\widetilde{y}(k)$ are both real, it follows from the relation $\widetilde{u}(-k) = [\widetilde{u}(k)]^*$, that $\widetilde{x}(-k) = \widetilde{x}(k)$ and $\widetilde{y}(-k) = -\widetilde{y}(k)$. Hence, the N variables $\widetilde{u}(k)$ may be replaced by $N/2$ pairs of independent variables, $\widetilde{x}$ and $\widetilde{y}$, whose argument k is positive (the limits on the values of k have to be chosen carefully, but the result remains the same: there are N independent harmonic oscillators).

Problem 5.13.

Derive the eigenmodes of the longitudinal vibrations in a finite one-dimensional lattice comprising N atoms, coupled together by nearest-neighbor interactions; assume rigid boundary conditions, for which $u_{N+1} = u_0 = 0$. How many eigenmodes are there in the first Brillouin zone?

Problem 5.14.

a. *Show that periodic boundary conditions in one dimension, Eq. (5.34), correspond to N atoms on a circular ring, like the carbon ions in the benzene molecule (Fig. 4.2), when the atoms move along the tangent to the ring.*

b. *Derive the equations of motion of the angles ϑ_n, which describe the angular deviations of the atoms away from their equilibrium sites. Use the harmonic approximation, and assume central-force interactions between nearest neighbors. Show that the vibrational modes are derived from the diagonalization of an $N \times N$ matrix. Find the eigenvalues and the eigenvectors of this matrix.*

Higher dimensions. The extension of the derivation above to dimensions higher than 1 is straightforward. Consider a single-atom lattice comprising N_m unit cells along the basis vector $\mathbf{a}_m$. The Born-von Karmann boundary condition along this direction is

$$\mathbf{u}(\mathbf{R}_i^0 + N_m\mathbf{a}_m) = \mathbf{u}(\mathbf{R}_i^0) \ . \tag{5.42}$$

This implies that $\exp[i\mathbf{k} \cdot N_m\mathbf{a}_m] = 1$. As for the one-dimensional lattice, the periodic boundary conditions turn the crystal into a three-dimensional torus. According to Eq. (3.17) the vector $N_m\mathbf{k}$ coincides with one of the reciprocal-lattice vectors and therefore in three dimensions it is a combination of the vectors $\{\mathbf{b}_\ell, \ell = 1, 2, 3\}$ with integer coefficients. Since it is only the vector $\mathbf{b}_m$ that survives the scalar product with $\mathbf{a}_m$ [from Eq. (3.20)], one chooses $\mathbf{k}_{\ell_m} = \ell_m\mathbf{b}_m/N_m$. As a displacement of this vector by a reciprocal-lattice vector does not change the wave function, i.e., $\exp[i\mathbf{k}_{\ell_m} \cdot N_m\mathbf{a}_m] = 1$, one may choose the independent values of the wave numbers within the first Brillouin zone, i.e., $-N_m/2 \le \ell_m < N_m/2 - 1$. The outcome is the basis functions $\{\exp[i\mathbf{k}_{\ell_1,\ell_2,\ell_3} \cdot \mathbf{R}_i^0]\}$, with the discrete wave vectors

$$\mathbf{k}_{\ell_1,\ell_2,\ell_3} = (\ell_1/N_1)\mathbf{b}_1 + (\ell_2/N_2)\mathbf{b}_2 + (\ell_3/N_3)\mathbf{b}_3 \ . \tag{5.43}$$

These wave vectors are all within the first Brillouin zone where they are distributed homogeneously, i.e., with a constant density that increases as the lattice is enlarged. The Fourier transform of $\mathbf{u}(\mathbf{R}_i^0)$, with this basis, is

$$\mathbf{u}(\mathbf{R}_i^0) = (N_1 N_2 N_3)^{-1/2} \sum_{\ell_1,\ell_2,\ell_3} \widetilde{\mathbf{u}}(\mathbf{k}_{\ell_1,\ell_2,\ell_3}) e^{i\mathbf{k}_{\ell_1,\ell_2,\ell_3} \cdot \mathbf{R}_i^0} \ . \tag{5.44}$$

Inserting it into the Hamiltonian yields

$$\mathcal{H} = \sum_{i,\mu} \frac{p_{i\mu}^2}{2M} + \frac{1}{2} \sum_{i \neq j} \sum_{\mu,\nu} u_{i\mu} K_{\mu\nu}(\mathbf{R}_{ij}^0) u_{j\nu} + \dots \ , \tag{5.45}$$

(where $p_{i\mu}$ is the momentum conjugate to the displacement $u_{i\mu}$). The Hamiltonian becomes a sum of $N_1 N_2 N_3$ Hamiltonians,

$$\mathcal{H} = \sum_{\mathbf{k}} \left\{ \frac{1}{2M} \sum_{\mu} |\widetilde{p}_\mu(\mathbf{k})|^2 + \frac{1}{2} \sum_{\mu,\nu} \widetilde{u}(\mathbf{k}) \widetilde{K}_{\mu\nu} \widetilde{u}_\nu(-\mathbf{k}) \right\} \ . \tag{5.46}$$

(The indices of $\mathbf{k}$ are omitted for brevity.) The three coupled equations of motion in Eq. (5.24) emerge from this Hamiltonian in the classical limit. Three vibrational eigenmodes for each wave vector are obtained from a diagonalization of the 3×3 matrix $\widetilde{K}_{\mu\nu}$. In symmetric configurations, those comprise a longitudinal wave and two transverse ones. When the unit cell, at d dimensions, contains n_B atoms, each of the discrete wave vectors in the first Brillouin zone corresponds to $n_B d$ harmonic oscillators, with the same frequencies as calculated in the previous paragraphs.

Though the derivations of the Hamiltonians (5.41) and (5.46) are based on classical mechanics, the transformations between the variables on the lattice, e.g., (u_n and p_n) and the variables in momentum space [$\widetilde{u}(k)$ and $\widetilde{p}(k)$ or $\widetilde{x}(k)$ and $\widetilde{y}(k)$, with the appropriate wave vectors] can be applied also on the quantum operators which correspond to these variables. It follows that the quantum description of lattice vibrations is derived from the Hamiltonians (5.41) or (5.46), upon replacing the coordinates and the momenta by the their quantum operators.

5.4   Density of states, specific heat of phonons

Energy levels, average energy, specific heat. Various thermal properties of crystals, e.g., the specific heat, follow from the thermodynamics of lattice vibrations. The thermodynamics of a single harmonic oscillator is discussed in Sec. 3.10. The extension to a system described by many vibrational eigenmodes is quite straightforward. As derived in the previous section, the Hamiltonian of the lattice vibrations in the harmonic approximation is a sum of Hamiltonians of one-dimensional harmonic oscillators,

$$\mathcal{H} = \sum_{\alpha} \left(\frac{1}{2M_\alpha} p_\alpha^2 + \frac{M_\alpha}{2} \omega_\alpha^2 u_\alpha^2 \right) \ , \tag{5.47}$$

where u_α and p_α denote the displacement and the momentum of the α–th eigen-mode [these replace $\widetilde{u}_\mu(\mathbf{k})$ and $\widetilde{p}_\mu(\mathbf{k})$ of Sec. 5.3]. The index α replaces the index μ (that has d values), the wave vector $\mathbf{k}$ (that has $N_1 N_2 \ldots N_d$ values in the first Brillouin zone) and the index of the atom in the unit cell (that has n_B values). As explained, each of these variables (that represent the eigenmodes) is a linear combination of the variables of a single atom, [see e.g., Eqs. (5.35) and (5.38), leading to Eq. (5.41)]. Each summand in Eq. (5.47) corresponds to a harmonic oscillator, of mass M_α, spatial coordinate u_α, and momentum p_α. Within a classical approach, it describes a periodic vibration whose frequency is ω_α. In a quantum-mechanical calculation, the energy levels of each oscillator are $\hbar\omega_\alpha(n_\alpha+1/2)$, where $n_\alpha = 0, 1, 2, \ldots, \infty$ are integers that classify the eigenstates. In addition to the ground-state energy, $\hbar\omega_\alpha/2$, the energy of a single vibrational mode contains an integer number of basic units (or quanta) of energy, $\hbar\omega_\alpha$. Very much like photons in quantum electrodynamics (e.g., black-body radiation) and as implied for quantum particles by the wave-particle duality principle of de Broglie, such a basic unit can be considered as a "particle" carrying this energy unit. In the present context these "particles" are termed **phonons**. In the quantum terminology, the classical wavy motion, given by the vibrational mode of frequency ω_α, is replaced by the motion of phonons, each carrying a discrete amount of energy $\hbar\omega_\alpha$ and momentum $\hbar k_\alpha$, related to that frequency by the respective dispersion relation. Substituting these eigenvalues in Eq. (5.47) yields the energy levels of the entire system,

$$E(\{n_\alpha\}) = \sum_\alpha \hbar\omega_\alpha(n_\alpha + 1/2) \ . \tag{5.48}$$

The thermal average of any physical quantity is carried out with the Boltzmann weights, $\exp[-\beta E]/Z$, where $\beta = 1/(k_\mathrm{B}T)$, and the normalization factor of the distribution (that ensures that the sum over all weights is 1) is the "partition function" Z, $Z = \sum_E \exp[-\beta E]$, where the sum comprises all quantum states of the system. In the present context,

$$Z \equiv \sum_E e^{-\beta E} = \sum_{\{n_1, n_2, n_3, \ldots\}} e^{-\beta \sum_\alpha \hbar\omega_\alpha(n_\alpha + 1/2)}$$

$$= \prod_\alpha \left[\sum_{n_\alpha} e^{-\beta\hbar\omega_\alpha(n_\alpha + 1/2)} \right] \equiv \prod_\alpha Z_\alpha \ . \tag{5.49}$$

(The sum over the states is replaced by a sum over all quantum numbers $\{n_\alpha\}$ in the second step.) As these sums are independent of each other, one may calculate each of them separately; the partition function is a product of the partition functions of each oscillator, Z_α. The latter is given by an infinite geometrical series,

$$Z_\alpha = \sum_{n_\alpha} e^{-\beta\omega_\alpha(n_\alpha + 1/2)} = \frac{e^{-\beta\hbar\omega_\alpha/2}}{1 - e^{-\beta\hbar\omega_\alpha}} = \frac{1}{2\sinh(\beta\hbar\omega_\alpha/2)} \ . \tag{5.50}$$

The free energy of the entire system is given by

$$F = -k_\mathrm{B}T \ln Z = -k_\mathrm{B}T \sum_\alpha \ln Z_\alpha = \sum_\alpha \hbar\omega_\alpha/2 + k_\mathrm{B}T \sum_\alpha \ln[1 - e^{-\beta\hbar\omega_\alpha}] \ . \tag{5.51}$$

The average energy of the system is obtained using the Boltzmann weights,

$$\langle E \rangle = \frac{\sum_E E e^{-\beta E}}{\sum_E e^{-\beta E}} = -\frac{\partial \ln Z}{\partial \beta} = -\sum_\alpha \frac{\partial \ln Z_\alpha}{\partial \beta}$$

$$= \sum_\alpha \left(\frac{\hbar \omega_\alpha}{2} + \frac{\hbar \omega_\alpha}{e^{\beta \hbar \omega_\alpha} - 1} \right) = \sum_\alpha \langle E_\alpha \rangle \, . \tag{5.52}$$

The last term is a sum over the average energy of each oscillator separately,

$$\langle E_\alpha \rangle = \hbar \omega_\alpha \left\langle \frac{1}{2} + n_\alpha \right\rangle = \hbar \omega_\alpha \left(\frac{1}{2} + \frac{1}{e^{\beta \hbar \omega_\alpha} - 1} \right) \, . \tag{5.53}$$

The right hand-side of Eq. (5.53) consists of two terms: the first, $\hbar \omega_\alpha / 2$, is the energy of the ground state of the oscillator. This is the **"zero-point energy"** of the oscillator, which does not affect the thermodynamic properties, being a number independent of the temperature (recall that energies are defined up to an additive constant). One does have to include this term in the total energy, as it might compensate for the "classical" binding energy and thus destroy the crystalline structure. As mentioned in Chapter 4, even in the ground state the location of the oscillator (i.e., its coordinate) is determined up to a certain uncertainty; when the amount of this uncertainty exceeds the lattice constant the crystal disintegrates. For this reason helium remains liquid down to ultra-low temperatures. The second term in Eq. (5.53) stems from the average over the quantum numbers n_α of the phonons that belong to an oscillator with frequency ω_α,

$$\langle n_\alpha \rangle = \frac{1}{e^{\beta \hbar \omega_\alpha} - 1} \, . \tag{5.54}$$

This number decays to zero at zero temperature ($\beta \to \infty$). In this limit there are no phonons, and the system is in its ground state. The number of phonons increases as the temperature is raised. Equation (5.54) also describes the number of photons in the black-body radiation, and in fact is the expression for the number of bosons (particles that obey the Bose-Einstein statistics) of a given energy.

Specific heat (at constant volume). The expression for the specific heat of a system is obtained by differentiating the energy with respect to the temperature,

$$C_V = \left(\frac{\partial \langle E \rangle}{\partial T} \right)_V = \sum_\alpha k_{\rm B} (\beta \hbar \omega_\alpha)^2 \frac{e^{\beta \hbar \omega_\alpha}}{(e^{\beta \hbar \omega_\alpha} - 1)^2} \, . \tag{5.55}$$

Figure 5.13 illustrates the temperature dependence of the specific heat due to a single vibrational mode. It is exponentially small at low temperatures, and approaches the Boltzmann constant, $k_{\rm B}$, at high ones (check!). The latter result is a manifestation of the **equipartition** law, which holds at high temperatures, i.e., in the classical limit: to each quadratic term in the Hamiltonian (for instance, the potential energy which is quadratic in the coordinate, or the kinetic energy, that is quadratic in the momentum) there corresponds an average energy $k_{\rm B} T / 2$. Thus, the total average energy of each oscillator (in the classical limit) is $k_{\rm B} T$, and its contribution to the specific heat is $k_{\rm B}$. This result also follows from the analysis above: when $k_B T \gg \hbar\omega_\alpha$, Eq. (5.54) yields $\langle n_\alpha \rangle \approx k_B T/(\hbar\omega_\alpha)$, and Eq. (5.53) gives $\langle E_\alpha \rangle \approx k_B T$. Inserting those into Eq. (5.55) multiples k_B by the number of terms in the sum, that is, by the number of vibrational modes. Each mole in a single-electron material contains $3N$ vibrational modes, where N is the Avogadro number. The specific heat per mole is $C_V = 3Nk_B = 3R$, where R is the gases' constant. This is the **Dulong and Petit** law, that is indeed observed in numerous high-temperature experiments. It is worthwhile to note that the function in Fig. 5.13 depends upon a single "scaled" variable $k_B T/(\hbar\omega_\alpha)$. The plot is **"universal"**: when the temperature is scaled by the frequency of the vibration (in energy units), and the specific heat is scaled by the Boltzmann constant, the plot describes the specific heat contributed by any vibrational mode.

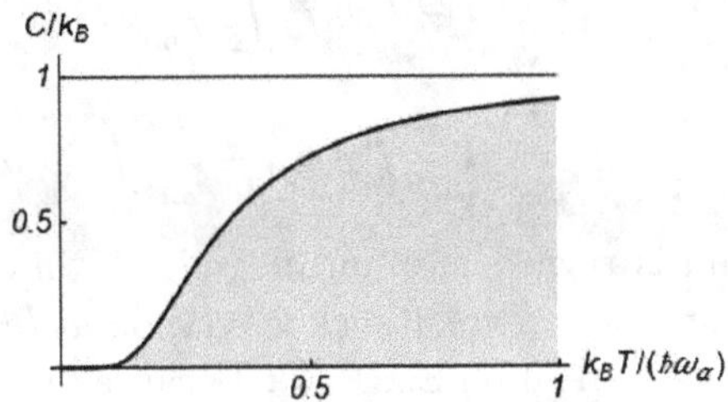

Fig. 5.13: The specific heat of a single vibrational mode of frequency ω_α, in units of k_B, as a function of the temperature measured in units of $\hbar\omega_\alpha/k_B$.

Problem 5.15.
In classical statistical mechanics, the partition function is an integral over the entire phase space. For a one-dimensional harmonic oscillator $Z = \prod_\alpha (1/h) \int dp_\alpha \int du_\alpha \exp[-\beta\mathcal{H}]$, where $\mathcal{H}$ is given in Eq. (5.47), and the Planck constant appears in order to make the partition function dimensionless. Derive the free energy, the energy, and the specific heat, and prove that the latter obeys the Dulong-Petit law.

Problem 5.16.
The area enclosed by the horizontal line and the curve of the specific heat in Fig. 5.13 represents the difference between the energy in the classical calculation and the one resulting from the quantum-mechanical one. Find the value of that difference.

Density of states. As explained in Sec. 5.4, the index α in Eq. (5.47) – and hence also in Eq. (5.55) – represents the $N_1 N_2 \ldots N_d$ wave vectors $\mathbf{k}$ in the first Brillouin zone; each of those has $n_B d$ vibrational modes that correspond to the d components of the displacement vector, and to the n_B atoms in the unit cell. The frequency of each of the $n_B d$ vibrational modes is related to the wave vector by the relevant dispersion law, $\omega_m(\mathbf{k})$, where $m = 1, 2, \ldots, n_B d$. Because the wave vectors cover densely the Brillouin zone, it is expedient at times to turn the sum over $\mathbf{k}$ into an integral. For instance, the $\mathbf{k}$–dependence of the contribution to the specific heat is solely via the dependence on the frequency $\omega_\alpha(\mathbf{k})$; by a variables' redefinition, the sum over $\mathbf{k}$ becomes an integral over ω.

The density of states in one dimension. The simplest configuration is that of the longitudinal oscillations of a single-atom, one-dimensional chain of length $L = Na$ (a is the lattice constant), built of N atoms with nearest-neighbor couplings. Here $d = n_B = 1$, and the sum runs over the wave vectors in the first Brillouin zone, $k = k_\ell = 2\pi\ell/(Na)$, with $\ell = -N/2, -N/2 + 1, -N/2 + 2, \ldots, N/2 - 1$ [see the discussion following Eq. (5.34)]. The sum over all these wave vectors of an arbitrary function $f[\omega(k)]$ is

$$\sum_k f[\omega(k)] = \sum_{\ell=-N/2}^{N/2-1} f[\omega(k_\ell)] = \int_{-N/2}^{N/2} d\ell\, f[\omega(k_\ell)]$$

$$= \int_{-\pi/a}^{\pi/a} dk\, \frac{d\ell}{dk} f[\omega(k)] = \frac{L}{2\pi} \int_{-\pi/a}^{\pi/a} dk\, f[\omega(k)] \,,$$

where the discrete sum is turned into an integral, of "integration step" $\Delta\ell = 1$, with the integration variable k (exploiting the relation $dk/d\ell = 2\pi/L$). In the next step one uses the dispersion relation Eq. (5.6). Because two vibrational modes with $\pm|k|$ correspond to each frequency, the integral over ω is multiplied by 2,

$$\sum_k f[\omega(k)] = \frac{L}{2\pi} \int_{-\pi/a}^{\pi/a} dk\, f[\omega(k)] = \frac{L}{a} \int_0^{\omega_0} d\omega\, \frac{dk}{d\omega} f(\omega) = \int_0^{\omega_0} d\omega\, g(\omega) f(\omega) \,,$$

$$(5.56)$$

where $g(\omega) = (L/\pi)(dk/d\omega)$ is the **density of states**. It is defined such that $g(\omega)d\omega$ is the number of vibrational modes with frequencies in the range $\{\omega, \omega + d\omega\}$. From Eqs. (5.6) and (5.10) one finds that the group velocity is

$$d\omega/dk = v_g = c\cos(ka/2) = c\sqrt{1 - (\omega/\omega_0)^2} \,,$$

(the sound velocity c is $c = \omega_0 a/2$) and thus the density of states can be written in the form

$$g(\omega) = (L/\pi)/v_g(\omega) = L/[\pi c\sqrt{1 - (\omega/\omega_0)^2}] \,. \qquad (5.57)$$

Figure 5.2 shows that as ω approaches ω_0, the group velocity decreases, thence the density of states, displayed in Fig. 5.14(a), increases. In particular, the density of states diverges at the upper edge of the Brillouin zone where $\omega = \omega_0$ and the group velocity vanishes. Such divergences are termed after **van Hove**.

The specific heat of a one-dimensional chain. The specific heat of the one-dimensional lattice discussed above is

$$C_V = k_B \int_0^{\omega_0} d\omega\, g(\omega) \frac{(\beta\hbar\omega)^2 e^{\beta\hbar\omega}}{(e^{\beta\hbar\omega} - 1)^2} = L \frac{k_B^2 T}{\hbar\pi c} \int_0^X dx\, \frac{x^2}{\sqrt{1 - (x/X)^2}} \frac{e^x}{(e^x - 1)^2} \,.$$

$$(5.58)$$

Here $X = \beta\hbar\omega_0$, and the integration variable is changed to $x = \beta\hbar\omega$. As expected, the specific heat, which is an extensive property, is proportional to the length of the lattice, L. At temperatures high enough compared to $\hbar\omega_0/k_B$ the frequencies are such that $\beta\hbar\omega < \beta\hbar\omega_0 \ll 1$, and consequently $(\beta\hbar\omega)^2 \exp[\beta\hbar\omega]/(\exp[\beta\hbar\omega]-1)^2 \approx 1$. In this limit, the integral is conveniently calculated using the second expression in Eq. (5.58). The result, $\int_0^{\omega_0} d\omega g(\omega) = N$, is the total number of vibrational modes in the range $0 < \omega < \omega_0$. [The equality stems from both the definition of the density of states, that counts the number of states per frequency unit, and from the explicit integration, see problem 5.17(a).] The specific heat at these temperatures is $C_V \approx k_B N$, and is a particular case of the Dulong-Petit law. In contrast, at low temperatures one has $X \gg 1$. Inserting into the integral on the right-hand side of Eq. (5.58) $X = \infty$, this integral converges, and its magnitude is

$$\int_0^\infty dx\, x^2 \frac{e^x}{(e^x - 1)^2} = 2\zeta(2) = \frac{\pi^2}{3}\,, \tag{5.59}$$

where $\zeta(2) = \sum_{m=1}^\infty m^{-2} = \pi^2/6$ is a particular case of the Riemann zeta function, $\zeta(s) = \sum_{m=1}^\infty m^{-s}$, see problem 5.17(b). It can be proven that the corrections to this result that arise from large but finite values of X are small (of the order of $X^2 \exp[-X]$, see also problem 5.19). It follows that at low temperatures the leading-order behavior of the specific heat of these vibrational modes (which are in fact acoustic phonons) in one dimension is linear in the temperature. This approximation fits quite well the low-temperature specific heat data, see Fig. 5.14(b) (the thin line there). This behavior differs significantly from the one depicted in Fig. 5.13: the sum over the small frequencies compensates for the exponential behavior, and the result is a much higher specific heat compared to the one of a single frequency, which increases much slower with the temperature.

The Debye approximation. At intermediate temperatures the integral on the right hand-side of Eq. (5.58) cannot be calculated analytically; a numerical computation validates the linear behavior of the specific heat at low temperatures, and the saturation at high temperatures [see Fig. 5.14(b)]. In order to simplify the calculations, and to obtain expressions that are insensitive to details of the density of states, **Debye** had proposed an approximation, named after him, for the density of states. The exact form, Eq. (5.57), represented by the thick line in Fig. 5.14(a), is replaced by a constant density of states, whose value is equal to that of the true density of states in the acoustic limit of low frequencies, i.e., $g(\omega) = g_0 = L/(\pi c)$. This is depicted by the horizontal thin line in Fig 5.14(a). As seen, the two lines are close to one another in the acoustic (low-frequency) range. Since the approximate density of states is smaller than the true one (in particular at frequencies close to ω_0), it yields less states. To compensate for this fault, Debye had proposed to extend the approximate density of states up to a new upper limit, termed the **Debye frequency** and marked by $\omega_D = k_B \Theta_D/\hbar$ (Θ_D is the **Debye temperature**; it expresses the same frequency in units of temperature). This frequency is chosen such that the total areas under the two curves in Fig. 5.14(a) are equal to one

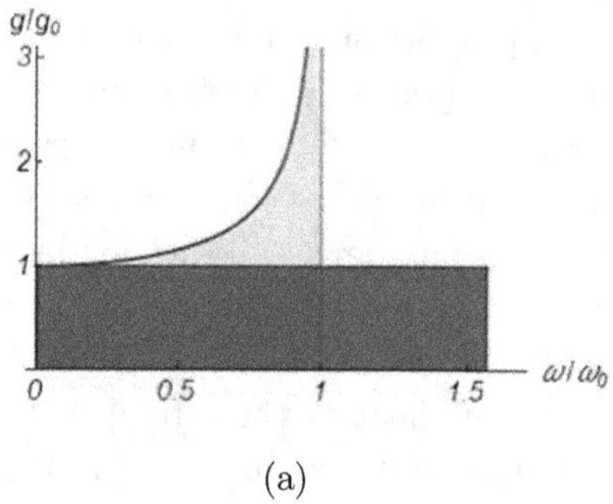

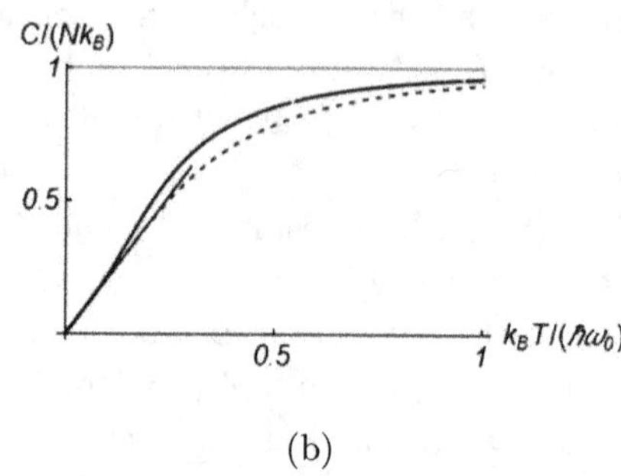

(a)   (b)

Fig. 5.14: (a) The thick upper line is the exact density of states of a single-atom, one-dimensional lattice with nearest-neighbor interactions, in units of $g_0 = L/(\pi c)$, as a function of ω/ω_0. The darker area below the horizontal line $g = g_0$, that extends up to $\omega_D/\omega_0 = \pi/2$, is identical to the light area below the (exact) density of states curve. (b) The specific heat of the same system in units of Nk_B, as a function of $k_B T/(\hbar\omega_0)$. The thick solid line: an exact computation of the integral. The dashed line: the Debye approximation. The thin solid line is obtained by approximating the integral on the right-hand side of Eq. (5.58) by $\pi^2/3$.

another, and the area is equal to the total number of vibrational modes. In the units used in the figure, the area under the curve of the exact density of states is $\int_0^1 dx/\sqrt{1 - x^2} = \pi/2$, while the new area [the darker one in Fig. 5.14(a)] is $\int_0^{x_D} dx = x_D$, hence one chooses $\omega_D/\omega_0 = x_D = \pi/2$. As $\int_0^{\omega_D} d\omega g(\omega) = N$, one finds $N = L\omega_D/(\pi c)$, i.e., $g_0 = N/\omega_D$. It follows that

$$C_V^{\text{Debye}} = \frac{k_B N}{X} \int_0^X dx x^2 \frac{e^x}{(e^x - 1)^2} , \quad X = \beta\hbar\omega_D . \tag{5.60}$$

Figure 5.14(b) compares the exact result for the specific heat [Eq. (5.58)] with the Debye approximation [Eq. (5.60)]. The two results are close to one another at the two edges: at low temperatures the specific heat is dominated by the low frequencies, where the density of states is indeed almost constant. At high temperatures the specific heat is proportional to the total number of vibrational modes and the Debye approximation is designed to comply with this number. At intermediate temperatures there is a difference between the Debye approximation and the exact result.

Typical values of the Debye temperatures are of the order of a few hundreds of degrees. As in the discussion of Fig. 5.13, the Debye approximation is **universal**: though the frequencies ω_0, and consequently the Debye temperatures, differ among materials, the coefficient in front of the integral on the right hand-side of Eq. (5.60) is $C_0 = k_B N/(\beta\hbar\omega_D) = k_B NT/\Theta_D$, and the integral itself depends solely on $X = \beta\hbar\omega_D \propto \Theta_D/T$. It follows that the specific heat depends on the temperature only via the ratio $T/\Theta_D \propto k_B T/(\hbar\omega_0)$. Hence, when the temperature and the specific heat of the acoustic vibrational modes of various one-dimensional materials are scaled as in Fig. 5.14(b), the scaled experimental results lie on the same curve.

The contribution of the optical branch to the specific heat. The optical branch appears when there are two atoms in the unit cell ($n_B = 2$). From the derivatives of the two dispersion relations in Eq. (5.15) (see Fig. 5.4), in conjunction with the expression $g(\omega) = (L/\pi)/(d\omega/dk)$ for the density of states, it is possible to deduce the density-of-states curves depicted in Fig. 5.15. As expected from Fig. 5.4, the density of states of the acoustic branch resembles that of the single-atom system [Fig. 5.14(a)]. In fact, the density of states is "flatter", implying that the Debye approximation is more reliable in this case. The optical branch possesses van Hove's singularities as well, since the group velocity vanishes at its both ends. It comprises relatively high frequencies within the range $\omega_+(\pi/a) < \omega < \omega_+(0)$. Its contribution to the specific heat is

$$C_V^{\text{optical}} = k_B \int_{\omega_+(\pi/a)}^{\omega_+(0)} d\omega\, g(\omega)(\beta\hbar\omega)^2 \frac{e^{\beta\hbar\omega}}{(e^{\beta\hbar\omega} - 1)^2} .$$

At low temperatures this contribution is bounded by a constant proportional to $\exp[-x_{\min}]$, where $x_{\min} = \beta\hbar\omega_+(\pi/a) \gg 1$, and hence it is rather tiny [see problem 5.17(c)], similarly to the illustration in Fig. 5.13. At high temperatures the integral approaches N (the number of optical vibrational modes), and then the specific heat is Nk_B, as expected from the Dulong-Petit law.

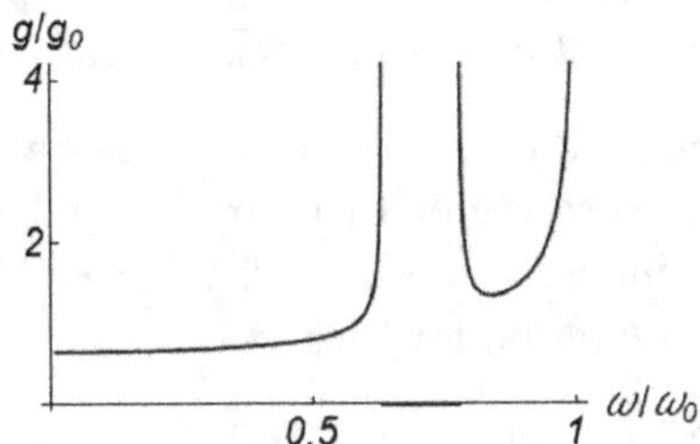

Fig. 5.15: The density of states of a one-dimensional crystal with two atoms in the unit cell, in units of $g_0 = 2L/(\omega_0 a)$, as a function of ω/ω_0. The mass ratio is 3/2, like in Fig. 5.4.

The Einstein approximation. Due to the exponential decay of the specific heat of the optical modes at low temperatures, and the approach to a constant value at high ones, and also since the integrals over complicated densities of states (like the one that displays the optical branch in Fig. 5.15) are quite cumbersome, it is customary to compute the contribution of the optical branch by the **Einstein approximation**. In this approximation, one chooses a typical optical frequency (for instance, the average frequency of the optical branch), finds the corresponding specific heat [a single term in Eq. (5.55)], and then multiplies it by the number of vibrational modes of this branch (N in the case of a one-dimensional lattice). The result resembles the one depicted in Fig. 5.13, multiplied by the number of vibrational modes of this branch. This approximation is in particular successful when the frequencies' range of the optical branch is narrow, a rather ubiquitous situation. Adding together the contributions of the two branches, yields a specific heat that increases linearly with the temperature at low temperatures (much lower than the maximal acoustic frequency or than the Debye temperature). At high temperatures the acoustic-branch specific heat reaches the value Nk_B, in accordance with the Dulong-Petit law. The contribution of the optical branch is negligible at temperatures smaller than the minimal optical frequency (in units of temperature), but it increases precipitately towards Nk_B at higher temperatures, so that asymptotically the specific heat tends to the value $2Nk_B$.

Problem 5.17.

a. *Prove, by a direct calculation of the integral, that the density of states given in Eq. (5.57) has the normalization $\int_0^{\omega_0} d\omega\, g(\omega) = N$.*

b. *Prove Eq. (5.59). [Hint: to prove the first step, expand the integrand in powers of $\exp[-x]$, and integrate by parts each term in the sum. To prove the second step compare the Taylor series of the function $f(x) = \sin x / x$ with the mathematical identity $f(x) = \prod_m (1 - x/x_m)$, where the product comprises all zeros of the function on the real axis.]*

c. *Prove that at low temperatures the contribution of the optical branch to the specific heat obeys the inequality $C_V^{\text{optical}} < k_B N x_{\min}^2 \exp[-x_{\min}]$, where $x_{\min} = \beta\hbar\omega_+(\pi/a) \gg 1$, and therefore it decays exponentially when β increases.*

Arbitrary dimensions. To derive the density of states at dimensions higher than 1, recall that for each wave vector there are $n_B d$ vibrational modes; the corresponding frequencies are indexed by $m = 1, 2, \ldots, n_B d$. The sum over the wave vectors is replaced by an integration for each value of m,

$$\sum_\alpha f(\omega_\alpha) = \sum_{m=1}^{n_B d} \sum_{\mathbf{k}} f[\omega_m(\mathbf{k})] = \sum_m \int n(\mathbf{k}) d^d k\, f[\omega_m(\mathbf{k})] , \qquad (5.61)$$

where $f(\omega)$ is an arbitrary function of the frequency, e.g., as in Eq. (5.56). In the last step, the discrete sum over $\mathbf{k}$ is transformed into an integral over the first Brillouin zone; $n(\mathbf{k}) d^d k$ is the number of (discrete) wave vectors in the volume element $d^d k$. Recall that, quite generally, the discrete wave vectors are given by $\mathbf{k}_{\ell_1, \ell_2, \ell_3} = (\ell_1/N_1)\mathbf{b}_1 + (\ell_2/N_2)\mathbf{b}_2 + (\ell_3/N_3)\mathbf{b}_3$, with $-N_i/2 \leq \ell_i < N_i/2 - 1$. The volume of the first Brillouin zone, which is the volume of a unit cell in reciprocal lattice, is $V_{\text{rec}} = (2\pi)^d/V$, where V is the volume of the unit cell in the original lattice (see problem 3.6). The Brillouin zone comprises $N = N_1 N_2 \ldots N_d$ discrete, uniformly distributed, wave vectors. Therefore, the **density of points in momentum space** (i.e., the number of points in the Brillouin zone) is

$$n(\mathbf{k}) = \frac{N}{V_{\text{rec}}} = \frac{NV}{(2\pi)^d} , \qquad (5.62)$$

where NV is the total volume of the crystal (the number of unit cells times the volume of one of them).

In the next step, the integration variables are changed, from $\mathbf{k}$ to ω,

$$\sum_m \int n(\mathbf{k}) d^d k f[\omega_m(\mathbf{k})] = \sum_m \frac{NV}{(2\pi)^d} \int d^d k f[\omega_m(\mathbf{k})] = \sum_m \int d\omega\, g_m(\omega) f(\omega) ,$$

(5.63)

where

$$g_m(\omega) = \frac{NV}{(2\pi)^d} \int d^d k \delta[\omega - \omega_m(\mathbf{k})]$$

(5.64)

is the **density of states** of the m-th vibrational mode [substituting Eq. (5.64) in the integral on the right hand-side of Eq. (5.63) gives exactly the terms that appear in the second equality there]. Phonons, in addition to their contribution to the specific heat, participate in many physical processes, e.g., heat conductance, scattering of electrons (that lead to a reduction of the electrical conductance), and more. These all involve the phonon density of states, given by Eq. (5.64).

In Eq. (5.64), $g_m(\omega) d\omega$ is the number of vibrational modes (of the m-th type), located in-between two equal-frequency surfaces in the Brillouin zone, of frequency ω and $\omega + d\omega$, respectively, see Fig. 5.16. As the density of the wave vectors in the Brillouin zone is uniform, it suffices to find the volume enclosed in-between the surfaces, and to multiply it by that density. Consider a point $\mathbf{k}$ located on the "lower" surface; the vector $\boldsymbol{\nabla}_{\mathbf{k}}\omega(\mathbf{k})$ is perpendicular to that surface, and points along the direction of increasing ω. The distance between the two surfaces in $\mathbf{k}$-space, at this point, $\Delta k_\perp$, is given by the relation $\Delta\omega = |\boldsymbol{\nabla}_{\mathbf{k}}\omega(\mathbf{k})|\Delta k_\perp$. Defining an area element on the surface by dS_ω, the volume in-between it and the area above on the neighboring surface is $dS_\omega \Delta k_\perp = \Delta\omega dS_\omega/|\boldsymbol{\nabla}_{\mathbf{k}}\omega(\mathbf{k})|$, leading to

$$g_m(\omega) = \frac{NV}{(2\pi)^d} \oint_S \frac{dS_\omega}{|\boldsymbol{\nabla}_{\mathbf{k}}\omega_m|} ,$$

(5.65)

where the integration is carried out over the entire surface (which is a closed one, encompassing the origin) on which the frequency is constant and equals ω. This result is a generalization of Eq. (5.57): the group velocity in arbitrary dimension is given by $\mathbf{v}_g = \boldsymbol{\nabla}_{\mathbf{k}}\omega|_{\mathbf{k}=\mathbf{k}_0}$. In one dimension, the coefficient in Eq. (5.65) is $Na/(2\pi) = L/(2\pi)$, and the "area" S encompassing the first Brillouin zone includes solely two points. Equation (5.65) can be also derived directly from Eq. (5.64): the identity $\delta f[(x)] = \delta[f'(x_0)(x - x_0)] = \delta(x - x_0)/f'(x_0)$, valid when $f(x_0) = 0$ [see part d of problem (B.2)] enables one to write

$$\int d^d k \delta[\omega - \omega_m(\mathbf{k})] = \oint_S dS_\omega \int dk_\perp \delta[k_\perp - k_\perp(\omega_m)]/(d\omega/dk_\perp) = \oint dS_\omega/|\boldsymbol{\nabla}_{\mathbf{k}}\omega_m| ,$$

(5.66)

where $k_\perp$ is the component of $\mathbf{k}$ normal to the surface.

As seen in Sec. 5.2, the dispersion relations in higher dimensions are quite complicated, and consequently the corresponding densities of states are complicated as

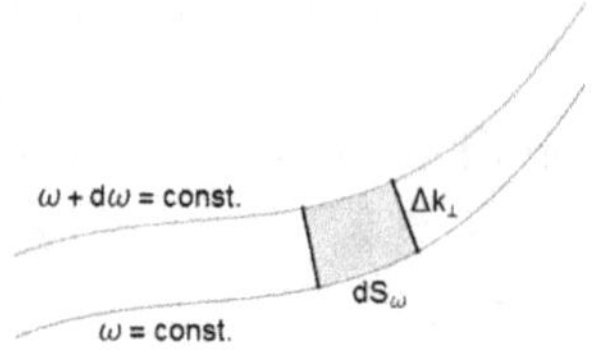

Fig. 5.16: Two equal-frequency surfaces in two dimensions. The dark area represents $dS_\omega dk_\perp$ (see text).

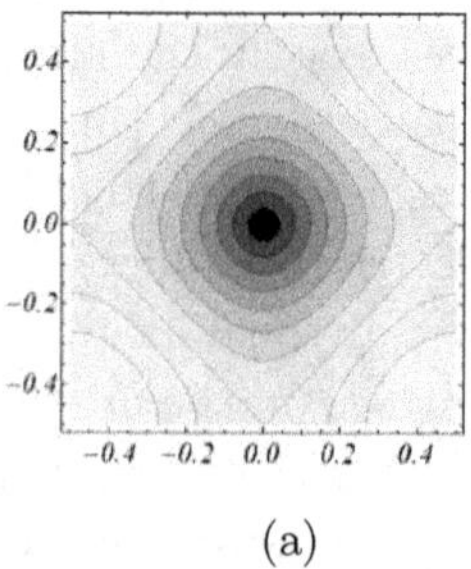

(a)

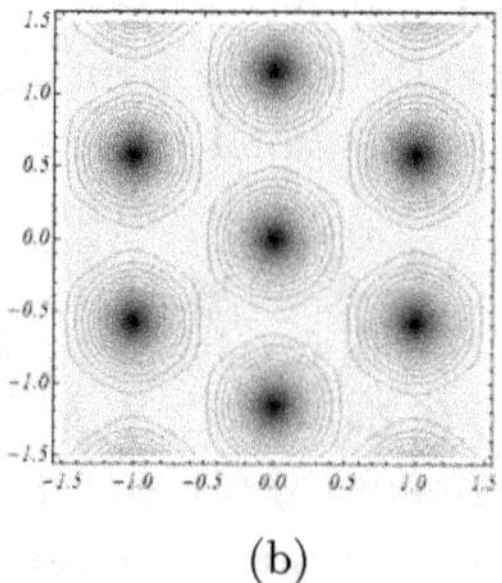

(b)

Fig. 5.17: Equal-frequency curves (in $k_x - k_y$ plane) for the dispersion relation of vibrations normal to the plane due to nearest-neighbor interactions. (a) The square lattice of Fig. 5.6(a). (b) The triangular lattice of Fig. 5.8(a).

well. Figure 5.17 displays the equal-frequency curves (corresponding to the "area" S) for the vibrations normal to the plane in the square and the triangular lattices, with nearest-neighbor interactions, according to Figs. 5.6 and 5.8. Around the center of the Brillouin zone, in the acoustic region, these lines are circles, the distances in-between them are identical along any direction, and the density of states takes a relatively simple form. However, farther away from the center the curves follow the symmetry of the lattice, and the distances between them depend on the direction of the wave vector. Moreover, for the vibrations within the plane of the square lattice the equal-frequency curves are not circular even in the acoustic range: the relation $\omega \propto a\sqrt{(D_L/M)[k_x^2 + (D_{T1}/D_L)k_y^2]}$ (Sec. 5.2) gives rise to elliptical equal-frequency lines. Nonetheless, given the explicit expression of the dispersion relation, the density of states can be derived from Eq. (5.66). Though the group velocity vanishes at special points in the Brillouin zone, the integration over the equal-frequency surfaces weakens the divergencies of the density of states, and thus the van Hove singularities of dimensions higher than one are weakened. As seen in problem 5.18, the divergence is logarithmic in two dimensions, and is not as strong as the one of one dimension, Eq. (5.57). In three dimensions, the density of states does not diverge at all, but its derivative does, see Fig. 5.19(b).

Problem 5.18.

a. *Find the group velocity of the vibrations normal to the plane in the single-atom square lattice with interactions between nearest-neighbors, whose dispersion relation, $M\omega^2 = \widetilde{K}_{zz}(\mathbf{k})$, is displayed in Eq. (5.26) and in Fig. 5.6(a). At which points in the first Brillouin zone does this velocity vanishes?*

b. *Give an expression for the dispersion relation at the corners of the Brillouin zone. Find the density of states at the vicinity of such a corner.*

c. *As seen in Fig. 5.17(a), the straight lines that join the points of the type M [i.e., $(\pi/a, 0) \to (0, \pi/a) \to (-\pi/a, 0) \to (0, -\pi/a) \to (\pi/a, 0]$ form an equal-frequency "surface" for this lattice. Prove this, and show that the density of states is infinite at the frequency corresponding to this curve, and diverges logarithmically at its vicinity.*

d. *Describe schematically the dependence of the density of states on the frequency.*

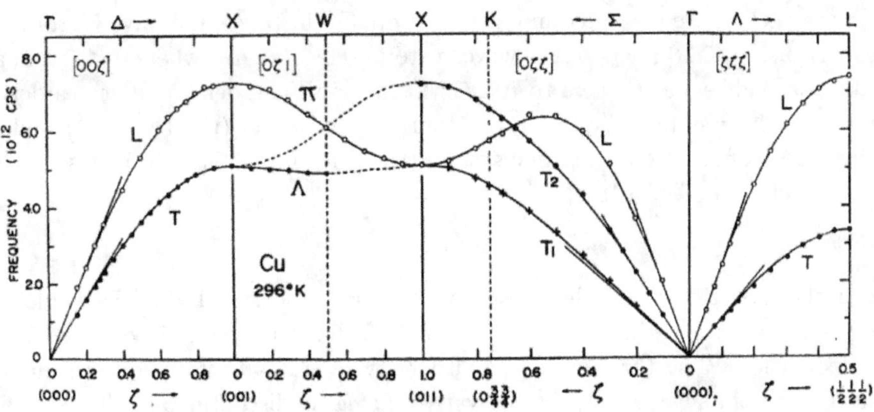

Fig. 5.18: The measured dispersion relations at various directions in the first Brillouin zone of copper. The frequency is $\nu = \omega/(2\pi)$. [E. C. Svensson, B. N. Brockhouse and J. M. Rowe, *Crystal dynamics of Copper*, Phys. Rev. **155**, 619 (1967)].

An example of a three-dimensional lattice. The complexity of the dispersion relations in three dimensions, in particular when there are many branches, is illustrated in Fig. 5.18, which displays those relations for copper (an **FCC** lattice). The curves in the figure are experimental data measured by **inelastic neutron scattering** (discussed below) for the wave-vector dependence of the frequency at various directions [see Fig. 5.19(a)] in the first Brillouin zone. Figure 5.19(b) illustrates the density of states as found from model calculations designed to fit the experimental data. Such a model can be based on educated "guesses" for the spring constants of the different interactions, which are adjusted until a plausible agreement with the data is reached [see also problem (s.5.6)]. At low frequencies,

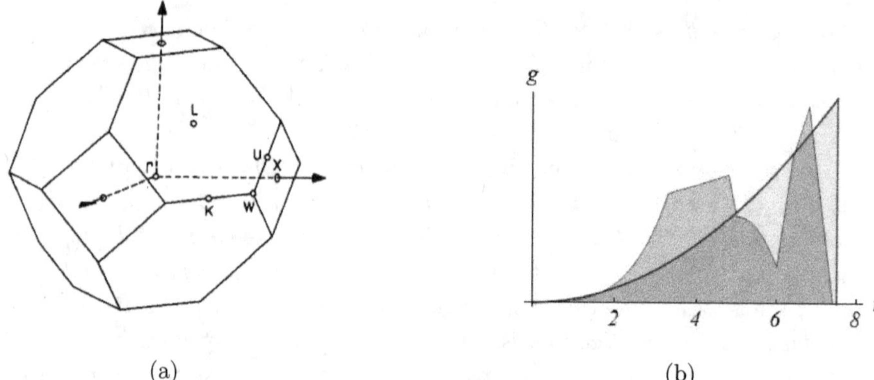

(a)   (b)

Fig. 5.19: (a) The first Brillouin zone of an **FCC** lattice. The coordinates of the special points Γ, X, W, K, and L are marked below the horizontal line in Fig. 5.18; the wave vectors are measured in units of $2\pi/a$. For instance, the wave vector in the left part of Fig. 5.18 refers to the wave vectors $(0\ 0\ \zeta)(2\pi/a)$, where $0 < \zeta < 1$, and that in second segment refers to $(0\ \zeta\ 1)(2\pi/a)$. (b) The density of states derived under certain assumptions about the "spring" constants (the thin line) and the Debye approximation for the same system (the thick line). Part (b) is based on the reference cited in Fig. 5.18.

i.e., in the acoustic branch, the density of states is proportional to ω^2, as would have been obtained from a linear dispersion relation, like the one derived in the Debye approximation. The three-dimensional density of states contains quite a number of singular points, at which the derivative $dg/d\omega$ is discontinuous. These points appear whenever an equal-frequency surface cuts the face of the first Brillouin zone. This behavior is related to the van Hove singularities: the group velocity vanishes at certain points near the corresponding surface in the Brillouin zone, but the two-dimensional integration [Eq. (5.65)] "mitigates" the divergence. Figure 5.19(b) shows the density of states found by the Debye approximation for this system.

The Debye and the Einstein approximations; arbitrary dimensions. The approximations of Debye and Einstein are also useful at dimensions higher than one. That of Einstein is the same in any dimension: each of the optical branches contributes to the specific heat a function similar to the one in Fig. 5.13, whose shape is dictated by the frequency typical to that branch. As opposed, the Debye approximation requires a re-examination. Recall that in one dimension, this approximation replaces the density of states of the material by a simple form, that captures the acoustic behavior at low frequencies, and extends that form up to the Debye frequency, which in turn is determined such that the total number of vibrational modes is conserved. This rather crude approximation produces results quite reliable for the specific heat. In the simplest version of this approximation the dispersion relation is assumed to be linear in the wave vector magnitude, i.e., it is independent of the direction of the wave vector, $\omega_m = c_m|\mathbf{k}|$. Consequently, the equal-frequency surfaces are spheres (in three dimensions) or circles (in two dimensions, as the ones near the origin in Fig. 5.17) in momentum space. The density of states of a single acoustic vibrational mode in Eq. (5.65) is

$$g_m(\omega) = \frac{NV}{(2\pi c_m)^d} S_d \omega^{d-1} , \qquad (5.67)$$

where S_d is the area of the sphere (or the perimeter of the circle) whose radius is 1 in d–dimensions $[S_3 = 4\pi,\ S_2 = 2\pi,\ S_1 = 2]$ and V is the volume of the d–dimensional crystal. The integrations over the frequency are carried out up to the Debye frequency, $\omega_{\mathrm{D}m}$, determined by the conservation of the total number of modes,

$$N = \int_0^{\omega_{\mathrm{D}m}} d\omega\, g_m(\omega) = \frac{NVS_d}{d}\left(\frac{\omega_{\mathrm{D}m}}{2\pi c_m}\right)^d , \qquad (5.68)$$

and hence $\omega_{\mathrm{D}m} = 2\pi c_m (VS_d/d)^{-1/d}$, and $g_m(\omega) = Nd\omega^{d-1}/\omega_{\mathrm{D}m}^d$. Inserting Eqs. (5.67) and (5.68) into Eqs. (5.55) and (5.61) yields the specific heat in the Debye approximation, for each of the acoustic branches,

$$C_V = k_{\mathrm{B}} dN (\beta\hbar\omega_{\mathrm{D}m})^{-d} \int_0^{\beta\hbar\omega_{\mathrm{D}m}} dx\, x^{d+1} e^x/(e^x - 1)^2 . \qquad (5.69)$$

At low temperatures, $k_{\mathrm{B}}T \ll \hbar\omega_{\mathrm{D}m}$, the integration may be carried out up to infinity [like in Eq. (5.59)]; the integral then converges to a constant value,

$$\int_0^{\infty} dx\, x^{d+1} e^x/(e^x - 1)^2 = I_d = (d+1)!\zeta(d+1) ,$$

where $I_1 = 2\zeta(2) = \pi^2/3 \approx 3.29$, $I_2 = 6\zeta(3) \approx 7.212$, and $I_3 = 24\zeta(4) = 4\pi^4/15 \approx 25.98$. $[\zeta(s) = \sum_{n=1}^{\infty} n^{-s}$ is the Riemann zeta function; its numerical values are derived in a method similar to that presented in connection with the one-dimensional case, problem 5.17(b).] The corrections to this approximation are exponentially small (see problem 5.19). The specific heat at low temperatures of the acoustic modes of the m–th branch is thus proportional to $(k_{\mathrm{B}}T/\hbar\omega_{\mathrm{D}m})^d$. As the density of states varies smoothly at low frequencies, this result pertains also to the full (exact) solution of the specific heat. It is worthwhile to emphasize again that the low-temperature specific heat is dominated by the acoustic modes, as the contributions of the optical ones decay there exponentially. At high temperatures, $k_{\mathrm{B}}T \gg \hbar\omega_{\mathrm{D}m}$, the integral encompasses all states, and gives the Dulong-Petit law, $C_V \to k_{\mathrm{B}}N$, for each of the $n_B d$ modes. The Debye temperatures in three dimensions, $\Theta_{\mathrm{D}m} = \hbar\omega_{\mathrm{D}m}/k_{\mathrm{B}}$, is of the order of a few hundreds degrees K.

Three dimensions. Equation (5.68) yields a different Debye temperature for each of the acoustic branches, i.e., the density of states of each branch is parabolic (in three dimensions) up to the corresponding Debye frequency, and then drops to zero. Accordingly, the specific heat is a sum of three different integrals, extending up to three different bounds. At times it is convenient to return to Eq. (5.67), and to write down the total density of states of the three vibrational modes (the longitudinal and the two transverse ones) in the small-frequency limit,

$$g(\omega) = \sum_{m=1}^{3} g_m(\omega) = \frac{NV}{(2\pi)^3}\left(\frac{1}{c_L^3} + \frac{2}{c_T^3}\right)S_3\omega^2 \equiv \frac{3NV}{2\pi^2\bar{c}^3}\omega^2 \, , \qquad (5.70)$$

where the average sound velocity $\bar{c}$ is defined by $3/\bar{c}^3 = [1/c_L^3] + [2/c_T^3]$. The Debye temperature is chosen so that the total number of vibrational modes is $3N$,

$$3N = \int_0^{\omega_D} d\omega g(\omega) = \frac{3NV}{2\pi^2}\left(\frac{\omega_D}{\bar{c}}\right)^3 \, , \qquad (5.71)$$

i.e., $g(\omega) = 9N\omega^2/\omega_D^3$, and $\omega_D = (6\pi^2/V)^{1/3}\bar{c}$. This expression is rather useful for explicit calculations, and gives the correct behavior of the specific heat at both low and high temperatures.

Problem 5.19.

The Debye function in d dimensions [appearing in Eq. (5.69)] is defined by

$$D_d(X) = \frac{d}{X^d}\int_0^X dx \frac{x^{d+1}e^x}{(e^x - 1)^2} \, .$$

a. *For $X \ll 1$ (which corresponds to high temperatures), show that $D_d(X) = 1 - dX^2/[12(d+2)] + \dots$. This is the leading correction to the Dulong-Petit law.*
b. *For $X \gg 1$ (corresponding to low temperatures), prove that $D_d(X) = [d/X^d]I_d - dX\exp[-X] + \mathcal{O}(\exp[-X])$. This is the leading correction to the T^d behavior of the vibrational modes' specific heat at low temperatures.*

Problem 5.20.

When the matrix $K_{\mu\nu}$ is diagonal, the dispersion relations are $M\omega_\mu^2 = \widetilde{K}_{\mu\mu}$, where the right hand-side is given by the Fourier transform of $K_{\mu\mu}(\mathbf{R})$ [Eq. (5.25)]. For long-range interactions, for which $K_{\mu\mu}(\mathbf{R}) = D_1/R^{\eta+d}$, with $\eta > 0$, the leading term in the Fourier transform is $\widetilde{K}_{\mu\mu} \propto |k|^\eta$ (under appropriate conditions). In the extended Debye approximation it is assumed that the dispersion relation $\omega^2 \propto |k|^\eta$ is valid at all frequencies, up to a generalized Debye frequency; the latter is chosen so that the total number of the vibrational modes is conserved. Give expressions for the density of states and for the Debye frequency in this model. Derive the low and high-temperature limits of the specific heat.

Problem 5.21.

Use the Debye approximation to obtain an expression for the acoustic specific heat of a system of the shape of box, of size $L_x \times L_y \times L$, with $L_x = L_y \gg L$. When the thickness L of the system is significantly shorter that the other dimensions, a two-dimensional behavior is expected. Use tetragonal symmetry with the "thin" direction normal to the plane of the system, and show that at low temperatures the temperature dependence of the specific heat crosses over from the two-dimensional behavior to the three-dimensional one; explain how this transition depends on L.

5.5 Crystal stability and the Debye-Waller factor

Mermin-Wagner theorem. The **Lindemann criterion** for the melting temperatures of solids is discussed in Sec. 3.10. There, it is assumed that each of the atoms in the lattice oscillates independently of the motions of all other; the estimates for the melting temperature are then based in turn on estimates of the average of the square of the oscillation's amplitude. In view of the discussions in the present chapter, that analysis can be improved. As seen from Eq. (5.44), The amplitude of the vibration of each atom, $\mathbf{u}(\mathbf{R}_i^0)$, is affected by the motions of all others. In fact, it is a sum over all $dn_B N$ vibrational modes (recall that d is the dimension, n_B is the number of atoms in the unit cell, and $N = N_1 N_2 \ldots N_d$ is the number of unit cells). The average of the square of this amplitude is $\langle u_m^2(\mathbf{R}_i^0)\rangle$, where $m = 1, 2, \ldots, n_B d$ counts the various vibrational modes. Due to the symmetry with respect to displacements of the lattice, the mean-square deviations of all atoms of the same type are identical. This allows for an average over all unit cells,

$$\langle u_m^2(\mathbf{R}_i^0)\rangle = \frac{1}{N}\sum_i \langle u_m^2(\mathbf{R}_i^0)\rangle = \frac{1}{N}\sum_{\mathbf{k}} \langle \widetilde{u}_m(\mathbf{k})\widetilde{u}_m(-\mathbf{k})\rangle \,, \qquad (5.72)$$

where the expansion in the eigenmodes, Eq. (5.44), is used; the rest of the derivation follows the same route as that leading to Eq. (5.46) (check!). The $\mathbf{k}$–sum runs over the discrete values of the wave vectors, as defined in Eq. (5.43).

The right hand-side of Eq. (5.72) is the average of the amplitude squared of a harmonic oscillator, whose frequency is $\omega_m(\mathbf{k})$. As seen in Sec. 3.10, one has first to calculate the quantum average, and then the thermal (statistical-mechanical) one, with the Boltzmann weights. The result is [see Eq. (3.59)]

$$\langle \widetilde{u}_m(\mathbf{k})\widetilde{u}_m(-\mathbf{k})\rangle = \frac{\hbar}{2M\omega_m(\mathbf{k})}\coth\!\Big(\frac{\beta\hbar\omega_m(\mathbf{k})}{2}\Big) \,, \qquad (5.73)$$

where the angular brackets represent the double average, over the quantum probability and over the thermal one. When this expression is inserted into Eq. (5.72), and the sum there is replaced by an integration over the frequency, one finds

$$\langle u_m^2(\mathbf{R}_i^0)\rangle = \frac{1}{N}\sum_{\mathbf{k}}\langle \widetilde{u}_m(\mathbf{k})\widetilde{u}_m(-\mathbf{k})\rangle = \frac{\hbar}{2MN}\int\frac{d\omega}{\omega}g_m(\omega)\coth\!\Big(\frac{\beta\hbar\omega}{2}\Big) \,. \qquad (5.74)$$

The integral on the right side is over a limited range of frequencies for the optical branches, and therefore it converges. This is in particular true when the Einstein approximation, which accounts for a single frequency, is used. In this approximation, the results derived in Sec. 3.10 are valid. As opposed, the same integral for the acoustic modes is not well-behaved: as seen from Eq. (5.67), at low frequencies $g(\omega) \propto \omega^{d-1}$. Consider first the limit of zero temperature. There, Eq. (5.74) yields $\langle \widetilde{u}_m(\mathbf{k})\widetilde{u}_m(-\mathbf{k})\rangle = \hbar/(2M\omega_m)$, and consequently

$$\langle u_m^2(\mathbf{R}_i^0)\rangle = \frac{\hbar}{2MN}\int\frac{d\omega}{\omega}g_m(\omega) \sim \int d\omega\,\omega^{d-2} \,.$$

As this expression is inversely proportional to the mass of the atoms, crystals comprising lighter atoms are less stable. The integral converges for $d > 1$, but diverges (due to the lower bound) at one dimension. This divergence obviously invalidates the Lindemann criterion: the average amplitude of each atom is by far longer than the lattice constant, and thus the crystal cannot remain stable, that is, in one dimension the harmonic vibrations destroy the crystalline structure even at zero temperature. At higher dimensions, however, the harmonic vibrations are stable down to zero temperature.

What happens at finite temperatures? At low frequencies, $\omega \ll k_B T/\hbar$, one finds $\langle \widetilde{u}_m(\mathbf{k})\widetilde{u}(-\mathbf{k})\rangle \approx k_B T/(M\omega_m^2)$, and therefore the integrand in Eq. (5.74) is proportional to ω^{d-3}. It thus diverges (at the lower bound) for $d \leq 2$, and the Lindemann criterion implies that a two-dimensional crystal disintegrates at any finite temperature. This has been already noted by **Landau** and **Peierls** and precisely phrased later by Mermin and Wagner. The **Mermin-Wagner theorem** states that **within the harmonic approximation, there are no stable one- and two-dimensional crystals at finite temperatures. At zero temperature there are no one-dimensional stable crystals.**

Does that mean that one should abandon the conclusions drawn from the previous discussions regarding one- and two-dimensional solids? Not necessarily. Contrary to the Mermin-Wagner theorem, experimental observations indicate that effective one-and two-dimensional materials do exist in Nature (recall, e.g., that graphene is a two-dimensional lattice). One has to understand why the Mermin-Wagner theorem does not pertain to them; several possible explanations are listed below.

Anharmonic effects. As mentioned, the Mermin-Wagner theorem is valid for harmonic vibrations alone, and does not pertain to anharmonic terms. Indeed, when the potential energy of a certain vibrational mode is of the form $U(u) = M\omega^2 u^2/2 + Bu^4/4$, with $B > 0$, then even though the frequency found within the harmonic approximation tends to zero, the quartic term confines the motion. Classically, in the $\omega \to 0$ limit, the energy of the particle is $E = p^2/(2M) + Bu^4/4$; the particle oscillates between the two points $u = \pm[4E/B]^{1/4}$, and the amplitude squared is of the order of $u^2 \propto \sqrt{4E/B}$. The classical Boltzmann average (discussed in problem 5.22) is

$$\langle u^2 \rangle = \frac{\int du\, u^2 \exp[-\beta B u^4/4]}{\int du \exp[-\beta B u^4/4]} = \sqrt{\frac{4k_B T}{B}} \frac{\int dx\, x^2 \exp[-x^4]}{\int dx \exp[-x^4]}$$

(the integration variable is changed in the second step). The integrals on the right hand-side converge [their values are tabulated, and yield $\Gamma(3/4)/\Gamma(5/4) \approx 1.35$]. It follows that the average of the amplitude squared is finite and proportional to $\sqrt{T}$. The quantum-mechanical calculation is subtle, as there is no analytical solution for the anharmonic oscillator; however, the motion is bounded, and the anharmonic terms stabilize the crystal.

Growth on a substrate and meta-stability. Often the low-dimensional system is attached to a three-dimensional substrate (examples are the absorbed layers described in Sec. 2.9 and Sec. 3.12). Even when the coupling to the substrate is weak, it suffices to stabilize the system. The two-dimensional layers fabricated in the lab are usually grown on a substrate, and then are peeled off. This is in particular true for graphene. Attempts to grow free crystals of graphene from isolated carbon atoms end up with small non-planar crystals, e.g., the Buckminster-Fuller spheres or the nano tubes (Fig. 1.9). Apparently during the peeling operation, the atoms keep the hexagonal structure they have had in the graphite, as this is a metastable state of the system (similar to the one of a glass): to reach a more stable state the system has to overcome numerous potential barriers, a process which is usually rather slow.

Finite size. As mentioned, samples of graphene are not in particular large; they are of the order of 100 microns. A possible explanation for their stability is based on their finite, rather small, size. The integral on the right hand-side of Eq. (5.74) is not continuous for a finite crystal; it is an approximation of a sum over discrete frequencies that correspond to discrete wave vectors (i.e., the central part of the equation). The latter, for a two-dimensional system, are $\mathbf{k}_{\ell_1,\ell_2} = (\ell_1/N_1)\mathbf{b}_1 + (\ell_2/N_2)\mathbf{b}_2$. The length of the shortest wave vector is of the order of $|\mathbf{b}_i|/N_i$. Assuming that the lengths of the crystal in both directions are similar, $L_1 = N_1 a_1$ and $L_2 = N_2 a_2$, one finds that $k_{\min} = 2\pi/\max\{L_1, L_2\} = 2\pi/L_{\max}$. At the small frequencies that belong to this wave number the dispersion relation is acoustic, and therefore $\omega_{\min} = ck_{\min} = 2\pi c_m/L_{\max}$. Even if the sum in Eq. (5.72) is approximated by an integral, the lower bound of the latter is (up to a constant) $\omega_{\min}$. Indeed, the Debye approximation (5.67) in conjunction with Eq. (5.74) for $d = 2$, $g_m(\omega) = NS\omega/(2\pi c_m^2)$, yields

$$\langle u_m^2(\mathbf{R}_i^0) \rangle = \frac{\hbar}{2MN} \int_{\omega_{\min}}^{\omega_D} \frac{d\omega}{\omega} g_m(\omega) \coth\left(\frac{\beta\hbar\omega}{2}\right) = \frac{Sk_{\mathrm{B}}T}{M 2\pi c_m^2} \ln\left(\frac{\sinh[\beta\hbar\omega_D/2]}{\sinh[\beta\hbar\omega_{\min}/2]}\right),$$

(5.75)

(check!) where S is the area of the two-dimensional unit cell. This expression diverges when applied to an infinite crystal, for which $L_{\max} \to \infty$. However, for a finite crystal for which $L_{\max} \gg \beta\hbar\pi c_m$, the logarithmic term on the right hand-side is dominated by $-\ln[\sinh(\beta\hbar\pi c_m/L_{\max})] \approx \ln[L_{\max}/(\beta\hbar\pi c_m)]$. As the logarithmic function varies slowly with its argument, the divergence of the amplitude squared as the crystal is increased is slow as well, and it is possible to achieve small displacements in relatively large samples. Moreover, the argument of the logarithmic function depends on the ratio $L_{\max}/\beta$, which decreases as the temperature is reduced. At sufficiently low temperatures, i.e., when $k_{\mathrm{B}}T \ll \hbar\omega_{\min} = \pi\hbar c_m/L_{\max}$, the logarithm on the right hand-side of Eq. (5.75) is approximately $\beta\hbar(\omega_D - \omega_{\min})/2$, and consequently the amplitude squared tends to a finite limit which is independent of the temperature. Thus, relative large stable crystals can exist within a range of non zero temperatures.

Problem 5.22.

Find the leading dependence of the average of the amplitude squared in a one-dimensional lattice (with a lattice constant a) on the sample's length, L. Is this lattice stable according to the Lindemann criterion?

Long-range forces. The denominator in Eq. (5.75) contains the sound velocity of the material. As shown in Sec. 5.1, the sound velocity increases when interactions with farther away neighbors are included. For instance, in problem 5.1 it is found that $c = a\sqrt{\sum_m m^2 D_{L,m}/M}$, which obviously becomes larger as more terms are added to the sum. It follows that when more couplings are taken into account, the average of the amplitude squared decreases, and larger crystals can be stabilized. This is even more so when the interaction decays slowly with the distance, e.g., like in problem 5.20: a spring constant that decays with the spring's length, $D(R) = D_1/|R|^{d+\eta}$ yields $\widetilde{K}(k) \propto k^\eta$, and in turn gives rise to $g(\omega) \propto \omega^{d/\eta-1}$. Inserting this form into Eq. (5.74) gives, in the zero-temperature limit, $\langle u^2(R_i^0)\rangle \sim \int_0^{\omega_D} d\omega \omega^{d/\eta-2}$; this integral converges for $\eta < d$. As opposed, at finite temperatures, the behavior near the lower bound of the integral is $\langle u^2(R_i^0)\rangle \sim \int_0^{\omega_D} d\omega \omega^{d/\eta-3}$, and this integral converges for $\eta < d/2$.

Topological phases. The heating of a solid crystalline causes the creation of crystalline imperfections whose number increases till the melting temperature is reached, and the crystal melts to become a liquid. As an example, consider a type of defects that is relevant to the issue of stability in two dimensions: **dislocations**. The dislocation displayed in Fig. 5.20(a) is a linear default, which is formed in a three-dimensional system when one of the lattice planes terminates along a line, while the distances between the other planes decrease, such that far away from the linear defect the ordered periodic structure is restored. In two dimensions [Fig. 5.20(b)] the dislocation is a point in the plane where one of the lattice lines terminates. The dislocation is a **topological phenomenon**: in an infinite lattice it cannot be eliminated even when the lattice is deformed continuously. (For instance, one can deform a cup with a handle into a ring, but one cannot eliminate the single hole that exists in the ring and in the handle of the cup; in fact, the number of holes, as the number of dislocations, is not changed by the deformation.) The dislocation in a finite lattice can be removed only by shifting it to the boundary, or when two "inverse" dislocations are joined so that they cancel one another. The energy "cost" of the creation of a dislocation in two dimensions increases logarithmically with the distance from the dislocation; as a result, a finite concentration of dislocations destroys the solid phase. On the other hand, when two dislocations are joined together (without cancelling one another), their joint effect at long distances is smaller. In fact, as discovered by Kosterlitz and Thouless and in parallel by Berezinskii, the periodic crystalline phase in two dimensions is replaced by another phase, in which all dislocations are paired. In this phase, called **hexatic**, the Bragg peaks are widened, which indicates the destruction of the periodic order. The relationships between the directions of the bonds, however, are maintained. For instance, the angles between the bonds are $0°$ or $60°$, as in the triangular lattice. The hexatic phase appears also in liquid crystals, see Fig. 7.2. At the melting temperature of the hexatic phase the pairs of dislocations disintegrate, and the material becomes a liquid. This is the Berezinskii-Kosterlitz-Thouless phase transition, that led to the discoveries of other **topological phase transitions.**

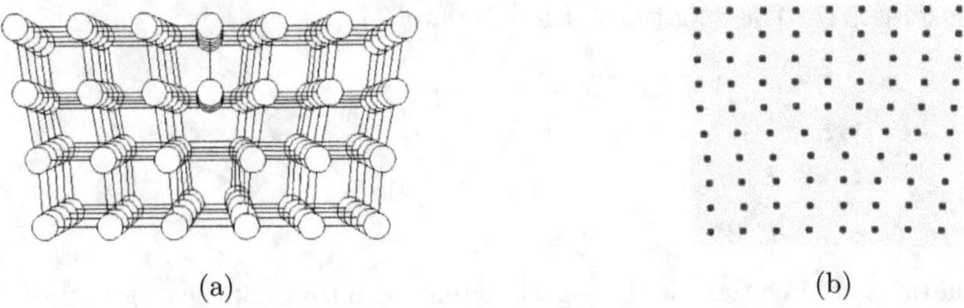

(a)   (b)

Fig. 5.20: (a) A dislocation in three dimensions. (b) A dislocation in two dimensions.

The Debye-Waller factor. The vibrational modes affect the form of the Debye-Waller factor, introduced in Sec. 3.10. The right hand-side of Eq. (3.57) is replaced by sums over the various modes that participate in the oscillations of the atoms. Using the Debye approximation for arbitrary dimension, Eqs. (5.67) and (5.74), one finds

$$\langle u_m^2(\mathbf{R}_i^0)\rangle\Big|_{\mathrm{D}} = \frac{\hbar d}{2M\omega_{\mathrm{D}}^d} \int_{\omega_{\min}}^{\omega_{\mathrm{D}}} d\omega\, \omega^{d-2} \coth\left(\frac{\beta\hbar\omega}{2}\right) \tag{5.76}$$

for each acoustic branch. Each optical mode, in the Einstein approximation, gives

$$\langle u_m^2(\mathbf{R}_i^0)\rangle\Big|_{\mathrm{E}} = \frac{\hbar}{2M\omega_{\mathrm{E}}^d} \coth\left(\frac{\beta\hbar\omega_{\mathrm{E}}}{2}\right) , \tag{5.77}$$

where ω_{E} is the frequency which represents the optical mode.

Problem 5.23.
Assuming that the melting temperature is higher than the Debye one, use the Lindemann criterion to express the melting temperature of a single-atom, $d-$dimensional lattice, in terms of its Debye temperature.

5.6 Impurities, localized modes, scattering of sound waves

In reality, crystals are not perfectly periodic; they contain a finite (though at times minute) concentration of "foreign" atoms, (**"impurities"**), or other flaws. The full description of disordered crystals, with randomly-distributed defects, is beyond the scope of this book. Instead, the effect of a small amount of impurities on the lattice vibrations of a simple system is studied. Consider for simplicity a single foreign atom that occupies one of the lattice sites of a one-dimensional lattice comprising $2N + 1$ atoms. Let this atom be located at $n = 0$, and all other atoms at

$n = \pm 1, \pm 2, \ldots, \pm N$. Though the following solution is obtained for rigid boundary conditions, $u_{\pm(N+1)} = 0$, similar solutions can be worked out for other boundaries (see problem 5.24), and for an infinite lattice as well. The atom at $n = 0$ is specified by its mass, m, which differs from the mass M of the host atoms. The constants of the springs that couple it to its nearest neighbors, D_0, also differ from all other spring constants, D. The equations of motion are

$$m\ddot{u}_0 = -D_0(2u_0 - u_1 - u_{-1}) \, ,$$
$$M\ddot{u}_{\pm 1} = -D_0(u_{\pm 1} - u_0) - D(u_{\pm 1} - u_{\pm 2}) \, ,$$
$$M\ddot{u}_{\pm n} = -D(2u_{\pm n} - u_{\pm(n-1)} - u_{\pm(n+1)}) \, , \quad 1 < n < N-1 \, ,$$
$$M\ddot{u}_{\pm N} = -D(2u_{\pm N} - u_{\pm(N-1)}) \, . \tag{5.78}$$

These equations are invariant with respect to reflection through the origin, $n \leftrightarrow -n$. Therefore the solutions are symmetric or anti-symmetric with respect to such a reflection. It is thus helpful to define symmetric and anti-symmetric functions, $u_n^{\pm} = u_n \pm u_{-n}$, for $1 \leq n \leq N$; the equations of motion then separate,

$$m\ddot{u}_0 = -D_0(2u_0 - u_1^+) \, , \quad M\ddot{u}_1^- = -D_0 u_1^- - D(u_1^- - u_2^-) \, ,$$
$$M\ddot{u}_1^+ = -D_0(u_1^+ - 2u_0) - D(u_1^+ - u_2^+) \, ,$$
$$M\ddot{u}_n^{\pm} = -D(2u_n^{\pm} - u_{n-1}^{\pm} - u_{n+1}^{\pm}) \, , \quad 2 \leq n \leq N-1 \, ,$$
$$M\ddot{u}_N^{\pm} = -D(2u_N^{\pm} - u_{N-1}^{\pm}) \, . \tag{5.79}$$

Thus there are N equations for $u_1^-, u_2^-, \ldots u_N^-$, and $N+1$ ones for $u_0, u_1^+, u_2^+, \ldots u_N^+$.

For $2 \leq n \leq N-1$, the equations are the "usual" ones for an homogeneous chain; the solutions are $u_n \propto C^{\pm n} \exp[-i\omega t]$ (this is implied by the Bloch theorem mentioned in Sec. 5.1). For $C = \exp[ika]$, where ka is real, these are propagating wavy solutions. Otherwise, these solutions grow or decay exponentially. Hence

$$u_n^{\pm} = (A^{\pm}C^n + B^{\pm}C^{-n})e^{-i\omega t} \, , \tag{5.80}$$

provided that the "usual" dispersion relation is obeyed,

$$M\omega^2 = D(2 - C - 1/C) \, . \tag{5.81}$$

Anti-symmetric solutions. These solutions obey two more equations, for $n = 1$ and for $n = N$. The symmetric case is discussed separately below. Indeed, inserting Eqs. (5.80) into the equations of motion yields

$$(M\omega^2 - D - D_0)(A^- C + B^-/C) + D(A^- C^2 + B^-/C^2) = 0 \, ,$$
$$(M\omega^2 - 2D)(A^- C^N + B^-/C^N) + D(A^- C^{n-1} + B^-/C^{N-1}) = 0 \, . \tag{5.82}$$

Using the dispersion relation (5.81), one obtains

$$-\frac{B^-}{A^-} = C^{2N+2} = \frac{(D - D_0)C - D}{(D - D_0)/C - D} \, . \tag{5.83}$$

When $D = D_0$, then $C^{2N+2} = 1$, which entails $C = \exp[ika]$, with $k = k_\ell = \pi\ell/[(N+1)a]$, $\ell = 1, 2, \ldots, N$. In this case $B^- = A^-$, and thus $u_n^- = 2iA^- \sin(nka)$.

These are just the wavy solutions that would have been obtained were the foreign atom identical to all others (see e.g., problem 5.13), since the symmetric vibrations' amplitudes, including u_0, vanish. As the oscillating atoms are not connected to the foreign atom, their amplitudes do not depend on its mass.

One may still look for wavy solutions when $D \neq D_0$. Indeed, using $C = \exp[ika]$ in Eq. (5.83), assuming that ka is real, turns that equation into $\exp[2i(N+1)ka] = \exp[2i\chi]$, with

$$\tan\chi = \frac{(D - D_0)\sin(ka)}{(D - D_0)\cos(ka) - D} .$$

It follows that $k = k_\ell + \chi(k)/[(N+1)a]$; this transcendental equation is solved graphically or numerically for each ℓ. Thus, there are N wavy solutions, with k_ℓ displaced from its previous value $\pi\ell/[(N+1)a]$. The corresponding frequencies, however, are still within the acoustic range found before.

Symmetric solutions. To derive these solutions, one first substitutes $u_0 = -D_0 u_1^+/(m\omega^2 - 2D_0)$ in the equation for u_1^+, and then finds that the coefficients A^+ and B^+ obey

$$\left(M\omega^2 - D - D_0 - \frac{2D_0^2}{m\omega^2 - 2D_0}\right)\left(A^+ C + \frac{B^+}{C}\right) + D\left(A^+ C^2 + \frac{B^+}{C^2}\right) = 0 ,$$

$$\left(M\omega^2 - 2D\right)\left(A^+ C^N + \frac{B^+}{C^N}\right) + D\left(A^+ C^{N-1} + \frac{B^+}{C^{N-1}}\right) = 0 . \tag{5.84}$$

The second equation is identical to that obtained in the solution for the anti-symmetric functions. All in all, one finds

$$-\frac{B^+}{A^+} = C^{2N+2} = \frac{(D - D_0 - X)C - D}{(D - D_0 - X)/C - D} , \tag{5.85}$$

where $X = 2D_0^2/(m\omega^2 - 2D_0)$ and $M\omega^2 = D(2 - C - 1/C)$. As before, one may insert $C = \exp[ika]$, to obtain N wavy solutions, with $k = k_\ell + \chi_1(k)/[(N+1)a]$, with

$$\tan\chi_1(k) = \frac{D - D_0 - X)\sin(ka)}{(D - D_0 - X)\cos(ka) - D} .$$

Localized vibrational mode. As all in all there are $2N + 1$ solutions, out of which $2N$ are wavy ones as found above, there is yet one more solution to be derived. One tries a solution for $|C| < 1$ (check that a trial solution with $|C| > 1$ gives rise to a similar result). For large enough N, the central part in Eq. (5.85) is very small. When it is completely ignored, one finds $(D - D_0 - X)C - D = 0$. With the dispersion relation (5.81), there emerges a cubic equation for C. The trivial value $C = 1$ is always a solution, and thus there remains a quadratic equation. Usually only one of its solutions complies with all the requirements. Consider for simplicity the case $D = D_0$. Then the afore-mentioned equation becomes

$$\left(1 - \frac{2M}{m}\right)C^2 - 2\left(1 - \frac{M}{m}\right)C + 1 = 0 ,$$

with the trivial solution $C = 1$ and the non trivial one $C = m/(m - 2M)$. The requirement $|C| < 1$ is equivalent to $m < M$, and then $-1 < C < 0$, implying that $C^n = (-|C|)^n$ and the wave function alternates its sign between nearest neighbors. In case $M < m$, then $|C| > 1$, and the whole procedure has to be repeated for that value.

Inserting the result into Eq. (5.81) yields

$$M\omega^2 = \frac{4DM^2}{m(2M - m)} \, .$$

(5.86)

One easily verifies that $\omega^2 > 4D/M = \omega_0^2$, where ω_0 is the maximal frequency of acoustic waves in one dimension (check!). When $m = M$ one finds that $\omega = \omega_0$, as expected. As seen in Sec. 5.1, frequencies higher than ω_0 correspond to exponential functions, which are not waves. Such solutions lead to unphysical consequences in the homogeneous system (as they diverge to infinity at one of the edges). However, here they decay on both sides of the impurity atom, and therefore they are acceptable. It is worth noting that for $2M < m$ the frequency becomes imaginary, and consequently the solutions decay with time.

Returning to Eq. (5.85), one finds that for sufficiently large N, $|B^+/A^+| = |C|^{2(N+1)} \ll 1$, and thus $u_n^+ \approx A^+C^n = A^+(-\exp[-\kappa a])^n$; that is, the amplitudes of the displacements decay with the distance from the foreign atom, with the decay coefficient $\kappa = -\ln|C|$. The coefficient is larger as the two masses become more disparate. This special mode is confined to the vicinity of the foreign atom; it is called a **localized vibrational mode**.

In summary, in the presence of a foreign atom, there are $2N + 1$ vibrational modes, as found also for a homogeneous system; the difference is that in the present case only $2N$ of them possess frequencies in the acoustic range. One vibrational mode is distinct, and has a **discrete frequency**. When there is a single foreign atom on the lattice, the contribution of its vibrations to the specific heat is smaller by a factor of $2N$ compared to the contributions of all other vibrations, making it difficult to be detected in experiments. In reality, however, there is a finite concentration, p, of foreign atoms. When this concentration is small, it is plausible to assume that the foreign atoms are far away from each other; consequently, the localized modes that are centered around each of the foreign atoms are similar to each other, and thus have the same frequency. In this case the contribution of the acoustic branch to the specific heat is multiplied by $1 - p$; the impurities' contribution is of the Einstein approximation type, with the frequency found above, and is proportional to the number of foreign atoms, i.e., $2Np$. Such a term in the specific heat indicates the existence of impurities in the crystal.

Problem 5.24.
Find the solutions of Eqs. (5.78) for periodic boundary conditions, i.e., $u_N = u_{-N}$.

The generalization of this discussion to more complex one-dimensional systems, e.g., with larger unit cells, yields similar results. In higher dimensions the equations turn out to be rather complicated. For instance, the localized states are not spherically symmetric (i.e., the decay coefficient depends on the direction); the analysis of those is beyond the scope of this book.

Scattering of phonons off an impurity. Impurities, beside causing localized vibrational modes, can also **scatter the waves** (i.e., scatter the phonons), and thus hinder the propagation of these waves in the crystal. Consider for concreteness a heavy foreign atom; the scattering off such an atom can be viewed as elastic – there is no energy exchange between the phonons and the scattering atom. To demonstrate this important phenomenon it is useful to consider a wave that impinges from the left on the one-dimensional crystal discussed above. The original wave propagates from left to right, and therefore its wave function is of the form $u_n = A\exp[i(kna - \omega t)]$. Once this wave hits the foreign atom, a reflected wave is created that moves to the left, with the same wave number and with the same frequency. Denoting the amplitude of the reflected wave by Ar (r is the **reflection amplitude** of the wave), its function is $u_n = Ar\exp[i(-kna - \omega t)]$. One therefore conjectures a solution of the form

$$u_n = A(e^{ikna} + re^{-ikna})e^{-i\omega t} , \quad \text{for} \;\; n \leq 0 . \tag{5.87}$$

On the other hand, the wave on the right side of the impurity moves entirely to the right; its origin is that part of the impinging wave that succeeded to pass the impurity. One therefore conjectures

$$u_n = Ate^{ikna}e^{-i\omega t} , \quad \text{for} \;\; n \geq 0 , \tag{5.88}$$

where t is the **transmission amplitude** (beware not to confuse it with the time!). The two functions, Eqs. (5.87) and (5.88), solve Eq. (5.79) for any $|n| > 0$, provided that the dispersion relation, $\omega(k) = \omega_0|\sin(ka/2)|$, inferred from Eq. (5.6) is fulfilled. That same equation, however, requires that $m\omega^2 t = K\{2t - t\exp[ika] - (\exp[-ika] + r\exp[ika])\}$. Using also the relation $u_0 = 1 + r = t$ (that follows from the equations for $n = \pm 1$) one obtains

$$t = \frac{i\sin(ka)}{e^{ika} - 1 + 2(m/M)\sin^2(ka/2)} . \tag{5.89}$$

For equal masses, $m = M$, the transmission amplitude is 1 (check!) since then the lattice is homogeneous and the wave is transmitted perfectly. In contrast, when the mass of the foreign atom differs from that of all other atoms, the transmission amplitude is a complex number, whose absolute value is smaller than 1. Denoting $t = \sqrt{T}\exp[i\alpha]$, the function of the wave on the right side of the impurity is $u_n = A\sqrt{T}\exp[i(kna + \alpha)]\exp[-i\omega t]$. The phase of the wave is shifted by α, and its amplitude is decreased by the factor $\sqrt{T}$. The dependence of T, which is termed the **transmission coefficient**, on the masses' ratio is displayed in Fig. 5.21, for several values of ka. All curves possess a maximum, $T = 1$, reached when the masses are equal; the transmission coefficient decreases as the wave number increases.

Generalizations. Three comments are in order. First, at higher dimensions, a plane wave, $\exp[i\mathbf{k} \cdot \mathbf{R}_i - i\omega t]$, impinging on a foreign atom, is scattered at all

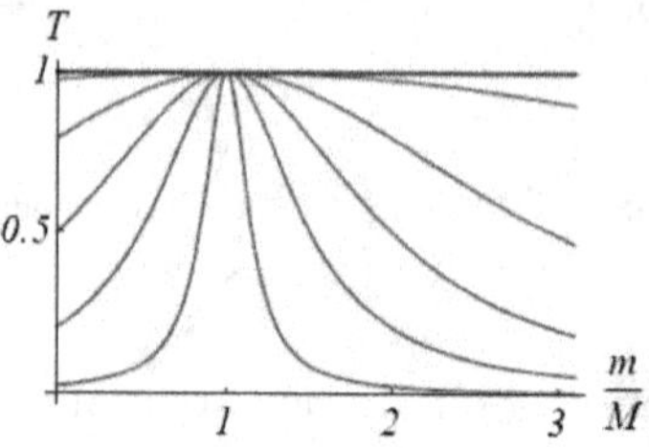

Fig. 5.21: The transmission coefficient, T, as a function of the masses' ratio m/M. It decreases as the original wave-number increases from $ka = 0.1\pi$ (upper curve) to $ka = 0.9\pi$ (lowest curve), in steps of 0.2π.

directions, with an amplitude that depends on the direction. Second, the classical treatment described above can be carried out quantum mechanically, e.g., see the arguments presented in Sec. 5.3, that lead to Eq. (5.41). Third, foreign atoms scatter other particles as well, predominantly electrons (see Chapter 6).

5.7 Lattice momentum and inelastic scattering of phonons

The lattice momentum of phonons. As mentioned in the context of Eq. (5.48), the description of vibrational modes in terms of waves can be translated into the "language" of particles: a vibrational mode of frequency ω, that possesses an average energy $\hbar\omega(1/2 + \langle n \rangle)$, is equivalent to a motion of $\langle n \rangle$ phonons, each with an energy $\hbar\omega$. In analogy with the de Broglie duality of particles and waves in quantum mechanics, each phonon may be also assigned a momentum, $\hbar\mathbf{k}$. In the previous section it is found that when a wave is scattered off an impurity elastically in one dimension, the frequency does not change but the momentum of the reflected wave is anti parallel to that of the impinging wave, and is $-\hbar\mathbf{k}$. The momentum lost by the scattered phonon is absorbed by the lattice. As the lattice is very large, the momentum absorbed from a single phonon is negligible (it is not measurable). In higher dimensions, the phonon can be scattered into various directions, and its momentum is changed accordingly. In the harmonic approximation, though, the energy is conserved.

In contrast with electrons or photons, whose momentum is uniquely defined by the de Broglie principle, $\mathbf{p} = \hbar\mathbf{k}$, the momentum of the phonon is different. As seen, it is enough to specify the wave vector of a vibration in the first Brillouin zone, as the properties of the wave are not changed upon displacing that vector by a reciprocal-lattice vector, $\mathbf{k} \Rightarrow \mathbf{k} + \mathbf{G}$. It follows that phonons whose momenta differ by $\hbar\mathbf{G}$ describe exactly the same physical phenomenon. For this reason, the phonon's momentum is termed **lattice momentum**, and is defined within the first Brillouin zone. In case this momentum, following a certain physical process (examples are given below) is augmented to lie beyond the first zone, one uses the translation $\mathbf{k} \Rightarrow \mathbf{k} + \mathbf{G}$ to bring it back into the first Brillouin zone. The physical outcome is not modified.

Scattering of radiation by phonons. The concept of lattice momentum is elucidated in the description of **scattering of radiation off solids**. This scattering may well be **inelastic**, i.e., the energy of the scattered particles differs from that of the impinging ones. To account for the inelasticity the real wave function is written in the form $\cos(\mathbf{k} \cdot \mathbf{r} - \Omega t) = \{\exp[i(\mathbf{k} \cdot \mathbf{r} - \Omega t)] + \exp[-i(\mathbf{k} \cdot \mathbf{r} - \Omega t)]\}/2$, where $\Omega = \Omega(\mathbf{k})$ is the frequency derived from the dispersion relation of the radiation, see Sec. 3.1. Assuming for simplicity point-like scatterers, then the generalization of Eq. (3.7) yields that the function of the scattered wave is

$$A \propto \sum_{\mathbf{R}} e^{i\mathbf{q}\cdot\mathbf{R} - i\Omega t} + \mathrm{cc} = \sum_{\mathbf{R}^0} e^{i\mathbf{q}\cdot[\mathbf{R}^0 + \mathbf{u}(\mathbf{R}^0, t)]} e^{-i\Omega t} + \mathrm{cc} , \qquad (5.90)$$

where cc denotes the complex conjugate of the other term and $\mathbf{q} = \mathbf{k} - \mathbf{k}'$ is the "momentum" transfer between the impinging particles (photons, neutrons, or electrons) and the scattered ones (up to the multiplicative factor $\hbar$).

Momentum and energy conservation. On the right hand-side of Eq. (5.90) appears the location of the nucleus, $\mathbf{R}$, expressed in terms of its equilibrium site on the lattice, $\mathbf{R}^0$, and its displacement, $\mathbf{u}(\mathbf{R}^0)$ (the index i is omitted here for brevity). As the displacement is small, the right hand-side of the equation may be expanded as a power series in $\mathbf{u}$,

$$A \propto \sum_{\mathbf{R}^0} e^{i\mathbf{q}\cdot\mathbf{R}^0} e^{-i\Omega t} [1 + i\mathbf{q} \cdot \mathbf{u}(\mathbf{R}^0) + \ldots] + \mathrm{cc} . \qquad (5.91)$$

When the lattice vibrations are ignored, this expression reproduces the result of Chapter 3, that gives rise to Bragg peaks whenever $\mathbf{q} = \mathbf{G}$. Then the frequency of the outgoing wave is also Ω, and the scattering is elastic, i.e., it conserves the energy of the impinging particle. As in Eq. (5.44), the oscillation' amplitude of each atom in the crystal is a sum over the contributions of all eigenmodes. Classically, the amplitude can be written in the form

$$\mathbf{u}(\mathbf{R}^0, t) = N^{-1/2} \sum_{\mathbf{K}} \widetilde{\mathbf{u}}(\mathbf{K}) e^{i\mathbf{K}\cdot\mathbf{R}^0 - i\omega(\mathbf{K})t} , \qquad (5.92)$$

where the sum contains only those discrete wave vectors which are allowed by the boundary conditions, $\mathbf{K} = (\ell_1/N_1)\mathbf{b}_1 + (\ell_2/N_2)\mathbf{b}_2 + (\ell_3/N_3)\mathbf{b}_3$, and where the explicit time dependence is kept, with the frequency that obeys the dispersion law, $\omega = \omega(\mathbf{K})$. Inserting Eq. (5.92) in each of the summands in Eq. (5.91) yields

$$\sum_{\mathbf{R}^0} e^{\pm i(\mathbf{q}\cdot\mathbf{R}^0 - \Omega t)} [1 \pm iN^{-1/2} \sum_{\mathbf{K}} \mathbf{q} \cdot \widetilde{\mathbf{u}}(\mathbf{K}) e^{i\mathbf{K}\cdot\mathbf{R}^0 - i\omega(\mathbf{K})t} + \ldots] . \qquad (5.93)$$

The second factor in the square brackets contains the sum $\sum_{\mathbf{R}^0} \exp[i(\mathbf{K} \pm \mathbf{q}) \cdot \mathbf{R}^0]$. As seen in the discussion following Eq. (3.11), this sum equals N when $\mathbf{K} \pm \mathbf{q} = \mathbf{G}$, for any reciprocal-lattice vector $\mathbf{G}$. Using the original wave numbers of the radiation thus results in an appreciable scattering amplitude only when

$$\hbar\mathbf{k} - \hbar\mathbf{k}' = \hbar\mathbf{q} = \pm\hbar(\mathbf{K} - \mathbf{G}) \ . \tag{5.94}$$

The left hand-side is the momentum transfer of the scattered particles, while the right hand-side comprises the lattice momentum of the relevant phonon. It follows that the radiation "emits" or "absorbs" a phonon (according to the sign on the right hand-side), and changes its momentum such that the new momentum is **conserved**, according to Eq. (5.94). The time dependence of the outgoing wave is $\exp[-i(\omega \pm \Omega)t]$; one may thus identify the frequency of the outgoing wave as $\Omega' = \Omega \pm \omega$; hence

$$\hbar\Omega - \hbar\Omega' = \pm\hbar\omega(\mathbf{K}) \ . \tag{5.95}$$

This relation expresses the **energy conservation** of the process. The upper sign corresponds to a "creation" of a new phonon in the system, while the radiation loses energy and momentum; the lower sign represents the "annihilation" of an existing phonon, with the radiation absorbing its energy and momentum. Recall that the dispersion relation of the phonons is periodic in momentum space, $\omega(\mathbf{K} - \mathbf{G}) = \omega(\mathbf{K})$, and thus the right hand-sides of Eqs. (5.94) and (5.95) represent the same vibrational mode. A quantum-mechanical treatment of this scattering event gives rise to Eqs. (5.94) and (5.95) as well; in such an analysis the Hamiltonian contains the interaction between the radiation and the lattice vibrations.

Bragg peaks of inelastic scattering. Once Eqs. (5.94) and (5.95) are used in Eq. (5.93), it is found that each Bragg peak, that results from elastic scattering, is accompanied by additional peaks, for which the momentum and the energy of the scattered particles obey the conservation laws of the energy and the lattice momentum. These new Bragg peaks represent **inelastic scattering events**. To detect the energy transfers in experiments, the energies of the scattered particles have to be of the same order of magnitude as those of the phonons. As the lattice constants are of the order of a few $\mathring{A}$'s, the relevant wave numbers are of the order of $\mathring{A}^{-1}$. Such wave numbers correspond to photons' energies of the order of a few thousands electron-volts, or neutrons' energies of the order of a few tens of milli-electron volts. The energies of the phonons are comparable to those of the neutrons, and therefore neutron radiation is the most suitable one for the research of phonons by inelastic scattering. Indeed, this radiation has been exploited in the experiments discussed in connection with Figs. 5.18 and 5.19.

Problem 5.25.

Neutrons of mass m_n and momentum $\hbar k$ impinge on a single-atom one-dimensional lattice (along the direction parallel to the lattice). The lattice consists of atoms of mass M coupled together by nearest-neighbor springs of a constant D. Find the possible values of the momentum and frequency of the scattered neutrons.

Inelastic scattering of light. As opposed to neutrons, the typical photon energy of the visible light is of the order of a few electron volts; the corresponding wave length is of the order of a few thousands Å. It follows that **inelastic scattering of light** off lattice vibrations might be realized only for wave vectors much shorter than the lengths of the reciprocal-lattice vectors, i.e., vectors at the vicinity of the center of the Brillouin zone. Inelastic scattering of light off phonons is called **Brillouin scattering**; when the scattering phonons are optical, it is referred to as **Raman scattering**. Equations (5.94) and (5.95) prevail for this type of scattering as well. The emission of a phonon is termed **Stokes scattering**, and the absorption process is the **anti-Stokes scattering**. As $c|\mathbf{k}| = \Omega \gg \omega(\mathbf{K})$, then approximately $|\mathbf{k}| = |\mathbf{k}|'$, and consequently $|\mathbf{K}| \approx 2k\sin\theta$ (see Fig. 3.7).

5.8   Phenomena related to anharmonic effects

The importance of anharmonic terms in the potential energy. The discussion above is focused on the outcome of the harmonic approximation, Eqs. (5.1) or (5.20), in which the potential between two atoms in the crystal is replaced by a paraboloid centered around around its minimal value. Figure 5.22 displays the Lennard-Johns potential, already discussed in previous chapters (see also Fig. 1.1), together with the parabola that passes through its minimum, such that the second derivative there is the same for both curves. Obviously the parabola is a faithful description of the potential close to the minimum, but it deviates from it farther away. At low temperatures the harmonic approximation is quite reliable, as only small deviations from the minimum are thence important. In the quantum language, the properties calculated within this approximation are determined mainly by the lower energy levels of each vibrational mode, i.e., by a small number of phonons. However, the harmonic potential misses several important aspects. First, the actual potential is steeper than the parabola to the left of the minimum, and smoother than it to the right. As a result, it is harder to compress the material than to stretch it. In contrary, the harmonic potential is symmetric and does not reproduce this feature.

Second, in the harmonic approximation the average distances between the atoms remain unchanged at all temperatures; the phenomenon of **thermal expansion** does not exist, whereas experiments show that crystals expand when heated. The volume **thermal expansion coefficient** of a material in d dimensions is defined by

$$\alpha_V = \frac{1}{V}\left(\frac{\partial V}{\partial T}\right)_p = d\alpha_L = \frac{d}{L}\left(\frac{\partial L}{\partial T}\right)_p, \tag{5.96}$$

where α_L is the longitudinal thermal expansion coefficient (the factor d results from the relation $V \propto L^d$, from which it follows that $\partial V/V = d\partial L/L$). It is found below that this coefficient vanishes in the harmonic approximation; it becomes finite only when anharmonic terms are included.

Third, in the harmonic approximation the vibrational waves extend over the entire lattice. In other words, in the absence of foreign atoms and other defects the

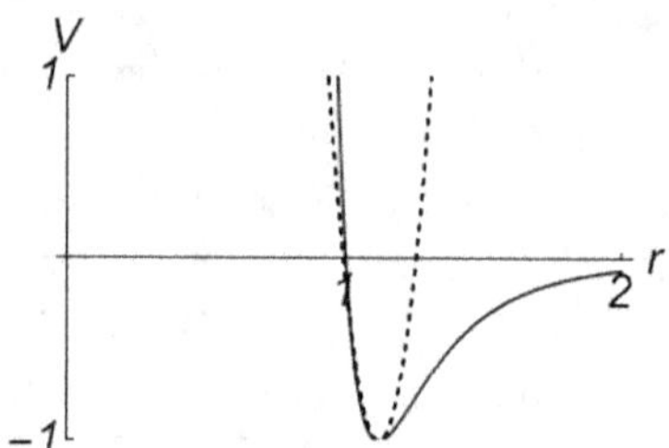

Fig. 5.22: A comparison between the Lennard-Johns potential, Eq. (4.33), and the harmonic approximation at the vicinity of its minimum. Distances are measured in units of σ and the energies in units of ϵ.

phonons move freely over the entire system. Phonons carry heat and contribute significantly to the **heat conduction** in solids. In experiments, the **heat conductance**, κ, is defined by the **Fourier law**, which relates the heat current, $\mathbf{j}_U$, to a temperature gradient imposed on the material, ∇T,

$$\mathbf{j}_U = -\kappa \nabla T \ . \tag{5.97}$$

The heat current is the amount of thermal energy that passes through a unit area per unit time; it is induced by the temperature gradient. (Note the analogy with the electric current in a metal, which is induced by an electric field, that is given by the gradient of the electrical potential.) When the phonons passing through the material are not hindered by any perturbation, there is no resistance to the heat conduction, that is, the heat conductance is infinite. However, as discussed, phonons are scattered by foreign atoms. The main cause for the finite value of the heat conductance in solids, though, are **collisions in-between phonons**. In the harmonic approximation, the Hamiltonian of the lattice vibrations is a sum over the separate Hamiltonians of each vibrational mode, and there are no interactions among the various modes; the modes do not evolve in time. The anharmonic terms, which exist beyond the harmonic approximation, give rise to interactions among the phonons; these cause collisions that lead in turn to a reduction in the heat conductance. As mentioned, other factors, e.g., the collisions of the phonons with impurities or with the sample's walls, or with other defects in the lattice, give rise to a further decrease in the heat conductance.

Thermal expansion. To understand the notion of thermal expansion it is useful to consider the simple anharmonic potential $U(x) = Cx^2/2 - g_3 x^3 + g_4 x^4 + \ldots$. The sign of the third-order term is chosen to make the leading-order correction to the parabola negative on the right of the maximum, and positive on its left, in accordance with Fig. 5.22. In the classical approximation, the thermal average of the particle's location (relative to its equilibrium site) is

$$\langle x \rangle = \frac{\int dx\, x e^{-\beta U(x)}}{\int dx\, e^{-\beta U(x)}} \approx \frac{\int dx\, x e^{-\beta C x^2/2}(1 + \beta g_3 x^3 - \beta g_4 x^4 + \ldots)}{\int dx\, e^{-\beta C x^2/2}(1 + \beta g_3 x^3 - \beta g_4 x^4 + \ldots)} \approx \frac{3 g_3}{C^2} k_{\mathrm{B}} T \; .$$

$$(5.98)$$

In the right-hand parts of the equation, both the numerator and the denominator are expanded in the anharmonic terms; the end result consists of the leading contribution (check!). As seen, the average position vanishes in the harmonic approximation, but increases linearly with the temperature once the first anharmonic correction is accounted for.

Problem 5.26.

At zero temperature the energy of the crystal can be written in the form $F_0 = U_{\mathrm{tot}}(V) \approx U_{\mathrm{tot}}(\overline{V}) + [B/(2\overline{V})](V - \overline{V})^2$, where $\overline{V}$ is the volume attained at the minimal value of the energy, and where $B = V(\partial^2 U/\partial V^2)$ is the coefficient of the volume expansion, see the discussion around Eq. (4.10). At finite temperatures the energy is augmented by the free energy of the phonons.

a. *Show that the total free energy is*

$$F_0(T, V) = U_{\mathrm{tot}}(\overline{V}) + \frac{B}{2\overline{V}}(V - \overline{V})^2 + k_{\mathrm{B}} T \sum_{\alpha} \log[2\sinh(\beta \hbar \omega_\alpha/2)] \; ,$$

where the sum encompasses all vibrational modes.

b. *The anharmonic effects can be accounted for by allowing the harmonic frequencies to depend on the volume. Assuming that the crystal is not under pressure, i.e., $p = -(\partial F_0/\partial V)_T = 0$, show that the change of the volume of the crystal caused by the phonons is*

$$V = \overline{V} + \frac{1}{B} \sum_{\alpha} \frac{\hbar \omega_\alpha}{2} \coth\left[\frac{\beta \hbar \omega_\alpha}{2}\right]\left(-\frac{\partial \log \omega_\alpha}{\partial \log V}\right) \; .$$

c. *To a certain extent, $\partial \ln \omega_\alpha / \partial \ln V = -\gamma$ is the same for all modes (i.e., it is independent of α). This quantity is termed the **Grüneisen constant**. Show that then $V = \overline{V} + \gamma U_{\mathrm{ph}}/B$, where $U_{\mathrm{ph}} = \langle E \rangle_{\mathrm{ph}}$ is the total average energy of the phonons.*

d. *Show that in this case $\alpha_V = [\gamma/(BV)]C_V$, where $C_V = (\partial \langle E \rangle_{\mathrm{ph}}/\partial T)_V$ is the phonon specific heat.*

Heat conductance. Heat conductance is mainly determined by the collisions among the phonons. These collisions can be treated in much the same way as done for those between the phonons and the photons, culminating in Eq. (5.94),

$$\mathbf{k}_1 - \mathbf{k}_2 = \pm(\mathbf{k}_3 - \mathbf{G}) \; ,$$

$$(5.99)$$

where $\mathbf{k}_1$ is the momentum of the phonon scattered into a state of momentum $\mathbf{k}_2$, once another phonon of momentum $\mathbf{k}_3$ is emitted or absorbed. Another way to present this relation is $\mathbf{k}_1 \pm \mathbf{k}_3 = \mathbf{k}_2 - \mathbf{G}$. The left hand-side is the total lattice momentum of the incoming phonons, and the right one contains the lattice momentum of outgoing phonon, up to a reciprocal-lattice vector. When the three momenta are small, then $\mathbf{G} = 0$, the momentum of the outgoing phonon equals the sum of the momenta of the two incoming ones; i.e., the total momentum of all phonons is not changed. On the other hand, when the sum of the two incoming momenta is a vector ending outside of the first Brillouin zone, the outgoing momentum is that sum minus a reciprocal-lattice vector. The outgoing phonon may thus propagate in a direction opposite to the original direction, and hence diminish the original phonons' stream. Such a process, that gives rise to an inverse momentum (relative to the incoming one) is termed **Umklapp**(the German word for inverse). Another example for such a process appears in Sec. 5.7, in connection with the scattering of a phonon off an impurity.

The kinetic theory. The simplest treatment of heat conductance is carried out in the context of the **kinetic theory of a phonon gas**, in which the phonons are viewed as particles moving in the material. Each phonon moves with a constant velocity $\mathbf{v}$ during a typical time interval τ, traversing a distance ℓ termed the "mean free-path", until it collides with a defect (or the vessel wall, or an electron, or another phonon). Imagine that the two opposite faces of the container are held at two different temperatures, such that the temperature decreases towards the positive $\hat{\mathbf{x}}$−axis. Assume further that the thermal-equilibrium energy of the phonon is determined by the local temperature, $T(x)$, i.e., $E(x) = E_{\mathrm{eq}}[T(x)]$. In the absence of a net current, half of the phonons moves to the right and the other half moves towards the left. The net number of phonons passing from left to right through a unit area normal to the $\hat{\mathbf{x}}$−axis per unit time is $nv_x/2$. (The number of phonons per unit volume is n, and it is assumed that $v_x > 0$). The phonons that impinge on this unit area from the left traveled a distance $v_x\tau$ and therefore carry the energy $E_{\mathrm{eq}}[T(x - v_x\tau)]$. Similarly, the phonons that come from the right carry the energy $E_{\mathrm{eq}}[T(x + v_x\tau)]$. Thus, the amount of energy passing through the unit area per unit time is

$$j_{U,x} = \frac{n}{2}v_x\{E_{\mathrm{eq}}[T(x - v_x\tau)] - E_{\mathrm{eq}}[T(x + v_x\tau)]\} \approx -\frac{n}{2}v_x\frac{\partial E_{\mathrm{eq}}}{\partial T}\frac{\partial T}{\partial x}2v_x\tau \ , \quad (5.100)$$

where both the energy and the temperature are expanded to leading order in the respective variable's difference. Recalling that the specific heat per unit volume is $C_V = n(\partial E_{\mathrm{eq}}/\partial T)$, one retrieves Eq. (5.97), with the heat conductance

$$\kappa = C_V\langle v_x^2\rangle\tau = \frac{1}{3}\langle \mathbf{v}^2\rangle C_V\tau \ . \quad (5.101)$$

The average is carried out over all velocity's directions (as there is no preference to the $\hat{\mathbf{x}}$−direction). When the phonons are acoustic, their velocity is the sound velocity, which is independent of the temperature. At temperatures higher than the Debye temperature the specific heat is close to the its value derived from the Dulong-Petit law, and thus it is also independent of the temperature. It remains to estimate the time in-between collisions, τ. It turns out that the most detrimental processes for the heat conduction are the Umklapp ones, in which two phonons are colliding with one another. The probability of a certain phonon to hit another one is proportional to the density of phonons in the crystal. From Eq. (5.54), one finds that at high temperatures $\langle n_\alpha \rangle = 1/(\exp[\beta\hbar\omega_\alpha] - 1) \approx k_\mathrm{B}T/(\hbar\omega_\alpha)$; the probability is proportional to the temperature and consequently the mean free-time in-between collisions is inversely proportional to the temperature, $\tau \propto 1/T$. In this limit $\kappa \propto 1/T$. Experiments find a somewhat higher power of $1/T$, probably because of collisions among more than two phonons. At low temperatures most of the phonons have small momenta; collisions among them leave the outgoing momentum within the first Brillouin zone, and consequently do not affect the heat conductance. To obtain Umklapp processes the two colliding phonons should have momenta larger than half of the reciprocal-lattice vector (in suitable units), and the probability to find such phonons at low temperatures is exponentially small, $\langle n_\alpha \rangle \approx \exp[-\beta\hbar\omega_\alpha]$, where ω_α is of the order of the Debye frequency; the mean free-path may then be even longer than the sample itself. It follows that the heat conductance at low temperatures is dominated by other processes, as collisions with the sample's walls or collisions with impurities. The dependence of the heat conductance on the temperature is then determined by the temperature dependence of the specific heat, $C_V \propto T^3$ in three dimensions. The outcome of these observations is that, roughly speaking, the heat conductance increases like T^3 at low temperatures, reaches a maximum, and then decays as T^{-1} at high temperatures.

5.9   Magnons

Magnetic excitations. In addition to phonons, there are other excitations in the crystal, some of them resembling phonons, for instance, **magnetic excitations** termed **magnons**. Recall that the magnetic behavior of quite a number of materials is represented by the Heisenberg model Eq. (4.41): $\hat{\mathcal{H}}_M = -\sum_{\langle nm \rangle} J_{nm} \mathbf{S}_n \cdot \mathbf{S}_m$, where $\boldsymbol{\mu}_n = -g\mu_\mathrm{B}\mathbf{S}_n$ is the magnetic moment located at the lattice site n, J_{nm} is the exchange coefficient (in units of energy), and the sum encompasses all pairs of sites $\langle nm \rangle$. At any given time, the spin (i.e., $\mathbf{S}_n$) at site n is subjected to a magnetic field, induced by the instantaneous magnetic moments of all its neighbors, $\mathbf{B}_n = \sum_m J_{nm} \mathbf{S}_m$ (as the "spins" are dimensionless, the magnetic field is in units of energy). Consider the **classical equations of motions** that emerge from this description. A magnetic field $\mathbf{B}$, applied on a classical magnetic moment $\boldsymbol{\mu}$, gives rise to a torque, such that the equation of motion is $\dot{\boldsymbol{\mu}} = \gamma\boldsymbol{\mu} \times \mathbf{B}$, where $\gamma = ge/(2mc) = \mu_\mathrm{B}/\hbar$. In the present case,

$$\dot{\boldsymbol{\mu}}_n = \boldsymbol{\mu}_n \times \sum_m \bar{J}_{nm}\boldsymbol{\mu}_m \, , \tag{5.102}$$

where $\bar{J}_{nm} = \gamma J_{nm}/(g\mu_\mathrm{B})^2 = J_{nm}/(g\mu_\mathrm{B}\hbar)$ (see problem 5.27). Equation (5.102) is a non-linear equation for the variables $\boldsymbol{\mu}_n$. However, at low temperatures the magnetic moments are ordered, that is, they mainly point along a certain direction, and it is plausible to consider only small deviations from this ordered state [in analogy with the lattice vibrations, where only small deviations (off the equilibrium sites) in the locations of the atoms are considered]. This situation is demonstrated by two examples.

Ferromagnets. In the fully-ordered ferromagnetic state, all magnetic moments are equal in magnitude and are parallel to each other, for instance along the $\hat{\mathbf{z}}$–axis. Thus, in the ground state $\boldsymbol{\mu}_n = \mu\hat{\mathbf{z}}$. Deviations from the ground state can be considered by assuming that the instantaneous components μ_{nx} and μ_{ny} are small. Up to first order in the x and y components, the length of the z–component is unchanged (its variation is quadratic in the lengths of the other components), $\mu_{nz}^2 = \mu^2 - \mu_{nx}^2 - \mu_{ny}^2 \approx \mu^2$. With these assumptions, the equation of motion (5.102) yields

$$\frac{d\mu_{nx}}{dt} = \sum_m \overline{J}_{nm}(\mu_{ny}\mu_{mz} - \mu_{nz}\mu_{my}) = \mu\sum_m \overline{J}_{nm}(\mu_{ny} - \mu_{my}) \ ,$$

$$\frac{d\mu_{ny}}{dt} = \sum_m \overline{J}_{nm}(\mu_{nz}\mu_{mx} - \mu_{nx}\mu_{mz}) = \mu\sum_m \overline{J}_{nm}(\mu_{mx} - \mu_{nx}) \ . \tag{5.103}$$

These are two linear equations in two variables, whose treatment is similar to the one carried out above, e.g., Eqs. (5.12), (5.23), or (5.32). Inserting a wavy solution, $\mu_{nx,y} = \mu_{x,y}\exp[i(\mathbf{k}\cdot\mathbf{R}_n^{(0)} - \omega t)]$, leads to

$$-i\omega\mu_y = -\mu[\widetilde{J}(0) - \widetilde{J}(\mathbf{k})]\mu_x \ , \quad -i\omega\mu_x = \mu[\widetilde{J}(0) - \widetilde{J}(\mathbf{k})]\mu_y \ , \tag{5.104}$$

where

$$\widetilde{J}(\mathbf{k}) = \sum_m \overline{J}_{nm}e^{-i\mathbf{k}\cdot\mathbf{R}_{nm}^0} \tag{5.105}$$

is the Fourier transform of the exchange coefficient, and $\mathbf{R}_{nm}^0 = \mathbf{R}_n^0 - \mathbf{R}_m^0$. For instance, for a nearest-neighbor interaction in one dimension $\widetilde{J}(k) = \overline{J}(\exp[-ika] + \exp[ika]) = 2\overline{J}\cos(ka)$. Inserting the second of Eqs. (5.104) into the first one yields $\omega^2\mu_y = \mu^2[\widetilde{J}(0) - \widetilde{J}(\mathbf{k})]^2\mu_y$; hence this is a valid solution provided that the dispersion relation $\omega^2 = \mu^2[\widetilde{J}(0) - \widetilde{J}(\mathbf{k})]^2$ is obeyed. The similarity with the dispersion relation of lattice vibrations is quite clear, for example, Eq. (5.5). Nonetheless, the dispersion relation is different,

$$\omega = \mu[\widetilde{J}(0) - \widetilde{J}(\mathbf{k})] \ , \tag{5.106}$$

where the positive value of the frequency is chosen. For instance, in one dimension with solely nearest-neighbor interactions

$$\omega = \mu[\widetilde{J}(0) - \widetilde{J}(k)] = \mu\overline{J}[2 - 2\cos(ka)] = 4\mu\overline{J}\sin^2(ka/2) \ . \tag{5.107}$$

The right hand-side may be expanded in k at small frequencies, yielding a quadratic dispersion relation, $\omega \approx \mu\overline{J}(ka)^2$. Note the disparity with the linear dispersion in the acoustic branch of lattice vibrations.

Problem 5.27.

a. *Show that the solutions of the equation $\dot{\boldsymbol{\mu}} = \gamma\boldsymbol{\mu} \times \mathbf{B}$ correspond to a precession of the vector $\boldsymbol{\mu}$ around the vector $\mathbf{B}$, with the Larmor frequency $\omega_{\mathrm{L}} = \gamma\mathrm{B}$.*
b. *Derive Eq. (5.102) from the equation in part (a).*

Spin waves. The waves derived in the calculation above are termed **"spin waves"**. Indeed, upon inserting Eq. (5.106) in Eqs. (5.104), one finds that $\mu_y = -i\mu_x$. This implies that there is a phase difference between the oscillations of the two spin components: the component of the magnetic moment in the XY plane rotates around the $\hat{\mathbf{z}}$–axis together with the propagation along the direction of the wave vector $\mathbf{k}$. Figure 5.23 depicts the spins (by the arrows) of a ferromagnet whose moment is perpendicular to the direction of the (one-dimensional) lattice, and hence is also normal to the direction of $\mathbf{k}$ itself. Has the magnetic moment been parallel to the lattice direction, it would have made a screw-like rotation around that direction.

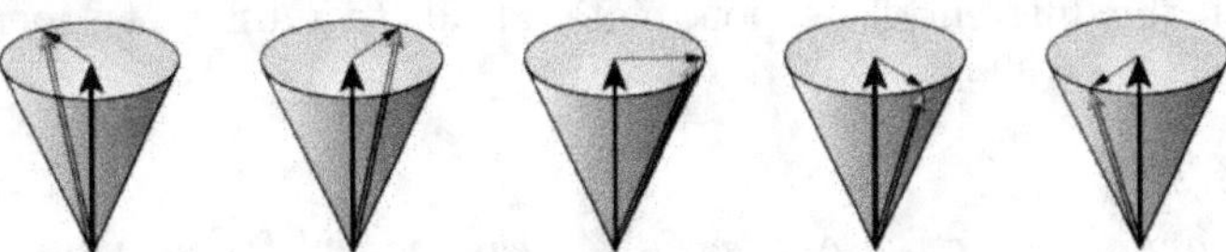

Fig. 5.23: Spin waves of a one-dimensional ferromagnet with a ferromagnetic moment (the vertical arrow) normal to the crystal, for a wave that propagates along the lattice (the horizontal direction in the figure). The component of the magnetic moment perpendicular to the ferromagnetic one rotates around the ferromagnetic moment along with the propagation of the wave.

Problem 5.28.
An alternative method to solve the equations of motion for spin waves of a ferromagnet is to construct equations for the functions $\mu_{n\pm} = \mu_{nx} \pm i\mu_{ny}$, and to solve them. Show that this method produces the same solutions as those obtained above.

Fourier transforms. An alternative way to obtain the spin-wave solutions is to Fourier transform the magnetic moments,

$$\widetilde{\boldsymbol{\mu}}(\mathbf{k}) = \sum_n e^{i\mathbf{k}\cdot\mathbf{R}_n^0}\boldsymbol{\mu}_n \ , \tag{5.108}$$

and then to insert the transforms into the equations of motion (5.103), which become

$$\frac{d\widetilde{\mu}_x}{dt} = \mu[\widetilde{J}(0) - \widetilde{J}(\mathbf{k})]\widetilde{\mu}_y \ ,$$

$$\frac{d\widetilde{\mu}_y}{dt} = -\mu[\widetilde{J}(0) - \widetilde{J}(\mathbf{k})]\widetilde{\mu}_x \ . \tag{5.109}$$

Differentiating the first equation with respect to the time t and then inserting the result into the second gives an equation for μ_x,

$$\frac{d^2\widetilde{\mu}_x}{dt^2} = -\mu^2[\widetilde{J}(0) - \widetilde{J}(\mathbf{k})]^2\widetilde{\mu}_x \ . \tag{5.110}$$

The same equation is obtained for $\widetilde{\mu}_y$. Equation (5.110) is identical to the one that describes the harmonic oscillator, with the frequency $\omega = \mu[\widetilde{J}(0) - \widetilde{J}(\mathbf{k})]$;

thus the previous solutions are reproduced. This mapping upon the harmonic oscillator also allows for the construction of a quantum Hamiltonian for the variable $\tilde{\mu}_x$. The energy levels of each such "oscillator" are of the form $\hbar\omega(n + 1/2)$, and their statistical mechanics is the same as that of phonons. In the present context the "particles" that represent spin waves are termed **"magnons"**. The only difference between (ferromagnetic) magnons and phonons is in the dispersion relation. Choosing in addition boundary conditions identical to the ones applied for phonons, i.e., periodic boundary conditions, yields discrete wave numbers, $\mathbf{k}_{\ell_1,\ell_2,\ell_3} = (\ell_1/N_1)\mathbf{b}_1 + (\ell_2/N_2)\mathbf{b}_2 + (\ell_3/N_3)\mathbf{b}_3$, where $-N_m/2 \le \ell_m < N_m/2$. The calculation of the specific heat of the magnons is very similar to that of the phonons, but the different dispersion relation leads to a different dependence on the temperature, see problem 5.29.

Problem 5.29.

As for phonons, one may also describe a Debye model for magnons, by extending the dispersion relation of long waves up to an appropriate Debye frequency.
a. *Find the dispersion relation of a cubic ferromagnet with nearest-neighbor interactions, and show that in the long-wave limit it can be written as $\omega \approx \Delta|\mathbf{k}|^2$. Determine the value of the coefficient Δ.*
b. *Find the leading-order dependence of the magnons' specific heat on the temperature, at low and at high temperatures.*

Antiferromagnets. For simplicity, the equations of motion for the spin waves of an **antiferromagnet** are derived for a simple configuration: a one-dimensional lattice, in which the exchange interactions are solely between nearest neighbors. The ground state of this system is depicted in part C of Fig. 2.42. The unit cell contains two magnetic moments, e.g., $\mu_1^{(1)}$ on the left and $\mu_1^{(2)}$ on the right; at equilibrium, $\boldsymbol{\mu}_n^{(1)} = \mu\hat{\mathbf{z}}$ and $\boldsymbol{\mu}_n^{(2)} = -\mu\hat{\mathbf{z}}$. The equations of motion are

$$\frac{d\boldsymbol{\mu}_n^{(1)}}{dt} = \boldsymbol{\mu}_n^{(1)} \times \overline{J}(\boldsymbol{\mu}_n^{(2)} + \boldsymbol{\mu}_{n-1}^{(2)}) \,,$$

$$\frac{d\boldsymbol{\mu}_n^{(2)}}{dt} = \boldsymbol{\mu}_n^{(2)} \times \overline{J}(\boldsymbol{\mu}_n^{(1)} + \boldsymbol{\mu}_{n+1}^{(1)}) \tag{5.111}$$

(check!). As again only linear-order terms in the deviations from the equilibrium configuration are accounted for, it is assumed that the $z-$component of the magnetic moment does not vary in time. Then, using the explicit forms of the vectors

$$\boldsymbol{\mu}_n^{(1)} = (\mu_{nx}^{(1)}, \mu_{ny}^{(1)}, \mu) \,,$$

$$\boldsymbol{\mu}_n^{(2)} = (\mu_{nx}^{(2)}, \mu_{ny}^{(2)}, \mu) \,,$$

one obtains four linear equations for the components of the magnetic moments in the XY plane (see problem 5.30). The solution of these equations yields spin waves with the dispersion relation

$$\omega = 2\mu\overline{J}|\sin(ka/2)| \,, \tag{5.112}$$

where a is the lattice constant.

Problem 5.30.
Write down the explicit equations of motion for spin waves that propagate in a one-dimensional antiferromagnet with nearest-neighbor exchange interactions, and prove Eq. (5.112).

Other excitations. In addition to phonons and magnons, there are several other types of excitations in condensed-matter systems. These are described classically by waves connected with the oscillations of a certain physical property, and quantum-mechanically are characterized by the appearance of "particles" moving in the material. Several such excitations are related to the motion of electrons in the crystal, discussed in Chapter 6. Examples are plasmons, polarons, and excitons. The **plasmon** refers to a collective excitation of the electron gas in a metal, relative to the lattice of ions. The origin of the **polaron** is the coupling between the electron and the vibrations of the crystal in which it resides. It describes a motion of an electron accompanied by a distortion of the surrounding ions' locations. The **exciton** results when an electron from the valence band is excited into the conduction band, leaving behind a "hole"; the electron and the hole remain connected, like the positive and negative charges in the hydrogen atom, and therefore move together. Detailed descriptions of these excitations are beyond the scope of the book.

5.10   Answers for the problems in the text

Answer 5.1.
a. Extending Eq. (5.3), the equations of motion for harmonic interactions between neighbors located at a distance ma from one another, are

$$M\ddot{u}_n = -\sum_m D_{L,m}(2u_n - u_{n-m} - u_{n+m}) \ .$$

Considerations similar to those presented at the beginning of Sec. 5.1 lead to a solution of the form $u_n = A_0 \exp[i(kna - \omega t)]$. This is indeed a solution provided that

$$M\omega^2 = \sum_m D_{L,m}(2 - e^{-ikma} - e^{ikma}) = 4\sum_m D_{L,m}\sin^2(mka/2) \ . \qquad (5.113)$$

This dispersion relation has the periodicity given in Eq. (5.7) and hence it suffices to investigate it in the first Brillouin zone. Due to the symmetry of k with respect to inversion, it is enough to consider the range $0 \leq k \leq \pi/a$. The group velocity is given by the derivative of the dispersion law, $v_{\mathrm{g}} = [a/(M\omega)]\sum_m D_{L,m}m\sin(mka)$. It vanishes at the edges of the Brillouin zone, $ka = \pi$. The expansion around $k = 0$ yields the sound velocity, $c = v_{\mathrm{g}} = a\sqrt{\sum_m m^2 D_{L,m}/M}$. This is valid as long as $mka \ll 1$ for each m in the sum.

b. The interaction is [see Eq. (4.33)]

$$U(R) = 4\epsilon\left[\left(\frac{\sigma}{R}\right)^{12} - \left(\frac{\sigma}{R}\right)^{6}\right] ,$$

and hence, using $D_{L,m} = \partial^2 U(R)/\partial R^2|_{R=ma}$, one finds

$$M\omega^2 = \frac{96\epsilon}{\sigma^2} \sum_{m} \left[26\left(\frac{\sigma}{ma}\right)^{14} - 7\left(\frac{\sigma}{ma}\right)^{8}\right] \sin^2(mka/2) .$$

Once the summation is carried out, the value of the lattice constant, $(a/\sigma)^6 = 2A_{12}/A_6$, is introduced [see discussion and notations after Eq. (4.34)]. In one dimension $A_n = \sum_{m=1}^{\infty} 1/m^n = \zeta(n)$, and the values of the zeta function, $\zeta(n)$, can be found in mathematical codes or treatises of mathematical functions. In particular, $\zeta(6) = \pi^6/945 \approx 1.0173$ and $\zeta(12) = 691\pi^{12}/638512875 \approx 1.000246$. In the long wave-length limit $\sin x$ can be approximated by x (though, strictly speaking, this approximation is valid only for $kam \ll 1$ and therefore fails at long distances). The result is

$$M\omega^2 = 24\epsilon\left(26A_{12}\left(\frac{\sigma}{a}\right)^{12} - 7A_6\left(\frac{\sigma}{a}\right)^{6}\right)k^2 = 72\epsilon\left(\frac{A_6^2}{A_{12}}\right)k^2 \approx 74.5\epsilon k^2 .$$

This leads to an acoustic dependence, with a sound velocity $c \approx 8.63\sqrt{\epsilon/M}$. Interestingly enough, a numerical computation of the sum (discarding the approximation of $\sin x$) gives rise to a similar behavior.

c. In this case

$$\omega^2 = \omega_0^2 \sin^2(ka/2)[1 + (4D_{L,2}/D_{L,1})\cos^2(ka/2)] ,$$

where $\omega_0 = 2\sqrt{D_{L,1}/M}$. Figure 5.24 displays the dispersion law. One observes that at the edge of the Brillouin zone the second term in the brackets vanishes and there appears a local minimum, $\omega = \omega_0$. Between the origin and this point there is a maximum, $\omega_{\max} = \omega_0(D_{L,1} + 2D_{L,2})/(4\sqrt{D_{L,1}D_{L,2}})$, at $\sin^2(k_{\max}a/2) = (D_{L,1} + 4D_{L,2})/(8D_{L,2})$. The group velocity vanishes at this maximum (like at any extremal point), leading to the appearance of standing waves there. For the parameters considered here, the maximum appears at $k_{\max}a = 2\pi/3$. As seen, there is a single wavy solution (for each sign of the wave number) for $\omega < \omega_0$. In the range $\omega_0 \leq \omega \leq \omega_{\max}$ there are two wavy solutions, and for $\omega \geq \omega_{\max}$ there are no wavy solutions at all. The number of solutions is doubled when one takes into account the negative k's.

Answer 5.2.

The vectors between the nth atom and its two neighbors are $\mathbf{R}_{n,n\pm1}$; the angle in-between the bonds is given by $\cos\gamma_n = \mathbf{R}_{n,n-1}\cdot\mathbf{R}_{n,n+1}/(R_{n,n-1}R_{n,n+1})$. Expanding $\mathbf{R}_{n,n\pm1} = \pm a\hat{\mathbf{x}} + \mathbf{u}_{n\pm1} - \mathbf{u}_n$ in a Taylor series in the displacements, results in $\cos\gamma_n + 1 = (2\mathbf{v}_n - \mathbf{v}_{n-1} - \mathbf{v}_{n+1})^2/(2a^2) + \ldots$, where $\mathbf{v}_n$ is the transverse component of the vector $\mathbf{R}_n$. Assuming that the energy is proportional to this expression, i.e.,

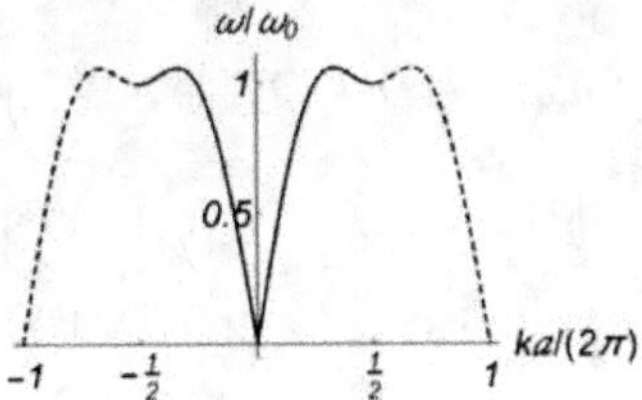

Fig. 5.24

$U = C \sum_n (1 + \cos\gamma_n)$, the equations of motion obeyed by the transverse components are

$$M\ddot{\mathbf{v}}_n = -\frac{C}{a^2}[2(2\mathbf{v}_n - \mathbf{v}_{n+1} - \mathbf{v}_{n-1})$$
$$+ (\mathbf{v}_n + \mathbf{v}_{n-2} - 2\mathbf{v}_{n-1}) + (\mathbf{v}_n + \mathbf{v}_{n+2} - 2\mathbf{v}_{n+1})] \ .$$

The dispersion relation for $\mathbf{v}_n = \mathbf{v}_0 \exp[i(kan - \omega_T t)]$ is

$$M\omega_T^2 = \frac{C}{a^2}[6 + 2\cos(2ka) - 8\cos(ka)] \approx Ca^2 k^4 + \dots \ .$$

Answer 5.3.

One may write Eq. (5.13) in the form

$$\omega^2 \begin{bmatrix} A \\ B \end{bmatrix} = D \begin{bmatrix} 2/M & -(e^{-ika}+1)/M \\ -(1+e^{ika})/m & 2/m \end{bmatrix} \begin{bmatrix} A \\ B \end{bmatrix} \ .$$

The vector whose components are A and B is an eigenvector of the matrix on the right hand-side, with the eigenvalue ω^2. Denoting this matrix by Ω, the eigenvalues are obtained by equating the determinant of $(\Omega - \hat{I}\omega)$ to zero, where $\hat{I}$ is the 2×2 unit matrix. This requirement is identical to Eq. (5.14). Inserting one of the eigenvalues into the equation yields the ratio between the components of the corresponding eigenvector, like in Eq. (5.18).

Answer 5.4.

The equations of motion are

$$M\ddot{u}_n = -D(u_n - v_{n-1}) - D(u_n - v_n) - e\mathcal{E}_0 e^{i(kna-\omega t)} \ ,$$
$$m\ddot{v}_n = -D(v_n - u_n) - D(v_n - u_{n+1}) + e\mathcal{E}_0 e^{i(kna-\omega t)} \ .$$

These describe a driven harmonic motion. One seeks for a particular solution of the equations: assuming that the atoms "follow" the oscillations of the electric field, one uses the forms $u_n = A\exp[ikna - i\omega t]$ and $v_n = B\exp[ikna - i\omega t]$, with the same wave vector and the same frequency as those of the electric field. Then

$$M\omega^2 A = D[2A - (1 + e^{-ika})B] + e\mathcal{E}_0 \ ,$$
$$m\omega^2 B = D[2B - (1 + e^{ika})A] - e\mathcal{E}_0 \ ,$$

and consequently

$$A = \frac{1}{\text{Det}}[m\omega^2 - 2D + D(1 + e^{-ika})]e\mathcal{E}_0 \;,$$

$$B = -\frac{1}{\text{Det}}[M\omega^2 - 2D + D(1 + e^{ika})]e\mathcal{E}_0 \;,$$

$$\text{Det} = (M\omega^2 - 2D)(m\omega^2 - 2D) - 4D^2\cos^2(ka/2) \;.$$

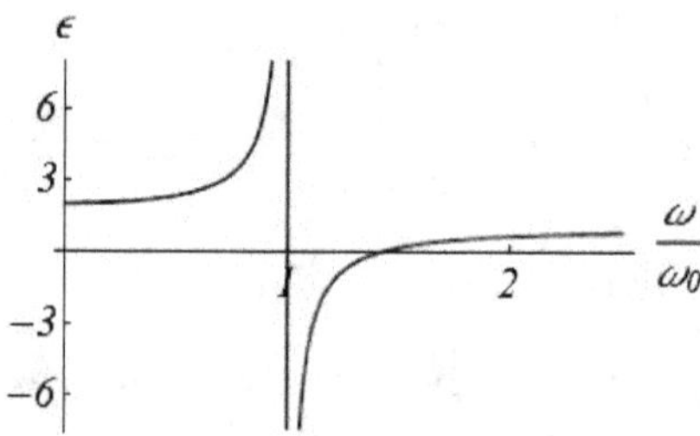

Fig. 5.25

The induced dipole moment in a unit cell is the charge times the distance between the negative and the positive ions, i.e.,

$$p(n,t) = e(v_n - u_n) = -e^2\mathcal{E}(n,t)[(m+M)\omega^2 - 4D\sin^2(ka/2)]/\text{Det}.$$

Because the speed of light is about 10^5 larger than the velocities of the acoustic and optical waves in the crystal, one may assume that $ka \ll 1$. Using strictly zero ka yields an electric field and an induced dipole moment which do not depend on the coordinate. The induced dipole moment per unit volume is then $P = \alpha\mathcal{E}$, with the polarizability $\alpha = \rho e^2/[\mu(\omega_0^2 - \omega^2)]$, where ρ is the density (the number of unit cells per unit volume), μ is the reduced mass, and $\omega_0 = \sqrt{2D/\mu}$ is the optical frequency in the $k = 0$ limit, Eq. (5.17). The polarizability is related to the dielectric function ε by $\varepsilon = 1 + 4\pi\alpha$ and hence $\varepsilon = 1 + 4\pi\rho e^2/[\mu(\omega_0^2 - \omega^2)]$.

The dielectric function is depicted in Fig. 5.25. The speed of light in the material is $c = c_{\text{light}}/\sqrt{\varepsilon}$, and the wave number is $k = \omega/c = \omega\sqrt{\varepsilon}/c_{\text{light}}$. The dielectric function is negative over a range of frequencies and there the wave number is imaginary, $k = i\kappa$. Then the light waves decay with the distance as $\exp[-\kappa na]$, and the material is opaque for radiation of frequencies in this range.

Answer 5.5.

The "new" equations of motion are

$$M\ddot{u}_n = -D(u_n - v_{n-1}) - D(u_n - v_n) - D_2(u_n - u_{n-1}) - D_2(u_n - u_{n+1}) \;,$$

$$m\ddot{v}_n = -D(v_n - u_n) - D(v_n - u_{n+1}) - D_2(v_n - v_{n-1}) - D_2(v_n - v_{n+1}) \;.$$

Inserting the wave functions $u_n = A \exp[i(kna - \omega t)]$ and $v_n = B \exp[i(kna - \omega t)]$ yields

$$M\omega^2 A = D[2A - (e^{-ika} + 1)B] - 2D_2[1 - \cos(ka)]A \;,$$
$$m\omega^2 B = D[2B - (e^{ika} + 1)A] - 2D_2[1 - \cos(ka)]B \;,$$

with the determinant

$$\mathrm{Det} = [M\omega^2 - 2D - 4D_2 \sin^2(ka/2)][m\omega^2 - 2D - 4D_2 \sin^2(ka/2)] - 4D^2 \cos^2(ka/2) \;.$$

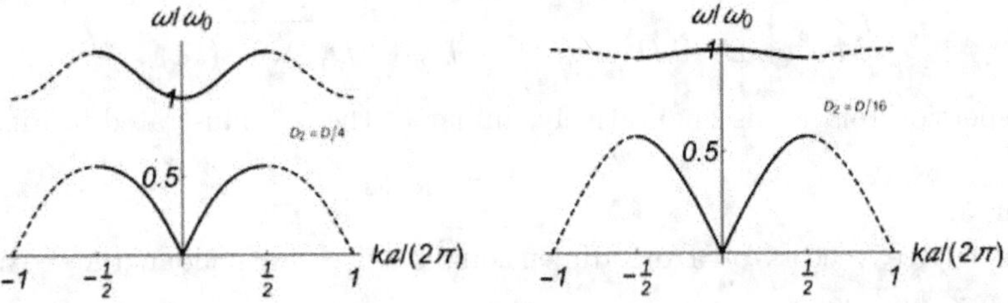

Fig. 5.26

Equating the determinant to zero gives two solutions,

$$\omega^2 = D\left(\frac{1}{M} + \frac{1}{m}\right)\left(1 + \frac{4D_2}{D}\sin^2(ka/2)\right)$$
$$\pm D\left[\left(\frac{1}{M} + \frac{1}{m}\right)^2\left(1 + \frac{4D_2}{D}\sin^2(ka/2)\right)^2\right.$$
$$\left. - \frac{4\sin^2(ka/2)}{Mm}\left(1 + \frac{4D_2}{D} + \frac{4D_2^2}{D^2}\sin^2(ka/2)\right)\right]^{1/2} \;.$$

These are shown in Fig. 5.26 for two values of D_2, with the same notations as in Fig. 5.4. Note in particular the qualitative modifications in the optical branch. When $D_2 = 0$, one retrieves Fig. 5.4. As the forces in-between next-nearest neighbors increase, this branch "flattens", and the maximum at $k = 0$ turns into a minimum which decreases as D_2 is increased. In the $D = 0$ limit, the only forces are those in-between identical atoms. The equations of motion then coincide with Eqs. (5.5), and give rise to the dispersion relations shown in Fig. 5.2 – but with different masses for each sublattice, and hence with different sound velocities and different maximal frequencies.

Answer 5.6.

The unit cell still contains two atoms, but their masses are identical. Denoting the spring constant between the atoms in the cell (i.e., the single bond) by D_1, and the one in-between nearest neighbors belonging to neighboring cells (i.e., the triple bond) by D_2 (these are the springs to the left and to the right of the solid circle in Fig. 5.3), the equations of motion are

$$M\ddot{v}_n = -D_1(v_n - u_n) - D_2(v_n - u_{n+1}) \ ,$$
$$M\ddot{u}_n = -D_2(u_n - v_{n-1}) - D_1(u_n - v_n) \ .$$

One then finds the two equations

$$M\omega^2 B = (D_1 + D_2)B - (D_1 + D_2 e^{ika})A \ ,$$
$$M\omega^2 A = (D_1 + D_2)A - (D_1 + D_2 e^{-ika})B \ ,$$

which have a non trivial solution when

$$M\omega^2 = D_1 + D_2 \pm \sqrt{D_1^2 + D_2^2 + 2D_1 D_2 \cos(ka)} \ .$$

This dispersion relation is qualitatively similar to the one illustrated in Fig. 5.4.

Answer 5.7.

a. The lattice vectors of a one-dimensional lattice lying along the $\hat{\mathbf{x}}$−axis are $\mathbf{R}_m^0 = ma\hat{\mathbf{x}}$, and hence

$$D_L(\mathbf{R}_{0m}^0) = D_{xx}(\mathbf{R}_m^0) = \left[\frac{U'(ma)}{ma} + U''(ma) - \frac{U'(ma)}{ma}\right] = U''(ma) \ .$$

On the other hand, the second term in Eq. (5.22) does not contribute to the transverse vibrations, and hence

$$D_T(\mathbf{R}_{0m}^0) = D_{yy}(\mathbf{R}_m^0) = D_{zz}(\mathbf{R}_m^0) = \frac{U'(ma)}{ma} \ .$$

b. The Lenard-Jones potential, when differentiated with respect to the distance between two atoms, gives

$$U'(ma) = \frac{24\epsilon}{\sigma}\left[\left(\frac{\sigma}{ma}\right)^7 - 2\left(\frac{\sigma}{ma}\right)^{13}\right], \text{ and } U''(ma) = \frac{24\epsilon}{\sigma^2}\left(26\left(\frac{\sigma}{ma}\right)^{14} - 7\left(\frac{\sigma}{ma}\right)^8\right) \ .$$

The remaining steps in the solution are rather straightforward.

Answer 5.8.

The function $u_{i\mu}$ depends on the coordinate $\mathbf{R}_i^0$. Denoting this function by $u_{i\mu} = U_\mu(\mathbf{R}_i^0)$, then $u_{i'\mu} = U_\mu(\mathbf{R}_i^0 + \mathbf{a}_1)$, and hence $U_\mu(\mathbf{R}_i^0 + \mathbf{a}_1) = C(\mathbf{a}_1)U(\mathbf{R}_i^0)$. Repeated displacements by the same lattice vector yield $U_\mu(\mathbf{R}_i^0 + 2\mathbf{a}_1) = C(2\mathbf{a}_1)U(\mathbf{R}_i^0) = [C(\mathbf{a}_1)]^2 U(\mathbf{R}_i^0)$, and consequently $C(n\mathbf{a}_1) = [C(\mathbf{a}_1)]^n$, or alternatively $\ln[C(n\mathbf{a}_1)] = n\ln[C(\mathbf{a}_1)]$. The only function of $n\mathbf{a}_1$ that increases linearly with n is the function proportional to this vector, i.e., $\ln[C(n\mathbf{a}_1)] = n\mathbf{a}_1 \cdot \mathbf{K}$, where $\mathbf{K}$ is an arbitrary vector. Since the absolute value of C is 1, the vector $\mathbf{K}$ has to be imaginary, i.e., $\mathbf{K} = i\mathbf{k}$, with $\mathbf{k}$ being a real vector. This completes the proof: $C(\mathbf{a}_1) = \exp[i\mathbf{k} \cdot \mathbf{a}_1]$. A similar argumentation with the other basis vectors yields $C(n_1\mathbf{a}_1 + n_2\mathbf{a}_2 + n_3\mathbf{a}_3) = \exp[i\mathbf{k} \cdot (n_1\mathbf{a}_1 + n_2\mathbf{a}_2 + n_3\mathbf{a}_3)]$. Choosing $\mathbf{R}_i^0 = 0$, i.e., $\mathbf{u}_i = \mathbf{A}\exp[-i\omega t]$, and $\mathbf{R}_{i'}^0 = n_1\mathbf{a}_1 + n_2\mathbf{a}_2 + n_3\mathbf{a}_3$, results in $\mathbf{u}_{i'} = C(n_1\mathbf{a}_1 + n_2\mathbf{a}_2 + n_3\mathbf{a}_3)\mathbf{u}_i = \mathbf{A}\exp[i(\mathbf{k} \cdot \mathbf{R}_i^0 - \omega t)]$.

Answer 5.9.

a. Following Eq. (5.22), the spring constants between second neighbors are denoted by $D_{\mu\nu}(\mathbf{R}^0) = D_3\delta_{\mu\nu} + (D_4 - D_3)\hat{R}^0_\mu\hat{R}^0_\nu$. This entails the following additions to the diagonal harmonic coefficients: $\Delta D_{zz} = D_3$, $\Delta D_{xx} = \Delta D_{yy} = D_3 + (D_4 - D_3)/2 = (D_3 + D_4)/2$, where $D_3 = U'(a\sqrt{2})/(a\sqrt{2})$ and $D_4 = U''(a\sqrt{2})$. Since $\hat{R}^0_{ijx}\hat{R}^0_{ijy} = \pm 1/2$, two additional non diagonal terms are added to the matrix of coefficients, $\Delta D_{xy} = \pm(D_4 - D_3)/2$, where the positive sign is for $\mathbf{R}^0_{ij} = \pm a(\hat{\mathbf{x}} + \hat{\mathbf{y}})$, and the negative one is for $\mathbf{R}^0_{ij} = \pm a(\hat{\mathbf{x}} - \hat{\mathbf{y}})$, and $\Delta D_{xz} = \Delta D_{yz} = 0$. Equation (5.25) then yields

$$\Delta\widetilde{K}_{xx}(\mathbf{k}) = \Delta\widetilde{K}_{yy}(\mathbf{k}) = (D_3 + D_4)[4 - 2\cos(k_x a + k_y a) - 2\cos(k_x a - k_y a)]/2$$
$$= 2(D_3 + D_4)[1 - \cos(k_x a)\cos(k_y a)] \ ,$$

$$\Delta\widetilde{K}_{zz}(\mathbf{k}) = 4D_3[1 - \cos(k_x a)\cos(k_y a)] \ ,$$

$$\Delta\widetilde{K}_{xy}(\mathbf{k}) = (D_3 - D_4)[\cos(k_x a + k_y a) - \cos(k_x a - k_y a)]$$
$$= -2(D_3 - D_4)\sin(k_x a)\sin(k_y a) \ .$$

The equations of motion are [see Eq. (5.24)]

$$M\omega^2 A_x = [\widetilde{K}_{xx}(\mathbf{k}) + \Delta\widetilde{K}_{xx}(\mathbf{k})]A_x + \Delta\widetilde{K}_{xy}A_y \ ,$$
$$M\omega^2 A_y = [\widetilde{K}_{yy}(\mathbf{k}) + \Delta\widetilde{K}_{yy}(\mathbf{k})]A_y + \Delta\widetilde{K}_{xy}A_x \ ,$$
$$M\omega^2 A_z = [\widetilde{K}_{zz}(\mathbf{k}) + \Delta\widetilde{K}_{zz}(\mathbf{k})]A_z \ .$$

Therefore, the frequency of the modes which vibrate along the normal to the plane is

$$M\omega^2_{T2} = \widetilde{K}_{zz}(\mathbf{k}) = 4D_{T2}[\sin^2(k_x a/2) + \sin^2(k_y a/2)] + 4D_3[1 - \cos(k_x a)\cos(k_y a)] \ .$$

The in-plane vibrations' frequencies are derived by solving two coupled linear equations, leading to a quadratic equation for the allowed frequencies, $(M\omega^2 - \widetilde{K}_{xx})(M\omega^2 - \widetilde{K}_{yy}) = (\widetilde{K}_{xy})^2$. Ignoring the relations between the spring constants implied by the equilibrium-statecondition of the central forces, leads to results not much different than those that pertain to the situation where only forces between nearest neighbors are involved. The three vibrational modes are presented in Fig. 5.27(b), for the three main directions in the Brillouin zone (with $D_4 = 0.5D_L$, $D_3 = 1.5D_L$, and $D_{T1} = 2D_L$, and in units of $a\sqrt{D_L/M}$). For a longitudinal wave propagating along one of the axes, say the $\hat{\mathbf{x}}-$ axis, the off diagonal element $\widetilde{K}_{xy}$ vanishes, and the dispersion relation is $M\omega^2_L = \widetilde{K}_{xx}(\mathbf{k}) = 4(D_L + D_3 + D_4)\sin^2(k_x a/2)$. This dispersion law is similar to the one obtained for a one-dimensional chain, and for the same wave in the case discussed above: atoms on a line normal to the direction of motion vibrate in unison. The sound velocity, though, is modified, $c = a\sqrt{(D_L + D_3 + D_4)/M}$. In a similar way, the frequency of the transverse wave vibrating along the $\hat{\mathbf{y}}-$axis and propagating along $\hat{\mathbf{x}}$ is given by $M\omega^2_{T1} = \widetilde{K}_{yy}(\mathbf{k}) = 4(D_{T1} + D_3 + D_4)\sin^2(k_x a/2)$. The one of the transverse wave oscillating along $\hat{\mathbf{z}}$ and moving along $\hat{\mathbf{x}}$ is $M\omega^2_{T2} = \widetilde{K}_{zz}(\mathbf{k}) = 4(D_{T2} + 2D_3)\sin^2(k_x a/2)$, with the sound velocity $c = a\sqrt{D_{T2} + 2D_3)/M}$. All in all, one obtains three distinct dispersion relations, with three different sound velocities. All degeneracies are removed, and there are three branches in each Brillouin zone.

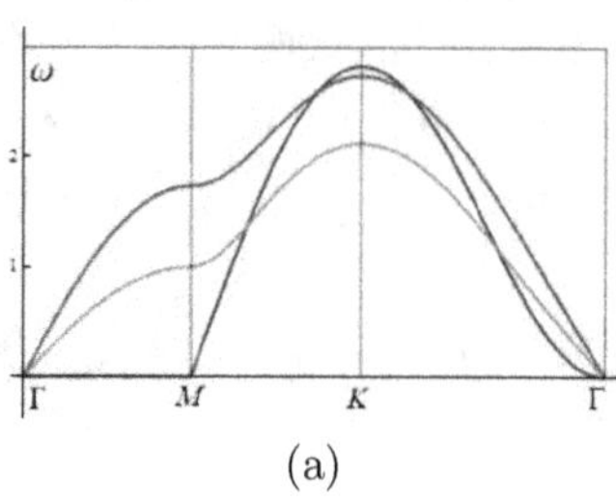

(a)

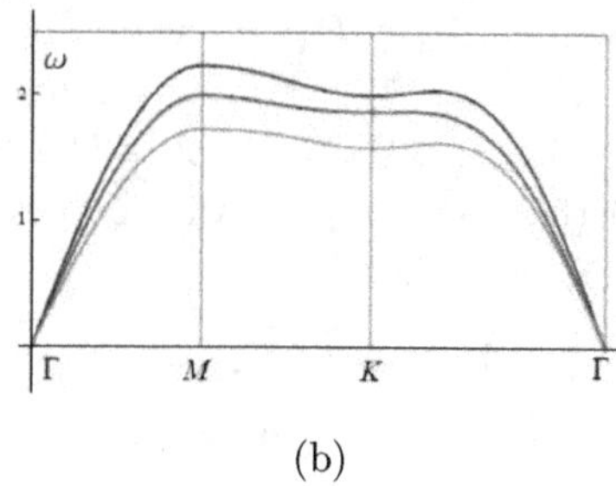

(b)

Fig. 5.27

As in one dimension, the constraint imposed by the condition of equilibrium modifies the picture. At equilibrium $U_{\text{tot}} = 2N[U(a) + U(\sqrt{2}a)]$, leading to $U'(\sqrt{2}a) = -U'(a)/\sqrt{2}$. It follows that $D_{T2} + 2D_3 = 0$, and then the sound velocity of vibrations normal to the plane vanishes. The dispersion law of these vibrations is $M\omega_{T2}^2 = \widetilde{K}_{zz}(\mathbf{k}) = 2D_{T2}[1 - \cos(k_x a)][1 - \cos(k_y a)]$ (check!), it vanishes on the axes in the Brillouin zone, and is approximately $M\omega_{T2}^2 \approx D_{T2}a^4 k_x^2 k_y^2$ at the vicinity of the Γ point. The behavior of the in-plane modes is qualitatively the same. Dispersion relations along chosen directions are depicted in Fig. 5.27(a). Note in particular the parabolic dependence of the frequency on the wave vector of the transverse oscillation normal to the plane around the Γ point in Fig. 5.27(b).

b. In the acoustic limit, the equations are

$$M\omega^2 A_x \approx a^2\{[(D_L + D_3 + D_4)k_x^2 + (D_{T1} + D_3 + D_4)k_y^2]A_x + 2(D_3 - D_4)k_x k_y A_y\},$$
$$M\omega^2 A_y \approx a^2\{[(D_L + D_3 + D_4)k_y^2 + (D_{T1} + D_3 + D_4)k_x^2]A_y + 2(D_3 - D_4)k_x k_y A_x\},$$
$$M\omega^2 A_z \approx a^2(D_{T2} + 2D_3)(k_x^2 + k_y^2)A_z.$$

The system has a square symmetry in the plane. A comparison with the equations in Appendix D enables the identifications $C_{xx,xx} = C_{yy,yy} = a(D_L + D_3 + D_4)$, $C_{xx,xy} + C_{xy,xy} = 2a(D_3 - D_4)$, and $C_{zx,zx} = C_{zy,zy} = a(D_{T2} + 2D_3)$. When the forces are central the last two coefficients are zero.

Answer 5.10.

a. From the expression following Eq. (5.33) one finds $\omega^2 = (3D/M)[1 \pm |\mathbf{U}|/3]$, where $\mathbf{U} = \exp[i\mathbf{k} \cdot \mathbf{a}_1] + \exp[i\mathbf{k} \cdot \mathbf{a}_2] + \exp[i\mathbf{k} \cdot (\mathbf{a}_1 + \mathbf{a}_2)]$. The wave vector at the K point is $\mathbf{k}_K = 4\pi\hat{\mathbf{x}}/(3a)$. Using $\mathbf{a}_1 = a\hat{\mathbf{x}}$ and $\mathbf{a}_2 = (\hat{\mathbf{x}} + \sqrt{3}\hat{\mathbf{y}})a/2$, shows that $\mathbf{U}$ vanishes, and hence the two solutions are degenerate.

b. Substituting $\mathbf{k} = \mathbf{k}_K + \mathbf{q}$ and expanding up to linear order in the components of $\mathbf{q}$, results in $\mathbf{U} = -ia(3i + \sqrt{3})(q_x + iq_y)/4 + \dots$. It follows that $|\mathbf{U}|^2 = 3a^2|\mathbf{q}|^2/4 + \dots$, and hence $\omega(\mathbf{k}_K + \mathbf{q}) \approx \sqrt{(3D/M)}[1 \pm (\sqrt{3}/6)a|\mathbf{q}|]$. The frequency is linear in the deviation of the vector from its value at the K point.

Answer 5.11.

a. The next-nearest neighbors of the atom whose displacements have been denoted by $u_{n,m}$ are precisely the atoms in the nearest cells that are of the same type; the contributions of the interactions with them are thus those listed in Eq. (5.28) (which pertains to the triangular lattice). The same argument holds for the atoms of the other type. Consequently, the equations of motion (5.33) become

$$M\omega^2 A = [3D + \Delta\widetilde{K}]A - D(e^{-i\mathbf{k}\cdot\mathbf{a}_1} + e^{-i\mathbf{k}\cdot\mathbf{a}_2} + e^{-i\mathbf{k}\cdot(\mathbf{a}_1+\mathbf{a}_2)})B \ ,$$

$$M\omega^2 B = [3D + \Delta\widetilde{K}]B - D(e^{i\mathbf{k}\cdot\mathbf{a}_1} + e^{i\mathbf{k}\cdot\mathbf{a}_2} + e^{i\mathbf{k}\cdot(\mathbf{a}_1+\mathbf{a}_2)})A \ ,$$

where

$$\Delta\widetilde{K} = 4D_2\left[\sin^2\left(\frac{\mathbf{k}\cdot\mathbf{a}_1}{2}\right) + \sin^2\left(\frac{\mathbf{k}\cdot\mathbf{a}_2}{2}\right) + \sin^2\left(\frac{\mathbf{k}\cdot(\mathbf{a}_2-\mathbf{a}_1)}{2}\right)\right] \ .$$

The frequencies (again derived by solving a quadratic equation) are $M\omega^2 = 3D + \Delta\widetilde{K} \pm D|\exp[i\mathbf{k}\cdot\mathbf{a}_1] + \exp[i\mathbf{k}\cdot\mathbf{a}_2] + \exp[i\mathbf{k}\cdot(\mathbf{a}_1+\mathbf{a}_2)]|$. Figure 5.28(a) shows the acoustic branch, and Fig. 5.28(b) the optical one, for $D_2 = 0.4D$. As seen, the competition between the two interactions creates a complex structure of the dispersion relations.

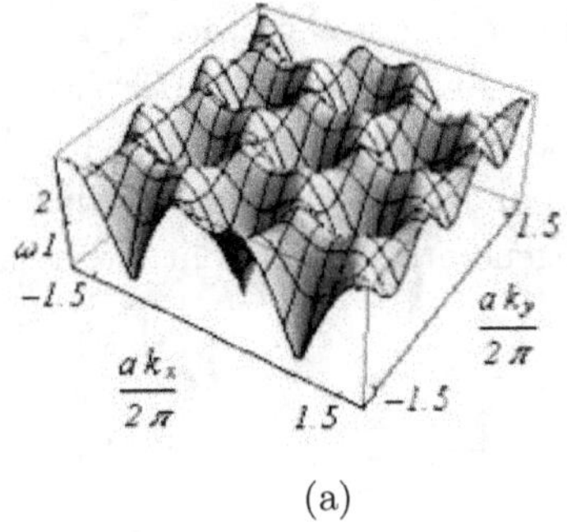

(a)

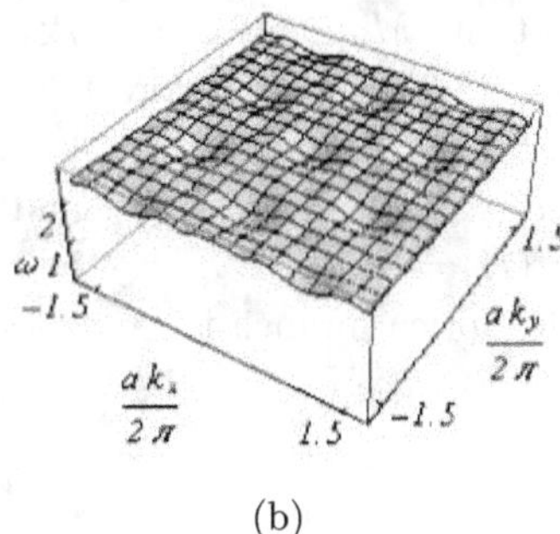

(b)

Fig. 5.28

b. The dependence of the expressions above on the wave vector $\mathbf{k}$ is via the terms $\exp[i\mathbf{k}\cdot\mathbf{a}_1]$ and $\exp[i\mathbf{k}\cdot\mathbf{a}_2]$. Since $\exp[i\mathbf{G}\cdot\mathbf{a}_1] = \exp[i\mathbf{G}\cdot\mathbf{a}_2] = 1$ for any reciprocal-lattice vector, the equations are invariant under the translation $\mathbf{k} \to \mathbf{k} + \mathbf{G}$, and the frequencies are not changed. In addition,

$$|\exp[i\mathbf{k}\cdot\mathbf{a}_1] + \exp[i\mathbf{k}\cdot\mathbf{a}_2] + \exp[i\mathbf{k}\cdot(\mathbf{a}_1+\mathbf{a}_2)]|^2 = 3 + 2\cos(\mathbf{k}\cdot\mathbf{a}_1) + 2\cos(\mathbf{k}\cdot\mathbf{a}_2)$$
$$+ 2\cos[\mathbf{k}\cdot(\mathbf{a}_2-\mathbf{a}_1)] \ .$$

Both this expression and the one for $\Delta\widetilde{K}$ are invariant under the transformation $\mathbf{a}_1 \to \mathbf{a}_2 \to \mathbf{a}_2 - \mathbf{a}_1 \to -\mathbf{a}_1$, which occurs when $\mathbf{k}$ is rotated by $60°$. The dispersion relations are thus invariant with respect to this rotation, as indeed can be seen in Figs. 5.28.

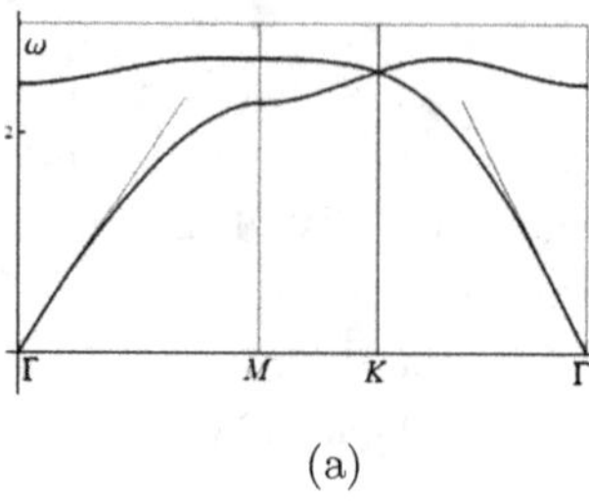

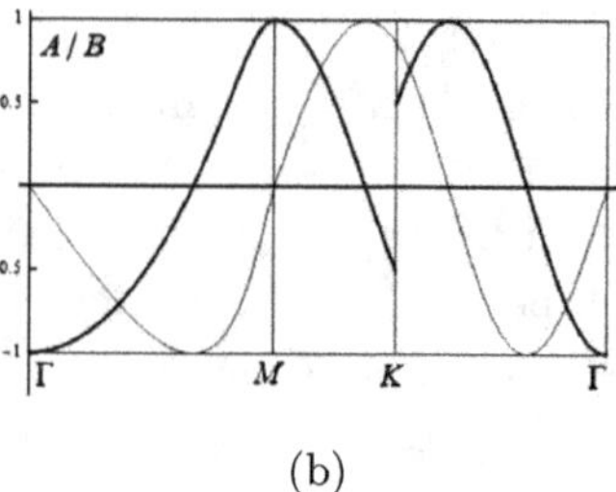

(a)   (b)

Fig. 5.29

c. The limit $|\mathbf{k}|a \ll 1$ allows one to expand the expressions obtained above. The frequency of the acoustic branch is given by $M\omega^2 = (D/4 + 3D_2/2)(|\mathbf{k}|a)^2$. It is independent of the direction of the wave vector; the corresponding sound velocity is $c = \omega/|\mathbf{k}| = a\sqrt{(D/4 + 3D_2/2)/M}$.

d. At $\mathbf{k} = 0$ one finds $M\omega^2(\mathbf{k} = 0) = 3D \pm 3D$, yielding $M\omega^2(\mathbf{k}) = 6D$ for the optical frequency. The equations of motion give $A = -B$, that is, the two sublattices are vibrating one against the other (as in the case of two different atoms in the unit cell of a one-dimensional lattice).

e. At the M point $\mathbf{k} = \mathbf{b}_1/2$ (where $\mathbf{b}_1$ is a reciprocal-lattice vector, Fig. 3.11 or Fig. 5.9), and therefore $\mathbf{k} \cdot \mathbf{a}_1 = \pi$ and $\mathbf{k} \cdot \mathbf{a}_2 = 0$. It follows that $M\omega^2(M) = (3 \pm 1)D + 8D_2$, and $A = \pm B$ (the upper sign pertains to the acoustic branch). At the K point $\mathbf{k} = 4\pi\hat{\mathbf{x}}/(3a)$, and then $\mathbf{k} \cdot \mathbf{a}_1 = 4\pi/3$ and $\mathbf{k} \cdot \mathbf{a}_2 = 2\pi/3$, leading to $M\omega^2(K) = 3D + 9D_2$, i.e., the two branches are degenerate. The amplitudes ratio in the optical branch is

$$\frac{A}{B} = -\frac{e^{i\mathbf{k}\cdot\mathbf{a}_1} + e^{i\mathbf{k}\cdot\mathbf{a}_2} + e^{i\mathbf{k}\cdot(\mathbf{a}_1+\mathbf{a}_2)}}{\left|e^{-i\mathbf{k}\cdot\mathbf{a}_1} + e^{-i\mathbf{k}\cdot\mathbf{a}_2} + e^{-i\mathbf{k}\cdot(\mathbf{a}_1+\mathbf{a}_2)}\right|},$$

as shown in Fig. 5.29(b) (the thick and thin lines correspond to the real and imaginary parts of the expression, respectively). The same expression but with the inverse sign describes the acoustic branch. Both numerator and denominator vanish at the K point leading to a discontinuity there: the real part is ± 0.5, and the imaginary one is $\sqrt{3}/2$. As all coefficients in Eqs. (5.33) vanish at that point, one may choose any values for A and B. In any event, the absolute values of the two amplitudes are equal; only their phases vary with the wave vector.

Answer 5.12.

The only difference between the two crystals is that in the present case the two ions in the unit cell are not identical, i.e., their masses differ. Hence, Eqs. (5.32) are modified to be

$$M\ddot{u}_{n,m} = -3Du_{n,m} + D(v_{n-1,m} + v_{n,m-1} + v_{n-1,m-1}) \,,$$

$$m\ddot{v}_{n,m} = -3Dv_{n,m} + D(u_{n+1,m} + u_{n,m+1} + u_{n+1,m+1}) \,.$$

Equations (5.33) are modified to be

$$M\omega^2 A = 3DA - D(e^{-i\mathbf{k}\cdot\mathbf{a}_1} + e^{-i\mathbf{k}\cdot\mathbf{a}_2} + e^{-i\mathbf{k}\cdot(\mathbf{a}_1+\mathbf{a}_2)})B \,,$$

$$m\omega^2 B = 3DB - D(e^{i\mathbf{k}\cdot\mathbf{a}_1} + e^{i\mathbf{k}\cdot\mathbf{a}_2} + e^{i\mathbf{k}\cdot(\mathbf{a}_1+\mathbf{a}_2)})A \,,$$

and the dispersion relations are the solutions of the quadratic equation

$$(M\omega^2 - 3D)(m\omega^2 - 3D) = D^2|\exp[i\mathbf{k}\cdot\mathbf{a}_1] + \exp[i\mathbf{k}\cdot\mathbf{a}_2] + \exp[i\mathbf{k}\cdot(\mathbf{a}_1 + \mathbf{a}_2)]|^2 \,.$$

Figure 5.30 displays the acoustic and optical branches for $M/m = 1.1$. The prime difference between it and the corresponding one for graphene is at the K point. Whereas in graphene the two branches merge together there, this degeneracy is lifted for the boron nitride. This is clear from the quadratic equation: at this point the right hand-side vanishes, and then the two roots are $M\omega_-^2 = 3D$ and $m\omega_+^2 = 3D$ (note in particular the similarity with the one-dimensional lattice whose unit cell contains two different atoms).

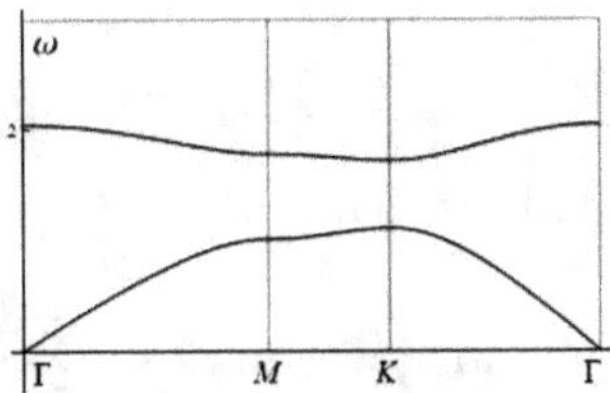

Fig. 5.30

Answer 5.13.
The equations of motion for the "inner" points $n = 2, 3, \ldots, N - 1$ are intact, and are given by Eq. (5.3). These are solved by the monochromatic wave functions $\exp[i(kna - \omega t)]$ or $\exp[i(-kna - \omega t)]$, and hence by any linear combination of them. The equations of motion for the edges, $n = 1$ and $n = N$, are again given by Eq. (5.3), but these should be augmented by boundary conditions. The condition $u_0 = 0$ is fulfilled by choosing the difference between the two wave functions, i.e., $u_n = A\sin(kna)\exp[-i\omega t]$. These functions, which describe a standing wave, indeed solve Eq. (5.3) (this can be shown by a direct substitution), with the same dispersion relations derived before, Eq. (5.5). It remains to consider the second boundary conditions, $u_{N+1} = \sin[k(N + 1)a] = 0$, which implies $k = k_\ell = \pi\ell/[(N + 1)a]$. Though it seems that the density of wave numbers is doubled (as compared to the solution based on periodic boundary conditions), the difference between subsequent

wave numbers is just $\pi/[(N+1)a]$ as compared to $2\pi/(Na)$. (The disparity between the two denominators is negligible for large N's.) Moreover, the solutions for k and $-k$ are identical up to a sign, and thus describe the **same** eigenmode. It follows that one has to include only the positive solutions within the (half) Brillouin zone, $\ell = 1, 2, \ldots, N$. All in all, there are N independent solutions, as before.

Answer 5.14.

a. At equilibrium, the locations of the atoms are $\mathbf{R}_n^0 = R(\cos[2n\pi/N], \sin[2n\pi/N])$; note that $\mathbf{R}_{n+N}^0 = \mathbf{R}_n^0$, which implies that the ring obeys Eq. (5.34). The same relation between the atoms' locations pertains also to the atoms vibrating around the equilibrium positions.

b. Denote the instantaneous locations of the atoms by $\mathbf{R}_n = R(\cos[2n\pi/N + \vartheta_n], \sin[2n\pi/N + \vartheta_n])$. The distance between two nearest neighbors is

$$R_{n,n+1} = 2R\sin\left[\frac{\pi}{N} + \frac{\delta\vartheta_{n+1,n}}{2}\right] \approx R\left(2\sin\left[\frac{\pi}{N}\right] + \cos\left[\frac{\pi}{N}\right]\delta\vartheta_{n+1,n} + \cdots\right),$$

where $\delta\vartheta_{n+1,n} = \vartheta_{n+1} - \vartheta_n$. Thus, the central potential between these two neighbors is

$$U(R_{n,n+1}) = U(X_0) + U'(X_0)Y_0\delta\vartheta_{n+1,n} + U''(X_0)(Y_0\delta\vartheta_{n+1,n})^2/2 + \cdots,$$

where

$$X_0 = 2R\sin(\pi/N) , \quad Y_0 = R\cos(\pi/N) , \quad \delta\vartheta_{N+1,N} = \vartheta_1 - \vartheta_N .$$

Using $\sum_{n=1}^N \delta\vartheta_{n+1,n} = 0$, it follows that

$$\sum_{n=1}^N U(R_{n,n+1}) = NU(X_0) + U''(X_0)(Y_0)^2 \sum_{n=1}^N (\delta\vartheta_{n+1,n})^2/2 + \cdots .$$

The equations of motion are

$$M\ddot{\vartheta}_n = -D(2\vartheta_n - \vartheta_{n-1} - \vartheta_{n+1}) , \quad n = 2, 3, \ldots, N-1 ,$$

$$M\ddot{\vartheta}_N = -D(2\vartheta_N - \vartheta_{N-1} - \vartheta_1) \quad M\ddot{\vartheta}_1 = -D(2\vartheta_1 - \vartheta_N - \vartheta_2) ,$$

where $D = U''(X_0)Y_0^2$. For $N = 6$, the equations can be written as

$$M\begin{bmatrix} \ddot{\vartheta}_1 \\ \ddot{\vartheta}_2 \\ \ddot{\vartheta}_3 \\ \ddot{\vartheta}_4 \\ \ddot{\vartheta}_5 \\ \ddot{\vartheta}_6 \end{bmatrix} = D\begin{bmatrix} -2 & 1 & 0 & 0 & 0 & 1 \\ 1 & -2 & 1 & 0 & 0 & 0 \\ 0 & 1 & -2 & 1 & 0 & 0 \\ 0 & 0 & 1 & -2 & 1 & 0 \\ 0 & 0 & 0 & 1 & -2 & 1 \\ 1 & 0 & 0 & 0 & 1 & -2 \end{bmatrix}\begin{bmatrix} \vartheta_1 \\ \vartheta_2 \\ \vartheta_3 \\ \vartheta_4 \\ \vartheta_5 \\ \vartheta_6 \end{bmatrix} .$$

The extension to other values of N is obvious. The column vector $(\exp[ik_\ell], \exp[2ik_\ell], \exp[3ik_\ell], \exp[4ik_\ell], \exp[5ik_\ell], \exp[6ik_\ell])$ is an eigenvector of the matrix on the right hand-side, with the eigenvalue $2[\cos k_\ell - 1]$, provided that $\exp[iNk_\ell] = 1$, which follows from the first and the last raws in the matrix. Therefore, there are just N allowed values of the "wave number", $k_\ell = \ell 2\pi/N$, with

$\ell = 1, 2, 3, \ldots, N$. The solution of the equations of motion can be presented as a sum of the eigenvectors, $\vartheta_n = \sum_{\ell=1}^{N} \widetilde{\vartheta}_\ell \exp[ik_\ell n]$, and then

$$M \sum_{\ell=1}^{N} \ddot{\widetilde{\vartheta}}_\ell \exp[ik_\ell n] = 2D \sum_{\ell=1}^{N} [\cos k_\ell - 1] \widetilde{\vartheta}_\ell \exp[ik_\ell n] ,$$

i.e.,

$$M \ddot{\widetilde{\vartheta}}_\ell = -2D[1 - \cos k_\ell] \widetilde{\vartheta}_\ell .$$

This is the equation of an harmonic oscillator, that gives the time dependence of the amplitudes $\widetilde{\vartheta}_\ell$, with the frequencies given in Eq. (5.6).

Answer 5.15.

The integrals, for instance for one of the vibrational modes, are computed using the identity $\int_{-\infty}^{\infty} dx \exp[-ax^2/2] = \sqrt{2\pi/a}$. One finds

$$Z_\alpha = (\sqrt{2\pi M/\beta}\sqrt{2\pi/(\beta M \omega_\alpha^2)}/h) = (\beta \hbar \omega_\alpha)^{-1} ,$$

and hence

$$\ln Z = \sum_\alpha \ln Z_\alpha = -[N \ln(\beta) + \sum_\alpha \ln(\hbar \omega_\alpha)] .$$

It follows that

$$\langle E \rangle = -\partial \ln Z/\partial \beta = N k_{\mathrm{B}} T ,$$

and

$$C_V = (\partial \langle E \rangle/\partial T)_V = N k_{\mathrm{B}} .$$

Answer 5.16.

The area below the dashed line in Fig. 5.13 is the total classical energy. The area up to a temperature T is $k_{\mathrm{B}}T$. The dark area in the figure is

$$\int_0^\infty dT C_V = \int_0^\infty (\partial \langle E \rangle/\partial T)_V = \langle E(T = \infty) \rangle - \langle E(T = 0) \rangle ,$$

where the expressions on the right-hand side are the average quantum energies, given by Eq. (5.52). This area is then

$$A = \int_0^\infty dT [k_{\mathrm{B}} - C_V(T)] = \lim_{T \to \infty} [k_{\mathrm{B}} T - \int_0^T dT C_V(T)$$

$$= \lim_{T \to \infty} [k_{\mathrm{B}} T - \langle E(T) \rangle + \langle E(T = 0) \rangle] .$$

Equation (5.52) implies that $\langle E(T = 0) \rangle = \hbar \omega_\alpha/2$. Expanding this equation at high temperatures yields

$$\langle E(T) \rangle = \hbar \omega_\alpha/2 + \hbar \omega_\alpha/[\beta \hbar \omega_\alpha + (\beta \hbar \omega_\alpha)^2/2 + \mathcal{O}((\beta \hbar \omega_\alpha)^3)] \approx k_{\mathrm{B}} T + \mathcal{O}(\beta \hbar \omega_\alpha) .$$

All in all,

$$A = \lim_{T \to \infty} [k_{\mathrm{B}}T - (k_{\mathrm{B}}T - \hbar\omega_\alpha/2)] = \hbar\omega_\alpha/2 \ ,$$

precisely the zero-point motion energy, which distinguishes the quantum-mechanical result from the classical one.

Answer 5.17.

a. Equation (5.6) implies that $\omega = \omega_0|\sin(ka/2)|$, which in the long-wave-length limit becomes $\omega = \omega_0|ka/2|$, and thus the sound velocity is $c = \omega_0 a/2$. The total number of states in the range $0 < \omega < \omega_0$, from Eq. (5.57), is

$$\int_0^{\omega_0} d\omega\, g(\omega) = \frac{L}{\pi c} \int_0^{\omega_0} \frac{d\omega}{\sqrt{1 - (\omega/\omega_0)}}$$

$$= \frac{L\omega_0}{\pi c} \int_0^1 \frac{dx}{\sqrt{1 - x^2}} = \frac{2L}{\pi a} \arcsin(x)\Big|_0^1 = \frac{L}{a} = N \ .$$

b. Expanding the integrand in Eq. (5.59) in $\exp[-x]$ yields

$$\frac{e^x}{(e^x - 1)^2} = e^{-x}(1 - e^{-x})^{-2} = -\frac{d}{dx}\frac{1}{1 - e^{-x}} = -\frac{d}{dx} \sum_{m=0}^{\infty} e^{-mx} = \sum_{m=1}^{\infty} m e^{-mx} \ .$$

Inserting this expression into the integral results in $\int_0^\infty dx\, x^2 \sum_{m=1}^\infty m \exp[-mx] = \sum_{m=1}^\infty 2/m^2 = 2\zeta(2)$ (integrating twice by parts). It remains to find $\zeta(2)$. This can be accomplished by expanding the function $\sin x/x$ up to second order in x. The function vanishes at the points $x = \pm m\pi$, and therefore

$$\frac{\sin x}{x} = \left(1 - \frac{x^2}{\pi^2}\right)\left(1 - \frac{x^2}{4\pi^2}\right)\left(1 - \frac{x^2}{9\pi^2}\right)\cdots = 1 - \frac{x^2}{\pi^2}\zeta(2) + \mathcal{O}(x^4) \ .$$

On the other hand, a straightforward Taylor expansion gives $\sin x/x = 1 - x^2/6 + \mathcal{O}(x^4)$, thus $\zeta(2) = \pi^2/6$.

c. At low temperatures, where $x_{\min} = \beta\hbar\omega_+(\pi/a) \gg 1$, the function $x^2 \exp[x]/(\exp[x] - 1)^2$ is bounded from above by $x_{\min}^2 \exp[x_{\min}]/(\exp[x_{\min}] - 1)^2 \approx x_{\min}^2 \exp[-x_{\min}]$. Replacing this function in the integrand by its bound (which can be taken out of the integral), it remains to calculate the integral that appears in part (a); in this way one derives an upper bound on the integral,

$$C_V^{\mathrm{optical}} < k_{\mathrm{B}} N x_{\min}^2 e^{x_{\min}}/(e^{x_{\min}} - 1)^2 \approx k_{\mathrm{B}} N x_{\min}^2 e^{-x_{\min}} \ .$$

As the upper bound decays exponentially when β increases, the specific heat follows this behavior (or even decays faster).

Answer 5.18.

a. The dispersion relation is

$$M\omega^2 = 2D_{T2}[1 - \cos(k_x a) + 1 - \cos(k_y a)] \ ,$$

and therefore the group velocity is

$$\mathbf{v}_g = \boldsymbol{\nabla}_{\mathbf{k}}(\omega) = [D_{T2}a/(M\omega)][\sin(k_x a)\hat{\mathbf{x}} + \sin(k_y a)\hat{\mathbf{y}}] \ .$$

This vanishes only when both components are zero, i.e., when $k_x a = 0, \pi$ and $k_y a = 0, \pi$; these are the points of the type M (the centers of the faces) and the points of the type K (the corners) in the Brillouin zone.

b. At the vicinity of the corner $K = (\pi/a, \pi/a)$ one may write $k_x a = \pi - k'_x a$ and $k_y a = \pi - k'_y a$, and obtain

$$M\omega^2 = 2D_{T2}[2 + \cos(k'_x a) + \cos(k'_y a)] \approx D_{T2}[8 - \{(k'_x)^2 + (k'_y)^2\}a^2] \ .$$

The equal-frequency surfaces are circles around the point K. The group velocity on a circle of radius k' is radial, and its magnitude is $|\mathbf{v}_g| = D_{T2}a^2 k'/(M\omega)$ (check!). The circumference is $2\pi k'$, and therefore Eq. (5.65) yields

$$g(\omega) = \frac{NM\omega}{2\pi D_{T2}} \approx \frac{N}{\pi}\sqrt{\frac{2M}{D_{T2}}}$$

(note that $V = a^2$). The density of states is independent of the frequency.

c. It suffices to calculate the contribution of k's from the upper right quarter of the Brillouin zone, as each quarter contributes the same amount. The line $(\pi/a, 0) \rightarrow (0, \pi/a)$ is $k_y a = \pi - k_x a$. Inserting it into the expression for the frequency confirms that it is indeed an equal-frequency curve, with $M\omega^2 = 4D_{T2}$. Substituting into the expression for the group velocity,

$$\mathbf{v}_g = [D_{T2}a/(M\omega)][\sin(k_x a)\hat{\mathbf{x}} + \sin(k_y a)\hat{\mathbf{y}}] \ ,$$

gives $|\mathbf{v}_g| = [D_{T2}a/(M\omega)]\sqrt{2}\sin(k_x a)$. Hence, the contribution of this segment to the integral in Eq. (5.66) is proportional to $\int_0^\pi dx/(\sin x) = \ln[\tan(x/2)]|_{x=0}^{x=\pi} = \infty$, where $x = k_x a$. This logarithmic dependence of the integral on its bounds implies that the same dependence still prevails upon shifting the frequency a bit below or above this value.

d. The line described in part (c) is the only equal-frequency line in the Brillouin zone that includes points where the group velocity vanishes. It follows that the density of states is finite at any other point in the Brillouin zone. It is a linear line near $\omega = 0$ [see Eq. (5.67)], diverges logarithmically at $\omega = \sqrt{4D_{T2}/M}$, and returns to the constant value calculated in part (b) around the maximal frequency $\omega = \sqrt{8D_{T2}/M}$. The density of states vanishes for frequencies above this value. Figure 5.31 illustrates this schematic behavior.

Answer 5.19.

a. For $X \ll 1$ the integrand may be expanded in powers of x, to yield

$$\int_0^X dx\, \frac{x^{d+1}e^x}{(e^x - 1)^2} = \int_0^X dx\, x^{d+1}(1 + x + \frac{x^2}{2} + \ldots)/(x + \frac{x^2}{2} + \frac{x^3}{6} + \ldots)^2$$

$$= \int_0^X dx\, x^{d-1}(1 - \frac{x^2}{12} + \ldots) = \frac{X^d}{d} - \frac{X^{d+2}}{12(d+2)} + \cdots ,$$

leading to the required answer.

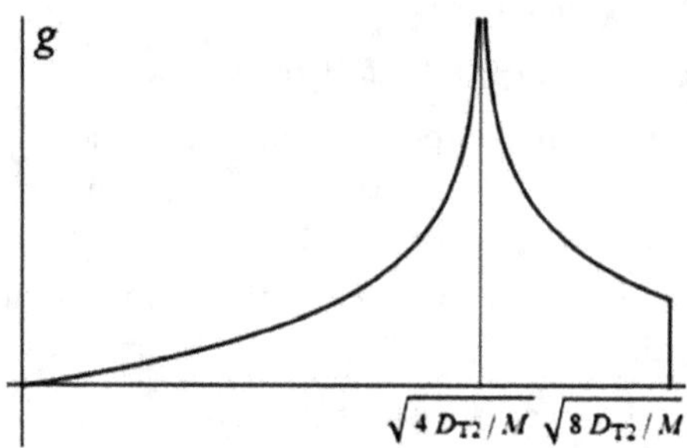

Fig. 5.31

b. It is possible to express the integral in the form

$$\int_0^X dx \frac{x^{d+1}e^x}{(e^x-1)^2} = \int_0^\infty dx \frac{x^{d+1}e^x}{(e^x-1)^2} - \int_X^\infty dx \frac{x^{d+1}e^x}{(e^x-1)^2} \ .$$

As mentioned after Eq. (5.69), the first integral is I_d, which attains finite values. Since $X \gg 1$, the integrand in the second term can be expanded in powers of $\exp[-x]$,

$$\int_X^\infty dx \frac{x^{d+1}e^{-x}}{(1-e^{-x})^2} = \int_X^\infty dx x^{d+1} e^{-x} [1 + \mathcal{O}(e^{-2x})] \ . \tag{5.114}$$

Integrating by parts,

$$J_d = \int_X^\infty dx x^{d+1} \exp[-x] = -x^{d+1}e^{-x}\Big|_X^\infty + (d+1)J_{d-1} = X^{d+1}e^{-X} + \mathcal{O}(X^d e^{-X}) \ .$$

This completes the proof.

Answer 5.20.
In spherical symmetry, Eq. (5.65) yields $g(\omega) = [NVS_d/(2\pi)^d][k^{d-1}/v_g]$. The group velocity, from the dispersion relation, is $v_g = d\omega/dk \propto k^{\eta/2-1}$, and hence $g(\omega) \propto \omega^{2d/\eta-1}$. As $\eta = 2$ for the acoustic dispersion, Eq. (5.67) is reproduced. The Debye frequency is derived from a generalized version of Eq. (5.68): $N = \int_0^{\omega_D} d\omega g(\omega) \propto \omega_D^{2d/\eta}$. Equation (5.69) yields

$$C_V = \int_0^{\omega_D} d\omega g(\omega)(\beta\hbar\omega)^2 \frac{e^{\beta\hbar\omega}}{(e^{\beta\hbar\omega}-1)^2}$$

$$= \frac{2k_B dN}{\eta}\left(\frac{k_B T}{\hbar\omega_D}\right)^{2d/\eta} \int_0^{\beta\hbar\omega_D} dx x^{\frac{2d}{\eta}+1} \frac{e^x}{(e^x-1)^2} \ .$$

When $k_B T \ll \hbar\omega_D$, the upper bound may be taken as infinity, to obtain $C_V \propto T^{2d/\eta}$. In the opposite limit one obtains the Dulong-Petit law, $C_V \approx k_B T$, which is independent of the details of the dispersion relation.

Answer 5.21.

The specific heat due to a single acoustic branch is given by Eq. (5.55),

$$C_V = \sum_{\mathbf{k}} k_{\mathrm{B}} [\beta\hbar\omega(\mathbf{k})]^2 \frac{e^{\beta\hbar\omega(\mathbf{k})}}{(e^{\beta\hbar\omega(\mathbf{k})} - 1)^2} \, ,$$

where the sum runs over the discrete values of the wave vector, $\mathbf{k}_{\ell_1,\ell_2,\ell_3} = (\ell_1/N_1)\mathbf{b}_1 + (\ell_2/N_2)\mathbf{b}_2 + (\ell_3/N_3)\mathbf{b}_3$, with $-N_m/2 \le \ell_m < N_m/2 - 1$. The three sums are turned into integrals in Eq. (5.63). However, the thickness along $\mathbf{a}_3$ is finite: $N_3 = L/a_3 \ll N_1, N_2$, and may be quite small; the corresponding sum cannot be turned into an integral. (The other two sums are still replaced by integrals.) It follows that Eq. (5.55) becomes

$$\sum_{\mathbf{k}} f[\omega(\mathbf{k})] = \sum_{\ell_3} \frac{N_1 N_2 a_1 a_2}{(2\pi)^2} \int d^2k \, f[\omega(\mathbf{k})] \, .$$

The Debye approximation, when applied to a system of tetragonal symmetry, implies that the dispersion relation is

$$\omega^2 = c^2(k_x^2 + k_y^2) + c_\perp^2 k_z^2 \, .$$

As the system is symmetric in the XY plane, the integrals are conveniently carried out using cylindrical coordinates, and thus $d^2k = 2\pi k\,dk = 2\pi\omega d\omega/c^2$, using the dispersion relation. Summing over all waves with frequency smaller than the Debye frequency, i.e., $\omega^2 < \omega_{\mathrm{D}}^2$, results in

$$\sum_{\mathbf{k}} f[\omega(\mathbf{k})] = 2 \sum_{k_z} \frac{N_1 N_2 a_1 a_2}{(2\pi)c^2} \int_{c_\perp k_z}^{\omega_{\mathrm{D}}} d\omega \, \omega f(\omega) \, , \qquad (5.115)$$

where the factor 2 is due to the restriction of the sum to positive values, $0 \le k_z = 2\pi\ell_3/L \le \pi N_3/L$. A positive contribution is obtained only when $c_\perp k_z < \omega_{\mathrm{D}}$. The total number of vibrational modes is $N = N_1 N_2 N_3 = \sum_{\mathbf{k}} 1$. One therefore inserts $f(\omega) = 1$ in Eq. (5.115) and finds

$$\frac{N}{Z} = \sum_{k_z \ge 0} \int_{c_\perp k_z}^{\omega_{\mathrm{D}}} \omega d\omega = \frac{1}{2} \sum_{k_z} (\omega_{\mathrm{D}}^2 - c_\perp^2 k_z^2)$$

$$= \frac{\omega_{\mathrm{D}}^2 N_3}{4} - \frac{1}{12} \left(\frac{\pi c_\perp}{L}\right)^2 N_3(N_3 + 1)(N_3 + 2) \, ,$$

where $Z = N_1 N_2 a_1 a_2/(\pi c^2)$, and $\sum_{\ell=0}^{N/2} \ell^2 = N(N+1)(N+2)/24$. When N_3 is very large, the sum over k_z can be turned into an integral; in this way the expressions for the three-dimensional system, Eq. (5.68) are reproduced, with $N_3 = \omega_{\mathrm{D}} L/(\pi c_\perp)$. In contrast, when the sum comprises only a small number of terms, the second term in small, and the expression is reduced to the formula for a two-dimensional crystal.

For the calculation of the specific heat one inserts

$$f(\omega) = k_{\mathrm{B}}(\beta\hbar\omega) \frac{e^{\beta\hbar\omega}}{e^{\beta\hbar\omega} - 1)^2} \, ,$$

into Eq. (5.115). Changing there the integration variable leads to

$$C_V = k_B Z \left(\frac{k_B T}{\hbar}\right)^2 \sum_{\ell_3=0}^{N_3/2} \int_{\alpha\ell_3}^{\beta\hbar\omega_D} dx\, x^3 \frac{e^x}{(e^x - 1)^2}\,, \quad \alpha = \beta\hbar c_\perp \frac{2\pi}{L}\,.$$

The two-dimensional expression for the specific heat is reproduced upon keeping only the first term, $\ell_3 = 0$, in the sum. For $\alpha > 1$, i.e., $\hbar c_\perp 2\pi/L > k_B T$, each of the other terms in the sum is in fact the specific heat of an optical mode: it is exponentially small at low temperatures and tends to k_B (the Dulong-Petit law) at higher ones. At low temperatures the sum is dominated by its first term, and the resulting specific heat is that of a two-dimensional system, $C_V \propto T^2$. In the intermediate range $\hbar c_\perp 2\pi/L < k_B T < \hbar\omega_D$ the lower bounds of the integrals for $\alpha\ell_3 < 1$ are relatively small, and their contributions are comparable to that of the first term. When this range is wide enough (i.e., L is large enough) the transformation to a (three-dimensional) integration results in a three-dimensional behavior, $C_V \propto T^3$. Figure 5.32 displays the specific heat, C_V/T^2 (scaled by the temperature) in arbitrary units, as a function of the temperature. The curves, in descending order, are for $N_3/2 = 1\,,3\,,10\,,1000$. The horizontal range at low temperatures corresponds to the two-dimensional dependence, while the region to its right ($N_3 > 2$) portrays an effective three-dimensional behavior.

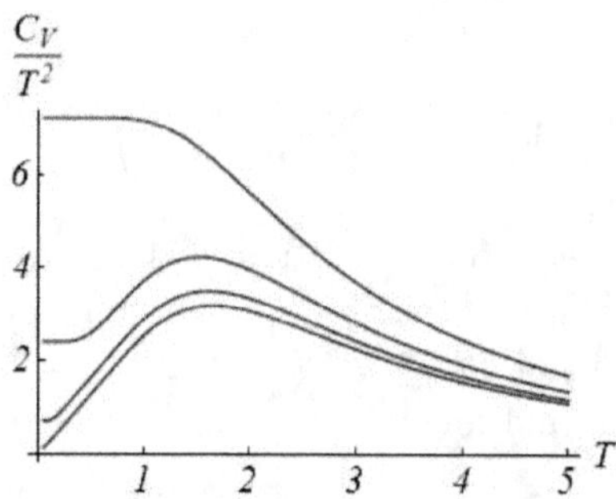

Fig. 5.32

Inter-dimensional crossovers like the one described here appear quite often in condensed matter. They occur also in one-dimensional systems, for instance, nano tubes covered with cylindrical layers, and quasi-two-dimensional systems (or quasi-one-dimensional ones), built of planar lattices (or wires) with very weak couplings in-between them.

Answer 5.22. The one-dimensional density of states in the Debye approximation, from Eq. (5.57), is $g = L/(\pi c)$. It follows that

$$\langle u_i^2(R_i^0)\rangle = \frac{\hbar a}{2\pi M c} \int_{x_1}^{x_2} \frac{dx}{x}\coth x \, , \quad x_1 = \frac{\beta\hbar\omega_{\min}}{2} \, , \quad x_2 = \frac{\beta\hbar\omega_{\mathrm{D}}}{2} \, .$$

The integral does not possess an analytical from, but when x is small, the lower limit implies that

$$\langle u_i^2(R_i^0)\rangle \approx \frac{\hbar a}{2\pi M c} \int_{x_1}^{x_2} \frac{dx}{x^2} \approx \frac{\hbar a}{2\pi M c x_1} = \frac{L a k_{\mathrm{B}} T}{2\pi^2 M c^2} \, .$$

According to the Lindemann criterion, the crystal is stable for $\langle u^2(R_i^0)\rangle < a^2/100$, and therefore $k_{\mathrm{B}} T < [\pi^2 M c^2/(50N)]$, where $N = L/a$.

Answer 5.23. At high temperatures, Eq. (5.76) yields

$$\langle \mathbf{u}^2(\mathbf{R}_i^0)\rangle \approx \frac{\hbar d}{2M\omega_{\mathrm{D}}^d} \int_{\omega_{\min}}^{\omega_{\mathrm{D}}} d\omega\,\omega^{d-2}\Big(\frac{\beta\hbar\omega}{2}\Big)^{-1} = \frac{d k_{\mathrm{B}} T}{(d-2) M \omega_{\mathrm{D}}^2} \, .$$

According to the Lindemann criterion, $\langle \mathbf{u}^2(\mathbf{R}_i^0)\rangle < a^2/100$, and therefore the melting temperature is $k_{\mathrm{B}} T_m = a^2 M (d-2)\omega_{\mathrm{D}}^2/(100d)$.

Answer 5.24. Periodic boundary conditions imply that $u_N = u_{-N}$. The last two equations of motion in (5.78) are thus modified to be

$$M\ddot{u}_{\pm n} = -D(2u_{\pm n} - u_{\pm(n-1)} - u_{\pm(n+1)}) \, , \quad 1 < n < N-2 \, ,$$
$$M\ddot{u}_{\pm(N-1)} = -D(2u_{\pm(N-1)} - u_{\pm(N-2)} - u_N) \, ,$$
$$M\ddot{u}_N = -D(2u_N - u_{N-1} - u_{-(N-1)}) \, .$$

The corresponding equations for the symmetric and anti-symmetric functions are modified as well,

$$M\ddot{u}_n^{\pm} = -D(2u_n^{\pm} - u_{n-1}^{\pm} - u_{n+1}^{\pm}) \, , \quad 1 < n < N-2 \, ,$$
$$M\ddot{u}_{N-1}^{+} = -D(2u_{N-1}^{+} - u_{N-2}^{+} - 2u_N) \, ,$$
$$M\ddot{u}_{N-1}^{-} = -D(2u_{N-1}^{-} - u_{N-2}^{-}) \, , \quad M\ddot{u}_N = -D(2u_N - u_{N-1}^{+}) \, .$$

The equations of motion for the anti-symmetric functions are similar to those obtained previously, upon interchanging N by $N-1$. It follows that there are $N-1$ anti-symmetric wavy solutions. The equations of motion for the symmetric functions are similar to the previously-derived ones as well; the difference is that one has to insert in the equation for u_{N-1}^{+} the expression $u_N = D u_{N-1}^{+}/(D - M\omega^2)$. With solutions of the type (5.80), one finds

$$\Big(M\omega^2 - D - D_0 - \frac{2D_0^2}{m\omega^2 - 2D_0}\Big)\Big(A^{+}C + \frac{B^{+}}{C}\Big) + D\Big(A^{+}C^2 + \frac{B^{+}}{C^2}\Big) = 0 \, ,$$

$$\Big(M\omega^2 - 2D - \frac{2D^2}{M\omega^2 - 2D}\Big)\Big(A^{+}C^{N-1} + \frac{B^{+}}{C^{N-1}}\Big) + D\Big(A^{+}C^{N-2} + \frac{B^{+}}{C^{N-2}}\Big) = 0 \, ,$$

modifying Eq. (5.85) to be

$$-\frac{B^+}{A^+} = \frac{C^2+3}{3+C^2}C^{2n} = \frac{(D-D_0-X)C-D}{(D-D_0-X)/C-D}\ .$$

For $C = \exp[ika]$ one obtains the same wavy solutions as found in the main text. For $|C| < 1$ and $N \gg 1$ the left hand-side may be again discarded, leading to precisely the same equation as found in the main text for the localized mode, $(D - D_0 - X)C - D = 0$. This could have been anticipated, as the localized state decays exponentially and thus is insensitive to conditions on boundaries that are situated far away from the foreign atom.

Answer 5.25. Denoting the momentum of the outgoing neutrons by $\hbar k'$, momentum conservation yields $k' = k \pm (G - K)$, where $G = 2\pi h/a$, and h is an arbitrary integer. Energy conservation implies

$$\hbar\omega_0|\sin(Ka/2)| = \hbar\omega(K) = \pm\frac{\hbar^2}{2m_n}[k^2 - (k')^2] = \pm\frac{\hbar^2}{2m_n}\Big[k^2 - [k \pm (G - K)]^2\Big]\ ,$$

$$(5.116)$$

where $\omega_0 = 2\sqrt{D/M}$. Figure 5.33 displays a graphical solution of this equation: the thin line is the left hand-side, and the thick solid line (or the thick dashed one) is the right hand-side for phonon emission (upper sign) or absorption (lower sign), as functions of $K - G$ in units of $G_0 = 2\pi/a$. As seen, the right hand-side is a parabola cutting the horizontal axis at the origin and at $\pm 2k$, with a maximum (minimum) which equals (up to a sign) the energy of the impinging particle, $\pm\hbar^2 k^2/(2m_n)$, at $K - G = \pm k$. Each crossing of the two curves gives a possible value of the energy and momentum of the phonon. The number of solutions varies with the value of k.

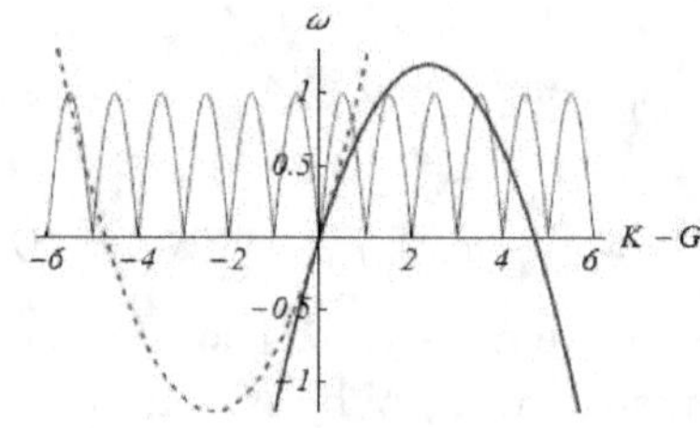

Fig. 5.33: The two sides of Eq. (5.116).

Answer 5.26.
a. Equation (5.51) is equivalent to the expression $F_{\mathrm{ph}} = k_{\mathrm B}T \sum_\alpha \ln[2\sinh(\beta\hbar\omega_\alpha/2)]$.
b. Equating the derivative of $F_0(T, V)$ from part (a) with respect to V to zero yields the required result.

c. Equation (5.52) is equivalent to $\langle E\rangle_{\rm ph} = \sum_\alpha (\hbar\omega_\alpha/2)\coth(\beta\hbar\omega_\alpha/2)$. Assuming that $\partial\ln\omega_\alpha/\partial\ln V$ is independent of α gives the required result.
d. Differentiation of the result of the previous part with respect to the temperature leads to the required result.

Answer 5.27.

a. Postulating that $\mathbf{B}$ is along $\hat{\mathbf{z}}$, $\mathbf{B} = B\hat{\mathbf{z}}$, then $\dot\mu_z = 0$, that is, the z–component of the magnetic moment does not vary in time. The equations for the two components normal to the field are $\dot\mu_x = \gamma B\mu_y$ and $\dot\mu_y = -\gamma B\mu_x$. An equation for μ_x is derived by differentiating the first and inserting the result into the second, $\ddot\mu_x = -(\gamma B)^2\mu_x$. The solution is $\mu_x = \mu^{(0)}\cos(\omega_{\rm L}t)$, with the Larmor frequency $\omega_{\rm L} = \gamma B$. Substituting in the first equation gives $\mu_y = \dot\mu_x/\omega_{\rm L} = -\mu^{(0)}\sin(\omega_{\rm L}t)$. Therefore, the component of the magnetic moment which is perpendicular to the field rotates around it with the Larmor frequency. The magnetic moment vector propagates at a fixed rate on a cone whose axis is parallel to the magnetic field.
b. The energy of the system, in terms of the magnetic moments, is

$$\hat{\mathcal{H}}_M = -\sum_{\langle nm\rangle} \overline{\overline{J}}_{nm}\boldsymbol{\mu}_n\cdot\boldsymbol{\mu}_m\ ,$$

where $\overline{\overline{J}}_{nm} = J_{nm}/(g\mu_{\rm B})^2$. It follows that $\hat{\mathcal{H}} = -\sum_n \boldsymbol{\mu}_n\cdot\mathbf{B}_n$, where $\mathbf{B}_n = \sum_m \overline{\overline{J}}_{nm}\boldsymbol{\mu}_m$. Hence,

$$\dot{\boldsymbol{\mu}}_n = \gamma\boldsymbol{\mu}_n\times\mathbf{B}_n = \boldsymbol{\mu}_n\times\sum_m \overline{J}_{nm}\boldsymbol{\mu}_m\ ,$$

$$\overline{J}_{nm} = \gamma\overline{\overline{J}}_{nm} = \gamma J_{nm}/(g\mu_{\rm B})^2 = J_{nm}/(g\mu_{\rm B}\hbar)\ .$$

Answer 5.28.

Inserting $\mu_{n\pm}$ in Eqs. (5.103) yields

$$\frac{d\mu_{n\pm}}{dt} = \pm i\mu\sum_m \overline{J}_{nm}(\mu_{n\pm} - \mu_{m\pm})\ .$$

With wavy solutions, $\mu_{n\pm} = \mu_\pm\exp[i(\mathbf{k}\cdot\mathbf{R}_n^0 - \omega t)]$, this equation yields $-i\omega\mu_\pm = \pm i\mu[\widetilde{J}(0) - \widetilde{J}(\mathbf{k})]\mu_\pm$, i.e., $\omega = \mp[\widetilde{J}(0) - \widetilde{J}(\mathbf{k})]$. It follows that the solution $\mu_{n-} = \mu_-\exp[i(\mathbf{k}\cdot\mathbf{R}_n^0 - |\omega|t)]$ propagates along the vector $\mathbf{k}$, while the other solution, $\mu_{n+} = \mu_+\exp[i(\mathbf{k}\cdot\mathbf{R}_n^0 + |\omega|t)]$, propagates in the opposite direction. Assuming that $\mu_+ = 0$ results in $\mu_x = \mu_-/2$ and $\mu_y = i\mu_-/2$, and thus there is a phase difference of $\pi/2$ between the spin components; the spins rotate in the XY plane.

Answer 5.29.

a. The equations of motion in three dimensions are identical to those presented in Eq. (5.102), but the sums run over nearest neighbors alone. For a cubic crystal of lattice constant a, whose nearest-neighbor exchange coefficient is $\overline{J}$, Eq. (5.105) yields

$$\widetilde{J}(\mathbf{k}) = \sum_m \overline{J}_{nm} e^{i\mathbf{k}\cdot\mathbf{R}_{nm}} = 2\overline{J}[\cos(k_x a) + \cos(k_y a) + \cos(k_z a)] ,$$

($\mathbf{R}_{nm}$ is a vector connecting two nearest neighbors) and Eq. (5.106) gives

$$\omega = 4\mu\overline{J}[\sin^2(k_x a/2) + \sin^2(k_y a/2) + \sin^2(k_z a/2)] \approx \Delta|\mathbf{k}|^2 ,$$

with $\Delta = \mu\overline{J}a^2$.

b. The dispersion relation of magnons in a ferromagnet is that found in problem 5.20, with $\eta = 4$. It follows from the results of that problem that $C_V(\text{FM}) \propto T^{d/2}$ at low temperatures. At high temperatures one obtains the Dulong-Petit law, $C_V(\text{FM}) = k_\text{B}N$, where N is the number of magnetic moments in the system.

Answer 5.30.

The equations of motion are

$$\frac{d\mu_{nx}^{(1)}}{dt} = -\mu\overline{J}(2\mu_{ny}^{(1)} + \mu_{ny}^{(2)} + \mu_{n-1y}^{(2)}) ,$$

$$\frac{d\mu_{ny}^{(1)}}{dt} = \mu\overline{J}(2\mu_{nx}^{(1)} + \mu_{nx}^{(2)} + \mu_{n-1x}^{(2)}) ,$$

$$\frac{d\mu_{nx}^{(2)}}{dt} = \mu\overline{J}(2\mu_{ny}^{(2)} + \mu_{ny}^{(1)} + \mu_{n+1y}^{(1)}) ,$$

$$\frac{d\mu_{ny}^{(2)}}{dt} = -\mu\overline{J}(2\mu_{nx}^{(2)} + \mu_{nx}^{(1)} + \mu_{n+1x}^{(1)}) .$$

Inserting the variables $\mu_{n\pm}^{(1,2)} = \mu_{nx}^{(1,2)} \pm i\mu_{ny}^{(1,2)}$ yields

$$\frac{d\mu_{n+}^{(1)}}{dt} = i\mu\overline{J}(2\mu_{n+}^{(1)} + \mu_{n+}^{2)} + \mu_{n-1+}^{(2)}) ,$$

$$\frac{d\mu_{n+}^{(2)}}{dt} = -i\mu\overline{J}(2\mu_{n+}^{(2)} + \mu_{n+}^{1)} + \mu_{n+1+}^{(1)}) .$$

Finally, exploiting the wavy form for the spin waves, one finds

$$-\omega\mu_+^{(1)} = \mu\overline{J}[2\mu_+^{(1)} + (1 + e^{-ika})\mu_+^{(2)}] ,$$

$$\omega\mu_+^{(2)} = \mu\overline{J}[2\mu_+^{(2)} + (1 + e^{ika})\mu_+^{(1)}] .$$

The determinant of the coefficients vanishes when $\omega = 2\mu\overline{J}|\sin(ka/2)|$, and therefore the dispersion relation is linear in the wave number for long waves, exactly like the one of phonons. The equations for $\mu_{n-}^{(1,2)}$ yield the same dispersion relation.

5.11 Problems for self-evaluation

s.5.1.

As explained in Sec. 4.4, the van der Waals interaction stems from displacements of the electrons' cloud (whose total mass is m) relative to the position of the nucleus (whose mass is M). Strictly speaking, there is no reason to assume identical displacements for both the electrons' center of mass and the nucleus' center of mass (these are located at the same place at equilibrium). Consider longitudinal vibrations in a single-atom, one-dimensional lattice, and denote the displacements of the nucleus and the electrons in the n—th atom by u_n and $u_{e,n}$, respectively. Assume further that the potential energy of the relative motion is described by the form $U = D_0(u_{e,n} - u_n)^2/2$, and that the harmonic interactions between nearest neighbors in the lattice result predominantly from the interaction between the electrons' clouds on these atoms; these harmonic interactions are characterized by a spring of constant D.

a. *Find the frequencies of the vibrational modes, and write down the expression for the dispersion relation of the acoustic branch in the $m \to 0$ limit.*

b. *Discuss the effects of the coupling between the nuclei and the electrons on the sound velocity, the maximal acoustic frequency, and the group velocity.*

s.5.2.

The lattice constant of a single-atom chain is 4.85Å, the mass of each atom is 6.81×10^{-26} Kg, and the maximal frequency of the lattice vibrations is 4.46×10^{13} sec^{-1}.

a. *Find the values of the spring constant and of the sound velocity.*

b. *One of the atoms is forced to oscillate at the frequency 5.75×10^{13} sec^{-1}, with amplitude U_0. Find the dependence of the displacement amplitudes of the other atoms on the distance from that atom. When is that amplitude equal to $U_0/10$?*

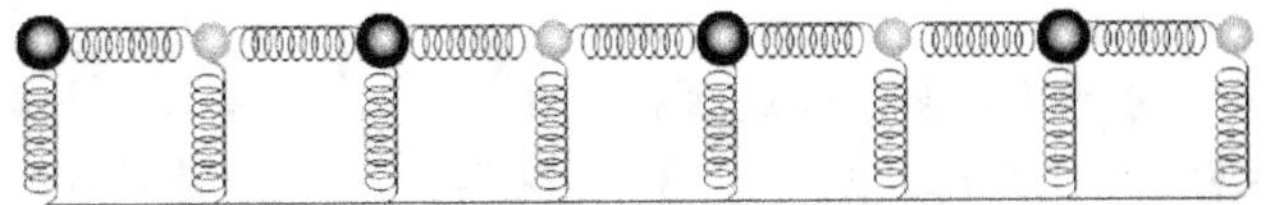

Fig. 5.34

s.5.3.

A chain that consists of two different masses m and M is coupled to the table on which it lies by springs (see Fig. 5.34). The constant of the spring joining two neighboring (different) masses is D, and that of the spring coupling each mass to the table is D_s. It is assumed that the masses oscillate only parallel to the table.

a. *Solve the classical equations of motion and find the dispersion law.*
b. *Find the frequencies for which the system is transparent. What can be concluded about the stability of this lattice against melting?*
c. *Plot schematically the density of states of the lattice vibrations as a function of the frequency, and the specific heat as a function of the temperature.*

s.5.4.

The unit cell of a one-dimensional lattice contains three atoms, two of mass m and one of mass M. The spring constants are all equal to D.
a. *Write down the equations of motion and derive the expression that determines the dispersion relations. How many branches are there?*
b. *Show that one of the branches is an acoustic one, and find the sound velocity.*
c. *Show that the solution approaches that of the single-atom lattice when the masses are equal, $M = m$.*
d. *Draw schematically the dispersion relations for all branches in the first Brillouin zone.*

s.5.5.

Consider a rectangular lattice with lattice constants a and b. The area of the lattice is A. The lattice sites contain two types of atoms, of masses M_1 and M_2, arranged such that the nearest neighbors of an atom of one type are all of the other type. The atoms oscillate only perpendicular to the plane, and the spring constants are D_a and D_b for the two lattice directions.
a. *Determine the lattice structure and find the primitive unit cell. How many atoms of each kind are there in the crystal?*
b. *Write down the classical equations of motion and derive the frequency spectrum of the lattice. Describe the frequency spectrum when $D_a = 0$ or $D_b = 0$. The crystal is perfectly symmetric when $D_a = D_b$ and $a = b$. Repeat part (a) for this case. Discuss in particular the spectrum of a wave along the (111) direction in the symmetric configuration.*
c. *Find the equal-frequency lines in momentum space in the low-frequency limit. Use them to derive the density of states.*
d. *Use the Debye approximation to obtain an expression for the contribution of the the acoustic modes to the specific heat, and find the specific heat at low temperatures.*
e. *Find the contribution of the high-frequencies modes to the specific heat. Which approximation is most suitable for its calculation? What is the relation between this contribution and the one of low temperatures, calculated in part (c)?*
f. *Give an expression for the high-temperature specific heat.*

s.5.6. *The nearest-neighbor interaction in an* **FCC** *lattice is described by "springs" whose constants are*

$$D_{\mu\nu}(\mathbf{R}^0) = D_1 \delta_{\mu\nu} + (D_2 - D_1)\hat{\mathbf{R}}^0_\mu \hat{\mathbf{R}}^0_\nu \ ,$$

where $D_1 = D_T$ and $D_2 = D_L$ are related to the transverse and longitudinal vibrations, respectively.

a. Derive the equations that relate the eigenfrequencies to the wave vector.

b. Obtain explicit expressions for these frequencies for waves corresponding to the following lines in the Brillouin zone: $\mathbf{k} = (k,0,0)$, $\mathbf{k} = (2\pi/a, k, 0)$, $\mathbf{k} = (k,k,0)$, and $\mathbf{k} = (k,k,k)$. Determine the range of $k-$values for each of those, needed to reproduce the path $\Gamma \to X \to W \to X \to \Gamma \to L$ in the Brillouin zone (Fig. 5.18). In particular, explain why the points $(2\pi/a, 0, 0)$ and $(2\pi/a, 2\pi/a, 0)$ are both marked by X.

c. Use the results of part (b) to reproduce Fig. 5.18 (choose for D_2/D_1 a value that yields approximately the ratio of the frequencies at the X point).

d. Deduce from Fig. 5.18 the sound velocities for a wave along the cube diagonal, and use them to estimate the Debye temperature of copper. The lattice constant of copper is $a = 3.61\text{Å}$.

e. Explain qualitatively (without an explicit calculation) the way the figure is modified when the crystal under consideration has two ions in the unit cell, e.g., table salt.

s.5.7.

a. The density of states depicted in Fig. 5.19(b) is singular at certain values of the frequency. Can these points be identified from Fig. 5.19(a)?

b. Some of these van Hove singularities pertain to points where the dispersion relation is maximal. Assuming that at the vicinity of such a maximum $\omega(\mathbf{k}) \approx \omega_0 - \omega_1(\mathbf{k} - \mathbf{k}_0)^2$, find the singularity of the density of states near the maximum.

c. What is the shape taken by the density of states of the vibrations in graphene near the frequency corresponding to the K point in the Brillouin zone (problem 5.10)?

s.5.8.

Consider the two-atom, one-dimensional system displayed in Fig. 5.3. Assume that the Einstein frequency ω_E coincides with the maximal optical frequency.

a. Find the Debye frequency ω_D, and the ratio ω_D/ω_E.

b. Copy Fig. 5.15, and add to it the density of states calculated within the Debye approximation and the Einstein one.

s.5.9.

Express the wave length that corresponds to the Debye frequency in terms of the volume of the unit cell. What is the ratio between this wave length and the distance between nearest neighbors in crystals of copper (**FCC**) and iron (**BCC**)?

s.5.10. The crystalline structures of NaCl and KCl are identical, though the lattice constants and the magnitudes of the "spring constants" differ. The sound velocities in the three acoustic branches of each crystal are rather comparable. The Debye *temperatures are 310 K and 230 K, respectively. The specific heat of KCl at 5 K is 0.038J/mol/K. What is the specific heat of NaCl at 2 K?*

s.5.11. *The dispersion relation of a two-dimensional lattice is $\omega(\mathbf{k}) = c(|k_x| + |k_y|)$. Derive the density of states of this lattice. Compare it with the result of the Debye approximation. Determine the Debye frequency. How does the specific heat vary with the temperature?*

s.5.12. *A beam of neutrons of wave length $\lambda = 3.5\text{Å}$ impinges on a single-atom* **SC** *lattice, of lattice constant $a = 4.25\text{Å}$, along the normal to one of its faces. Part of the neutrons leave the lattice along its principal diagonal, [111], with wave length $\lambda' = 2.33\text{Å}$. What are the frequency and the wave length of the phonon involved in this inelastic scattering? Is the phonon absorbed, or is it emitted?*

s.5.13. *Find the frequencies of the spin waves in a one-dimensional ferri-magnet (whose unit cell contains two antiparallel moments of unequal magnitudes), in which the exchange interactions are only between nearest-neighbors.*

5.12 Answers for the self-evaluation problems

Answer s.5.1.

a. The equations of motion are

$$M\omega^2 u_n = D_0(u_n - u_{e,n}) \,,$$
$$m\omega^2 u_{e,n} = D_0(u_{e,n} - u_n) + D(2u_{e,n} - u_{e,n-1} - u_{e,n+1}) \,,$$

where ω is the frequency of the motion. Assuming $u_n = A \exp[ikna]$ and $u_{n,e} = B \exp[ikna]$, one finds

$$M\omega^2 A = D_0(A - B) \,,$$
$$m\omega^2 B = D_0(B - A) + 4D\sin^2(ka/2)B \,.$$

The frequencies are the roots of a quadratic equation, as is also obtained for the two-atom chain. Hence there are two branches, acoustic and optical. However, in the limit where the electronic mass is extremely small compared to the mass of the nucleus, the optical branch comprises very high frequencies, approaching infinity. Only the acoustic branch survives, with the dispersion relation

$$M\omega^2 = 4D\sin^2(ka/2)/[1 + (4D/D_0)\sin^2(ka/2)] \,.$$

This dispersion is depicted in Fig. 5.35, for $4D/D_0 = 0, .5, 1, 5, 10$ (the smaller is the ratio, the higher is the frequency).

b. In the acoustic limit one expands the dispersion relation in powers of the wave number k; to leading order, $M\omega^2 \approx 4D(ka/2)^2$. This expression is identical to the dispersion relation of a chain in which the electrons and the nuclei move in unison. The sound velocity is thus unaffected by the internal motion within each ion. Likewise, the specific heat in the Debye approximation is not affected. In contrast, the maximal frequency, attained for $ka = \pi$, is $M\omega^2_{\max} = 4D/[1 + (4D/D_0)]$, and it decreases as D_0 is reduced. When the latter is small, it dominates the dispersion, and the dispersion relation becomes $M\omega^2 \approx D_0$. It follows that the dependence of the frequency on the wave number is gradually weakened, and so is the group velocity. In this limit most of the motion takes place within each atom, and the relative displacements of neighboring atoms have almost no effect.

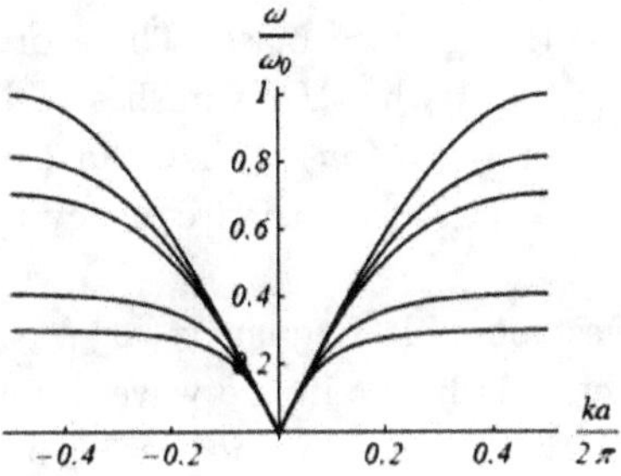

Fig. 5.35

Answer s.5.2.

a. Using Eq. (5.6), one finds $D = M\omega_0^2/4 = 3.39 \times 10^4$ dyn/cm. The sound velocity is $c = a\omega_0/2 = 1.08 \times 10^6$ cm/sec.

b. The forced frequency exceeds the maximal frequency of the dispersion relation, thence the amplitudes of the vibrations decay with the distance away from the special atom, $u_n = U_0 \exp[-\kappa a n]$. The coefficient that characterizes the decay is derived from the relation $\omega/\omega_0 = \cosh(\kappa a/2)$. It follows that $\kappa a = 2\mathrm{arccosh}(\omega/\omega_0) \approx 1.5$. A decay by a factor of 10 requires $10 = \exp[\kappa n a] \approx \exp[1.5n]$, i.e., $n \approx 1.55$ or $na \approx 7.5\text{Å}$. Since n turned out to be non integer, the amplitude of the vibration is smaller than $U_0/10$ already at the third atom in the chain.

Answer s.5.3.

a. A generalization of Eqs. (5.12) gives that the equations of motion are

$$M\ddot{u}_n = -D(2u_n - v_{n-1} - v_n) - D_s u_n \,,$$
$$m\ddot{v}_n = -D(2v_n - u_{n+1} - u_n) - D_s v_n \,,$$

where the notations are defined in Fig. 5.3. Assuming the forms $u_n = A\exp[i(kna-$

$\omega t)]$ and $v_n = B \exp[i(kna - \omega t)]$ extends Eqs. (5.13), to yield

$$M\omega^2 A = D[2A - (1 + e^{-ika})B] + D_s A \,,$$
$$m\omega^2 B = D[2B - (1 + e^{ika})A] + D_s B \,.$$

The determinant of the coefficients vanishes for [*cf.* Eq. (5.14)]

$$(M\omega^2 - 2D - D_s)(m\omega^2 - 2D - D_s) - 4D^2 \cos^2(ka/2) = 0 \,,$$

and thus

$$\omega^2 = \frac{2D + D_s}{2\mu}\left\{1 \pm \sqrt{1 - \frac{4\mu}{M + m}\left[1 - \frac{\cos^2(ka/2)}{[1 + D_s/(2D)]^2}\right]}\right\} \,,$$

where $\mu = Mm/(M + m)$ is the reduced mass. The acoustic and optical branches shown in Fig. 5.4 are reproduced when D_s vanishes. They are presented in Fig. 5.36 by the dashed curves (for $M = 2m$). The solid curves in Fig. 5.36 display the dispersion relations for $D_s = D$, with the frequency measured in units of $\omega_0 = \sqrt{2D/\mu}$.

b. When $D_s \neq 0$, the lowest possible frequency (for $k = 0$) is finite; the acoustic regime in which the frequency is linear in the wave number disappears. The two branches are both "optical"; this means that the average of the displacement squared does not diverge, and the instability problem discussed in Sec. 5.5 does not arise. As mentioned, the **coupling to the substrate stabilizes the crystal**. At the edge of the Brillouin zone, where $ka = \pi$, the frequencies are $\omega_2^2 = (2D + D_s)/M$ or $\omega_3^2 = (2D + D_s)/m$. In the middle of the Brillouin zone one finds

$$\omega_\pm^2 = \frac{2D + D_s}{2\mu}\left\{1 \pm \sqrt{1 - \frac{4\mu}{M + m}\left[1 - \frac{1}{[1 + D_s/(2D)]^2}\right]}\right\} \,.$$

The material is transparent to radiation in the ranges $\omega < \omega_-$, $\omega_2 < \omega < \omega_3$, and $\omega_+ < \omega$.

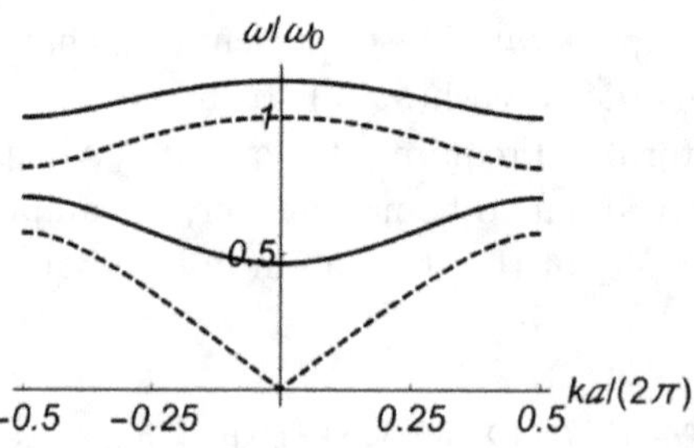

Fig. 5.36

c. In one dimension, the density of states is $g(\omega) = (L/\pi)/(d\omega/dk)$. Figure 5.36 shows that the density of states in both branches diverges at the edges of the Brillouin zone. The plot portrayed in Fig. 5.37(a) is obtained by differentiating the dispersion law, with the same parameters as used in Fig. 5.36 [the frequency is in units of ω_0, and the density of states is in units of $g_0 = 2L/(\omega_0 a)$]. The contribution to the specific heat of each of the frequencies has the same form as the one resulting from the Einstein model, Fig. 5.13. Figure 5.37(b) shows the contributions of the two branches, for a mass ratio of $M/m = 10$; such a ratio allows for two distinct branches. The two dotted curves display the contribution to

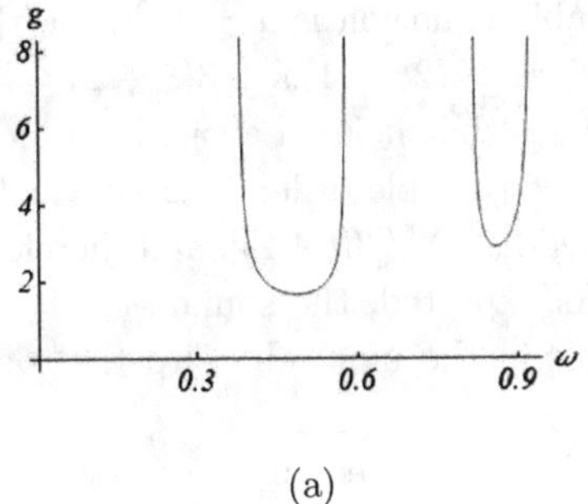

(a)

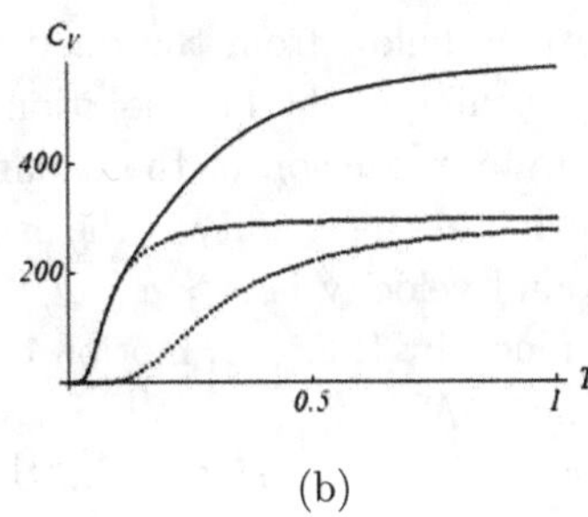

(b)

Fig. 5.37

the specific heat of each branch separately (those are computed from a summation over 300 frequency values, that correspond to 300 wave numbers in the first Brillouin zone). The specific heat of each branch is exponentially small at temperatures lower than its threshold frequency; as the temperature is raised the specific heat saturates towards the Dulong-Petit constant ($300k_{\mathrm{B}}$ in the plots). The solid line is the total contribution. As seen, at low temperatures the specific heat is predominantly given by the lower branch; when the temperature approaches the threshold frequency of the upper branch, the latter begins to contribute as well, such that at high temperatures the total specific heat reaches the Dulong-Petit value $600k_{\mathrm{B}}$.

Answer s.5.4.

a. Denoting the displacements of the three atoms in the cell (from left to right) by $u_n = A\exp[i(kna - \omega t)]$, $v_n = B\exp[i(kna - \omega t)]$, and $w_n = C\exp[i(kna - \omega t)]$, the equations of motion are

$$m\omega^2 A = D(2A - B - C\exp[-ika])\,,$$
$$m\omega^2 B = D(2B - A - C)\,,$$
$$M\omega^2 A = D(2C - B - A\exp[ika])\,.$$

When the determinant of the coefficients' matrix vanishes,

$$\det \begin{bmatrix} m\omega^2 - 2D & D & De^{-ika} \\ D & m\omega^2 - 2D & D \\ De^{ika} & D & M\omega^2 - 2D \end{bmatrix} = 0 \ ,$$

the solutions are given by

$$(x-2)^2(\xi x - 2) - (2+\xi)x + 6 + 2\cos(ka) = 0 \ , \quad x \equiv \frac{m\omega^2}{D} \ , \quad \xi = \frac{M}{m} \ .$$

The three roots of the cubic equation give rise to three branches of frequencies, as expected for a one-dimensional lattice with three atoms in the unit cell.

b. When $ka = 0$, one of the solutions of the cubic equation is $x = 0$, while the two others follow from the quadratic equation $\xi x^2 - 2(2\xi + 1)x + 3(\xi + 2) = 0$. The vanishing of the first solution at $ka = 0$ implies that it is an acoustic branch. Indeed, an expansion of the equation to the lowest possible order in x and in ka yields that at the vicinity of the origin $x = m\omega^2/D \approx (ka)^2/[3(\xi+2)]$, and therefore the sound velocity is $c \approx a\sqrt{D/[3(2m + M)]}$. As expected, the sound velocity is determined by the joint motion of all three atoms in the unit cell, with the total mass $2m + M$.

c. Figure 5.38(a) displays the three roots for the case of equal masses, $\xi = 1$. Exploiting the symmetry of the solution with respect to a displacement by a reciprocal lattice-vector, and the one with respect to reflections, shows that the three branches represent three parts of the Brillouin zone of a single-atom lattice, whose lattice constant is $b = a/3$, and its Brillouin zone extends up to $\pi/(a/3) = 3\pi/a$.

d. Figure 5.38(b) displays the results for $\xi = 2$ (obtained by an explicit solution of the cubic equation for an arbitrary wave number). An expansion around the origin shows that the successive branches have a sequence of minimum–maximum– minimum at the origin. One may therefore "guess" this form, and draw it schematically. One may also derive the result starting from the plots in Fig. 5.38(a), and then opening gaps in-between the branches at the locations where they coincide in Fig. 5.38(a).

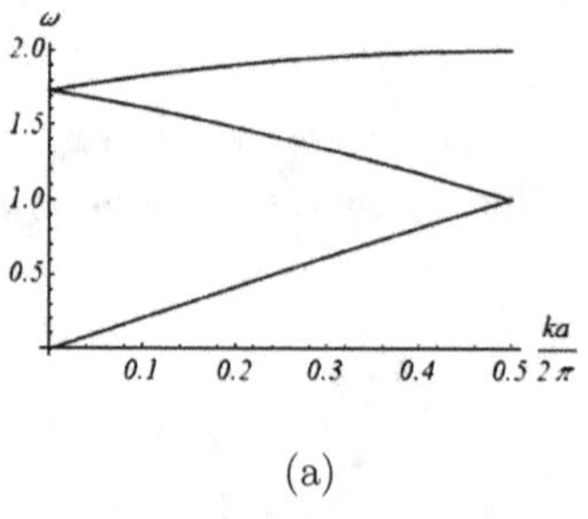

(a)

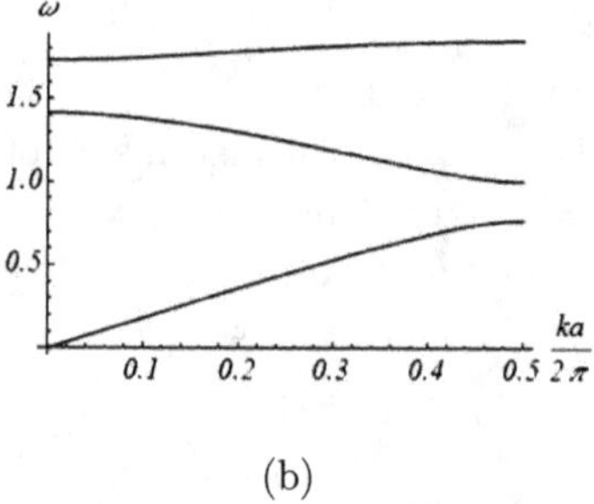

(b)

Fig. 5.38

Answer s.5.5.

a. The lattice is a rectangular centered one; its primitive unit cell, encompassed in Fig. 5.39 by the thick-line rhombus, comprises a single atom of each type. The lattice vectors are $\mathbf{a}_1 = a\hat{\mathbf{x}} - b\hat{\mathbf{y}}$ and $\mathbf{a}_2 = a\hat{\mathbf{x}} + b\hat{\mathbf{y}}$, and the locations of the atoms in this unit cell are $\mathbf{R}_{n,m}^{(1)} = n\mathbf{a}_1 + m\mathbf{a}_2 + u_{nm}\hat{\mathbf{z}}$ and $\mathbf{R}_{n,m}^{(2)} = n\mathbf{a}_1 + m\mathbf{a}_2 + a\hat{\mathbf{x}} + v_{n,m}\hat{\mathbf{z}}$. The displacements away from the equilibrium locations are marked in Fig. 5.39. The area of the unit cell is $2ab$, and the number of cells in the given area is $N = A/(2ab)$.

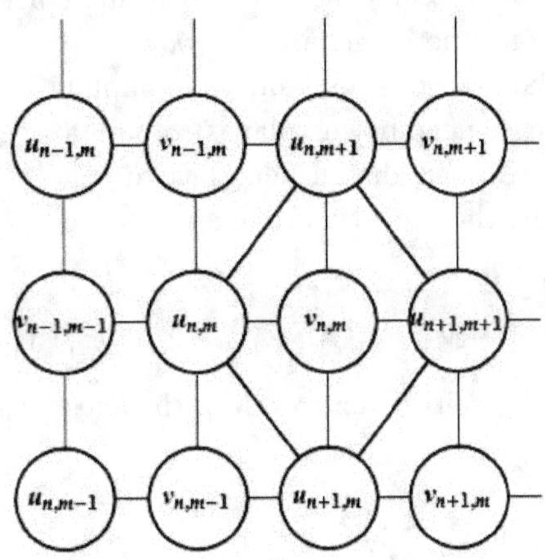

Fig. 5.39

b. The classical equations of motion are

$$M_1\ddot{u}_{n,m} = -D_a(2u_{n,m} - v_{n,m} - v_{n-1,m-1}) - D_b(2u_{n,m} - v_{n,m-1} - v_{n-1,m}) \,,$$

$$M_2\ddot{v}_{n,m} = -D_a(2v_{n,m} - u_{n,m} - u_{n+1,m+1}) - D_b(2v_{n,m} - u_{n,m+1} - u_{n+1,m}) \,.$$

Substituting $u_{n,m} = A\exp[i\mathbf{k}\cdot(n\mathbf{a}_1 + m\mathbf{a}_2) - i\omega t]$ and $v_{n,m} = B\exp[i\mathbf{k}\cdot(n\mathbf{a}_1 + m\mathbf{a}_2) - i\omega t]$ yields

$$M_1\omega^2 A = D_a(2A - B - Be^{-i\mathbf{k}\cdot(\mathbf{a}_1+\mathbf{a}_2)}) + D_b(2A - Be^{-i\mathbf{k}\cdot\mathbf{a}_1} - Be^{-i\mathbf{k}\cdot\mathbf{a}_2}) \,,$$

$$M_2\omega^2 B = D_a(2B - A - Ae^{i\mathbf{k}\cdot(\mathbf{a}_1+\mathbf{a}_2)}) + D_b(2B - Ae^{i\mathbf{k}\cdot\mathbf{a}_1} - Ae^{i\mathbf{k}\cdot\mathbf{a}_2}) \,,$$

and thus

$$\begin{bmatrix} M_1\omega^2 - 2(D_a + D_b) & 2e^{-ik_x a}[D_a\cos(k_x a) + D_b\cos(k_y b)] \\ 2e^{ik_x a}[D_a\cos(k_x a) + D_b\cos(k_y b)] & M_2\omega^2 - 2(D_a + D_b) \end{bmatrix}\begin{bmatrix} A \\ B \end{bmatrix} = 0 \,.$$

A non trivial solution is obtained when the determinant vanishes,

$$M_1 M_2\omega^4 - 2(D_a + D_b)(M_1 + M_2)\omega^2 + 4(D_a + D_b)^2$$
$$- 4[D_a\cos(k_x a) + D_b\cos(k_y b)]^2 = 0 \,.$$

There are therefore two dispersion branches,

$$\omega_\pm^2 = \frac{D_a + D_b}{\mu} \pm \left[\frac{(D_a + D_b)^2}{\mu^2} - \frac{16}{M_1 M_2} \left(D_a \sin^2(k_x a/2) + D_b \sin^2(k_y b/2) \right) \right.$$

$$\left. \times \left(D_a \cos^2(k_x a/2) + D_b \cos^2(k_y b/2) \right) \right]^{1/2} ,$$

where $\mu = M_1 M_2/(M_1 + M_2)$ is the reduced mass. In the limits $D_a = 0$ or $D_b = 0$ this spectrum coincides with that of a one-dimensional lattice with two atoms in the unit cell, Eq. (5.15), with lattice constants $2a$ or $2b$, respectively. In the symmetric configuration, where $D_a = D_b$ and $a = b$, the result corresponds to that of the table salt, Fig. 4.4. Along the diagonal, $k_x = k_y$, the expression resembles again Eq. (5.15) (once the lattice constant and the coupling coefficients are properly adjusted), since in this case neighboring planes contain atoms of the same type; the system becomes similar to a one-dimensional lattice.

c. In the long-wave limit, the acoustic branch is

$$\omega_-^2 \approx (c_x k_x)^2 + (c_y k_y)^2 ,$$
$$c_x^2 = 2\mu D_a a^2/(M_1 + M_2) , \quad c_y^2 = 2\mu D_b b^2/(M_1 + M_2) .$$

c_x and c_y are the sound velocities along each of the axes. It follows that the equal-frequency curves are ellipses,

$$\left(\frac{k_x}{\omega/c_x} \right)^2 + \left(\frac{k_y}{\omega/c_y} \right)^2 = 1 .$$

The area of such an ellipse is $\pi(\omega/c_x)(\omega/c_y)$. The area of the unit cell in the reciprocal lattice (which equals the area of the first Brillouin zone) is $(2\pi)^2/(2ab)$ (see problem 3.6), and it contains N points; therefore the area per site is $2\pi^2/(Nab) = 4\pi^2/A$. Consequently, the total number of points within the ellipse is $[\pi\omega^2/(c_x c_y)]/[4\pi^2/A] = A\omega^2/(4\pi c_x c_y)$, and the density of states is

$$g(\omega) = \frac{d[A\omega^2/(4\pi c_x c_y)]}{d\omega} = \frac{A\omega}{2\pi c_x c_y} .$$

d. The Debye approximation assumes that the acoustic density of states, obtained in the previous part, is valid up to the Debye frequency. The total number of acoustic vibrational modes is

$$N = \frac{A}{2ab} = \int_0^{\omega_D} g(\omega) d\omega = A \frac{\omega_D^2}{4\pi c_x c_y} ,$$

and hence $\omega_D^2 = 2\pi c_x c_y/(ab)$. The contribution of the acoustic modes to the specific heat is

$$C_V = k_B \int_0^{\omega_D} d\omega g(\omega)(\beta\hbar\omega)^2 \frac{e^{\beta\hbar\omega}}{(e^{\beta\hbar\omega} - 1)^2} = 2Nk_B \frac{T^2}{\Theta_D^2} \int_0^{\Theta_D/T} dx x^2 \frac{e^x}{(e^x - 1)^2} ,$$

and at low temperatures, $C_V \approx 2Nk_B I_2(T/\Theta_D)^2$, where I_2 is defined after Eq. (5.69).

e. The frequencies in the optical branch are of the order of $\omega_+ \approx \sqrt{2(D_a + D_b)/\mu}$. In the Einstein approximation, their contribution to the specific heat is

$$C_V \approx N k_{\mathrm{B}} (\beta\hbar\omega_+)^2 \exp[\beta\hbar\omega_+]/(\exp[\beta\hbar\omega_+] - 1)^2 \ ,$$

yielding at low temperatures $C_V \simeq N k_{\mathrm{B}} (\beta\hbar\omega_+)^2 \exp[-\beta\hbar\omega_+]$. This contribution is exponentially smaller than that of the acoustic modes.

f. At sufficiently high temperatures the specific heat obeys the Dulong-Petit law, $C_V \approx 2N k_{\mathrm{B}}$.

Answer s.5.6.

a. Each atom in the **FCC** lattice has 12 nearest neighbors, at the points $(\pm 1, \pm 1, 0)a/2$, $(\pm 1, 0, \pm 1)a/2$, and $(0, \pm 1, \pm 1)a/2$. Inserting these into the expression for the spring constant yields

$$D_{zz}[(\pm 1, \pm 1, 0)/2] = D_1 \ , \quad D_{xx}[(\pm 1, \pm 1, 0)/2] = D_{yy}[(\pm 1, \pm 1, 0)/2] = (D_1 + D_2)/2 \ ,$$
$$D_{yy}[(\pm 1, 0, \pm 1)/2] = D_1 \ , \quad D_{xx}[(\pm 1, 0, \pm 1)/2] = D_{zz}[(\pm 1, 0, \pm 1)/2] = (D_1 + D_2)/2 \ ,$$
$$D_{xx}[(0, \pm 1, \pm 1)/2] = D_1 \ , \quad D_{yy}[(0, \pm 1, \pm 1)/2] = D_{zz}[(0, \pm 1, \pm 1)/2] = (D_1 + D_2)/2 \ ,$$
$$D_{xy}[\pm(1, 1, 0)/2] = -D_{xy}[\pm(1, -1, 0)/2] = (D_2 - D_1)/2 \ ,$$
$$D_{yz}[\pm(0, 1, 1)/2] = -D_{yz}[\pm(0, 1, -1)/2] = (D_2 - D_1)/2 \ ,$$
$$D_{xz}[\pm(1, 0, 1)/2] = -D_{xz}[\pm(1, 0, -1)/2] = (D_2 - D_1)/2 \ .$$

All other non diagonal spring constants vanish. (The lattice constants are omitted on the left hand-sides for brevity.) Equation (5.25) then yields

$$\begin{aligned}
\widetilde{K}_{xx} &= (D_1 + D_2)\{4 - \cos[(k_x + k_y)a/2] - \cos[(k_x - k_y)a/2] - \cos[(k_x + k_z)a/2] \\
&\quad - \cos[(k_x - k_z)/2]\} + 2D_1\{2 - \cos[(k_y + k_z)a/2] - \cos[(k_y - k_z)a/2]\} \ , \\
\widetilde{K}_{yy} &= (D_1 + D_2)\{4 - \cos[(k_x + k_y)a/2] - \cos[(k_x - k_y)a/2] - \cos[(k_y + k_z)a/2] \\
&\quad - \cos[(k_y - k_z)/2]\} + 2D_1\{2 - \cos[(k_x + k_z)a/2] - \cos[(k_x - k_z)a/2]\} \ , \\
\widetilde{K}_{zz} &= (D_1 + D_2)\{4 - \cos[(k_x + k_z)a/2] - \cos[(k_x - k_z)a/2] - \cos[(k_y + k_z)a/2] \\
&\quad - \cos[(k_y - k_z)/2]\} + 2D_1\{2 - \cos[(k_y + k_x)a/2] - \cos[(k_y - k_x)a/2]\} \ , \\
\widetilde{K}_{xy} &= (D_2 - D_1)\{-\cos[(k_x + k_y)a/2] + \cos[(k_x - k_y)a/2]\} \ , \\
\widetilde{K}_{xz} &= (D_2 - D_1)\{-\cos[(k_x + k_z)a/2] + \cos[(k_x - k_z)a/2]\} \ , \\
\widetilde{K}_{yz} &= (D_2 - D_1)\{-\cos[(k_z + k_y)a/2] + \cos[(k_z - k_y)a/2]\} \ .
\end{aligned}$$

The values of $M\omega^2$ for the eigenfrequencies are derived from a diagonalization of this matrix.

b. For $\mathbf{k} = (k, 0, 0)$ the matrix is diagonal,

$$\widetilde{K}_{xx} = 8(D_1 + D_2)\sin^2(ka/4) \ , \quad \widetilde{K}_{yy} = \widetilde{K}_{zz} = 4(3D_1 + D_2)\sin^2(ka/4) \ .$$

As the wave propagates along the $\hat{\mathbf{x}}$–direction, the longitudinal motion has the frequency $\omega_L^2 = \widetilde{K}_{xx}(k00)/M$, while the frequency of the two transverse modes is

$\omega_T^2 = \widetilde{K}_{yy}(k00)/M$. The path $\Gamma \to X$ in the Brillouin zone is given by $0 < k < 2\pi/a$. The matrix is again diagonal for $\mathbf{k} = (2\pi/a, k, 0)$,

$$\widetilde{K}_{xx} = 4(D_1 + D_2)[2 - \sin^2(ka/4)] + 8D_1 \sin^2(ka/4) ,$$
$$\widetilde{K}_{yy} = 4(3D_1 + D_2) ,$$
$$\widetilde{K}_{zz} = 4(D_1 + D_2)[1 + \sin^2(ka/4)] + 8D_1 \cos^2(ka/4) .$$

Hence, there are three different eigenfrequencies, $\omega_\mu^2 = \widetilde{K}_{\mu\mu}/M$. The path $X \to W \to X$ is given by $0 < k < 2\pi/a$. The end point of this path, $(2\pi/a, 2\pi/a, 0)$ is marked in Fig. 5.18 by X, much as the starting point, $(2\pi/a, 0, 0)$. Indeed, when the point $(2\pi/a, 2\pi/a, 0)$ is displaced by the reciprocal-lattice vector $(2\pi/a, 2\pi/a, 2\pi/a)$, it becomes the point $(0, 0, -2\pi/a)$, which is equivalent to the point $(2\pi/a, 0, 0)$, but along the $\hat{\mathbf{z}}$–direction. Recall that around the origin of the first Brillouin zone there are six points of the type X, along all axes. In fact, Fig. 5.18 and Fig. 5.40 demonstrate that the dispersion relation on the path $X \to W \to X$ is symmetric under a reflection through the point W, since such an operation transforms the Brillouin zone back to itself. For $\mathbf{k} = (k, k, 0)$ one finds

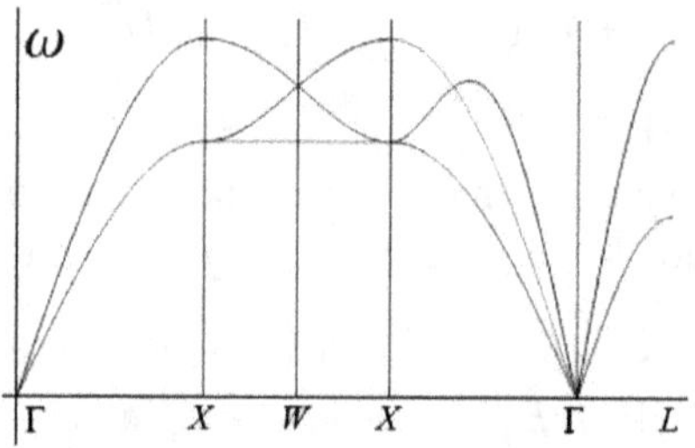

Fig. 5.40

$$\widetilde{K}_{xy} = 2(D_2 - D_1)\sin^2(ka/2) , \quad \widetilde{K}_{xz} = \widetilde{K}_{yz} = 0 ,$$
$$\widetilde{K}_{xx} = \widetilde{K}_{yy} = 4(3D_1 + D_2)\sin^2(ka/4) + 2(D_1 + D_2)\sin^2(ka/2) ,$$
$$\widetilde{K}_{zz} = 8(D_1 + D_2)\sin^2(ka/4) + 4D_1\sin^2(ka/2) ,$$

and therefore the frequencies are $M\omega_\pm^2 = \widetilde{K}_{xx} \pm \widetilde{K}_{xy}$ and $M\omega_{T2}^2 = \widetilde{K}_{zz}$. The corresponding eigenvectors are $(1, \pm 1, 0)$ and $(0, 0, 1)$, which can be identified as belonging to a longitudinal vibration (parallel to the wave vector, with ω_+), and two transverse modes, with $\omega_{T1} = \omega_-$ and ω_{T2}. The path $X \to \Gamma$ pertains to $0 < k < 2\pi/a$. Finally, for $\mathbf{k} = (k, k, k)$,

$$\widetilde{K}_{xx} = \widetilde{K}_{yy} = \widetilde{K}_{zz} = 4(2D_1 + D_2)\sin^2(ka/2) ,$$
$$\widetilde{K}_{xy} = \widetilde{K}_{yz} = \widetilde{K}_{xz} = 2(D_2 - D_1)\sin^2(ka/2) ,$$

which constitutes a 3×3 matrix,

$$\begin{bmatrix} \widetilde{K}_{xx} & \widetilde{K}_{xy} & \widetilde{K}_{xy} \\ \widetilde{K}_{xy} & \widetilde{K}_{xx} & \widetilde{K}_{xy} \\ \widetilde{K}_{xy} & \widetilde{K}_{xy} & \widetilde{K}_{xx} \end{bmatrix} , \text{ whose eigenvectors are } \begin{bmatrix} 1 \\ 1 \\ 1 \end{bmatrix} , \begin{bmatrix} 1 \\ -1 \\ 0 \end{bmatrix} , \begin{bmatrix} 1 \\ 1 \\ -2 \end{bmatrix} ,$$

(the eigenvectors are not normalized). The corresponding eigenvalues are $\widetilde{K}_{xx} + 2\widetilde{K}_{xy}$ (for the left vector) and $\widetilde{K}_{xx} - \widetilde{K}_{yy}$ (for the other two). The first is the longitudinal frequency, as its eigenvector is parallel to the wave vector; the other two are transverse, as their vectors are perpendicular to the first one. In the latter case, the path $\Gamma \to L$ corresponds to $0 < k < \pi/a$.

c. At the point X, the expressions found in part (b) yield $\omega_L/\omega_T = \sqrt{2(D_1 + D_2)/(3D_1 + D_2)}$. This ratio, as judged from Fig. 5.18, is approximately $1.4 \approx \sqrt{2}$; it implies that $D_2 \gg D_1$. Figure 5.40 displays the results of part (b), for $D_2 = 100D_1$. The plots are quite similar to those portrayed in Fig. 5.18. This indicates that accounting only for nearest-neighbor interactions, with forces which are predominantly longitudinal, is quite a reliable approximation.

d. The left part of Fig. 5.18 shows that the slopes are $c_L \approx 4.4 \times 10^5$ cm/sec, and $c_T \approx 2.8 \times 10^5$ cm/sec (note that the figure refers to frequencies and not to angular frequencies). It follows that $\bar{c} \approx 2.9 \times 10^5$ cm/sec. Equation (5.71) then yields $\omega_D = (24\pi^2/a^3)^{1/3}\bar{c} \approx 5 \times 10^{13}$ sec^{-1}, where $V = a^3/4$. Consequently, $\Theta_D = \hbar\omega_D/k_B \approx 360$ K.

e. The acoustic branches of table salt are qualitatively similar to those calculated above, as they pertain to the joint vibrations of the two ions in the unit cell. Because the unit cell comprises two atoms, there are also three optical branches with higher frequencies, which describe motions within the unit cell. For the acoustic branches, and for specific directions in the Brillouin zone, the two transverse modes are degenerate. The degeneracies are implied by the symmetries with respect to rotations around the longitudinal direction (and do not result from the approximations). One therefore expects two optical frequencies for the paths $\Gamma \to X$ (due to the symmetry with respect to rotations around the edge of the cube) and $\Gamma \to L$ (due to symmetries with respect to rotations around the cube's diagonal), and three optical frequencies along other directions.

Answer s.5.7.

a. The van Hove singularities appear when the group velocity vanishes; this happens when the frequency is maximal, minimal, or at a saddle point, as a function of the wave numbers. Indeed, a vanishing slope of the frequency can be detected in Fig. 5.18 at points X (two frequencies), W (a single frequency), and L (two frequencies), and also at a point in-between X and Γ. The values of these frequencies agree with those at which the derivatives in Fig. 5.19 are discontinuous.

b. The equal-frequency surfaces are spheres centered around the point $\mathbf{k}_0$ in the Brillouin zone, $\omega = \omega_0 - \omega_1|\mathbf{k} - \mathbf{k}_0|^2$. The group velocity is $\boldsymbol{\nabla}_{\mathbf{k}}\omega = -2\omega_1(\mathbf{k} - \mathbf{k}_0)$.

For a sphere of radius $\mathbf{q} = |\mathbf{k} - \mathbf{k}_0| = [(\omega_0 - \omega)/\omega_1]^{1/2}$, Eq. (5.65) yields

$$g(\omega) = \frac{NV}{(2\pi)^3} \oint_S \frac{dS}{|\nabla_{\mathbf{k}}\omega|} = \frac{NV}{(2\pi)^3} \frac{4\pi q^2}{2\omega_1 q} = \frac{NV}{4\pi^2 \omega_1^{3/2}} (\omega_0 - \omega)^{1/2} \ , \quad \omega < \omega_0 \ ,$$

$$g(\omega) = 0 \ , \quad \text{else} \ .$$

The density of states of this branch tends continuously to zero as ω approaches ω_0 from below, but the derivative diverges.

c. At the vicinity of point K in the Brillouin zone of graphene $\omega = \omega_0 \pm \omega_1|\mathbf{q}|$, where $\mathbf{q} = \mathbf{k} - \mathbf{k}_K$. The group velocity is $\nabla_{\mathbf{k}}\omega = \pm\omega_1\mathbf{q}/q$, and therefore $|\nabla_{\mathbf{k}}\omega| = \omega_1$. For this two-dimensional case, Eq. (5.65) gives

$$g(\omega) = \frac{2NV}{(2\pi)^2} \oint_S \frac{dS}{|\nabla_{\mathbf{k}}\omega|} = \frac{2NV}{(2\pi)^2} \frac{2\pi q}{\omega_1} = \frac{NV}{\omega_1^2}|\omega_0 - \omega| \ ,$$

where the factor of 2 reflects the existence of two points of the type K in each Brillouin zone. The density of states vanishes at $\omega = \omega_0$, and increases linearly as the frequency moves away from it along the two directions. The derivative flips its sign at the singular point.

Answer s.5.8.

a. Using Eq. (5.17), one finds that $\omega_E = \sqrt{2D/\mu}$. Equation (5.16) yields that the sound velocity is $c = a\sqrt{D/[2(M+m)]}$. The density of states in the Debye approximation is

$$g(\omega) = L/(\pi c) = Na/(\pi c) \ .$$

Hence,

$$N = \int_0^{\omega_D} g(\omega)d\omega = Na\omega_D/(\pi c) \ ,$$

i.e., $\omega_D = \pi c/a = \pi\sqrt{D/[2(M+m)]}$. It follows that

$$\omega_D/\omega_E = (\pi/2)\sqrt{m/M}/(1 + m/M) \ .$$

The ratio of the frequencies depends only on the masses' ratio. It increases from 0 to $\pi/4$ when this ratio varies between 0 and 1.

b. Figure 5.41 illustrates the approximate density of states: a constant density between 0 and ω_D, and a delta function at ω_E.

Answer s.5.9.

The wave length corresponding to the Debye frequency is obtained from the relation $\omega_D = \bar{c}k_D = 2\pi\bar{c}/\lambda_D$. Equation (5.71) implies that $V = 6\pi^2(\bar{c}/\omega_D)^3 = 6\pi^2[\lambda_D/(2\pi)]^3$, and therefore $\lambda_D = (4\pi V/3)^{1/3}$. The volume of the unit cell in an **FCC** lattice with a lattice constant a is $V = a^3/4$, and the distance between nearest-neighbor atoms is $d = a/\sqrt{2}$. Consequently, $(\lambda_D/d)|_{\mathbf{FCC}} = (2\pi\sqrt{2}/3)^{1/3} \approx 1.44$. The volume of the unit cell of a **BCC** lattice is $V = a^3/2$, and the distance between

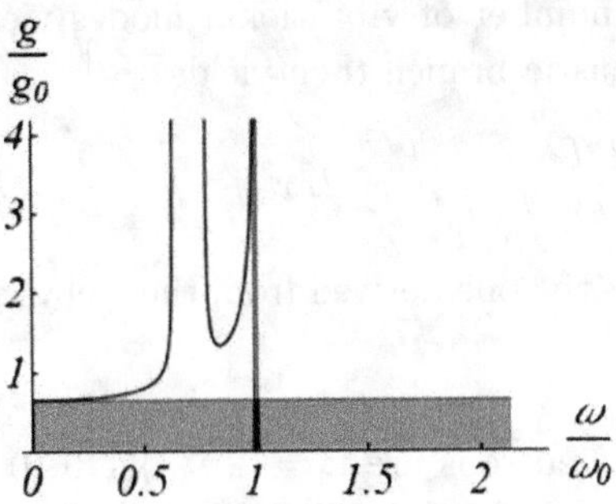

Fig. 5.41

nearest-neighbor atoms is $d = a\sqrt{3}/2$. Thus, $(\lambda_D/d)|_{BCC} = [16\pi/(9\sqrt{3})]^{1/3} \approx 1.48$. In both cases the wave length is longer than the nearest-neighbor distance, and therefore the description of the vibrations in terms of waves is plausible. Such a description is meaningless once the wave length is shorter than the distance between nearest-neighbor atoms.

Answer s.5.10.

The temperatures at which the specific heat is measured are much lower than the Debye temperature, and therefore Eq. (5.69), with the integral there replaced by I_3, suffices for the calculation of the specific heat. These two crystals possess two atoms in the unit cell, and consequently they have also optical branches. However, the specific heat contributed by the latter is exponentially small and may be ignored. Assuming that the sound velocities are the same along all directions, one finds $C_V \approx C_0(\beta\hbar\omega_D)^{-3}$, where the coefficient C_0 is the same for both crystals. Using $\hbar\omega_D = k_B\Theta_D$ yields $C_V \approx C_0(\Theta_D/T)^3$, and consequently $C_0 = C_V(\text{KCl,5K})(5/230)^3$. It follows that

$$C_V(\text{NaCl, 2K}) = C_V(\text{KCl, 5K})\left(\frac{310}{2}\frac{5}{230}\right)^3 = 1.45 \text{ J/mol/K} .$$

Answer s.5.11.

The equal-frequency lines, at frequency ω, are squares of edge $\sqrt{2}\omega/c$, and hence their circumference is $S = 4\sqrt{2}\omega/c$. The group velocity (which is normal to the equal-frequency curves) is $\mathbf{v}_g = \nabla_k\omega = c(\text{sgn}[k_x], \text{sgn}[k_y])$ and thus $|\mathbf{v}_g| = c\sqrt{2}$. Equation (5.65) yields

$$g(\omega) = \frac{Na^2}{(2\pi)^2}\frac{4\sqrt{2}\omega/c}{c\sqrt{2}} = \frac{Na^2\omega}{(2\pi)^2} .$$

As the density of states is linear in the frequency, all thermodynamic properties calculated with it are identical to those obtained in the Debye approximation at two dimensions, except for a multiplicative factor that determines the scale. A calculation of the total number of vibrational modes yields the Debye frequency, $\omega_{\mathrm{D}} = \pi c\sqrt{2}/a$. Each acoustic branch then yields

$$C_V = 2k_{\mathrm{B}}N\left(\frac{T}{\Theta_{\mathrm{D}}}\right)^2 \int_0^{\Theta_{\mathrm{D}}/T} dx\, x^3 \frac{e^x}{(e^x-1)^2} \,, \qquad k_{\mathrm{B}}\Theta_{\mathrm{D}} = \hbar\omega_{\mathrm{D}} \,.$$

This result is identical to the one derived from the Debye model in two dimensions.

Answer s.5.12.

The wave vectors of the neutrons are $\mathbf{k} = (2\pi/\lambda)(1,0,0) = (1.795(\text{Å})^{-1},0,0)$ and $\mathbf{k}' = (2\pi/\lambda')(1,1,1)/\sqrt{3} = (1.5571,1.557,1.557)(\text{Å})^{-1}$. For neutrons, $\hbar^2/(2m_n) = 1.965\times 10^{-3}$ eV $(\text{Å})^2$, which implies that the energy change of the scattered neutron is $\Omega - \Omega' = \hbar^2(k^2 - k'^2)/(2m_n) = -7.96 \times 10^{-3}$ eV. This is a negative change, i.e., the neutron emits a phonon. The momentum the neutron loses is $\mathbf{K} - \mathbf{G} = \mathbf{k} - \mathbf{k}' = (0.238, -1.557, -1.557)(\text{Å})^{-1}$. The boundaries of the first Brillouin zone along each of the axes are $|K_\alpha| < \pi/a = 0.739(\text{Å})^{-1}$. One therefore concludes that the $y-$ and $z-$components of the momentum exchange vector are outside the first Brillouin zone. Hence, the reciprocal-lattice vector $\mathbf{G} = (0, 2\pi/a, 2\pi/a)$ is used, and then $\mathbf{K} = \mathbf{k} - \mathbf{k}' = (0.238, -0.079, -0.079)(\text{Å})^{-1}$.

Answer s.5.13.

The magnetic moments in the ordered state (i.e., the ground state) are denoted $\boldsymbol{\mu}^{(1)} = \mu_1\hat{\mathbf{z}}$ and $\boldsymbol{\mu}^{(2)} = -\mu_2\hat{\mathbf{z}}$. The equations of motion given in problem 5.29 are then

$$\frac{d\mu_{nx}^{(1)}}{dt} = -\bar{J}[2\mu_2\mu_{ny}^{(1)} + \mu_1(\mu_{ny}^{(2)} + \mu_{n-1y}^{(2)})] \,,$$

$$\frac{d\mu_{ny}^{(1)}}{dt} = \bar{J}[2\mu_2\mu_{nx}^{(1)} + \mu_1(\mu_{nx}^{(2)} + \mu_{n-1x}^{(2)})] \,,$$

$$\frac{d\mu_{nx}^{(2)}}{dt} = \bar{J}[2\mu_1\mu_{ny}^{(2)} + \mu_2(\mu_{ny}^{(1)} + \mu_{n+1y}^{(1)})] \,,$$

$$\frac{d\mu_{ny}^{(2)}}{dt} = -\bar{J}[2\mu_1\mu_{nx}^{(2)} + \mu_2(\mu_{nx}^{(1)} + \mu_{n+1x}^{(1)})] \,.$$

Introducing the variables $\mu_{n\pm}^{(1,2)} = \mu_{nx}^{(1,2)} \pm i\mu_{ny}^{(1,2)}$ changes these equations to be

$$\frac{d\mu_{n+}^{(1)}}{dt} = i\bar{J}[2\mu_2\mu_{n+}^{(1)} + \mu_1(\mu_{n+}^{(2)} + \mu_{n-1+}^{(2)})] \,,$$

$$\frac{d\mu_{n+}^{(2)}}{dt} = -i\bar{J}[2\mu_1\mu_{n+}^{(2)} + \mu_2(\mu_{n+}^{(1)} + \mu_{n+1+}^{(1)})] \,.$$

Finally, with the wavy-solution form of spin waves, one finds

$$-\omega\mu_+^{(1)} = \bar{J}[2\mu_2\mu_+^{(1)} + \mu_1(1 + e^{-ika})\mu_+^{(2)}] \,,$$

$$\omega\mu_+^{(2)} = \bar{J}[2\mu_1\mu_+^{(2)} + \mu_2(1 + e^{ika})\mu_+^{(1)}] \,.$$

The determinant vanishes for

$$\omega = \overline{J}[(\mu_1 - \mu_2) \pm \sqrt{(\mu_1 - \mu_2)^2 + 4\mu_1\mu_2 \sin^2(ka/2)]} \ ,$$

(a is the lattice constant). The two solutions for the frequency, in units of $\mu_1\overline{J}$, are presented in Fig. 5.42 for $\mu_2 = 2\mu_1$. One of the modes is similar to that found for a ferromagnet, with a dispersion relation quadratic in the wave number (at long waves). The other mode is "optical". The sign does not matter, as it just determines the direction of the propagation. When the two moments are equal, the linear dispersion law of long waves, as for an antiferromagnet, is reproduced.

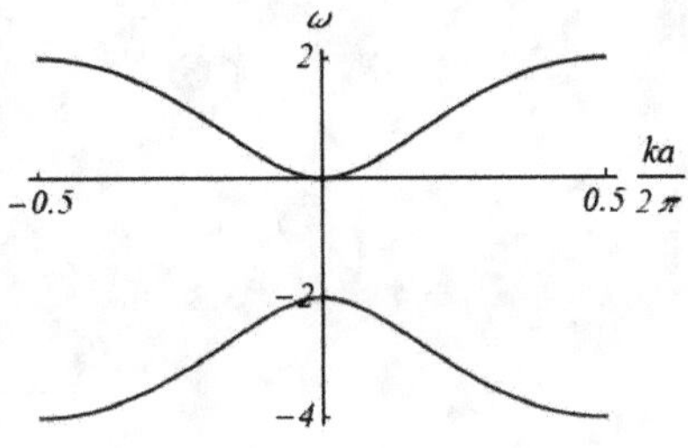

Fig. 5.42

Appendix D

The continuum limit

Equation (5.8) shows that the equations of motion approach the continuum limit when the lattice constant is significantly shorter than the wave length of the acoustic wave. The theory of elasticity yields these equations of motion, irrespective of the lattice structure of the material. Here are listed certain basic concepts of elasticity theory, from which are derived the equations governing acoustic waves in the continuum.

Consider first the one-dimensional system, lying on the $\hat{\mathbf{x}}$–axis. Denoting the longitudinal displacement of a point x by $u_x(x)$, the **strain** is defined by $\varepsilon_{xx} = \partial u_x/\partial x$. According to Hooke's law, the **stress**, σ_{xx} (that is, the force per unit area normal to the $\hat{\mathbf{x}}$–axis) is linear in the strain, i.e., $\sigma_{xx} = Y\varepsilon_{xx} = Y(\partial u_x/\partial x)$, where Y is a material constant called the **Young modulus**. For a cylinder lying along the $\hat{\mathbf{x}}$–axis, whose height is dx and whose base area is A, Newton's second law takes the form

$$(\rho A dx)\frac{\partial^2 u_x(x,t)}{\partial t^2} = A[\sigma_{xx}(x+dx) - \sigma_{xx}(x)] = A\frac{\partial \sigma_{xx}}{\partial x}dx = YA\frac{\partial^2 u_x(x,t)}{\partial x^2}dx \ ,$$

where ρ is the mass density. The result is the wave equation,

$$\frac{\partial^2 u_x(x,t)}{\partial t^2} = \left(\frac{Y}{\rho}\right)\frac{\partial^2 u_x(x,t)}{\partial x^2} \ .$$

A monochromatic wave, $u_x \propto \exp[i(kx - \omega t)]$, of wave vector k and frequency ω, has a linear dispersion law, $\omega = ck$, with the sound velocity $c = \sqrt{Y/\rho}$.

In three dimensions, the displacement vector $\mathbf{u}$ has three components, and the **strain tensor** is defined by

$$\varepsilon_{\mu\nu} = \frac{1}{2}\left(\frac{\partial u_\mu}{\partial x_\nu} + \frac{\partial u_\nu}{\partial x_\mu}\right) \ .$$

The stress is a tensor as well, $\sigma_{\mu\nu}$ (i.e., the ν–component of the force on a unit area normal to the $\hat{\boldsymbol{\mu}}$–axis), and Hooke's law becomes $\sigma_{\mu\nu} = \sum_{\alpha,\beta} C_{\mu\nu,\alpha\beta}\varepsilon_{\alpha\beta}$. The tensor $C_{\mu\nu,\alpha\beta}$ contains the **elastic constants** (or elastic moduli) of the material. By definition, the strain tensor is symmetric, $\varepsilon_{\mu\nu} = \varepsilon_{\nu\mu}$. When no moment is exerted on the material so that its angular velocity is zero, the stress tensor is symmetric

411

as well, $\sigma_{\mu\nu} = \sigma_{\nu\mu}$. It follows that the tensor comprises six independent quantities, conventionally denoted by

$$e_1 = \varepsilon_{xx} \ , \ e_2 = \varepsilon_{yy} \ , \ e_3 = \varepsilon_{zz} \ , \ e_4 = 2\varepsilon_{yz} \ , \ e_5 = 2\varepsilon_{zx} \ , \ e_6 = 2\varepsilon_{xy} \ ,$$

(at times, these are defined without the factor 2 in the last three). Similarly, the **elastic constants' tensor**, $C_{ij} \equiv C_{\mu\nu,\alpha\beta}$ is of order 6×6 and contains 21 independent quantities (check!). The free energy connected with the elasticity of the material is given by

$$F = \frac{1}{2} \int d^3r \sum_{i,j=1}^{6} e_i C_{ij} e_j \ . \tag{D.1}$$

The elastic tensor determines the coefficients which relate the response of the material to various forces. All 21 components of the triclinic lattice differ from each other. The number of non-zero coefficients of more symmetric crystals is smaller. For instance, in cubic symmetry there are four rotation axes of order 3 (around the cube's diagonals). Rotations by $120°$ around these axes give rise to the transformations $x \to y \to z \to x$, $x \to z \to -y \to x$, $-x \to z \to -y \to -x$, $-x \to y \to z \to -x$. Because of these mappings, the free energy (D.1) is invariant under the transformations $\varepsilon_{xy} \to \varepsilon_{yz} \to \varepsilon_{z(-x)} \to \varepsilon_{x(-y)} = -\varepsilon_{xy}$. As a result, the free energy contains only three independent coefficients,

$$C_{12} = C_{23} = C_{31}, \ C_{44} = C_{55} = C_{66} \ , \ C_{11} = C_{22} = C_{33} \ , \tag{D.2}$$

and all other coefficients vanish, e.g., $C_{45} = C_{56} = C_{64} = 0$ and $C_{14} = C_{15} = C_{16} = 0$. By similar considerations one finds that the monoclinic lattice has 13 elastic constants, the orthorhombic one has 9, 6 or 7 coefficients describe the tetragonal and rhombohedral lattices, and 5 are needed for the hexagonal lattice.

The three-dimensional wave equation in cubic symmetry is obtained in a similar way. For a cube whose edges are dx, dy, and dz, along the directions of the three axes, the force acting on this cube along the $\hat{\mathbf{x}}$−direction is

$$(\rho dx dy dz)\frac{\partial^2 u_x(\mathbf{r},t)}{\partial t^2} = \left[\frac{\partial \sigma_{xx}}{\partial x} + \frac{\partial \sigma_{yx}}{\partial y} + \frac{\partial \sigma_{zx}}{\partial z}\right] dx dy dz \ .$$

In cubic symmetry, Hooke's law yields

$$\sigma_{xx} = C_{11}\varepsilon_{xx} + C_{12}(\varepsilon_{yy} + \varepsilon_{zz}) = C_{11}\frac{\partial u_x}{\partial x} + C_{12}\left(\frac{\partial u_y}{\partial y} + \frac{\partial u_z}{\partial z}\right) \ ,$$

$$\sigma_{xy} = 2C_{44}\varepsilon_{xy} = C_{44}\left(\frac{\partial u_x}{\partial y} + \frac{\partial u_y}{\partial x}\right) \ , \quad \sigma_{xz} = 2C_{44}\varepsilon_{xz} = C_{44}\left(\frac{\partial u_x}{\partial z} + \frac{\partial u_z}{\partial x}\right) \ .$$

Collecting all terms, one obtains the wave equation in the form

$$\rho\frac{\partial^2 u_x(\mathbf{r},t)}{\partial t^2} = C_{11}\frac{\partial u_x^2(\mathbf{r},t)}{\partial x^2} + C_{12}\left(\frac{\partial u_x^2(\mathbf{r},t)}{\partial y \partial x} + \frac{\partial u_x^2(\mathbf{r},t)}{\partial x \partial z}\right)$$

$$+ C_{44}\left(\frac{\partial u_x^2(\mathbf{r},t)}{\partial y^2} + \frac{\partial u_y^2(\mathbf{r},t)}{\partial y \partial x} + \frac{\partial u_x^2(\mathbf{r},t)}{\partial z^2} + \frac{\partial u_z^2(\mathbf{r},t)}{\partial z \partial x}\right) \ ,$$

and two more equations for the forces along the $\hat{\mathbf{y}}$ and $\hat{\mathbf{z}}$ directions. Inserting the plane-wave function, $\mathbf{u} = \mathbf{A}\exp[i(\mathbf{k}\cdot\mathbf{r} - \omega t)]$, results in three coupled linear equations,

$$\rho\omega^2 A_x = [C_{11}k_x^2 + C_{44}(k_y^2 + k_z^2)]A_x + (C_{12} + C_{44})(k_x k_y A_y + k_z k_x A_z) \,,$$
$$\rho\omega^2 A_y = [C_{11}k_y^2 + C_{44}(k_x^2 + k_z^2)]A_y + (C_{12} + C_{44})(k_x k_y A_x + k_z k_y A_z) \,,$$
$$\rho\omega^2 A_z = [C_{11}k_z^2 + C_{44}(k_y^2 + k_x^2)]A_z + (C_{12} + C_{44})(k_x k_z A_x + k_y k_x A_y) \,.$$

For instance, for a wave along the $\hat{\mathbf{x}}$–direction, there is a longitudinal wave with the dispersion $\omega_L = \sqrt{C_{11}/\rho}\,k_x$, and two transverse waves with the dispersion law $\omega_T = \sqrt{C_{44}/\rho}\,k_x$. In the more general case $\rho\omega^2 A_\mu = \sum_{\sigma,\nu,\tau} C_{\mu\sigma,\nu\tau}k_\sigma k_\nu A_\tau$.

Problem D.1.

a. *Derive the dispersion relations for acoustic waves in cubic symmetry, for a plane wave along an arbitrary direction in the XY–plane.*
b. *Repeat part (a) for a plane wave along (111).*

D.1 Answer

Answer D.1.

a. When $k_z = 0$, the wave equations are

$$\rho\omega^2 A_x = [C_{11}k_x^2 + C_{44}k_y]A_x + (C_{12} + C_{44})k_x k_y A_y \,,$$
$$\rho\omega^2 A_y = [C_{11}k_y^2 + C_{44}k_x]A_y + (C_{12} + C_{44})k_x k_y A_x \,,$$
$$\rho\omega^2 A_z = C_{44}[k_x^2 + k_y^2]A_z \,.$$

Hence, there are transverse vibrations normal to the plane, whose dispersion is $\omega_{T2} = \sqrt{C_{44}/\rho}\,|\mathbf{k}|$. To obtain the in-plane vibrations one has to solve the first two equations; the requirement that the determinant vanishes gives

$$[C_{11}k_x^2 + C_{44}k_y^2 - \rho\omega^2][C_{11}k_y^2 + C_{44}k_x^2 - \rho\omega^2] - (C_{12} + C_{44})^2 k_x^2 k_y^2 = 0 \,.$$

It follows that

$$\rho\omega^2 = \frac{1}{2}(C_{11} + C_{44})\mathbf{k}^2 \pm \sqrt{\frac{1}{4}(C_{11} - C_{44})^2\mathbf{k}^4 + [C_{12}^2 - C_{11}^2 + 2C_{44}(C_{11} + C_{12})]k_x^2 k_y^2} \,.$$

Generally, there is no special relation between the eigenvectors and the direction of $\mathbf{k}$. Such a relation pertains to symmetry directions, like (100) or (110). For a wave along the latter direction, $\rho\omega_L^2 = (C_{11} + C_{12} + 2C_{44})\mathbf{k}^2/2$, with a longitudinal vibration along (110), or $\rho\omega_{T1}^2 = (C_{11} - C_{12})\mathbf{k}^2/2$, with a transverse motion normal to that direction.

b. For a wave along (111), i.e., $k_x = k_y = k_z = |\mathbf{k}|/\sqrt{3}$, the wave equations are

$$3\rho\omega^2 A_x = [(C_{11} + 2C_{44})A_x + (C_{12} + C_{44})(A_y + A_z)]k^2 \,,$$
$$3\rho\omega^2 A_y = [(C_{11} + 2C_{44})A_y + (C_{12} + C_{44})(A_x + A_z)]k^2 \,,$$
$$3\rho\omega^2 A_z = [(C_{11} + 2C_{44})A_z + (C_{12} + C_{44})(A_y + A_x)]k^2 \,.$$

The eigenfrequencies are $\rho\omega_L^2 = (C_{11} + 2C_{12} + 4C_{44})\mathbf{k}^2/3$, corresponding to a longitudinal motion along (111), and two transverse modes with $\rho\omega_T^2 = (C_{11} - C_{12} + C_{44})\mathbf{k}^2/3$. Problem 5.9 compares these equations with those of the acoustic branch of the square lattice.

Chapter 6

Electrons in solids

This chapter discusses the thermal and electric properties of electrons in solids. Certain phenomena can be explained in terms of the classical theory of Drude, while others are analyzed within the framework of the quantum theory of free electrons, as formulated by Sommerfeld. The main part of the chapter focuses on phenomena that result from, or that are related to, the periodic potential of the lattice of ions in which the electrons move. This potential gives rise to bands of energies accessible for the electrons, that in turn determine the electronic properties of solids, and the classification of those into metals, insulators and semiconductors. The fascinating quantum behavior of electrons subjected to magnetic fields, e.g., the quantum Hall effect, is discussed towards the end of the chapter.

6.1 Preface

Metals are characterized by their high **electrical conductance**. The electric current is carried by the electrons, that move relatively freely through the solid. Each atom in the metal "releases" its **valence electrons** (usually one or two electrons per atom), generally from the outer "shells" of the free-atom electronic states. These electrons are termed **"conduction electrons"**. In the classical picture of Drude (called below the "Drude theory") these electrons are treated as particles whose motion is governed by Newton's laws, until they collide with a certain defect. (In the original formulation of the theory it was assumed that the collisions are with the ions forming the metal, but in fact other types of collisions dominate the conductance.) The collisions change the direction of the electrons' propagation, and may slow down the flow of electrons along the electric field. This is the physical origin of the electrical resistance (the inverse of the conductance), R, and the explanation of **Ohm's law**, which relates the electrical current, I, to the voltage difference across the sample, V: $I = V/R$.

In the Drude theory the electrons are treated as classical particles, and their quantum nature is ignored. As found in Sec. 6.2, even this naïve picture captures many of the phenomena observed in experiments. However, there exist phenomena

415

whose origin is in the quantum aspect of the electrons and in the Pauli principle. These are included in the Sommerfeld theory, described in Sec. 6.3. Both the Sommerfeld and the Drude theories ignore the periodic potential of the crystal in which the electrons reside, as well as the Coulomb interactions among them. Though there are properties that cannot be explained without invoking the periodic potential, surprisingly enough, many other phenomena can be described in terms of free electrons.

The central theorem that dictates the form of the solutions of the Schrödinger equation with a periodic potential is the **Bloch theorem**. It is presented and proven in Sec. 6.4. The next three sections contain examples of calculations of the wave functions: the exact solution of the **Kronig and Penney potential** in Sec. 6.5, the **nearly-free electron approximation** in Sec. 6.6, and the **tight-bindingapproximation** in Sec. 6.7. These calculations reveal the existence of **energy bands**, (comprising energies which the electrons can assume) with **energy gaps** in between them, of forbidden energies. The mere existence of such bands explains the disparities between insulators, conductors, and semiconductors, as explained in Sec. 6.8. The electronic **density of states** is analyzed in Sec. 6.9, together with the occupations of the various bands and the specific heat, and Sec. 6.10 discusses the equations of motion of an electron moving in a periodic potential; these enable, e.g., the calculation of the charge conductance. Section 6.11 explains the motion of an electron in a magnetic field and the quantum Hall effect. Section 6.12 surveys the Anderson and the Mott insulators.

6.2　The Drude theory

As mentioned, the Drude model (that preceded the appearance of quantum mechanics) disregards the Coulomb interactions among the electrons. Rather, the electrons are viewed as a gas of **free particles**. The electron gas is described by the **kinetic theory of gases**, the same as the one exploited for gases of freely-moving molecules.

Time in-between collisions (the relaxation time) and the electrons' velocity. The kinetic theory presumes that the probability per unit time for the electron to collide with a defect is given by $1/\tau$. Problem 6.1 demonstrates that τ is in fact the **mean time in-between collisions**. It is also shown below that the average velocity of the electron decays during a time interval τ to a stationary value (that is independent of time), rendering τ to be also the **"relaxation time"**. Both terms are used below. Drude conjectured that the electrons collide with the ions situated at the lattice sites. As the ions are far heavier than the electrons, he assumed that the collisions are instantaneous and elastic, such that the direction of the electron's velocity is changed randomly, but not its magnitude. In reality the collisions that lead to relaxation are not with the ions in the lattice. Nonetheless, the existence of defects that scatter elastically the electrons suffices for the derivation of Ohm's law (without specifying the nature of those defaults).

Problem 6.1.

Assuming that the probability per unit time for a collision between the electron and a certain defect is $1/\tau$, prove that:
a. *the probability that the electron will not collide during a time interval t is $P(t) = \exp[-t/\tau]$;*
b. *the mean time in-between collisions is τ.*

In-between collisions, the motion of the electron (whose mass is m and whose velocity is $\mathbf{v}$) is determined by Newton's second law, $\dot{\mathbf{p}} = m\dot{\mathbf{v}} = m\mathbf{a} = \mathbf{F}$, where $\mathbf{p} = m\mathbf{v}$ is the momentum of the electron and $\mathbf{F}$ is the force that acts on it. In case the electron at time t is subjected to an electric field $\boldsymbol{\mathcal{E}}(t)$, then $\mathbf{F} = -e\boldsymbol{\mathcal{E}}$, where $-e$ is the electron's charge. It follows that during a very short time interval dt $\mathbf{p}(t + dt) = \mathbf{p}(t) - e\boldsymbol{\mathcal{E}}(t)dt$. As the collisions the electron performs are random, and since eventually the result is summed over many electrons, it follows that the momentum is averaged over all events that can occur. The probability for the electron to undergo a collision during the time interval dt and to lose the velocity along the direction of propagation is dt/τ. The electron will then attain a new random velocity $\mathbf{v}'$, or a new random momentum $\mathbf{p}' = m\mathbf{v}'$. On the other hand, it may not collide at all and then its momentum is not changed. The probability for the latter scenario is $1 - dt/\tau$. The average over all possible events during dt yields the new momentum, $\langle\mathbf{p}(t + dt)\rangle = (1 - dt/\tau)[\langle\mathbf{p}(t)\rangle - e\boldsymbol{\mathcal{E}}dt] + \langle\mathbf{p}'\rangle dt/\tau$, and thus

$$\frac{d}{dt}\langle\mathbf{p}(t)\rangle = -\frac{\langle\mathbf{p}(t)\rangle}{\tau} - e\boldsymbol{\mathcal{E}}(t) . \tag{6.1}$$

(It is assumed that the random velocity vanishes upon averaging, $\langle\mathbf{v}'\rangle = 0$.) Equation (6.1) resembles the equation of motion of a particle moving in a viscous fluid, with $1/\tau$ playing the role of the friction coefficient (in appropriate units). The solution of Eq. (6.1), for an electric field that does not vary in time, is

$$\langle\mathbf{p}(t)\rangle = -e\boldsymbol{\mathcal{E}}\tau + \mathbf{A}e^{-t/\tau} , \tag{6.2}$$

where $\mathbf{A} = \langle\mathbf{p}(0)\rangle + e\boldsymbol{\mathcal{E}}\tau$ is determined by the initial conditions. At times longer than the relaxation time, $t \gg \tau$, the second term on the right hand-side decays and becomes exponentially small; the solution is then stationary (independent of time),

$$\langle\mathbf{v}(t)\rangle = \langle\mathbf{p}(t)\rangle/m \Rightarrow -e\boldsymbol{\mathcal{E}}\tau/m = -\mu\boldsymbol{\mathcal{E}} , \quad \mu = e\tau/m , \tag{6.3}$$

where μ is the electronic **mobility**. The relaxation time is in fact the time interval during which there still remains a certain memory of the initial conditions. Once time longer than the relaxation has elapsed, $t \gg \tau$, the average velocity is stationary, with its magnitude determined by the electric field.

Another point of view on Eq. (6.3) is the following. In-between collisions the electron moves under the effect of the electric field with a constant acceleration, and therefore its velocity is $\mathbf{v}(t) = \mathbf{v}' - e\boldsymbol{\mathcal{E}}t/m$, where $\mathbf{v}' \equiv \mathbf{v}(0)$ is the initial velocity (i.e., the one it had immediately after the previous collision). This expression is valid on the average, at times shorter than τ; after a time τ has elapsed, the electron collides again and attains a new velocity. Upon averaging over all possible collisions, the average $\langle \mathbf{v} \rangle$ vanishes, as after each scattering event the velocity is directed at random. It follows that since the typical time between collisions is τ, the average velocity is $\langle \mathbf{v}(t) \rangle = \langle \mathbf{v}' \rangle - e\boldsymbol{\mathcal{E}}\langle t \rangle/m = -e\boldsymbol{\mathcal{E}}\tau/m$, as in Eq. (6.3).

Problem 6.2.

So far, it is assumed that after each collision the electron loses its original velocity, and attains a new one, $\mathbf{v}'$, at a completely random direction. Find the velocity of the electron in a stationary state when $\mathbf{v}'$ is not randomly oriented, but is changed at most within an angle β.

Ohm's law. Consider a gas of electrons whose density (the number of electrons per unit volume) is n, and where each electron is moving with the average velocity $\langle \mathbf{v} \rangle$. The number of electrons that pass through an area A normal to this average velocity during a unit time is $nA|\langle \mathbf{v} \rangle|$ (this is the number of electrons located within a volume whose base area is A and whose height is the velocity). The **charge current-density**, $\mathbf{j}$ (a vector), is defined as the amount of charge that is transported through a unit area orthogonal to $\mathbf{j}$, per unit time, i.e.,

$$\mathbf{j} = -ne\langle \mathbf{v} \rangle = -\frac{ne}{m}\langle \mathbf{p} \rangle \ . \tag{6.4}$$

Using here Eq. (6.3) yields

$$\mathbf{j} = ne^2\tau\boldsymbol{\mathcal{E}}/m = \sigma\boldsymbol{\mathcal{E}} = \frac{1}{\rho}\boldsymbol{\mathcal{E}} \ , \tag{6.5}$$

where

$$\sigma = 1/\rho = ne^2\tau/m = ne\mu \tag{6.6}$$

is the **specific conductivity** of the material, while ρ is the corresponding **specific resistivity**. Once a voltage drop of magnitude V is applied across a sample whose length, along the direction of the current, is L and the cross section normal to it is A, then the electric field is $\mathcal{E} = V/L$. Denoting the total current by I, the current per unit area is $j = I/A$. One then finds **Ohm's law**, $V = IR$, with the resistance

$$R = L\rho/A = L/(\sigma A) \ . \tag{6.7}$$

Equation (6.6) relates the macroscopic specific conductivity of the material to the microscopic processes of electrons' scattering in the sample. Equation (6.5) represents the **local Ohms's law**.

Characteristic values of the specific resistivity of metals are of the of the order of micro-Ohm cm, $\rho \approx 1 \ \mu\Omega$ cm. Typical electrons' densities are those of the ions' densities, i.e., $n \approx 10^{23}$ cm$^{-3} = 10^{29}$ m^{-3}. Using $m = 9.1 \times 10^{-31}$ Kg and $e = 1.6 \times 10^{-19}$ Coul leads to $\tau \approx 10^{-14}$ sec. Drude assumed that the average velocity of a particle in a gas can be estimated from the equipartition law,

$m\mathbf{v}^2/2 = 3k_{\mathrm{B}}T/2$. At room temperature, $T = 300K$, this relation yields $|\mathbf{v}| \approx 10^7$ cm/sec. It follows that the **mean free-path** of an electron, i.e., the mean distance it traverses in-between successive collisions, is $\ell = |\mathbf{v}|\tau \approx 10\mathring{A}$. This is a rather short distance, which fits Drude's conjecture that the electrons collide with the ions, and consequently the length of their mean free-path is of the order of the lattice constant. However, (see Sec. 6.3), the estimate of the velocity magnitude given above is not correct, and in fact the mean free-path is considerably longer. Moreover, using the velocity derived from the equipartition law implies that the mean free-path is proportional to $\sqrt{T}$, whereas the lattice constant is almost independent of the temperature. Nonetheless, the wrong estimate of the velocity is the only fault in the argument above; the assumption that there exists a relaxation time that determines the electrical resistance is benign. Once its correct value is used in Eq. (6.6), the result is a faithful description of the resistances of many metals. This topic is resumed below; for the time-being, the Drude picture is used to address three more phenomena.

The Hall effect. Adding a constant magnetic field (that does not vary neither in time nor in space) leads to the **Hall effect**, discovered by Hall in 1879. Such a field, when directed normal to the electric field, gives rise to a voltage drop normal to its direction and to that of the electric current. The measurement of this voltage enables the determination of the sign of the charge carriers, and also their density. The experimental setup is displayed in Fig. 6.1, which illustrates the situation where two fields affect the electron's motion: an electric field $\boldsymbol{\mathcal{E}}$, and a magnetic one, $\mathbf{B}$, with $\mathbf{B} \perp \boldsymbol{\mathcal{E}}$; both fields are constant in time and in space. The classical equation of motion, obeyed by the electron in-between collisions, is dominated by the Lorentz force, $\dot{\mathbf{p}} = -e(\boldsymbol{\mathcal{E}} + \mathbf{v} \times \mathbf{B}/c)$ where $\mathbf{p} = m\mathbf{v}$ and c is the speed of light (this is the Lorentz force in CGS units; in the SI ones the speed of light does not appear). The constant magnetic field does not affect the velocity component parallel to it. When one adds the probability for collisions, and averages over all possible scattering events that occur within a unit time, one obtains a generalization of Eq. (6.1),

$$\langle \dot{\mathbf{p}}(t) \rangle = -\frac{\langle \mathbf{p}(t) \rangle}{\tau} - e\left(\boldsymbol{\mathcal{E}} + \frac{\langle \mathbf{p}(t) \rangle}{mc} \times \mathbf{B} \right) . \tag{6.8}$$

When the magnetic field is along the $\hat{\mathbf{z}}$−axis and the electric field and the velocity are in the XY plane, the equations of motion for each component of the momentum are

$$\langle \dot{p}_x \rangle = -e\mathcal{E}_x - \omega_c \langle p_y \rangle - \langle p_x \rangle / \tau ,$$
$$\langle \dot{p}_y \rangle = -e\mathcal{E}_y + \omega_c \langle p_x \rangle - \langle p_y \rangle / \tau .$$

At a stationary state these equations become [see Eq. (6.2)],

$$- e\mathcal{E}_x - \omega_c \langle p_y \rangle - \langle p_x \rangle / \tau = 0 ,$$
$$- e\mathcal{E}_y + \omega_c \langle p_x \rangle - \langle p_y \rangle / \tau = 0 , \tag{6.9}$$

where $\omega_c = eB/(mc)$ is the **"cyclotron frequency"** of the electron. (This is the angular frequency of the circular motion in the plane normal to the magnetic field in the absence of an electric field, and when collisions are ignored, see problem 6.3.) Replacing the momentum by the current density from Eq. (6.4), $\mathbf{j} = -ne\langle \mathbf{v}\rangle = -ne\langle \mathbf{p}\rangle/m$, leads to

$$\mathcal{E}_y = \rho(-\omega_c\tau j_x + j_y) \,, \quad \mathcal{E}_x = \rho(\omega_c\tau j_y + j_x) \tag{6.10}$$

where the specific resistivity ρ is given in Eq. (6.6). Using the latter allows one to introduce the notation $\rho\omega_c\tau = B/(nec) = -R_{\mathrm{H}}B$, where the **Hall coefficient** R_{H} is

$$R_{\mathrm{H}} = -\frac{1}{nec} \,, \tag{6.11}$$

which brings the equations into the compact form

$$\begin{bmatrix} \mathcal{E}_x \\ \mathcal{E}_y \end{bmatrix} = \begin{bmatrix} \rho_{xx} & \rho_{xy} \\ \rho_{yx} & \rho_{yy} \end{bmatrix}\begin{bmatrix} j_x \\ j_y \end{bmatrix} = \begin{bmatrix} \rho & -R_{\mathrm{H}}B \\ -R_{\mathrm{H}}B & \rho \end{bmatrix}\begin{bmatrix} j_x \\ j_y \end{bmatrix} \,. \tag{6.12}$$

In the presence of the magnetic field the specific resistivity turns out to be a tensor, that relates the components of the current density with the components of the electric field, in the general case where these two vectors are not parallel to one another. The minus sign in Eq. (6.11) is designed such the sign of R_{H} reflects that of the charge carriers, i.e., R_{H} is negative when the charge carriers are electrons. It happens at times that the Hall coefficient is positive; the positive sign of the charge carriers needs then to be explained.

Inverting the matrix in Eq. (6.12) yields

$$\begin{bmatrix} j_x \\ j_y \end{bmatrix} = \begin{bmatrix} \sigma_{xx} & \sigma_{xy} \\ \sigma_{yx} & \sigma_{yy} \end{bmatrix}\begin{bmatrix} \mathcal{E}_x \\ \mathcal{E}_y \end{bmatrix} \,, \tag{6.13}$$

which is an extension of Eq. (6.5), with the specific conductivity replaced by a 2×2 tensor, whose components are

$$\sigma_{xx} = \sigma_{yy} = \frac{\sigma}{1 + (\tau\omega_c)^2} = \frac{\sigma}{1 + (\sigma R_{\mathrm{H}}B)^2} \,,$$

$$\sigma_{xy} = -\sigma_{yx} = \frac{\sigma\tau\omega_c}{1 + (\tau\omega_c)^2} = \frac{\sigma^2 R_{\mathrm{H}}B}{1 + (\sigma R_{\mathrm{H}}B)^2} \,. \tag{6.14}$$

Problem 6.3.
The motion of an electron in a magnetic field. Prove that in the absence of an electric field, and when collisions are ignored, the electron moves with a constant velocity along the direction of the magnetic field, and performs a circular motion of period $2\pi/\omega_c$ in the plane perpendicular to the magnetic field.

Measuring the Hall coefficient. Consider a sample of the shape of a box, as illustrated in Fig. 6.1, and assume that a current flows only along the $\hat{\mathbf{x}}$–direction (i.e., $j_y = 0$). The electric field along the current direction follows Ohm's law [Eq.

(6.5)], $\mathcal{E}_x = \rho j_x$. The current flowing along the $\hat{\mathbf{x}}$−direction creates an electric field along the $\hat{\mathbf{y}}$−axis,

$$\mathcal{E}_y = \rho_{yx} j_x = -\rho \tau \omega_c j_x = -\frac{B}{nec} j_x \equiv R_{\mathrm{H}} B j_x \ . \tag{6.15}$$

As no current flows along the $\hat{\mathbf{y}}$−direction, negative charges are accumulated on the lower face, and positive charges are collected on the upper one.

It is interesting to note that several of the coefficients specific to a certain system (e.g., the relaxation time) disappear from the Hall coefficient, Eq. (6.11). This coefficient, detected in experiment by monitoring the voltage created between the opposite faces of the sample (along the $\hat{\mathbf{y}}$−direction in the example above) depends only on the charge of the carriers ($-e$ in the present discussion) and their density n. When it is known that the charge carriers are electrons, as assumed here, the measurement yields their density. There are quite a number of materials for which this description is faithful enough. However, as mentioned, there are materials with positive Hall coefficients; these deserve a special consideration.

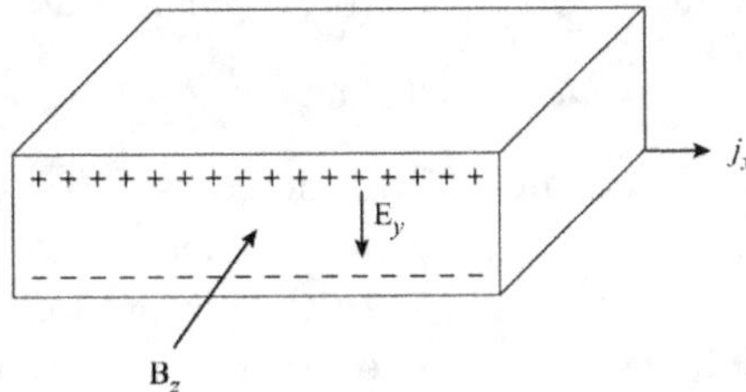

Fig. 6.1: The Hall effect geometry: the magnetic field is along the $\hat{\mathbf{z}}$−axis, the current flows along the $\hat{\mathbf{x}}$−axis, creating an electric field along the $\hat{\mathbf{y}}$−axis (in addition to the external electric field pointing along the $\hat{\mathbf{x}}$−axis).

Problem 6.4.
*Compute the Hall coefficients of lithium (a **BCC** lattice with a lattice constant $a = 3.49\mathring{A}$) and copper (an **FCC** lattice with a lattice constant $a = 3.61\mathring{A}$). Each atom donates a single electron to the total conductance.*

ac transport. When the electric field depends periodically on time, it gives rise to an alternating current (ac). To find the corresponding **ac conductivity**, it is convenient to present the field in the form $\boldsymbol{\mathcal{E}}(t) = \mathrm{Re}\{\boldsymbol{\mathcal{E}}(\omega)\exp[-i\omega t]\}$. The equation of motion (6.1) then requires the use of an analogous expression for the average momentum, $\langle \mathbf{p}(t)\rangle = \mathrm{Re}\{\mathbf{p}(\omega)\exp[-i\omega t]\}$. Hence

$$-i\omega \mathbf{p}(\omega) = -\mathbf{p}(\omega)/\tau - e\boldsymbol{\mathcal{E}}(\omega) \ .$$

The resulting expression for $\mathbf{p}(\omega)$ yields the **local Ohm's law for alternating (ac) current**,

$$\mathbf{j}(\omega) = -ne\mathbf{p}(\omega)/m = \sigma(\omega)\boldsymbol{\mathcal{E}}(\omega) \ , \tag{6.16}$$

where

$$\sigma(\omega) = \frac{\sigma}{1 - i\omega\tau} \tag{6.17}$$

is the **complex conductivity for an alternating current**.

The Maxwell equations for electromagnetic fields that oscillate with a frequency ω, $\boldsymbol{\mathcal{E}}(t) = \text{Re}\{\boldsymbol{\mathcal{E}}(\omega)\exp[-i\omega t]\}$, in the presence of an electric current, are

$$\boldsymbol{\nabla} \times \mathbf{B}(\omega) = \frac{4\pi}{c}\mathbf{j}(\omega) - \frac{i\omega}{c}\boldsymbol{\mathcal{E}}(\omega) \,, \quad \nabla^2\boldsymbol{\mathcal{E}}(\omega) = \frac{i\omega}{c}\boldsymbol{\nabla} \times \mathbf{B}(\omega) \,. \tag{6.18}$$

Combining the two and using Eq. (6.16) leads to

$$-\nabla^2\boldsymbol{\mathcal{E}}(\omega) = \frac{\omega^2}{c^2}\epsilon(\omega)\boldsymbol{\mathcal{E}}(\omega) \,, \tag{6.19}$$

with the complex **dielectric function**

$$\epsilon(\omega) = 1 + 4\pi i\sigma(\omega)/\omega \,. \tag{6.20}$$

In the high-frequency limit, $\omega\tau \gg 1$, Eq. (6.17) yields $\sigma(\omega) \approx i\sigma/(\omega\tau)$; hence

$$\epsilon(\omega) \approx 1 - \omega_p^2/\omega^2 \,, \tag{6.21}$$

where ω_p is the **plasma frequency** of the electrons, $\omega_p^2 = 4\pi ne^2/m$. In SI units, the plasma frequency reads $\omega_p^2 = ne^2/(m\epsilon_0)$, where ϵ_0 is the dielectric constant of the vacuum. At frequencies smaller than the plasma frequency (but still higher than $1/\tau$), the dielectric function $\epsilon(\omega)$ is negative, and therefore the solutions of Eq. (6.19) decay exponentially with the distance, e.g., like $\exp[-\kappa x]$, with $\kappa = \omega\sqrt{|\epsilon|}/c$ (check!). The fields decay, and waves of those frequencies cannot penetrate into the material; instead, they are reflected from the surface. This is the reason that metals reflect light and behave like mirrors. At frequencies higher than the plasma frequency the solutions are waves that propagate through the material. The metal is thus transparent to fields at such frequencies.

Problem 6.5.

A metal is subjected to a constant magnetic field, $\mathbf{B} = B\hat{\mathbf{z}}$, and to an oscillating electric field, $\boldsymbol{\mathcal{E}}(t) = \text{Re}[\boldsymbol{\mathcal{E}}_0\exp[-i\omega t]]$ oriented at an arbitrary direction. The oscillating magnetic field connected to the latter is ignored.
a. *Find the tensor of the electrical resistivity and that of the electrical conductivity.*
b. *Find the conductivity tensor in the limit of high frequencies, $\omega\tau \gg 1$. What happens when $\omega \approx \omega_c = eB/(mc)$?*

Thermal conductivity. The contribution of the electrons to the **thermal conductivity** of metals is described in the Drude picture within the kinetic theory, (as is also argued for the thermal conductivity of phonons, Sec. 5.8). By Eq. (5.101), the heat current is

$$\mathbf{j}_U = -\kappa\boldsymbol{\nabla}T \,, \tag{6.22}$$

where

$$\kappa = \frac{n\mathbf{v}^2 \tau c}{3} = \frac{\mathbf{v}^2 \tau c_V}{3} = \frac{3}{2m} n k_B^2 \tau T \tag{6.23}$$

is the **thermal conductivity** of the electrons. The last expression on the right hand-side follows from a double use of the equipartition theorem that assigns to each electron the energy $u = m\mathbf{v}^2/2 = 3k_B T/2$, and hence a specific heat, $c = 3k_B/2$ (per electron). The central expression in Eq. (6.23) exploits the specific heat per unit volume, $c_V = nc$.

Note the similarity between Eq. (6.22) and Eq. (6.5), keeping in mind that the electric field is (minus) the gradient of the electric potential, $\mathcal{E} = -\nabla\phi$. One may obtain Eq. (6.5) from arguments similar to those that lead to Eq. (6.22). In both cases the carriers are the same, carrying charge in the first case, and energy in the second. This similarity gives rise to the **Lorenz ratio**,

$$L = \kappa/(\sigma T) . \tag{6.24}$$

Indeed, using Eqs. (6.6) and (6.23), results in

$$L = \frac{3}{2}\left(\frac{k_B}{e}\right)^2 . \tag{6.25}$$

Thus, the Lorenz ratio depends solely upon the physical constants e and k_B; it is independent of any material coefficient. Equation (6.25) is the **Wiedemann-Franz law**. Its numerical value is $L = 1.11 \times 10^{-8}$ Watt Ω/K^2 (K represents Kelvin's scale). Using experimental values (for the charge conductivity and for the thermal conductivity of the electrons), one finds that indeed the Lorenz ratio is approximately the same constant for quite a number of metals. Nonetheless, that constant is about twice the one derived form the Drude theory. Improved estimates of both conductivities, as done in the next section, leads to a ratio much closer to the experimental values.

The thermoelectric (Seebeck) effect. As seen, a temperature gradient causes the electrons on the warmer side of a sample to be faster than those on the colder side. Such a gradient creates an average velocity in the opposite direction of the gradient. This nonzero average velocity gives rise to an electric current, induced by a temperature gradient (without an electric field). This is the **thermoelectric effect**; the electric current is proportional to the temperature gradient that causes it. Alternatively, when no current is flowing (e.g., because the circuit is not closed) the temperature gradient creates a voltage drop due to the accumulation of electrons, i.e. charge, on the edge of the sample. The voltage drop that forms is termed the **Seebeck effect**. This voltage cancels the electric current.

Problem 6.6.
Use the Drude model to obtain an expression for
a. *the electric current created by a temperature gradient (the thermoelectric effect), and*
b. *the electric field created when there is no current flowing (the Seebeck effect). Can the magnitude of this field be estimated from the equipartition law for the specific heat of the electrons?*

6.3   The Sommerfeld theory

Flaws in the Drude theory. The Drude theory is based on the classical picture. For instance, one of the assumptions of this theory is that the specific heat contributed by each electron is given by the equipartition theorem, $c = 3k_B/2$, and therefore the total contribution of the electrons to the specific heat is $C_{V,e} = N_e c = 3N_e k_B/2$, where N_e is the number of conduction electrons. Experiment finds that the specific heat of metals at room temperature is essentially given by the Dulong-Petit law [see the discussion after Eq. (5.55)] that includes solely the phonons' contribution, $C_{V,\text{tot}} \approx 3Nk_B$. The actual contribution of the electrons to the specific heat is hence much smaller than $3Nk_B$. It follows that the equipartition theorem does not pertain to electrons at room temperature. The use of the equipartition theorem leads to other discrepancies with experiment, e.g., the coefficient L of the Wiedemann-Franz law. These caveats of the Drude theory are removed by using **quantum mechanics** for the electrons. That is, the electrons are still considered as a gas of free particles (i.e., the electronic interactions, as well as the interactions of the electrons with the ions, are ignored), but the particles are treated quantum-mechanically. The theory that describes electrons as a gas of free quantum particles is named after **Sommerfeld.**

    Free quantum electrons. Quantum mechanics modifies the calculations in two ways. First, the free electron is described by the Schrödinger equation,

$$-\frac{\hbar^2}{2m}\nabla^2\psi(\mathbf{r}) = E\psi(\mathbf{r}) , \tag{6.26}$$

and its energy levels are given by the eigenvalues of this equation. Assuming that the electron is confined in a box whose dimensions are $L_x \times L_y \times L_z$, then by a variable separation the eigenfunctions are $\psi(\mathbf{r}) = \psi_x(x)\psi_y(y)\psi_z(z)$, where each multiplier is a solution of the one-dimensional Schrödinger equation. In most of the following considerations it is assumed that the wave function ψ obeys **periodic boundary conditions,** as the ones exploited in Sec. 5.3 for the lattice vibrations. Thus, e.g., for $\psi_x(x) = \psi_x(x + L_x)$, the normalized solution is a wavy one, $\psi_x(x) = L_x^{-1/2}\exp[ik_x x]$, with discrete values of the wave number, $k_x = 2\pi n_x/L_x$, $n_x = 0, \pm 1, \pm 2, \ldots$. Similarly, $k_y = 2\pi n_y/L_y$ and $k_z = 2\pi n_z/L_z$, such that $\psi(\mathbf{r}) = \exp[i\mathbf{k} \cdot \mathbf{r}]/\sqrt{V}$, where $V = L_x L_y L_z$ is the volume of the box. Inserting this solution into the Schrödinger equation (6.26) yields discrete energy levels,

$$E_{n_x,n_y,n_z} = \frac{\hbar^2}{2m}\mathbf{k}^2 = \frac{2\pi^2\hbar^2}{m}\left[\left(\frac{n_x}{L_x}\right)^2 + \left(\frac{n_y}{L_y}\right)^2 + \left(\frac{n_z}{L_z}\right)^2\right] . \tag{6.27}$$

Because the wave number is related to the momentum of the electron by $\mathbf{p} = \hbar\mathbf{k}$, $\mathbf{k}$ is referred to as either a "wave vector" or a "momentum".

    As mentioned, other boundary conditions give rise to different discrete energies; e.g., the vanishing of the wave functions on the sample's edges yields $\psi_x(x) = (2/L_x)^{1/2}\sin(k_x x)$, with $k_x = \pi n_x/L_x$, $n_x = 1, 2, \ldots$, and similar expressions for the other two directions (see problem 5.13). The physical quantities of a macroscopic sample are not affected by these minor changes. The energy levels given in Eq. (6.27) are discrete, but they become very dense as the size of the sample grows. For instance, the interval between two levels corresponding to successive values of n_x is proportional to $1/L_x^2$, and as such it tends to zero for an infinite system. Thus, the infinite system possesses a continuum of energy levels. In most cases the focus is on large but finite systems; one has at hand a **discrete continuum** of levels. A similar discrete continuum appears in Chapter 5, in the context of the dispersion relations of lattice vibrations.

The Pauli principle and the Fermi energy. The second modification introduced by quantum mechanics is much more significant. As electrons are fermions, they obey the **Pauli principle**, a basic concept of quantum mechanics: like electronic levels in atoms, each of the single-particle energies in Eq. (6.27) can accommodate solely two electrons, with opposite spins. Thus, not only the thermal fluctuations prevent the electrons from being in their lowest energy level [out of all those given in Eq. (6.27)]; it is also the Pauli principle. The most conspicuous manifestation of the Pauli principle is at zero temperature, where there appears a **degenerate electron gas**: the electrons fill in pairs the lowest $N_e/2$ energy levels, such that their total energy is minimal. This is the **ground state of the free electron gas**.

The topmost occupied energy level is termed the **Fermi level**, and its energy (the Fermi energy) is denoted E_F. For free particles one may write $E_F = \hbar^2 k_F^2/(2m)$, where $\hbar k_F$ is the **Fermi momentum** (the maximal momentum an electron can have in the degenerate gas at zero temperature). k_F is the **Fermi wave number**, corresponding to the Fermi momentum. From the discussion that leads to Eq. (6.27) it can be deduced that the momentum component (in units of $\hbar$) along the i–th axis, $i = x, y, z$, assumes discrete values in steps of $\Delta k_i = 2\pi/L_i$; in the three-dimensional space of the momentum there is a single "allowed" value of the quantum momentum in each basic cube, whose volume (in momentum space) is $\Delta k_x \Delta k_y \Delta k_z = (2\pi)^3/V$, where $V = L_x L_y L_z$ is the volume of the sample. The energy levels below the Fermi energy, for which

$$E_{n_x,n_y,n_z} = \frac{\hbar^2 \mathbf{k}^2}{2m} < E_F = \frac{\hbar^2 k_F^2}{2m} \ ,$$

are located within a sphere in momentum space (the **"Fermi sphere"**), whose radius is $\hbar k_F$ [an example of such a "sphere" with the discrete points it contains, for a (two-dimensional) square system with $L_x = L_y$, is depicted in Fig. 6.3(a)]. The number of points contained in the Fermi sphere equals its volume divided by the "basic" volume of each point,

$$N_e = 2[4\pi k_F^3/3]/[(2\pi)^3/V] = k_F^3 V/(3\pi^2) = nV \ , \tag{6.28}$$

where $n = N_e/V = k_F^3/(3\pi^2)$ is the electrons' density, and the factor of 2 counts for the two spin directions. As explained below, the electrons that can change their states at zero temperature are those located on the surface of the Fermi sphere, since the other electrons are surrounded by occupied states which are blocked for them due to the Pauli principle. Consequently, the typical momentum of the electrons that are free to move in the quantum gas is $\hbar k_F$, and the corresponding velocity is the **Fermi velocity**, $v_F = \hbar k_F/m$. At low enough temperatures (as found shortly, room temperature is very low in this context), this is the velocity that has to be used in the Drude formula, replacing there the one derived from the equipartition law. Recall that $n \approx 10^{23}$ cm^{-3}, and thus $k_F \approx \text{Å}^{-1}$, implying $v_F \approx 10^8$ cm/sec (check!) This value is ten times larger than the one used by Drude, and therefore the mean free-pathis significantly longer than the lattice constant. This finding invalidates the conjecture that most of the electron's collisions are with the ions in the lattice.

Problem 6.7.

The Fermi temperature is defined by the relation $k_B T_F = E_F$. Estimate the Fermi temperature of a three-dimensional gas of free electrons whose density is $n \approx 10^{23}$ cm^{-3}. [$\hbar = 1.05 \times 10^{-27}$ erg sec, $m = 9.1 \times 10^{-28}$ g, $k_B = 1.38 \times 10^{-16}$ erg/K].

The energy of the electron gas at zero temperature. Given the electronic population of the single-particle energy levels, the total kinetic energy of the electron gas at zero temperature is given by

$$E_{\text{tot}} = 2 \sum_{E < E_F} E_{n_x, n_y, n_z} \, , \tag{6.29}$$

where the sum runs over all discrete values of the energy within the Fermi sphere. For a large-volume system, the discrete points in momentum space are dense (they form a "discrete continuum"), and the sum may be replaced by an integral,

$$E_{\text{tot}} = V \int_0^{E_F} E g(E) dE \, , \tag{6.30}$$

of which $V g(E) dE$ is the number of discrete energy levels with energies in-between E and $E + dE$, multiplied by 2 due to the two spin states. As $E = (\hbar k)^2/(2m)$, this number is given by the volume of the spherical shell with wave number in-between k and $k + dk$, divided by the volume assigned to a point in momentum space,

$$V g(E) dE = \frac{8\pi k^2 dk}{(2\pi)^3/V} = V \frac{m}{\hbar^3 \pi^2} \sqrt{2mE} dE \tag{6.31}$$

(check!). The **density of states** per unit energy and per unit volume, is given by [using Eq. (6.28)],

$$g(E) = \frac{3n}{2E_F^{3/2}} \sqrt{E} \, . \tag{6.32}$$

It follows that the kinetic energy of the degenerate electron gas, at zero temperature, is

$$E_{\text{tot}} = V \int_0^{E_F} E g(E) dE = (3/5) V n E_F \, ,$$

from which follows the average kinetic energy per an electron,

$$\overline{E} = E_{\text{tot}}/N_e = E_{\text{tot}}/(Vn) = 3E_{\text{F}}/5 \ . \tag{6.33}$$

With $k_{\text{F}} = (3\pi^2 n)^{1/3}$ [Eq. (6.28)] this becomes

$$E_{\text{tot}} = N_e \frac{3}{5}\frac{\hbar^2}{2m}(3\pi^2 n)^{2/3} = \frac{3\hbar^2 N_e}{10m}\left(3\pi^2 \frac{N_e}{V}\right)^{2/3} , \tag{6.34}$$

and therefore the pressure of the electron gas is $p = -(\partial E_{\text{tot}}/\partial V)_{N_e} = 2E_{\text{tot}}/(3V)$, and the bulk modulus is $B = -V(\partial p/\partial V)_{N_e} = 5p/3 = 10E_{\text{tot}}/(9V)$ (check!).

The binding energy of metals. When each atom in the system contributes a single electron, then V/N_e is the volume of the unit cell, and is of the order of R^3, where R is the distance between nearest-neighbor ions. Hence, Eq. (6.34) implies that $E_{\text{tot}}/N_e \propto 1/R^2$. This result could have been derived also from arguments based on dimensional analysis: when the only length in the problem is R, then the only momentum possible is proportional to $\hbar/R$, and the average kinetic energy is proportional to $(\hbar/R)^2/(2m)$. By similar dimensionality arguments, the Coulomb potential energy of the interaction between the electrons and the ions, and the interaction among the electrons themselves, is of the order of $-e^2/R$. The total energy of the electron gas takes the form $E_{\text{tot}}/N_e = A/R^2 - B/R$, with the positive constants, $A, B > 0$. Its minimization with respect to R provides an estimate for the size of the unit cell and for the binding energy of the metal: $R = 2A/B$ and $(E_{\text{tot}}/N_e) = -B^2/(2A)$. Thus, the minimization of the energy of the electron gas determines the average distance between neighboring ions in the metal, i.e., the lattice constant.

The Fermi-Dirac distribution. At finite temperatures, there is a finite probability for the electrons from the degenerate gas to go into energy levels higher than the Fermi one. According to the Pauli principle, each single-electron quantum state of energy E_i is occupied by n_i electrons (or more generally, fermions), where n_i can be assigned only the values 0 or 1 (ignoring the spin degree of freedom). The average value of n_i is given by the **Fermi-Dirac distribution**,

$$\langle n_i \rangle = f(E_i) = \frac{1}{1 + e^{\beta(E_i - \mu)}} \ , \tag{6.35}$$

where $\beta = 1/(k_{\text{B}}T)$ and μ, which depends on the temperature, is the **chemical potential**. The chemical potential is the free energy that is added to the system with each extra electron. Ignoring the spin degree of freedom, the total number of electrons is $N = \sum_i n_i$. The chemical potential is determined by the requirement that the average number of electrons which occupy all possible states equals the total number of electrons in the system,

$$N_e = 2\langle N \rangle = 2\sum_i \langle n_i \rangle = \sum_i 2/[1 + e^{\beta(E_i - \mu)}] \ . \tag{6.36}$$

Equation (6.35) resembles in certain aspects Eq. (5.54) for the average number of phonons in a given vibrational mode. As mentioned there, Eq. (5.54) is a particular case of the Bose-Einstein distribution, which pertains to particles of integer-value spins. In contrast, the Fermi-Dirac distribution describes particles of half-integer spin values, e.g., electrons. In the case of bosons, n_i can have any integer value between 0 and ∞, while for fermions n_i is 0 or 1.

As for fermions $0 \leq n_i \leq 1$, the function $f(E_i)$, Eq. (6.35), can be regarded as the average probability to find a particle in a state of energy E_i. A simple way to derive that expression is by conjecturing that the ratio of this probability to the probability of that state to be empty, $[1 - f(E_i)]$, is proportional to the Boltzmann distribution, i.e., $f(E_i)/[1 - f(E_i)] = z \exp[-\beta E_i]$. Indeed, when $z = \exp[\beta\mu]$ Eq. (6.35) is reproduced.

A more technical proof of Eq. (6.35) is based on statistical physics (see also problem 6.8). The thermodynamic properties at finite temperatures are derived from the partition function of the electron gas, i.e., its free energy. It is convenient to calculate the latter for a fixed value of the **chemical potential** μ (i.e., for an undetermined value of the number of electrons), and then to compute the average number of electrons which corresponds to that chemical potential, i.e., to use Eq. (6.36). For a certain configuration $\{n_i\}$ of the states' occupations, the energy of the electron gas (omitting the spin) is $E_{\text{tot}} = \sum_i n_i E_i$, and their number is $N = \sum_i n_i$. The Boltzmann's weight of such a configuration is $\exp[-\beta(E_{\text{tot}} - \mu N)]$, and the partition function is the sum of all weights over all possible configurations, that is, over all possible occupations $\{n_i\}$,

$$Z = \sum_{E_{\text{tot}},N} e^{-\beta(E_{\text{tot}}-\mu N)} = \sum_{\{n_i\}} e^{-\beta \sum_i n_i (E_i-\mu)}$$

$$= \prod_i \sum_{n_i=0,1} e^{-\beta n_i (E_i-\mu)} = \prod_i [1 + e^{-\beta(E_i-\mu)}] . \qquad (6.37)$$

The average number of electrons at energy E_i (excluding spin) is obtained upon averaging n_i with the Boltzmann weights,

$$\langle n_i \rangle = f(E_i) = \frac{\sum_{\{n_j\}} n_i e^{-\beta \sum_j n_j (E_j-\mu)}}{\sum_{\{n_j\}} e^{-\beta \sum_j n_j (E_j-\mu)}} = -\frac{\partial \ln Z}{\partial(\beta E_i)} = \frac{1}{1 + e^{\beta(E_i-\mu)}} , \qquad (6.38)$$

which is the result given in Eq. (6.35). The Fermi-Dirac distribution replaces the Maxwell-Boltzmann one, $f_{\text{MB}}(E_i) = \exp[-\beta(E_i - \mu)]$, in the quantum regime. The latter gives rise to the equipartition theorem used in the classical Drude theory. The Fermi-Dirac distribution is identical to the Maxwell-Boltzmann one only in the limit of energies much higher than $\beta^{-1} = k_{\text{B}}T$, or at temperatures that largely exceed the Fermi temperature (see problem 6.7), where the exponential dominates the denominator in the right-hand side of Eq. (6.35).

Problem 6.8.
Another way to derive Eq. (6.35) is based on combinatorics. One assumes that each quantum state may contain g_i electrons at most (including a possible degeneracy of

the energy level E_i, but ignoring the spin), and is occupied by n_i electrons. The total energy and the total number of electrons in the gas are given by $E_{\text{tot}} = \sum_i n_i E_i$ and $N = \sum_i n_i$, respectively.

a. *Show that the number of occupations' possibilities is $W = \prod_i \{g_i! / [n_i!(g_i - n_i)!]\}$.*

b. *Use the Stirling formula, $\ln n! \approx n \ln n - n$, to derive expressions for the entropy, defined by $S = k_B \ln W$, and for the free energy, $F = E_{\text{tot}} - TS$.*

c. *Assuming that the free energy is not changed when δn electrons are transferred from the state m to the state ℓ, show that $\partial F / \partial n_m = \partial F / \partial n_\ell$, and therefore this derivative is identical for all states, and can be denoted as the chemical potential, $\partial F / \partial n_m = \mu$.*

d. *Use the expression for F derived in part (b) to find μ, calculate the average $\langle n_i \rangle$ and show that the result reproduces Eq. (6.35).*

At zero temperature, i.e., when the electron gas is in the ground state, the function $f(E)$ becomes a step function: it is 1 for $E < \mu$ and zero otherwise (Fig. 6.2). Only the levels below μ are occupied, and the chemical potential coincides with the Fermi level, $\mu(T = 0) = E_F$. At finite temperatures, as seen in Fig. 6.2, $f(E)$ is close to 1 as long as $E - \mu \ll k_B T$, and is close to zero in the opposite limit, $E - \mu \gg k_B T$. In the latter case $f(E) \approx \exp[-\beta(E - \mu)]$; the "tail" of $f(E)$ decays to zero like the Maxwell-Boltzmann distribution [see also part (c) in problem 6.10].

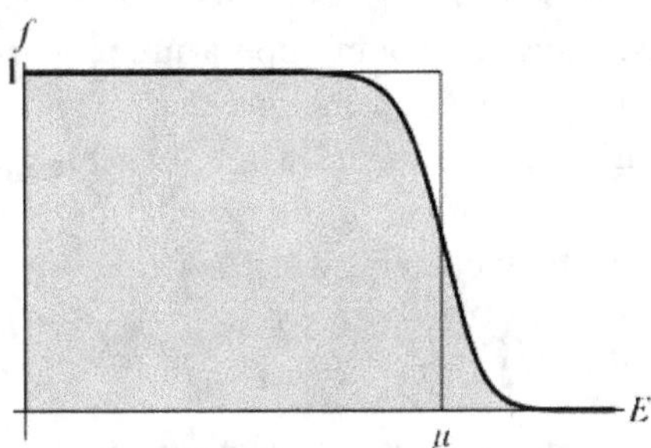

Fig. 6.2: The Fermi-Dirac distribution, Eq. (6.35). The thin line is for $T = 0$, the thick one (with the shadowed area below) is for $k_B T / \mu = 0.05$.

The specific heat. At any finite temperature, $f(\mu) = 1/2$. The exponential factor $\exp[\beta(E - \mu)]$ differs significantly from 1 or from 0 only when $\beta|E - \mu| \leq 1$, and therefore the gradual decrease of the function from 1 to zero takes place at energies in the range $|E - \mu| \leq k_B T$. As found in problem 6.7, at room temperature $k_B T \ll E_F = k_B T_F$, and so this width of this energy range is much smaller than the Fermi energy, $|E - \mu| \ll E_F$. The probability of an electron to pass from an energy within the Fermi sphere to a state of energy above that sphere is the ratio of the area in-between the two lines (which is not shadowed) below μ and the total area below the thin line in the figure. The former equals the shadowed area above μ, which then represents the probability density of the electrons that are moved from below to above the Fermi sphere. A rough estimate yields that each of these areas is of the order of $k_\mathrm{B}T$ (the area of a triangle whose base is $k_\mathrm{B}T$ and whose height is 1/2). A more careful calculation yields for the shadowed area above μ

$$\int_\mu^\infty dE e^{-\beta(E-\mu)}/[1 + e^{-\beta(E-\mu)}] = -k_\mathrm{B}T \ln[1 + e^{-\beta(E-\mu)}]\Big|_\mu^\infty = k_\mathrm{B}T \ln 2$$

(check that the area in-between the two lines below μ is indeed the same as this result). The area of the (big) rectangular in Fig. 6.2 is μ, and at low enough temperatures $\mu \approx E_\mathrm{F}$ (as the temperature-dependent corrections to the chemical potential of the degenerate gas are small). It follows that the probability of an electron to ascend above the Fermi level is of the order of $k_\mathrm{B}T/E_\mathrm{F}$. As the total number of electrons is N_e, the total number of electrons that move above the Fermi level is $\Delta N \approx N_e(k_\mathrm{B}T/E_\mathrm{F})$ (up to a coefficient of order 1). The total energy of the gas then increases as compared to its zero-temperature value, since some of the electrons have changed their states from below to above the Fermi sphere. As each of those changes the energy by $|E - \mu| \approx k_\mathrm{B}T$, the total change in the energy is $\Delta E_\mathrm{tot} \approx \Delta N k_\mathrm{B}T \approx N_e(k_\mathrm{B}T)^2/E_\mathrm{F}$, and hence the specific heat is (approximately) linear in the temperature, $C_V \approx 2N_e k_\mathrm{B}(k_\mathrm{B}T/E_\mathrm{F})$. As found in problem 6.7, the Fermi temperature of a typical electron gas is $T_\mathrm{F} = E_\mathrm{F}/k_\mathrm{B} \approx 90,000$ K. (In fact, it is $T_\mathrm{F} = 136,000$ K for aluminium, and $T_\mathrm{F} = 18,400$ K for cesium.) Hence, at room temperature, $k_\mathrm{B}T/E_\mathrm{F} \approx 1/300$, and the electronic specific heat is much smaller than that of the phonons, which at room temperature follows the Dulong-Petit law, $C_V(\text{phonons}) = 3N k_\mathrm{B}$, in agreement with experiment.

The above calculation is an approximate one; in reality the changes in the energies of the electrons are not identical; a full calculation requires the computation of the total energy at finite temperature, which averages over all those energies

$$\langle E_\mathrm{tot} \rangle = V \int_0^\infty E g(E) f(E) dE = \frac{3N_e}{2E_\mathrm{F}^{3/2}} \int_0^\infty dE \frac{E^{3/2}}{1 + e^{\beta(E-\mu)}} . \tag{6.39}$$

The lower bound is set to zero because the density of states vanishes below zero energy. The electronic specific heat is

$$C_V(\text{electrons}) = \frac{\partial \langle E_\mathrm{tot} \rangle}{\partial T} = V \int_0^\infty E g(E) \frac{\partial f(E)}{\partial T} dE . \tag{6.40}$$

An expansion to leading order in the temperature (see problem 6.9) yields

$$C_V(\text{electrons}) \approx V(3\pi^2 n)^{1/3} m k_\mathrm{B}^2 T/(3\hbar^2) = \frac{\pi^2}{2} N k_\mathrm{B} \frac{T}{T_\mathrm{F}} \equiv \gamma T . \tag{6.41}$$

The result is indeed linear in the temperature, but the coefficient slightly differs from the one derived in the crude approximation used above. At temperatures lower than the Debye temperature of the phonons, the phonons' contribution to the specific heat, at three dimensions, is $C_V(\text{phonons}) \approx AT^3$ [see the discussion after Eq. (5.69)]; the total specific heat is thus given by

$$C_V \approx \gamma T + AT^3 . \tag{6.42}$$

Plotting the experimental values for C_V/T as a function of T^2, one usually obtains a straight line. The measured value of γ is not the one found from the expression given in Eq. (6.41). As γ is proportional to the mass of the electron [Eq. (6.41)], the discrepancy is resolved by using an effective mass m^*, different from that of free electrons. This issue is resumed in the following.

Problem 6.9.

a. *Plot the function* $-f'(E) \equiv -\partial f/\partial E$ *for several values of the temperature. Show that the integral of this function is 1, and that at low temperatures it has a symmetric narrow peak around the Fermi level. Estimate the dependence of the peak's width on the temperature. Show that in the* $T \to 0$ *limit, the function tends to a delta function,* $-\partial f/\partial E \to \delta(E - \mu)$.

b. The Sommerfeld expansion. *Use the results of part (a) to show that the leading correction to average quantities at low temperatures is proportional to* T^2. *[Hint: integrate by parts the integral that yields the average, say* $\int dE f(E) A(E)$, *and zoom in on the contributions around the Fermi level.]*

Problem 6.10.

Denote the surface of the unit sphere in d dimensions by S_d *[e.g.,* $S_2 = 2\pi$ *and* $S_3 = 4\pi$*].*

a. *Find the relation between the electrons' number and the Fermi energy at zero temperature. Express the zero-temperature Fermi energy and the zero-temperature Fermi momentum in terms of the density of electrons at arbitrary dimensions.*

b. *Find the electrons' density of states at arbitrary dimensions, d. Find the ratio of the total energy of the gas to the Fermi energy.*

c. *Given the number of electrons in the system, compare it with the average number found from the Fermi-Dirac distribution at arbitrary temperatures, and derive thence an equation which yields the dependence of the chemical potential* μ *on the temperature and on the Fermi energy. Use this equation to prove that in the high-temperature limit, the chemical potential of a system with a finite number of electrons tends to* $\mu \approx -(d/2)(k_BT)\ln(k_BT) \to -\infty$. *Show that in this limit all thermodynamic averages can be obtained from the Maxwell-Boltzmann distribution, and thus the equipartition theorem and the Dulong-Petit law are valid at high temperatures.*

d. *Show that the chemical potential of a quantum electron gas at two dimensions is given by* $\mu(T) = k_BT \ln\{\exp[E_F/(k_BT)] - 1\}$.

e. *Find the exponents in Eq. (6.42) at arbitrary dimensions.*

The electrical conductivity. The quantum description of a free electron, according to the de Broglie principle, is by a wave of frequency $\omega = E/\hbar$ and of wave vector $\mathbf{k} = \mathbf{p}/\hbar$. ($E$ is the energy of the electron and $\mathbf{p}$ is the momentum.) The plane-wave eigenfunctions $\psi_{\mathbf{k}}(\mathbf{r}) = \exp[i\mathbf{k} \cdot \mathbf{r} - i\omega(\mathbf{k})t]/\sqrt{V}$ are spread over the entire space. An electron confined to a finite volume is described by a wave packet. The (average) velocity of the electron is then the group velocity of the wave packet, $\mathbf{v} = \boldsymbol{\nabla}_{\mathbf{k}}\omega(\mathbf{k})$. For a free particle, with $E = \hbar^2\mathbf{k}^2/2m$, this velocity is $\mathbf{v} = \hbar\mathbf{k}/m$.

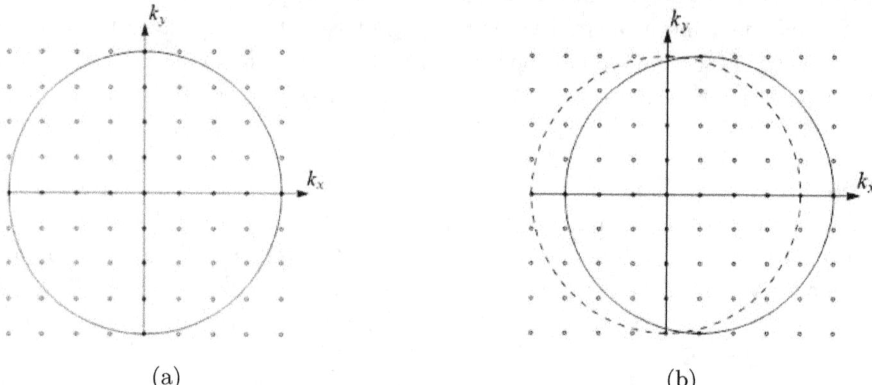

(a)   (b)

Fig. 6.3: The dots represent discrete values of the momentum for the electron's eigenstates in a finite square sample, $L_x = L_y$. The states within the solid circles are occupied by electrons (at zero temperature). (a) Without an electric field. (b) In the presence of an electric field pointing along the negative $\hat{\mathbf{x}}$–axis.

At zero temperature, when the electron gas is in its ground state, all states within the Fermi sphere are occupied. Consequently the total current density vanishes,

$$\mathbf{J} = -\frac{e}{V} \sum_i \mathbf{v}_i = 0 \,, \tag{6.43}$$

as for each point $\mathbf{k}_i$ in the sphere there is also the point $\mathbf{k}'_i = -\mathbf{k}_i$, see Fig. 6.3(a). When the electron gas is subjected to an electric field, $\mathcal{E}$, the picture is changed. At distances long compared with the wave length of the electron the quantum averages of the location and of the momentum fulfill the classical equations of motion; that is, in-between collisions, the momentum, and hence the wave vector, obeys

$$\mathbf{k}_i(t) - \mathbf{k}_i(0) = -e\mathcal{E}t/\hbar \,. \tag{6.44}$$

The Pauli principle, though, forbids the electron to move into a state already occupied by two other electrons. Assuming that the electric field is along the negative $\hat{\mathbf{x}}$–axis, the electrons move along the positive $\hat{\mathbf{x}}$–direction. When the field is weak the change in the momentum of each electron is small, and therefore the electron moves to a close-by momentum state. The only electrons that are capable of doing so are those represented by points on the right side of the Fermi sphere. These move to the right according to Eq. (6.44), vacating their original locations for electrons on their left, which then also move to the right according to the same equation, i.e., the entire Fermi sphere is shifted towards the right, as illustrated in Fig. 6.3(b). The states in-between the two spheres there, on the left ($k_x < 0$), remain empty, and the states in-between the two spheres on the right become occupied.

Reproducing the Drude formula. Recall that the electrical resistivity originates in **collisions** of the electrons with various flaws in the sample. As room temperature is always significantly lower than the Fermi temperature, the only electrons that can collide and as a result change their states are those located near empty states, i.e., the electrons close to the surface of the Fermi sphere. The electrons that reside deep in the sphere have no place to move to after the collision. Therefore, the typical velocity of the colliding electrons is the Fermi velocity, and not the thermal one that appears in the Drude theory. Equation (6.6) of the Drude theory does not contain the velocity; hence, the relaxation time is still of the order of $\tau \sim 10^{-14}$ sec. With the value $v_{\mathrm{F}} \sim 10^8$ cm/sec, the mean free-path is $\ell = v_{\mathrm{F}}\tau \sim 100\mathrm{\AA}$, implying that the collisions cannot be with the ions of the lattice, whose distances from each other are of the order of the lattice constant. The periodic array does not scatter the electrons for reasons similar to those discussed in Chapter 3. The quantum electron is described by a wave, and that wave can be scattered only when its wave vector is changed by a reciprocal-lattice vector, as in the case of the Bragg peaks of waves scattered off a periodic structure. As discussed below, a change by a reciprocal-lattice vector does not alter the state of the electron, and hence cannot contribute to the electrical resistivity.

In a stationary situation, the Fermi sphere is shifted by the amount $m\Delta\mathbf{v}/\hbar = \Delta\mathbf{k} = -e\boldsymbol{\mathcal{E}}\tau/\hbar$. Most of the electrons still occupy pairs of states with opposite momenta. Nonetheless, Fig. 6.3(b) illustrates that the points in-between the solid and dashed circles do not have a counterpart. The number of these points is of the order of $N_e|\Delta v|/v_{\mathrm{F}} = N_e|e\tau\mathcal{E}/m|/v_{\mathrm{F}}$. The velocity of each such electron is v_{F}, and thus the resultant current density is

$$|J| \approx n(|\Delta v|/v_{\mathrm{F}})ev_{\mathrm{F}} = ne\Delta v = \frac{ne^2\tau}{m}|\mathcal{E}| \ . \tag{6.45}$$

This is the Drude formula; the relaxation time, however, is that of electrons that move with the Fermi velocity.

The heat conductivity and the Seebeck coefficient. The expression found above for the heat conductivity is $\kappa = nv^2\tau c/3$ [Eq. (6.23)]; however, in this expression one has to use the Fermi velocity and the specific heat from Eq. (6.41). Substituting in the Lorenz ratio yields

$$L = \frac{\kappa}{\sigma T} = \frac{\pi^2}{3}\left(\frac{k_{\mathrm{B}}}{e}\right)^2 \ . \tag{6.46}$$

With the numerical values of the physical constants, this becomes $L = 2.45 \times 10^{-8}$ Watt Ω/K^2. In contrast with the result of the Drude model, this value of L is in fair agreement with experimental values of metals.

It is found in problem 6.6 that the Seebeck coefficient is $Q = -c/(3e)$. With the equipartition-law value for the specific heat this gives a result higher by a factor of 100 as compared with experimental measurements. Equation (6.41) implies that the ratio of the electronic specific heat calculated within the Sommerfeld theory to its counterpart resulting from the Drude model is $(\pi^2/3)(T/T_{\mathrm{F}})$. As according to problem 6.7 $T/T_{\mathrm{F}} \approx 1/300$, the Seebeck coefficient within the Sommerfeld theory is smaller by a factor of 100 as compared with its Drude's analogue, and thus is in agreement with experiment.

The collisions that cause the electrical resistivity. As mentioned, the collisions leading to the electrical resistivity cannot be among the electrons and the ions forming the periodic lattice. In fact, the electrons collide with various types of impurities, that might be foreign atoms or other flaws in the lattice structure, or they are scattered by phonons or other electrons. As $1/\tau$ is the probability for a collision to occur, it can be presented as a sum over the probabilities pertaining to the various collision processes,

$$\frac{1}{\tau} = \frac{1}{\tau_{\text{imp}}} + \frac{1}{\tau_{\text{ph}}} + \frac{1}{\tau_{\text{ee}}} + \dots \,, \tag{6.47}$$

where the terms on the right hand-side refer to the three scattering processes mentioned above. The first of them is estimated as follows. Denoting the impurities' concentration (their number per unit volume) by N_{imp}, and the scattering cross section of each of them (the effective area normal to the propagation direction from which the electron is scattered) by σ_{imp}, then the mean free-path of the electron in-between such collisions, ℓ_{imp}, is determined by the relation $\ell_{\text{imp}}\sigma_{\text{imp}}N_{\text{imp}} = 1$. Here, ℓ_{imp} is the distance the electron travels between collisions with the impurities, i.e., it is the average distance between the impurities. (The left hand-side is the product of the impurities' density by the volume that contains one of them.) It follows that $1/\tau_{\text{imp}} = \langle v\rangle/\ell_{\text{imp}} = \langle v\rangle\sigma_{\text{imp}}N_{\text{imp}}$, proportional to the impurities' concentration. This density is independent of the temperature, and therefore comprises the only contribution to the resistivity at zero temperature. This is called the **residual resistivity**, and is given by

$$\rho_{\text{res}} = \frac{m}{ne^2\tau_{\text{imp}}} = \frac{m\langle v\rangle\sigma_{\text{imp}}}{ne^2}N_{\text{imp}} \,. \tag{6.48}$$

Next consider the probability of **collisions with phonons**. At temperatures higher than the Debye one, the number of phonons in the $\alpha-$vibrational mode is [from Eq. (5.54)] $\langle n_\alpha\rangle \approx k_{\text{B}}T/(\hbar\omega_\alpha)$. Some of these phonons have momenta of the order of the Fermi momentum, and so might upon colliding with an electron cancel its average momentum along the direction of the current. Their number is proportional to $k_{\text{B}}T$, and therefore the contribution to the electrical resistivity is proportional to $k_{\text{B}}T$ as well. At lower temperatures the typical phonon momentum is proportional to the temperature; consequently the change in the momentum of an electron after colliding with a phonon is of the order of $k_{\text{ph}}^2/k_{\text{F}} \propto T^2$ (this is the momentum component along the direction of the motion after losing a momentum k_{ph} along the normal direction, see the discussion of scattering into small angles in problem 6.2). In d dimensions, and within the Debye approximation [Eqs. (5.54) and (5.67)], the number of such phonons, at low temperatures, is

$$N_{\text{ph}} = \sum_\alpha \langle n_\alpha\rangle = \int_0^{\omega_{\text{D}}} d\omega \frac{g_{\text{ph}}(\omega)}{e^{\beta\hbar\omega} - 1} \propto T^d \int_0^{\beta\hbar\omega_{\text{D}}} dx \frac{x^{d-1}}{e^x - 1} \propto T^d \,,$$

where in the last step the upper limit of the integration is set to infinity. The phonon contribution to the resistivity of a three-dimensional material is proportional to T^5 at low temperatures, and to T at higher ones.

The probability of an electron to **collide with other electrons** is proportional to the square of the density of the electrons that take part in the transport (this is the probability for two electrons to be close to one another in space). The two electrons have to be at an (energy) distance of the order of $k_\mathrm{B}T$ from the Fermi energy (to have empty states to move to), and therefore the contribution to the resistivity is proportional to T^2. At low temperature, this contribution is greater than that arising from collisions with phonons. Figure 6.4 illustrates the resistivity as a function of the temperature: the curve begins from the value ρ_res close to $T = 0$, then ascends and becomes linear with the temperature at T higher than the Debye temperature, T_D. This behavior of $\rho(T)$ is termed **"Matthiessen's law"**,

$$\rho(T) = \begin{cases} \rho_\mathrm{res} + AT^2 \,, & T \ll T_\mathrm{D} \,, \\ B\,T\,, & T \gg T_\mathrm{D} \,, \end{cases} \quad A, B > 0 \,. \tag{6.49}$$

Since ρ_res depends on the impurities' concentration, it varies from sample to sample, as opposed to the temperature-dependent part which is not, because it is determined by the properties of the material. Indeed, data taken on different samples fall one on top of the other once the resistivity curves are shifted vertically to account for the differences in ρ_res. To illustrate this point, Fig. 6.4 presents two such curves. It is also customary to define the **resistivity ratio**, the ratio between the room-temperature resistivity and the very-low-temperature one. This ratio increases as the material becomes cleaner.

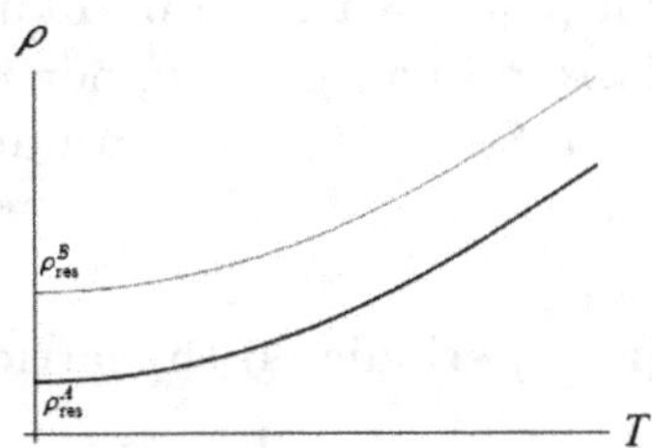

Fig. 6.4: Schematic dependence of the resistivity on the temperature according to the Matthiessen law, for two samples with different residual resistivities.

Problem 6.11.

a. *The mass density of copper is $\rho_m = 8.95$ g/cm^3, and its specific resistivity at room temperature is $\rho = 1.55$ $\mu\Omega$ cm (beware of the notation ρ, for the mass density and for the resistivity). Assuming that each atom donates a single electron, whose mass is that of a free electron, find the concentration of conduction electrons, the relaxation time, the Fermi velocity, the Fermi energy, the Fermi temperature, and the mean free-path of the electrons.*

b. *The residual resistivity of copper that contains one atomic percent of arsenic is $\rho = 6.8$ $\mu\Omega$ cm. Find the cross section for the scattering of an electron by an arsenic atom in the copper.*

6.4   Electrons in a periodic potential: the Bloch theorem

Preface. The free-electron theories discussed above, that ignore the periodic potential in which the electrons reside, fail to explain various phenomena. For instance, certain materials are insulators though the Sommerfeld theory predicts a metallic behavior for them; another example is the Hall coefficient that can be positive. Recall that in the presence of a periodic potential the Schrödinger equation for a single electron [Eq. (6.26)] is replaced by

$$\hat{\mathcal{H}}(\mathbf{r})\psi(\mathbf{r}) = \left[-\frac{\hbar^2}{2m}\boldsymbol{\nabla}^2 + U(\mathbf{r})\right]\psi(\mathbf{r}) = E\psi(\mathbf{r}) \,, \tag{6.50}$$

where the potential U has the periodicity of the lattice,

$$U(\mathbf{r}) = U(\mathbf{r} + \mathbf{R}) \,, \tag{6.51}$$

for **any displacement vector in the lattice,** $\mathbf{R} = n_1\mathbf{a}_1 + n_2\mathbf{a}_2 + n_3\mathbf{a}_3$: the potential "seen" by an electron at each point in the unit cell coincides with that felt at the same point in any other unit cell. The wave functions which solve Eq. (6.50) in conjunction with Eq. (6.51), possess special properties, that are described by the **Bloch theorem.** This theorem is mentioned also in the context of lattice vibrations, see the discussion following Eq. (5.4).

    The Bloch theorem.[1] **Any wave function** $\psi_{n\mathbf{k}}(\mathbf{r})$ **that is a solution for the Schrödinger equation (6.50) with the periodic potential Eq. (6.51), is characterized by a discrete quantum number** n**, and by a vector k of quantum numbers, and can be written as a product,**

$$\psi_{n\mathbf{k}}(\mathbf{r}) = e^{i\mathbf{k}\cdot\mathbf{r}}u_{n\mathbf{k}}(\mathbf{r}) \,, \tag{6.52}$$

where the function $u_{n\mathbf{k}}(\mathbf{r})$ **is periodic on the lattice,**

$$u_{n\mathbf{k}}(\mathbf{r}) = u_{n\mathbf{k}}(\mathbf{r} + \mathbf{R}) \tag{6.53}$$

with an arbitrary displacement vector on the lattice, R. For an infinite system, the values of k form a continuum, while for a finite system they are discrete and are given by the dense values listed in Eq. (5.43). The solutions of the Schrödinger equation for a certain vector k are indexed by the quantum number n**; the eigenenergies are denoted by** $E_n(\mathbf{k})$**.**

    The second version of the Bloch theorem. Upon inserting Eq. (6.53) into Eq. (6.52), one finds

$$\psi_{n\mathbf{k}}(\mathbf{r} + \mathbf{R}) = \exp[i\mathbf{k}\cdot(\mathbf{r} + \mathbf{R})]u_{n\mathbf{k}}(\mathbf{r} + \mathbf{R}) = \exp[i\mathbf{k}\cdot\mathbf{R}]\{\exp[i\mathbf{k}\cdot\mathbf{r}]u_{n\mathbf{k}}(\mathbf{r})\} \,,$$

i.e.,

$$\psi_{n\mathbf{k}}(\mathbf{r} + \mathbf{R}) = e^{i\mathbf{k}\cdot\mathbf{R}}\psi_{n\mathbf{k}}(\mathbf{r}) \,. \tag{6.54}$$

---

[1]Bloch's paper appeared in the same year, 1928, when he had been awarded his PhD degree.

This means that a function which complies with the Bloch theorem obeys Eq. (6.54). The opposite is true as well: if $\psi_{n\mathbf{k}}(\mathbf{r})$ obeys Eq. (6.54), one may define another function,

$$u_{n\mathbf{k}}(\mathbf{r}) = e^{-i\mathbf{k}\cdot\mathbf{r}}\psi_{n\mathbf{k}}(\mathbf{r}) \ . \tag{6.55}$$

Displacing the coordinate by $\mathbf{R}$ and using Eq. (6.54) yield

$$\begin{aligned}
u_{n\mathbf{k}}(\mathbf{r} + \mathbf{R}) &= e^{-i\mathbf{k}\cdot(\mathbf{r}+\mathbf{R})}\psi_{n\mathbf{k}}(\mathbf{r} + \mathbf{R}) \\
&= e^{-i\mathbf{k}\cdot\mathbf{r}}[e^{-i\mathbf{k}\cdot\mathbf{R}}\psi_{n\mathbf{k}}(\mathbf{r} + \mathbf{R})] = e^{-i\mathbf{k}\cdot\mathbf{r}}\psi_{n\mathbf{k}}(\mathbf{r}) = u_{n\mathbf{k}}(\mathbf{r}) \ . \tag{6.56}
\end{aligned}$$

It follows that the function $u_{n\mathbf{k}}(\mathbf{r})$ is periodic with the lattice vectors, as in Eq. (6.53). **Equation (6.54) is identical to the Bloch theorem** (this equation is referred to as "the second version of the Bloch theorem"). Though the quantum numbers $\mathbf{k}$ in Eq. (5.43) coincide with the discrete values of the momentum of a free particle (with periodic boundary conditions on the surface of an arbitrary triclinic box), they do **not coincide with the momentum of the electron**. However, as shown in Sec. 6.10, they do possess properties similar to those of the momentum, and therefore $\mathbf{k}$ is referred to as the **"lattice momentum"** of the electron.

A qualitative proof. Suppose that $\psi(\mathbf{r})$ is a solution of the Schrödinger equation in a certain cell, say the one at the origin, and denote the wave function in a neighboring cell, displaced from the first one by the lattice vector $\mathbf{a}_1$, by $\psi(\mathbf{r} + \mathbf{a}_1)$. Due to the periodicity, the surroundings of $\mathbf{r}$ in the first cell and $(\mathbf{r} + \mathbf{a}_1)$ in the second are identical; it is hence expected that the probability densities to find the electron at these two points are also identical, i.e., $|\psi(\mathbf{r})|^2 = |\psi(\mathbf{r} + \mathbf{a}_1)|^2$. It follows that the ratio $A = \psi(\mathbf{r} + \mathbf{a}_1)/\psi(\mathbf{r})$ (whose absolute value is 1) must be of the form of $A = \exp[i\alpha_1]$, where α_1 is a real phase. Consider next the wave function in the next cell, $\psi(\mathbf{r} + 2\mathbf{a}_1)$. Again due to the periodicity, it is expected that this cell does not differ from the first and the second ones, and hence $A = \psi(\mathbf{r}+2\mathbf{a}_1)/\psi(\mathbf{r}+\mathbf{a}_1)$. Likewise, $A = \psi(\mathbf{r}+(n_1+1)\mathbf{a}_1)/\psi(\mathbf{r}+n_1\mathbf{a}_1)$, and therefore $\psi(\mathbf{r} + n_1\mathbf{a}_1) = A^{n_1}\psi(\mathbf{r})$. Periodic boundary conditions on the face of the crystal along $\mathbf{a}_1$ imply that $\psi(\mathbf{r} + N_1\mathbf{a}_1) = \psi(\mathbf{r})$; consequently $A^{N_1} = \exp[iN_1\alpha_1] = 1$, i.e., $\alpha_1 = 2\pi h/N_1$, where h is an arbitrary integer. Hence α_1 is independent of $\mathbf{r}$. Nonetheless, $A = \exp[i\alpha_1]$ may depend on $\mathbf{a}_1$. This discussion leads to the conclusion that $A(n_1\mathbf{a}_1) = [A(\mathbf{a}_1)]^{n_1}$ for any integer n_1, i.e., $\alpha(n_1\mathbf{a}_1) = n_1\alpha(\mathbf{a}_1)$. The only possible solution for this equation is a linear function of $\mathbf{a}_1$, $\alpha(n_1\mathbf{a}_1) = n_1\mathbf{k}\cdot\mathbf{a}_1$, where $\mathbf{k}$ is a real vector. The boundary condition $\alpha(N_1\mathbf{a}_1) = N_1\mathbf{k} \cdot \mathbf{a}_1 = 2\pi h$ is fulfilled only when $\mathbf{k} = h\mathbf{b}_1$, where $\mathbf{b}_1$ is a reciprocal-lattice vector [Eq. (3.20)]. Hence $\psi(\mathbf{r} + n\mathbf{a}_1) = \exp[i\mathbf{k} \cdot n_1\mathbf{a}_1]\psi(\mathbf{r})$. Repeating this consideration for the other directions of the lattice vectors results in Eq. (6.54), where $\mathbf{k}$ attains one of the values listed in Eq. (5.43). As the change of the function between one of the unit cells and another depends on $\mathbf{k}$, the function itself has to depend on $\mathbf{k}$ as well. The Schrödinger equation possesses many solutions for Eq. (6.54); the index n is added to distinguish among them, and hence the notation $\psi_{n\mathbf{k}}(\mathbf{r})$. This completes the proof of the second version of the Bloch theorem, which is equivalent to the original Bloch theorem.

Translation operators. The proof given above is based on plausibility arguments and hence a more mathematical derivation is called for. Such a proof is presented here; another one is given in Sec. 6.6. To this end, one needs to introduce **translation operators on the lattice.** The translation operator $\hat{T}(\mathbf{R})$ shifts the coordinate $\mathbf{r}$ of an arbitrary function $f(\mathbf{r})$ by the vector $\mathbf{R} = n_a\mathbf{a}_1 + n_2\mathbf{a}_2 + n_3\mathbf{a}_3$,

$$\hat{T}(\mathbf{R})f(\mathbf{r}) = f(\mathbf{r} + \mathbf{R}) \ . \tag{6.57}$$

Obviously

$$\hat{T}(\mathbf{R})\hat{T}(\mathbf{R}') = \hat{T}(\mathbf{R} + \mathbf{R}') = \hat{T}(\mathbf{R}')\hat{T}(\mathbf{R}) \ , \tag{6.58}$$

and therefore all translation operators commute with each other [it is worthy to note that according to Eq. (6.58) the collection of all translation operators forms a group in the algebraic sense. That is, the members of this group are in one-to-one correspondence with those of the translation group discussed at the beginning of Sec. 2.7]. According to the theorem proved at the end of Appendix C, this implies that there is a basis of the Hilbert space in which each of the basis states ψ is **simultaneously an eigenstate of all translation operators,** $\hat{T}(\mathbf{R})\psi(\mathbf{r}) = t(\mathbf{R})\psi(\mathbf{r})$, where the eigenvalue $t(\mathbf{R})$ is a number.

Applying Eq. (6.58) on the eigenfunction $\psi(\mathbf{r})$ yields $t(\mathbf{R})t(\mathbf{R}') = t(\mathbf{R} + \mathbf{R}')$, where the eigenvalue $t(\mathbf{R})$ is a number. The logarithm of this equation is $\ln t(\mathbf{R}) + \ln t(\mathbf{R}') = \ln t(\mathbf{R} + \mathbf{R}')$. The only function that obeys this additive relation for **any** pair of vectors $\mathbf{R}$ and $\mathbf{R}'$ is linear in $\mathbf{R}$, i.e., $\ln t(\mathbf{R}) = \mathbf{c} \cdot \mathbf{R}$ [one may add a constant, but that constant is zero as $t(0) = 1$], leading to $t(\mathbf{R}) = \exp[\mathbf{c} \cdot \mathbf{R}]$. The normalization condition requires the absolute value of $t(\mathbf{R})$ to be 1, and therefore $\mathbf{c}$ is imaginary, $\mathbf{c} = i\mathbf{k}$, with a real $\mathbf{k}$. Thus, the result is $t(\mathbf{R}) = \exp[i\mathbf{k} \cdot \mathbf{R}]$. Periodic boundary conditions on an arbitrary lattice with lattice vectors $\{\mathbf{a}_m, \ m = 1, 2, 3\}$, comprising N_m unit cells along the lattice vector $\mathbf{a}_m$, imply that $f(\mathbf{r} + N_m\mathbf{a}_m) = f(\mathbf{r})$, that is, $t(N_m\mathbf{a}_m) = t(\mathbf{a}_m)^{N_m} = \exp[i\mathbf{k} \cdot N_m\mathbf{a}_m] = 1$. One may repeat the arguments in Sec. 5.3 (also used in the "qualitative proof" of the Bloch theorem), to obtain that the independent quantum numbers that characterize the eigenvalues of the translation operators are those given in Eq. (5.43), $\mathbf{k}_{\ell_1,\ell_2,\ell_3} = (\ell_1/N_1)\mathbf{b}_1 + (\ell_2/N_2)\mathbf{b}_2 + (\ell_3/N_3)\mathbf{b}_3$, with $-N_m/2 \le \ell_m < N_m/2 - 1$. In summary, an eigenstate of any translation operator obeys

$$\hat{T}(\mathbf{R})\psi_\mathbf{k}(\mathbf{r}) = e^{i\mathbf{k}\cdot\mathbf{R}}\psi_\mathbf{k}(\mathbf{r}) = t_\mathbf{k}(\mathbf{R})\psi_{|\mathbf{k}}(\mathbf{r}) \ , \quad \mathbf{k} = \mathbf{k}_{\ell_1,\ell_2,\ell_3} \ . \tag{6.59}$$

A simple example is that of a free particle, for which the potential energy vanishes, $U(\mathbf{r}) = 0$. The eigenfunctions of Eq. (6.59) are then those of a free particle in a box, $\Psi_\mathbf{k}^{(0)}(\mathbf{r}) = \exp[i\mathbf{k} \cdot \mathbf{r}]/\sqrt{V}$, for which

$$\hat{T}(\mathbf{R})\Psi_\mathbf{k}^{(0)}(\mathbf{r}) = \Psi_\mathbf{k}^{(0)}(\mathbf{r} + \mathbf{R}) = \frac{1}{\sqrt{V}}e^{i\mathbf{k}\cdot(\mathbf{r}+\mathbf{R})} = e^{i\mathbf{k}\cdot\mathbf{R}}\Psi_\mathbf{k}^{(0)}(\mathbf{r}) \ .$$

Proof of the Bloch theorem. Equation (6.51) implies that the Hamiltonian $\hat{\mathcal{H}}(\mathbf{r}) = -(\hbar\boldsymbol{\nabla})^2/(2m) + U(\mathbf{r})$ is such that $\hat{\mathcal{H}}(\mathbf{r}+\mathbf{R}) = \hat{\mathcal{H}}(\mathbf{r})$ [the kinetic energy is invariant to translations, as, e.g., $d(x+a) = dx$]. Hence, any function obeys the relations

$$\hat{T}(\mathbf{R})[\hat{\mathcal{H}}(\mathbf{r})f(\mathbf{r})] = \hat{\mathcal{H}}(\mathbf{r}+\mathbf{R})f(\mathbf{r}+\mathbf{R}) = \hat{\mathcal{H}}(\mathbf{r})f(\mathbf{r}+\mathbf{R}) = \hat{\mathcal{H}}(\mathbf{r})[\hat{T}(\mathbf{R})f(\mathbf{r})] \,,$$

i.e.,

$$\hat{\mathcal{H}}(\mathbf{r})\hat{T}(\mathbf{R}) = \hat{T}(\mathbf{R})\hat{\mathcal{H}}(\mathbf{r}) \,,$$

the Hamiltonian $\hat{\mathcal{H}}(\mathbf{r})$ commutes with all the translation operators of the lattice, $\{\hat{T}(\mathbf{R})\}$. The theorem proven at the end of Appendix C implies that there is a basis of the Hilbert space comprising states that are simultaneously eigenstates of all translation operators and of the Hamiltonian. Each of these states has to obey Eq. (6.59); such a state is also a solution of the Schrödinger equation,

$$\hat{\mathcal{H}}(\mathbf{r})\psi_{n\mathbf{k}}(\mathbf{r}) = E_n(\mathbf{k})\psi_{n\mathbf{k}}(\mathbf{r}) \,, \tag{6.60}$$

where the index n distinguishes among the various solutions of the equation, all belonging to the subspace of the Hilbert space with the eigenvalue $\exp[i\mathbf{k}\cdot\mathbf{R}]$ of $\hat{T}(\mathbf{R})$ (for any $\mathbf{R}$). As all functions within this subspace depend on $\mathbf{k}$, the eigenvalues of the energy, $E_n(\mathbf{k})$, depend on it as well. In other words, one may choose the eigenfunctions of the Schrödinger equation with a periodic potential such that they obey both Eq. (6.59) and Eq. (6.60). As Eq. (6.59) coincides with Eq. (6.54), this is a proof of the Bloch theorem in its second version. As shown in Eq. (6.56), it also proves the first version of the Bloch theorem.

Conclusions drawn from the Bloch theorem. The Bloch theorem has quite an impact on the possible solutions of the Schrödinger equation with a periodic potential. Inserting Eq. (6.52) into Eq. (6.60) yields

$$\left(-\frac{\hbar^2}{2m}\boldsymbol{\nabla}^2 + U(\mathbf{r})\right)\left(e^{i\mathbf{k}\cdot\mathbf{r}}u_{n\mathbf{k}}(\mathbf{r})\right) = E_n(\mathbf{k})\left(e^{i\mathbf{k}\cdot\mathbf{r}}u_{n\mathbf{k}}(\mathbf{r})\right) \,. \tag{6.61}$$

By differentiating (in the left hand-side) and then eliminating $\exp[i\mathbf{k}\cdot\mathbf{r}]$, this becomes

$$\left(\frac{\hbar^2}{2m}(-i\boldsymbol{\nabla}+\mathbf{k})^2 + U(\mathbf{r})\right)u_{n\mathbf{k}}(\mathbf{r}) = E_n(\mathbf{k})u_{n\mathbf{k}}(\mathbf{r}) \,. \tag{6.62}$$

According to the Bloch theorem [Eq. (6.53)], the function $u_{n\mathbf{k}}(\mathbf{r})$ is periodic, and hence it **suffices to solve this differential equation in a single unit cell**, with periodic boundary conditions on its faces, and then to exploit Eq. (6.52) to obtain the solution in the entire space.

The eigenfunctions $u_{n\mathbf{k}}(\mathbf{r})$ and the eigenenergies $E_n(\mathbf{k})$ depend on $\mathbf{k}$, as it appears explicitly in the left hand-side of Eq. (6.62). However, Eq. (6.62) is a second-order differential equation within the unit cell, and as such it may have many solutions with different eigenenergies and different eigenfunctions (recall the examples of an infinite potential well, a harmonic oscillator, or the hydrogen atom, whose respective Schrödinger equations have many discrete energy levels). As mentioned, the index n distinguishes among these various solutions. It is worthwhile to note that in three dimensions a single discrete index usually does not suffice to characterize the eigenfunctions. For instance, in the case of the hydrogen atom there are the three indices n, ℓ, and m, while for a free particle in a box these are the indices n_x, n_y, and n_z [Eq. (6.27)]. For brevity, a single index n is kept below.

Periodicity in the reciprocal lattice. The Bloch theorem is phrased for **k**'s whose values are confined to the first Brillouin zone, Eq. (5.43). This equation follows from the periodic boundary conditions, $\exp[i\mathbf{k}\cdot N_m\mathbf{a}_m] = 1$; these are obeyed also by vectors shifted into other Brillouin zones. In other words, the eigenvalues of Eq. (6.59) obey $t_{\mathbf{k}+\mathbf{G}}(\mathbf{R}) = \exp[i(\mathbf{k}+\mathbf{G})\cdot\mathbf{R}] = \exp[i\mathbf{k}\cdot\mathbf{R}] = t_{\mathbf{k}}(\mathbf{R})$, where **G** is an arbitrary reciprocal-lattice vector. The central step in the equation results from Eq. (3.17), $\exp[i\mathbf{G}\cdot\mathbf{R}] = 1$. Consequently, there is no difference between the solutions in the first Brillouin zone, and those in any other zone shifted by a reciprocal-lattice vector **G**: one may replace in the previous equations **k** by $\mathbf{k}+\mathbf{G}$, to obtain

$$\psi_{n,\mathbf{k}+\mathbf{G}}(\mathbf{r}) = \psi_{n,\mathbf{k}}(\mathbf{r}) \ , \tag{6.63}$$

and

$$E_n(\mathbf{k}+\mathbf{G}) = E_n(\mathbf{k}) \ . \tag{6.64}$$

Note the similarity between Eq. (6.64) and Eq. (5.7): the spectrum of the lattice vibrations is also periodic in reciprocal lattice; in both cases it suffices to obtain the wave functions and the eigenvalues in the first Brillouin zone.

Problem 6.12.

a. *Prove that $\psi^*_{n,\mathbf{k}}(\mathbf{r}) = \psi_{n,-\mathbf{k}}(\mathbf{r})$, and hence $E_n(-\mathbf{k}) = E_n(\mathbf{k})$. [Complex conjugating the time-dependent Schrödinger equation, $\hat{\mathcal{H}}(\mathbf{r})\psi(\mathbf{r},t) = i\hbar(\partial/\partial t)\psi(\mathbf{r},t)$, is equivalent to time inversion, $t \to -t$. Therefore this result proves that the Hamiltonian is invariant under time reversal.]*
b. *Repeat the proof for an orthorhombic lattice in three dimensions, and show that $E_n(-k_x, k_y, k_z) = E_n(k_x, k_y, k_z)$.*

Energy bands. As mentioned, the values of **k** are rather dense (they form a "discrete continuum"); they become a continuum in the infinite-lattice limit. As opposed, the values of n are discrete. Thus, for each n there is, in the continuum limit, a separate continuous function of **k**, that repeats itself in the various Brillouin zones (or among the unit cells in reciprocal lattice). Several examples for the calculation of such functions are presented in the next three sections. They all give rise to the same qualitative picture as illustrated in Fig. 6.5: each of the sinusoidal lines in Fig. 6.5 belongs to a different energy function $E_n(\mathbf{k})$, with a different index n. For a specific value of n the function $E_n(\mathbf{k})$ is periodic in **k**. In part (a) of the figure, the period is $G_0 = 2\pi/a$ (a is the lattice constant, and the horizontal coordinate is k/G_0). The vertical lines in that figure are the boundaries of three unit cells in reciprocal lattice (that are equivalent to three Brillouin zones).

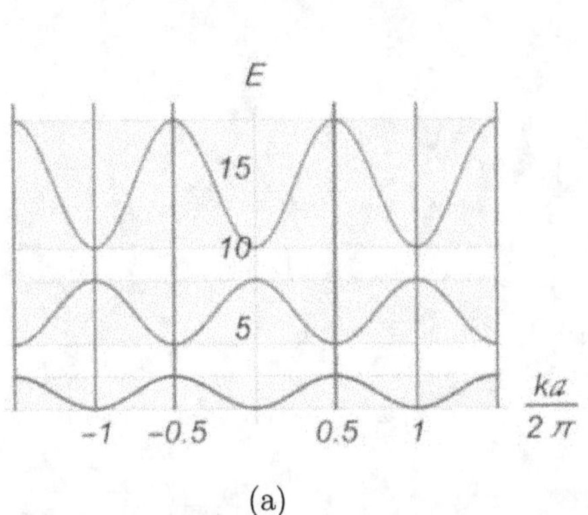

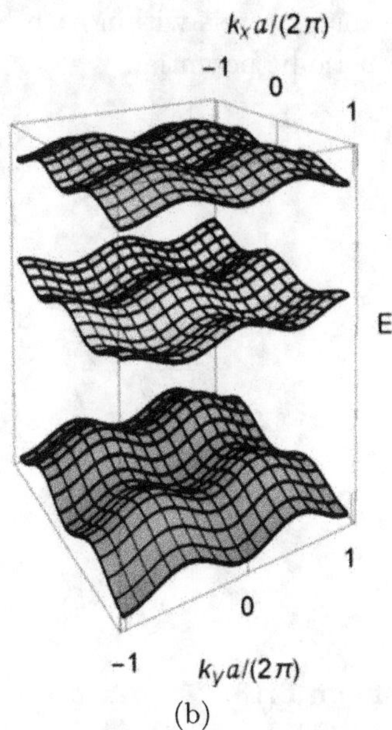

(a) (b)

Fig. 6.5: (a) Energy bands of an electron in a one-dimensional periodic potential, computed in the tight-binding approximation (Sec. 6.7). The thick sinusoidal curves are the energies as functions of the lattice momentum, and the shadowed areas indicate the range of energies in each band. The vertical lines are the boundaries between the Brillouin zones in reciprocal lattice. The central cell is the first Brillouin zone. (b) An example of energy bands of a two-dimensional periodic potential on a square lattice, computed within the same approximation.

Each of the functions $E_n(\mathbf{k})$ comprises energy values confined to a certain range. This range, shadowed in Fig. 6.5(a), is termed an **"energy band"**. In simple cases, the bands are separated by **energy gaps**, in which the Schödinger equation has no eigenvalues (see also problem 6.13). Two types of states are frequently encountered in quantum mechanics: bound states, e.g., those of an electron in the hydrogen atom, or the ones of a harmonic oscillator; these belong to discrete energy levels separated from each other by finite energies. Alternatively, the energy levels of a free particle confined to a large box are very dense, see for instance those given in Eq. (6.27), and then they form a "discrete continuum"; in the limit of an infinite system those levels become a continuum of positive energies. The spectrum of the periodic potential joins together the two types of states: each band comprises many dense energy levels, but the bands are separated by gaps in which there exist no levels, similar to the gaps between the energies of the bound states in the atom. These gaps remain finite also in the limit of an infinite system. It is worthwhile to note the similarity between the energy gaps of the periodic potential and the gap separating the acoustic and optical modes of the lattice vibrations, see Chapter 5.

The energy levels within each band are the available energies for a single electron in a periodic potential.

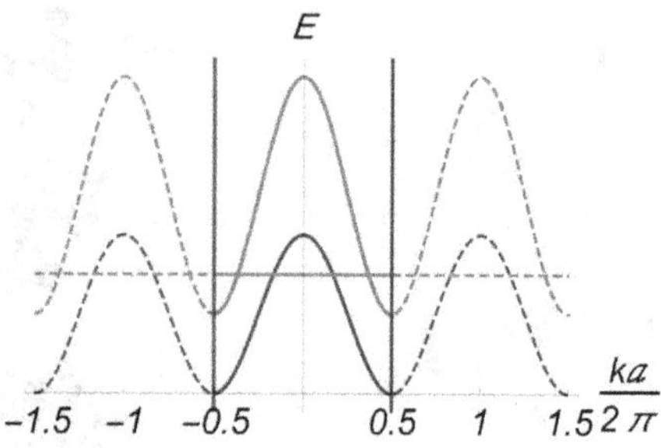

Fig. 6.6

Problem 6.13.

Figure 6.6 exhibits bands that "penetrate" into each other. Prove that such a penetration is impossible in one dimension. Hence, in one dimension the successive bands are always separated by energy gaps, and at most they may touch each other. Why is a penetration possible at higher dimensions?

Note that the energy levels are derived for a single electron. These levels group themselves into bands that are filled with electrons according to the Pauli principle.

6.5   The Kronig-Penney potential

A square-shape periodic potential. A simple example that illustrates the band structure of the energy levels is based on the **one-dimensional potential of Kronig and Penney**, built of periodic steps of square-shape potential,

$$U(x) = \begin{cases} 0, & na \leq x \leq na + b \,, \\ U_0 \,, & na + b \leq x < (n+1)a \,, \end{cases} \tag{6.65}$$

for any integer n ($b < a$). As seen in Fig. 6.7, the lattice constant is a. The segments with the lower potential (0 in the present choice) represent regions to which the electron is attracted, similarly to the Coulomb potential of the positive ions arranged periodically in a crystal. Since the potential is periodic, the solutions of the Schrödinger equation obey the Bloch theorem, and can be represented by Eq. (6.52), $\psi_k = \exp[ikx]u_k(x)$. The periodic function $u_k(x)$ is the solution of the one-dimensional version of Eq. (6.62),

$$\left(\frac{\hbar^2}{2m}(-i\frac{d}{dx} + k)^2 + U(x)\right)u_k(x) = E(k)u_k(x) \,, \tag{6.66}$$

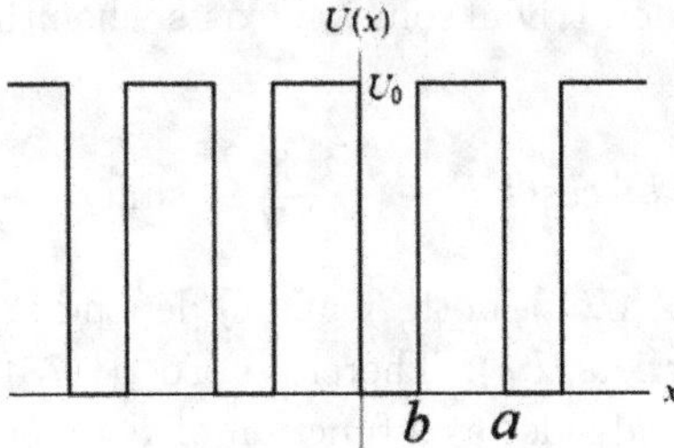

Fig. 6.7: The periodic Kronig-Penney potential. The width of each "well" is b, and the width of each potential barrier (whose height is U_0) is $a - b$.

within a single unit cell, $0 \le x \le a$, with periodic boundary conditions, $u_k(a) = u_k(0)$ and $u'_k(a) = u'_k(0)$.

The wave function. Consider the solutions for $0 < U_0 < E$; the treatment of the case $0 < E < U_0$ is quite similar, see below. As can be verified by a direct substitution, the solutions of Eq. (6.66) are

$$u_k(x) = e^{-ikx}[Ae^{iKx} + Be^{-iKx}] , \quad \text{for } 0 < x < b ,$$
$$u_k(x) = e^{-ikx}[Ce^{iQx} + De^{-iQx}] , \quad \text{for } b < x < a , \tag{6.67}$$

where

$$K^2 = 2mE/\hbar^2 , \quad Q^2 = 2m(E - U_0)/\hbar^2 . \tag{6.68}$$

By continuity, $u_k(b^+) = u_k(b^-)$ and $u'_k(b^+) = u'_k(b^-)$, and hence

$$Ae^{iKb} + Be^{-iKb} = Ce^{iQb} + De^{-iQb} ,$$
$$K[Ae^{iKb} - Be^{-iKb}] = Q[Ce^{iQb} - De^{-iQb}] . \tag{6.69}$$

The periodicity of $u_k(x)$ implies that $u_k(a) = u_k(0)$ and $u'_k(a) = u'_k(0)$, and so

$$A + B = e^{-ika}[Ce^{iQa} + De^{-iQa}] ,$$
$$K(A - B) = Qe^{-ika}[Ce^{iQa} - De^{-iQa}] . \tag{6.70}$$

Adding and subtracting the two equations in (6.70) give

$$A = \frac{1}{2}e^{-ika}[(1 + Q/K)Ce^{iQa} + (1 - Q/K)De^{-iQa}] ,$$

$$B = \frac{1}{2}e^{-ika}[(1 - Q/K)Ce^{iQa} + (1 + Q/K)De^{-iQa}] . \tag{6.71}$$

In a similar fashion one finds from Eqs. (6.69)

$$A = \frac{1}{2}e^{-iKb}[(1 + Q/K)Ce^{iQb} + (1 - Q/K)De^{-iQb}] ,$$

$$B = \frac{1}{2}e^{iKb}[(1 - Q/K)Ce^{iQb} + (1 + Q/K)De^{-iQb}] . \tag{6.72}$$

A comparison of the two pairs of expressions for A and B yields two homogeneous equations for C and D. A non-trivial solution exists when the determinant vanishes, i.e., when

$$\cos(ka) = \cos[Q(a-b)]\cos(Kb) - \frac{K^2+Q^2}{2KQ}\sin[Q(a-b)]\sin(Kb) \qquad (6.73)$$

(check!) As seen from Eqs. (6.68), both K and Q depend on the energy (and on the height of the potential barriers, U_0). Therefore, Eq. (6.73) gives explicitly k (that appears only on the left hand-side) as a function of E (contained solely in the right hand-side). A plot of k as a function of E, together with a rotation of the resulting figure by $90°$, produces the energy levels as functions of k, i.e., the energy bands. As the same value of k pertains for numerous energies, there exist numerous bands.

The eigenenergies. Introducing the dimensionless variables,

$$\epsilon = \frac{E}{U_0}\ , \quad z = \frac{a}{b} - 1\ , \quad W = \frac{2mU_0}{\hbar^2}b^2\ , \qquad (6.74)$$

transforms Eq. (6.73) into the form

$$\cos(ka) = F(\epsilon) = \cos[z\sqrt{W(\epsilon-1)}]\cos(\sqrt{W\epsilon})$$
$$- \frac{2\epsilon - 1}{2\sqrt{\epsilon(\epsilon-1)}}\sin[z\sqrt{W(\epsilon-1)}]\sin(\sqrt{W\epsilon})\ . \qquad (6.75)$$

When $\epsilon > 1$ then $Q = \sqrt{W(\epsilon-1)}/b$ is real, and consequently all quantities on the right hand-side are real. On the other hand, when $\epsilon < 1$ then Q is purely imaginary; using the identities $\sqrt{\epsilon - 1} = i\sqrt{1-\epsilon}$, $\cos(i\alpha) = \cosh(\alpha)$, and $\sin(i\alpha) = i\sinh(\alpha)$, one obtains again real quantities in the right-hand side. The function $F(\epsilon)$ is depicted in Fig. 6.8(a), for $W = 200$ and $z = 0.1$. (Other values give similar results.) As seen, there are energy segments for which $1 < |F(\epsilon)|$. In those, Eq. (6.75) yields complex values for (ka); these are unphysical as the Bloch theorem requires k to be real (the normalization implies that $|\exp[ika]| = 1$). Hence, the allowed energy levels are in the segments for which $|F(\epsilon)| < 1$. Each such segment (shadowed in the figure) corresponds to a band of allowed energies. In-between these segments there appear energy gaps, in which the Schrödinger equation has no acceptable solutions. For each allowed value of the energy, Eq. (6.75) produces definite values for $\cos(ka)$, and then the solutions repeat themselves with the period $ka \to ka + 2\pi\ell$, i.e., under a translation of k by a reciprocal-lattice vector, $G = 2\pi\ell/a = \ell G_0$. These solutions, as computed from Eq. (6.75), are displayed in Fig. 6.8(b). Indeed, the energies that correspond to a certain value of k, in the various Brillouin zones, coincide with each other. Each such value allows for many values of the energy: the horizontal line [that represents $\cos(ka)$] has many cuts with the branches of $F(\epsilon)$ in Fig. 6.8(a).

As seen in Fig. 6.8(a), the function $F(\epsilon)$ is monotonic in each shadowed area; this implies that the energy is minimal or maximal at the ends of the segment, i.e., when $\cos(ka) = 1$ or $\cos(ka) = -1$. In particular, there are minima or maxima for

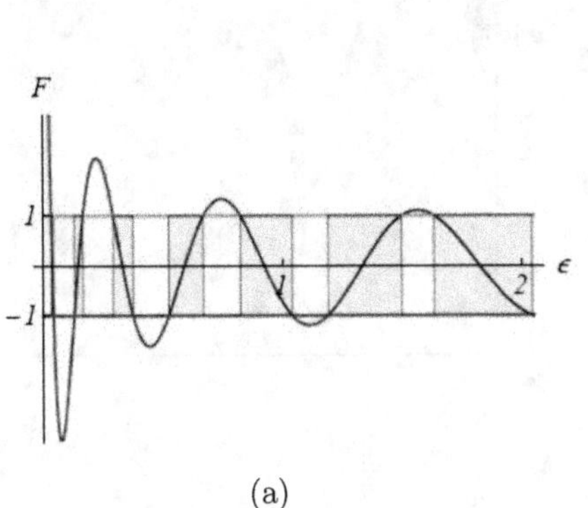

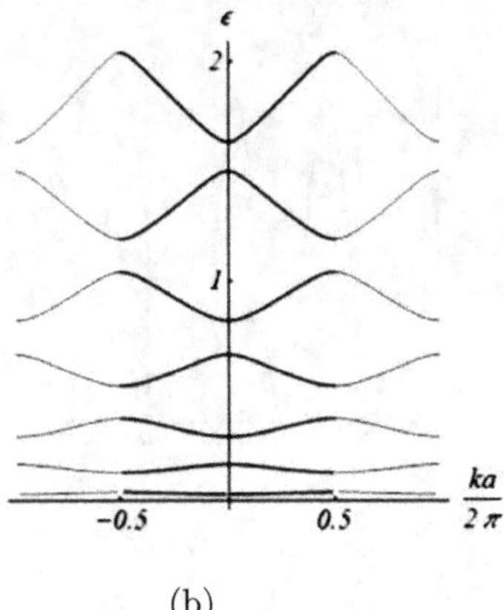

(a)   (b)

Fig. 6.8: The right hand-side of Eq. (6.75), for $W = 200$ and $z = .1$. The shadowed areas contain the "allowed values" of the energy, i.e., the energy bands. (b) The solutions of Eq. (6.75), for the same parameters. The thick lines mark the first Brillouin zone.

$ka = 0$, that is, at the center of the Brillouin zone, or for $ka = \pm\pi$, i.e., at the ends of the first Brillouin zone, as indeed seen in Fig. 6.8(b). Figure 6.8(a) also shows that the slopes of $F(\epsilon)$ change sign between successive bands, and therefore the locations of the minima and the maxima alternate between such bands. This observation is confirmed in Fig. 6.8(b).

The wave function. Figure 6.9 illustrates the absolute value squared of the wave function, that is, the probability density to find the electron, for parameters given in the caption, with the energies (in units of U_0) $\epsilon = .6$ and $\epsilon = 2$. In both cases the probability is oscillatory in the ranges where the potential barrier vanishes. In Fig. 6.9(a) the energy is lower than the height of the potential barrier and consequently Q is purely imaginary; the probability to find the electron in the range $b < x < a$ is exponentially decaying on the two sides of this segment. Figure 6.9(b) displays the case where the energy is above the potential step, and therefore the probability to find the electron in that range is considerably higher.

The "nearly-free" electron. Equation (6.73) has simple solutions in two limits. When $U_0 = 0$ then $Q = K = \sqrt{2mE}/\hbar$, and the right hand-side becomes $\cos(Ka)$. In that case $ka = Ka + 2\ell\pi$ for any integer ℓ, yielding $E_\ell = \hbar^2(k - G_\ell)^2/(2m)$, with $G_\ell = \ell G_0 = 2\pi\ell/a$. One might have expected that when the periodic potential vanishes the solutions of the Schrödinger equation are those of the free particle, $E_0(k) = \hbar^2 k^2/(2m)$. This parabola is indeed obtained in the present case for $\ell = 0$. Nonetheless, even when the periodic potential is infinitesimally small, almost unnoticed in the energy levels, the system does "identify" its existence, and then the Bloch theorem yields Eq. (6.64). This equation in turn dictates that the energy is periodic in reciprocal lattice, and therefore the parabola $E_0(k)$ can be

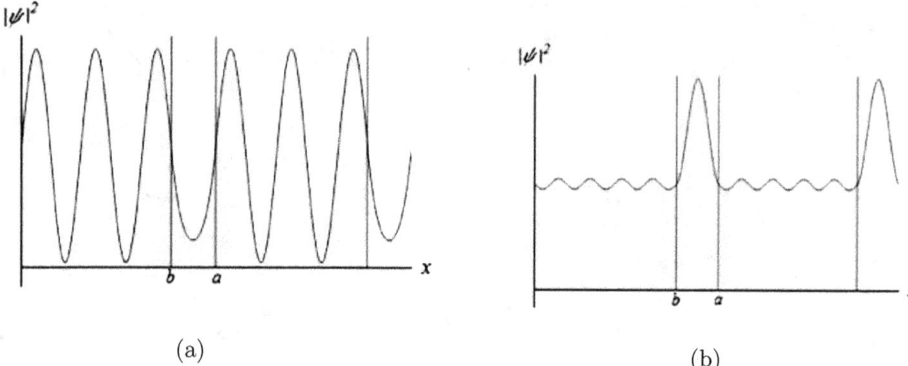

Fig. 6.9: The probability density to find the electron within two neighboring unit cells, for $W = 100$, $z = -.3$, and (a) $\epsilon = .6$, (b) $\epsilon = 2$.

shifted to the left or to the right by an arbitrary reciprocal-lattice vector, giving rise to an infinite ensemble of parabolae, $E_\ell(k) = E_0(k - G_\ell) = \hbar^2(k - G_\ell)^2/(2m)$. This ensemble, as derived from the limiting solution of Eq. (6.73), is displayed in Fig. 6.10(a). It is discussed in more detail in the next section.

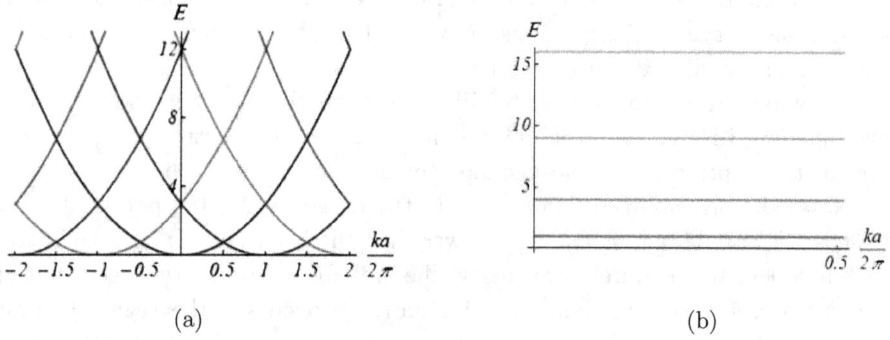

Fig. 6.10: The energy levels of the Kronig-Penney potential. (a) The free-electron limit, $U_0 = 0$. [The energy is in units of $2\pi^2\hbar^2/(6ma^2)$.] (b) The limit of infinite potential wells, $U_0 \to \infty$. [The energy is in units of $\pi^2\hbar^2/(2mb^2)$.]

The tight-binding limit. Another simple case is obtained when the potential barrier is quite high, $U_0 \to \infty$. Then $Q^2 \to \infty$ and the right-side member of Eq. (6.73) is considerably larger than the other two. As a result, the equality exists only when $\sin(Kb) = 0$, i.e., $E = \hbar^2K^2/(2m) = \hbar^2\pi^2n^2/(2mb^2)$, where n is an arbitrary integer. These are precisely the discrete energy levels in a single infinite potential well of width b (check!). The energy levels do not depend on the wave number k, and therefore each of the energy bands becomes an horizontal line at one of the discrete levels, see Fig. 6.10(b). As opposed to the single-well case, there are now N identical wells, and therefore each energy level is N-fold degenerate. Decreasing gradually U_0 widens these horizontal lines and narrows the gaps in-between them, until the free-particle limit is reached, and then neighboring bands touch one another. Figures 6.8(b) and 6.11 display intermediate situations between these two limits.

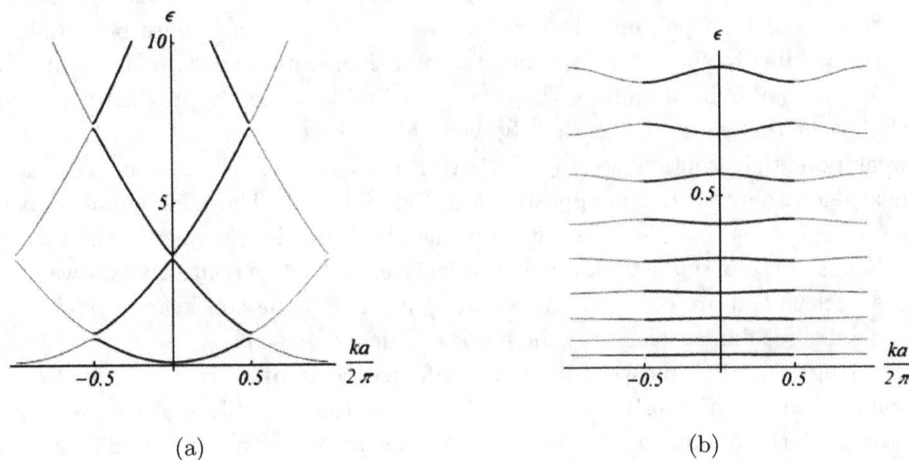

(a) (b)

Fig. 6.11: Energy bands of the Kronig-Penney potential, with $z = .1$ (a) Weak potential, $W = 10$. (b) Strong potential, $W = 1000$. Energy in units of U_0.

As seen in Fig. 6.8(a), the amplitude of the oscillations of the function $F(\epsilon)$ decreases when the energy is increased for finite values of U_0. Therefore, the regions where $|F(\epsilon)| > 1$ (that represent the energy gaps in-between the energy bands) are narrowed down for high-energy bands. Indeed, when $\epsilon \gg 1$, one finds $F(\epsilon) \approx \cos[(z+1)\sqrt{W\epsilon}] = \cos(Ka)$, which reproduces the free-particle solutions (check!). In other words, electrons of high energies do not "feel" the potential.

Problem 6.14.
a. *Find the limit of Eq. (6.73) for $U_0 \to \infty$ and $(a-b) \to 0$ under the condition that $u_0 = U_0(a-b)$ remains constant. This is the limit in which each step becomes a delta function, and the potential has the shape of a "comb", $U(x) = u_0 \sum_n \delta(x-na)$. Show that in this limit the value $Ka = n\pi$ is always the border between an energy band and an energy gap.*
b. *Calculate the energy bands for large values of the constant $\phi = u_0 am/\hbar^2 = -Q^2(a-b)a/2$.*
c. *Find the energy gaps between the bands that correspond to small values of ϕ. What is then the energy at the bottom of the lowest band?*

6.6   Nearly-free electrons

Weak and strong potentials. The Kronig-Penney potential is a special case, since it allows for an exact solution of the Schrödinger equation. As in most cases an exact solution does not exist, approximations or numerical methods are called for. Figure 6.10 illustrates the energy levels in two extreme limits, when the potential is vanishingly small and when it is very strong. In the first case, the solution of the Schrödinger equation is that of a free particle; in the second the solution coincides with that of a single "unit", described in Sec. 6.5 by an infinite potential well corresponding to the attractive potential of a single ion in the lattice. It turns out that one may build on any of these extreme limits, to find approximate solutions that yield the energy levels slightly away from those limits. The analysis of the weak-potential situation [see Fig. 6.11(a)] is carried out in the present section; the next one is devoted to the opposite case [Fig. 6.11(b)]. The approximation based on a weak potential describes faithfully the alkali metals (the ones in the first two columns of the periodic table), whose valence electrons are relatively far away from the nucleus and are only loosely bound to it. The discussion below pertains to a general three-dimensional periodic potential, denoted $U(\mathbf{r})$.

Bragg conditions for scattering of electrons off a periodic lattice. A comparison of Figs. 6.10(a) and 6.11(a) shows that the effect of the weak periodic potential on the energy levels is conspicuous close to the boundaries of the Brillouin zones, that is, when k is near an integral multiple of π/a. A qualitative understanding of this feature is gained by considering the discussion in Chapter 3. It is shown there that a plane wave that impinges on a periodic lattice is significantly scattered only when the difference between the incoming and outgoing wave vectors equals one of the reciprocal-lattice vectors, Eq. (3.9). This condition leads to Eq. (3.26), $\mathbf{k} \cdot \hat{\mathbf{G}} = -G/2$, that defines the boundaries of the Brillouin zones. Indeed, in one dimension this equation yields $k = G/2 = \ell\pi/a$. In the absence of the periodic potential the wave function of the electron is that of a free particle, $\Psi_{\mathbf{k}}^{(0)}(\mathbf{r}) = \exp[i\mathbf{k} \cdot \mathbf{r}]/\sqrt{V}$ (V is the total volume of the crystal), and this is a plane wave of the same type as discussed in Chapter 3. A wave with $\mathbf{k}$ far away from the boundaries of the Brillouin zone traverses the lattice almost unscattered. The energy corresponding to such a wave vector is hence very close to that of the free particle, $E(\mathbf{k}) \approx E^{(0)}(\mathbf{k}) = \hbar^2 \mathbf{k}^2/(2m)$. The probability density of the particle is constant in space, $|\Psi_{\mathbf{k}}^{(0)}(\mathbf{r})|^2 = 1/V$. As opposed, when $\mathbf{k}$ complies with the condition for the Bragg scattering, Eq. (3.26), the wave (that describes the electron) is scattered; the wave function contains the original wave together with all the scattered ones.

In one dimension the incident wave is $\exp[ikx]/\sqrt{L}$, and the only possible scattered wave is $\exp[-ikx]/\sqrt{L}$. The wave function is a linear combination of these two waves,

$$\Psi \to \frac{1}{\sqrt{L}}\left(\alpha e^{ikx} + \beta e^{-ikx}\right) = \frac{1}{\sqrt{L}}\left((\alpha + \beta)\cos(kx) + i(\alpha - \beta)\sin(kx)\right) ,$$

$$|\alpha|^2 + |\beta|^2 = 1 .$$

The probability density of the electron's location is thence a function of the coordinate. Assuming that the positive ions are located at $x = (n + 1/2)a$, i.e., at the centers of the unit cells, then the potential there is attractive, tending to bind the electron to the ion. It is thus plausible that the energy of the electron is lowest when the probability density to find it at these locations is maximal. This is particularly prominent at the boundary of the Brillouin zone, $k = \pi/a$. The probability density of the electron at the ions' locations is maximal when its wave function is $\Psi^-(x) = \sin(\pi x/a)/\sqrt{2L}$. As opposed, the energy of the electron is highest in the other extreme case, when the wave function is $\Psi^+(x) = \cos(\pi x/a)/\sqrt{2L}$, because then the probability density at the vicinity of the ions is the lowest. One therefore may expect that when $k = \pi/a$ there is an energy gap between the two states Ψ^+ and Ψ^-, as can indeed be observed in Fig. 6.11(a). Interestingly enough, both these states represent standing waves, with equal weights for the incident and reflected waves, $|\alpha|^2 = |\beta|^2$. This feature of the wave functions stems from the total reflection that is realized when the wave vector obeys the Bragg condition (3.26). Recall that standing waves also characterize lattice vibrations at the vicinity of the Brillouin-zone boundaries [see the discussion following Eq. (5.8)]. A systematic derivation of this result, as well as the magnitude of the energy gap, is presented in the following.

Bragg scattering does not contribute to the electrical resistance. In one dimension, the wave function of the scattered particle comprises the incident wave, with wave vector k, and the scattered wave, with the wave vector $-k = k+G$. In three dimensions there are many scattered waves, and the wave function of the electron is a linear combination of the incident wave of wave vector $\mathbf{k}$, and all scattered waves, each with a wave vector $\mathbf{k} \to \mathbf{k} + \mathbf{G}$. The coefficients of this linear combination are computed below. Note though that by Eqs. 6.63) and (6.64) a shift of the lattice momentum by a reciprocal-lattice vector does not change the wave function and the corresponding energy. This is the reason why **Bragg scattering does not contribute to the electrical resistance of the metal**, contrary to the original conjecture of Drude.

The Schrödinger equation in momentum space. The analysis in this section provides yet another **proof of the Bloch theorem**. Consider the Fourier transforms of the periodic potential and of the electron's wave function. As seen in Sec. 3.8, the Fourier transform of a function that is periodic on a lattice contains only components with wave vectors belonging to the reciprocal lattice [Eq. (3.34)], and therefore the potential energy can be written as

$$U(\mathbf{r}) = \sum_{\mathbf{G}} \tilde{U}(\mathbf{G})e^{i\mathbf{G}\cdot\mathbf{r}} , \tag{6.76}$$

where the sum runs over all N reciprocal-lattice vectors (N is the number of unit cells in the original lattice). The coefficients in the sum are given by the Fourier transforms,

$$\widetilde{U}(\mathbf{G}) = \frac{1}{\mathrm{v}} \int_{\mathrm{v}} d^3 r \, e^{-i\mathbf{G}\cdot\mathbf{r}} U(\mathbf{r}) \,, \tag{6.77}$$

where the integration is carried out over a single unit cell, $\mathrm{v} = V/N$. For $\mathbf{G} = 0$, the Fourier transform is $\widetilde{U}(0) = (1/\mathrm{v}) \int_{\mathrm{v}} d^3 r \, U(\mathbf{r})$, and is a constant that represents the choice of the coordinate origin. This constant is subtracted from all energies, so that the sum over $\mathbf{G}$ in Eq. (6.76) comprises only nonzero reciprocal-lattice vectors [i.e., it is practically assumed that $\widetilde{U}(0) = 0$]. The potential $U(\mathbf{r})$ is real, and hence $\widetilde{U}(-\mathbf{G}) = \widetilde{U}^*(\mathbf{G})$. When the potential is invariant under reflection (as is usually the case for Bravais lattices), then $U(\mathbf{r}) = U(-\mathbf{r})$, and consequently $\widetilde{U}^*(\mathbf{G}) = \widetilde{U}(-\mathbf{G}) = \widetilde{U}(\mathbf{G})$; it follows that $\widetilde{U}(\mathbf{G})$ is real [see also problem (B.1)].

The Fourier transform of the wave function is

$$\psi(\mathbf{r}) = \sum_{\mathbf{k}} \widetilde{\psi}(\mathbf{k}) e^{i\mathbf{k}\cdot\mathbf{r}} \,, \tag{6.78}$$

where periodic boundary conditions along each of the basis vectors imply that $\exp[i\mathbf{k} \cdot N_i \mathbf{a}_i] = 1$, and hence $\mathbf{k} = \sum_i (m_i/N)\mathbf{b}_i$. Here, $\mathbf{b}_i$ are the reciprocal-lattice basis vectors, and the sum encompasses the vectors $\mathbf{k}$ in the first Brillouin zone. Equation (6.78) presents the wave function in a periodic potential as a linear combination of the free-particle wave functions, that by themselves constitute a basis of the Hilbert space of all wave functions (this is the Fourier theorem, see App. B).

The Schrödinger equation. Applying the Hamiltonian on Eq. (6.78) yields

$$-\frac{\hbar^2}{2m} \nabla^2 \Big(\sum_{\mathbf{k}} \widetilde{\psi}(\mathbf{k}) e^{i\mathbf{k}\cdot\mathbf{r}} \Big) = \sum_{\mathbf{k}} \frac{\hbar^2 k^2}{2m} \widetilde{\psi}(\mathbf{k}) e^{i\mathbf{k}\cdot\mathbf{r}} \,, \tag{6.79}$$

and

$$U(\mathbf{r})\psi(\mathbf{r}) = \sum_{\mathbf{G}} \widetilde{U}(\mathbf{G}) e^{i\mathbf{G}\cdot\mathbf{r}} \sum_{\mathbf{k}'} \widetilde{\psi}(\mathbf{k}') e^{i\mathbf{k}'\cdot\mathbf{r}} = \sum_{\mathbf{k}} \Big(\sum_{\mathbf{G}} \widetilde{U}(\mathbf{G}) \widetilde{\psi}(\mathbf{k} - \mathbf{G}) \Big) e^{i\mathbf{k}\cdot\mathbf{r}} \,. \tag{6.80}$$

In the second equality of Eq. (6.80) the summation index is changed, $\mathbf{k}' \to \mathbf{k} = \mathbf{k}' + \mathbf{G}$. Inserting Eqs. (6.79) and (6.80) into the Schrödinger equation (6.50) results in

$$\sum_{\mathbf{k}} \Big(\Big[\frac{\hbar^2 k^2}{2m} - E \Big] \widetilde{\psi}(\mathbf{k}) + \sum_{\mathbf{G}} \widetilde{U}(\mathbf{G}) \widetilde{\psi}(\mathbf{k} - \mathbf{G}) \Big) e^{i\mathbf{k}\cdot\mathbf{r}} = 0 \,. \tag{6.81}$$

This sum vanishes only when all the coefficients (in the circular brackets) vanish, i.e.,

$$\Big(\frac{\hbar^2 k^2}{2m} - E \Big) \widetilde{\psi}(\mathbf{k}) + \sum_{\mathbf{G}} \widetilde{U}(\mathbf{G}) \widetilde{\psi}(\mathbf{k} - \mathbf{G}) = 0 \,. \tag{6.82}$$

Equation (6.82) is a set of linear homogeneous equations, involving N unknowns $\{\widetilde{\psi}(\mathbf{k} - \mathbf{G})\}$, where $\mathbf{G}$ is one of the reciprocal-lattice vectors (including $\mathbf{G} = 0$). Each of the $\mathbf{k}$ vectors in the first Brillouin zone possesses such a set. Except for certain particular values of $\mathbf{k}$ (discussed below) these sets do not mix coefficients related to different $\mathbf{k}$'s, and therefore each set can be treated separately. Equation (6.82) can be presented in a matrix form, with an $N \times N$ matrix multiplying the vectors comprising the N unknowns. In the one-dimensional example discussed at the beginning of this section, the matrix equation is

$$\begin{bmatrix} \cdots & \cdots & \cdots & \cdots & \cdots & \cdots \\ \cdots & E^{(0)}(k-G_0)-E & \widetilde{U}(-G_0) & \widetilde{U}(-2G_0) & \widetilde{U}(-3G_0) & \cdots \\ \cdots & \widetilde{U}(G_0) & E^{(0)}(k)-E & \widetilde{U}(-G_0) & \widetilde{U}(-2G_0) & \cdots \\ \cdots & \widetilde{U}(2G_0) & \widetilde{U}(G_0) & E^{(0}(k+G_0)-E & \widetilde{U}(-G_0) & \cdots \\ \cdots & \widetilde{U}(3G_0) & \widetilde{U}(2G_0) & \widetilde{U}(G_0) & E^{(0)}(k+2G_0)-E & \cdots \\ \cdots & \cdots & \cdots & \cdots & \cdots & \cdots \end{bmatrix}$$

$$\times \begin{bmatrix} \cdots \\ \widetilde{\psi}(k-G_0) \\ \widetilde{\psi}(k) \\ \widetilde{\psi}(k+G_0) \\ \widetilde{\psi}(k+2G_0) \\ \cdots \end{bmatrix} = 0 \, , \tag{6.83}$$

where $G_0 = 2\pi/a$ is the basic reciprocal-lattice vector, and where k is in the first Brillouin zone. Note that the number of rows and columns equals the number of reciprocal-lattice vectors, that is, the number of lattice sites, N. Equation (6.83) has nontrivial solutions when the determinant vanishes; this requirement determines the values of the energy E for each k. These values are termed $E_n(k)$; the corresponding wave functions, $\psi_{nk}(x)$, are obtained by inserting the solutions of Eq. (6.83) in Eq. (6.78).

Another proof of the Bloch theorem. The solutions of the Schrödinger equation, as obtained from Eq. (6.82) for a given $\mathbf{k}$, are

$$\psi_{n\mathbf{k}}(\mathbf{r}) = \sum_{\mathbf{G}} \widetilde{\psi}_n(\mathbf{k} - \mathbf{G})e^{i(\mathbf{k}-\mathbf{G})\cdot\mathbf{r}} \, . \tag{6.84}$$

They may be written in the form

$$\psi_{n\mathbf{k}}(\mathbf{r}) = e^{i\mathbf{k}\cdot\mathbf{r}} \sum_{\mathbf{G}} \widetilde{\psi}_n(\mathbf{k} - \mathbf{G})e^{-i\mathbf{G}\cdot\mathbf{r}} \equiv e^{i\mathbf{k}\cdot\mathbf{r}} u_{n\mathbf{k}}(\mathbf{r}) \, . \tag{6.85}$$

The function $u_{n\mathbf{k}}(\mathbf{r})$ defined in the right hand-side obeys Eq. (6.53), $u_{n\mathbf{k}}(\mathbf{r}) = u_{n\mathbf{k}}(\mathbf{r} + \mathbf{R})$, which validates the Bloch theorem. It can be verified that

$$\psi_{n,\mathbf{k}+\mathbf{G}}(\mathbf{r}) = \sum_{\mathbf{G}'} \widetilde{\psi}_n(\mathbf{k} + \mathbf{G} - \mathbf{G}')e^{-i(\mathbf{k}+\mathbf{G}-\mathbf{G}')\cdot\mathbf{r}}$$

$$= \sum_{\mathbf{G}''} \widetilde{\psi}_n(\mathbf{k} - \mathbf{G}'')e^{i(\mathbf{k}-\mathbf{G}'')\cdot\mathbf{r}} = \psi_{n\mathbf{k}}(\mathbf{r}) \tag{6.86}$$

(the summation index is changed, $\mathbf{G}' \to \mathbf{G}'' = \mathbf{G}' - \mathbf{G}$); this proves Eq. (6.63). Equation (6.64) follows from the latter. Thus, it suffices to investigate the solutions in the first Brillouin zone.

The weak-binding approximation. The complete solution of the Schrödinger equation can be derived, in principle, from Eqs. (6.82). It is a formidable task to solve those because of the huge number of equations; hence approximation methods are called for. When the lattice is "empty", i.e., $\widetilde{U}(\mathbf{G}) = 0$, Eq. (6.82) yields $[\hbar^2 k^2/(2m) - E]\widetilde{\psi}(\mathbf{k}) = 0$; consequently $\widetilde{\psi}(\mathbf{k}) \neq 0$ only when the energy is $E = E^{(0)}(\mathbf{k}) = \hbar^2 k^2/(2m)$. Then all other coefficients in Eq. (6.84), $\widetilde{\psi}(\mathbf{k} - \mathbf{G})$ with $\mathbf{G} \neq 0$, have to vanish, and so the sum in Eq. (6.84) comprises a single term, $\psi_{\mathbf{k}}(\mathbf{r}) = \widetilde{\psi}^{(0)}(\mathbf{k}) \exp[i\mathbf{k} \cdot \mathbf{r}]$. This is the wave function of the free electron, $\Psi_{\mathbf{k}}^{(0)}(\mathbf{r}) = \exp[i\mathbf{k} \cdot \mathbf{r}]/\sqrt{V}$, implying that $\widetilde{\psi}^{(0)}(\mathbf{k}) = 1/\sqrt{V}$. Suppose that the periodic potential is very weak, such that it is a small fraction of the free-particle energy. Specifically, $|\widetilde{U}(\mathbf{G})| \leq |\widetilde{U}(\mathbf{G}_0)| = \lambda E^{(0)}(\mathbf{G}_0)$, where λ is a small dimensionless parameter, $\lambda \ll 1$, which characterizes the weakness of the potential, and $\mathbf{G}_0$ is the shortest reciprocal-lattice vector. [The first inequality follows from the integral in Eq. (6.77): it is smaller when $|\mathbf{G}|$ is larger, due to the oscillations of the factor $\exp[i\mathbf{G} \cdot \mathbf{r}]$ in the integrand.] Barring exceptional cases, to be discussed separately, the other coefficients in the sum are such that $|\widetilde{\psi}(\mathbf{k} - \mathbf{G})/\widetilde{\psi}(\mathbf{k})| = \mathcal{O}(\lambda)$, and hence can be considered as small corrections to the wave function: both the wave function $\psi(\mathbf{r})$ and the energy $E(\mathbf{k})$ are quite close to their free-particle values, $\Psi_{\mathbf{k}}^{(0)}(\mathbf{r})$ and $E^{(0)}(\mathbf{k})$. For one of these other coefficients, $\widetilde{\psi}(\mathbf{k} - \mathbf{G})$ with $\mathbf{G} \neq 0$, Eq. (6.82) reads

$$[E^{(0)}(\mathbf{k} - \mathbf{G}) - E(\mathbf{k})]\widetilde{\psi}(\mathbf{k} - \mathbf{G}) = -\sum_{\mathbf{G}'} \widetilde{U}(\mathbf{G}')\widetilde{\psi}(\mathbf{k} - \mathbf{G} - \mathbf{G}') \ .$$

The right hand-side includes the term $\widetilde{U}(-\mathbf{G})\widetilde{\psi}(\mathbf{k})$, that comes from $\mathbf{G}' = -\mathbf{G}$. This is the largest term in the sum; all other terms are smaller as they contain the "other" coefficients of the solution, $\widetilde{\psi}(\mathbf{k} - \mathbf{G} - \mathbf{G}')$, with $\mathbf{G} + \mathbf{G}' \neq 0$. They thus can be discarded, leading to

$$\widetilde{\psi}(\mathbf{k} - \mathbf{G}) \approx -\widetilde{U}(-\mathbf{G})\widetilde{\psi}(\mathbf{k})/[E^{(0)}(\mathbf{k} - \mathbf{G}) - E^{(0)}(\mathbf{k})] \ , \qquad (6.87)$$

where $E(\mathbf{k}) \approx E^{(0)}(\mathbf{k})$ is used. Assuming that the denominator is not too small, then $\widetilde{\psi}(\mathbf{k} - \mathbf{G})/\widetilde{\psi}(\mathbf{k}) = \mathcal{O}(\lambda)$, making the discarded terms of the order of λ^2. Inserting Eq. (6.87) in Eq. (6.82) yields

$$E(\mathbf{k}) = E^{(0)}(\mathbf{k}) - \sum_{\mathbf{G}} \frac{|\widetilde{U}(\mathbf{G})|^2}{E^{(0)}(\mathbf{k} - \mathbf{G}) - E^{(0)}(\mathbf{k})} + \dots \ , \qquad (6.88)$$

where $\dots$ represent higher-order terms in λ. Finally, the assumption that none of the denominators in the sum is too small implies that the correction to the energy due to the periodic potential is of the order of λ^2; thus it is indeed small. Higher-order corrections are even smaller.

The weak-binding approximation near Brillouin-zone boundaries. The expansion in Eq. (6.88) is valid as long as $|\widetilde{U}(\mathbf{G})/[E^{(0)}(\mathbf{k} - \mathbf{G}) - E^{(0)}(\mathbf{k})]| < \lambda$, but it fails for pairs of wave vectors for which the difference $E^{(0)}(\mathbf{k} - \mathbf{G}) - E^{(0)}(\mathbf{k})$ is small [e.g., of the order of $\lambda E^{(0)}(\mathbf{G}_0)$], because then the corrections to $E(\mathbf{k})$ might be larger than the intervals between the original energies. This energy difference vanishes for

$$E^{(0)}(\mathbf{k} - \mathbf{G}) - E^{(0)}(\mathbf{k}) = \frac{\hbar^2}{2m}[(\mathbf{k} - \mathbf{G})^2 - \mathbf{k}^2] = 0 \;,$$

i.e., when $2\mathbf{k} \cdot \mathbf{G} = \mathbf{G}^2$. This is, up to a sign, the condition for Bragg scattering, Eq. (3.26), discussed also at the beginning of this section. The condition implies that the energies of the states corresponding to $\mathbf{k}$ and to $(\mathbf{k} - \mathbf{G})$ are degenerate whenever the tip of $\mathbf{k}$ is located on the plane perpendicular to a reciprocal-lattice vector at its mid point, i.e., it is located on the border between two Brillouin zones.

The energy gaps in one dimension. In a one-dimensional lattice with a constant a Eq. (3.26) is fulfilled at a single point for each reciprocal-lattice vector, i.e., when $k = G/2 = \ell\pi/a$, where the "free states" with k and $(k - G)$ are degenerate. For such states, the energy $E(k)$ is close to both $E^{(0)}(k)$ and $E^{(0)}(k - G)$, but is rather far away from all other energies $E^{(0)}(k - G')$, with $G' \neq G, 0$. Considerations similar to those leading to Eq. (6.87) yield that the coefficients $\widetilde{\psi}(k - G')$, with $G' \neq G$, are still rather small. Nonetheless, in the present situation there are two coefficients that are not necessarily small, $\widetilde{\psi}(k)$ and $\widetilde{\psi}(k - G)$. For these two special coefficients, Eq. (6.82) yields

$$[E^{(0)}(k) - E(k)]\widetilde{\psi}(k) + \widetilde{U}(G)\widetilde{\psi}(k - G) = 0 \;,$$
$$[E^{(0)}(k - G) - E(k)]\widetilde{\psi}(k - G) + \widetilde{U}(-G)\widetilde{\psi}(k) = 0 \;. \tag{6.89}$$

All other, much smaller, coefficients are discarded. Equations (6.89) may be obtained from Eq. (6.83) once only the first two rows and the first two columns there are retained. In general, keeping only the rows and the columns related to degenerate energies gives a reliable approximation for the energy bands. [A similar result is derived in perturbation theory for degenerate states; see Eq. (C.4) and problem 6.15.]

The two linear homogeneous equations (6.89) have a nontrivial solution when the determinant of their coefficients vanishes. On the border of the Brillouin zone $E^{(0)}(k) = E^{(0)}(k - G) = E^{0)}(G/2)$; then this requirement yields two possible energies

$$E_{\pm}^{(1)}(G/2) = E^{(0)}(G/2) \pm |\widetilde{U}(G)| \;, \tag{6.90}$$

with a gap in-between them, $2|\widetilde{U}(G)|$. In contrast to Eq. (6.88), which gives corrections of the order λ^2 to the energy, the corrections obtained here are far larger, of order λ.

As the potential energy is real, Eq. (6.77) implies that $\widetilde{U}(G) = \widetilde{U}^*(-G)$. When it is also even, $U(x) = U(-x)$, then $\widetilde{U}(G) = \widetilde{U}(-G)$ [see Problem (B.1)]. Hence, $\widetilde{U}(G) = \widetilde{U}_0(G)$ is real, and the eigenvectors of Eq. (6.89) are such that $\widetilde{\psi}(-G/2) = \pm\widetilde{\psi}(G/2)$. The eigenfunctions are $\Psi_{G/2}^{\pm} = [\Psi_{G/2}^{(0)} \pm \Psi_{-G/2}^{(0)}]/\sqrt{2L}$; these are precisely the two standing-wave functions obtained qualitatively in the discussion of Bragg reflections. Indeed, the potential energy "responsible" for these wave functions, for $G = G_0 = 2\pi/a$, is

$$U(x) = \widetilde{U}_0(G)(e^{iGx} + e^{-iGx}) = 2\widetilde{U}_0(G)\cos(Gx) = 2\widetilde{U}_0(G)\cos(2\pi x/a) \ .$$

Assuming that $\widetilde{U}_0(G) > 0$, the minima of the potential are at the points $x = (n + 1/2)a$; therefore the function $\Psi^-_{G/2}(x) = \sin(\pi x/a)/\sqrt{2L}$ has the lower energy.

Energies near Brillouin-zone boundaries. In the general case where k is close to the border of a Brillouin zone, the eigenvalues of Eq. (6.89) are

$$E^{(1)}_\pm(k) = \left[E^{(0)}(k) + E^{(0)}(k - G) \pm \sqrt{[E^{(0)}(k) - E^{(0)}(k - G)]^2 + 4|\widetilde{U}(G)|^2} \right]/2 \ . \tag{6.91}$$

Denoting $k = G/2 + \delta$ an expansion in δ up to second order gives

$$\begin{aligned}
E^{(1)}_\pm &= \frac{\hbar^2}{2m}\left(\frac{G^2}{4} + \delta^2 \right) \pm \sqrt{|\widetilde{U}(G)|^2 + \left(\frac{\hbar^2 G\delta}{2m} \right)^2} \\
&\approx \frac{\hbar^2 G^2}{8m} \pm |\widetilde{U}(G)| + \frac{\hbar^2}{2m}\left(1 \pm \frac{\hbar^2 G^2}{2m|\widetilde{U}(G)|} \right)\delta^2 + \dots \ . \tag{6.92}
\end{aligned}$$

(Show that the approximation is valid only when

$$\hbar^2\delta^2/(2m) \ll \widetilde{U}(G)|^2/E^{(0)}(G) = \mathcal{O}[\lambda^2 E^{(0)}(G)] \ ,$$

i.e., when $\delta \ll \lambda G$, in the proximity of the point $k = G/2$.) Thus, an energy gap of width $2|\widetilde{U}(G)|$ opens at $k = G/2$. Around each of the "new" energies $E^{(1)}_\pm(G/2) = \hbar^2 G^2/(8m) \pm \widetilde{U}_0(G)$ the energy depends quadratically on δ. Equation (6.92) gives an approximate expression for the energies near each border of the Brillouin zones, that is, at the vicinity of the points $k = G/2 = \ell\pi/a$. Away from these points, the degeneracy may be ignored, and a good approximation for the energies is the form (6.88).

Problem 6.15.
Show that Eqs. (6.88) and (6.89) are derived directly from the perturbation theory, as summarized in Appendix C.

Problem 6.16.
An alternative way to solve problem 6.14, with the potential $U(x) = u_0 \sum_n \delta(x - na)$: calculate the Fourier transform of the wave function, $\widetilde{\psi}(k)$, using Eq. (6.82). [Hint: Derive an equation for the function $f(k) = \sum_G \widetilde{\psi}(k - G)$, and exploit the mathematical identity $\cot z = \sum_{n=-\infty}^{\infty}(z - n\pi)^{-1}$, documented in mathematical-formulae books.]

Problem 6.17.
Use the weak-binding potential to derive the energy gaps for the Kronig-Penney potential, Eq. (6.65). Confining yourself to the delta-function limit, compare the results with those of problem 6.14.

Figure 6.12(a) compares the energy of a free particle with the energy corresponding to a periodic potential described by the shortest reciprocal-lattice vector $G_0 = 2\pi/a$. The figure displays the jump between the two energies on the boundary of the first Brillouin zone, Eq. (6.90). Similar jumps appear at the other boundaries of the Brillouin zones, as seen in Fig. 6.12(b). In each case, the behavior near $k = G/2$ is derived from Eq. (6.92). For the sake of clarity, the parameters are chosen such that all gaps are equal, $2\widetilde{U}_0(G_0) = 2\widetilde{U}(2G_0) = 2\widetilde{U}(3G_0) = 0.6$ [in energy units of $E^{(0}(G_0/2) = \hbar^2\pi^2/(2ma^2)$]. In general one expects $\widetilde{U}_0(G)$ to decrease as G increases, due to the oscillations of the factor $\exp[iGx]$ in the integrand of the Fourier transform, Eq. (6.77).

Figure 6.12(b) illustrates the modifications in the parabolic energy of the free particle stemming from the weak periodic potential, for **all** k values $-\infty \ll k \ll \infty$. For each value of k (except the boundaries of the Brillouin zones) there is a single value of the energy, derived from the original free-particle energy with an additional small correction. This description is termed the **"extended spectrum"**.

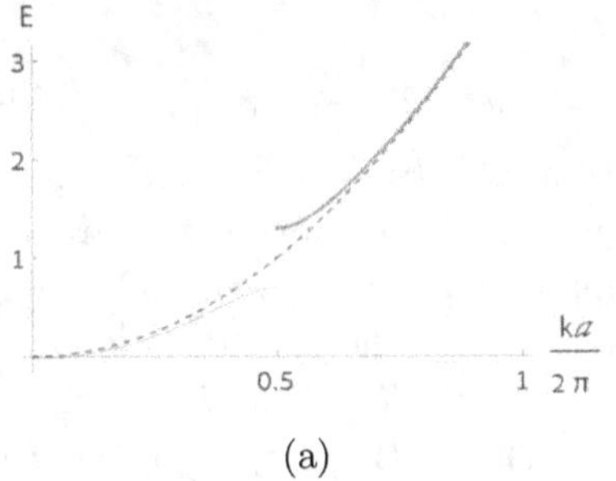
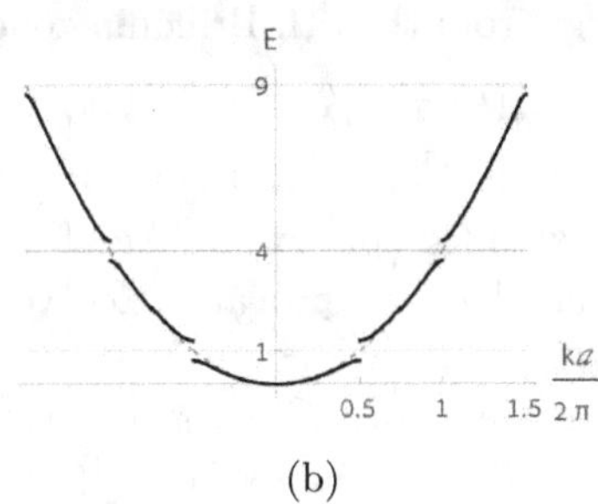

(a)   (b)

Fig. 6.12: (a) The dashed line is the parabolic energy of a free particle in one dimension, as a function of k, $E^{(0)}(k) = \hbar^2 k^2/(2m)$, in units of $E^{(0)}(G_0/2) = \hbar^2\pi^2/(2ma^2)$. The solid lines are the energies of a particle moving in a periodic potential described by the reciprocal-lattice vector $G_0 = 2\pi/a$, with $\widetilde{U}_0(G_0) = 0.3$ (in the same units). These lines are computed from Eq. (6.92), $E_-^{(1)}(k)$ for $k < \pi/a$, and $E_+^{(1)}(k)$ for $k > \pi/a$. (b) The dashed line illustrates the energy of the free particle, while the solid lines display the energies close to additional boundaries of Brillouin zones, due to periodic potentials that correspond to other reciprocal-lattice vectors, with $\widetilde{U}_0(2G_0) = \widetilde{U}_0(3G_0) = 0.3$. The behavior at $k < 0$ is obtained by reflection, $E(-k) = E(k)$.

The reduced spectrum. The Bloch theorem states that it is impossible to distinguish between the energies in the different Brillouin zones, see Eq. (6.64). Indeed, once the energies in Fig. 6.12(b) are displaced by a reciprocal-lattice vector to the right or to the left, the resulting figure resembles Fig. 6.11(a). In the limit of a very weak periodic potential the figure resembles Fig. 6.10(a), which at times is referred to as the **empty-lattice approximation**. Figure 6.13(a) reproduces Fig. 6.10(a) for a very weak potential. As mentioned in the context of Fig. 6.10(a), this approximation is plausible provided that one keeps in mind that it is an approximation of Fig. 6.11(a), which displays the spectrum for a weak periodic potential.

Figures as Fig. 6.10(a) or Fig. 6.11(a) comprise numerous energy levels for each value of k. Such figures contain many repetitions that convey the same information numerous times. To avoid this extensive redundancy one may employ the **extended spectrum**, Fig. 6.12(b). Another option in which each energy level appears only once is presented in Fig. 6.13(b) (showing the spectrum in the empty-lattice approximation). In this figure all energies of the extended spectrum are moved into the first Brillouin zone. The energy levels of the extended spectrum are in one-to-one correspondence with the shifted ones in the first Brillouin zone. The latter description is called the **reduced spectrum**. For each value of k, the first Brillouin zone contains many energy levels, shifted from other Brillouin zones. These are distinguished by the index n which counts the bands, and are denoted $E_n(k)$. The function corresponding to the nth energy band is obtained upon displacing the energy from the nth Brillouin zone into the first one,

$$E_n(k) = E[k - 2(n-1)\pi/a] , \quad \text{or} \quad E_n(k) = E[k + 2(n-1)\pi/a] . \tag{6.93}$$

The dashed curve in Fig. 6.13(b) is the parabola $E^{(0)}(k) = \hbar^2 k^2/(2m)$, that is, the extended spectrum of the free particle. Figure 6.13(a) comprises all parabolas displaced by reciprocal-lattice vectors (as compared to the original one). When $E^{(0)}(k) < \hbar^2(\pi/a)^2/(2m)$, the wave number k is within the first Brillouin zone, and the corresponding energy is $E_1^{(0)} = E^{(0)}(k) = \hbar^2 k^2/(2m)$, again within the first Brillouin zone. When $\hbar^2(\pi/a)^2/(2m) < E^{(0)}(k) < \hbar^2(2\pi/a)^2/(2m)$, the wave number k is in the second Brillouin zone, and consists of two segments, $\pi/a < k < 2\pi/a$ and $-2\pi/a < k < -\pi/a$. These segments are moved into the first Brillouin zone by $k \Rightarrow k - 2\pi/a$ and $k \Rightarrow k + 2\pi/a$, respectively. The corresponding shifted energies are $E_2^{(0)}(k) = \hbar^2(k - 2\pi/a)^2/(2m)$ and $E_2^{(0)}(k) = \hbar^2(k + 2\pi/a)^2/(2m)$, respectively. In a similar fashion, the k values in the nth Brillouin zone are displaced: $k \Rightarrow k - 2(n-1)\pi/a$ and $k \Rightarrow k + 2(n-1)\pi/a$, giving rise to the energies $E_n^{(0)}(k) = \hbar^2[k \pm 2(n-1)\pi/a]^2/(2m)$ in the reduced spectrum. The energies shifted into the first Brillouin zone are presented by the thick curves in Fig. 6.13(b), and the corresponding displacements of the wave numbers are marked by the arrows there. The boundaries of the first Brillouin zone are the vertical lines at $k = \pm\pi/a$, and the borders between the various n values are the vertical dashed lines. The figure illustrates that the spectrum can be presented in two equivalent forms: by the original parabola $E^{(0)}(k)$ or by the energy levels $E_n^{(0)}(k)$ in the first Brillouin zone, with a one-to-one correspondence between the two. Figure 6.13(a) emphasizes the periodicity of the energy levels in the reciprocal lattice; the information on the electronic energies is contained within the first Brillouin zone. One therefore confines oneself to the energies $E_n^{(0)}(k)$ in the first Brillouin zone, and examines the corrections caused to them by the periodic potential. These corrections are crucial when the original levels are degenerate, i.e., when the curves $E_n^{(0)}(k)$ cut each other. In Fig. 6.13(b) such cuts occur at the origin or at the edges of the Brillouin zone. At each of them the periodic potential causes the appearance of an energy gap, with the energies nearby varying quadratically as a function of the distance from the cut.

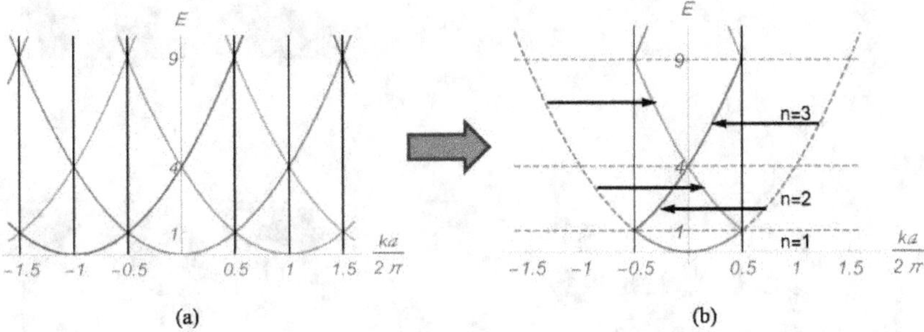

Fig. 6.13: (a) The periodic functions $E_n^{(0)}(k)$ in several Brillouin zones in the empty-lattice approximation [see also Fig. 6.10(a)]. The vertical lines are the zone boundaries [the wave number k is measured in units of the shortest reciprocal-lattice vector, $2\pi/a$, and the energy is measured in units of $E^{(0)}(G_0/2) = \hbar^2\pi^2/(2ma^2)$]. (b) The dashed parabola [$E^{(0)}(k) = \hbar^2 k^2/(2m)$ in suitable units] is the free-particle energy, in the absence of the periodic potential. This is the extended spectrum of the particle in the empty-lattice approximation. The thick lines display the reduced spectrum: these are the energies displaced by reciprocal-lattice vectors, from the various parabolas of part (a). The index n "counts" the energy levels $E_n^{(0)}(k)$ for a given k value in the first Brillouin zone.

The energy bands in the empty-lattice approximation in two dimensions. At higher dimensions the description of the spectrum is more involved, as the free-particle energy, the paraboloid $E^{(0)}(\mathbf{k}) = \hbar^2\mathbf{k}^2/(2m)$, cuts the boundaries of the Brillouin zones at various energies for various direction of $\mathbf{k}$. Consider for instance the square lattice in the plane. Figure 6.14(a) shows this paraboloid in the first Brillouin zone. As seen, the paraboloid meets the center of the face, $\mathbf{k}_M = (\pi/a, 0)$ at the energy $E^{(0)}(\mathbf{k}_M) = \hbar^2(\pi/a)^2/(2m)$, but touches the corner, $\mathbf{k}_K = (\pi/a, \pi/a)$ at twice that energy, $E^{(0)}(\mathbf{k}_K) = 2\hbar^2(\pi/a)^2/(2m)$. The notations for these special points of the Brillouin zone are defined in Fig. 5.7(a). Figure 6.14(b) displays the equal-energy curves corresponding to this paraboloid. At the center of the zone these curves are perfect circles, but as the radius becomes larger than π/a, part of the circle penetrates into the second Brillouin zone. As noticed in the one-dimensional case, the energies located within the second zone are part of the second energy band. Indeed, exploiting the periodicity of the energies in reciprocal

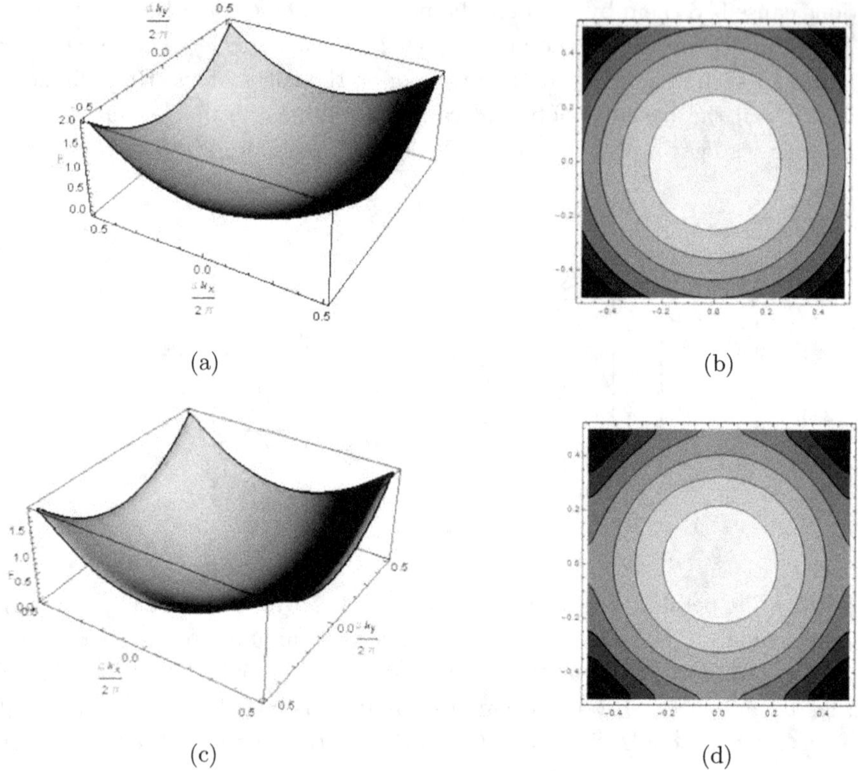

Fig. 6.14: The energy in the first Brillouin zone [measured in units of $E^{(0)}(\mathbf{k}_M)$ and $\mathbf{k}$ is measured in units of the shortest reciprocal-lattice vector, $2\pi/a$]. (a) The free-particle energy as a function of the two components of $\mathbf{k}$. (b) Equal-energy curves for part (a): the brighter areas represent lower energies. (c) The energy of a particle in a weak periodic potential, $\widetilde{U}(\mathbf{k}_M) = 0.2$, $\widetilde{U}(\mathbf{k}_K) = 0.15$. Note in particular the modifications near the faces of the Brillouin zone: the energy surface is orthogonal to the faces. (d) Equal-energy curves for part (c). Again notice that the equal-energy curves are normal to the faces at the faces [see also Fig. 6.18(b)].

lattice enables one to bring them all into the first Brillouin zone. Figure 6.15(a) reproduces Fig. 3.14(a), with different notations, ($2a$, $2b$, $2c$, $2d$), for the various parts of the second Brillouin zone. Displacements by basic reciprocal-lattice vectors along the two axes transfer these parts into the first zone, as illustrated in Fig. 6.15(b). Figure 6.16(a) shows those parts of the original free-particle paraboloid that belong to the second zone, and Fig. 6.16(b) exhibits them after being displaced into the first zone, according to the scheme of Figs. 6.15. As seen, the energies of the second band are minimal at the centers of the faces, and are maximal at the center of the zone. Figure 6.17 portrays the same construction of the second band in the first zone, in terms of the equal-energy curves. It is convenient

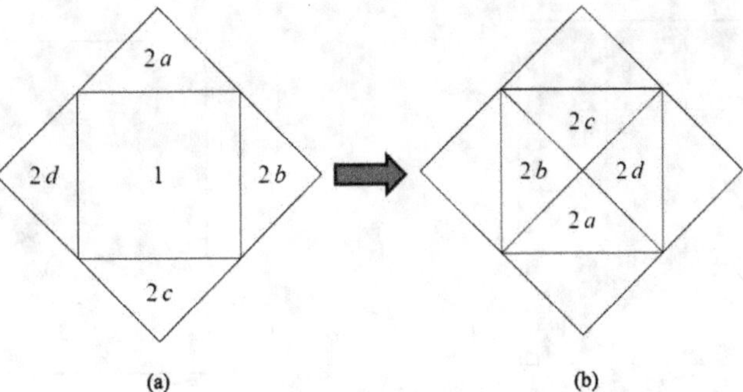

Fig. 6.15: Mapping of the second Brillouin zone [panel (a)] into the first Brillouin zone [in panel (b)].

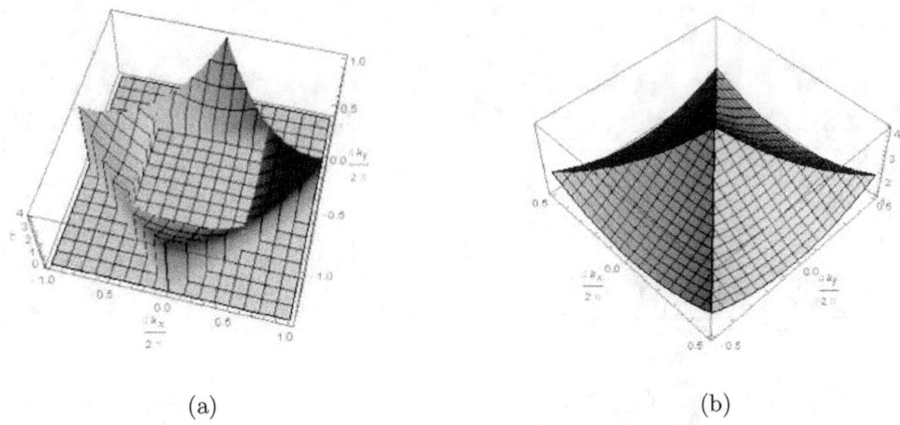

(a)   (b)

Fig. 6.16: (a) Parts of the original paraboloid in the second Brillouin zone (that belong to the extended spectrum). (b) The same parts after being shifted into the first Brillouin zone (these form the second band in the reduced spectrum).

at times to present the energy bands as a function of the wave vector along certain lines in the Brillouin zone (similarly to the figures showing the frequencies of the normal modes in Chapter 5). Such a path, pertaining to the square lattice, is displayed in Fig. 5.7. The energies of the first two bands along this path are shown in Fig. 6.18(a).

Problem 6.18.

Find the criteria for mapping the third Brillouin zone into the first one for a free particle on a square lattice. Display the equal-energy curves that belong to the third band in the first Brillouin zone.

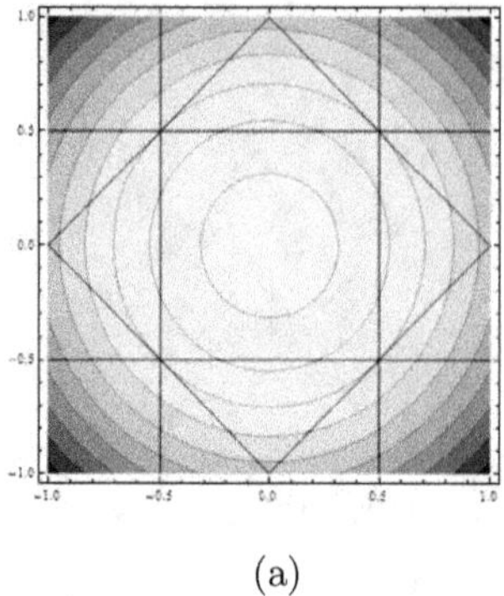

(a)

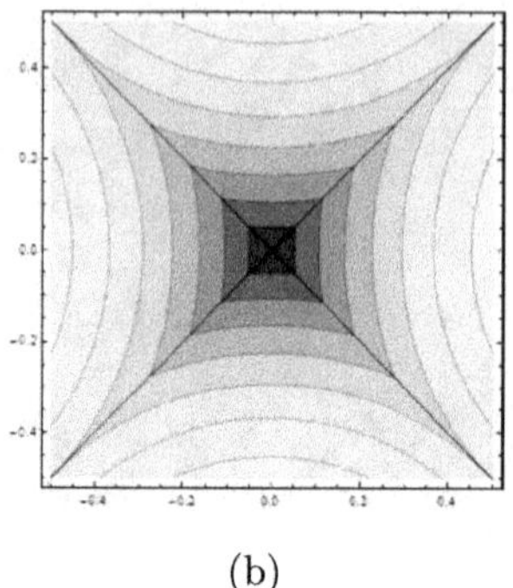

(b)

Fig. 6.17: (a) Equal-energy curves for the paraboloid of a free particle. The straight lines are the boundaries between the Brillouin zones. (b) Equal-energy curves from the second band after being shifted into the first Brillouin zone, as illustrated in Fig. 6.15.

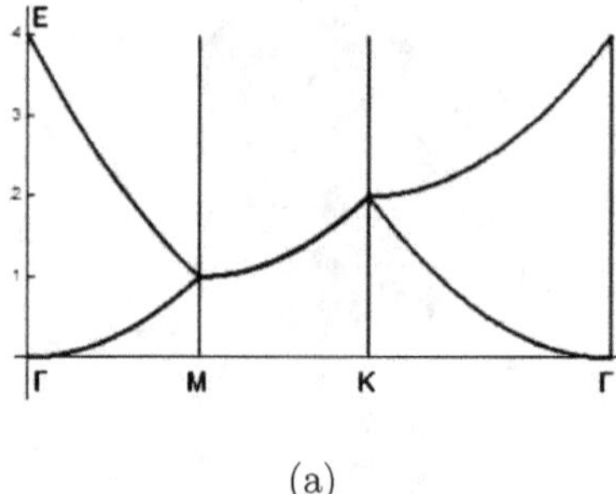

(a)

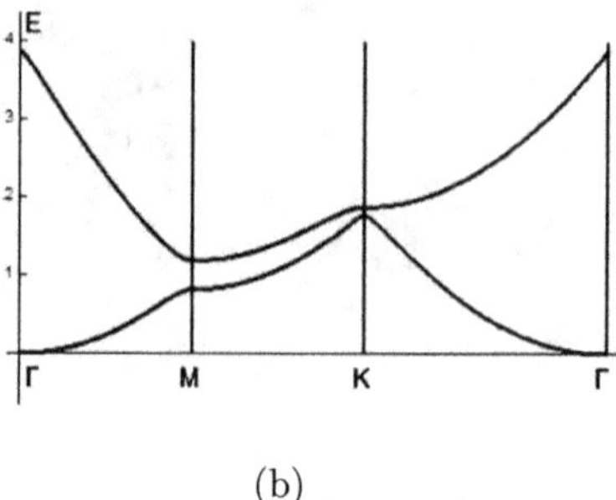

(b)

Fig. 6.18: The first two energy bands of a square lattice, on special lines in the first Brillouin zone, as defined in Fig. 5.7. (a) A free particle. (b) Particle in a weak periodic potential [as in Fig. 6.14(c)].

Band structure for a weak potential. Once the first two bands of the free particle are constructed, one can "switch-on" a weak periodic potential, and explore where are there degenerate states that cause energy gaps between these bands. A comparison of Fig. 6.14(a) (that illustrates the first energy band in the first zone) with Fig. 6.16(b) (which displays the second energy band in that zone) shows that the energies in the two bands are equal only on the perimeter of the zone. As noticed, this is a general property of neighboring bands (that have the same energies of the free particle on the zone boundaries). Indeed, the energies of the bands coincide along the line connecting points M and K in Fig. 6.18(a). The displacements of the two bands against one another created upon the opening of the gaps is shown schematically in Fig. 6.18(b). These gaps are calculated in problem 6.19. When the potential is weak the gaps are quite narrow, and then the minima of the second band, located at the centers of the first zone's faces, are lower than the maxima of the first band, located at the corners. Hence, as opposed to the situation in one dimension, in dimensions higher than 1 a band may penetrate into its neighbors (problem 6.13).

Figure 6.14(c) portrays the energy of the lowest band for a weak periodic potential (calculated in problem 6.19), and Fig. 6.14(d) displays the corresponding equal-energy curves. Note the differences between these two figures and Figs. 6.14(a) and (b) that present the bands for a free particle. These are in particular conspicuous near the zone boundaries. Due to the opening of the gaps, the energy of the first band is maximal there, and hence its derivative with respect to the component of the lattice momentum normal to this curve vanishes. For the same reason, the equal-energy curves are always normal to this boundary. This property of the spectrum is in particular visible in Fig. 6.18(b), that illustrates the energy along certain lines in the Brillouin zone.

Problem 6.19.

a. *Obtain an approximate equation for the energies near the zone boundaries by exploiting the nearly-free electron model, for an arbitrary lattice in arbitrary dimensions, when there are solely two degenerate states on each such boundary. Use the result to compute the energy gap created at a point $\mathbf{q}_0$ on one of these boundaries.*

b. *Calculate the energies at the two sides of the gap found in part (a) at $\mathbf{q} = \mathbf{q}_0 + \boldsymbol{\delta}$, for small $|\boldsymbol{\delta}|$. Discuss in particular the dependence of the energies on the normal and tangential components (with respect to the boundary surface) of the wave vector.*

6.7   *The tight-binding approximation*

c. *Use the results to obtain roughly Figs. 6.14(c) and (d), and explain the differences between your plots and Figs. 6.14(a) and (b).*

d. *Optional: the results above do not pertain to the corners of the Brillouin zone. Explain why, and describe how should the bands be calculated there. Apply this calculation to a square lattice, for which the only nonzero Fourier components are $\widetilde{U}(\mathbf{k}_M) = 0.2$ and $\widetilde{U}(\mathbf{k}_K) = 0.15$ (together with all other points obtained from those by symmetry operations).*

States built of atomic wave functions. Section 6.6 zeroes on the limit of a weak periodic potential, such that the eigenenergies remain close to those of a free particle, with minor corrections. When the binding potential-energy of the electron to one of the ions is rather strong, the energy, to a good approximation, is found by ignoring all other ions to which the electron is only loosely bound. That is, the calculation is based on the state in which the electron is bound solely to its parent atom, as in the isolated atom. This description is in particular useful when the ions are relatively far away from each other. It is also suitable for electrons located in the inner electronic shells, as these are quite close to the nucleus. Generally, the potential energy is the sum of the potential energies arising from the attraction of the electron to all ions; the Hamiltonian of a single electron is thus

$$\hat{\mathcal{H}} = -\frac{\hbar^2}{2m}\nabla^2 + \sum_{\mathbf{R}} u(\mathbf{r} - \mathbf{R}) , \qquad (6.94)$$

where $u(\mathbf{r} - \mathbf{R}) = -e^2/|\mathbf{r} - \mathbf{R}|$ is the potential energy of the Coulomb attraction between the electron and the ion at the lattice site $\mathbf{R}$. (For simplicity, we ignore here the screening of these potentials by the other electrons in the system.) This Hamiltonian is an extension of the one of the molecule H_2^+, Eqs. (4.12) and (4.13), that contains two ions. Strictly speaking, the various states of the electron on a single ion (or on all ions in the unit cell) have to be considered. This point is elaborated upon at the end of the section. Here is analyzed the simplest configuration in which there is a single ion possessing a single state in each unit cell, $\varphi^{(0)}(\mathbf{r})$. (In a similar fashion to the treatment of the hydrogen molecule, where only the ground state of each ion is included.) Assuming also that the corresponding atomic energy $E^{(0)}$ is not degenerate (save for the trivial degeneracy of the spin states), then when the ions are far away from each other there are N degenerate states on the lattice. The electron's wave function is the atomic $\varphi^{(0)}(\mathbf{r} - \mathbf{R})$ on each of the ions located at each of the N lattice sites $\{\mathbf{R}\}$, and the energy is the discrete energy of the electron on that ion, $E^{(0)}$,

$$\left[-\frac{\hbar^2}{2m}\nabla^2 + u(\mathbf{r} - \mathbf{R})\right]\varphi^{(0)}(\mathbf{r} - \mathbf{R}) = E^{(0)}\varphi^{(0)}(\mathbf{r} - \mathbf{R}) . \qquad (6.95)$$

The treatment of the H_2^+ molecule exploits the degenerate states of the electron on each of the hydrogen ions, see Sec. 4.3. As in that case, one aims to find a variational approximation for the wave function of the electron in the lattice, by using an **arbitrary linear combination of the "atomic" wave functions**,

$$\varphi(\mathbf{r}) = \sum_{\mathbf{R}} a(\mathbf{R})\varphi^{(0)}(\mathbf{r} - \mathbf{R}) . \qquad (6.96)$$

This wave function is termed **LCAO**, which abbreviates "Linear Combination of Atomic Orbitals". A similar combination is encountered in the context of electrons in the benzene molecule, Eq. (4.27).

The conditions for the Bloch theorem. The coefficients $a(\mathbf{R})$ have to fulfill certain requirements for the function (6.96) to comply with the Bloch theorem, Eq. (6.54), $\varphi(\mathbf{r} + \mathbf{R}) = \exp[i\mathbf{k} \cdot \mathbf{R}]\varphi(\mathbf{r})$. Displacing $\mathbf{r}$ by $\mathbf{R}$, and changing the summation index $\mathbf{R}' \to \mathbf{R} + \mathbf{R}'$ in Eq. (6.96) give

$$\varphi(\mathbf{r} + \mathbf{R}) = \sum_{\mathbf{R}'} a(\mathbf{R}')\varphi^{(0)}(\mathbf{r} + \mathbf{R} - \mathbf{R}') = \sum_{\mathbf{R}'} a(\mathbf{R} + \mathbf{R}')\varphi^{(0)}(\mathbf{r} - \mathbf{R}') . \qquad (6.97)$$

The change of summation index does not cause any harm, as (upon employing periodic boundary conditions) the sum comprises N displacement vectors on the lattice. On the other hand, the second version of the Bloch theorem requires that any solution of the Schrödinger equation with a periodic potential is described by the quantum number $\mathbf{k}$, such that

$$\sum_{\mathbf{R}'} a(\mathbf{R} + \mathbf{R}')\varphi^{(0)}(\mathbf{r} - \mathbf{R}') = \varphi(\mathbf{r} + \mathbf{R}) = e^{i\mathbf{k}\cdot\mathbf{R}}\varphi(\mathbf{r}) = e^{i\mathbf{k}\cdot\mathbf{R}}\sum_{\mathbf{R}'} a(\mathbf{R}')\varphi^{(0)}(\mathbf{r} - \mathbf{R}') \ .$$

$$(6.98)$$

Adding $\mathbf{k}$ as an index to the function, $\varphi_{\mathbf{k}}(\mathbf{r})$, and to the coefficients, $a_{\mathbf{k}}(\mathbf{R})$, in all previous equations, and comparing the coefficients on both sides of Eq. (6.98) yield the relation

$$a_{\mathbf{k}}(\mathbf{R} + \mathbf{R}') = e^{i\mathbf{k}\cdot\mathbf{R}}a_{\mathbf{k}}(\mathbf{R}') \ . \tag{6.99}$$

For $\mathbf{R}' = 0$ one finds

$$a_{\mathbf{k}}(\mathbf{R}) = e^{i\mathbf{k}\cdot\mathbf{R}}a_{\mathbf{k}}(0) \equiv A_{\mathbf{k}}e^{i\mathbf{k}\cdot\mathbf{R}} \ , \tag{6.100}$$

where $A_{\mathbf{k}} = a_{\mathbf{k}}(0)$ is a constant, and therefore

$$\varphi_{\mathbf{k}}(\mathbf{r}) = A_{\mathbf{k}}\sum_{\mathbf{R}} e^{i\mathbf{k}\cdot\mathbf{R}}\varphi^{(0)}(\mathbf{r} - \mathbf{R}) \ . \tag{6.101}$$

Normalization. The coefficient $A_{\mathbf{k}}$ is determined by the normalization of the wave function,

$$1 = \langle\varphi_{\mathbf{k}}|\varphi_{\mathbf{k}}\rangle = |A_{\mathbf{k}}|^2 \sum_{\mathbf{R},\mathbf{R}'} e^{i\mathbf{k}\cdot(\mathbf{R}-\mathbf{R}')}\langle\varphi^{(0)}(\mathbf{r} - \mathbf{R}')|\varphi^{(0)}(\mathbf{r} - \mathbf{R})\rangle \ . \tag{6.102}$$

The expression on the right-hand side includes the overlap integral between the two wave functions at the two sites,

$$\langle\varphi^{(0)}(\mathbf{r} - \mathbf{R}')|\varphi^{(0)}(\mathbf{r} - \mathbf{R})\rangle = \int d^3r [\varphi^{(0)}(\mathbf{r} - \mathbf{R}')]^* \varphi^{(0)}(\mathbf{r} - \mathbf{R}) \equiv \alpha(\mathbf{R}' - \mathbf{R}) \ ,$$

$$(6.103)$$

which depends only on the distance between the two ions, $|\mathbf{R} - \mathbf{R}'|$, since the origin within the integral can be shifted to the point $\mathbf{R}$. This integral decays exponentially with $|\mathbf{R} - \mathbf{R}'|$, because $|\varphi^{(0)}(\mathbf{r})|$ decays with $\mathbf{r}$. For instance, for the ground-state wave function of the hydrogen atom, $\varphi^{(0)}(\mathbf{r}) = \psi_{100}(r) = \exp[-r/a_{\mathrm{B}}]/\sqrt{\pi a_{\mathrm{B}}^3}$, Eq. (4.16) yields

$$\alpha(\mathbf{R}) = e^{-R/a_{\mathrm{B}}}[1 + R/a_{\mathrm{B}} + (R/a_{\mathrm{B}})^2/3] \ . \tag{6.104}$$

Replacing the dummy index $\mathbf{R}'$ in the sum in the right hand-side of Eq. (6.102) by $\mathbf{R} - \mathbf{R}'$ gives N identical sums, and hence

$$|A_{\mathbf{k}}|^{-2} = N\left(1 + \sum_{\mathbf{R}\neq 0} e^{i\mathbf{k}\cdot\mathbf{R}}\alpha(\mathbf{R})\right) \ , \tag{6.105}$$

where it is assumed that each of the original functions $\varphi^{(0)}(\mathbf{r})$ is normalized, $\alpha(0) = \langle\varphi^{(0)}|\varphi^{(0)}\rangle = 1$. When the ions are far away from each other the overlap of wave functions localized around two different ions may be ignored, and

$|a_{\mathbf{k}}(\mathbf{R})|^2 = |A_{\mathbf{k}}|^2 \approx 1/N$, is then roughly the probability to find the electron around the ion located at $\mathbf{R}$. Equation (6.100) then implies that these probabilities are all equal, as can indeed be expected since the lattice is invariant under displacements by lattice vectors.

Problem 6.20.

Use Eq. (6.101) to prove that the function $\varphi_{\mathbf{k}}(\mathbf{r})$ obeys the two versions of the Bloch theorem, Eqs. (6.53) and (6.54). Prove that this function is periodic in reciprocal space, Eq. (6.63).

In the case of a one-dimensional lattice with periodic boundary conditions, the quantum numbers that characterize the wave functions are $k = k_m = 2\pi m/L$ [Eq. (6.60)], where $L = Na$. For $N = 2$ one finds $k_m = 0, \pi/a$, and hence there exist two solutions, $\varphi_0 = A_0[\varphi^{(0)}(x) + \varphi^{(0)}(x - a)]$ and $\varphi_{\pi/a} = A_{\pi/a}[\varphi^{(0)}(x) - \varphi^{(0)}(x - a)]$. These solutions resemble the symmetric (bonding) and the antisymmetric (anti bonding) ones obtained for the molecule H_2^+, Eq. (4.14). Another relevant example is that of the benzene molecule, also discussed in Sec. 4.3: each of the six carbons there has four valence states. The electronic states in the molecule's plane create local covalent bonds, called sp^2. At low temperatures these are fully occupied and do not contribute, neither to the electrical conductance around the ring, nor to the specific heat. Each of the carbons has in addition an electron in the p_z state, whose wave function is "normal" to the molecule's plane. These can serve as a basis for single-electron molecular states as the one used in Eq. (6.101) [or Eq. (4.27)]. These six electrons occupy six states similar to those in Eq. (6.101), with equal probability to occupy each of the carbon atoms. Each such state can accommodate two electrons; there are, therefore, empty states to which the electrons can move under the effect of external fields. The electrons in the p_z states are capable of moving around the ring and hence contribute, e.g., to the magnetic susceptibility of the molecule.

The energy bands in the variational approximation. As argued in the discussion of the hydrogen molecule, the average energy in the state (6.101), $E(\mathbf{k}) = \langle \varphi_{\mathbf{k}} | \hat{\mathcal{H}} | \varphi_{\mathbf{k}} \rangle$ is an upper bound on the ground-state energy of the system; in the lowest variational approximation it is the energy of the electron in the ground-state of the periodic lattice. This energy is

$$E(\mathbf{k}) = |A_{\mathbf{k}}|^2 \sum_{\mathbf{R},\mathbf{R}'} e^{i\mathbf{k}\cdot(\mathbf{R}'-\mathbf{R})} \langle \varphi^{(0)}(\mathbf{r}-\mathbf{R}) | \hat{\mathcal{H}} | \varphi^{(0)}(\mathbf{r}-\mathbf{R}') \rangle$$

$$= N|A_{\mathbf{k}}|^2 \sum_{\mathbf{R}} \langle \varphi^{(0)}(\mathbf{r}-\mathbf{R}) | \hat{\mathcal{H}} | \varphi^{(0)}(\mathbf{r}) \rangle . \tag{6.106}$$

The same manipulations exploited in treating Eq. (6.105) are used: in the second line, the origin in the integral is shifted,

$$\langle \varphi^{(0)}(\mathbf{r}-\mathbf{R}) | \hat{\mathcal{H}} | \varphi^{(0)}(\mathbf{r}-\mathbf{R}') \rangle = \langle \varphi^{(0)}(\mathbf{r}-\mathbf{R}+\mathbf{R}') | \hat{\mathcal{H}} | \varphi^{(0)}(\mathbf{r}) \rangle ,$$

and the summation index is changed $\mathbf{R} \to (\mathbf{R} - \mathbf{R}')$. The terms in the sum are independent of $\mathbf{R}'$, which then comprises N identical summands.

Next, the Hamiltonian is written in the form $\hat{\mathcal{H}} = \hat{\mathcal{H}}_0 + \Delta U$, where

$$\hat{\mathcal{H}}_0 = -\frac{\hbar^2}{2m}\nabla^2 + u(\mathbf{r}) \tag{6.107}$$

is the Hamiltonian of an electron bound solely to the ion located at the origin, and

$$\Delta U = \sum_{\mathbf{R} \neq 0} [u(\mathbf{r} - \mathbf{R}) + e^2/|\mathbf{R}|] \tag{6.108}$$

is the potential energy of an electron near $\mathbf{R}$ due to all other ions (save the one at the origin), to which the repulsive energy between each such ion and the ion at $\mathbf{R}$ is added. The wave function $\varphi^{(0)}(\mathbf{r})$ is a solution of the atomic Schrödinger equation (6.95). Multiplying that equation from the left by $[\varphi^{(0)}(\mathbf{r} - \mathbf{R}')]^*$ and then integrating gives

$$\langle \varphi^{(0)}(\mathbf{r} - \mathbf{R})|\hat{\mathcal{H}}_0|\varphi^{(0)}(\mathbf{r})\rangle = E^{(0)}\alpha(\mathbf{R}) , \tag{6.109}$$

where α is defined in Eq. (6.103); hence,

$$E(\mathbf{k}) = E^{(0)} - \frac{\beta + \sum_{\mathbf{R} \neq 0} e^{i\mathbf{k}\cdot\mathbf{R}}\gamma(\mathbf{R})}{1 + \sum_{\mathbf{R} \neq 0} e^{i\mathbf{k}\cdot\mathbf{R}}\alpha(\mathbf{R})} , \tag{6.110}$$

with

$$\gamma(\mathbf{R}) = -\langle \varphi^{(0)}(\mathbf{r} - \mathbf{R})|\Delta U(\mathbf{r})|\varphi^{(0)}(\mathbf{r})\rangle , \tag{6.111}$$

and

$$\beta = -\langle \varphi^{(0)}(\mathbf{r})|\Delta U(\mathbf{r})|\varphi^{(0)}(\mathbf{r})\rangle . \tag{6.112}$$

The ground state of hydrogen is $\varphi^{(0)} = \psi_{100}$, and Eqs. (4.18) and (4.19) show that

$$\gamma(R) = \frac{e^2}{a_{\mathrm{B}}}\left(1 + \frac{R}{a_{\mathrm{B}}}\right)e^{-R/a_{\mathrm{B}}} , \tag{6.113}$$

and

$$\beta = -\sum_{\mathbf{R} \neq 0} \frac{e^2}{R}\left(1 + \frac{R}{a_{\mathrm{B}}}\right)e^{-2R/a_{\mathrm{B}}} . \tag{6.114}$$

When R exceeds the Bohr radius, $R \gg a_{\mathrm{B}}$, then $\beta \ll \gamma$. This small ratio is found also for heavier atoms. As mentioned, for sufficient long $|\mathbf{R}|$ the integrals $\alpha(\mathbf{R})$ and $\gamma(\mathbf{R})$ decay exponentially with $|\mathbf{R}|$, due to the small overlap of wave functions localized around different ions, in addition to the exponential decay of the atomic wave functions themselves. The constant β decreases as the lattice constant is increased, and therefore in the limit of a very large lattice constant the energy bands are extremely narrow, with $E(\mathbf{k}) \approx E^{(0)}$. As the distance between nearest neighbors is diminished, the correction to this value gradually increases and the bands broaden. Since the integrals $\alpha(\mathbf{R})$ decay exponentially with the distance $|\mathbf{R}|$, one may expand the denominator in Eq. (6.110); retaining only the first-order,

$$E(\mathbf{k}) \approx E^{(0)} - \beta - \sum_{\mathbf{R}\neq 0} e^{i\mathbf{k}\cdot\mathbf{R}}\Gamma(\mathbf{R}) \,, \tag{6.115}$$

where $\Gamma(\mathbf{R}) = \gamma(\mathbf{R}) - \beta\alpha(\mathbf{R}) \approx \gamma(\mathbf{R})$. For a Bravais lattice $\alpha(\mathbf{R}) = \alpha(-\mathbf{R})$ and $\gamma(\mathbf{R}) = \gamma(-\mathbf{R})$ (check!). The summations in Eqs. (6.110) or (6.115) can then be carried out pairwise, e.g.,

$$\sum_{\mathbf{R}\neq 0} e^{i\mathbf{k}\cdot\mathbf{R}}\Gamma(\mathbf{R}) = \sum_{\mathbf{R}\neq 0} \cos(\mathbf{k}\cdot\mathbf{R})\Gamma(\mathbf{R}) \tag{6.116}$$

(the sum includes both $\mathbf{R}$ and $-\mathbf{R}$, thus each term appears twice). Note the resemblance of Eqs. (6.115) and (5.35). In both cases there appear sums of the type displayed in Eq. (6.116). The difference is that Eq. (5.35) yields the frequency-squared of the lattice vibrations, while Eq. (6.115) gives the energy of the electron. Nonetheless, this similarity might be useful: details of the calculations presented in Chapter 5 can be exploited to obtain energy bands in thetight-bindingapproximation.

Many bands. In the calculation presented above, it is assumed that each ion has a ingle atomic energy level; upon adding the other ions in the crystal this level creates an energy band. The isolated atom has many discrete energy levels, and the calculation can be repeated for each of them. Since the wave functions of different atomic states are orthogonal to each other the variational approximation gives rise to separate energy bands for each atomic level, with wave functions built of the atomic wave functions of that same level. Below, the energies within the energy band obtained from the atomic level $E_n^{(0)}$ are denoted $E_n(\mathbf{k})$. These bands are very narrow when the lattice constant is large (that is, the overlap of wave functions of neighboring ions is small, which leads to a small $|\Gamma_n(\mathbf{R})|$), and they broaden gradually as the ions approach each other. In many cases the sign of the coefficient $\Gamma_n(\mathbf{R})$ alternates for successive states of the ions. As a result, the energies of neighboring bands have maxima or minima at the center of the first Brillouin zone, as seen in Fig. 6.5.

Example: one-dimensional lattice. For a single-atom, one-dimensional chain, with lattice constant a, Eq. (6.115) gives

$$E_n(k) = E_n^{(0)} - \beta_n - 2 \sum_{m=1}^{\infty} \Gamma_n(ma) \cos(mka) \tag{6.117}$$

for the nth band (the factor 2 results from $m = \pm|m|$). As $\Gamma_n(ma)$ decays sharply with the distance between the ions ma, the summands in the sum may all be ignored save those of the nearest neighbors. Then, denoting $\Gamma_n(a) = \Gamma_n$,

$$E_n(k) \simeq E_n^{(0)} - \beta_n - 2\Gamma_n \cos(ka) = \overline{E}_n + 2\Gamma_n[1 - \cos(ka)] \,. \tag{6.118}$$

Except for the constant $\overline{E}_n = E_n^{(0)} - \beta_n - 2\Gamma_n$, that shifts the band's location on the energy axis, the last term in Eq. (6.118) is the same as the one that appears in Eq. (5.5) for the dispersion relation of a one-dimensional, single-atom chain with nearest-neighbor spring constants. Equation (5.5) pertains to the square of the frequency and therefore Fig. 5.2 displays its root. In the present case there is no such root; hence, leaving aside the constant $\overline{E}_n$, the energy in each band is proportional to $1 - \cos(ka) = 2\sin^2(ka/2)$. Figure 6.5(a) shows three bands described by this expression. Γ_n of the first and the third bands is positive, and that of the second is negative. It is worth noting that near the bottom of the lower band the energy is approximately $E_1(k) \approx \overline{E}_1 + \Gamma_1 k^2 a^2$. It is parabolic in k, as is the energy of a free particle $[E = (\hbar k)^2/(2m)]$, but with a coefficient that may be very different from that of a free particle. Nonetheless, it is customary to denote $\Gamma_1 = \hbar^2/(2m^* a^2)$, where m^* is the **effective mass** of the electron. The meaning of this "mass" is discussed in the following.

Wave functions. It is of interest to examine the wave function that corresponds to each of the energies in the band. From Eq. (6.101), $\varphi_k(\mathbf{r}) = A_k \sum_m \exp[ikma]\varphi^{(0)}(\mathbf{r} - ma)$, where the sum runs over the unit cells, $\mathbf{R} = m\mathbf{a}$. An example is a linear chain of hydrogen ions for which it is assumed that $\varphi^{(0)}(\mathbf{r}) = \psi_{100}$. This function appears in the sum multiplied by an oscillating phase factor, and therefore the sum is larger for smaller wave numbers. Figure 6.19 shows the real part of the wave function and its absolute value squared on the $x-$axis (on which the ions are located), for several values of $k_\ell = 2\pi\ell/(Na)$, for a lattice of 10 ions. For $\ell = 0$ the wave function is real, and oscillates between its value on the ions and the one in-between them. For $\ell \neq 0$ the real and the imaginary parts of the wave function alternate in sign, but the absolute value (and hence the probability to find the electron) oscillates periodically. As seen, the probability to find the electron at the vicinity of any of the ions is equal (about 0.1 in the figure). The probability to find it at the mid-point between two neighbors is maximal in the ground state, $\ell = 0$. This probability decreases as ℓ increases, and vanishes at the middle of the band ($\ell = 5$), where the wave function changes sign at nearest neighbors (compare with Fig. 4.9, which is similar to the case with two ions, $N = 2$).

Square lattice. When only the overlaps of the wave functions on nearest neighbors are included, Eq. (6.115) for a square lattice reads

$$E_n(\mathbf{k}) \approx \overline{E}_n + 4\Gamma_n(a)[\sin^2(k_x a/2) + \sin^2(k_y a/2)] , \qquad (6.119)$$

where $\overline{E}_n = E_n^{(0)} - \beta_n - 4\Gamma_n(a)$ (check!) Again it is useful to compare Eq. (6.119) with the last equation in (5.26), which gives the dispersion relation of vibrations normal to the plane in a square lattice. Figure 6.5(b) displays three energy bands described by Eq. (6.119), as a function of the two components of the vector $\mathbf{k}$. A parabolic behavior is observed near the center of the Brillouin zone for $\Gamma_1 > 0$, $E_1(k) = \overline{E}_1 + \Gamma_1 \mathbf{k}^2 a^2 \equiv \overline{E}_1 + (\hbar k)^2/(2m^*)$. The equal-energy curves for each band are similar to those presented in Fig. 5.17(a): circles near the center of the Brillouin zone and its corners, and a square connecting the mid-points on the faces of the zone in-between these two ranges, see Fig. 6.20(b). [In the middle of the band $\cos(k_x a) + \cos(k_y a) = 0$, which implies that $k_y a = \pm\pi - k_x a$. These are

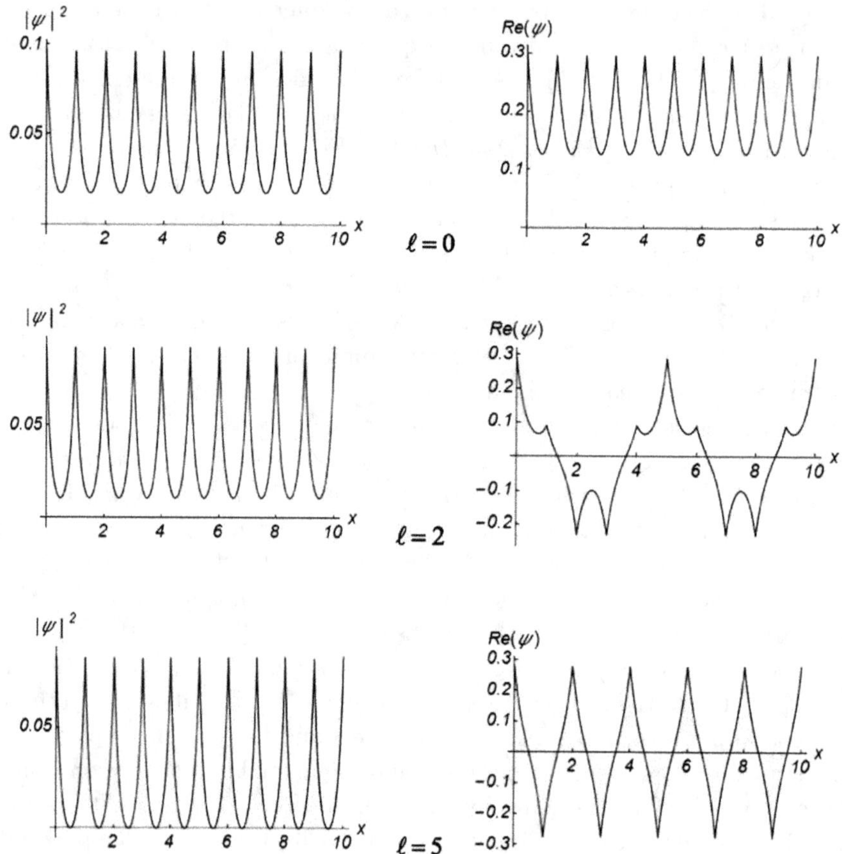

Fig. 6.19: Wave functions (their real parts and their absolute values squared) on the lattice axis for a few values of the wave number $k_\ell = 2\pi\ell/(Na)$. The lattice comprises $N = 10$ hydrogen ions (with periodic boundary conditions) and the lattice constant, for the sake of the demonstration, is $a = 3a_{\rm B}$.

the equations of the straight lines that build the square whose edges connect the mid-points on the faces of the Brillouin zone.] Like in Fig. 6.18, the energy along the path $\Gamma \to M \to K \to \Gamma$ is presented in Fig. 6.20(c). It is illuminating to compare Fig. 6.20(b) with Fig. 6.14(b), that displays the lower energy band of a free particle. The equal-energy curve is deformed away from the center of the zone, so that it gradually attains the square symmetry of the lattice. A similar deformation is found in the nearly-free electron model, as calculated in problem 6.19: there, the sharp maxima at the corners of the Brillouin zone are diminished and become smoother due to the periodic potential, leading to a paraboloid behavior near the maxima. The spectrum of the lattice vibrations behaves similarly, see Sec. 5.2. In particular, the dependence of the energy on the vector **k** is sensitive to its direction.

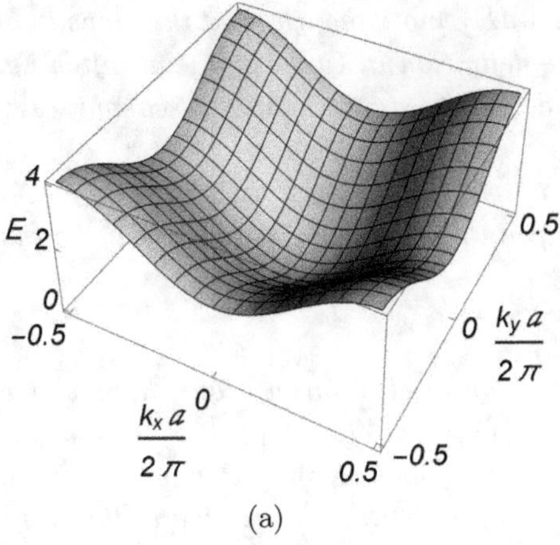
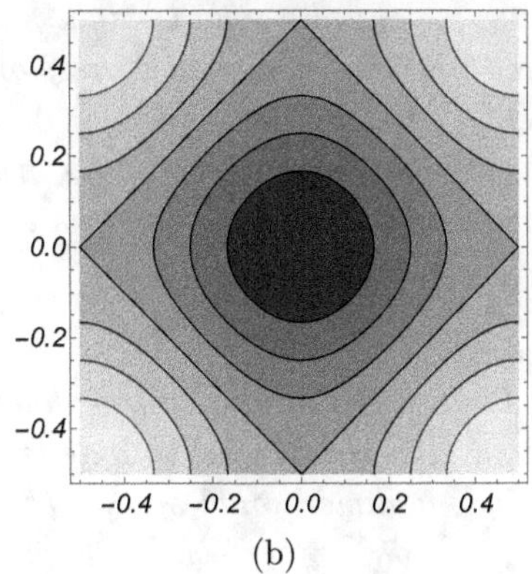

(a)   (b)

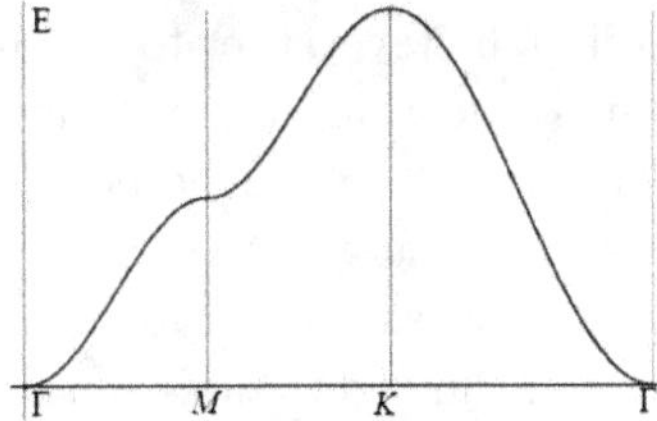

Fig. 6.20: (a) An energy band in the square lattice, according to Eq. (6.119) (with overlaps between nearest neighbors alone). (b) Equal-energy curves for the same band. (c) The energies along the path $\Gamma \to M \to K \to \Gamma$ in the Brillouin zone.

Figure 6.20 illustrates the following features. (1) The derivative of the energy with respect to the component of $\mathbf{k}$ normal to the face of the Brillouin zone vanishes on that face. (2) At the center of the face the energy has a saddle point: it is minimal parallel to the face, and maximal normal to it. (3) The energy is maximal at the corners of the Brillouin zone, and can be presented there in the quadratic form $E \approx E_{\mathrm{max}} - \Gamma_1(\mathbf{k} - \mathbf{G}_{11})^2 a^2$ where $\mathbf{G}_{11} = (\pm\pi/a, \pm\pi/a)$. Similar results are found in the nearly-free electron model, see problem 6.19.

Problem 6.21.

Consider a one-dimensional lattice of lattice constant a, with overlap energy $\Gamma_1 > 0$ for nearest neighbors and $\Gamma_2 = -\zeta\Gamma_1$ for the next nearest-neighbors. The bottom of the band is at the energy $\overline{E}_0 = \Gamma_1$.

a. *Plot schematically the energies in the first Brillouin zone for $\zeta = 1/8$ and for $\zeta = 1/4$.*

b. *As in Eq. (6.31), the density of states is defined such that $Lg(E)dE$ equals the* *number of discrete levels in the range $\{E, E + dE\}$ [note though, that this density of states is defined per unit length, not like the definition in Chapter 5, e.g., after Eq. (5.56)]. Derive a general expression for the density of states, and draw schematically $g(E)$ for the present case.*

c. *Assume that the Fermi level of the electron gas, E_F, is such that $E_F - \Gamma_1 \ll \Gamma_1$. Find the total energy of the gas at zero temperature, and compare it with the result for a free electron gas.*

Problem 6.22.

a. *Find the wave function in the ground state of an electron in a one-dimensional potential well described by a delta function, $U(x) = -u_0 \delta(x)$, $u_0 > 0$.*

b. *A one-dimensional system is described by a periodic Kronig-Penney potential, $U(x) = -u_0 \sum_m \delta(x - ma)$. Exploit the result of part (a) to obtain the width of the lowest energy band in the tight-binding approximation.*

Several states in the unit cell. The calculation above produces a single energy band because it assumes a single quantum state on each atom, and a single atom in each unit cell. Quite often, however, the atomic energy levels are degenerate; for instance, three orbital states, according to the quantum number m, belong to the $2p$ atomic shell, and each such state can be occupied by two electrons with opposite spins. It follows that the energy of the $2p$ state is six-fold degenerate. Moreover, as encountered in chapter 4, the $2s$ states are quite close to the $2p$ ones, and may hybridize with them to form eight states in the second shell. In addition, the unit cell may well contain more than one atom.

In order to account for such configurations, the wave functions required for the description of all the states within a single unit cell centered around the lattice site $\mathbf{R}$ are denoted as $\varphi_{i\lambda}^{(0)}(\mathbf{r} - \mathbf{R})$, where the index λ represents the degenerate (or almost degenerate) states on the atom located at $\mathbf{r}_i$ in the unit cell. The corresponding eigenenergies are $E_{i\lambda}^{(0)}$. The solutions of the Schrödinger equation are written as [extending Eq. (6.96)]

$$\varphi(\mathbf{r}) = \sum_{\mathbf{R}} \sum_{i,\lambda} a_{i\lambda}(\mathbf{R}) \varphi_{i\lambda}^{(0)}(\mathbf{r} - \mathbf{R}) \ . \tag{6.120}$$

Inserting this form into the Schrödinger equation, multiplying it on the left by $[\varphi_{i'\lambda'}^{(0)}(\mathbf{r} - \mathbf{R}')]^*$ and integrating the result over $\mathbf{r}$ yields

$$(E - E_{i'\lambda'}^{(0)}) \left[\sum_{\mathbf{R}} \sum_{i,\lambda} \alpha_{i'\lambda',i\lambda}(\mathbf{R}' - \mathbf{R}) a_{i\lambda}(\mathbf{R}) \right] = - \sum_{\mathbf{R}} \sum_{i,\lambda} \gamma_{i'\lambda',i\lambda}(\mathbf{R}' - \mathbf{R}) a_{i\lambda}(\mathbf{R}) \ ,$$

$$\tag{6.121}$$

where

$$-\gamma_{i'\lambda',i\lambda}(\mathbf{R}' - \mathbf{R}) = \langle \varphi_{i'\lambda'}^{(0)}(\mathbf{r} - \mathbf{R}') | \Delta U_i(\mathbf{r}) | \varphi_{i\lambda}^{(0)}(\mathbf{r} - \mathbf{R}) \rangle \ ,$$

with $\Delta U_i(\mathbf{r})$ comprising all interactions with other ions, and where $\alpha_{i'\lambda',i\lambda}(\mathbf{R}' - \mathbf{R}) = \langle \varphi_{i'\lambda'}^{(0)}(\mathbf{r} - \mathbf{R}') | \varphi_{i\lambda}^{(0)}(\mathbf{r} - \mathbf{R}) \rangle$ are the overlap integrals (including for brevity the term $\mathbf{R} = \mathbf{R}'$, treated separately before). For n_B atoms in the unit cell, each possessing n_S states, there are Nn_Bn_S linear equations for all the coefficients $\{a_{i\lambda}(\mathbf{R})\}$, and these have a nontrivial solution when the determinant vanishes. Again generalizing the previous discussion, it is noted that according to the Bloch theorem $a_{i\lambda}(\mathbf{R}) = A_{i\lambda}\exp[i\mathbf{k}\cdot\mathbf{R}]$. Then

$$(E - E_{i'\lambda'}^{(0)})\left[\sum_{\mathbf{R}}\sum_{i,\lambda}\alpha_{i'\lambda',i\lambda}(\mathbf{R}' - \mathbf{R})A_{i\lambda}e^{i\mathbf{k}\cdot(\mathbf{R}'-\mathbf{R})}\right]$$
$$= -\sum_{\mathbf{R}}\sum_{i,\lambda}\gamma_{i'\lambda',i\lambda}(\mathbf{R}' - \mathbf{R})A_{i\lambda}e^{i\mathbf{k}\cdot(\mathbf{R}'-\mathbf{R})} . \tag{6.122}$$

In this way, separate equations are obtained for the n_Bn_S coefficients $\{A_{i\lambda}\}$. For $n_B = n_S = 1$ there is solely a single equation, that reproduces Eq. (6.110).

Problem 6.23.

Calculate the energy bands of a **one-dimensional lattice with two atoms in the unit cell**, *A at the origin and B at the point $x = a/2$. The lattice constant is a. Each atom when "free" has a single energy level; the wave function of an electron on A is $\varphi_A^{(0)}(x - ma)$ and the energy is $E_A^{(0)}$, and the wave function on B is $\varphi_B^{(0)}(x - ma - a/2)$ and the energy is $E_B^{(0)}$. Ignore the overlap integrals α between different atoms, and keep only the overlaps γ of nearest neighbors.*

Problem 6.24.

a. *Calculate the energy bands of graphene which are formed from atomic states of π type (wave functions that are normal to the plane of the lattice). Neglect the overlap integrals α and take into account only the overlaps γ between nearest neighbors.*

b. *Show that at the vicinity of each corner in the Brillouin zone, e.g., the point marked K in Fig. 5.9(a), the energies of the two bands appear as two cones joined at their apexes.*

Problem 6.25.

Consider the following way to derive Eq. (6.115). The Hamiltonian of a single electron in the lattice is given by a matrix whose diagonal elements are $\langle\mathbf{R}|\hat{\mathcal{H}}|\mathbf{R}\rangle = \overline{E}^{(0)} = E^{(0)} - \beta$ and the nondiagonal ones are real and depend on the distance between the atoms, $\langle\mathbf{R}|\hat{\mathcal{H}}|\mathbf{R}'\rangle = \langle\mathbf{R}'|\hat{\mathcal{H}}|\mathbf{R}\rangle = -\gamma(|\mathbf{R} - \mathbf{R}'|)$, for $\mathbf{R} \neq \mathbf{R}'$.

a. *Under which assumptions does this matrix reflect the tight-binding approximation?*

b. *A one-dimensional chain contains N atoms at the sites $\mathbf{R} = na\hat{\mathbf{x}}$, $n = 1, 2, 3, \ldots, N$; $\gamma(n - m) = \gamma\delta_{n-m,1}$. Assuming periodic boundary conditions, $\psi(n) = \psi(n + N)$, show that the eigenstates and the eigenvalues of the matrix are $\psi_\ell(n) = \exp[ink_\ell a]/\sqrt{Na}$, $E(k_\ell) = \overline{E}^{(0)} - 2\gamma\cos(k_\ell a)$, where $k_\ell = 2\pi\ell/(Na)$ and $\ell = 0, 1, 2, \ldots, N - 1$. This result coincides with Eq. (6.118).*

c. *A foreign atom is located at the origin ($n = 0$) of the one-dimensional lattice discussed in part (b). Assume that the chain is infinite, $N \to \infty$ (see also Sec.*

5.6). The matrix elements of the Hamiltonian are those of part (b), except for the elements $\langle 0|\hat{\mathcal{H}}|0\rangle = \overline{E}_{\mathrm{imp}}^{(0)}$, $\langle 0|\hat{\mathcal{H}}|1\rangle = \langle 1|\hat{\mathcal{H}}|0\rangle = -\gamma_{\mathrm{imp}}$. Show that in addition to the wavy solutions, the Schrödinger equation for this Hamiltonian can possess a localized solution, of the form $\psi(n) = A\exp[-\kappa|n|]$ for $|n| \geq 1$, where $|\exp[\kappa]| > 1$, and find the conditions for the existence of such a solution.

6.8   Metals, insulators, and semiconductors

Introduction. The previous three sections include various examples of electronic band-structure calculations. An exact solution of the Kronig-Penney model is presented in Sec. 6.5; the main advantage of this model is the possibility to derive energy bands for **any** strength of the potential that binds the electron to a certain ion in the lattice, U_0. The band structure in the empty-lattice limit is displayed in Fig. 6.10(a): a structure with no energy gaps, i.e., the bands are touching each other. A small U_0 gives rise to the energy structure shown in Fig. 6.11(a) with narrow gaps in-between the bands. A similar result is depicted in Fig. 6.12(b), calculated within the nearly-free electron approximation, or the weak-binding approximation. As U_0 increases, the gaps widen while the bands' widths narrow down. An intermediate configuration is illustrated in Fig. 6.8(b). Very narrow bands are found for large values of U_0, as in Fig. 6.11(b). Such bands are also derived in the tight-binding approximation, in Sec. 6.7. As U_0 approaches infinity, the bands are infinitesimally narrow, see Fig. 6.10(b).

The calculations within the weak-binding approximation produce expressions for the energy gaps, and clarify the reason for the widening of those gaps with the increasing strength of the attractive potential. As opposed, calculations within the tight-binding model focus on the bands themselves and show that the bands widen as the overlap of wave functions on nearby neighbors increases. This width is defined as the difference between the maximal and the minimal energy of the band, $W_n = E_n(\mathrm{max}) - E_n(\mathrm{min})$. In one-dimension, and when only the overlap of nearest neighbors is accounted for, Eq. (6.118) gives $W_n = 4\Gamma_n$; as shown, Γ_n decays exponentially with the distance between nearest-neighbor ions. In two dimensions Eq. (6.119) yields $W_n = 8\Gamma_n$, and in the general case [Eq. (6.115)] the band widens as overlaps with more neighbors are included, and narrows down as a function of the distance between the neighbors. Figure 6.21(a) illustrates the edges of the band for a one-dimensional chain of hydrogen atoms, Eq. (6.118), with $ka = 0$ and $ka = \pi$ [the figure exploits the explicit expressions given in Eqs. (6.104) and (6.113)]: the energy band shrinks as the distance between nearest-neighbor ions is increased.

In higher dimensions the energies involve all components of $\mathbf{k}$, see e.g., Fig. 6.5(b). However, even there the energies are independent of $\mathbf{k}$ when there are a few atoms located very far away from each other, and the bands widen gradually until touching each other or even penetrating into one another [see e.g., the square lattice with a weak potential, discussed at the end of Sec. 6.6]. Figure 6.21(b) illustrates the bands of neon and sodium (in three dimensions) as a function of the distance in-between the ions. As seen, the bands are narrow and are well separated when the ions are far away from each other; as these distances are shortened, certain gaps are closed, and there is even an overlap between the bands. Recall that such overlaps are impossible in one dimension, but may well appear in higher dimensions. In particular, the overlap between the $2s$ band and the $2p$ one is related to the hybridization between these two types of states, as discussed in length in Sec. 4.3.

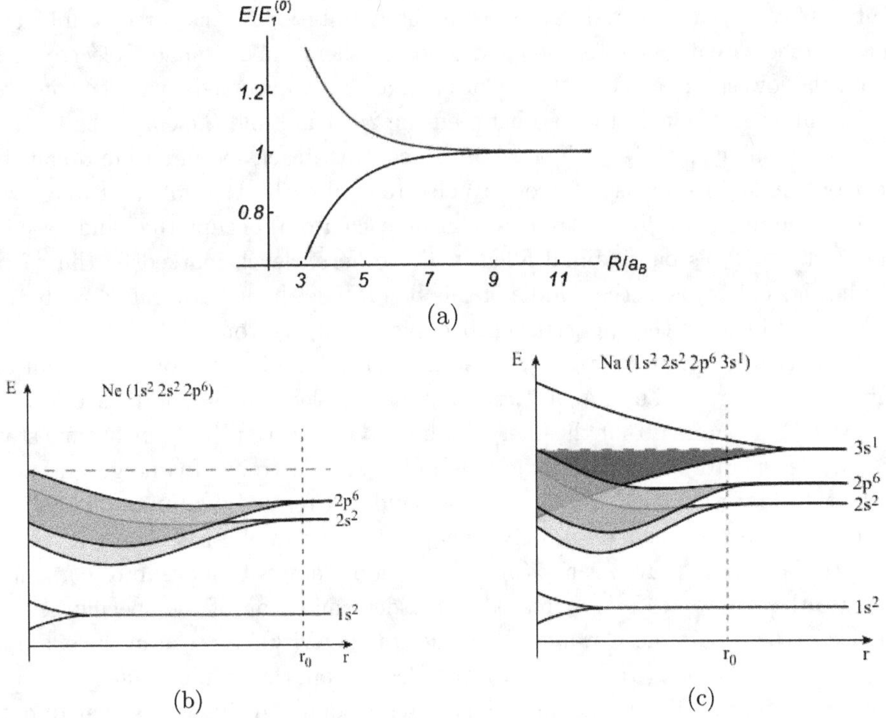

Fig. 6.21: (a) The boundaries of the energy bands of a one-dimensional chain, comprising hydrogen atoms in their ground state, with overlap between nearest neighbors, as a function of the distance between those neighbors (distances in units of the Bohr radius, energy in units of the ground-state energy of the hydrogen atom). (b and c) A schematic illustration of the energy bands of neon (b) and sodium (c). The vertical dashed lines indicate the measured distance in the materials, and the horizontal ones are located at the Fermi energies.

Filling of energy bands. At zero temperature, when the electron gas is in its ground state, the electrons occupy all energy levels (two of them in each level), starting at the lowest one, until their number is exhausted. The upmost filled level is termed the **Fermi level**. In contrast with the Sommerfeld picture (Sec. 6.3), in which the energy of the levels increases with the wave number, in the periodic solid the energy levels form bands separated by gaps. As seen in Eq. (6.27), there are precisely $N = N_1 N_2 N_3$ discrete values of lattice momentum $\mathbf{k}$ in the first Brillouin zone. The number of $\mathbf{k}$ values is the number of quantum states in each band: there is a wave function and an energy level for each discrete value of $\mathbf{k}$. As a quantum state can accommodate two electrons, each band (whose "parent" is a single atomic level) can accommodate at most $2N$ electrons. In the example of neon illustrated in Fig. 6.21(b), each atom donates 10 electrons. The lowest band, marked by $1s^2$, indeed contains $2N$ electrons, and hence is full. Similarly, the band denoted $2s^2$ contains $2N$ electrons as well, and is completely full too. As the three atomic levels in the $2p$ shell are degenerate, the band based on them can contain $6N$ electrons. All in all, the lowest three energy bands of neon are fully occupied with electrons, and the Fermi level is thus in the gap between the $2p^6$ band and the empty band above it, corresponding to the atomic level $3s$. In contrast, each sodium atom donates 11 electrons. Like in the case of neon, 10 electrons of each atom fill the three lowest bands. The remaining N electrons (one from each atom) occupy the band based on the $3s$ states. This band is half filled, and the Fermi level, marked by the dashed line in Fig. 6.21(c) is at the middle of this band. The shaded areas in Figs. 6.21(b) and 6.21(c) indicate the energy levels occupied with electrons.

Metals and Insulators. When each atom in the lattice contributes an odd number of electrons, i.e., $2M + 1$, the number of electrons in the lattice is $N_e = N(2M+1)$. Assuming that the energy bands do not overlap (i.e., the atomic states are not degenerate), each band contains up to $2N$ electrons. In the ground state, each of the lowest M bands is filled. These bands are termed **"valence bands"**; the electrons residing in them essentially do not contribute to physical processes, mainly because they do not have nearby (in energy) empty states to move into (e.g., either because the temperature is increased, for a determination of the specific heat, or an electric field is applied for a study of the conductivity). As seen in the examples displayed in Figs. 6.21(b) and 6.21(c), the electrons in the valence bands are bound mainly to inner atomic shells, and thus remain bound to the atoms that donated them. The overlap of the "inner" wave functions is small, and the corresponding bands are narrow. The remaining N electrons occupy the $(M + 1)$th band. This band is only half filled, as it can accommodate $2N$ electrons. This is the situation in sodium, shown in Fig. 6.21(c). The upmost band, which is not fully occupied, is called the **"conduction band"**. The electrons in it usually come from the outmost atomic shell. As the band is half filled, there are empty states nearby (in energy) to the occupied ones; this implies that the behavior of the specific heat and the electrical conductance of the material are similar to those analyzed in Sec. 6.3. The calculations have to be modified, though, to include the dispersion relation of the electrons, $E(\mathbf{k})$, and the relevant density of states that are determined by the band structure. This task is carried out in Sec. 6.9. However, as explained below, the electrons in the conduction band are relatively free to move over the lattice, rendering the material to be a **metal**.

A different result emerges when the number of electrons in each atom in the lattice is even, $2M$. The total number of electrons is then $N_e = 2MN$, and they fully occupy M valence bands, leaving an energy gap in-between the electrons in the upmost band and the next one, which is the conduction band. This is the configuration of neon, Fig. 6.21(b). As at zero temperature there are no nearby (in energy) empty states to which the electrons can move, such materials are usually **insulators**: their electrical conductance is zero at zero temperature. The energy bands of an insulator and a metal are illustrated in the right and left panels of Fig. 6.22.

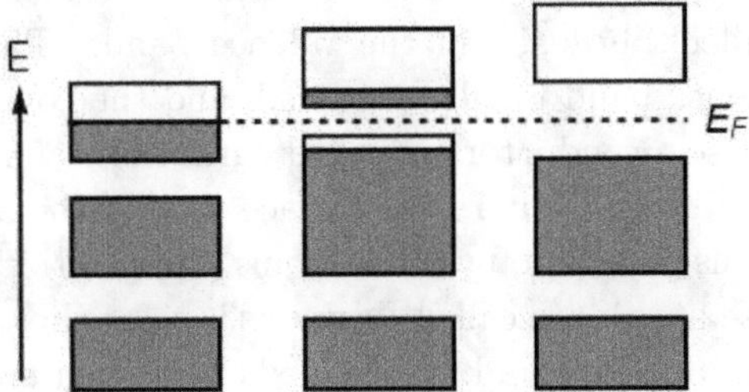

Fig. 6.22: Schematic illustration of energy bands' occupations, depicted by the rectangles, at low temperatures. The grey areas are filled with electrons, and the white ones are empty. The energy increases upwards. The upmost rectangles show the conduction band, and the lower ones the valence bands. The dashed line is at the Fermi energy. Right: insulator; middle: semiconductor; left: metal.

Semiconductors. Assuming that at zero temperature the upper valence band is fully occupied and the conducion band above it is empty, as in the right panel of Fig. 6.22, there is a finite probability to excite electrons from the upper part of the upmost valence band to the bottom of the conduction band; these electrons become conducting. Since they evacuate states in the valence band, the electrons remaining there can also move around and hence become conducting as well. This latter motion can be described as that of **holes**, which represent the empty states. At low temperatures the number of electrons that pass between the bands is quite small; it is proportional to $\exp[-E_g/(2k_\mathrm{B}T)]$, where $E_g = E_c - E_v$ is the energy gap between the bottom of the conduction band, E_c, and the top of the valence one, E_v. For typical insulators E_g is of the order of several electron volts, and therefore is much larger than the thermal energy at room temperature; the number of electrons in the conduction band is minute, leading to a negligible conductance. The material is still termed an "insulator" even at finite temperatures. **Semiconductors**, as opposed , are insulators for which E_g is relatively small (parts of electron volts as compared to several electron volts). At room temperature there are electrons at the bottom of the conduction band, that moved there from the top of the valence band, as in the central panel of Fig. 6.22. The details of the electrical conductance in this configuration are discussed in Sec. 6.9. Contrary to the Matthiessen rule [Eq. (6.49)] that predicts a decrease of the conductance as the temperature is raised, the one of a semiconductor increases with the temperature, since there are more free charge carriers at higher temperatures. The distinction between an insulator and a semiconductor is not clear cut: the electrical conductance of both materials vanishes at zero temperature and increases as the temperature is raised; but in the insulator this increase is rather slow, such that the conductance is tiny even at room temperature.

Intrinsic and extrinsic semiconductors. In a semiconductor as described above, the number of electrons in the conduction band balances precisely the number of empty states (called **"holes"**) in the valence band. This is so since at zero temperature the valence band is fully occupied, and the conducion band is totally empty. This type of semiconductors is called **"intrinsic"**. Another way to inject electrons into the conduction band is to replace part of the atoms in an insulator, that have $2M$ electrons per atom, by other atoms ("impurities"), which have $2M+1$ electrons each. These atoms are called **"donors"**, since they add surplus electrons to the states in the conduction band. The added charge carriers are negative; for this reason such a semiconductor is called **n-type**. Alternatively, part of the atoms may be replaced by atoms with $2M-1$ electrons. In this case, electrons from the upmost valence band can move to these impurities, leaving behind them empty states in the valence band. These foreign atoms are called **"acceptors"**, as they capture electrons and create "holes" in the valence band. The holes behave as positive charge carriers; the semiconductor is then termed **a p-type semiconductor**. Both the n-type and the p-type semiconductors are called **extrinsic**. As shown in Sec. 6.9, the energy levels of the donors (or acceptors) are close to the conduction band (or the valence one), and electronic transitions between them and the bands significantly enhance the conductance.

Double-valence metals. The above description suffices to explain the metallicity of crystals built of single-valence atoms, like the alkali metals (sodium, potassium), copper or silver. Each atom donates a single electron to the conduction band, which is then half filled (see the left panel of Figs. 6.22). The electrons can reach quite easily empty states and thus contribute to, e.g., the electrical conductance and the specific heat. This picture remains valid also for numerous elements that contribute an even number of electrons, like diamond, silicon, or germanium, in which the conduction band is empty at zero temperature, and is separated by a finite gap from the valence band below. Such materials are insulators (or semiconductors). Nonetheless, solids comprising double-valence atoms are at times more complicated. Figure 6.22 illustrates the simplest configuration, where the bands do not overlap (as indeed happens in one-dimensional lattices). At higher dimensions the bands widen as the distance between atoms is diminished (as occurs, e.g., in the tight-binding approximation); it is hence plausible that two neighboring bands overlap with one another, as shown in Fig. 6.21(b). The gap in-between them closes, and there appear degenerate states in the two bands that may hybridize. This situation occurs, for example, in certain systems containing carbon, due to the hybridization of $2s$ states with $2p$ ones (see Sec. 4.3). Moreover, in dimensions higher than 1 the atomic states are frequently degenerate (e.g., due to different values of the quantum number of the angular momentum). In both cases, the number of electrons available to populate the joint band is still even, and is $2Ng_M$, where g_M is the number of degenerate states of each original atomic energy level, or the number of overlapping bands. For $g_M > 1$ the $2N$ electrons that can occupy the conduction band from double-valence atoms, like the alkaline earths (calcium, barium, and strontium) or iron, fill the conduction band only partially, leading to a metallic character of these materials.

At dimensions higher than one it may happen that two neighboring bands, which do not touch one another, contain states of identical energies, as illustrated in Fig. 6.23(a). A hypothetical case with this configuration is discussed in problem 6.13. It can occur, e.g., in a square lattice, for which the lower band is separated from the upper one at the middle of the face of the first Brillouin zone, e.g., at $(\pi/a, 0)$; the lower band though, possesses states on the diagonal of the Brillouin zone; an energy gap is created only around the corner $(\pi/a, \pi/a)$, see Figs. 6.14, 6.16, and 6.18. Within the weak-potential approximation, the second band, illustrated in Fig. 6.16(b), digs a "dip" in the first band along each edge of the Brillouin zone. The energy levels along the edge are displayed schematically in Fig. 6.23(b). With the Fermi energy as given by the horizontal line in the figure, electrons occupy states in the two bands, leaving empty states in the upper parts of both bands. Such a material, though metallic, is called **semimetal**.

The periodic table. Figure 6.24 presents the periodic table; the various colors indicate the different types of crystals built of the corresponding elements. The alkali atoms in the first column have a valence of 1 and create metallic crystals (save hydrogen, which under standard ambient conditions forms a gas comprising diatomic molecules, and is an insulator). The alkaline earths in the second column have a valence of 2, but their bands overlap significantly, and therefore they create metallic crystals as well. The same is true for the transition metals in which the inner electronic shell nd is full, but they donate electrons from the $(n + 1)$th electronic shell to the conduction band. At the other end of the table, the lattices formed by the noble gases are insulators, as each of the atoms possesses an even number of electrons, that remain relatively close to their parent nuclei. It follows that the overlap of their wave functions is small [see the example of neon in Fig. 6.21(b)] and the energy gaps are wide. The valence bands are completely full with an even number of electrons and the conduction one is empty. According to the table, the atoms in the upper right triangle are insulators. This fact is usually related to details of the crystalline structure, and not only to the number of electrons in the outmost atomic shell. For instance, the halogens are insulators since the unit cell of their crystalline structure contains an even number of atoms, and consequently also an even number of electrons. Carbon is a borderline case: it has four electrons in the outmost shell, and therefore should be an insulator, provided that the bands do not overlap. Indeed, diamond is an insulator as it has a large energy gap in-between the valence and the conduction bands. However, graphite is a semimetal, since its valence bands touch one another (as is also the case for graphene, see problem 6.24). The elements on the diagonal below the triangle (marked in blue) are **metalloids**; their properties are partially metallic and partially semiconducting (e.g., germanium and silicon). The other elements in the table crystalize in metallic structures, for the aforementioned reasons.

The Fermi surface. Possible band structures are illustrated in Fig. 6.22. In practice, however, the specific heat and the electrical conductivity are determined by the number of electrons that can pass among the states near the Fermi level. In lattice-momentum space, these states are at the vicinity of an equal-energy surface whose energy is the Fermi one. It is called the **Fermi surface.** In contrast with the Sommerfeld model, in which the electrons are free and therefore the Fermi surface is spherical (referred to in Sec. 3.3 as the "Fermi sphere"), the periodic potential entails a replacement of that sphere by a surface whose geometry may be rather complicated (it is determined by the energy dependence on the lattice momentum). As the energy bands are customarily pictured within the reduced spectrum, the Fermi surface is also presented within the first Brillouin zone.

The Fermi surface of free electrons. In three dimensions, the Fermi momentum of a free particle is given by $n = k_{\mathrm{F}}^3/(3\pi^2)$ [Eq. (6.28)] while in two dimensions

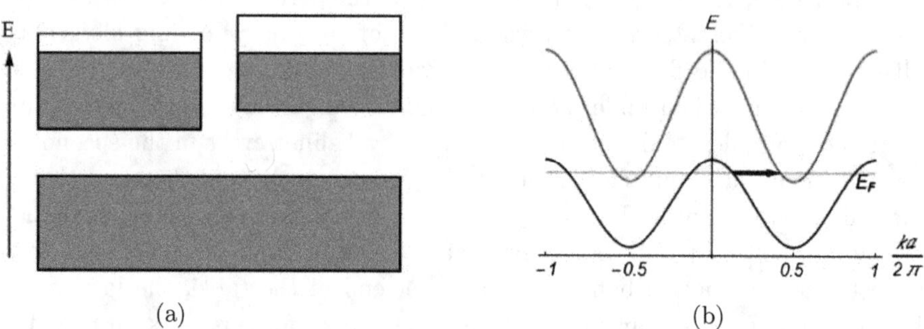

(a)   (b)

Fig. 6.23: A schematic description of the energy bands in a "semimetal". (a) A generalization of Figs. 6.22: the two upper bands are partially overlapping, and both are not full. (b) The energy levels along an edge of the Brillouin zone of a square lattice, in which a relatively weak periodic potential is active. The arrow indicates a transition of an electron from the top of the lower band to the bottom of the upper band, that penetrates into the lower one.

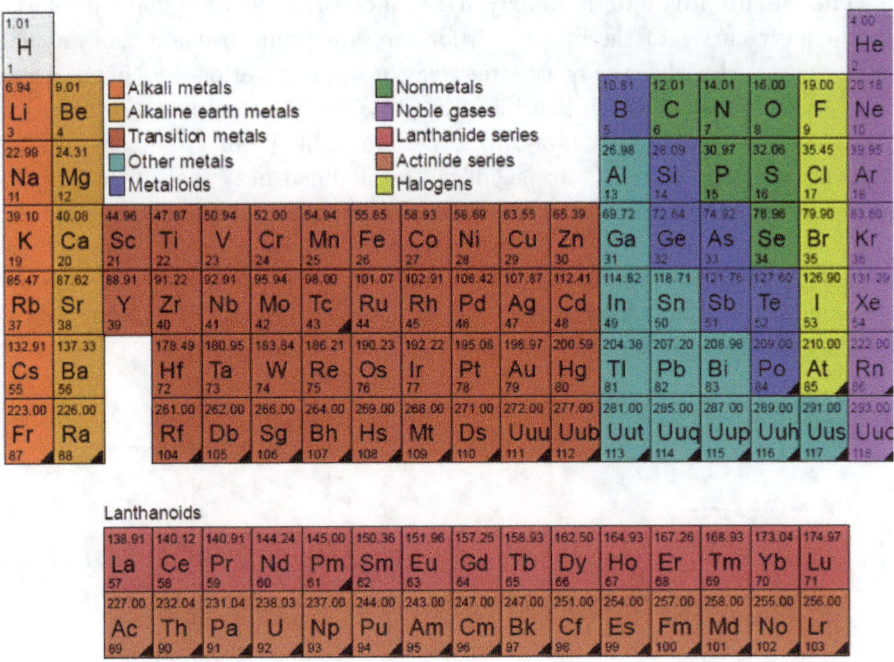

Fig. 6.24: The periodic table. The colors distinguish between the various metals and insulators. Created by the software Atomic PC.

(problem 6.10) $n = k_{\mathrm{F}}^2/(2\pi)$. It follows that the radius of the Fermi sphere (or circle) increases with the electrons' density n. When the density is small enough, the Fermi sphere is contained within the first Brillouin zone; as it increases, the sphere cuts the borders of the zone and penetrates into the second Brillouin zone. This is illustrated in Fig. 6.25(a) for the square lattice: when the valence of each atom is ℓ (i.e., it donates ℓ electrons) the electrons' density is $n = \ell/a^2$, and hence $k_{\mathrm{F}}a = \sqrt{2\pi\ell}$; that is, $k_{\mathrm{F}}a = \sqrt{2\pi} \approx 0.798\pi$ for $\ell = 1$ (curve A in the figure) and $k_{\mathrm{F}}a = \sqrt{4\pi} \approx 1.128\pi$ for $\ell = 2$ (curve C). The Fermi circle cuts the border of the first Brillouin zone once $k_{\mathrm{F}}a$ exceeds π (curve B). Thus curve A, corresponding to single-valence atoms, is within the Brillouin zone, while curve C, representing a double-valence material, "overflows" into the second zone. In the first case the Fermi line is a circle, like for a free particle. In the second case only the arcs that remained in the first zone represent states there; the ones belonging to states in the second zone are discussed below. When the neighboring unit cells (in reciprocal lattice) are drawn as well, the Fermi curve that belongs to the parts of C in the first zone creates a closed line encircling the corner of the zone, like the thick line in Fig. 6.25(b). The perimeter of this curve shrinks as the Fermi energy rises, until it becomes a single point, and the band is completely full.

The Fermi level of a nearly-free electron. Figure 6.25(c) displays the equal-energy curves of the first band, for the same four Brillouin zones as above, when calculated within the nearly-free electron approximation, i.e., in the presence of a weak periodic potential [see Fig. 6.14(d)]. The periodic potential deforms the curves: line A, which is far away from the border of the first zone is still roughly circular, but lines B and C are significantly modified near the boundaries of the zone. As explained following Fig. 6.14, the periodic potential "smoothens" the sharp corners of curve C in Fig. 6.25(b), since the equal-energy curves must be normal to the borderline between the Brillouin zones. A further increase of the periodic potential, as shown in Fig. 6.25(d), causes curve A to penetrate into the second zone as well.

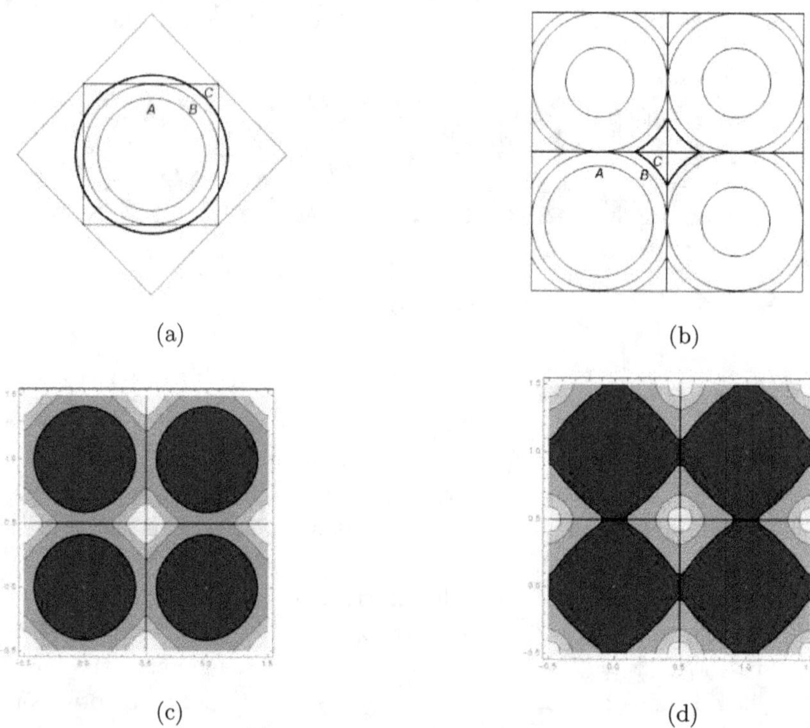

(a)   (b)

(c)   (d)

Fig. 6.25: (a) Three Fermi circles (in the $k_x - k_y$ plane) of a free particle on a square lattice. A – each atom donates a single electron, C – each atom donates two electrons, B – a Fermi circle that touches the boundaries of the first Brillouin-zone (whose edge length is $2\pi/a$). (b) The four Brillouin zones derived upon displacing part (a) by reciprocal lattice vectors. (c) Like (b), for energies calculated within the nearly-free electron model. (d) Like (c), but with a stronger periodic potential.

The Fermi surface of the second energy band. The circle C in Fig. 6.25(a), is reconstructed in Fig. 6.26 (right panel). The parts of the circle that penetrate into the second zone contain states belonging to the second energy band. Indeed, a displacement of the circle by reciprocal-lattice vectors, as in the mid panel of Fig. 6.26, creates the islands marked by X; the perimeters of these islands are parts of the Fermi surface. A weak periodic potential displaces these curves, like in the left panel of Fig. 6.26, and causes the equal-energy curves to cross smoothly the boundaries of the Brillouin zone, along a direction normal to the faces. As a result, the Fermi surface consists of the curves marked by C in Fig. 6.25(b) in the first band, and those appearing in the left panel of Fig. 6.26 in the second band.

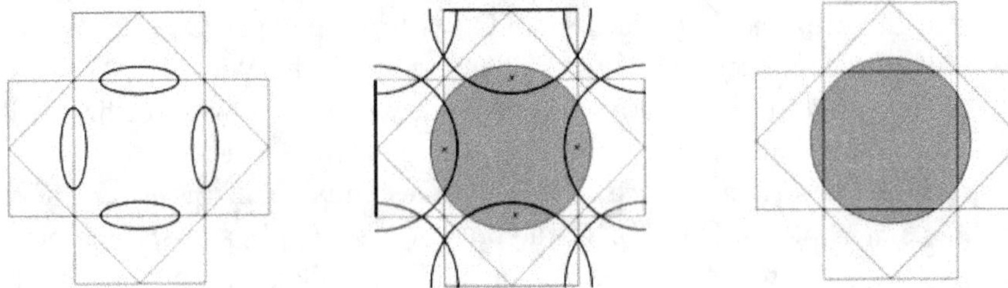

Fig. 6.26: Right panel: the dark circle contains the points in lattice-momentum space that are occupied by free electrons, for double-valence atoms on a square lattice. The perimeter consists of the arcs C in the first Brillouin zone, pertaining to the first energy band, and from the arcs that penetrate into the second zone, and thus refer to the second band. Mid panel: the thick circles represent the circle of part (a) displaced into the neighboring Brillouin zones. The areas marked by X contain occupied states in the second energy band; their circumferences are part of the Fermi surface. Left panel: whereas the right and middle panels describe the spectrum of free electrons, this panel is drawn for a weak periodic potential. The equal-energy curves, and in particular the Fermi curve, are slightly shifted. The potential is mainly effective close to the faces of the Brillouin zone, where these curves are normal to the faces.

Within the nearly-free electron approximation, the second energy band of the square lattice penetrates into the first one, and therefore the Fermi surface is contained in the two overlapping bands; as the number of the electrons in the lattice increases, the electrons occupy both bands, and therefore such a material behaves as a **semimetal**. When the periodic potential is strong, the gaps in-between the bands widen, the lower band is fully occupied for double-valence materials and the solid is an insulator. Thus, double-valence materials are metals when the electrons interact weakly with the ions, and are insulators when that interaction is strong. The equal-energy curves (and in particular the Fermi one), within the tight-binding approximation and for a single band, are illustrated in Fig. 6.20.

Fermi surfaces in three dimensions. The configuration of the band structure, and consequently of that of the Fermi surface, is quite complex in three dimensions. A general discussion is beyond the scope of this volume; instead, several explicit examples based on the nearly-free electron approximation are described. The **alkali metals** in the first column of the periodic table are all of the **BCC** type, see Fig. 2.22. There are two atoms in the unit cell of the **BCC** and hence the electrons' density is $n = 2/a^3$. By Eq. (6.28) $n = k_{\rm F}^3/(3\pi^2)$, and thus $k_{\rm F} = (3\pi^2 n)^{1/3} = (6\pi^2)^{1/3}/a \approx 1.24\pi/a$. The reciprocal lattice of the **BCC** is an **FCC**, with a lattice constant $4\pi/a$, and reciprocal-lattice vectors $\mathbf{G} = (2\pi/a)[h(\hat{\mathbf{y}} + \hat{\mathbf{z}}) + k(\hat{\mathbf{x}} + \hat{\mathbf{z}}) + \ell(\hat{\mathbf{x}} + \hat{\mathbf{y}})]$, as given in problem 3.8. The reciprocal-lattice vectors connecting the center of the Brillouin zone with the 12 nearest neighbors in reciprocal lattice are $\mathbf{G} = (2\pi/a)(\pm\hat{\mathbf{y}}\pm\hat{\mathbf{z}})$, $\mathbf{G} = (2\pi/a)(\pm\hat{\mathbf{x}}\pm\hat{\mathbf{z}})$, or $\mathbf{G} = (2\pi/a)(\pm\hat{\mathbf{y}} \pm \hat{\mathbf{x}})$. The first Brillouin zone is the Wigner-Seitz cell of the **FCC**, as displayed in Fig. 2.19(a). The boundary planes between the first Brillouin zone and its neighbors are the normals to those vectors, at their mid points. The distance from the origin to each of these planes equals half the length of such a vector, i.e., $k_0 = \sqrt{2}\pi/a$ (check!). In the present case, $k_{\rm F}/k_0 \approx 0.877$ and hence the Fermi surface is entirely contained within the first Brillouin zone, not approaching its borders. In the nearly-free electron model this surface is roughly spherical (that of potassium is presented in Fig. 6.27). Indeed, the Fermi surfaces of the alkali metals are almost spheres, rendering the Sommerfeld model an adequate description of their electronic properties. Apparently, this implies that the periodic potential in these materials is rather weak. On the other hand, the **alkaline earths** in the second column of the periodic table have a double valence. For instance, barium has the **BCC** structure, with $k_{\rm F} = (12\pi^2)^{1/3}/a \approx 1.56\pi/a$. The Fermi sphere of the free particles penetrates into the neighboring zones, and the Fermi surfaces for both bands encircle small islands in lattice-momentum space, as in Figs. 6.25 and 6.26 (see Fig. 6.27). Hence the elements in the second column are semimetals.

Another example is the single-valence transition metals, copper, silver, and gold. They are all of the **FCC** type and have similar electronic configurations. That of copper, for instance, is $[\text{Ar}]3d^{10}4s^1$ (where $[\text{Ar}]=1s^2 2s^2 2p^6 3s^2 3p^6$ denotes the electronic configuration of argon); the energy bands are constructed from hybridizations of s-states with d-ones. One of those bands is above the others, and therefore copper is treated as a single-valence metal, in which this band is half-filled. In the **FCC** lattice there are four atoms in the cubic unit cell, and hence $n = 4/a^3$. Within the nearly-free electron model the corresponding Fermi momentum is $k_{\rm F} = (3\pi^2 n)^{1/3} = (12\pi^2)^{1/3}/a \approx 1.56\pi/a$. The reciprocal lattice is the **BCC** one with a lattice constant $4\pi/a$; the Brillouin zone is depicted in Fig. 2.19(b) [and also in Fig. 5.17(b)]. As seen, the surface of the Brillouin zone contains hexagons and squares. The distance from the center of the zone to a hexagonal surface plane is a quarter of the cube's diagonal, $k_0 = \sqrt{3}\pi/a$, leading to $k_{\rm F}/k_0 \approx 1.084$; the Fermi sphere cuts these planes and penetrates into the neighboring zones. In contrast, the distance form the center of the Brillouin zone to a square part of the surface is $k_0' = 2\pi/a$, giving $k_F/k_0' \approx 0.78$. All in all, the Fermi surface is close to being spherical save several narrow "sleeves" that appear along the eight directions from the center of the zone to the corners of the cubic unit cell, connecting neighboring zones, see Fig. 6.27.

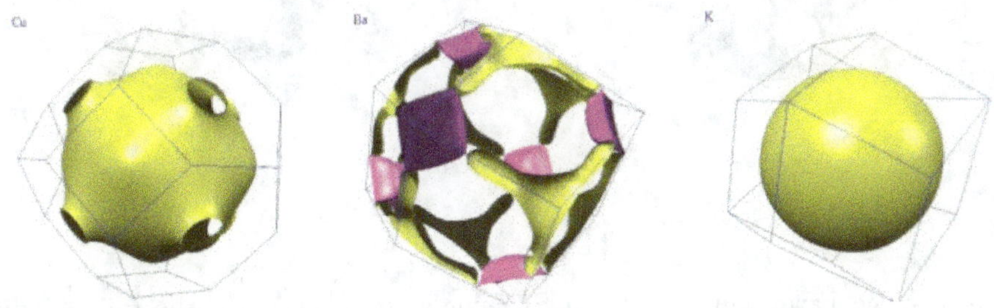

Fig. 6.27: The Fermi surfaces of copper, barium, and potassium. (Taken from `http://www.phys.ufl.edu/fermisurface`.)

Fermi surfaces of other materials may be more complex. Consider for example the simple-cubic lattice, in which–in the tight-binding approximation–only overlaps of the wave functions between nearest neighbors are taken into account. Extending Eqs. (6.115) and (6.119), the energies in the tight-binding approximation are given by

$$E_n(\mathbf{k}) \approx \overline{E}_n + 2\Gamma_n(a)[3 - \cos(k_x a) - \cos(k_y a) - \cos(k_z a)] . \qquad (6.123)$$

The Fermi surface of a single-valence metal is at the middle of the band, i.e., at $E_n = \overline{E}_n + 6\Gamma_n$, which implies $\cos(k_x a) + \cos(k_y a) + \cos(k_z a) = 0$. This surface is shown in Fig. 6.28. In contrast with the nearly-free electron model, for which the Fermi surface is entirely within the first Brillouin zone [problem 6.27(a)] the tight-binding approximation produces a surface that penetrates into the second zone, similar to the surface displayed in Fig. 6.25(d), where also a strong potential opens "sleeves" in-between the zones.

Problem 6.26.
Double-valence atoms are situated on the sites of a square lattice.
a. Assume a very weak periodic potential. Find the electrons' densities for which the electrons occupy states within the first Brillouin zone alone. At which density they occupy states in the third zone?
b. Exploit the nearly-free electron model, the results of problem 6.19(a) and Fig. 6.18(b), to derive the criterion for the transition from a semimetallic behavior to an insulating one.

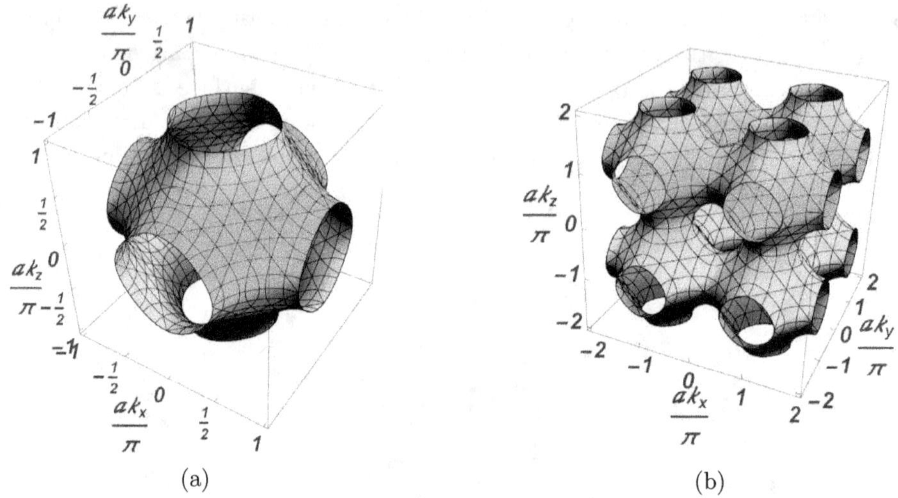

Fig. 6.28: The Fermi surfaces of an **SC** single-atom metal, in the tight-binding approximation. (a) In the first Brillouin zone. (b) In eight Brillouin zones.

c. *Which form do the two lowest bands take in the tight-binding approximation? (assume overlaps of wave functions only on nearest-neighbors). Is semimetallic behavior possible here?*

Problem 6.27.

a. *Consider a single-valence metal on an* **SC** *lattice in the presence a very weak periodic potential. Find the percentage of double-valence atoms that has to be added to the system in order for the Fermi surface to cross the boundary of the first Brillouin zone.*

b. *The band's energies of an electron on a cubic lattice are given in Eq. (6.123). What is the upmost energy to be filled such that the Fermi surface is still within the first Brillouin zone?*

c. *What is the answer of part (a) for a* **BCC** *lattice?*

6.9 Density of states, charge-carriers' density, specific heat

Preface. The occupation of the energy levels is determined by the Fermi-Dirac distribution, $f(E)$ [Eq. (6.35)], which depends on the temperature. Most electronic properties are determined by the location of the Fermi energy, i.e., whether it is within an energy band or in an energy gap. For example, when at zero temperature the electrons fill all states within a certain band then at a finite temperature they have to fill states above that band; however, the only available ones are those in the next band. The energy distance to those can be quite significant, making the corresponding Fermi-Dirac distribution exponentially small, $f(E) \approx \exp[-\beta(E - \mu)]$, and consequently decreasing considerably the electrical conductivity, the electronic specific heat, *etc.* On the other hand, when at zero temperature the electrons occupy only part of the band and have easily-available states near the Fermi level, the finite-temperature behavior becomes similar to that emerging from the Sommerfeld model.

Density of states. As discussed in Sec. 6.3, the total energy of the electron gas, from Eq. (6.39), is

$$\langle E_{\text{tot}} \rangle = V \int_0^\infty dE \, E g(E) f(E) , \tag{6.124}$$

where $g(E)$ is the density of the electrons' energy levels: $V g(E) dE$ is the number of eigenenergies of the single-particle Schrödinger equation in the energy range $\{E, E + dE\}$. As opposed to the density of states of a free particle [e.g., Eq. (6.32)], a simple monotonic function of the energy, the band structure has gaps in which there are no states at all, implying zero density of states in the gaps. The density of states is no more a monotonic function of the energy.

The density of states in one dimension. The number of states with wave numbers in the range $\{k, k + dk\}$ in one dimension is $L g(k) = 2 \times 2 \times dk/(2\pi/L)$, where $(2\pi/L)$ is the length of the segment on the $k-$axis that contains a single discrete value of $k_\ell = 2\pi\ell/L$; consequently, $dk/(2\pi/L)$ is the number of states in the segment dk. The density of states as a function of energy, taking into account only positive energies, is multiplied by 2 since to each state of a wave number k there corresponds a state with $-k$. Another factor of 2 accounts for the two spin states. Save this last factor, the calculation is identical to the one following Eq. (5.56). Note, though, that here the density of states is defined **per unit length** and so the final result is not multiplied by L. Upon transforming a $k-$integration into an energy one, one obtains

$$g(E) = \frac{2}{\pi} \frac{1}{(dE/dk)} . \tag{6.125}$$

As an example, consider the density of states of the bands shown in Fig. 6.5(a). These bands are derived within the tight-binding approximation, Eq. (6.118). Exploiting those, one finds that $dE/dk = 2a\Gamma_n \sin(ka)$ in the range $0 < E - \overline{E}_n < 4\Gamma_n$. Using the energy as a variable gives

$$dE/dk = a\sqrt{(E - \overline{E}_n)(4\Gamma_n - E + \overline{E}_n)} .$$

In-between the bands the density of states vanishes. Inserting these relations into Eq. (6.125) produces Fig. 6.29. As in the case of phonons in one dimension, there appear **van Hove singularities** at the bands' edges.

The density of states in arbitrary dimensions. The extension of the calculation of the density of states in one dimension to higher dimensions is similar to the derivation in Chapter 5, Eqs. (5.61)-(5.66). The density of states is derived by counting the number of states in the range $\{E, E + dE\}$ between equal-energy

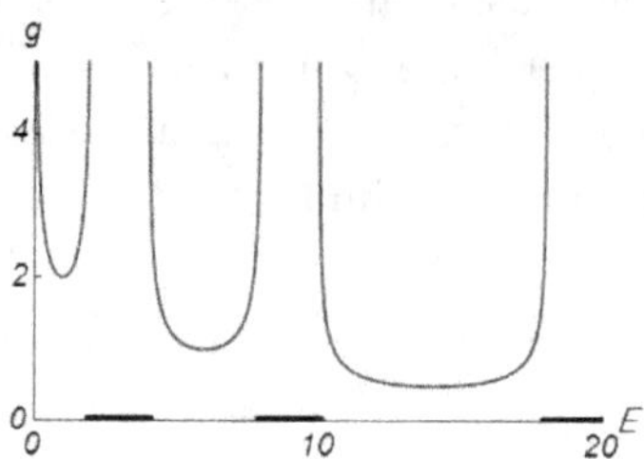

Fig. 6.29: The density of states [in units of $1/(\pi a \Gamma_1)$] of the energy bands displayed in Fig. 6.5(a) (energy in units of $2\Gamma_1$).

surfaces in $d-1$ dimensions. The number of states in a volume element $d^d k$ in $\mathbf{k}$−space is $2d^d k/[(2\pi)^d/V]$, where the factor of 2 counts the spin states. Hence, for an arbitrary function $F[E_n(\mathbf{k})]$,

$$2\sum_{n,\mathbf{k}} F[E_n(\mathbf{k})] = 2\frac{V}{(2\pi)^d}\sum_n \int d^d k F[E_n(\mathbf{k})]$$

$$= \frac{V}{2^{d-1}\pi^d}\int d^d k \int dE \sum_n \delta[E - E_n(\mathbf{k})]F(E) ,$$

and thus

$$2\sum_{n,\mathbf{k}} F[E_n(\mathbf{k})] = V \int dE g(E) F(E) , \qquad (6.126)$$

with

$$g(E) = \frac{1}{2^{d-1}\pi^d}\int d^d k \sum_n \delta[E - E_n(\mathbf{k})] = \frac{1}{2^{d-1}\pi^d}\sum_n \int \frac{dS}{|\boldsymbol{\nabla}_\mathbf{k} E_n(\mathbf{k})|} . \qquad (6.127)$$

Here, $\boldsymbol{\nabla}_\mathbf{k} E_n(\mathbf{k})$ is the gradient of the energy in the n−band with respect to $\mathbf{k}$, such that $|\boldsymbol{\nabla}_\mathbf{k} E_n(\mathbf{k})|$ is the derivative of the energy with respect to the $\mathbf{k}$−component perpendicular to the equal-energy surface whose energy is E; the integration is over this surface (see Fig. 5.16). The final result is obtained upon transforming the integration to the one over the volume in-between the two equal-energy surfaces, and using $d^d k = dS\Delta k_\perp$, where $\Delta k_\perp$ is the local distance between the two surfaces. Equation (6.125) is a particular case of Eq. (6.127).

In one dimension, the van Hove singularities near the bands' edges are caused by the vanishing of the derivative dE/dk. To explore the analogous behavior at higher dimensions, consider, e.g., the square lattice. The point $\mathbf{k}_0 = (\pi/a, \delta)$ is on the border between the first two Brillouin zones, that is, on the midperpendicular of the reciprocal-lattice vector $\mathbf{G} = (2\pi/a, 0)$. Since $E_n(\mathbf{k}) = E_n(-k_x, k_y)$ (see problem 6.12), it follows that $\partial_{k_x} E_n(\mathbf{k}) = -\partial_{k_x} E_n(-k_x, k_y)$. On the other hand, Eq. (6.64) yields $\boldsymbol{\nabla}_\mathbf{k} E_n(\mathbf{k}) = \boldsymbol{\nabla}_\mathbf{k} E_n(\mathbf{k} + \mathbf{G})$, and hence

$$\frac{\partial}{\partial k_x} E_n(-\pi/a, \delta) = \frac{\partial}{\partial k_x} E_n(\pi/a, \delta) = -\frac{\partial}{\partial k_x} E_n(-\pi/a, \delta) .$$

It follows that

$$\frac{\partial}{\partial k_x} E_n(-\pi/a, \delta) = \frac{\partial}{\partial k_x} E_n(\mathbf{k}_0) = 0 \ . \tag{6.128}$$

A parabolic behavior of the energies as the surface separating the two Brillouin zones is crossed normally to it is indeed observed in the solution of problem 6.19. In such a case $\nabla_{\mathbf{k}} E_n$ is parallel to this surface, and the equal-energy surfaces are normal to it. Similarly, Eq. (6.119) yields that the component of the gradient parallel to the vector $\mathbf{G} = (2\pi/a, 0)$, i.e., the component normal to the border line between the two Brillouin zones, is $\partial_{k_x} E_n(\mathbf{k}) = 2a\Gamma_n \sin(k_x a)$, that vanishes at all points on this border line, on which $k_x = G/2 = \pi/a$.

The density of states of a square lattice. The gradient of the energy, which vanishes on the zone boundaries, is still in the denominator for dimensions higher than 1. However, it appears within an integral over the equal-energy surface, see Eq. (6.127). The integration usually smoothens the divergence, and leaves a weaker singularity in the density of states (as also happens for the density of states of the phonon modes). Consider for example the electronic density of states on the square lattice, derived within the tight-binding approximation. According to Eq. (6.119), the equal-energy surfaces are given by

$$E_n(\mathbf{k}) \approx \overline{E}_n + 4\Gamma_n[\sin^2(k_x a/2) + \sin^2(k_y a/2)] \ ,$$

yielding for the denominator in Eq. (6.127)

$$|\nabla_{\mathbf{k}} E_n| = |2\Gamma_n a[\sin(k_x a)\hat{\mathbf{x}} + \sin(k_y a)\hat{\mathbf{y}}]| = 2\Gamma_n a\sqrt{\sin^2(k_x a) + \sin^2(k_y a)} \ .$$

Around the origin $E_n(\mathbf{k}) \approx \overline{E}_n + 2\Gamma_n(a)(ak)^2$, and therefore the density of states jumps from zero to the constant $g(E) \approx 1/(2\pi\Gamma_n a^2)$ as the energy rises above $E_{\min} = \overline{E}_n$. Similarly, near the maximum of the band energy, at the K point (the corner of the Brillouin zone), the energy is $E_n(\mathbf{k}) \approx \overline{E}_n + 4\Gamma_n(a) - \Gamma_n(a)a^2(\mathbf{k} - \mathbf{G}_{11})^2$, with $\mathbf{G}_{11} = (\pi/a, \pi/a)$. The density of states there jumps from 0 to the same constant as the energy is reduced through the upper edge of the band, $E_{\max} = \overline{E}_n + 4\Gamma_n(a)$. When the energy is raised from the bottom of the band, or when it is decreased away from the top of the band, the equal-energy curves deform gradually, changing from circles to the square obtained at the middle of the band [see the discussion after Eq. (6.119)]. As a result, the density of states increases gradually until the middle of the band is reached; there $\cos(k_x a) + \cos(k_y a) = 0$, i.e., $k_x a = \pi - k_y a$. The equal-energy curve is a straight line, and in such a case the integral of Eq. (6.127) is

$$8 \int_0^{\pi/(2a)} \frac{d\ell}{\sqrt{2\sin^2(k_x a)}} = 8 \int_0^{\pi/(2a)} \frac{dk_x}{\sin(k_x a)} = \frac{8\ln[\tan(k_x a/2)]}{a}\Big|_{k_x=0}^{k_x=\frac{\pi}{2a}} \ ,$$

where the symmetry of the equal-energy curves under rotations by $45°$ is used (each integral covers just $1/8$ of the Fermi line, i.e., half of the line in a single quarter); note that $d\ell^2 = dk_x^2 + dk_y^2$ for a length element on the equal-energy curve.

The integral diverges logarithmically near its lower bound, i.e., near the cut of the middle of the face with the Fermi curve. A somewhat more complicated calculation yields that the density of states diverges logarithmically near the middle of the band, as $\ln[|E - \overline{E}_n - 2\Gamma_n(a)|]$; this singularity is far weaker than that found in one dimension (see problem 6.28). Figure 6.30(a) illustrates the density of states of the square lattice in the tight-binding approximation.

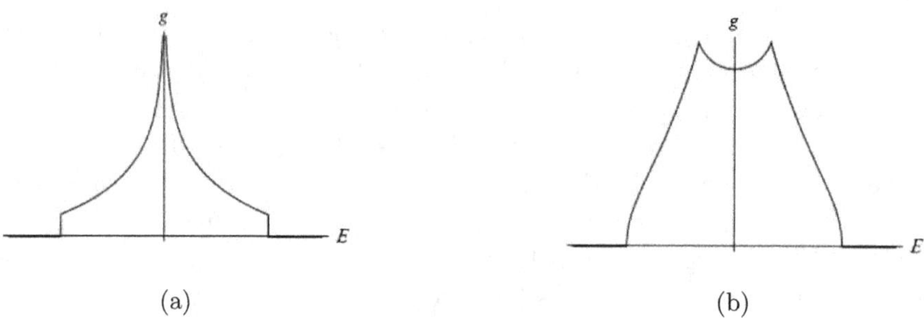

(a)   (b)

Fig. 6.30: Schematic illustration of the density of states in the tight-binding approximation. (a) Square lattice. (b) cubic lattice.

Problem 6.28.
The density of states of energies at the vicinity of the middle of the energy band of a square lattice is governed by the integral Eq. (6.127) around the point $(0, \pi/a)$. Use a Taylor expansion around this point to show that for such energies the density of states is roughly given by $g(E) \propto \ln[|E - \overline{E}_n - 2\Gamma_n(a)|]$.

The density of states in a cubic crystal. The three-dimensional density of states does not diverge; its derivative, however, does. The parabolic form of the energy as a function of the crystal momentum at the vicinity of the band's edges yields a density of states proportional to $\sqrt{|E - E_{\min}|}$ or to $\sqrt{|E - E_{\max}|}$, with van Hove singularities that appear at certain planes in the Brillouin zone. For instance, the density of states of the lower band in a cubic lattice, illustrated in Fig. 6.30(b), exhibits weak singularities at the reciprocal-lattice vectors $(\pi/a, 0, 0)$ (the maxima in the figure) and $(\pi/a, \pi/a, 0)$ (the edges of the band).

Problem 6.29.
The energy of the electrons at the bottom of the band may be often (e.g., for semi-conductors) approximated by

$$E(\mathbf{k}) = E_0 + \hbar^2 \left(\frac{k_x^2}{2m_x^*} + \frac{k_y^2}{2m_y^*} + \frac{k_z^2}{2m_z^*} \right).$$

This relation between the energy and the crystal momentum is termed "ellipsoid" dispersion, due to the ellipsoid form of the equal-energy surfaces. The quantities $\{m^*\}$ are called "effective masses"; these are discussed in the next section.

a. *Show that this is the dispersion relation at the bottom of the band of an orthorhombic crystal, calculated within the tight-binding approximation, with overlaps between nearest neighbors.*

b. *Find the density of states for this case.*

c. *Find the number of states in the conduction band in the regimes where the parabolic approximation is valid. Express the density of states in terms of this number.*

d. *At the top of the valence band (where the holes in semiconductors are located) the energy again has an ellipsoid dependence on the crystal momentum,*

$$E(\mathbf{k}) = E_0 - \hbar^2 \left(\frac{k_x^2}{2m_x^*} + \frac{k_y^2}{2m_y^*} + \frac{k_z^2}{2m_z^*} \right) .$$

The effective masses that appear here are generally different from those used for the description of the conduction band. Find the modifications in the answers for parts (a) and (b) in this region.

The specific heat of metals. The Fermi level of metals lies within the conduction band, and hence the density of states at the Fermi level, $g(E_\mathrm{F})$, differs from zero. The specific heat of a metal can be found by following the arguments leading to Eq. (6.41): at low temperatures, the energy change relative to that of the ground state comes from electrons displaced from the energy range $E_\mathrm{F} - k_\mathrm{B}T < E < E_\mathrm{F}$ to the range $E_\mathrm{F} < E < E_\mathrm{F} + k_\mathrm{B}T$. Assuming that the density of states is continuous and is almost independent of the energy over this narrow range, one concludes that the number of states there is roughly $V g(E_\mathrm{F}) k_\mathrm{B}T$. As each electron carries the energy $k_\mathrm{B}T$, the total change in the energy is proportional to $V g(E_\mathrm{F})(k_\mathrm{B}T)^2$, and hence the electronic specific heat is linear in the temperature. A more accurate calculation, which extends the one leading to Eq. (6.41), (see also problem 6.9) yields

$$C_V(\text{electrons}) \approx \frac{V g(E_\mathrm{F})}{T} \int_0^\infty (E - \mu)^2 \left[-\frac{\partial f(E)}{\partial E} \right] dE = \frac{\pi^2}{3} V g(E_\mathrm{F}) k_\mathrm{B}^2 T \equiv \gamma T .$$

$$(6.129)$$

The electronic specific heat of metals is indeed linear in the temperature, but the coefficient γ can be very different from that derived within the Sommerfeld model. This coefficient carries the information on the density of states at the Fermi level. In particular, much like the density of states, the coefficient γ is maximal wherever the Fermi surface crosses the boundary of the Brillouin zone (see Fig. 6.30). When electrons reside in several bands, like in Fig. 6.23, the measured coefficient represents the sum of the densities of states of the bands that contribute at the Fermi level. Varying the number of electrons, e.g., by replacing single-valence atoms with double-valence ones (as described in problem 6.27) facilitates a controlled variation of the Fermi energy, and the ensuing variation of γ. Thus, measuring this coefficient provides information on the dependence of the density of states upon the energy.

At low temperatures the electronic specific heat is smaller than that of the phonons, and the total specific heat is still given by Eq. (6.42).

There is a pronounced disparity between the specific heat of electrons subjected to a periodic potential and the one computed within the Sommerfeld model, even when the energy is approximated by a parabolic form, like in problem 6.29. In this approximation one finds

$$\gamma = (3\pi^2 n m_x^* m_y^* m_z^*)^{1/3} k_{\rm B}^2/(3\hbar^2) \ .$$

This expression resembles the one in Eq. (6.41), with the mass there replaced here by the geometrical average of the three effective masses, $(m_x^* m_y^* m_z^*)^{1/3}$, see problem 6.29. Since these effective masses reflect the dependence of the energy on the lattice-momentum vector near the bottom of the band, they are mostly not related to the mass of a free electron.

The concentration of charge carriers in semiconductors. Assuming that the electrons' concentration in the conduction band is low, and that they are mostly located near the band's bottom, their energy is generally quadratic in the lattice momentum, and thus their density of states is given by the answer to problem 6.29, $g(E) = C_e(E-E_c)^{1/2}$, where E_c denotes the energy at the bottom of the conduction band, and $C_e = \sqrt{2m_x^* m_y^* m_z^*}/(\pi^2\hbar^3)$. Exploiting the Fermi-Dirac distribution, and the expression for N_e, Eq. (6.36), yields

$$n_e = \int_{E_c}^{\infty} dE g(E)f(E) = C_e\sqrt{\pi}(k_{\rm B}T)^{3/2}e^{-\beta(E_c-\mu)}/2 \equiv n_{0e}e^{-\beta(E_c-\mu)} \ , \quad (6.130)$$

for the electrons' concentration in the conduction band. [This derivation makes use of the relation $\int_0^{\infty} dx\sqrt{x}\exp[-\beta x] = \sqrt{\pi}/(2\beta^{3/2})$.] An analogous calculation gives the number of electrons removed from the valence band, leaving empty states (called "holes"). The probability to find an empty state of energy E in the valence band is

$$1 - f(E) = 1 - \frac{1}{1+e^{\beta(E-\mu)}} = \frac{1}{1+e^{-\beta(E-\mu)}} \approx e^{\beta(E-\mu)} \ , \quad (6.131)$$

where E is below the top of the upmost valence band, E_v, and it is assumed that $\mu - E \gg k_{\rm B}T$. Using part (c) of problem 6.29, the density of states is $g(E) = C_h(E_v-E)^{1/2}$, (C_h includes the effective masses of the holes), and the concentration of holes is

$$n_h = n_{0h}e^{\beta(E_v-\mu)} \ . \quad (6.132)$$

Note that by multiplying the electrons' concentration with that of the holes, one finds

$$n_e n_h = n_{0e}n_{0h}e^{-\beta E_g} = \frac{\pi}{4}C_e C_h (k_{\rm B}T)^3 e^{-\beta E_g} \equiv n_i^2 \ , \quad (6.133)$$

where $E_g = E_c - E_v$ is the energy gap between the two bands. This result is a particular case of the **law of mass action**, which relates the concentrations of different particles held at equilibrium. It is independent of the chemical potential or of the electrons' source, and thus enables the detection of the energy gap.

In an intrinsic semiconductor the electrons' concentration coincides with that of the holes,

$$n_e = n_h = n_i = \sqrt{n_{0e}n_{0h}}\,e^{-\beta E_g/2} = \sqrt{C_e C_h}\,\sqrt{\pi}(k_{\mathrm{B}}T)^{3/2}e^{-\beta E_g/2}/2 \ . \qquad (6.134)$$

Using here Eq. (6.130) yields

$$\mu = (E_v + E_c)/2 + k_{\mathrm{B}}T \ln[C_h/C_e]/2 = E_{\mathrm{F}} + \mathcal{O}(k_{\mathrm{B}}T) \ . \qquad (6.135)$$

At low temperatures the Fermi energy is located at the center of the gap between the two bands. The ratio C_h/C_e is determined by the ratio of the effective masses of the holes and of the electrons.

The electrons' energy levels in donors and acceptors. Quite generally, the electron "contributed" by a donor atom is in the outer shell of the latter. This electron is bound to the nucleus and to the other electrons in the donor. As that atom is within the lattice, one may assume that the Coulomb energy of the electron is roughly given by $U_{\mathrm{D}} \approx -ze^2/(\epsilon r)$, where ze is the charge of the nucleus and all other nearby electrons, and ϵ is the dielectric constant of the material (this approximation is valid as long as the electron is relatively far from the nucleus, such that it "samples" the average dielectric effect of many atoms; it is justified below). The Schrödinger equation of this electron is

$$[E_c - \hbar^2 \nabla^2/(2m^*) + U_{\mathrm{D}}]\psi = E\psi$$

(assuming that the effective mass is isotropic); extending the solution of the hydrogen atom yields the energy levels,

$$E(n) - E_c = -z^2 R_y m^*/(m\epsilon^2 n^2) \ .$$

Here, $R_y = 13.6$ eV is the Rydberg constant. The energy is measured relative to the bottom of the conduction band. The consideration below refers to the ground state of the donated electron, with the binding energy $E_{\mathrm{D}} = E_c - E(1)$. In the solution of the hydrogen atom, where $z = 1$, the probability to find the electron at a distance r from the donor is proportional to $\exp[-r/r_0]$, where $r_0 = a_{\mathrm{B}}\epsilon m/m^*$. The dielectric constant of semiconductors is usually quite high, making this distance much longer than the Bohr radius $a_{\mathrm{B}} = 0.53\text{Å}$ (and also longer than the lattice constant, validating the use in the average dielectric constant). The binding energy E_{D} is usually quite small (of the order of 0.001-0.005 meV); it is thus relatively easy to ionize the atom and release the electron. Similar conclusions can be drawn regarding the energies of the electrons in acceptors, measured relative to E_v.

The density of charge carriers in extrinsic semiconductors. Equation (6.135) implies that the Fermi energy of an intrinsic semiconductor at zero temperature is at the middle of the gap between the highest occupied level and the lowest empty one [Fig. 6.31, right panel]. In the more general case, the chemical potential is determined by requiring **charge neutrality**: the number of free negative charges should compensate the number of positive charges on the donors that contribute them. As seen, the "outer" electron of the donor has energy within the gap between the two bands, at a distance E_D below the bottom of the conduction band (so that these electrons are easily excited into the conduction band). The probability to find an electron on the donor is $f(E_c - E_D)$. Denoting the concentration of donors by x_D and assuming that each of them donates a single electron, then the number of electrons added to the conduction band is

$$x_D[1 - f(E_c - E_D)] = x_D\left[1 - \frac{1}{1 + e^{\beta(E_c - E_D - \mu)}}\right] = \frac{x_D}{1 + e^{-\beta(E_c - E_D - \mu)}} \ .$$

Very similar considerations yield that the energy of the "outer" electron in an acceptor is $E_v + E_A$, and that it is located slightly above the top of the valence band, such that electrons from that band are easily excited into the levels on the acceptors. For a concentration x_A of acceptors, each accommodating an additional single electron, the concentration of electrons located on the acceptors is

$$x_A f(E_v + E_A) = x_A / (1 + \exp[\beta(E_v + E_A - \mu)]) \ .$$

The total number of electrons obeys

$$n_e + \frac{x_A}{1 + \exp[\beta(E_v + E_A - \mu)]} = n_h + \frac{x_D}{1 + \exp[-\beta(E_c - E_D - \mu)]} \ , \qquad (6.136)$$

where n_e and n_h are given in Eqs. (6.130) and (6.132).

The chemical potential μ of an **extrinsic semiconductor** is determined by Eq. (6.136). Consider for instance an n-type semiconductor. At low temperatures, such that all energy differences are larger than $k_B T$, the second term on the right-hand side may be expanded to obtain

$$n_{0e} \exp[-\beta(E_c - \mu)] = n_e \approx n_{0h} \exp[\beta(E_v - \mu)] + x_D \exp[\beta(E_c - E_D - \mu)] \ .$$

Ignoring completely the holes' concentration, then

$$\mu \approx E_c - E_D/2 + k_B T \ln[x_D/n_{0e}]/2 \ . \qquad (6.137)$$

Inserting this result into Eq. (6.132) yields $n_h \approx n_{0h} \exp[\beta(E_g - E_D/2)]$, which justifies its aforementioned neglect (the gap energy is larger than $k_B T$). The Fermi energy, $E_F = E_c - E_D/2$ is at the mid point between the filled level of the donors and the empty level at the bottom of the conduction band, see the mid panel in Fig. 6.31. As the energy E_D (measured relative to the bottom of the conduction band) is generally smaller than the energy gap, the Fermi level is close to the threshold of the conduction band, and then the approximations used above are valid as long as $E_D \gg k_B T$. Otherwise, one has to solve a quadratic equation in the unknown variable $\exp[\beta(\mu - E_c)]$ see problem 6.30. Inserting the result (6.137) into Eq. (6.130) yields

$$n_e \simeq \sqrt{n_{0e} x_D}\, e^{-\beta E_D/2} \ . \qquad (6.138)$$

Analogous results are obtained for p-type semiconductors, left panel in Fig. 6.31. Here again, a number of electrons reside on the acceptors, leaving holes in the valence band that contribute to the conductance (check!).

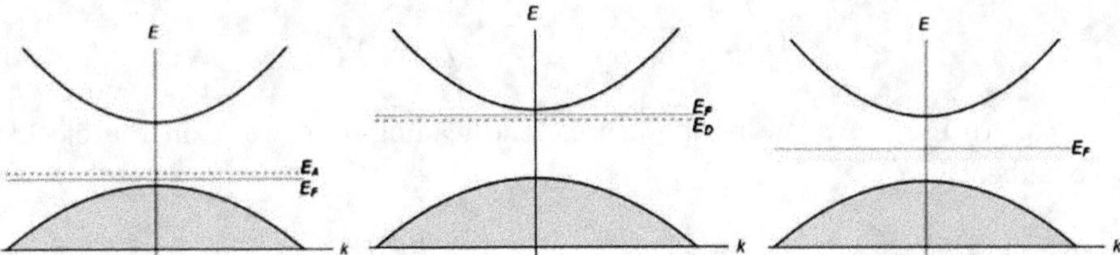

Fig. 6.31: Right panel: the Fermi energy at the middle of the gap between the full valence band and the empty conduction one. Mid panel: the Fermi energy at the mid energy between the level of the electrons in the donors and the bottom of the conduction band. Left panel: the Fermi energy at the mid energy between the level of the electrons in the acceptors and the top of the valence band.

Problem 6.30.

a. *Find the chemical potential which results from Eq. (6.136), when there are only donors, without applying any approximations except that $k_{\rm B}T \ll E_g$. Discuss the temperature dependence of the solution.*

b. *What is the answer when there are only acceptors?*

Problem 6.31.

The following data have been obtained for silicone at $T = 300K$: $n_{0e} = 2.8 \times 10^{19}$ cm^{-3}, $n_{0h} = 1.04 \times 10^{19}$ cm^{-3}. The energy gap is 1.14 eV.

a. *What is the chemical potential when the concentration of electrons in the conduction band is $n_e = 10^{17}$ cm^{-3}? What is the concentration of holes in the valence band in this case?*

b. *What is the chemical potential when the concentration of holes in the valence band is $n_h = 10^{14}$ cm^{-3}? What is the concentration of electrons in the conduction band?*

The specific heat of semiconductors. In **intrinsic semiconductors**, the number of electrons that cross over from the valence to the conduction band is proportional to $\exp[-\beta E_g/2]$ at low temperatures, see Eq. (6.134). Each of these carries an energy of the order of the gap, E_g, between the bands. It follows that the electronic specific heat of an intrinsic semiconductor at low temperatures is also proportional to $\exp[-\beta E_g/2]$, and hence is much smaller than the phonons' specific heat. A more accurate computation yields that the average energy of the electrons in the conduction band of an intrinsic semiconductor is

$$\frac{\langle E \rangle}{V} = \int_{E_c}^{\infty} dE\, g(E) f(E) E = E_c n_e + C_e e^{-\beta(E_c - \mu)} \int_0^{\infty} dx\, x^{3/2} e^{-\beta x}$$

$$= (E_c + 3k_{\rm B}T/2) n_e \,, \tag{6.139}$$

using the relation

$$\int dx\, x^{3/2} \exp[-\beta x] = 3\sqrt{\pi}/(4\beta^{5/2}) \ , \qquad (6.140)$$

and Eq. (6.130). At low temperatures, the leading-order term in the electronic specific heat is

$$C_V \approx E_c(\partial n_e/\partial T) \propto \beta^{1/2}\exp[-\beta E_g/2] \ .$$

A similar term is contributed by the holes in the valence band, but both contributions are much smaller than the phonons' specific heat. They can though be measured when the energy gap between the valence and the conduction bands is not too wide.

The electronic specific heat of an **extrinsic semiconductor** is also not very significant. In an n-type semiconductor, the density of states is

$$g(E) = \begin{cases} x_{\rm D}\sum_\ell g_\ell \delta[E - E(\ell)] \ , & E < E_c \ , \\[2mm] C_e\sqrt{E - E_c} \ , & E > E_c \ , \end{cases} \qquad (6.141)$$

where g_ℓ is the degeneracy of the energy level E_ℓ, located within the gap. The average energy is hence

$$\frac{\langle E\rangle}{V} = \int dE\, g(E) f(E) E = \sum_\ell g_\ell f[E(\ell)]E(\ell) + E_c n_c + C_e e^{-\beta(E_c-\mu)}\int_0^\infty dx\, x^{3/2}e^{-\beta x}$$

$$= \sum_\ell g_\ell f[E(\ell)]E(\ell) + (E_c + 3k_{\rm B}T/2)n_e \ , \qquad (6.142)$$

where the second step exploits Eq. (6.130) and the integral Eq. (6.140). This result is identical to the one of an intrinsic semiconductor, augmented by contributions from the electrons on the donors.

6.10 The semiclassical equations of motion for electrons in a periodic potential

Lattice momentum, electrons' velocity. The band structure induces modifications in the description of the electrons' motion as compared to the simplified calculations within the Sommerfeld model, Sec. 6.3. A conspicuous example is the average velocity of the electrons. In quantum mechanics, the operator that represents the momentum is $\hat{\mathbf{p}} = -i\hbar\boldsymbol{\nabla}$. Its expectation value, in a quantum state that obeys the Bloch theorem, $\psi_{n\mathbf{k}} = \exp[i\mathbf{k}\cdot\mathbf{r}]u_{n\mathbf{k}}(\mathbf{r})$ [Eq. (6.52)], is

$$\langle\mathbf{p}\rangle_{n\mathbf{k}} = \langle\psi_{n\mathbf{k}}| - i\hbar\boldsymbol{\nabla}|\psi_{n\mathbf{k}}\rangle = \int d^3r\, [e^{i\mathbf{k}\cdot\mathbf{r}}u_{n\mathbf{k}}(\mathbf{r})]^*(-i\hbar\boldsymbol{\nabla})[e^{i\mathbf{k}\cdot\mathbf{r}}u_{n\mathbf{k}}(\mathbf{r})] \ . \qquad (6.143)$$

As

$$(-i\hbar\boldsymbol{\nabla})\{\exp[i\mathbf{k}\cdot\mathbf{r}]u_{n\mathbf{k}}(\mathbf{r})\} = \exp[i\mathbf{k}\cdot\mathbf{r}][\hbar\mathbf{k} - i\hbar\boldsymbol{\nabla}]u_{n\mathbf{k}}(\mathbf{r}) \ ,$$

one finds

$$\langle \mathbf{p} \rangle_{n\mathbf{k}} = \hbar \mathbf{k} + \int d^3 r\, u_{n\mathbf{k}}^*(\mathbf{r})(-i\hbar \boldsymbol{\nabla}) u_{n\mathbf{k}}(\mathbf{r}) \,, \tag{6.144}$$

using the normalization of the wave functions. Thus, $\hbar \mathbf{k}$ **is not the momentum of the particle** (except for a free particle, for which the second term vanishes). Nonetheless, in many aspects the quantity $\hbar \mathbf{k}$ does play the role of a momentum, and therefore it is called the **"lattice momentum"**.

To derive an expression for the average momentum of the electron in a Bloch state, one begins from Eq. (6.62) and considers the variation of its solution as the vector $\mathbf{k}$ is slightly changed,

$$E_n(\mathbf{k} + \boldsymbol{\delta}) = E_n(\mathbf{k}) + \boldsymbol{\delta} \cdot \boldsymbol{\nabla}_\mathbf{k} E_n(\mathbf{k}) + \mathcal{O}(\delta^2) \,. \tag{6.145}$$

A similar variation of the Hamiltonian operator in the left hand-side of Eq. (6.62)) yields

$$\hat{\mathcal{H}}(\mathbf{k} + \boldsymbol{\delta}) = \hat{\mathcal{H}}(\mathbf{k}) + \boldsymbol{\delta} \cdot \frac{\hbar^2}{m}(-i\boldsymbol{\nabla} + \mathbf{k}) + \frac{\hbar^2 \delta^2}{2m} \,. \tag{6.146}$$

Treating the second term as a perturbation, Eq. (C.2) for the eigenenergies of $\hat{\mathcal{H}}(\mathbf{k} + \boldsymbol{\delta})$ shows that the first-order correction of the energy is the expectation value of the perturbation,

$$E_n(\mathbf{k} + \boldsymbol{\delta}) = E_n(\mathbf{k}) + \boldsymbol{\delta} \cdot \frac{\hbar^2}{m} \langle u_{n\mathbf{k}} | - i\boldsymbol{\nabla} + \mathbf{k} | u_{n\mathbf{k}} \rangle + \mathcal{O}(\delta^2) \,. \tag{6.147}$$

A comparison of the two expressions, Eqs. (6.146) and (6.147), reveals that

$$\frac{\hbar^2}{m} \langle u_{n\mathbf{k}} | - i\boldsymbol{\nabla} + \mathbf{k} | u_{n\mathbf{k}} \rangle = \boldsymbol{\nabla}_\mathbf{k} E_n(\mathbf{k}) \,,$$

and upon inserting into Eq. (6.144), one obtains the average velocity of the electron,

$$\langle \mathbf{v} \rangle_{n\mathbf{k}} = \frac{1}{m} \langle \mathbf{p} \rangle_{n\mathbf{k}} = \frac{1}{\hbar} \boldsymbol{\nabla}_\mathbf{k} E_n(\mathbf{k}) \,. \tag{6.148}$$

This expression is identical to that for the **group velocity** of a wave packet in quantum mechanics [see also Eq. (5.10) and the comments preceding Eq. (6.43)].

Equation (6.148) reproduces the result $\mathbf{p} = \hbar \mathbf{k}$ for a free particle [see also Eq. (6.144)]. In this case the momentum does not vary in time and is conserved. When the state of the electron belongs to a band structure, Eq. (6.148) can produce very different results. In particular, the momentum is not a "good quantum number", and therefore is not conserved. As seen in Eq. (6.128), the component of $\boldsymbol{\nabla}_\mathbf{k} E_n(\mathbf{k})$ normal to the faces of the Brillouin zones vanishes at the zones' boundaries, and therefore the velocity of the electron is zero there as well, though the lattice momentum is not. This conclusion stems from the fact that on the faces of the Brillouin zones the electron's wave functions correspond to standing waves. This, in turn, originates from the Bragg reflection off the periodic potential created by the ions (Sec. 6.6). The second term in Eq. (6.144) represents the momentum exchange between the electron and the lattice in this scattering event. [Indeed, substituting $u_{n\mathbf{k}}$ from Eq. (6.85) into Eq. (6.144) gives a weighted sum of reciprocal-lattice vectors.]

Equations of motion. In quantum mechanics the electron is described by a **wave packet**, as in Eq. (5.9). Roughly speaking, this wave packet is concentrated around the average momentum in momentum space, and around the average location in real space. The Ehrenfest theorem relates the time derivative of the expectation values of the position and momentum operators to the classical equations of motion (this is the correspondence principle). The wave packets of an electron in a periodic potential are composed of linear combinations of Bloch functions (that are the eigenfunctions of the Hamiltonian with a periodic potential). In the presence of solely the periodic potential, the energy of the electron is $E_n(\mathbf{k})$ of the conduction band; the lattice momentum $\hbar\mathbf{k}$ is a good quantum number, and hence it is conserved. The average velocity of the electron, Eq. (6.148), is conserved as well, and does not vary with time. The only way to change this situation is to apply additional forces on the electron, e.g., an external electric field. When this force is weak enough so that the width $\delta\mathbf{k}$ of the wave packet that describes the electron is narrower than the width of the Brillouin zone, the wave packet remains within a single energy band, and does not include contributions of other bands. The index n is then a good quantum number, and is not changed, e.g., by collisions. In real space the wave packet extends over distances of the order of $1/\delta\mathbf{k}$, comprising many unit cells, and the location of the electron, denoted $\mathbf{r}$, is within this region in space. Under these circumstances, the time dependence of $\mathbf{k}$ and $\mathbf{r}$ is derived from the classical equations of motion, with the kinetic energy replaced by $E_n(\mathbf{k})$. This approximation is termed **semiclassical**, since the effect of the periodic potential due to the ions, which essentially extends over short ranges, is found by a fully-quantum calculation, while that of the weaker forces, effective over larger distances, is accounted for classically. The success of this approximation is only partial: For example, the explanation of the quantum Hall effect (in the next section) requires quantum corrections.

In the presence of a static electric potential ϕ, the energy of the electron in the semiclassical approximation reads

$$E = E_n(\mathbf{k}) - e\phi \ , \tag{6.149}$$

where the electric field is $\boldsymbol{\mathcal{E}} = -\boldsymbol{\nabla}_{\mathbf{r}}\phi$. Assuming that the wave packet is narrow enough, the weak electric potential does not modify significantly the energy distribution among the constituents of the wave packet. Then, differentiating with respect to the time, and using the (approximate) relation $\mathbf{v} = \dot{\mathbf{r}}$ and Eq. (6.148), one finds

$$\dot{E} = \dot{E}_n(\mathbf{k}) - e\dot{\phi} = \boldsymbol{\nabla}_{\mathbf{k}}E_n(\mathbf{k}) \cdot \dot{\mathbf{k}} - e\boldsymbol{\nabla}_{\mathbf{r}}\phi \cdot \dot{\mathbf{r}} = \langle\mathbf{v}\rangle_{n\mathbf{k}} \cdot (\hbar\dot{\mathbf{k}} - e\boldsymbol{\nabla}_{\mathbf{r}}\phi) \ . \tag{6.150}$$

As the energy does not vary in time, the left hand-side vanishes, and hence

$$\hbar\dot{\mathbf{k}} = e\boldsymbol{\nabla}_{\mathbf{r}}\phi = -e\boldsymbol{\mathcal{E}} \ . \tag{6.151}$$

This equation of motion resembles the classical one, $\dot{\mathbf{p}} = -e\boldsymbol{\mathcal{E}}$, which is the basis of Eq. (6.3) in the Drude theory, or Eq. (6.44) of the Sommerfeld model.

The only difference is that in Eq. (6.151) the **momentum is replaced by the lattice momentum, $\hbar\mathbf{k}$.** Though the latter is not the true momentum of the electron, it plays the role of momentum in the semiclassical equation of motion, and consequently in all equations derived from momentum conservation, e.g., those that describe collisions of the electrons with impurities. However, as opposed to the classical conservation law, the lattice momentum is defined up to a reciprocal-lattice vector, and hence it may "jump" by such a vector after a collision. This process is called Umklapp [see also the discussion following Eq. (5.99)]. The excess momentum is absorbed by the entire lattice.

Ignoring collisions, then in the presence of a constant electric field, the lattice momentum increases linearly with time, $\hbar\mathbf{k}(t) = \hbar\mathbf{k}(0) - e\mathcal{E}t$, precisely as in Eq. (6.44). Indeed, substituting this time-dependent lattice-momentum in the right-hand side of Eq. (6.148) yields

$$\langle \mathbf{v}(t)\rangle = \frac{1}{\hbar}\boldsymbol{\nabla}_{\mathbf{k}}E_n[\mathbf{k}(0) - e\mathcal{E}t/\hbar] \ . \tag{6.152}$$

Since the band energy oscillates periodically as a function of $\mathbf{k}$, the electron velocity oscillates as a function of time. Thus, a constant electric field causes the electrons to move periodically in time. This periodic motion is named after Bloch, and is called **Bloch oscillations.** In particular, when $\mathbf{k}(t)$ approaches the boundary between two Brillouin zones, the velocity in Eq. (6.152) tends to zero [Eq. (6.128)] and changes its sign as a function of time. This surprising behavior, very different from the linear increase with time of a free particle, stems directly from the band structure and the interference of the waves describing the electrons upon their scattering off lattice points. Combining this result with the one derived from the Drude theory, where the electron moves freely only during a relaxation time τ, gives rise at equilibrium to an average velocity,

$$\langle \mathbf{v}\rangle = \boldsymbol{\nabla}_{\mathbf{k}}E_n[\mathbf{k}(0) - e\mathcal{E}\tau/\hbar]/\hbar \ .$$

This velocity may have a sign opposite to that of the electric field when the relaxation time is long enough. The relaxation times, however, are usually considerably shorter than the period of the Bloch oscillations (see problem 6.32), and hence this behavior is not realized.

Equation (6.152) implies that a constant electric field induces a periodic-in-time motion of the electrons. As the positive charges on the ions (almost) do not move, it follows that a constant electric field creates an oscillating dipole moment (for sufficiently long relaxation times). In reality the dipole moment is a sum over all electrons' dipole moments in the conduction band. As electrons move in all directions, the total moment vanishes for a completely filled band. A detailed calculation is given in problem 6.32.

Problem 6.32.

a. *Show that when the electric field $\mathcal{E}$ is oriented along a basis vector $\mathbf{b}$ in reciprocal lattice, the electron moves periodically in space, with the period $\mathbf{T} = \hbar b/(e|\mathcal{E}|)$.*

b. *Estimate the period in a* **SC** *lattice of lattice constant* $a = 1\text{Å}$, *in the presence of an electric field* $|\mathcal{E}| = 10^4$ *eV/cm. Compare the result with the typical value for the relaxation time, as estimated in Secs. 6.2 and 6.3. Are there conditions under which the Bloch oscillations can be detected experimentally?*

c. *In a one-dimensional semiconductor of length L and lattice constant a, the energy of the conduction band is given by* $E(k) = E_0 - \Gamma\cos(ka)$. *At zero temperature the band is filled up to the Fermi level. Find the dipole moment induced by a constant electric field along the lattice direction.*

Acceleration and effective mass. The average acceleration of the electron, Eq. (6.148), is

$$\langle \mathbf{a} \rangle_{n\mathbf{k}} = \frac{d}{dt}\langle \mathbf{v} \rangle_{n\mathbf{k}} = \frac{1}{\hbar}\frac{d}{dt}[\boldsymbol{\nabla}_{\mathbf{k}}E_n(\mathbf{k})] \ . \tag{6.153}$$

Using Eq. (6.151), the α−component of the acceleration is

$$\langle a_\alpha \rangle_{n\mathbf{k}} = \frac{1}{\hbar}\sum_\beta \frac{\partial}{\partial k_\beta}\left[\frac{\partial}{\partial k_\alpha}E_n(\mathbf{k})\right]\frac{d}{dt}k_\beta = -\sum_\beta \left(\frac{1}{m^*}\right)_{\alpha\beta}e\mathcal{E}_\beta \ , \tag{6.154}$$

where the tensor of coefficients,

$$\left(\frac{1}{m^*}\right)_{\alpha\beta} = \frac{1}{\hbar^2}\frac{\partial^2 E_n(\mathbf{k})}{\partial k_\alpha \partial k_\beta} \ , \tag{6.155}$$

replaces the mass in Newton's second law, $\mathbf{a} = -(1/m)e\mathcal{E}$. Admittedly, this notation is confusing: the tensor m^* is the inverse of the matrix in the right hand-side of Eq. (6.155). This is the tensor of the **effective mass of the electron.** In the nearly-free electron model, and at the bottom of the conduction band, the energy is roughly quadratic in k, $E(k) \approx E_c + \hbar^2 k^2/(2m)$; the effective mass tensor is then $(m^*)_{\alpha\beta} \approx m\delta_{\alpha\beta}$, as in classical mechanics. In other parts of the band structure the effective mass is not necessarily related to the mass of a free electron. Equation (6.154) is counterintuitive to classical mechanics: when the tensor is not diagonal, an electric field may accelerate an electron along a direction normal to the field. Even when the tensor is diagonal, it may happen that different components of the electric field induce different accelerations. Note though, that in many cases the tensor is diagonal, leading in turn to the dispersion relation discussed in problem 6.29.

Problem 6.33.

Exploit second-order perturbation theory [Eq. (C.3)] to express the inverse tensor of the effective mass in terms of integrals containing the periodic Bloch functions $u_{n\mathbf{k}}$.

Problem 6.34.

Use the results of the nearly-free electron model for the energies of the electron close to the energy gap [e.g., Eq. (6.80)] to obtain the tensor of the effective mass.

The conductance of electrons and holes.   Within the Drude model, the electrons' velocity at constant electric fields is [Eq. (6.154)]

$$\langle v_\alpha \rangle_{n\mathbf{k}} = -e \sum_{\beta} \left(\frac{1}{m^*_{n\mathbf{k}}} \right)_{\alpha\beta} \mathcal{E}_\beta \tau_{n\mathbf{k}} \, . \tag{6.156}$$

This relation assumes that the relaxation time of the electron depends on the band it occupies, and on its lattice momentum. The total electric current density is the sum over all occupied states, in all bands,

$$\mathbf{j} = -\frac{e}{V} \sum_{n,\mathbf{k}} \langle \mathbf{v} \rangle_{n\mathbf{k}} \, , \tag{6.157}$$

where V is the volume of the sample. [The sum over the states divided by the volume replaces the electronic concentration n in Eq. (6.4); note the different meaning of the notation n in Eq. (6.4) and in Eq. (6.157).] The valence bands are fully occupied by electrons, but then they do not contribute to the current, because for each electron with a certain velocity there is in the band another electron with the opposite velocity (due to the $\mathbf{k} \Leftrightarrow -\mathbf{k}$ symmetry), and therefore the sum over such a band vanishes. It follows that the current is carried only by electrons in partially-filled bands, i.e., by electrons from the conduction bands. One thus finds

$$\mathbf{j} = \boldsymbol{\sigma} \cdot \boldsymbol{\mathcal{E}} \, , \tag{6.158}$$

where the **conductivity tensor** is

$$\sigma_{\alpha\beta} = \frac{e^2}{V} \sum_{n,\mathbf{k}} \left(\frac{1}{m^*_{n\mathbf{k}}} \right)_{\alpha\beta} \tau_{n\mathbf{k}} \, , \tag{6.159}$$

and the sum runs over the partially-filled bands. This equation extends Eqs. (6.5) and (6.6) of the Drude model for an electron in a periodic potential.

At the bottom of the conduction band the energy is minimal as a function of the lattice momentum; denoting this point by $\mathbf{k}_c$ and the corresponding energy by E_c, then for an isotropic diagonal effective-mass tensor, the energy around this point is

$$E(\mathbf{k}) \approx E_c + (\hbar^2/[2m^*_c])(\mathbf{k} - \mathbf{k}_c)^2 \, ,$$

from which one finds $\langle \mathbf{v} \rangle_c = -e\boldsymbol{\mathcal{E}}\tau_c/m^*_c$. The electrons thus move in the opposite direction to the field, as expected for negative charges. On the other hand, at the upper part of the valence band the energy has a maximum, E_v, at the lattice momentum $\mathbf{k}_v$. Assuming that the maximal energy is isotropic, then

$$E(\mathbf{k}) \approx E_v - (\hbar^2/[2m^*_v])(\mathbf{k} - \mathbf{k}_v)^2 \, ,$$

where $m^*_v > 0$. The effective mass, as defined in Eq. (6.155) is negative, and equals to $-m^*_v$. The equation of motion gives

$$\langle \mathbf{v} \rangle_v = -(\hbar/m^*_v)[\mathbf{k}(\tau_v) - \mathbf{k}_v] = e\boldsymbol{\mathcal{E}}\tau_v/m^*_v \, .$$

The electrons thus move in the opposite direction to that of the lattice momentum, and along the same direction as that of the electric field. This means that they behave like positively-charged particles. For this reason the charge carriers in the valence band are treated as holes with a positive charge. This may be understood also intuitively: an electron moving due to an applied electric field occupies an empty place in the valence band (i.e., it fills a hole). The place left behind is now occupied by a hole. Thus, the electron and the hole move at opposite directions.

The current in intrinsic semiconductors is carried by the electrons in the conduction band and the holes in the upper part of the upmost valence band. The effective mass of the electrons at the bottom of the conduction band is positive; their contribution to the conductivity tensor can be roughly written as $n_e \tau_e e^2 (1/m_c^*)_{\alpha\beta}$. In contrast, the effective mass of the electrons in the upper part of of the valence band is negative; they appear as positively-charged particles. This also results from the following argument. Were the band completely full, then the current carried by the electrons in it would have vanished. One may therefore express the electrons' current in the partially-filled valence band as

$$\mathbf{j} = \mathbf{j}_{\text{full}} - \mathbf{j}_{\text{edge}} = -\mathbf{j}_{\text{edge}} = n_h e \langle \mathbf{v} \rangle \, , \tag{6.160}$$

where $\mathbf{j} = (-e/V) \sum_{\text{electrons}} \langle \mathbf{v} \rangle$ is the current of the electrons in the band, $\mathbf{j}_{\text{full}} = 0$ is the current of the fully-occupied band, and $\mathbf{j}_{\text{edge}} = -n_h e \langle \mathbf{v} \rangle$ is the current of the missing electrons in the upper edge of the band. The number of these electrons is the number of states out of which electrons are removed into the conduction band; each of them carries the current $-e \langle \mathbf{v} \rangle$. The right hand-side of Eq. (6.160) appears as a current of particles that carry the charge $+e$ and move with a velocity $\langle \mathbf{v} \rangle$. Altogether,

$$\sigma_{\alpha\beta} = \frac{e^2}{V} \sum_{n\mathbf{k}} \left(\frac{1}{m_{n\mathbf{k}}^*} \right)_{\alpha\beta} \tau_{n\mathbf{k}} = e^2 \left[n_e \tau_e \left(\frac{1}{m_c^*} \right)_{\alpha\beta} + n_h \tau_h \left(\frac{1}{m_v^*} \right)_{\alpha\beta} \right] \, , \tag{6.161}$$

is the conductivity tensor; the sign of the effective masses of the holes is reversed, so that they are positive.

Recall that the numbers of the electrons and holes, n_e and n_h, are exponential functions of the temperature, see e.g., the discussion in Sec. 6.8, or Eqs. (6.134) and (6.138). Ignoring the temperature dependence of the relaxation times, the conductivity of intrinsic semiconductors is such that $\sigma \propto T^{3/2} \exp[-\beta E_g/2]$. It thus increases with the temperature, as opposed to that of metals, that decreases as the temperature is raised, due to the decrease of the relaxation time [Matthiessen rule, Eq. (6.49)]. In extrinsic semiconductors, the number of charge carriers reaches the number of donors (or acceptors) at relatively low temperatures and then it is almost independent of the temperature; the corresponding conductivity decreases as the temperature is raised due to the decrease of the relaxation times.

6.11 Electrons in a magnetic field

Semiclassical equations of motion. The discussion following Eq. (6.151) indicates that $\hbar\mathbf{k}$ plays the role of the momentum in the semiclassical equations of motion. Indeed, in the presence of a uniform constant magnetic field $\mathbf{B}$, and when collisions of the electron with various defects are ignored, the semiclassical equation of motion (6.8) is

$$\frac{d}{dt}\hbar\mathbf{k}(t) = -\frac{e}{c}[\mathbf{v} \times \mathbf{B}] = -\frac{e}{\hbar c}[\boldsymbol{\nabla}_{\mathbf{k}}E(\mathbf{k}) \times \mathbf{B}] \,, \tag{6.162}$$

[making use of Eq. (6.148)]. The increment in time of the vector $\mathbf{k}$ lies in a plane perpendicular to the magnetic field. In the same plane, that increment is also normal to $\boldsymbol{\nabla}_{\mathbf{k}}E$. As the gradient of the energy with respect to $\mathbf{k}$ is perpendicular to an equal-energy surface (which is a surface in the three-dimensional momentum space), the vector $\mathbf{k}$ moves on a curve created by the cut of the equal-energy surface and the plane normal to the magnetic field. For a magnetic field along $\hat{\mathbf{z}}$, the vector $\mathbf{k}$ moves in the XY plane. Since at low temperatures the electrons are confined to move on the Fermi surface, it follows that the electron moves on **cuts of that surface with planes perpendicular to the magnetic field.**

The periodic motion on the Fermi surface. As seen in Sec. 6.8, the Fermi surface is roughly a sphere at the vicinity of the extremal points of the bands; the cuts with a plane are thus approximately circles. Electrons are moving at the bottom of the band, and holes are moving near the top; in any event, when the cut is a closed orbit, the motion of a particle around it is periodic. According to Eq. (6.162), the time dependence of the component of the lattice momentum which is tangential to the orbit is $dk_\parallel/dt = (e\mathrm{B}/[\hbar^2 c])|\boldsymbol{\nabla}_{\mathbf{k}}E|$, and thus the period is

$$\mathbf{T} = \int_0^T dt = \frac{\hbar^2 c}{e\mathrm{B}} \oint \frac{dk_\parallel}{|\boldsymbol{\nabla}_{\mathbf{k}}E|} = \frac{\hbar^2 c}{e\mathrm{B}} \frac{dA(E, k_\perp)}{dE} \,, \tag{6.163}$$

where the integration marked by $\oint$ is over a scalar function, along the closed orbit created by the cut of the Fermi surface with a plane perpendicular to the magnetic field. This integration gives the difference between the two areas encircled by the equal-energy curves at E and at $E + dE$, denoted dA/dE. The component of the lattice momentum parallel to the magnetic field (and thus perpendicular to the plane of motion) remains constant.

Problem 6.35.
Derive the result (6.163) directly from the equations of motion

$$\dot{k}_x = \frac{e\mathrm{B}}{\hbar^2 c}\frac{\partial E}{\partial k_y} \,, \qquad \dot{k}_y = -\frac{e\mathrm{B}}{\hbar^2 c}\frac{\partial E}{\partial k_x} \,,$$

(the magnetic field is along $\hat{\mathbf{z}}$).

In a **two-dimensional lattice** the electrons move on the Fermi curve. When the energy is a paraboloid, $E = \hbar^2 \mathbf{k}^2/(2m^*)$, this curve (close to the extremal points) is a circle and the velocity is a constant, $|\boldsymbol{\nabla}_{\mathbf{k}}E| = \hbar v_{\mathrm{F}} = \hbar^2 k_{\mathrm{F}}/m^*$; the perimeter of the circle is $2\pi k_{\mathrm{F}}$ and hence the period is $\mathbf{T} = 2\pi/\omega_c$, where $\omega_c = e\mathrm{B}/(m^* c)$ is the **cyclotron frequency**, known from the motion of a free particle in a constant magnetic field [Eq. (6.9) and problem 6.3], with the mass replaced by the effective mass. As the effective mass of the holes has the opposite sign, it follows that they move around the same perimeter in the opposite direction. Problem 6.36 shows that for the ellipsoidal energy introduced in problem 6.29 and a magnetic field along an arbitrary direction, $\mathbf{B} = \hat{n}\mathrm{B}$, the cyclotron frequency is

$$\omega_c = \frac{e\mathrm{B}}{c} \frac{\sqrt{m_x^* \hat{n}_x^2 + m_y^* \hat{n}_y^2 + m_z^* \hat{n}_z^2}}{\sqrt{m_x^* m_y^* m_z^*}} \ . \tag{6.164}$$

Thus, measuring the cyclotron frequency of the electrons and the holes on a periodic lattice (for instance, by monitoring the resonant absorption of radiation at that frequency) at various directions of the magnetic field, provides valuable information on the Fermi surface and on the effective masses.

In three dimensions, the cuts of planes with the Fermi surface may attain diverse shapes. When the band is almost empty (or almost full) the Fermi surfaces are approximately spheres, and the cuts are closed orbits (circles or ellipses). In such cases the motion is periodic; however, as the band is filled up and the Fermi surfaces reach the boundaries of Brillouin zones this simple picture breaks down. Consider for instance the Fermi surface depicted in Fig. 6.28, pertaining to a single-valence metal of a **SC** structure. The upper row in Fig. 6.32 displays cuts of this surface with planes normal to a magnetic field oriented along $\hat{z}$. As seen, all cuts are closed orbits, on which the motion is periodic. On the other hand, the lower row in the figure illustrates the situation where the magnetic field is rotated around the $\hat{x}$–axis, in the YZ plane. At small inclinations the orbits are still closed; as the angle increases there appear open orbits, on which the motion does not repeat itself.

Problem 6.36.
Prove Eq. (6.164).

Problem 6.37.

Equation (6.162) yields the orbit of an electron in lattice-momentum space. Exploit it to show that the orbit in real space is

$$\mathbf{r}(t) = \mathbf{r}(0) - \ell_B^2 \hat{\mathbf{B}} \times [\mathbf{k}(t) - \mathbf{k}(0)] + v_B t \hat{\mathbf{B}} \ ,$$

where $\hat{\mathbf{B}} = \mathbf{B}/|\mathbf{B}|$, $\ell_B^2 = \hbar c/(e\mathrm{B})$ is the square of the "magnetic length", and $v_B = \dot{\mathbf{r}} \cdot \hat{\mathbf{B}}$ is the component of the velocity parallel to the magnetic field (this component does not vary in time). Show that the projection of this orbit on a plane normal to the magnetic field is produced by rotating the orbit in lattice-momentum space, and scaling the coordinates by the multiplicative factor ℓ_B^2.

Problem 6.38.
a. *Find the Hall coefficient when the charge is carried by electrons as well as by holes. Keep only terms linear in the magnetic field.*

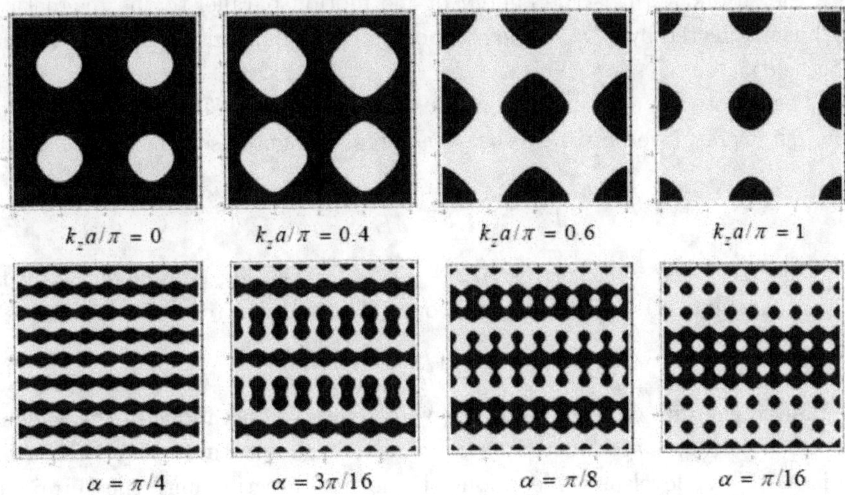

Fig. 6.32: Upper panels: the boundaries between the dark and light areas represent cuts of the Fermi surface from Fig. 6.28 with planes normal to the $\hat{\mathbf{z}}$–axis, for several values of $k_z a/\pi$. Lower panels: cuts of the same surface with planes passing through the $\hat{\mathbf{x}}$–axis, making an angle α with the $\hat{\mathbf{y}}$–axis.

b. *Silicon is an intrinsic semiconductor, with $n_i = 1.5 \times 10^{10}\ cm^{-3}$ and $\mu_e = 3\mu_h = 1450\ cm^2\ V^{-1}\ sec^{-1}$. Plot the dependence of the Hall coefficient on the electrons' concentration in the conduction band. For what value of this concentration does the Hall coefficient vanish? At what concentration is it extremal?*
c. *Incorporate terms quadratic in the magnetic field and show that – in contradiction with Eq. (6.12) – the specific resistivity ρ_{xx} does depend on the magnetic field. This dependence is termed* **magneto-resistance.** *Discuss in particular the cases $n_e = n_h$ and $n_e \neq n_h$.*

Landau levels. Without loss of generality, the magnetic field is chosen to point along $\hat{\mathbf{z}}$. The semiclassical picture gives that a free particle of mass m^* moves on a circle in the plane normal to the field, with a period given by the cyclotron frequency, $\omega_c = eB/(m^*c)$. As in other cases, quantum mechanics "translates" a periodic motion into quantized energy levels corresponding to this motion, such that the level spacing between consecutive levels is $E_{\ell+1} - E_\ell = \hbar\omega_c$. The quantum-mechanical derivation (see Appendix E) results in a motion similar to that of a harmonic oscillator, with the discrete energy levels

$$E_\ell = \hbar\omega_c(\ell + \frac{1}{2})\ . \tag{6.165}$$

These are the **Landau levels.** In three dimensions, the total energy of the electron is thus

$$E(\ell, k_z) = E_\ell + \hbar^2 k_z^2/(2m^*)\ ,$$

where the second term is the energy of the motion parallel to the magnetic field (which is unaffected by it; k_z is the component of the momentum along the direction of the field).

Let us denote the area in momentum space, enclosed by the orbit corresponding to the ℓth energy level, by A_ℓ. When ℓ is large, one may use Eq. (6.163), to find

$$\frac{A_{\ell+1} - A_\ell}{E_{\ell+1} - E_\ell} = \frac{dA}{dE} = \frac{TeB}{\hbar^2 c} \, ,$$

and then $E_{\ell+1} - E_\ell = \hbar\omega_c$ gives $A_{\ell+1} - A_\ell = 2\pi eB/(\hbar c)$; consequently, up to an additive constant, $A_\ell = \ell 2\pi eB/(\hbar c) + \text{const}$. The solution to problem 6.39 yields

$$A_\ell = (\ell + 1/2) 2\pi eB/(\hbar c) \, . \tag{6.166}$$

The allowed energies of a free particle (whose energy is parabolic in the wave numbers) for a certain ℓ are located on a cylinder in momentum space, whose base's area is A_ℓ. These levels are filled along the normal direction until the Fermi energy is reached; there the cylinder cuts the Fermi sphere. Its height there is $k_{z\mathrm{F}}$, as determined by $E(\ell, k_{z\mathrm{F}}) = E_\mathrm{F}$. Figure 6.33(a) illustrates those cylinders within the Fermi sphere (that in the absence of the magnetic field corresponds to a free electron). As the Landau level is increased, the radius of the cylinder increases as well but its height diminishes. The area encompassed by the periodic orbit of the electron on the Fermi surface is

$$A_\mathrm{F} = (\ell + 1/2) 2\pi eB/(\hbar c) = E_\mathrm{F} 2\pi m^*/\hbar^2$$

(check!).

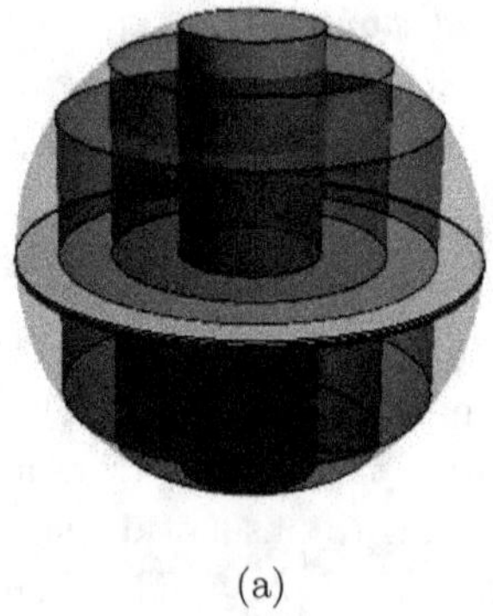

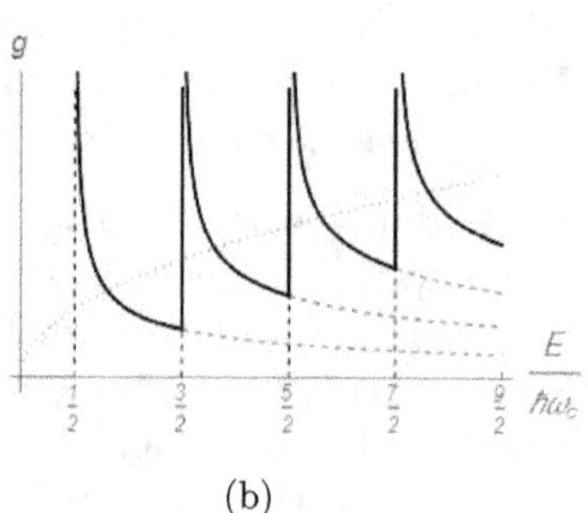

(a)   (b)

Fig. 6.33: (a) The energy levels of a free electron in a magnetic field along the normal direction. Each Landau level encompasses the states on one of the cylinders; the circular cut of the cylinder with the original Fermi sphere gives the corresponding Fermi energy. (b) The density of the states of (a), Eq. (6.167). The dashed curve represents the density of states of a free particle in three dimensions, in the absence of the field, $g \propto \sqrt{E}$.

Problem 6.39.

Consider a free electron in a magnetic field. The radius of the ℓth cylinder in Fig. 6.33, K_ℓ, is determined by $E_\ell = \hbar^2 K_\ell^2/(2m^*)$. Find the area A_ℓ in terms of K_ℓ and prove Eq. (6.166).

The density of states in three dimensions. Since the system is finite along the direction parallel to the field, k_z attains discrete values, $k_z = 2\pi n_z/L_z$ (where n_z is an integer, and L_z is the length of the sample along the field). These values form a discrete continuum for a long sample. As the motion along this direction is a free one in one dimension, the corresponding density of states is $C[E-(\ell+1/2)\hbar\omega_c]^{-1/2}$ (problem 6.10). Whenever ℓ is augmented by 1, another Landau level is "open" for electrons. It therefore follows that the total density of states is

$$g(E) = \sum_{\ell=0}^{\infty} C[E - (\ell+1/2)\hbar\omega_c]^{-1/2}\Theta[E - (\ell+1/2)\hbar\omega_c] , \qquad (6.167)$$

where $\Theta(x)$ is the step function, $\Theta(x) = 0$ for $x < 0$ and $\Theta(x) = 1$ for $x > 0$. This density of states is depicted in Fig. 6.33(b). All occupied Landau levels are such that $(\ell+1/2)\hbar\omega_c \leq E_F$ (E_F is the Fermi energy). As the magnetic field is increased, the spacing between the Landau levels broadens, and the number of filled Landau levels diminishes (the cylinder whose radius is the largest moves out of the Fermi sphere). Whenever the value of

$$\ell_F + 1/2 = E_F/(\hbar\omega_c) = E_F m^* c/(\hbar e B)$$

is reduced by 1, the upmost Landau level, previously occupied, is depopulated, and the density of states "jumps" from its upmost branch to the branch below [on one of the vertical lines in Fig. 6.33(b)].

de Haas and van Alphen effect. The magnetic susceptibility of metallic bismuth was measured by de Haas and van Alphen. It was found that at strong magnetic fields and low temperatures, the susceptibility is a periodic function in $1/B$, where B is the field. This result can be explained from an examination of the density of states. As seen, the density of states is determined by the geometry of the cylinders in Fig. 6.33(a): equal areas encompassed by the electron's orbits on the Fermi surface in a plane normal to the field give rise to identical densities of states, and hence to identical results for the physical properties that are dictated by the density of states. Denoting by B_ℓ the field that obeys $A_F = (\ell + 1/2)2\pi e B_\ell/(\hbar c)$, and by $B_{\ell+1}$ the one that is given by $A_F = (\ell + 3/2)2\pi e B_{\ell+1}/(\hbar c)$ (with the same area) then the properties measured at these two fields should be the same, and therefore they are periodic functions of $1/B$, with the period

$$\Delta(1/B) = 1/B_{\ell+1} - 1/B_\ell = 2\pi e/(\hbar c A_F) = e\hbar/(m^* c E_F) .$$

A similar periodicity was found in measurements of the electrical resistance in the presence of a magnetic field, a phenomenon termed **Shubnikov-de Haas effect**. This periodicity is obtained also when the Fermi surfaces are not spherical, e.g., those in Fig. 6.27. As the direction of the magnetic field is changed, the area A_{F} is changed as well, according to the geometry of the Fermi surface. In this way the entire surface can be mapped.

Modifications due to the Zeeman interaction. The description given above is modified when the spin degree of freedom is taken into account. The energy of an electron in a magnetic field is affected by the Zeeman interaction: each Landau level splits into two,

$$E_\ell - \mathbf{B} \cdot \boldsymbol{\mu} = E_\ell \pm g^* \mu_{\mathrm{B}} \mathrm{B} \ ,$$

where $\boldsymbol{\mu} = \pm g^* \mu_{\mathrm{B}} \hat{\mathbf{z}}$ are the two possible values of the magnetic moment of the electron along the magnetic field, μ_{B} is the Bohr magneton, and g^* is the gyromagnetic ratio (see Secs. 2.10 and 4.7). The corresponding density of states, that extends the one in Fig. 6.33(b), is

$$g(E) = \sum_{\ell=0}^{\infty} C[(E - E_\ell + g^* \mu_{\mathrm{B}} \mathrm{B})^{-1/2} + (E - E_\ell - g^* \mu_{\mathrm{B}} \mathrm{B})^{-1/2}] \ . \qquad (6.168)$$

Each of the vertical lines in Fig. 6.33(b) splits into two, whose distance from one another increases as the magnetic field becomes stronger. In a periodic solid, the mass of the electron is replaced by m^*, and consequently $\omega_c = e\mathrm{B}/(m^* c)$. As in semiconductors m^* is usually much smaller than m, it turns out that $\hbar\omega_c \gg \mu_{\mathrm{B}}\mathrm{B}$. The Zeeman splitting can be detected only at strong magnetic fields.

The density of states in two dimensions. As opposed to the spectrum in three dimensions, that of a two-dimensional system, when the magnetic field is perpendicular to the sample, is discrete. Taking into account also the Zeeman interaction, the density of states is

$$g(E) = \sum_{\ell=0}^{\infty} g_\ell [\delta(E - E_\ell + g^* \mu_{\mathrm{B}}\mathrm{B}) + \delta(E - E_\ell - g^* \mu_{\mathrm{B}}\mathrm{B})] \ , \qquad (6.169)$$

where g_ℓ denotes the degeneracy (per unit area) of the ℓth Landau level (without the spin degree of freedom). This density of states is illustrated in Fig. 6.34(a). In the absence of the magnetic field, the density of states per unit area (not accounting for the spin degree of freedom) is $g^{(2d)} = m^*/(2\pi\hbar^2)$, problem 6.10. The N_e occupied single-particle energy levels are represented by the dense points that fill the Fermi circle (N_e is the number of electrons). In the presence of the field, there are only a finite number, say ν, of occupied Landau levels up to the Fermi energy. As the total number of states up to that energy cannot be changed by the magnetic field, each of the Landau levels should be highly (macroscopically) degenerate. In other words, the states of the free particle that, in the absence of the field, have (continuous) energies in the range $E_\ell - \hbar\omega_c/2$ and $E_\ell + \hbar\omega_c/2$ are all confined in the presence of the field to the single level E_ℓ. The number of these states (per unit area, and without the spin degree of freedom) is the product of the two-dimensional density of states with this energy range, i.e., $g_\ell = g^{(2d)} \hbar\omega_c = e\mathrm{B}/(hc)$. Denoting the area of the two-dimensional sample by A, then the number of states belonging to each level (per spin) is

$$N_\Phi = g_\ell A = (AB)/(hc/e) = \Phi/\Phi_0 \ ,$$

where $\Phi = AB$ is the **magnetic flux** through the sample, and $\Phi_0 = hc/e$ is the **magnetic flux quantum**. Thus, the number of states belonging to each level is the number of flux quanta contained within the total magnetic flux, N_Φ. Using the magnetic length from problem 6.37 gives $N_\Phi = A/(2\pi\ell_B^2)$, and therefore the flux quantum corresponds to the area $2\pi\ell_B^2$. From the expression in problem 6.37, one finds

$$\ell_B = \sqrt{\hbar c/(2B)} = 257\text{Å}/\sqrt{\text{B(Tesla)}} \ ,$$

(check!). Another derivation of the degeneracy of each Landau level appears in Appendix E. In conclusion, when the sample contains N_e electrons, the number of levels occupied at zero temperature is

$$\nu \equiv \frac{N_e}{N_\Phi} = \frac{N_e}{g_\ell A} = \frac{n_e}{g_\ell} \ ,$$

where $n_e = N_e/A$ is the density of electrons in the plane. The number ν is termed the **filling factor**.

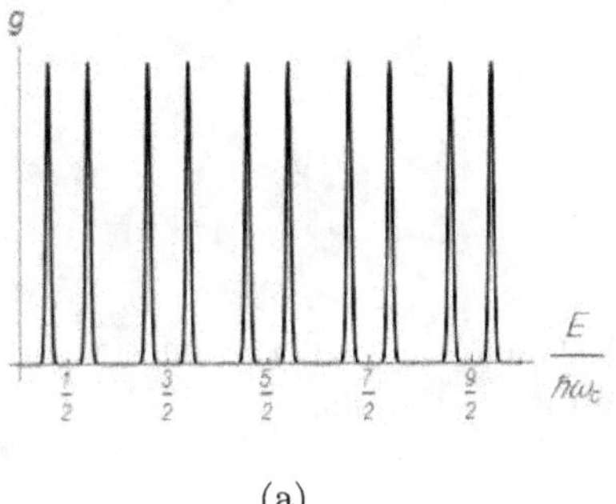

(a)

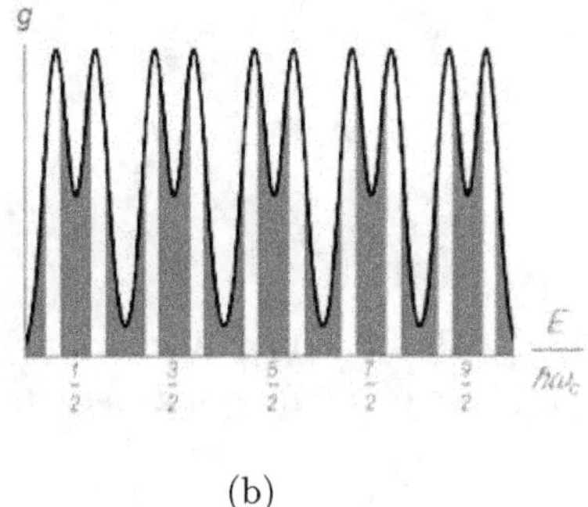

(b)

Fig. 6.34: The density of states of a free electron in a magnetic field normal to the plane, at two dimensions. (a) The Landau levels, each bifurcated due to the Zeeman interaction. (b) The same as in (a), but in the presence of disorder. The dark areas contain localized states, that do not contribute to the conductance; the light ones contain extended states.

The degeneracy of the Landau levels, surface states, and topological insulators. Within the semiclassical picture, all eigenstates that belong to the same Landau level (save one) describe each a periodic circular motion, that encompasses the area $2\pi\ell_B^2$. Figure 6.35 illustrates two examples for such orbits (in the center of the figure). Due to the Pauli principle, each orbit is centered around a different point on the plane, and therefore the plane is covered by circular paths, whose number is the degeneracy of the level. An electron occupying such an orbit is localized around the center, and is not free to move away; consequently, it does not contribute to the electrical transport. However, in addition to these states there are others, near the edges of the sample. An electron that moves along a circular path centered on the boundary eventually collides with the surface, and then is reflected, i.e., its velocity is reversed. It then launches on a new circular path (see the figure). This motion along the boundary is called a **skipping orbit**; it is this motion that carries the electric current along the edges of the sample. The quantum-mechanical treatment of these **edge states** is detailed in Appendix E. The electrons in the skipping orbits, or in the edge states, propagate along a single direction (shown by the arrows in Fig. 6.35). They cannot be scattered into the opposite direction and therefore the corresponding electric resistance vanishes. These are **topological** states: they extend between the edges of the sample and the vacuum around it, and their number is independent of the shape of the boundary [for instance, measuring the transverse resistance of a rectangular sample in quantum-Hall regime (see below) yields the same result as that of a circular system]. Materials for which the bulk is an insulator while the surface conducts electricity are called **topological insulators**.

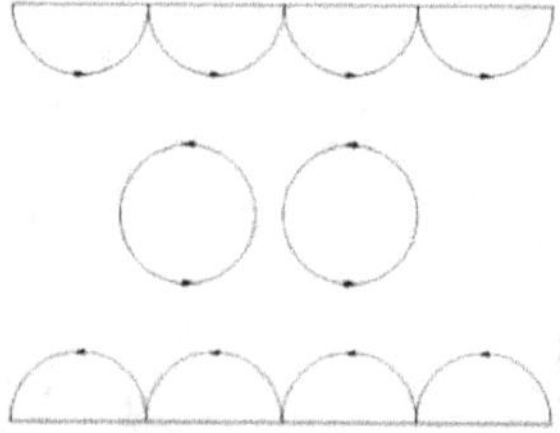

Fig. 6.35: Circular and skipping orbits of electrons at the Fermi energy, in the presence of a strong magnetic field directed normal to the plane.

The quantum Hall effect. The "classical" Hall resistivity, Eq. (6.12), is linear in the magnetic field, $\rho_{xy} = -R_H B$. Surprisingly enough, measurements of the Hall effect of electrons in a thin layer of GaAs [realized in a heterostructure, Sec. 2.9, which includes a thin layer of GaAs grown epitaxially between two layers of $Al_x Ga_{1-x} As$; the latter is an n-type semiconductor and therefore donates electrons to the thin layer] reveal a different behavior. At high magnetic fields, of the order of 1-40 Tesla (i.e., 10-400 kGauss), the Hall resistance has a staircase structure, at discrete values $|\rho_{xy}| = h/(\nu e^2)$, where ν is an integer ($h = 2\pi\hbar$ is the Planck constant). These measurements are presented in the upper part of Fig. 6.36. The Hall resistance remains constant on the plateaux, while the "usual" (longitudinal) resistance there is very small, as seen in the lower part of the figure. In the narrow space in-between steps this resistance is very large. This phenomenon is termed the **quantum Hall effect**; it is rather amazing, since it enables the detection of quantum effects on macroscopic length scales.

There are many other measurements of the quantum Hall resistance on other materials, including graphene. The results are **universal**, in the sense that they are independent of the geometry of the sample or its composition. In fact, the quantization of the Hall resistance reflects that of the Landau levels. In a thin layer the width along the direction normal to the plane, L_z, is small and hence the corresponding wave vector, $k_z = 2\pi n_z/L_z$, is large, implying that at low temperatures (the measurements shown in Fig. 6.36 for $n_z \neq 0$ are taken at 8 mK) only the states with $k_z = 0$ are occupied (and thus the system is effectively two dimensional).

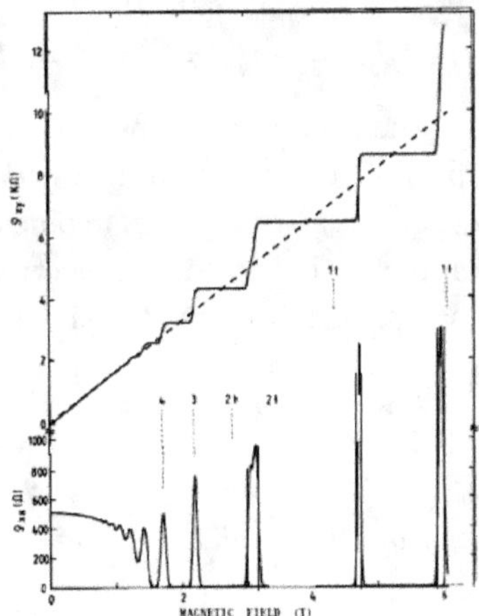

Fig. 6.36: Measurements of the quantum Hall effect on a planar layer of GaAs, grown on $Al_x Ga_{1-x} As$. The upper plot shows the Hall resistance, $R_H B = |\rho_{xy}|$; the dashed line depicts the classical Hall resistance. The lower part portrays the electric resistance, $R = \rho_{xx}$. [K. von Klitzing, *The quantum Hall effect*, Rev. Mod. Phys. **58**, 519 (1986).]

According to Ohm's law [Eq. (6.7)], the resistance of a $d-$dimensional cube of length L is $R = \rho L^{2-d}$. Thus, in two dimensions, $R = \rho$. When the Fermi level is located in-between two Landau levels, that is, between two maxima in Fig. 6.34(a) then all levels below it are populated, and the filling factor ν is an integer. When ν is an even number, then the upmost Landau level is completely full; when it is odd, it is only half filled (because of the spin index of the electrons). In both cases, the density of electrons is $n_e = \nu g_\ell = \nu Be/(hc)$. Exploiting Eqs. (6.11-6.12) yields

$$\rho_{xy} = -R_H B = -B/(n_e ec) = -h/(\nu e^2) \ . \tag{6.170}$$

For a constant number of electrons in the sample, the νth Landau level is filled when $B_\nu = n_e hc/(e\nu)$, and then there appears a step in Fig. 6.36. The next stair occurs when the field reaches the value $B_{\nu+1} = n_e hc/[e(\nu + 1)]$. The difference, $1/B_{\nu+1} - 1/B_\nu = e/(n_e hc)$, is a constant independent of the field, similar to the situation in the Shubnikov-de Haas effect.

When the chemical potential (i.e., the Fermi energy) is located in one of the gaps between the Landau levels the material is an insulator, as indeed confirmed in the experiment [Fig. 6.22(a)]. The longitudinal conductance vanishes, $\sigma_{xx} = \sigma_{yy} = 0$. Problem 6.40 shows that by inverting the conductance tensor one finds $\rho_{xx} = \rho_{yy} = 0$. Surprisingly – but again confirmed in the experiment – the diagonal elements in the conductance tensor and the diagonal elements of the resistance tensor are all zero, and off-diagonal ones are $\rho_{xy} = -\rho_{yx} = 1/\sigma_{yx}$, so that

$$\sigma_{xx} = \sigma_{yy} = 0 , \quad \sigma_{xy} = -\sigma_{yx} = \nu(e^2/h) ,$$
$$\rho_{xx} = \rho_{yy} = 0 , \quad \rho_{xy} = -\rho_{yx} = 1/\sigma_{yz} = -h/(\nu e^2) . \tag{6.171}$$

Within each of the energy gaps the Hall conductance, σ_{xy}, attains discrete constant values that are integer multiples of the basic conductance unit, e^2/h. As opposed, when the chemical potential (i.e., the Fermi energy) coincides with one of the Landau levels, that level is not completely occupied, the electrons there are free to carry charge, and consequently $\sigma_{xx} = \sigma_{yy}$ is finite; its value is proportional to the number of states in that level. These are the narrow peaks displayed in the lower part of Fig. 6.36.

Problem 6.40.

Calculate the resistance tensor from the conductance tensor, and verify Eq. (6.171).

In the earlier experiments on the quantum Hall effect, the strength of the magnetic field was kept constant, but the number of the electrons was varied by the potential applied to the sample [so that electrons are injected into (depleted from) the system, from (into) the neighboring layers]. At low temperature, this is equivalent to a modification of the Fermi energy. When one of the Landau levels in Fig. 6.34(a) is full, and the one above it is still empty, then by varying the number of electrons the Fermi energy may jump into the gap in-between the levels, as in Fig. 6.31(a). When the electrons' density attains the appropriate value, the Hall resistance reaches the discrete value given in Eq. (6.170), and the "conventional" resistance vanishes (as happens in the experiment). However, in clean systems, the addition of a single electron may bring the Fermi energy into the next discrete Landau level, and then the discrete value is achieved only for a single value of the density. In the experiment whose data are given in Fig. 6.36 the density of the electrons was fixed and the strength of the magnetic field was varied. One expects that as the strength of the magnetic field is increased, the energy of each of the Landau levels will grow as well (and also the Zeeman bifurcation); the upmost filled level can cross the Fermi energy and depopulate into the lower levels. In this way the discrete values of the Hall resistance are reached at discrete values of the magnetic field. However, in the experiment the discrete values of the Hall resistance pertain to finite ranges of the magnetic field (these are the stairs in the upper part of Fig. 6.36). A possible solution to this conundrum is to add quantum states whose energies are in-between the Landau levels. When these states are localized, the electrons that occupy them do not contribute to the longitudinal conductance (at zero temperature they cannot move under the effect of an electric field) that remains very low. The density of the electrons that participate in the Hall conductance is not changed, and the Hall resistance is a constant, as in Eq. (6.170).

What happens when, in addition to the magnetic field, the electrons are also subjected to a random potential, that is not homogeneous in space? Such a potential represents "disorder" in the system [see, e.g., part (c) of problem 6.25 and also Sec. 6.12]. In the presence of disorder each Landau level is broadened and becomes a discrete continuum, see the curve of the density of states in Fig. 6.34(a). The energies close to the original Landau level (the light areas in the figure) correspond to states which are extended over the plane; the electrons residing in them contribute to the electrical conductance. The states belonging to the dark areas are localized near minima of the random potential; the electrons occupying those do not contribute to the conductance (at zero temperature), as they cannot move under the effect of an electric field. Thus, the longitudinal conductance is still zero and the Hall conductance is fixed at one of its discrete values. The next stair of the Hall resistance occurs when the Fermi energy is increased and the region of extended states belonging to the next Landau level is reached. When the magnetic field is reduced (in the shadowed areas) electrons are transferred from the extended to the localized states, but the longitudinal conductance is still very low and the Hall conductance has still the discrete values. Once the extended states' region is reached, near the consecutive Landau level, the longitudinal conductance is increased, and the Hall conductance is gradually moving to the next step.

The measurements of the quantum Hall effect yield the most accurate value of the quantum unit of the conductance,

$$\frac{h}{e^2} = 25812.807557(18)\ \Omega \sim 26\mathrm{k}\Omega\ .$$

This quantity, sometimes called **Klitzing** for obvious reasons, is nowadays the **international standard of the resistance unit**. Such standards are the basis of the field of **metrology**, which determines the units of physical quantities. The very accurate number (the accuracy is 10 digits!) is also relevant for the scrutinization of the **fine-structure constant** in quantum electrodynamics, $\alpha = e^2/(\hbar c)$.

The fractional Hall effect. The quantum Hall effect as discussed so far is also known as the **integer quantum Hall effect, IQHE**. There exist two-dimensional systems whose Hall resistance contains stairs at non-integer, rational values of the filling factor, e.g., $\nu = 3/7,\ 2/5,\ 1/3$, etc. This phenomenon is termed the **fraction quantum Hall effect, FQHE**. Its interpretation is based on the electronic interactions, that lead to the creation of quasiparticles of fractional charge, e.g., $e/3$. These quasiparticles are neither bosons nor fermions, and are termed **anyons**. Their discussion is beyond the scope of this volume.

6.12   A metal or an insulator?

When only **single-electron states of an electron in a periodic lattice** are
considered, a certain material is an insulator if the valence band is fully populated
and the conduction band is empty (at zero temperature), with a finite energy gap
in-between the two bands. While this picture describes faithfully quite a number of
solids, there are cases in which it fails. That is, a material can be an insulator though
its single-electron upmost band is not fully populated. Such cases are surveyed
tersely below.

Anderson insulator. The electrons' wave functions do not obey the Bloch
theorem when the potential acting on them is not periodic. In particular, when
that potential is random in space with large variations of its magnitude at various
sites in the lattice (e.g., due to the presence of impurities – foreign atoms located
randomly in space) there appear electronic states centered around locations where
the potential is strongly attractive. Such states are termed **localized**; the probability to find an electron in them decays exponentially with the distance from the
local minimum of the potential. A similar situation is described in Sec. 5.6, for
localized states of the lattice vibrations. The similarity is further explored in part
(c) of problem 6.25.

At zero temperature, a localized electron cannot move under the effect of an
electric field. In case there are only localized electronic states around the Fermi
level, the material is an insulator. This type of insulators is named after Anderson.
In less extreme situations, the states whose energies are close to the band's edges
are localized, while those belonging to energies around the center of the band are
extended. Once the Fermi energy passes from an "extended region" to a "localized"
one, the system changes from being an insulator to being a conductor. This is the
zero-temperature Anderson transition, and it is precisely the transition at the stairs
of the quantum Hall effect. Crossing over from a light region to a shadowed one in
Fig. 6.34(b) illustrates this transition.

The Anderson transition depends on the dimensionality of the system. In the
absence of a magnetic field, any amount of disorder (i.e., randomness) will cause a
total localization of all electrons at one and two dimensions. Thus all such materials
are insulators at zero temperature. In three dimensions, the states that correspond
to energies around the center of the band are extended, provided that the amount
of the disorder is not too high. As the level of randomness is increased, that region
is narrowed down, until all states become localized.

Mott insulatorinsulator. The description in terms of single-electron wave
functions ignores the **Coulomb interactions** among the electrons. These interactions can be viewed as turning the electron gas into a liquid, termed the **Fermi
liquid**. The motion of each electron within this liquid affects the electrons around,
in such a way that those **screen** its electric potential. In other words, each electron is replaced by a **quasiparticle**, an entity composed of the electron and the

"screening cloud" around it. At low energies the quasiparticles behave as independent fermions, and their wave functions can be classified according to the Bloch theorem, with parabolic energies near the band boundaries.

Consider the configuration in which each atom contributes a single electron to the conduction band. When the atoms in the lattice are far away from each other, that is, the lattice constant is quite large, the preferred situation is that in which each electron remains in the atomic state belonging to a certain atom (whose energy is ϵ_0), as on the left side in Fig. 4.30. This situation corresponds to a very narrow band, as in Fig. 6.10(b), or as in the limit of long inter-atomic distance in Fig. 6.21. Another electron added to this system has to join an electron already residing on a certain atom. The cost in energy is the short-range repulsion energy between two electrons, denoted U (also called the **Hubbard energy**). When the lattice constant is large, the energy of the second electron is $\epsilon_0 + U$. In this way a new band is created, at an energy distance U from the original band (see the left side of Fig. 6.37). The original band, that can contain $2N$ electrons, is bifurcated into two, each of which can contain N electrons. It follows that in its ground state, the material is an insulator, as each electron remains localized near one of the atoms in the lattice. This is the **Mott insulators insulator**. In certain configurations these localized states also give rise to the **exchange interaction** (Sec. 4.7), that causes their magnetic moments to be arranged in an antiferromagnetic pattern.

As the inter-atomic distance is decreased (e.g., due to a pressure applied on the sample) the overlap energy between wave functions on nearest neighbors increases and the two bands become wider. This is because the energy width, W, is proportional to the overlap energy between nearest neighbors, Γ, as in Fig. 6.21. As a result, the gap in-between them becomes narrower, and the two bands may even overlap. This implies a transition between the insulating state to the conducting one, called the **Mott transition**. The dependence of the two bands on the overlap energy is illustrated in Fig. 6.37.

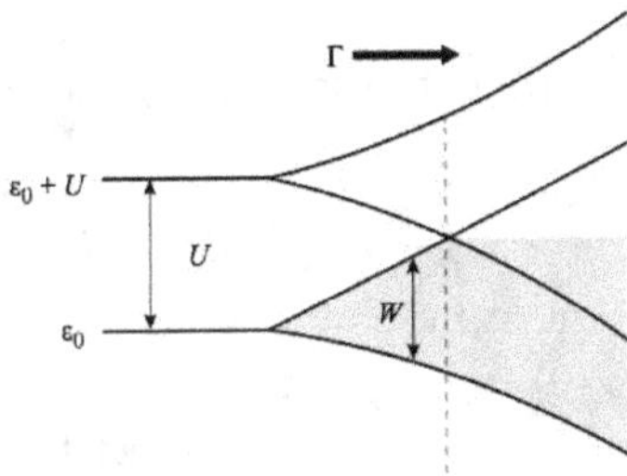

Fig. 6.37: The two energy bands formed from a single band by the repulsion interaction between the electrons on the same atom in the lattice. As the lattice constant becomes smaller, the overlap of the wave functions on neighboring atoms increases, causing the bands to broaden, as illustrated in the figure (from left to right).

6.13   Answers for the problems in the text

Answer 6.1.

a. The probability for the electron to undergo a collision during the time interval dt is dt/τ. Hence, the probability that it will not collide during the time $t + dt$ is the product of the two probabilities, $P(t)$ and $(1 - t/\tau)$, $P(t + dt) = P(t)(1 - dt/\tau)$. It follows that $dP/dt = -P/\tau$, and hence $P(t) = \exp[-t/\tau]$ [it is assumed here that the initial condition is $P(0) = 1$].

b. The probability for a collision to occur precisely within the time interval between t and $t + dt$ is $P(t)(dt/\tau)$. Consequently, the mean time in-between two successive collisions is

$$\langle t \rangle = \frac{\int_0^\infty t P(t) dt/\tau}{\int_0^\infty P(t) dt/\tau} = \tau \ .$$

Answer 6.2.

The component of the new velocity $\mathbf{v}'$ normal to the direction of the original one is still averaged to zero; its component along the direction of the original velocity is the projection along that direction, that obeys $\langle v'_\parallel \rangle \geq \langle v(\tau) \rangle \cos \beta$, as the angle between the old and the new velocities is smaller or equal to β. By inserting this relation into

$$\langle \mathbf{v} \rangle = \langle \mathbf{v}' \rangle - e\mathcal{E}\tau/m \ ,$$

one finds that

$$\langle \mathbf{v} \rangle (1 - \cos \beta) \geq -e\mathcal{E}\tau/m \ .$$

This result might be interpreted as yielding a generalized relaxation time, obeying $\tau' \geq \tau/(1 - \cos \beta)$. This implies that the effective relaxation time becomes longer as the scattering angle is decreased, and diverges to infinity when this angle vanishes and no scattering takes place.

Answer 6.3.

When the magnetic field is parallel to the $\hat{\mathbf{z}}$−axis, the classical equations of motion are

$$\dot{p}_x = -\omega_c p_y \ , \quad \dot{p}_y = \omega_c p_x \ , \quad \dot{p}_z = 0 \ .$$

It follows that p_z is a constant of motion, and the motion along the $\hat{\mathbf{z}}$−axis is with a constant velocity, $z(t) = z(0) + p_z(0)t/m$.

Differentiating the second equation with respect to time and inserting the first one leads to

$$\ddot{p}_y = \omega_c \dot{p}_x = -\omega_c^2 p_y \ ,$$

with the solution

$$p_y(t) = p_y(0) \cos(\omega_c t) \ .$$

As

$$\dot{x} = p_x/m = \dot{p}_y/(m\omega_c) = -[p_y(0)/m]\sin(\omega_c t) \; ,$$

one finds

$$x(t) = x(0) + [p_y(0)/(m\omega_c)][\cos(\omega_c t) - 1] \; .$$

In a similar way,

$$\dot{y} = p_y/m \Rightarrow y(t) = y(0) + [p_y(0)/m\omega_c]\sin(\omega_c t) \; .$$

The last two equations correspond to a circular motion in the XY plane with the cyclotron frequency ω_c, and with the radius $p_y(0)/(m\omega_c)$.

Answer 6.4.

The density of lithium and copper atoms (in CGS units) is $n_{\mathrm{Li}} = 2/(3.49 \times 10^{-8}\ \mathrm{cm})^3 = 4.7 \times 10^{22}\ \mathrm{cm}^{-3}$, $n_{\mathrm{Cu}} = 4/(3.61 \times 10^{-8}\ \mathrm{cm})^3 = 8.45 \times 10^{22}\ \mathrm{cm}^{-3}$. Using $e = 4.8 \times 10^{-10}$ esu and $c = 3 \times 10^{10}$ cm/sec yields $R_{\mathrm{H}}(\mathrm{Li}) = -1.48 \times 10^{-24}$ (CGS units) and $R_{\mathrm{H}}(\mathrm{Cu}) = -0.82 \times 10^{-24}$ (CGS units).

Answer 6.5.

a. From the equations of motion (6.8) it follows that the equations for the in-plane momentum components are identical to those analyzed following Eq. (6.8); the one for the normal component is

$$\langle \dot{p}_z \rangle = -e\mathcal{E}_z - \langle p_z \rangle/\tau \; .$$

Using $\langle \mathbf{p}(t) \rangle = \mathbf{p}_0 \exp[-i\omega t]$ yields

$$-i\omega p_{0x} = -e\mathcal{E}_{0x} - \omega_c p_{0y} - p_{0x}/\tau \; ,$$
$$-i\omega p_{0y} = -e\mathcal{E}_{0y} + \omega_c p_{0x} - p_{0y}/\tau \; ,$$
$$-i\omega p_{0z} = -e\mathcal{E}_{0z} - p_{0z}/\tau \; . \tag{6.172}$$

The current density, from Eq. (6.4), is

$$\langle \mathbf{j}(t) \rangle = \mathbf{j}_0 \exp[-i\omega t] \; , \quad \mathbf{j}_0 = -(en/m)\mathbf{p}_0 \; .$$

Using these relations in Eqs. (6.172) yields the required result,

$$\begin{bmatrix} \mathcal{E}_{0x} \\ \mathcal{E}_{0y} \\ \mathcal{E}_{0z} \end{bmatrix} = \frac{1}{\sigma_0} \begin{bmatrix} 1 - i\omega\tau & \omega_c\tau & 0 \\ -\omega_c\tau & 1 - i\omega\tau & 0 \\ 0 & 0 & 1 - i\omega\tau \end{bmatrix} \begin{bmatrix} j_{0x} \\ j_{0y} \\ j_{0z} \end{bmatrix} \; ,$$

where $\sigma_0 = ne^2\tau/m$. The matrix is the specific-resistivity tensor ρ; its inverse is the specific-conductivity one,

$$\boldsymbol{\sigma} = \boldsymbol{\rho}^{-1} = \sigma_0 \begin{bmatrix} \dfrac{1-i\omega\tau}{(1-i\omega\tau)^2+(\omega_c\tau)^2} & \dfrac{-\omega_c\tau}{(1-i\omega\tau)^2+(\omega_c\tau)^2} & 0 \\ \dfrac{\omega_c\tau}{(1-i\omega\tau)^2+(\omega_c\tau)^2} & \dfrac{1-i\omega\tau}{(1-i\omega\tau)^2+(\omega_c\tau)^2} & 0 \\ 0 & 0 & \dfrac{1}{1-i\omega\tau} \end{bmatrix} \; .$$

b. At high frequencies,

$$\sigma \Rightarrow \sigma_0 \begin{bmatrix} \frac{i\omega}{\tau(\omega^2-\omega_c^2)} & \frac{\omega_c}{\tau(\omega^2-\omega_c^2)} & 0 \\ \frac{-\omega_c}{\tau(\omega^2-\omega_c^2)} & \frac{i\omega}{\tau(\omega^2-\omega_c^2)} & 0 \\ 0 & 0 & \frac{i}{\omega\tau} \end{bmatrix} .$$

At $\omega \approx \omega_c$, the currents induced in the plane normal to the magnetic field are rather high, even though the electric field magnitude is not. The system approaches resonance conditions, which enable an efficient conversion of electromagnetic energy into kinetic energy of the electrons.

Answer 6.6.

a. Assuming that the temperature gradient is along the positive $\hat{\mathbf{x}}$–axis, then electrons impinging on the origin from the left, from the point $-|v_x|\tau$, have the "thermal velocity" $|v_x[T(-|v_x|\tau)]|$. The velocity of the electrons coming from the opposite direction is $-|v_x[T(|v_x|\tau)]|$. It follows that for a small gradient, the average thermal velocity is

$$\begin{aligned}
\langle v_{Ex}\rangle &= \{|v_x[T(-|v_x|\tau)]| - |v_x[T(|v_x|\tau)]|\}/2 \approx -|v_x|\tau(d|v_x|/dT)(dT/dx) \\
&= -(\tau/m)[d(mv_x^2/2)/dT](dT/dx) = -(\tau/[3m])(du/dT)(dT/dx) \\
&= -(\tau c/[3m])(dT/dx) .
\end{aligned}$$

Inserting this expression into Eq. (6.4) yields

$$\mathbf{j}_Q = -ne\langle \mathbf{v}_E\rangle = (\sigma c/(3e)\boldsymbol{\nabla}T ,$$

for the charge current.

b. The absence of an electric current implies an electric field giving rise to a current flowing along the opposite direction to that computed above, such that the resultant vanishes, $\mathbf{j}_E + \mathbf{j}_Q = 0$. Exploiting Eq. (6.5) yields the current induced by the electric field, $\mathbf{j}_E = \sigma\boldsymbol{\mathcal{E}}$. One therefore finds $\boldsymbol{\mathcal{E}} = Q\boldsymbol{\nabla}T$, with the **Seebeck coefficient** given by $Q = -c/(3e)$. The equipartition theorem yields $c = 3k_B/2$, and hence $Q = -k_B/(2e)$. This estimate exceeds by a factor of 100 the experimentally-measured values; it is corrected by introducing modifications into the Drude theory.

Answer 6.7.

By Eq. (6.28), $k_F = (3\pi^2 n)^{1/3} = 1.4 \text{Å}^{-1}$. It follows that the Fermi energy is $E_F = (\hbar k_F)^2/(2m) = 1.25 \times 10^{-11}$ erg, and the Fermi temperature is $T_F = 90500$ K. Room temperature is far lower than the Fermi temperature.

Answer 6.8.

a. As the particles are identical, all possible arrangements of them among the g_i degenerate states yield the same result. The total number of these possibilities amounts to the number of permutations of n_i electrons among g_i states, i.e., $\{g_i!/[n_i!(g_i - n_i)!]\}$. The total number of possibilities, W, is the product of the numbers of all configurations.

b. It follows that

$$\ln W = \sum_i \{\ln[g_i!] - \ln[n_i!] - \ln[(g_i - n_i)!]\} \ .$$

By the Stirling formula,

$$\ln W = \sum_i \{g_i \ln g_i - g_i - n_i \ln n_i + n_i - (g_i - n_i)\ln(g_i - n_i) + (g_i - n_i)\} \ ,$$

and

$$F = E - TS = \sum_i \{n_i \epsilon_i + k_{\mathrm{B}} T[n_i \ln n_i + (g_i - n_i)\ln(g_i - n_i) - g_i \ln g_i]\} \ .$$

c. Transferring δn particles from state m to state ℓ changes the free energy by

$$\delta F = (\partial F/\partial n_\ell - \partial F/\partial n_m)\delta n \ .$$

It follows from the condition $\delta F = 0$ that

$$\partial F/\partial n_\ell = \partial F/\partial n_m \ ,$$

that is, this derivative is the same for all states, and can be denoted by $\mu = \partial F/\partial n_m$.
d. Differentiating the expression for F with respect to n_i gives

$$\mu = \frac{\partial F}{\partial n_i} = \epsilon_i + k_{\mathrm{B}} T \ln\left[\frac{n_i}{g_i - n_i}\right] \ ,$$

that is

$$n_i = g_i/\{1 + \exp[\beta(\epsilon_i - \mu)]\} \ ,$$

as in Eq. (6.35) for $g_i = 1$.

Answer 6.9.
a. The derivative of the function in Eq. (6.35) is

$$-f' = -\partial f(E)/\partial E = \beta e^{\beta(E-\mu)}/[1 + e^{\beta(E-\mu)}]^2 \ .$$

This derivative is displayed in Fig. 6.38 for $\beta = 10,\ 20,\ 40$, where energies are measured in units of μ. The function is concentrated symmetrically around the value $E = \mu$,

$$-f'(\mu - E) = \beta e^{-\beta(E-\mu)}/[1 + e^{-\beta(E-\mu)}]^2 = -f'(E - \mu) \ ,$$

where it takes the value $-f'(E = \mu) = \beta/4$. As the temperature is lowered this height increases while the width decreases. In particular, at zero temperature the height tends to infinity while the width shrinks to zero. There are several options to estimate the width. For instance, the energy at which the height is halved, i.e., the solution of

$$f'(E_{1/2})/f'(\mu) = 1/2 \ , \text{ i.e. } 4x/(1+x)^2 = 1/2 \ , \text{ where } x = \exp[\beta(E_{1/2} - \mu)] \ .$$

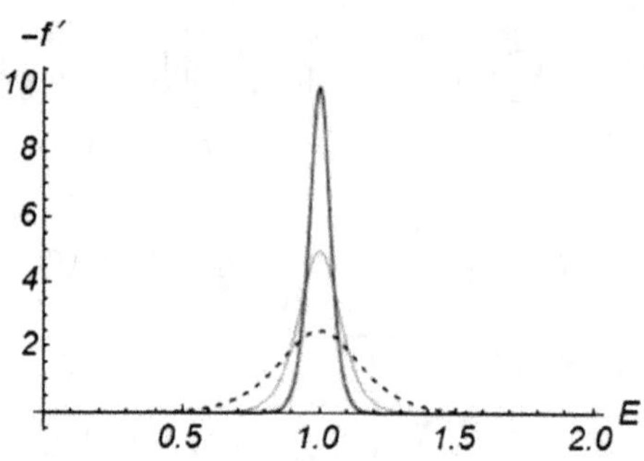

Fig. 6.38

One finds $x = 3 \pm 2\sqrt{2}$, and consequently $E_{1/2} - \mu = k_{\mathrm{B}}T \ln[3 \pm 2\sqrt{2}] = \pm 1.76 k_{\mathrm{B}}T$. The width $\Delta E_{1/2} = 3.52 k_{\mathrm{B}}T$ is proportional to the temperature, and indeed decreases with the temperature. The integral over this function is nonetheless a constant,

$$\int_{-\infty}^{\infty} dE[-f'(E)] = -f(E)\Big|_{-\infty}^{\infty} = f(-\infty) - f(\infty) = 1 ,$$

independent of the temperature. In the $T \to 0$ limit, $f'(E) = 0$ for any $E \neq \mu$, while $f'(E = \mu) = \infty$. As the integral equals 1, the function $-f'(E)$ obeys the same properties as a delta function; hence $-f'(E) \to \delta(E - \mu)$.

b. The average (per particle) of a certain property $A(E)$ of the electron gas at an arbitrary temperature T is

$$\langle A(E) \rangle = V \int_{-\infty}^{\infty} dE g(E) A(E) f(E) / \langle N_e \rangle ,$$

$$\langle N_e \rangle = V \int_{-\infty}^{\infty} dE g(E) f(E) . \tag{6.173}$$

We introduce the function

$$B(E) = V \int_{-\infty}^{E} dE' g(E') A(E') ,$$

and integrate by parts the numerator in Eq. (6.173),

$$\int_{-\infty}^{\infty} dE g(E) A(E) f(E) = B(E) f(E)\Big|_{-\infty}^{\infty} - \int_{-\infty}^{\infty} dE B(E) f'(E)$$

$$= \int_{-\infty}^{\infty} dE B(E) [-f'(E)] .$$

The first term in the intermediate step vanishes in both limits [$B(-\infty) = 0$ and $f(\infty) = 0$]. As the function $-f'(E)$ is narrow, and is concentrated around the Fermi level, the integral on the right hand-side is significant mainly at energies for which $|E - \mu| \leq k_{\mathrm{B}}T$. Therefore, at low enough temperatures one may use the expansion

$$B(E) = B(\mu) + (E - \mu)B'(\mu) + (E - \mu)^2 B''(\mu)/2 + \dots ,$$

to obtain

$$\int_{-\infty}^{\infty} dE\, B(E)[-f'(E)] = B(\mu) + B''(\mu)(\delta E)^2 + \dots .$$

The first term here is the zero-temperature result, and the second contains $(\delta E)^2 = \int_{-\infty}^{\infty} dE (E-\mu)^2 [-f'(E)]$, which is the standard deviation of the energy off the chemical potential with the weight function $-f'(E)$. As seen in part (a), the width of this function is proportional to T, and therefore the leading-order correction at low temperatures is proportional to T^2 [and indeed, a precise computation of the integral yields $(\delta E)^2 = (\pi^2/3)(k_{\rm B}T)^2$]. The same correction appears in the integral in the expression for $\langle N_e \rangle$; it follows that the average of any quantity calculated with the Fermi-Dirac distribution contains corrections to the zero-temperature value which are quadratic in the temperature. The higher terms include higher powers of the temperature.

Answer 6.10.

a. The volume of the Fermi sphere in d dimensions is $S_d k_{\rm F}^d/d$ and the volume that corresponds to a discrete "allowed" value of the momentum is $(2\pi)^d/V$. Each such state accommodates two electrons. Hence, an extension of Eq. (6.28) is

$$N_e = \frac{2S_d k_{\rm F}^d/d}{(2\pi)^d/V} = \frac{2S_d(2mE_{\rm F})^{d/2}}{d(2\pi\hbar)^d}V = nV .$$

It follows that $k_{\rm F} = 2\pi[nd/(2S_d)]^{1/d}$, and the Fermi energy is

$$E_{\rm F} = (2\pi\hbar)^2[nd/(2S_d)]^{2/d}/(2m) .$$

The electrons' density is

$$n = 2S_d[k_{\rm F}/(2\pi)]^2/d = 2S_d(2mE_{\rm F})^{d/2}/[d(2\pi\hbar)^2] .$$

b. Similarly, an extension of Eq. (6.31) to an arbitrary dimension is

$$V g(E)dE = \frac{2S_d k^{d-1}dk}{(2\pi)^d/V} = V\frac{nd}{2(E_{\rm F})^{d/2}}E^{\frac{d}{2}-1}dE .$$

The total energy of the electron gas at zero temperature is hence

$$E_{\rm tot} = V\int_0^{E_{\rm F}} E g(E)dE = \frac{d}{d+2}VnE_{\rm F} .$$

c. The average number of the electrons is

$$\langle N_e \rangle = V\int_0^{\infty} dE\, g(E)f(E) .$$

Equating it to $N = Vn$, and using for $g(E)$ the expression from the previous part, yield

$$1 = \frac{d}{2(E_{\rm F})^{d/2}}\int_0^{\infty}\frac{E^{\frac{d}{2}-1}}{1+e^{\beta(E-\mu)}}dE = \frac{d(k_{\rm B}T)^{d/2}}{2(E_{\rm F})^{d/2}}\int_0^{\infty}\frac{y^{\frac{d}{2}-1}}{1+e^{y-\beta\mu}}dy . \qquad (6.174)$$

The solution of this equation is the chemical potential μ as a function of β and E_{F}. At very high temperatures the coefficient in front of the last integral tends to infinity, and consequently the integral should tend to zero, i.e., $\exp[-\beta\mu] \to \infty$. The 1 in the denominator may then be discarded, and the distribution function becomes approximately $f(E) \to \exp[-\beta(E-\mu)]$, with the proper normalization. The integral in Eq. (6.174) becomes

$$\int_0^\infty \frac{y^{\frac{d}{2}-1}}{e^{y-\beta\mu}} = e^{\beta\mu}\Gamma(d/2) ,$$

where $\Gamma(z)$ is the Gamma function, $\Gamma(3/2) = \sqrt{\pi/2}$, $\Gamma(1) = 1$, and $\Gamma(1/2) = \sqrt{\pi}$. The equation for the chemical potential gives that the product $(k_{\mathrm{B}}T)^{d/2}\exp[\beta\mu]$ is independent of the temperature, and thus the leading-order dependence on the temperature, at high temperatures, is $\mu \approx -(d/2)(k_{\mathrm{B}}T)\ln(k_{\mathrm{B}}T)$. The factor $\exp[\beta\mu]$ in the distribution function is cancelled in the calculation of any weighed average; therefore in the high-temperature limit, the results are those of the Maxwell-Boltzmann distribution, e.g.,

$$\frac{\langle E_{\mathrm{tot}}\rangle}{\langle N_e \rangle} = \frac{\int_0^\infty E^{d/2} f(E) dE}{\int_0^\infty E^{d/2-1} f(E) dE} = k_{\mathrm{B}}T \frac{\int_0^\infty x^{d/2} e^{-x} dx}{\int_0^\infty x^{d/2-1} e^{-x} dx} = d \frac{k_{\mathrm{B}}T}{2} .$$

This is the equipartition theorem, that also leads to the Dulong-Petit law.

d. In two dimensions, the equation for the chemical potential (from the previous part) is

$$1 = \int_0^\infty dE g(E) f(E) = \frac{1}{E_{\mathrm{F}}} \int_0^\infty dE \frac{1}{e^{\beta(E-\mu)} + 1} = \frac{1}{\beta E_{\mathrm{F}}} \ln[1 + e^{\beta\mu}] ,$$

and consequently $\mu = \ln[e^{\beta E_{\mathrm{F}}} - 1]/\beta$.

e. The arguments used in the estimation of the specific heat still hold: the number of electrons excited above the Fermi level is proportional to T; each of them carries an energy of the order of $k_{\mathrm{B}}T$. The specific heat is then proportional to T. However, as seen after Eq. (5.69), the phonon specific heat is proportional to T^d. Hence, the extension of Eq. (6.42), upon taking into account the contribution of the phonons, is

$$C_V \approx \gamma_d T + A_d T^d .$$

Answer 6.11.

a. The density of the electrons is given by $n = 6.022 \times 10^{23} \, z \, \rho_m/A$, where z is the number of conduction electrons "donated" by each atom, A is the atomic mass (in grams) and the coefficient is the Avogadro number (the number of atoms in one mole). For copper, $z = 1$ and $A \approx 64$, and hence $n = 0.84 \times 10^{29}$ m^{-3}. From the expressions following Eq. (6.6) one finds $\tau = m/(\rho n e^2) = 2.7 \times 10^{-14}$ sec. Equation (6.28) implies that $k_{\mathrm{F}} = (3\pi^2 n)^{1/3} = 1.35 \times 10^8$ cm^{-1}, and then $v_{\mathrm{F}} = \hbar k_{\mathrm{F}}/m = 1.56 \times 10^8$ cm/sec, $E_{\mathrm{F}} = m v_{\mathrm{F}}^2/2 = 13.8$ eV, $T_{\mathrm{F}} = E_{\mathrm{F}}/k_{\mathrm{B}} = 1.6 \times 10^5$ K, and $\ell = v_{\mathrm{F}}\tau = 4.2 \times 10^{-6}$ cm$= 420$Å.

b. The relaxation time is $\tau_{\mathrm{imp}} = m/(\rho m e^2) = 0.6 \times 10^{-14}$ sec, and then using the expression before Eq. (6.48), $\sigma_{\mathrm{imp}} = 1/(\tau_{\mathrm{imp}} v_F N_{\mathrm{imp}}) = 12.5(\mathring{A})^2$.

Answer 6.12.

a. The function $\psi_{n\mathbf{k}}(\mathbf{r})$ obeys Eq. (6.59),

$$\psi_{n\mathbf{k}}(\mathbf{r} + \mathbf{R}) = \hat{T}(\mathbf{R})\psi_{n\mathbf{k}}(\mathbf{r}) = \exp[i\mathbf{k} \cdot \mathbf{r}]\psi_{n\mathbf{k}}(\mathbf{r}) \ .$$

Complex conjugating yields

$$\hat{T}(\mathbf{R})\psi_{n\mathbf{k}}^*(\mathbf{r}) = \psi_{n\mathbf{k}}^*(\mathbf{r} + \mathbf{R}) = \exp[-i\mathbf{k} \cdot \mathbf{r}]\psi_{n\mathbf{k}}^*(\mathbf{r}) \ .$$

Hence, $\psi_{n\mathbf{k}}^*(\mathbf{r})$ is an eigenfunction of $\hat{T}(\mathbf{R})$, with the eigenvalue $\exp[-i\mathbf{k} \cdot \mathbf{r}]$. The function $\psi_{n-\mathbf{k}}(\mathbf{r})$ is an eigenfunction of $\hat{T}(\mathbf{R})$ too, with the same eigenvalue. As the eigenvalues of $\hat{T}(\mathbf{R})$ for a certain value of n are non-degenerate, it follows that $\psi_{n\mathbf{k}}^*(\mathbf{r}) = \psi_{n-\mathbf{k}}(\mathbf{r})$. Complex-conjugating the Schrödinger equation yields $\hat{\mathcal{H}}(\mathbf{r})\psi_{n\mathbf{k}}^*(\mathbf{r}) = E_n(\mathbf{k})\psi_{n\mathbf{k}}^*(\mathbf{r})$. The replacement $\mathbf{k} \to -\mathbf{k}$ in the Schrödinger equation gives $\hat{\mathcal{H}}(\mathbf{r})\psi_{n-\mathbf{k}}(\mathbf{r}) = E_n(-\mathbf{k})\psi_{n-\mathbf{k}}(\mathbf{r})$. Comparing the two equations implies that $E_n(\mathbf{k}) = E_n(-\mathbf{k})$.

b. Denote $\mathbf{k}' = (-k_x, k_y, k_z)$. From Eq. (6.59),

$$\psi_{n\mathbf{k}}(\mathbf{r} + \mathbf{a}_1) = \hat{T}(\mathbf{a}_1)\psi_{n\mathbf{k}}(\mathbf{r}) = \exp[ik_x a_1]\psi_{n\mathbf{k}}(\mathbf{r}) \ ,$$

and upon complex-conjugating,

$$\psi_{n\mathbf{k}}^*(\mathbf{r} + \mathbf{a}_1) = \hat{T}(\mathbf{a}_1)\psi_{n\mathbf{k}}^*(\mathbf{r}) = \exp[-ik_x a_1]\psi_{n\mathbf{k}}^*(\mathbf{r}) \ .$$

On the other hand,

$$\psi_{n\mathbf{k}'}(\mathbf{r} + \mathbf{a}_1) = \hat{T}(\mathbf{a}_1)\psi_{n\mathbf{k}'}(\mathbf{r}) = \exp[-ik_x a_1]\psi_{n\mathbf{k}'}(\mathbf{r}) \ .$$

As the eigenvalues of $\hat{T}(\mathbf{a}_1)$ are not degenerate, it follows that $\psi_{n\mathbf{k}}^*(\mathbf{r}) = \psi_{n\mathbf{k}'}(\mathbf{r})$. Finally, complex-conjugating the Schrödinger equation for $\psi_{n\mathbf{k}}(\mathbf{r})$ and a comparison of the same equation for $\psi_{n\mathbf{k}'}(\mathbf{r})$ yields that $E_n(\mathbf{k}) = E_n(\mathbf{k}')$.

Answer 6.13.

According to the Bloch theorem, it suffices to consider the solutions of the Schrödinger equation in the first Brillouin zone, $-\pi/a \le k < \pi/a$ (in-between the vertical lines in the figure). The one-dimensional Schrödinger equation with periodic boundary conditions has solely two independent solutions for each energy $E_n(k)$. It is found in problem 6.12 that these two solutions are $\psi_{nk}(x)$ and $\psi_{n-k}(x)$ [related to one another by $\psi_{n-k}^*(x) = \psi_{nk}(x)$]. Figure 6.6 shows that for the energy represented by the horizontal line there are four independent solutions, in contradiction with the requirement that there are only two independent solutions. As explained following Eq. (6.62), in higher dimensions there are more independent solutions, and consequently the bands may well overlap each other.

Answer 6.14.

a. In the limit specified in the problem, each potential barrier becomes very high and very narrow; the area it covers is $\int_a^b dx U(x) = U_0(a-b) = u_0$. This step has precisely the same properties as the function $u_0\delta(x-a)$, from which follows the comb's shape. Assuming that the energy of the electron remains finite, i.e., $E \ll U_0$, one may neglect K^2 as compared to $Q^2 \approx -2mU_0/\hbar^2$, turning Eq. (6.73) into

$$\cos(ka) = \cos(Ka) - [Q^2(a-b)/2K]\sin(Ka) ,$$

and hence

$$\cos(ka) = F(K) = \cos(Ka) + \phi\sin(Ka)/(Ka) .$$

This reproduces the solutions of a free particle for $\phi = 0$, i.e., $ka = Ka + 2\ell\pi$. The borders of the bands are obtained from the relation $F(K) = \pm 1$; since $F(K) = (-1)^n$ for $Ka = n\pi$ where n is an integer, these points are always borders of energy bands.

b. In the $\phi \gg 1$ limit, the energy bands are narrow because $F(K)$ varies steeply as a function of K [and therefore also as a function of $E = \hbar^2 K^2/(2m)$] within each band. Consequently, close to each border found in part (a) there is another one, where the function takes the value $(-1)^{n+1}$. Assuming that these borders are close to each other, one defines $Ka = n\pi + \kappa$, and expands in the small variable κ. The equation obtained for the second border is then

$$(-1)^{n+1} = \cos(n\pi + \kappa) + \phi\sin(n\pi + \kappa)/(n\pi + \kappa) = (-1)^n[\cos\kappa + \phi\sin\kappa/(n\pi + \kappa)]$$
$$\approx (-1)^n[1 - \kappa^2/2 + \ldots + \phi(\kappa - \kappa^3/6 + \ldots)/(n\pi + \kappa)] \approx (-1)^n[1 + \phi\kappa/(n\pi)] .$$
$$(6.175)$$

To leading order $\kappa \approx -2n\pi/\phi$, and the bands are in the ranges

$$n\pi(1 - 2/\phi) < Ka < n\pi , \quad \text{for} \quad n = 1, 2, 3, \ldots .$$

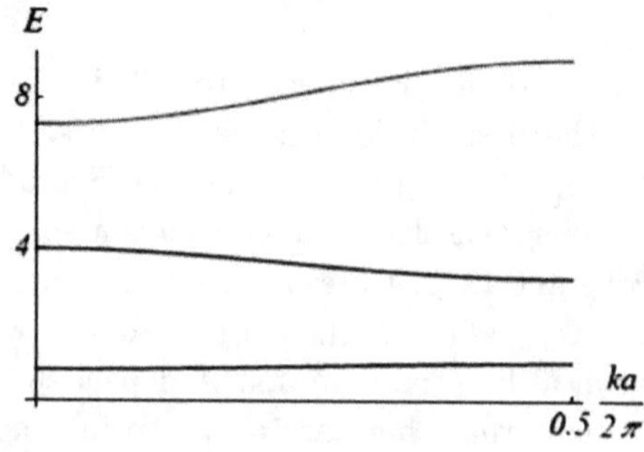

Fig. 6.39

Denoting $K_{n,\max} = n\pi/a$ and $K_{n,\min} = n\pi(1 - 2/\phi)/a$, one finds that the width of the n–th band is

$$\Delta E_n = \hbar^2(K_{n,\max}^2 - K_{n,\min}^2)/(2m) \approx 2\hbar^2 n^2\pi^2/(m\phi) ,$$

and hence the bands widen as the energy increases. Within each band $Ka = n\pi(1 - x)$, where $0 < x < 2/\phi$. Inserting this into the equation for $F(K)$, $\cos(ka) = F(K) \approx (-1)^n(1 - \phi x)$, yields $x \approx [1 - (-1)^n \cos(ka)]/\phi$, and hence the energy in the band is

$$E = \hbar^2 K^2/(2m) = \hbar^2(n\pi)^2(1 - x)^2/(2ma^2) .$$

These energies are plotted in Fig. 6.39, for $\phi = 20$. In the $\phi \to \infty$ limit these are the discrete energies of an infinite potential well of width a; the leading-order correction to those is proportional to x,

$$E = \frac{\hbar^2 K^2}{2m} = \frac{\hbar^2(n\pi)^2}{2ma^2}\left[1 - \frac{2}{\phi}\left(1 - (-1)^n \cos(ka)\right)\right] .$$

c. When $\phi \ll 1$ the second term in Eq. (6.175) is a minor correction to the free-particle solutions, for which $ka = Ka + 2\ell\pi$. As it is found that there exists a border of a band at each point $Ka = n\pi$, when the right hand-side equals $(-1)^n$, one considers another point close to this one, at which the right hand-side has the same value. That is, denoting $Ka = n\pi + y$ and requiring that

$$(-1)^n = \cos(n\pi + y) + \phi\sin(n\pi + y)/(n\pi + y) \approx (-1)^n[1 - y^2/2 + \phi y/(n\pi)] ,$$

it follows that $y \approx 2\phi/(n\pi)$. Thus, the difference between the two energies on the opposite sides of the energy gap is

$$\Delta E = \hbar^2[(Ka + y)^2 - (Ka)^2]/(2ma^2) \approx 2\hbar^2\phi/(ma^2) = 2u_0/a .$$

The bottom of the lowest band is obtained from the solution of the equation

$$1 = \cos(Ka) + \phi\sin(Ka)/(Ka) \approx 1 - (Ka)^2/2 + \phi[1 - (Ka)^2/6] ,$$

and therefore $(Ka)^2 \approx 2\phi$, which yields

$$E_{\min} \approx \hbar^2(Ka)^2/(2ma^2) \approx \hbar^2\phi/(ma^2) = u_0/a .$$

Answer 6.15.

When the periodic potential $U(\mathbf{r})$ is weak, it can be regarded as a small perturbation. In the absence of the potential the Hamiltonian describes a free particle, $\hat{\mathcal{H}}_0 = -\hbar^2\nabla^2/(2m)$, whose solution is discussed in Sec. 6.3. Using periodic boundary conditions, the eigenfunctions of $\hat{\mathcal{H}}_0$ are $\Psi_{\mathbf{k}}^{(0)}(\mathbf{r}) = \exp[i\mathbf{k}\cdot\mathbf{r}]/\sqrt{V}$, and the corresponding eigenenergies are $E^{(0}(\mathbf{k}) = \hbar^2 k^2/(2m)$ [with the discrete values of the wave vectors from Eq. (5.43) as reproduced in Eq. (6.27)]. As explained in App. C, the corrections to the energy levels and the wave functions are obtained from the matrix of the perturbation Hamiltonian, $\langle\Psi_{\mathbf{k}'}^{(0)}|U|\Psi_{\mathbf{k}}^{(0)}\rangle$. When the wave vector of the electron is far away from Brillouin-zone boundaries the "original" states (without the perturbation) are not degenerate (see Sec. 6.6), and then Eq. (C.3) yields

$$E(\mathbf{k}) = E^{(0)}(\mathbf{k}) + \langle\Psi_{\mathbf{k}}^{(0)}|U|\Psi_{\mathbf{k}}^{(0)}\rangle - \sum_{\mathbf{k}'\neq\mathbf{k}} \frac{|\langle\Psi_{\mathbf{k}'}^{(0)}|U|\Psi_{\mathbf{k}}^{(0)}\rangle|^2}{E^{(0)}(\mathbf{k}') - E^{(0)}(\mathbf{k})} \; . \qquad (6.176)$$

The matrix elements of the perturbation are given by

$$\langle\Psi_{\mathbf{k}'}^{(0)}|U|\Psi_{\mathbf{k}}^{(0)}\rangle = \frac{1}{V}\int d^3 r\, e^{i(\mathbf{k}-\mathbf{k}')\cdot\mathbf{r}} U(\mathbf{r}) \equiv \widetilde{U}(\mathbf{k}'-\mathbf{k}) \; .$$

The first-order correction is given by the diagonal term, $\langle\Psi_{\mathbf{k}}^{(0)}|U|\Psi_{\mathbf{k}}^{(0)}\rangle = \int d^3 r\, U(\mathbf{r})/V$. This is just the average of the potential over the whole crystal; it is a constant independent of $\mathbf{k}$, and as such can be subtracted from all energy levels (i.e., it shifts the origin of the energy) and then ignored. As explained in Sec. 3.8, the only nonzero Fourier components of a periodic function are those for which the wave vector equals one of the reciprocal-lattice vectors, $\mathbf{G}$. Hence, the sum in Eq. (6.176) comprises only terms for which $\mathbf{k}' - \mathbf{k} = \mathbf{G}$,

$$E(\mathbf{k}) = E^{(0)}(\mathbf{k}) - \sum_{\mathbf{G}\neq 0} \frac{|\widetilde{U}(\mathbf{G})|^2}{E^{(0)}(\mathbf{k}-\mathbf{G}) - E^{(0)}(\mathbf{k})} + \cdots \; ,$$

reproducing Eq. (6.88).

In a similar fashion App. C gives the correction to the wave function,

$$\Psi_{\mathbf{k}}^{(1)} = \sum_{\mathbf{G}}\langle\Psi_{\mathbf{k}-\mathbf{G}}^{(0)}|\Psi_{\mathbf{k}}^{(1)}\rangle\Psi_{\mathbf{k}-\mathbf{G}}^{(0)} = \sum_{\mathbf{G}\neq 0} \frac{\widetilde{U}(\mathbf{G})}{E^{(0)}(\mathbf{k}-\mathbf{G}) - E^{(0)}(\mathbf{k})}\Psi_{\mathbf{k}-\mathbf{G}}^{(0)} + \cdots \; .$$

When the unperturbed states are degenerate (or when the difference between two "original" energies is not large enough), App. C indicates that the degenerate states have to be treated separately, choosing linear combinations of them which are eigenstates of the perturbation Hamiltonian. This calculation yields the first-order correction to the energy. In the one-dimensional configuration analysed in Sec. 6.6 there is double degeneracy only near each reciprocal-lattice vector, with the perturbation matrix

$$\begin{bmatrix} E^{(0)}(k) - E^{(0)}(G/2) & \widetilde{U}(G) \\ \widetilde{U}(-G) & E^{(0)}(k-G) - E^{(0)}(G/2) \end{bmatrix} \; .$$

The eigenvalue equation of this matrix is identical to Eq. (6.89).

Answer 6.16.

The Fourier transform of the potential is $\widetilde{U}(G) = (1/a) \int_{-a/2}^{a/2} dx \exp[-iGx]U(x) = u_0/a$. Hence, the equation for the Fourier coefficients of the wave function is

$$\left(\frac{\hbar^2 k^2}{2m} - E\right)\widetilde{\psi}(k) + \frac{u_0}{a}\sum_G \widetilde{\psi}(k - G) = 0 \ .$$

The second term contains the function $f(k) = \sum_G \widetilde{\psi}(k - G)$. It follows that

$$\widetilde{\psi}(k) = -\frac{u_0}{a}\frac{f(k)}{\hbar^2 k^2/(2m) - E} \ ,$$

which, for $k \to k - G$ yields

$$\widetilde{\psi}(k - G) = -\frac{u_0}{a}\frac{f(k - G)}{\hbar^2(k - G)^2/(2m) - E} \ .$$

As the sum representing $f(k)$ contains all reciprocal-lattice vectors, one may shift the summation index and prove that

$$f(k - G') = \sum_G \widetilde{\psi}(k - G - G') = \sum_{G''} \widetilde{\psi}(k - G'') = f(k) \ ,$$

and therefore

$$\widetilde{\psi}(k - G) = -\frac{u_0}{a}\frac{f(k)}{\hbar^2(k - G)^2/(2m) - E} \ .$$

It remains to sum this expression over G to obtain

$$f(k) = -\frac{u_0}{a}f(k)\sum_G \frac{1}{\hbar^2(k - G)^2/(2m) - E} \ ,$$

which has a solution provided that

$$\sum_G \frac{1}{(k - G)^2 - K^2} = -\frac{\hbar^2 a}{2mu_0} \ , \qquad E = \hbar^2 K^2/(2m) \ .$$

The last equation relates K, i.e., the energy, to k.

When u_0 is very small, the right hand-side is rather large, and then the sum on the left side is dominated by the terms with the smallest denominators. In the $u_0 \to 0$ limit this yields $E = \hbar^2(k - G)^2/(2m)$, reproducing the reduced spectrum of a free particle. When ka is away from the zone boundaries, an expansion in powers of u_0 reproduces the second-order correction, which is also obtained from perturbation theory (Sec. 6.6). On the border of a Brillouin zone, e.g., at $ka = \pi$, the sum is dominated by two terms, with $G = 0$ and $G = 2\pi/a$; then the result reproduces the energy gap, proportional to u_0. In the opposite limit, when $u_0 \to \infty$, the right hand-side vanishes, and a plot of the left side, as a function of K, shows that it is zero for any value of k when $Ka = 2\pi n$; this yields the discrete energy levels of an infinite potential well. (Check!)

The equation derived in problem 6.14 is obtained as follows. The denominator in the left hand-side is $(k-G)^2 - K^2 = (k-G-K)(k-G+K)$, leading to

$$\sum_G \frac{1}{(k-G)^2 - K^2} = \frac{a}{2K}\sum_G \left[\frac{1}{ka - Ka - Ga} - \frac{1}{ka + Ka - Ga}\right].$$

$$(6.177)$$

Utilizing the identity given in the problem and other trigonometrical relations, and $G = 2\pi n/a$, gives

$$\sum_G \frac{1}{(k-G)^2 - K^2} = \frac{a}{4K}\sum_n \left[\frac{1}{a(k-K)/2 - n\pi} - \frac{1}{a(k+K)/2 - n\pi}\right]$$

$$= \frac{a}{4K}\left[\cot(\frac{ka - Ka}{2}) - \cot(\frac{ka + Ka}{2})\right] = \frac{a}{2K}\frac{\sin(Ka)}{\cos(Ka) - \cos(ka)}.$$

The last equation is identical to the result of problem 6.14.

Answer 6.17.

The energy gaps are determined by the Fourier transforms of the potential, Eq. (6.77), $\widetilde{U}(G) = \int_0^a dx \exp[-iGx]U(x)/a$, where the integration is carried out over the unit cell. Inserting here Eq. (6.65) gives

$$\widetilde{U}(G) = \frac{U_0}{a}\int_b^a dx e^{-iGx} = \frac{U_0}{iGa}\left(e^{-iGb} - e^{-iGa}\right).$$

The energy gap at $k = G/2$ is

$$2|\widetilde{U}(G)| = \frac{2U_0}{|G|a}\sqrt{2(1 - \cos[G(a-b)])} = \frac{4U_0}{|G|a}|\sin[G(a-b)/2]| ,$$

where the identity $|\exp[i\alpha] - \exp[i\beta]|^2 = 2 - 2\cos(\alpha - \beta) = 4\sin^2[(\alpha - \beta)/2]$ is used. As expected, and indeed found in Sec. 6.5, the gap decreases as $|G|$ increases, that is, in-between higher-energy bands. In the delta-function limit, $U_0(a-b) \to u_0$, one finds

$$2|\widetilde{U}(G)| \approx \frac{4U_0}{|G|a}|[G(a-b)/2]| \to \frac{2u_0}{a} ,$$

in agreement with the results of problem 6.14(a).

Answer 6.18.

Figure 6.40(a) reproduces Fig. 3.14; the parts of the third Brillouin zone are indicated. Displacing those by reciprocal-lattice vectors shifts them all into the first Brillouin zone, as shown in Fig. 6.40(b).

Figure 6.41(a) displays the dependence of the energy on the components of $\mathbf{k}$, once the energies are shifted into the first Brillouin zone. Figure 6.41(b) shows the equal-energy curves. The minima are located at the corners of the zone, and the maxima are at the centers of the faces.

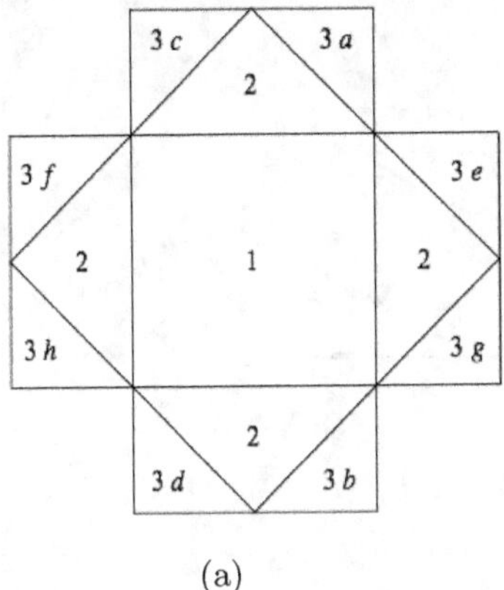

(a)

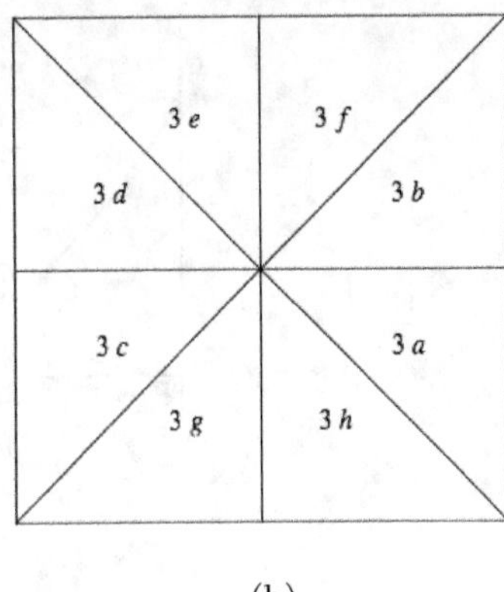

(b)

Fig. 6.40

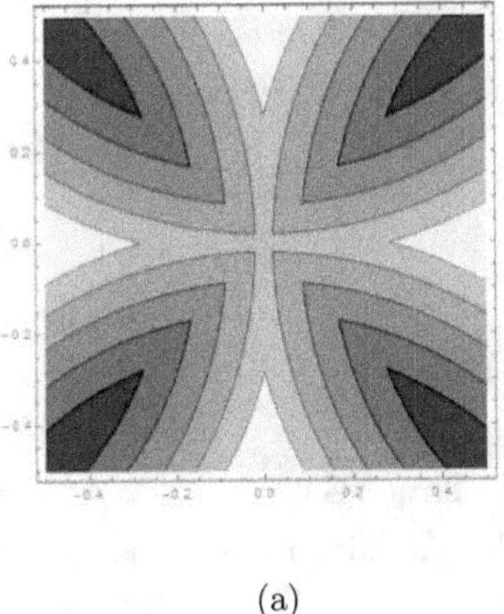

(a)

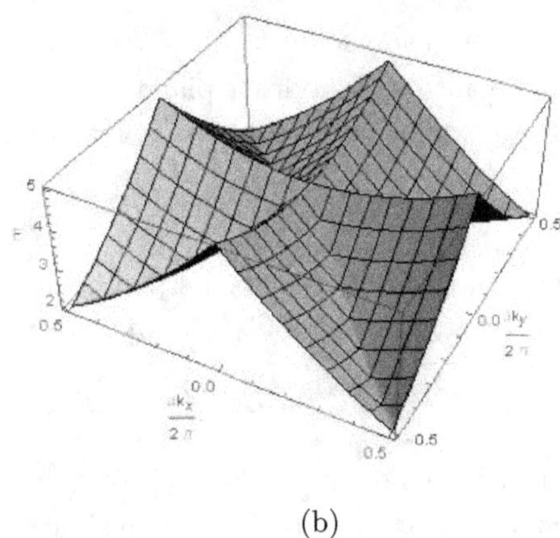

(b)

Fig. 6.41

Figure 6.42 summarizes the results for the three bands, along a special path in the first Brillouin zone. As expected, the third and the second bands overlap over the diagonal of the zone (in much the same way as the second and the first bands overlap on the faces).

Answer 6.19.

a. Assume that the boundary between two Brillouin zones is given by the vectors $\mathbf{q}_0$ that obey Eq. (3.26) [see also Fig. 3.13], $2\mathbf{q}_0 \cdot \mathbf{G} = \mathbf{G}^2$ where $\mathbf{G}$ is a reciprocal-lattice

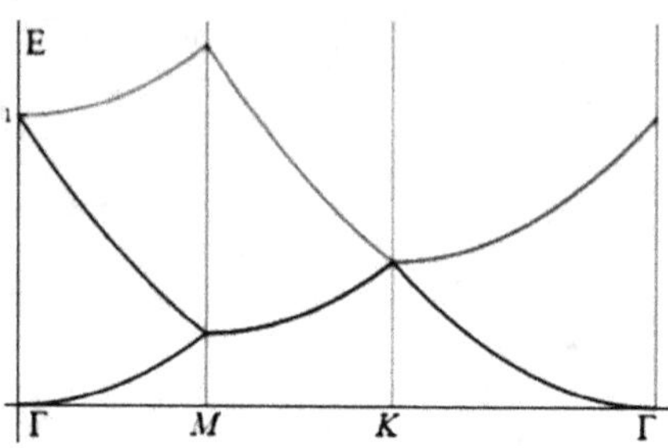

Fig. 6.42

vector. In a three-dimensional system, Eq. (6.91) becomes

$$E_{\mathbf{q}\pm}^{(1)} = \left[E^{(0)}(\mathbf{q}) + E^{(0)}(\mathbf{q} - \mathbf{G}) \pm \sqrt{[E^{(0)}(\mathbf{q}) - E^{(0)}(\mathbf{q} - \mathbf{G})]^2 + 4|\widetilde{U}(\mathbf{G})|^2} \right]/2 \ . \tag{6.178}$$

For $\mathbf{q} = \mathbf{q}_0$ it yields $E_{\mathbf{q}_0\pm}^{(1)} = E^{(0)}(\mathbf{q}_0) \pm |\widetilde{U}(\mathbf{G})|$, and therefore the energy gap at each point on the border plane (or line) is constant, equals $2|\widetilde{U}(\mathbf{G})|$.

b. An expansion around the point $\mathbf{q}_0$, $\mathbf{q} = \mathbf{q}_0 + \boldsymbol{\delta}$, gives

$$E_{\mathbf{q}\pm}^{(1)} \approx E_{\mathbf{q}_0\pm}^{(1)}(\mathbf{q}_0) \pm |\widetilde{U}(\mathbf{G})| + + \frac{\hbar^2}{2m}\left[\boldsymbol{\delta}^2 + \boldsymbol{\delta} \cdot (2\mathbf{q}_0 - \mathbf{G}) \pm \frac{\hbar^2(\boldsymbol{\delta} \cdot \mathbf{G})^2}{4m|\widetilde{U}(\mathbf{G})|}\right] \ .$$

When $\boldsymbol{\delta} \parallel \mathbf{G}$ then it is perpendicular to the plane between the two zones. In particular, $\boldsymbol{\delta} \cdot (2\mathbf{q}_0 - \mathbf{G}) = 0$, and consequently

$$E_{\mathbf{q}\pm}^{(1)} \approx E_{\mathbf{q}_0\pm}^{(1)}(\mathbf{q}_0) \pm |\widetilde{U}(\mathbf{G})| + \frac{\hbar^2}{2m}\left[1 \pm \frac{\hbar^2\mathbf{G}^2}{2m|\widetilde{U}(\mathbf{G})|}\right]\boldsymbol{\delta}^2 \ ,$$

which is analogous to Eq. (6.92) pertaining to the one-dimensional case. For a weak enough potential it is plausible that $1 < \hbar^2\mathbf{G}^2/(2m|\widetilde{U}(\mathbf{G})|)$. Therefore the coefficient of the quadratic term is positive for the upper band and negative for the lower one. As the dependence is quadratic, the energy of the lower band possesses a maximum on the plane between the two zones, and an equal-energy surface intersects that plane normally (i.e., the energy is not changed when the boundary is crossed), as indeed seen in Figs. 6.14(c) and (d). The energy of the upper band is minimal along the same direction, and there again the equal-energy curves are perpendicular to the boundary plane.

On the other hand, when $\boldsymbol{\delta}$ is on the border plane then $\mathbf{q} = \mathbf{q}_0 + \boldsymbol{\delta}$ lies there as well, and the energies are given by $E^{(0)}(\mathbf{q}) \pm |\widetilde{U}(\mathbf{G})|$. These are two paraboloids shifted one against the other by the energy gap. It follows that at the middle of the zone's face the upper band is minimal along all directions; the coefficients of the parabolas, however, are different along the different directions. The lower band has a saddle point: a maximum perpendicular to the plane and a minimum within it.

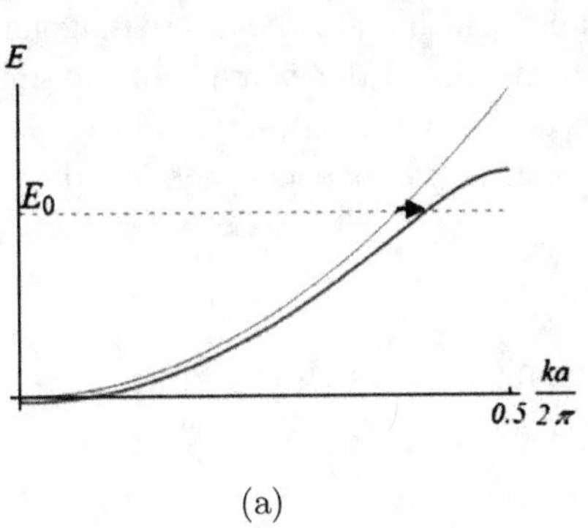

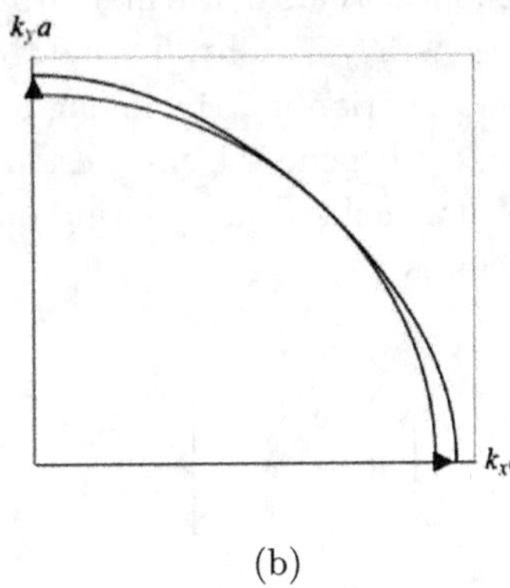

(a)   (b)

Fig. 6.43

c. Figures 6.14(c) and (d) can be deduced from Eq. (6.178) by using an approximation: in each quarter of the Brillouin zone [Fig. 6.15(b)] one uses the vector $\mathbf{q}_0$ corresponding to the boundary separating that quarter from the closest zone. As expected, away from the zone boundaries the plots are quite similar to those in Figs. 6.14(a) and (b). The dissimilarities appear near the zone boundaries. In the vicinity of the zone faces the normal derivative of the energy vanishes. As the energy of the lower band is smaller than the one of a free electron, the equal-energy curves are displaced outwards, intersecting perpendicularly the face (as explained quantitatively in the previous part of this solution).

In the absence of the periodic potential the energy is $E^{(0)} = \hbar^2\mathbf{k}^2/(2m)$, and hence the equal-energy curves are planar circles. A weak periodic potential leaves them almost unchanged, except close to the zone boundaries. Figure 6.43(a) shows the energy along the line connecting the center of the Brillouin zone with the center of one of the faces, e.g., M. As this face is approached, the energy is affected by the gap there, and then the lower band (parabolic for a free electron) is shifted downwards, from the thin to the thick line. The intersection of the horizontal line in the figure, E_0, with the function $E(k)$ indicates the location of the equal-energy curve whose energy is E_0. The figure shows that the intersection point moves to the right upon passing from the original parabola to the new energy, that is affected by the gap. Hence, the points on the equal-energy curves, which are circular for a free particle, move towards the centers of the faces of the zone, as illustrated by the arrows in Fig. 6.43(b). The circular curves are gradually deformed, until the boundary curve turns into a square rotated by 45° relative to the Brillouin zone. Increasing further the energy, the equal-energy curve reaches the maximum of the thick line where it is normal to the face of the zone.

d. At the corners of the Brillouin zone there is an additional degeneracy, since there the energy is unchanged with respect to displacements by more reciprocal-lattice vectors. For example, in a square lattice the energy of the free particle is degenerate at the four corners of the first Brillouin zone, while at the corners of the cubic lattice the degeneracy is eight. To find the energy gaps in such cases one has to diagonalize the perturbation matrix among **all** degenerate states. In the example of the square lattice, the degenerate energies on the corners are $e_0 = \hbar^2\pi^2/(ma^2)$, and the matrix to be diagonalized, accounting only for the Fourier components mentioned in the problem, is

$$\begin{bmatrix} e_0 & u & u & v \\ u & e_0 & v & u \\ u & v & e_0 & u \\ v & u & u & e_0 \end{bmatrix} \ , \ \ \text{with} \ \ u = \widetilde{U}(\mathbf{k}_M) = 0.2 \ , \ \ v = \widetilde{U}(\mathbf{k}_K) = 0.15 \ .$$

The rows and the columns represent the corners, $\mathbf{k} = (\pi/a)(1,1)$, $(\pi/a)(-1,1)$, $(\pi/a)(1,-1)$ and $(\pi/a)(-1,-1)$. This matrix has two eigenvectors with eigenvalue $e_0 - v$, $(-1,0,0,1)$ and $(0,-1,1,0)$, an eigenvector with the eigenvalue $e_0 - 2u + v$, $(1,-1,-1,1)$, and another one with eigenvalue $e_0 + 2u + v$, $(1,1,1,1)$. It follows that out of the four degenerate states two remain degenerate, with no change in their energy, while the other two are shifted upwards and downwards. Assuming that $u > v$, the lower band moves further down, to the energy $e_0 - 2u + v$, and the next band to the energy $e_0 - v$. Thus the energy gap at the point K is $2(u - v)$. This gap is smaller than the one at M, that is equal to $2u$. Figures 6.14 (c) and (d) and 6.18(b) are computed by generalizing the above determinant, with the energies in the diagonal elements that correspond to an arbitrary point in the Brillouin zone. For instance, when $k_x > 0$ and $k_y > 0$, the energies on the diagonal are

$$E^{(0)}(k_x, k_y) \ , \ \ E^{(0)}(k_x - \frac{2\pi}{a}, k_y) \ , \ \ E^{(0)}(k_x, k_y - \frac{2\pi}{a}) \ , \ \ E^{(0)}(k_x - \frac{2\pi}{a}, k_y - \frac{2\pi}{a}) \ .$$

Near the center of the Brillouin zone it is enough to keep the first row; near the M point it suffices to keep the first two rows (as done in the previous parts of the problem), and near the K point all four rows have to be used.

Answer 6.20.
Displacing $\mathbf{r}$ by $\mathbf{R}'$ in Eq. (6.101) yields

$$\varphi_{\mathbf{k}}(\mathbf{r} + \mathbf{R}') = A_{\mathbf{k}} \sum_{\mathbf{R}} \exp[i\mathbf{k} \cdot \mathbf{R}]\varphi^{(0)}(\mathbf{r} + \mathbf{R}' - \mathbf{R}) \ .$$

Setting the dummy index of the sum to be $\mathbf{R}'' = \mathbf{R} - \mathbf{R}'$ gives

$$\varphi_{\mathbf{k}}(\mathbf{r} + \mathbf{R}') = \exp[i\mathbf{k} \cdot \mathbf{R}']A_{\mathbf{k}} \sum_{\mathbf{R}''} \exp[i\mathbf{k} \cdot \mathbf{R}'']\varphi^{(0)}(\mathbf{r} - \mathbf{R}'') \ ,$$

as required by Eq. (6.54). The function

$$u_{\mathbf{k}}(\mathbf{r}) = \exp[-i\mathbf{k} \cdot \mathbf{r}]\varphi_{\mathbf{k}}(\mathbf{r}) = A_{\mathbf{k}} \sum_{\mathbf{R}} \exp[-i\mathbf{k} \cdot (\mathbf{r} - \mathbf{R})]\varphi^{(0)}(\mathbf{r} - \mathbf{R}) \ ,$$

is periodic, since a displacement of $\mathbf{r}$ by $\mathbf{R}'$ and a change of the dummy index from $\mathbf{R}$ to $\mathbf{R} - \mathbf{R}'$ give

$$u_{\mathbf{k}}(\mathbf{r} + \mathbf{R}') = e^{-i\mathbf{k}\cdot(\mathbf{r}+\mathbf{R}')}\varphi_{\mathbf{k}}(\mathbf{r} + \mathbf{R}')$$
$$= A_{\mathbf{k}}\sum_{\mathbf{R}} e^{-i\mathbf{k}\cdot(\mathbf{r}+\mathbf{R}'-\mathbf{R})}\varphi^{(0)}(\mathbf{r} + \mathbf{R}' - \mathbf{R}) = e^{-i\mathbf{k}\cdot\mathbf{r}}\varphi_{\mathbf{k}}(\mathbf{r}) = u_{\mathbf{k}}(\mathbf{r}) \, ,$$

as required in Eq. (6.53). Displacing $\mathbf{k}$ of Eq. (6.101) by a reciprocal-lattice vector $\mathbf{G}$, and using Eq. (3.17) give

$$\varphi_{\mathbf{k}+\mathbf{G}}(\mathbf{r}) = A_{\mathbf{k}+\mathbf{G}}\sum_{\mathbf{R}} e^{i(\mathbf{k}+\mathbf{G})\cdot\mathbf{R}}\varphi^{(0}(\mathbf{r} - \mathbf{R}) = A_{\mathbf{k}+\mathbf{G}}\sum_{\mathbf{R}} e^{i\mathbf{k}\cdot\mathbf{R}}\varphi^{(0)}(\mathbf{r} - \mathbf{R}) \, .$$

By the normalization of the wave function $|A_{\mathbf{k}+\mathbf{G}}|^2 = |A_{\mathbf{k}}|^2$. The wave function is determined up to a phase (that does not change any quantum average); thus $\varphi_{\mathbf{k}+\mathbf{G}}(\mathbf{r}) = \varphi_{\mathbf{k}}(\mathbf{r})$.

Answer 6.21.

a. The energies in the band are

$$E(k) = \Gamma_1\{1 + 2[1 - \cos(ka)] - 2\zeta[1 - \cos(2ka)]\}$$
$$= \Gamma_1\{1 + 4\sin^2(ka/2) - 4\zeta\sin^2(ka)\} \, .$$

Figure 6.44(a) portrays these energies in the first Brillouin zone (in unit of Γ_1), for $\zeta = 1/8$ (the thick line) and for $\zeta = 1/4$ (the thin line). The leading-order behavior at the bottom of the band is

$$E(k) - \Gamma_1 \approx \Gamma_1(1 - 4\zeta)(ka)^2 + \mathcal{O}(k^4) \, .$$

It is parabolic for $\zeta = 1/8$, but $E(k) - \Gamma_1 \propto k^4$ for $\zeta = 1/4$. Larger negative values of Γ_2 give rise to four solutions for energies close to zero, which is impossible [see problem 6.13]. This means that the tight-binding approximation breaks down for such values.

b. Similar to Eq. (5.56), the density of states is $g(E) = (1/\pi)dk/dE$. In the present case the derivative is

$$\frac{dE}{dk} = \Gamma_1\sin(ka)[1 + 4(1 - \zeta) - 32\zeta\sin^2(ka/2)] \, .$$

The density of states diverges near the center of the Brillouin zone and at its boundaries. This van Hove singularity emerges from the vanishing of the derivative of the energy with respect to the wave number at these points. The result, in units of $1/(\Gamma_1 a)$ is exhibited in Fig. 6.44(b) for $\zeta = 0.1$.

c. At zero temperature all levels up to the Fermi energy are filled, and hence $E_{\text{tot}} = L\int_{\Gamma_1}^{E_F} dE g(E)E$. From the condition $E_F - \Gamma_1 \ll \Gamma_1$ it follows that $E - \Gamma_1 \ll \Gamma_1$ for all filled energies, and thus the parabolic approximation, $E(k) - \Gamma_1 \approx \Gamma_1(1 - 4\zeta)(ka)^2$, can be used. As a result,

$$g(E) = \frac{1}{\pi}\frac{dk}{dE} \approx \frac{1}{2\pi\Gamma_1(1 - 4\zeta)ka^2} \, ,$$

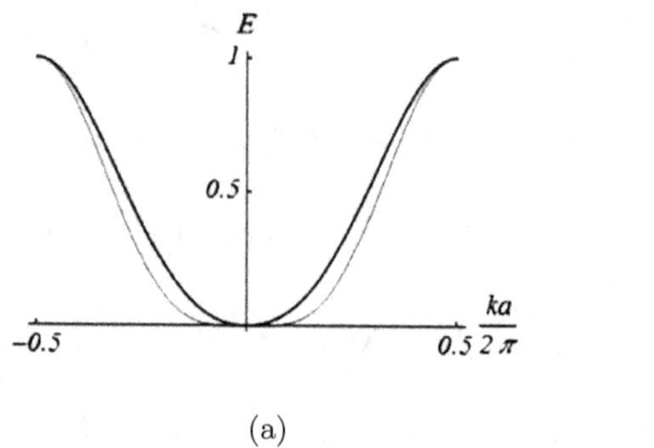

(a)

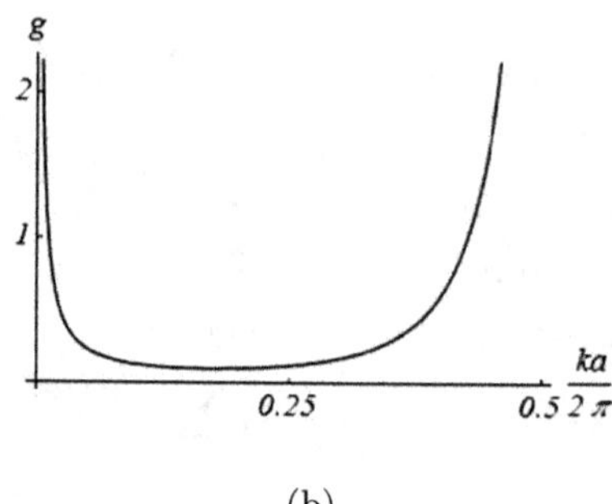

(b)

Fig. 6.44

leading to

$$E_{\text{tot}} = N\Gamma_1 + L \int_{\Gamma_1}^{E_{\text{F}}} dE \frac{\sqrt{E - \Gamma_1}}{2\pi a \sqrt{\Gamma_1(1 - 4\zeta)}} = N\Gamma_1 + N \frac{(E_{\text{F}} - \Gamma_1)^{3/2}}{3\pi \sqrt{\Gamma_1(1 - 4\zeta)}} .$$

The resemblance to the density of states of a free particle is conspicuous: the density of states is inversely proportional to the square root of the energy (relative to the bottom of the band). The difference is in the numerical coefficient.

Answer 6.22.

a. Inserting the function $\varphi^{(0)}(x) = A \exp[-\kappa|x|]$, with $\kappa > 0$, into the Schrödinger equation shows that it is indeed a solution for any $x \neq 0$, with the energy $E^{(0)} = -\hbar^2 \kappa^2/(2m)$. At the vicinity of the point $x = 0$, the Schrödinger equation is

$$-\frac{\hbar^2}{2m} \frac{d^2\varphi^{(0)}}{dx^2} - u_0 \delta(x)\varphi^{(0)} = E^{(0)}\varphi^{(0)} .$$

Integrating it between $-\epsilon$ to ϵ yields, in the $\epsilon \to 0$ limit,

$$-\frac{\hbar^2}{2m} \left[\frac{d\varphi^{(0)}}{dx} \bigg|_{x=0+} - \frac{d\varphi^{(0)}}{dx} \bigg|_{x=0-} \right] - u_0 \varphi^{(0)}(0) = 0 ,$$

i.e., $-[\hbar^2/(2m)][-\kappa A - \kappa A] - u_0 A = 0$, and hence $\kappa = u_0 m/\hbar^2$. The normalization of the wave function implies that

$$1 = \int_{-\infty}^{\infty} dx |\varphi^{(0)}(x)|^2 = |A|^2/\kappa ,$$

and consequently $A = \sqrt{\kappa}$. In this way a unique solution is obtained, and since its energy is negative, $E^{(0)} = -mu_0^2/(2\hbar^2)$, it is the ground state.

b. The overlap integral between nearest neighbors is

$$\alpha = \int dx [\varphi^{(0)}(x)]^* \varphi^{(0)}(x - a) = \kappa \left[\int_{-\infty}^{0} dx e^{\kappa(2x-a)} + \int_{0}^{a} dx e^{-\kappa a} \right.$$

$$\left. + \int_{a}^{\infty} dx e^{-\kappa(2x-a)} \right] = \kappa a e^{-\kappa a} + \cosh(\kappa a) .$$

The integral yielding γ, for nearest neighbors, is

$$\gamma = -\int dx\,[\varphi^{(0)}(x)]^* \Delta U \varphi^{(0)}(x-a) = u_0 \sum_{m\neq 0} \int dx\,[\varphi^{(0)}(x)]^* \delta(x-ma)\varphi^{(0)}(x-a)$$

$$= -u_0\kappa \sum_{m\neq 0} e^{-\kappa|ma|}e^{-\kappa|m-1|a} = -2u_0\kappa[e^{-\kappa a} + 2e^{-3\kappa a} + 2e^{-5\kappa a} + \ldots]$$

$$\approx -2u_0\kappa e^{-\kappa a} \ .$$

Likewise,

$$\beta = -\int dx\,[\varphi^{(0)}(x)]^* \varphi^{(0)}(x) = -u_0 \sum_{m\neq 0} |\varphi^{(0)}(x)|^2 \delta(x-ma)$$

$$= -u_0\kappa \sum_{m\neq 0} e^{-2\kappa|ma|} \approx -2u_0\kappa e^{-2\kappa a} \ .$$

In the tight-binding approximation $\kappa a = u_0 a m/\hbar^2 \gg 1$ and therefore $|\beta| \ll |\gamma|$. Discarding the correction to the normalization in the denominator of Eq. (6.110) gives $E(k) \approx E^{(0)} - \beta - 2\gamma\cos(ka)$. The width of this band is $2\gamma = 4u_0\kappa\exp[-\kappa a]$.

Answer 6.23.

For the configuration described in this problem, Eq. (6.122) gives rise to two equations,

$$(E - E_A^{(0)})A_A = -\tilde{\gamma}_{AA}(\mathbf{k})A_A - \tilde{\gamma}_{BA}(\mathbf{k})A_B \ ,$$

$$(E - E_B^{(0)})A_B = -\tilde{\gamma}_{BB}(\mathbf{k})A_B - \tilde{\gamma}_{AB}(\mathbf{k})A_A \ ,$$

where $\tilde{\gamma}_{ij}(\mathbf{k}) = \sum_{\mathbf{R}} \gamma_{ij}(\mathbf{R})\exp[-i\mathbf{k}\cdot\mathbf{R}]$, and when all overlap integrals save $\alpha_{ii}(0) = 1$ are ignored. The two equations possess a nontrivial solution provided that their determinant vanishes. Hence,

$$E = \left[E_A^{(0)} + E_B^{(0)} - \tilde{\gamma}_{AA} - \tilde{\gamma}_{BB} \pm \sqrt{(E_A^{(0)} + E_B^{(0)} - \tilde{\gamma}_{AA} - \tilde{\gamma}_{BB})^2 + 4|\tilde{\gamma}_{AB}|^2}\right]/2 \ .$$

When only nearest-neighbors hopping are retained, as stated in the problem, then $\tilde{\gamma}_{AB} = 2\gamma_{AB}\cos(ka/2)$, where $\gamma_{AB} = \langle\varphi_A^{(0)}(x)|\Delta U|\varphi_B^{(0)}(x-a/2)\rangle$, and $\tilde{\gamma}_{AA} = \tilde{\gamma}_{BB} = 0$. With these approximations,

$$E = \left[E_A^{(0)} + E_B^{(0)} \pm \sqrt{(E_A^{(0)} - E_B^{(0)})^2 + 16|\gamma_{AB}|^2\cos^2(ka/2)}\right]/2 \ .$$

Figure 6.45(a) displays the two bands for different atoms; it resembles qualitatively Fig. 5.9 that portrays the acoustic and optical bands of lattice vibrations in a crystal with two atoms in the unit cell. When the two atoms are identical, the unit cell is twice smaller, and the Brillouin zone is doubled, precisely as in the case of lattice vibrations, see Fig. 6.45(b).

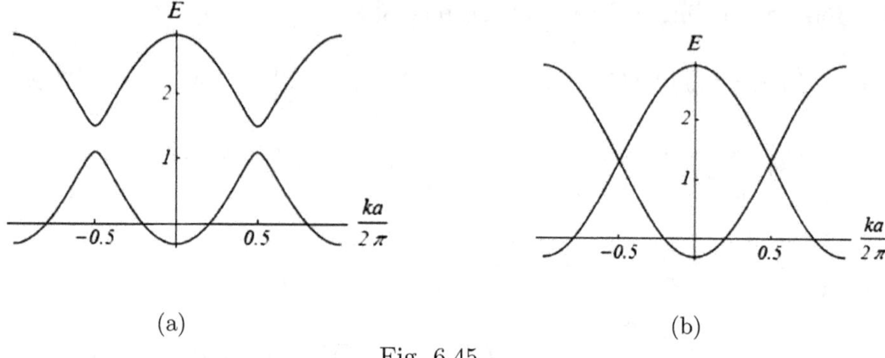

(a)   (b)

Fig. 6.45

Answer 6.24.

a. Recall that the graphene crystal is represented by a triangular lattice with two atoms in the unit cell. The wave functions on the different atoms can be indexed as in Fig. 5.10, with u_{nm} and v_{nm} representing the wave functions in the two points in the $nm-$unit cell. As in problem 6.23, there are two linear equations for the wave functions on the two sites in the unit cell. Since the atoms are all identical, $E_A^{(0)} = E_B^{(0)}$, and the energies are given by $E_\pm = E^{(0)} \pm |\widetilde{\gamma}_{AB}|$, with

$$\widetilde{\gamma}_{AB}(\mathbf{k}) = \sum_{nn} \gamma_{AB} e^{-i\mathbf{k}\cdot\mathbf{r}_{nn}} = \gamma_{AB}(e^{-i\mathbf{k}\cdot\mathbf{a}_1} + e^{-i\mathbf{k}\cdot\mathbf{a}_2} + e^{-i\mathbf{k}\cdot(\mathbf{a}_1+\mathbf{a}_2)}) ,$$

where the sum is over nearest neighbors; $\mathbf{a}_1 = (a,0)$ and $\mathbf{a}_2 = (a,\sqrt{3}a)/2$. [The expression resembles the one in Eq. (5.33), due to the similarity of the tight-binding equations and those for lattice vibrations.] A three-dimensional calculation of the two energy bands is depicted in Figs. 6.46. Two view points are shown, in order to emphasize the triangular symmetry and the linear dispersion relations near the point K in the Brillouin zone.

b. As seen in Fig. 6.46, and similar to the result in Chapter 5, the two bands touch one another at the six points marked by K in Fig. 5.12. For example, around one of them, say $(4\pi/(3a),0)$, $\mathbf{k} = (4\pi/(3a) + \overline{k}_x, \overline{k}_y)$ and hence

$$\widetilde{\gamma}_{AB}(\mathbf{k}) = \gamma_{AB}(e^{-i(4\pi/3+\overline{k}_x a)} + e^{-i(2\pi/3+\overline{k}_x a/2+\sqrt{3}\,\overline{k}_y a/2)} + e^{-i(3\overline{k}_x a/2+\sqrt{3}\,\overline{k}_y a/2)}) .$$

Expanding in the deviation away from this point gives

$$\widetilde{\gamma}_{AB}(\mathbf{k}) = \gamma_{AB}(\sqrt{3} - 3i)(\overline{k}_x - i\overline{k}_y)a/4 + \mathcal{O}(\mathbf{k}^2) .$$

The energy is thus linear in the deviation vector, $E = \pm|\widetilde{\gamma}_{AB}(\mathbf{k})| \approx \pm(\gamma_{AB}\sqrt{3}/2)|\overline{\mathbf{k}}|a$. Each of the signs gives an equation of a cone, symmetric with respect to rotations around the $\hat{\mathbf{z}}-$axis, as indeed seen in the figure. Such a linear dispersion is typical to solutions of the **Dirac equation**, that describes a particle and an antiparticle with zero mass. This unique behavior facilitates experimental explorations of the solutions of the Dirac equation, and entails a very peculiar electrical behavior of graphene. There is currently much interest in this special material.

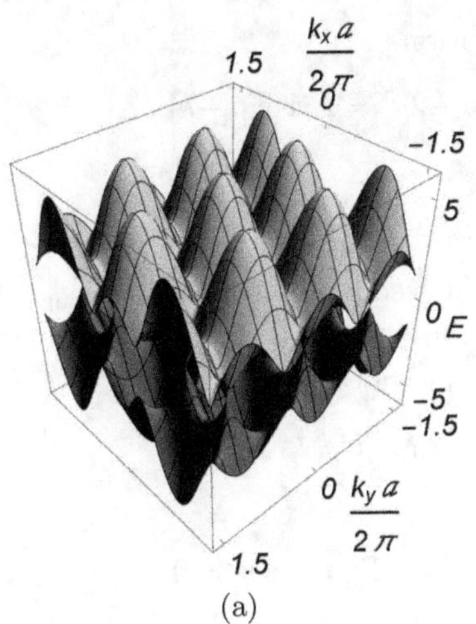

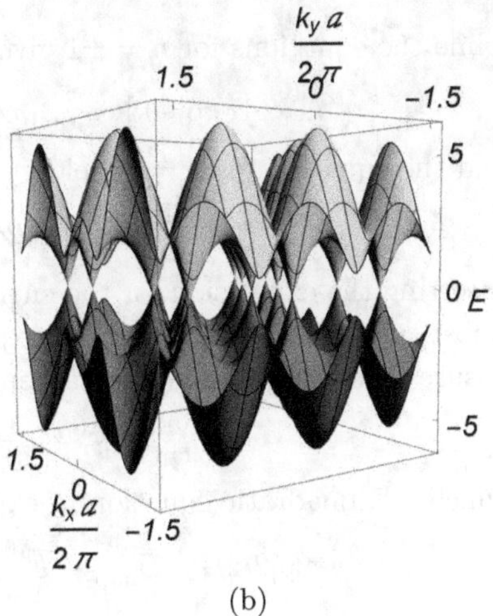

Fig. 6.46

Answer 6.25.

a. The matrix in the problem corresponds to an Hamiltonian in which the overlap of wave functions on different atoms is absent, $\alpha(\mathbf{R}-\mathbf{R}') = \langle \varphi^{(0)}(\mathbf{r}-\mathbf{R})|\varphi^{(0)}(\mathbf{r}-\mathbf{R}')\rangle = \delta(\mathbf{R}-\mathbf{R}')$; β, Eq. (6.112), is "absorbed" in the diagonal term, $\overline{E}^{(0)} = E^{(0)} - \beta$.

b. The Schrödinger equation $\hat{\mathcal{H}}\psi = E\psi$ takes the form $\sum_m \langle n|\hat{\mathcal{H}}|m\rangle \psi(m) = E\psi(n)$. In the present case, it becomes N linear homogeneous equations in N variables,

$$\overline{E}^{(0)}\psi(n) - \gamma\psi(n-1) - \gamma\psi(n+1) = E\psi(n) \ .$$

The similarity between this equation and Eq. (5.4) allows one to build on the arguments used to solve the latter. In particular, using $\psi(n) = A\exp[ikna]$, and eliminating $A\exp[ikna]$ on both sides proves that the wavy solution is an accepted one, provided that

$$E = \overline{E}^{(0)} - \gamma(e^{ika} + e^{-ika}) \ .$$

The periodic boundary conditions, $\psi(n) = \psi(n+N)$ imply that $\exp[ikNa] = 1$, so that the discrete wave vectors are $k_\ell = 2\pi\ell/(Na)$, $\ell = 0, 1, 2, \ldots, N-1$. This result reproduces Eq. (6.118).

c. In this case, the equations for $n = 0$ and $n = \pm 1$ have to be handled separately,

$$\overline{E}^{(0)}_{\text{imp}}\psi(0) - \gamma_{\text{imp}}\psi(-1) - \gamma_{\text{imp}}\psi(1) = E\psi(0) \ ,$$

$$\overline{E}^{(0)}\psi(\pm 1) - \gamma_{\text{imp}}\psi(0) - \gamma\psi(\pm 2) = E\psi(\pm 1) \ .$$

All other equations, for $|n| > 1$, remain as in part (b). Is $\psi(n) = A\exp[-\kappa|n|]$, for $|n| \geq 1$, an accepted solution? The equations of part (a) imply

$$E = \overline{E}^{(0)} - \gamma(\exp[\kappa] + \exp[-\kappa])] \ ,$$

while the equations for $n = \pm 1$ give for this solution

$$\overline{E}^{(0)} A \exp[-\kappa] - \gamma_{\mathrm{imp}} \psi(0) - \gamma A \exp[-2\kappa] = E A \exp[-\kappa] \ ,$$

and the equation for $n = 0$ yields

$$\overline{E}^{(0)}_{\mathrm{mp}} \psi(0) - 2\gamma_{\mathrm{imp}} A \exp[-\kappa] = E\psi(0) \ .$$

Inserting the expression for the energy into the first equation gives $\gamma A = \gamma_{\mathrm{imp}} \psi(0)$. Substituting the latter and the expression for the energy in the second equation ensures that this is indeed an accepted solution, provided that

$$\overline{E}^{(0)}_{\mathrm{imp}} - \overline{E}^{(0)} + \gamma(e^{-\kappa} + e^{\kappa}) = 2\gamma^2_{\mathrm{imp}} e^{-\kappa}/\gamma \ ,$$

which is a quadratic equation in $\exp[\kappa]$,

$$\exp[2\kappa] + [(\overline{E}^{(0)}_{\mathrm{imp}} - \overline{E}^{(0)})/\gamma] \exp[\kappa] + 1 - 2(\gamma_{\mathrm{imp}}/\gamma)^2 = 0 \ .$$

The solutions are

$$\exp[\kappa] = b \pm \sqrt{b^2 + 2c - 1} \ , \quad b = (\overline{E}^{(0)}_{\mathrm{imp}} - \overline{E}^{(0)})/(2\gamma) \ , \quad c = (\gamma_{\mathrm{imp}}/\gamma)^2 > 0 \ .$$

The solutions are complex when $b^2 + 2c - 1 < 0$, but then $|\exp[\kappa]|^2 = b^2 + (1 - 2c - b^2) = 1 - 2c < 1$; such solutions are not physical. For $b^2 + 2c - 1 > 0$ the two solutions are real. They cross the values $\exp[\kappa] = \pm 1$ when $c = 1 \pm b$ (substitute $\exp[\kappa] = \pm 1$ in the quadratic equation). The condition $|\exp[\kappa]| > 1$ is fulfilled when there is a single solution, $c = 1 - |b|$, and for the two solutions when $c > 1 + |b|$.

Answer 6.26.

a. As shown in the solution of problem 6.10, the Fermi wave vector in two dimensions is $k_{\mathrm{F}} = \sqrt{2\pi n}$, where n is the planar electrons' density. The second Brillouin zone is first filled when the Fermi circle touches the face of the first zone [see Fig. 6.25(a)]. Therefore, it is only the first zone which is occupied as long as $k_{\mathrm{F}} < \pi/a$, i.e., $n < \pi/(2a^2)$. As the density can be written in the form $n = \ell/a^2$, one finds $\ell < \pi/2 \approx 1.57$. The third Brillouin zone is first occupied [see Fig. 6.25(a)] when the Fermi circle crosses the corner of the first zone, i.e., when $k_{\mathrm{F}} > \sqrt{2}\pi/a$, or $n > \pi/a^2$, which implies $\ell > \pi \approx 3.14$.

b. As found in the solution of problem 6.19(a) the energy gap in-between the two bands on the face of the Brillouin zone is $2\Delta = 2|\widetilde{U}(\mathbf{G})|$. Ignoring the correction found in problem 6.19(d), the energy at the top of the lower band (at point K, the corner of the Brillouin zone) is about $\hbar^2\pi^2/(ma^2) - \Delta$; the energy at the center of the face (point M) is approximately $\hbar^2\pi^2/(2ma^2) - \Delta$. The bottom of the second band is also at M, with the energy $\hbar^2\pi^2/(2ma^2) + \Delta$. The difference between the bottom of the upper band and the top of the lower one is

$$[\hbar^2\pi^2/(ma^2) - \Delta] - [\hbar^2\pi^2/(2ma^2) + \Delta] = \hbar^2\pi^2/(2ma^2) - 2\Delta \ .$$

The transition from the semimetal behavior, which is characterized by the configuration of two overlapping bands, to an insulating one (with a finite gap in-between the bands) occurs for $\hbar^2\pi^2/(2ma^2) = 2\Delta$.

c. The band energy, in the tight-binding approximation, Eq. (6.119), is

$$E_n(\mathbf{k}) \approx \overline{E}_n + 4\Gamma_n(a)[\sin^2(k_x a/2) + \sin^2(k_y a/2)] \ .$$

The form of the bands is dictated by the sign of Γ_n. When the signs of both Γ_1 and Γ_2 are identical (positive), the two bands reach their maximal and minimal points at the same locations. The lower part of the upper band is then also at the center of the Brillouin zone (as opposed to the situation for a free particle!). The material is thus a semimetal as long as that minimal energy, $E_2(0,0) \approx \overline{E}_2$, is lower than the maximal energy of the lower band,

$$E_1(\pi/a, \pi/a) \approx \overline{E}_1 + 8\Gamma_1(a) \ .$$

When the signs of Γ_1 and Γ_2 are opposite to one another, say $\Gamma_2 < 0$, the minimum of the second band is above the maximum of the first one, and the configuration is similar to that found for a one-dimensional solid: the material is an insulator unless the bands overlap near the corners of the Brillouin zone. The condition for this configuration to occur is

$$\overline{E}_1 + 8\Gamma_1 \approx E_1(\pi/a, \pi/a) < E_2(\pi/a, \pi/a) \approx \overline{E}_2 - 8|\Gamma_2| \ .$$

Answer 6.27.

a. The Fermi sphere touches the face of the Brillouin zone when $k_F = \pi/a$. At three dimensions and for free electrons, $n = k_F^3/(3\pi^2)$ [Eq. (6.28)]. The electrons' density is then $n = \pi/(3a^3)$. Denoting the concentration of the double-valence atoms by x, the average number of electrons in each unit cell is $\ell = 2x + 1(1 - x) = 1 + x$, and the spatial electrons' density is $n = (1+x)/a^3$. It follows that $(1+x)/a^3 = \pi/(3a^3)$, and $x = \pi/3 - 1 \approx 4.72\%$.

b. As seen in Fig. 6.28, the curve formed by the cut of the Fermi surface of an **SC** lattice with the face of the Brillouin zone is roughly circular. The face is characterized by, e.g., $k_x = \pi/a$, and hence the energy is

$$E_n(\mathbf{k}) \approx \overline{E}_n + 2\Gamma_n(a)[4 - \cos(k_y a) - \cos(k_z a)] \ .$$

In the middle of the band, from Eq. (6.123), $0 = \cos(k_x a) + \cos(k_y a) + \cos(k_z a)$, which becomes $\cos(k_y a) + \cos(k_z a) = 1$ on the face, and this indeed appears like a circle. The radius is gradually diminished as $\cos(k_y a) + \cos(k_z a)$ increases, and is reduced to a single point precisely at the center of the face when this sum becomes 2. For larger values of the sum there are no points on the face that obey this condition, and the Fermi surface is contained within the first Brillouin zone. Indeed, inserting 2 into Eq. (6.123) results in $E_n(\mathbf{k}) \approx \overline{E}_n + 4\Gamma_n(a)$, i.e., an energy located at the third of the band width.

c. In a **BCC** lattice there are two atoms in the unit cell, and therefore the electrons' density is $n = 2(1+x)/a^3$. The reciprocal lattice is an **FCC**, whose lattice constant is $4\pi/a$, and whose lattice vectors are given in Eq. (3.21),

$$\mathbf{G} = (2\pi/a)[h(\hat{\mathbf{y}} + \hat{\mathbf{z}}) + k(\hat{\mathbf{x}} + \hat{\mathbf{z}}) + \ell(\hat{\mathbf{x}} + \hat{\mathbf{y}})] \ .$$

The center of the zone is connected with its 12 nearest neighbors by the reciprocal-lattice vectors

$$\mathbf{G} = (2\pi/a)(\pm\hat{\mathbf{y}} \pm \hat{\mathbf{z}}) , \quad \mathbf{G} = (2\pi/a)(\pm\hat{\mathbf{z}} \pm \hat{\mathbf{x}}) , \quad \text{or} \quad \mathbf{G} = (2\pi/a)(\pm\hat{\mathbf{x}} \pm \hat{\mathbf{y}}) .$$

The surfaces separating the first Brillouin zone from its neighbors cut perpendicularly these vectors at their mid points; the distance from the origin to each of these planes is $\sqrt{2}\pi/a$. It follows that $n = k_{\mathrm{F}}^3/(3\pi^2) = (\sqrt{2}\pi/a)^3/(3\pi^2)$, and $x = \sqrt{2}\pi/3 - 1 \approx 48\%$.

Answer 6.28.

For convenience, denote $k_x a = \xi$ and $k_y a = \pi - \eta$, and expand in the variables ξ and η,

$$E_n(\mathbf{k}) \approx E_n^{(0)} - \beta_n - 2\gamma_n[\cos\xi + \cos(\pi - \eta)] \approx E_0 + \gamma_n(\xi^2 - \eta^2) ,$$
$$E_0 = E_n^{(0)} - \beta_n .$$

The point $(0, \pi/a)$ is a saddle point. On the equal-energy curve $E_n(\mathbf{k}) \equiv \overline{E}$, $dE_n = 2\gamma_n(\xi d\xi - \eta d\eta) = 0$, and $|\boldsymbol{\nabla} E_n| \approx 2\gamma_n a\sqrt{\xi^2 + \eta^2}$. The length segment on this curve obeys

$$(d\ell)^2 = (dk_x)^2 + (dk_y)^2 = a^2[(d\xi)^2 + (d\eta)^2] = a^2[1 + (\xi/\eta)^2](d\xi)^2 ,$$

and therefore the integrand is

$$\frac{d\ell}{2\gamma_n a\sqrt{\xi^2 + \eta^2}} = \frac{d\xi}{2\gamma_n \eta} = \frac{d\xi}{2\gamma_n \sqrt{\xi^2 - (\overline{E} - E_0)/\gamma_n}} .$$

Since $\xi^2 + \eta^2 > 0$, it follows that $\xi > \xi_0 = \sqrt{(\overline{E} - E_0)/\gamma}$. As the integral is dominated by the surrounding of $\xi \approx \xi_0$, the other bound of the integration is of no importance, and the integral can be evaluated in the range $\xi_0 < \xi < \pi/2$. An equivalent contribution comes from the two sides of the four points $(0, \pm\pi/a)$, $(\pm\pi/a, 0)$, leading to

$$g(\overline{E}) \approx 8 \int_{\xi_0}^{\pi/2} \frac{d\xi}{2\gamma_n\sqrt{\xi^2 - (\overline{E} - E_0)/\gamma_n}} \approx -\frac{4}{\gamma_n} \ln\left[\frac{\overline{E} - E_0}{\gamma_n \pi}\right] ,$$

where the relation $\int dx/\sqrt{x^2 - a} = \ln[x + \sqrt{x^2 - a}]$ is used; it is also assumed that $\overline{E}$ is very close to the middle of the band, $\overline{E} - E_0 \ll \gamma_n$.

Answer 6.29.

a. Denote the lattice constants of the orthorhombic lattice by a_x, a_y, and a_z, and the overlap integrals between nearest neighbors along the corresponding directions by Γ_x, Γ_y, and Γ_z. Then

$$E_n(\mathbf{k}) \approx \overline{E}_n + 4\Gamma_x \sin^2(k_x a_x/2) + 4\Gamma_y \sin^2(k_y a_y/2) + 4\Gamma_z \sin^2(k_z a_z/2) .$$

A Taylor expansion around the origin yields

$$E_n(\mathbf{k}) \approx \overline{E}_n + \Gamma_x(k_x a_x)^2 + \Gamma_y(k_y a_y)^2 + \Gamma_z(k_z a_z)^2 \ .$$

Introducing the definition $\hbar^2/(2m_i^*) = \Gamma_i a_i^2$ results in the expression given in the problem.

b. Equation (6.127) gives $g(E) = [1/(4\pi^3)] \int d^3k \sum_n \delta[E - E_n(\mathbf{k})]$. Changing variables, $\hbar k_\alpha / \sqrt{m_\alpha^*} = \overline{k}_\alpha$, yields

$$g(E) = \frac{\sqrt{m_x^* m_y^* m_z^*}}{4(\pi\hbar)^3} \int d\overline{k}_x d\overline{k}_y d\overline{k}_z \delta[E - E_0 - (\overline{k}_x^2 + \overline{k}_y^2 + \overline{k}_z^2)/2] \ .$$

Transforming back to spherical coordinates,

$$g(E) = \frac{\sqrt{m_x^* m_y^* m_z^*}}{4(\pi\hbar)^3} 4\pi \int d\overline{k}^2 d\overline{k} \delta[E - E_0 - \overline{k}^2/2] = \frac{\sqrt{2m_x^* m_y^* m_z^*}}{\pi^2 \hbar^3} \sqrt{E - E_0} \ .$$

This result resembles the one given in Eq. (6.31), where the mass there is replaced here by the geometrical mean of the three effective masses that appear in the expression for the energy.

c. The density of the electrons in the conduction band is

$$n = \frac{N_e}{V} = \int_{E_0}^{E_F} dE g(E) = \frac{\sqrt{2m_x^* m_y^* m_z^*}}{\pi^2 \hbar^3} \frac{2}{3} (E_F - E_0)^{3/2} \ ,$$

and thus, generalizing Eq. (6.32),

$$g(E) = \frac{3n(E - E_0)^{1/2}}{2(E_F - E_0)^{3/2}} \ .$$

As mentioned, this result pertains to holes in the upper part of the valence band.

d. Similar manipulations yield $g(E) = \sqrt{2m_x^* m_y^* m_z^*} \sqrt{E_{\max} - E}/(\pi^2 \hbar^3)$, and hence

$$n = \frac{N_e}{V} = \int_{E_0}^{E_F} dE g(E) = \frac{\sqrt{2m_x^* m_y^* m_z^*}}{\pi^2 \hbar^3} \frac{2}{3} (E_{\max} - E_F)^{3/2} \ .$$

Answer 6.30.

a. The required equation is

$$n_{0e} e^{-\beta(E_c - \mu)} = n_{0h} e^{\beta(E_v - \mu)} + \frac{x_D}{1 + \exp^{-\beta(E_c - E_D - \mu)}} \ .$$

Denoting $x = \exp[-\beta(E_c - \mu)]$ yields

$$n_{0e} x = \frac{n_{0h}}{x} e^{-\beta E_g} + \frac{x_D}{1 + x \exp[\beta E_D]} \ .$$

The assumption $k_B T \ll E_g$ validates the neglect of the first term on the right-hand side. This results in a quadratic equation, whose only positive solution is

$$x = \{-1 + \sqrt{1 + 4e^{\beta E_D} x_D / n_{0e}}\}/(2\exp[\beta E_D]) \ .$$

For $k_{\rm B}T \ll E_{\rm D}$ it becomes $x \approx \exp[-\beta E_{\rm D}/2]\sqrt{x_{\rm D}/n_{0e}}$, and thus

$$\mu \approx E_c - E_{\rm D}/2 + k_{\rm B}T \ln[x_{\rm D}/n_{0e}]/2 \ ,$$

in accordance with the solution in the text. On the other hand, for $E_{\rm D}/\ln[n_{0e}/x_{\rm D}] \ll k_{\rm B}T \ll E_g$, one finds $x \approx x_{\rm D}/n_{0e}$, and then $\mu \approx E_c + k_{\rm B}T \ln[x_{\rm D}/n_{0e}]$. In this limit $n_e \approx x_{\rm D}$, i.e., all electrons are transferred from the donors to the conduction band. At sufficiently low temperatures $x_{\rm D} > n_{0e}$, since $n_{0e} \propto T^{3/2}$. Thus, as the temperature rises, the chemical potential rises as well, from $\mu(T = 0) = E_{\rm F} = E_c - E_{\rm D}/2$ towards E_c. As long as $x_{\rm D} > n_{0e}$, the chemical potential may even surpass E_c. At higher temperatures $x_{\rm D} < n_{0e}$, and the chemical potential decreases as the temperature is increased. In any case, since $E_{\rm D} \ll E_g$, the chemical potential is always around the threshold of the conduction band, and the neglect of the holes' concentration (the first term on the right hand-side of the original equation) is justified.

b. For $x_{\rm D} = 0$, Eq. (6.136) yields

$$n_{0e}\exp[-\beta E_g]/x + x_{\rm A}/(1 + x\exp[\beta E_{\rm A}]) = n_{0h}x \ , \quad x = \exp[\beta(E_v - \mu)] \ .$$

Discarding the first term on the left produces a quadratic equation with the solution

$$x = \{-1 + \sqrt{1 + 4\exp[\beta E_{\rm A}]x_{\rm A}/n_{0h}}\}/(2\exp[\beta E_{\rm A}]) \ .$$

It follows that

$$\mu \approx E_v + E_{\rm A}/2 - k_{\rm B}T \ln[x_{\rm A}/n_{0h}]/2 \ , \quad \text{for} \ \ k_{\rm B}T \ll E_{\rm A}$$
$$\mu \approx E_v - k_{\rm B}T \ln[x_{\rm A}/n_{0h}], \quad \text{for} \ \ E_{\rm A}/\ln[n_{0h}/x_{\rm A}] \ll k_{\rm B}T \ll E_g \ .$$

The chemical potential is always around the top of the valence band.

Answer 6.31.

a. The chemical potential is found from Eq. (6.130),

$$E_c - \mu = k_{\rm B}T \ln[n_{0e}/n_e] = 0.026 \ (\text{eV}) \ln[2.8 \times 10^{19}/10^{17}] = 0.146 \ \text{eV} \ .$$

As by Eq. (6.133)

$$n_i^2 = n_e n_h = n_{0e}n_{0h}e^{-\beta E_g}$$
$$= 2.8 \times 10^{19} \times 1.04 \times 10^{19}e^{-1.14/0.026} \ \text{cm}^{-6} = 2.6 \times 10^{19} \ \text{cm}^{-6} \ ,$$

one finds $n_h = n_i^2/n_e \approx 260 \ \text{cm}^{-3}$. The same result is obtained from a direct calculation, $n_h = n_{0h}\exp[-\beta(\mu - E_v)]$.

b. In this case

$$\mu - E_v = k_{\rm B}T \ln[n_{0h}/n_h] = 0.026 \ (\text{eV}) \ln[1.04 \times 10^{19}/10^{14}] = 0.31 \ \text{eV} \ ,$$

and therefore $n_e = n_i^2/n_h \approx 2.6 \times 10^5 \ \text{cm}^{-3}$.

Answer 6.32.

a. As the motion is solely along the $\mathbf{b}$–axis, it suffices to consider the vectors' components along this direction. From Eq. (6.148), $v_b = \partial E_n/(\hbar \partial k_b)$ and hence

$$r_b(t) = r_b(0) + (1/\hbar) \int_0^t [\partial E_n(k)/\partial k_b] dt \ .$$

According to Eq. (6.151) one may write $dt = -\hbar dk_b/(e\mathcal{E})$, and hence

$$r_b(t) = r_b(0) - \frac{1}{e\mathcal{E}} \int_{k_b(0)}^{k_b(t)} \frac{\partial E_n(k)}{\partial k_b} dk_b = r_b(0) - \frac{E_n[k_b(t)] - E_n[k_b(0)]}{e\mathcal{E}} \ .$$

(The components of $\mathbf{k}$ other than k_b remain unchanged.) The energy is periodic in the reciprocal-lattice vectors, and thus returns to the same value once $k_b(t) = k_b(0) + b$. The period is found by integrating,

$$\mathbf{T} = \left| \int_0^{\mathbf{T}} dt \right| = \int_{k_b(0)}^{k_b(\mathbf{T})} \hbar dk_b/(e\mathcal{E}) = \frac{\hbar b}{e\mathcal{E}} \ .$$

b. In a cubic lattice $b = 2\pi/a$, and thus $\mathbf{T} = h/(e E a)$. Using $\mathcal{E} = 10^4$ V/cm and $a = 1 \overset{\circ}{A}$ gives $\mathbf{T} = 10^{-10}$ sec. This implies that $\mathbf{T} \gg \tau \sim 10^{-14}$ sec, and the electron is scattered before it reaches the boundary of the Brillouin zone. In order to detect Bloch oscillations, one has to increase considerably the lattice constant , e.g., by utilizing superlattices, increase the magnitude of the electric field, and extend the relaxation time, e.g., by using high quality crystals at very low temperatures.

c. The lattice momentum of each electron, in the presence of a constant electric field, is $k(t) = k(0) - e\mathcal{E}t$; the induced dipole moment is

$$P = -e \sum_{k(0)=-k_{\mathrm{F}}}^{k_{\mathrm{F}}} x[k(0), t] = -e(L/\pi) \int_{-k_{\mathrm{F}}}^{k_{\mathrm{F}}} dk(0) x[k(0), t] \ ,$$

where $x[k(0), t] = -(E[k(t)] - E[k(0)])/(e\mathcal{E})$ is the location at time t of an electron with an initial momentum $k(0)$, and the sum runs over all electrons in the conduction band (two electrons in each state). It follows that

$$P = \frac{\Gamma L}{\pi a \mathcal{E}} [\sin(k_{\mathrm{F}} a - e\mathcal{E} a t/\hbar) + \sin(k_{\mathrm{F}} a + e\mathcal{E} a t/\hbar) - 2\sin(k_{\mathrm{F}} a)]$$

$$= -\frac{4\Gamma L}{\pi a \mathcal{E}} \sin(k_{\mathrm{F}} a) \sin^2[e\mathcal{E} a t/(2\hbar)] \ .$$

A constant electric field induces an oscillating dipole moment! this moment vanishes when the band is fully occupied, i.e., $k_{\mathrm{F}} a = \pi$.

Answer 6.33.

One begins from Eq. (6.146), treating $(\hbar^2/m)(-i\nabla + \mathbf{k}) \cdot \boldsymbol{\delta}$ as a perturbation. Then Eq. (C.3) yields a generalization of Eq. (6.147): up to second order in the perturbation,

$$E_n(\mathbf{k} + \boldsymbol{\delta}) = E_n(\mathbf{k}) + \boldsymbol{\delta} \cdot \frac{\hbar^2}{m} \langle u_{n\mathbf{k}}| -i\nabla + \mathbf{k}|u_{n\mathbf{k}}\rangle$$

$$+ \frac{\hbar^2 \delta^2}{2m} + \sum_{n \neq n'} \frac{|\langle u_{n\mathbf{k}}|\frac{\hbar^2}{2m}(-i\nabla + \mathbf{k}) \cdot \boldsymbol{\delta}|u_{n'\mathbf{k}}\rangle|^2}{E_n(\mathbf{k}) - E_{n'}(\mathbf{k})} \ . \tag{6.179}$$

Next, one notes that a continuation of the expansion of Eq. (6.155) gives

$$E_n(\mathbf{k} + \boldsymbol{\delta}) = E_n(\mathbf{k}) + \boldsymbol{\delta} \cdot \boldsymbol{\nabla}_\mathbf{k} E_n(\mathbf{k}) + \frac{1}{2} \sum_{\alpha,\beta} \frac{\partial^2 E_n(\mathbf{k})}{\partial k_\alpha \partial k_\beta} \delta_\alpha \delta_\beta + \dots \ . \tag{6.180}$$

Comparing Eqs. (6.179) and (6.180) and using Eq. (6.155) yields

$$\left(\frac{1}{m^*}\right)_{\alpha\beta} \equiv \frac{1}{\hbar^2} \frac{\partial^2 E_n(\mathbf{k})}{\partial k_\alpha \partial k_\beta}$$

$$= \frac{1}{m} \delta_{\alpha\beta} + \sum_{n \neq n'} \frac{\langle u_{n\mathbf{k}} | - i\boldsymbol{\nabla}_\alpha | u_{n'\mathbf{k}} \rangle \langle u_{n'\mathbf{k}} | - i\boldsymbol{\nabla}_\beta | u_{n\mathbf{k}} \rangle + \{\alpha \Leftrightarrow \beta\}}{E_n(\mathbf{k}) - E_{n'}(\mathbf{k})} \ ,$$

where the orthogonality of the wave functions is exploited in the constant terms with $\boldsymbol{\delta} \cdot \mathbf{k}$, and where $\{\alpha \Leftrightarrow \beta\}$ indicates the previous term in the numerator, with interchanged cartesian components.

Answer 6.34.

The solution of problem 6.19 gives that for the wave vector $\mathbf{q} = \mathbf{q}_0 + \boldsymbol{\delta}$, where $\mathbf{q}_0$ is on the boundary between two Brillouin zones [i.e., $2\mathbf{q}_0 \cdot \mathbf{G} = \mathbf{G}^2$], the energies are

$$E_{\mathbf{q}\pm}^{(1)} \approx E_{\mathbf{q}_0\pm}^{(1)}(\mathbf{q}_0) \pm |\widetilde{U}(\mathbf{G})| + \frac{\hbar^2}{2m}\left(\delta^2 + \boldsymbol{\delta} \cdot (2\mathbf{q}_0 - \mathbf{G}) \pm \hbar^2(\boldsymbol{\delta} \cdot \mathbf{G})^2/[4m|\widetilde{U}(\mathbf{G})|]\right) \ .$$

This expression is quadratic in the momentum only at the points $\mathbf{q}_0 = \mathbf{G}/2$. Differentiating it there, one obtains

$$\left(\frac{1}{m^*}\right)_{\alpha\beta} \equiv \frac{1}{\hbar^2} \frac{\partial^2 E_\pm(\mathbf{q})}{\partial q_\alpha \partial q_\beta} = \frac{1}{m}\left(\delta_{\alpha\beta} \pm \hbar^2 G_\alpha G_\beta (1 + \delta_{\alpha\beta})/(8m|\widetilde{U}(\mathbf{G})|)\right) \ ,$$

where the two signs refer to the two bands.

Answer 6.35.

Using $dt = dk_x/\dot{k}_x$ and

$$\dot{k}_x = \frac{e\mathrm{B}}{\hbar^2 c} \frac{\partial E}{\partial k_y} \ ,$$

yields

$$\mathbf{T} = \int_0^{\mathbf{T}} dt = \frac{\hbar^2 c}{e\mathrm{B}} \oint \frac{dk_x}{|\partial E/\partial k_y|} = \frac{\hbar^2 c}{e\mathrm{B}} \oint \frac{\partial k_y}{\partial E} dk_x = \frac{\hbar^2 c}{e\mathrm{B}} \frac{d}{dE} \oint k_y dk_x \ ,$$

and the integral is over the area encircled by the orbit.

Answer 6.36.

The energy is

$$E(\mathbf{k}) = E_0 + \hbar^2 \left[\frac{k_x^2}{2m_x^*} + \frac{k_y^2}{2m_y^*} + \frac{k_z^2}{2m_z^*}\right] \ . \tag{6.181}$$

The momentum component parallel to the field does not vary with time, $\mathbf{k}\cdot\hat{\mathbf{n}} = k_0$. Equation (6.181) describes a cone whose apex is at the origin; the area of its base is A and its volume is $\Omega = Ak_0/3$. Transforming to the coordinates $\hbar k_\alpha/\sqrt{m_\alpha^*} = \overline{k}_\alpha$, the equal-energy surface becomes the paraboloid

$$E(\mathbf{k}) = E_0 + \frac{1}{2}[\overline{k}_x^2 + \overline{k}_y^2 + \overline{k}_z^2] \,,$$

and the cone equation is

$$\overline{\mathbf{k}}\cdot\overline{\mathbf{n}} = k_0 \,, \quad \overline{n}_\alpha = n_\alpha\sqrt{m_\alpha^*}/\hbar \,.$$

In these coordinates $\Omega = \Omega_c\sqrt{m_x^* m_y^* m_z^*}/\hbar^3$, where Ω_c is the volume of the cone in the spherical coordinates. The base of the cone is the cut of a plane with a sphere of radius $R = \sqrt{2(E - E_0)}$, and its height is

$$h = k_0/|\overline{\mathbf{n}}| = k_0\hbar/\sqrt{m_x^*\hat{n}_x^2 + m_y^*\hat{n}_y^2 + m_z^*\hat{n}_z^2} \,.$$

It follows that the radius of the cone's base is $\sqrt{R^2 - h^2}$, and its volume is $\Omega_c = \pi(R^2 - h^2)h/3$. Thus,

$$A = 3\Omega/k_0 = \pi\sqrt{m_x^* m_y^* m_z^*}[2(E - E_0) - h^2](h/k_0)/\hbar^3 \,,$$

and hence

$$\frac{\partial A}{\partial E} = 2\pi\frac{\sqrt{m_x^* m_y^* m_z^*}}{\hbar^2\sqrt{m_x^*\hat{n}_x^2 + m_y^*\hat{n}_y^2 + m_z^*\hat{n}_z^2}} \,.$$

Equation (6.163) yields the period, as given in Eq. (6.164). For instance, for a magnetic field along the $\hat{\mathbf{z}}$–axis it yields $\omega_c = e\mathrm{B}/[c\sqrt{m_x^* m_y^*}]$. For an isotropic effective mass one obtains for the cyclotron frequency an expression similar to that of a free particle; in the general case the cyclotron frequency depends on the weighed effective mass.

The result above may be obtained by an alternative method. For a field along the $\hat{\mathbf{z}}$–axis, the equations of motion are

$$\dot{k}_x = (e\mathrm{B}/[m_y^*c])k_y \,, \quad \dot{k}_y = -e\mathrm{B}/([m_x^*c])k_x \,.$$

It follows that $\ddot{k}_x = -[(e\mathrm{B})^2/(m_x^* m_y^* c^2)]k_x$; this equation describes a periodic motion with the cyclotron frequency $\omega_c = e\mathrm{B}/[c\sqrt{m_x^* m_y^*}]$.

Answer 6.37.
The vector product of Eq. (6.162), $\hbar\dot{\mathbf{k}} = -(e/c)\dot{\mathbf{r}}\times\mathbf{B}$, with $\mathbf{B}$ yields

$$\hbar\mathbf{B}\times\dot{\mathbf{k}} = -\frac{e}{c}\mathbf{B}\times[\dot{\mathbf{r}}\times\mathbf{B}] = -\frac{e}{c}[\dot{\mathbf{r}}\mathbf{B}^2 - \mathbf{B}(\mathbf{B}\cdot\dot{\mathbf{r}})] \,.$$

Hence $\dot{\mathbf{r}} = -[\hbar c/(e\mathrm{B})]\hat{\mathbf{B}}\times\dot{\mathbf{k}} + v_B\hat{\mathbf{B}}$, and upon carrying out the temporal integration one obtains the required result. It shows that the vector

$$\mathbf{r}_\perp(t) - \mathbf{r}_\perp(0) = -\frac{\hbar c}{e\mathrm{B}}\hat{\mathbf{B}}\times[\mathbf{k}(t) - \mathbf{k}(0)]$$

lies in a plane perpendicular to $\mathbf{B}$. It is also normal to the vector $[\mathbf{k}(t) - \mathbf{k}(0)]$, and its length is

$$|\mathbf{r}_\perp(t) - \mathbf{r}_\perp(0)| = \ell_B^2 |\mathbf{k}(t) - \mathbf{k}(0)| \, ,$$

Thus, the orbit $[\mathbf{r}_\perp(t) - \mathbf{r}_\perp(0)]$ can be derived from the orbit $[\mathbf{k}(t) - \mathbf{k}(0)]$ by a $90°-$ rotation, and a multiplication by ℓ_B^2. This conclusion coincides with the result of problem 6.3 (check!).

Answer 6.38.

a. One begins by extending Eq. (6.8) to account for particles of charge q_j (i.e., $q_e = -e$ for electrons and $q_h = e$ for holes), a positive effective mass m_j^*, a relaxation time τ_j, a density n_j, and an average velocity $\mathbf{v}_j$:

$$m_j^*(\dot{\mathbf{v}}_j + \mathbf{v}_j/\tau_j) = q_j(\boldsymbol{\mathcal{E}} + \mathbf{v}_j \times \mathbf{B}/c) \, .$$

The magnetic field is along the $\hat{\mathbf{z}}-$direction, while the electric field and the particles are assumed to be in the XY plane. In a stationary state

$$m_j^* \frac{v_{jx}}{\tau_j} = q_j\left(\mathcal{E}_x + v_{jy}\frac{\mathrm{B}}{c}\right) \, , \quad m_j^* \frac{v_{jy}}{\tau_j} = q_j\left(\mathcal{E}_y - v_{jx}\frac{\mathrm{B}}{c}\right) \, .$$

Solving the two equations (discarding quadratic terms in the magnetic field) yields

$$v_{jy} \approx \frac{q_j\tau_j}{m_j^*}\left(\mathcal{E}_y - \frac{q_j\tau_j\mathrm{B}}{m_j^*c}\mathcal{E}_x\right) \, , \quad v_{jx} \approx \frac{q_j\tau_j}{m_j^*}\left(\mathcal{E}_x + \frac{q_j\tau_j\mathrm{B}}{m_j^*c}\mathcal{E}_y\right) \, .$$

The current density is given by

$$\mathbf{j} = \sum_j n_j q_j \mathbf{v}_j \, .$$

The component j_y vanishes in the geometry of Fig. 6.1, leading to

$$\sum_j n_j q_j^2 \frac{\tau_j}{m_j^*}\mathcal{E}_y - \sum_j n_j q_j^3 \left(\frac{\tau_j}{m_j^*}\right)^2 \frac{\mathrm{B}\mathcal{E}_x}{c} = 0 \, .$$

Similarly, the current along the $\hat{\mathbf{x}}-$direction is

$$j_x = \sum_j n_j q_j^2 \frac{\tau_j}{m_j^*}\mathcal{E}_x + \sum_j n_j q_j^3 \left(\frac{\tau_j}{m_j^*}\right)^2 \frac{\mathrm{B}\mathcal{E}_y}{c} \, .$$

As $\mathcal{E}_y \propto \mathcal{E}_x\mathrm{B}$, the second term in this expression is quadratic in the magnetic field and thus is ignored; then

$$R_H = \frac{\mathcal{E}_y}{j_x\mathrm{B}} \approx \frac{\sum_j n_j q_j(q_j\tau_j/m_j^*)^2}{c[\sum_j n_j(q_j^2\tau_j/m_j^*)]^2} = \frac{n_h\mu_h^2 - n_e\mu_e^2}{ce(n_e\mu_e + n_h\mu_h)^2} \, ,$$

where $\mu_j = e\tau_j/m_j^*$ are the mobilities of the two types of charge carriers. This result reproduces Eq. (6.11), $R_H = -1/(n_e ec)$, when the density of the holes vanishes. In contrast, when the charge carriers are holes, then $R_H = 1/(n_h ec)$, as can be expected for positively-charged particles. In an intrinsic semiconductor $n_e = n_h$, and the sign of the result depends on the sign of $\mu_h - \mu_e$. In the more general case, the sign, as well as the numerical factor, can be arbitrary. This removes one of the caveats of the Drude theory.

b. Introducing the notations $b = \mu_e/\mu_h$ and $x = n_e/n_i = \sqrt{n_e/n_h}$, part (a) yields

$$R_H = \frac{x}{cen_i}\frac{1 - b^2x^2}{(1 + bx^2)^2} \; .$$

This function is plotted in Fig. 6.47 (in units of $1/(cen_i)$). As seen, $R_H = 0$ when $n_e/n_i = x = 1/b = 1/3$, and hence $n_e = n_i/3 = 5 \times 10^9$ cm^{-3}. The extremal points of R_H are derived from $dR_H/dx = 0$, i.e.,

$$b^3x^4 - 3b(1 + b)x^2 + 1 = 0 \;\; \Rightarrow \;\; x^2 = \frac{3b(1 + b) \pm \sqrt{9b^2(1 + b)^2 - 4b^3}}{2b^3} \; .$$

The two solutions are $n_e/n_i = x = 1.14, \; 0.17$, and $n_h/n_i = 1/x = 0.88, \; 2.5$, which correspond to the maximal and minimal R_H.

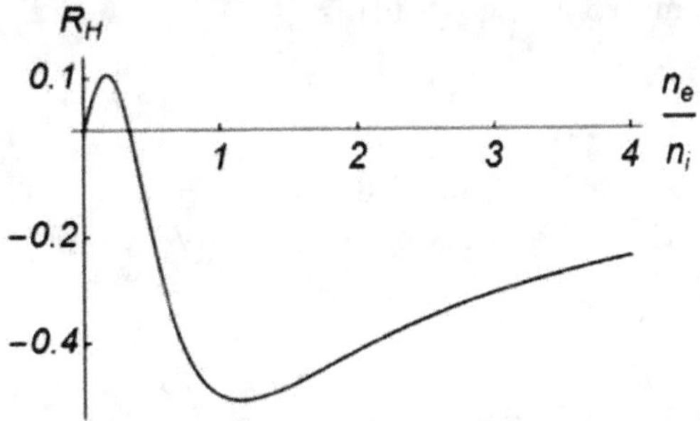

Fig. 6.47

c. When higher orders in the magnetic field are taken into account Eq. (6.13) pertains separately for each type of particles. The current is the sum over the currents carried by each type of particles, and therefore Eq. (6.14) is generalized to be

$$\sigma_{xy} = -\sigma_{yx} = \sum_j \frac{\sigma_j \tau_j \omega_{cj}}{1 + (\tau_j \omega_{cj})^2} \; , \quad \sigma_{xx} = \sigma_{yy} = \sum_j \frac{\sigma_j}{1 + (\tau_j \omega_{cj})^2} \; ,$$

where $\sigma_j = n_j q_j^2 \tau_j/m_j^*$ and $\omega_{cj} = q_j B/(m_j^* c)$. The resistivity tensor is the inverse of the conductivity one,

$$\rho_{yx} = -\rho_{xy} = \sigma_{xy}/[\sigma_{xx}^2 + \sigma_{xy}^2] \; , \quad \rho_{xx} = \rho_{yy} = \sigma_{xx}/[\sigma_{xx}^2 + \sigma_{xy}^2] \; .$$

Equation (6.12) is reproduced upon assuming that there is only one type of particles, and then the diagonal element, $\rho_{xx} = \rho$ is independent of the magnetic field. Otherwise, the diagonal elements do depend on the magnetic field. In particular, for very strong fields, with $q_e = -e$ for the electrons and $q_e = e$ for the holes, one finds

$$\sigma_{xx} = \sigma_{yy} = (n_e/\mu_e + n_h/\mu_h)ec^2/B^2 \ , \quad \sigma_{xy} = -\sigma_{yx} = (n_e - n_h)ec/B \ .$$

In intrinsic semiconductors, for which $n_e = n_h$, the diagonal resistivity is quadratic in the magnetic field, $\rho_{xx} \propto B^2$, producing very large magneto-resistance. As opposed, for $n_e \neq n_h$ the magnetic field is cancelled out, and ρ_{xx} approaches a finite value.

Answer 6.39.

The cylinder cuts the Fermi sphere at a circle of radius K_ℓ, whose area is $A_\ell = \pi K_\ell^2$. Since $K_\ell^2 = 2m^* E_\ell/\hbar^2$, one finds

$$A_\ell = \frac{2\pi m^* E_\ell}{\hbar^2} = \frac{2\pi m^* \omega_c}{\hbar}(\ell + \frac{1}{2}) = \frac{2\pi eB}{\hbar c}(\ell + \frac{1}{2}) \ .$$

Answer 6.40.

The inverse of the matrix in Eq. (6.13) is

$$\begin{bmatrix} \rho_{xx} & \rho_{xy} \\ \rho_{yx} & \rho_{yy} \end{bmatrix} = \frac{1}{\sigma_{xx}\sigma_{yy} - \sigma_{yx}\sigma_{xy}} \begin{bmatrix} \sigma_{yy} & -\sigma_{xy} \\ -\sigma_{yx} & \sigma_{xx} \end{bmatrix} = \begin{bmatrix} 0 & 1/\sigma_{yx} \\ 1/\sigma_{xy} & 0 \end{bmatrix} \ .$$

The last step stems from the substitution $\sigma_{xx} = \sigma_{yy} = 0$. Comparing the left hand-side with the right one yields $\rho_{xy} = \rho_{yx} = 1/\sigma_{yx}$ and $\rho_{xx} = \rho_{yy} = 0$.

6.14   Problems for self-evaluation

s.6.1.

Exploit the Drude model to compute the average energy that the electron loses within a unit time, and use the result to prove the Joule law for the energy loss per unit volume per unit time, $W = \sigma \mathcal{E}^2$.

s.6.2.

Consider a two-dimensional hexagonal lattice, as the one displayed in Fig. 6.48, with a lattice constant $a = 3\text{Å}$. Each atom in the lattice donates a single electron of mass m, that is free to move in the plane. The total area of the sample is A. The measured conductivity of this sample is $\sigma = 10^{-2}/\Omega$.

a. How many free electrons are there in the sample?
b. What is the density of states?
c. Determine the Fermi energy.
d. Find the average energy of an electron in the ground state of this system.
e. Find the specific heat at low temperatures and at high temperatures.

f. *Determine the mean free-path of the electron.*

g. *How many Brillouin zones are occupied by the electrons? Suppose that each atom releases ℓ electrons. At which ℓ value will the second Brillouin zone begin to be occupied?*

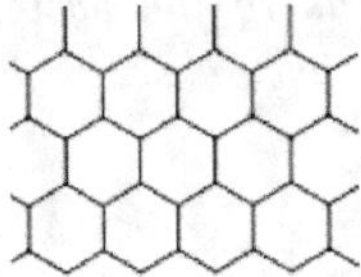

Fig. 6.48

s.6.3.

A three-dimensional gas of free electrons, contained in a volume V is subjected to a constant magnetic field, $\mathbf{B}$. The energies of electrons whose moment is parallel (antiparallel) to the field are $E - \mu_{\rm B}B$ $(E + \mu_{\rm B}B)$, where $\mu_{\rm B}$ is the magnetic moment resulting from the spin of the electron.

a. *Exploit the Sommerfeld theory to write down expressions for the number of electrons whose magnetic moment is parallel or is antiparallel to the field.*

b. *Prove that at low temperatures, and for a weak magnetic field, the total magnetic moment of the system is $M = 3N\mu_{\rm B}^2 B/(2E_{\rm F})$. [Hint: use the results of problem 6.9].*

s.6.4.

Suppose that the potential steps of the Kronig-Penney model are replaced by potential wells. This is accomplished by assuming that U_0 of Eq. (6.65) is negative.

a. *Describe the wave functions for energies in the range $U_0 < E < 0$. What is the relation between the energy and the wave number in this case?*

b. *Find the width of the lowest band in the delta-function limit, $U_0 \to -\infty$, $(a-b) \to 0$, $U_0(a-b) \to u_0 < 0$, and compare it with the calculation in the tight-binding limit, problem 6.22.*

s.6.5.

A material is described by a one-dimensional lattice with three atoms in each primitive unit cell, whose length is a. The potential that the atoms exert on the electrons is

$$U(x) = U_0 a \sum_n [2\delta(x - na) - \delta(x - (n + 1/4)a) - \delta(x - (n + 3/4)a)] \ .$$

It is assumed that $U_0 \ll \hbar^2 \pi^2/(2ma^2)$, where m is the free electron mass.

a. *Find the value of the energy gap at the end of the first Brillouin zone, close to the energy $\hbar^2 \pi^2/(2ma^2)$.*

b. *Determine the effective mass of the electrons at the bottom of the band located slightly above $\hbar^2\pi^2/(2ma^2)$. What is the answer for the holes at the top of the band located slightly below this energy?*

c. *Find the value of the energy gap at the center of the first Brillouin zone, at the vicinity of $2\hbar^2\pi^2/(ma^2)$.*

d. *Each atom donates ℓ electrons to the system. For which values of ℓ is the material an insulator? and for which ones is it conducting?*

s.6.6.

A free electron of mass m is moving in a square lattice of lattice constant a.

a. *Calculate the energy of the electron at the corners and at the faces' centers of the first Brillouin zone.*

b. *Assuming that each atom donates ℓ electrons, calculate the radius of the Fermi circle. How many Brillouin zones are occupied when $\ell = 1$? when $\ell = 2$? in these two cases, are the charge carriers electrons or holes?*

c. *The electron is subjected to a weak two-dimensional potential of the form*

$$U(x,y) = -4U_0 \cos \frac{2\pi x}{a} \cos \frac{2\pi y}{a} \ ,$$

as shown in Fig. 6.49. This potential can be artificially created in the lab, by combining oscillating electric fields produced by lasers. The experimentalists can capture atoms in such a potential, and thus explore the behavior of electrons in a periodic potential. Estimate the energy gaps created at the corners and at the mid points of the edges of the first Brillouin zone.

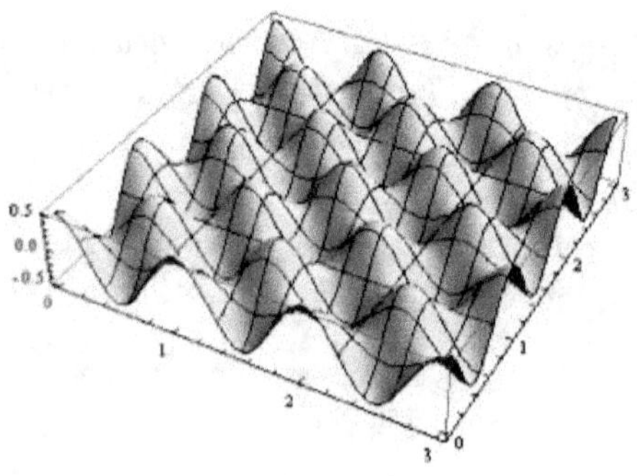

Fig. 6.49

s.6.7.

*Consider an electron gas in a single-atom lattice of a **BCC** structure.*

a. *Exploit the tight-binding approximation to obtain the energies in a single band, as functions of the lattice momentum. The overlap of the wave functions between* *nearest neighbors (nn) is parametrized by the energy γ_1, and between next-nearest neighbors (nnn) by the energy γ_2.*

b. *Find the acceleration along the $\hat{\mathbf{y}}$−direction caused by an external electric field along the $\hat{\mathbf{x}}$−direction.*

c. *Calculate the effective mass tensor at the bottom of this band.*

d. *Assuming that the band is the conduction band, and that all bands below are completely full, derive the electric conductivity tensor in the Drude approximation. The (low) electrons' density in the band is n and their relaxation time is τ.*

s.6.8.

The energy of electrons in a three-dimensional crystal, at the bottom of the conduction band, is

$$E = E_0 + \frac{\hbar^2}{2m^*}\mathbf{k}^2 + \frac{\hbar\alpha}{4}\mathbf{k}^4 \ ,$$

where the effective mss m^* and the coefficient α are positive. The material is placed in an electric field of frequency ω, $\mathcal{E} = \mathcal{E}_0 \cos(\omega t)$. At time $t = 0$ there is an electron at the bottom of the band, with $\mathbf{k} = 0$. Exploit the Semiclassical picture for the electron motion to find

a. *the location of the electron at time t;*

b. *the frequencies and the amplitudes of the electron's motion;*

c. *the amplitude of the field $\mathcal{E}_0$ required to make the effect of the quartic term in the energy on the electron motion at frequency ω comparable to that of the quadratic one.*

s.6.9.

The energy gap of crystalline silicon (whose valence is 4) is 1.14 eV, and the effective masses of the electrons and of the holes are 0.2 m and 0.3 m, respectively (m is the free-electron mass). What is the concentration x_D of arsenic atoms, which are five-valence donors, that has to be added to the crystal such that at room temperature the extrinsic conductivity exceeds the intrinsic one by a factor of 10^4? Assume that the binding energy of the electron, E_D, is small, $E_D \ll k_{\mathrm{B}}T$.

s.6.10.

A material is described by a square lattice of lattice constant a. The dispersion relation in the first energy band is $E(\mathbf{k}) = -\gamma \cos[(|k_x|+|k_y|)a/2]$, with $\gamma > 0$. This system is placed in a magnetic field $\mathbf{B} = B\hat{\mathbf{z}}$. The Fermi energy of the electrons is $E_{\mathrm{F}} = a\gamma$.

a. *Find the Fermi curve.*

b. *Find the velocity of an electron at the Fermi energy within the semiclassical approximation.*

c. *Find the force exerted on the electron by a magnetic field along an arbitrary direction.*

d. *What is the time needed for the electron to complete a full circle around the Fermi curve?*

s.6.11.

a. *Find the effective-mass tensor of an electron in the two energy bands of graphene. Exploit the tight-binding approximation, confining yourselves to overlap couplings between nearest neighbors. In particular, determine the effective mass at the center of the Brillouin zone (the Γ point) and at the center of the hexagonal face (the M point).*

b. *What is the density of states in graphene around the Γ and the K points of the Brillouin zone? (the latter is at the corner of the hexagonal face).*

s.6.12.

Figure 6.50 displays the density of states in a material, in which the density of electrons is 5×10^{28} m^{-3}.

a. *Determine the Fermi energy. Is this material a metal, an insulator, or a semiconductor? Explain your choice.*

b. *Determine the specific heat of this material at low temperatures.*

c. *How are the answers modified when the electrons' density is 7×10^{28} m^{-3}?*

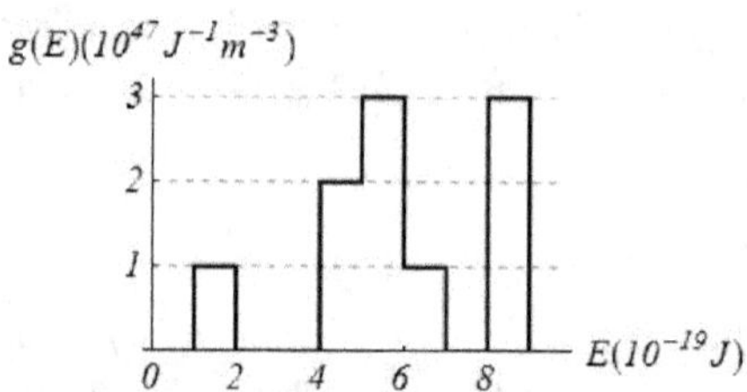

Fig. 6.50

s.6.13.

The wave functions of electrons in the d-shell decay strongly with the distance from their parent atom; consequently, the overlap between wave functions localized on neighboring atoms is small, and the energy band (in the tight-binding approximation) is narrow. Assume that the center of the band is on the atomic energy E_d, and that the density of states can be taken as a rectangular function, of width W.

a. *Determine the height of the rectangular function.*

b. *Find the total energy, at zero temperature, of the electrons in the band, when each atom contains N d-electrons (ignore the other bands).*

c. *Assume that the energy found in part (b) is the binding energy of the lattice. Find the conditions for this energy to be maximal.*

s.6.14.

The model of Su, Schrieffer, and Heeger is a generalization of problem 6.23: electrons in a one-dimensional lattice. The atoms on the lattice sites are identical and it is assumed that the energies on the atoms vanish. The coupling energies of the tight-binding model between nearest neighbors alternate,

$$\gamma_{2n-1,2n} = \gamma_{2n,2n-1} = \gamma_- = \gamma_0(1-\Delta) \ ,$$

$$\gamma_{2n+1,2n} = \gamma_{2n,2n+1} = \gamma_+ = \gamma_0(1+\Delta) \ .$$

Such a model can describe, e.g., the lattice of polyyne, Fig. 1.7(b), which comprises two types of bonds. Su, Schrieffer, and Heeger designed their model to describe polyethylene, $[C_2H_2]_n$, where each carbon on the chain is coupled also to a hydrogen, and the bonds between the carbons alternate between single and double bonds; see the discussion concerning polyyne, Sec. 4.3. The shift Δ can result also from elastic longitudinal deformation of the lattice, of the type $u_m = u_0 \exp[i\pi m]$. When u_0 is positive, the atoms at $m = 2n$ move slightly to the right, and those with $m = 2n-1$ move to the left, thus forming a static phonon at the edge of the Brillouin zone. The basis for this model is the dependence of γ on the interatomic distance. The pairs of atoms that move closer together are called **dimers**.

a. *Ignore the overlap of the atomic wave functions, and show that there is an energy gap when Δ is finite.*

b. *Assume that the gain in elastic energy due to the deformation is $\Delta E_d = Nx\Delta^2$, where the lattice contains $2N$ atoms. Show that when each atom donates a single electron the material is an insulator, and the energy is minimal for both possibilities $\Delta = \pm\Delta_0$. It follows that the state with $\Delta = 0$, where the material is a metal, is unstable: the material is "dimerized" into the insulating phase. This instability is named after Peierls, and is known as the* **Peierls instability**.

c. *What are the two states of the infinite system when $|\Delta| = 1$? What happens in a finite system comprising $2N$ atoms?*

6.15 Answers for the self-evaluation problems

Answer s.6.1.

In-between collisions, the equation of motion of the electron is $\mathbf{p}(t) = -e\mathcal{E}t$, and consequently its energy at time t is $(e\mathcal{E}t)^2/(2m)$. The probability to reach the time t without undergoing an additional collision is $\exp[-t/\tau]$, and therefore the probability to collide and lose all the energy between t and $t + dt$ is $\exp[-t/\tau]dt/\tau$. Hence, the average energy the electron loses at each collision is

$$\int_0^\infty [(e\mathcal{E}t)^2/(2m)]e^{-t/\tau} dt/\tau = (e\mathcal{E}\tau)^2/m \ .$$

The energy loss per unit time, per unit volume, is found by multiplying this result by the electron density n and dividing it by the mean free-time τ; then $W = (ne^2\tau/m)\mathcal{E}^2 = \sigma\mathcal{E}^2$.

Answer s.6.2.

a. The area of each hexagon is $S = 6 \times (\sqrt{3}a^2/4) = 3\sqrt{3}a^2/2$. Each hexagon contributes two electrons (as each atom around it belongs to three hexagons, there are two atoms per each hexagon). Hence, the planar density of the electrons is

$$n = 2/S = 4/(3\sqrt{3}a^2) \ .$$

The total number of electrons in the sample is

$$nA = 2A/S = 4A/(3\sqrt{3}a^2) \ .$$

Another option to obtain this result is to note that the unit cell of the hexagonal lattice is a rhombus whose area is $a_\triangle^2\sqrt{3}/2$, where $a_\triangle = a\sqrt{3}$ is the lattice constant of the triangular lattice. This unit cell contains two atoms, and hence

$$n = 2/(a_\triangle^2\sqrt{3}/2) = 4/(3\sqrt{3}a^2) \ .$$

b. Consider a circle of radius k. Its area is πk^2, and therefore the number of states within it (including the spin degree of freedom) is $2 \times (\pi k^2)/[(2\pi)^2/A] = Ak^2/(2\pi)$. Using $E = \hbar^2 k^2/(2m)$ results in $AmE/(\pi\hbar^2)$ for the number of states. It follows that the number of states per unit area between the energies E and $E + dE$ is

$$g(E)dE = mdE/(\pi\hbar^2) \ .$$

c. The Fermi energy is given by the relation

$$n = \int_0^{E_F} g(E)dE \ ,$$

and therefore

$$n = mE_F/(\pi\hbar^2) \ , \quad \Rightarrow \quad E_F = n\pi\hbar^2/m \ .$$

Inserting this into the result of part (b) gives $g(E) = n/E_F$, in accordance with the solution of problem 6.10.

d. The average energy per particle is

$$\langle E \rangle = A \int_0^{E_F} Eg(E)dE/N = (1/n)(n/E_F)E_F^2/2 = E_F/2 \ ,$$

again in accordance with the solution of problem 6.10.

e. According to the solution of problem 6.9, the specific heat at low temperatures is linear in the temperature (in any dimension). At high temperatures it obeys the Dulong-Petit law, $C_A = 2Nk_B$.

f. From Eq. (6.6), $\tau = m\sigma/(ne^2)$. The mean free-path is

$$\ell = v_F\tau = (\hbar k_F/m)\tau = \hbar k_F\sigma/(ne^2) \ .$$

Using $\hbar k_F = \sqrt{2mE_F}$ and $E_F = n\pi\hbar^2/m$ yields

$$\ell = \sqrt{2\pi\hbar^2/n}(\sigma/e^2) \ .$$

As $n = 2/S = 4/(3\sqrt{3}a^2)$, one finds

$$\ell = \sqrt{3\pi\sqrt{3}/2}(\hbar a\sigma/e^2) \ ,$$

and with the numerical value of the conductivity, this becomes $\ell \approx 354\text{Å}$.

g. Exploiting the expressions above yields $k_F = \sqrt{2\pi n} = \sqrt{8\pi/(3\sqrt{3})}/a$. The closest point to the center of the Brillouin zone is located at the mid point of the reciprocal-lattice basis vector. The length of the reciprocal-lattice vector in the triangular lattice is $b = 4\pi/(a\sqrt{3})$ (see, e.g., Fig. 3.11). Hence, all occupied states are within the first Brillouin zone provided that $k_F < b/2$, which is indeed the case here, since $2k_F/b = \sqrt{2/(\pi\sqrt{3})} < 1$. When each atom contributes ℓ electrons, then $2k_F/b = \sqrt{2\ell/(\pi\sqrt{3})}$; the Fermi circle cuts the boundary of the Brillouin zone for $\ell = \pi\sqrt{3}/2 \approx 2.72$.

Answer s.6.3.

a. The energies of the electrons are shifted by the magnetic field, and thus the Fermi-Dirac distribution [Eq. (6.35)] for the two types of electrons, are $f(E \mp \mu_B B)$, where E is the kinetic energy, Eq. (6.27). As the total number of electrons is N, the density of states of each type is

$$g(E)/2 = 3n\sqrt{E}/(4E_F^{3/2}) \ ,$$

[Eq. (6.32), where $n = N/V$]. The average number of electron of each type is

$$N_\pm = (V/2) \int_0^\infty dE\, g(E) f(E \mp \mu_B B) \ .$$

b. When the magnetic field is weak, one may expand the distributions, $f(E \mp \mu_B B) \approx f(E) \mp f'(E)\mu_B B$. At low temperatures $f'(E) \approx -\delta(E - E_F)$ [problem 6.9]. Hence,

$$N_\pm = \frac{V}{2} \int_0^\infty dE\, g(E)[f(E) \mp f'(E)\mu_B B] \approx \frac{N}{2} \pm \frac{V}{2} g(E_F)\mu_B B \ .$$

The total moment is

$$M = (N_+ - N_-)\mu_B \approx V g(E_F)\mu_B^2 B = \frac{3N}{2E_F}\mu_B^2 B \ ,$$

where in the last step use has been made of Eq. (6.32). To leading order, the magnetic moment is linear in the field; this phenomenon is called the **paramagnetism of the electron gas.**

Answer s.6.4.

a. Using Eq. (6.67), one finds that in the present case $Q^2 > 0$, but

$$K^2 = 2mE/\hbar^2 = -\kappa^2 < 0 \ .$$

It follows that the wave functions oscillate within the potential wells, but decay (or increase) exponentially, $A\exp[-\kappa x] + B\exp[\kappa x]$ in-between the wells. Equation (6.73) still pertains, but one has to insert there $K = i\kappa$; consequently

$$\cos(ka) = \cos[Q(a-b)]\cosh(\kappa b) - \frac{-\kappa^2 + Q^2}{2\kappa Q}\sin[Q(a-b)]\sinh(\kappa b) \ .$$

b. In the limit specified in the problem,

$$\cos(ka) \approx \cosh(\kappa b) - \frac{Q^2}{2\kappa}(a-b)\sinh(\kappa b) \to \cosh(\kappa a) + \frac{mu_0 a}{\hbar^2}\frac{\sinh(\kappa a)}{\kappa a} \ .$$

In the tight-binding limit one may approximate $\cosh(\kappa a) \approx \sinh(\kappa a) \approx \exp[\kappa a]/2$. The band boundaries are at the points where $\cos(ka) = \pm 1$. There

$$\pm 1 \approx \frac{1}{2}e^{\kappa_\pm a}\left(1 + \frac{mu_0}{\hbar^2 \kappa_\pm}\right) \ ,$$

that is,

$$\kappa_\pm \approx \kappa_0(1 \pm 2\exp[-\kappa_\pm a]) \ , \quad \kappa_0 = -mu_0/\hbar^2 \ .$$

The band width is the difference between the corresponding energies, i.e.,

$$\Delta E = \frac{\hbar^2}{2m}(\kappa_+^2 - \kappa_-^2) \approx \frac{\hbar^2 \kappa_0}{m}(\kappa_+ - \kappa_-) \approx \frac{\hbar^2 \kappa_0}{m}4\kappa_0 e^{-\kappa_0 a} = \frac{4\hbar^2 \kappa_0^2}{m}e^{-\kappa_0 a} \ .$$

Inserting here $\kappa_0 = m|u_0|/\hbar^2$ shows that this is precisely the gap obtained in the solution of problem 6.22.

Answer s.6.5.

The Fourier transform of the potential is

$$\widetilde{U}(\mathbf{G}) = \frac{1}{a}\int_{-a}^{a} dx U(x)e^{-iGx}$$

$$= U_0 \int_{-a}^{a} dx\left(\sum_n [2\delta(x - na) - \delta[x - (n+1/4)a] - \delta[x - (n+3/4)a]e^{-iGx}\right) \ .$$

The first two terms in the integrand contribute solely for $n = 0$, while the third one necessitates $n = -1$. It follows that

$$\widetilde{U}(\mathbf{G}) = U_0[2 - e^{-iGa/4} - e^{iGa/4}] = 2U_0[1 - \cos(Ga/4)] \ .$$

a. The required gap separates the first two bands, and is due to the degeneracy $E^{(0)}(G_1/2) = E^{(0)}(-G_1/2)$, where $G_1 = 2\pi/a$. The energy gap is $2|\widetilde{U}(G_1)| = 4U_0$ [see Eq. (6.90)].

b. The energies near the boundary of the first Brillouin zone, by Eq. (6.92), are

$$E_\pm^{(1)}(k) \approx \frac{\hbar^2 G_1^2}{8m} \pm |\widetilde{U}(G_1)| + \frac{\hbar^2}{2m}\left(1 \pm \frac{\hbar^2 G_1^2}{2m|\widetilde{U}(G_1)|}\right)\delta^2 + \dots.$$

The coefficient of $(\hbar\delta)^2/2$ in the last term defines the inverse effective mass, $1/m^*$,

$$m^*_\pm = m/\left(1 \pm \frac{\hbar^2 G_1^2}{8mU_0}\right).$$

The upper sign corresponds to electrons in the conduction band, and the lower one corresponds to holes in the valence band. The effective mass in the valence band is negative because of the assumption that $U_0 < \hbar^2 G_1^2/(4m)$.

c. The required gap is in-between the second and the third bands, owing to the degeneracy $E^{(0)}(G_2/2) = E^{(0)}(-G_2/2)$, with $G_2 = 4\pi/a$. Its value is $2|\widetilde{U}(G_2| = 8U_0$.

d. The electrons' density is $n = N_e/L = 3\ell/a$. The total number of electrons is related to the Fermi wave vector,

$$N_e = 2(2k_{\rm F})/[2\pi/L] = 2k_{\rm F}L/\pi , \quad \Rightarrow \quad k_{\rm F} = \pi n/2 = 3\pi\ell/(2a)$$

When $\ell = 1$ then $k_{\rm F} = 3\pi/(2a)$. This value is at the middle of the second Brillouin zone, between $G_1/2$ and $G_2/2$. Therefore, this band is half-full, and the material is metallic. For $\ell = 2$ $k_{\rm F} = 3\pi/a$, exactly on the border between the third and the fourth Brillouin zones. Hence the material is an insulator. Generally speaking, the material is metallic for odd ℓ's and is an insulator for even ones.

Answer s.6.6.
a. The mid point of the edge of the first Brillouin zone is, e.g., at $(\pi/a, 0)$, and the corner is at $(\pi/a, \pi/a)$. The corresponding energies are $E(\pi/a, 0) = (\hbar\pi/a)^2/(2m)$ and $E(\pi/a, \pi/a) = (\hbar\pi/a)^2/m$.
b. The Fermi wave vector in two dimensions is

$$k_{\rm F} = \sqrt{2\pi n} = \sqrt{2\pi\ell}/a .$$

When $\ell = 1$, then $k_{\rm F}/(\pi/a) = \sqrt{2/\pi} < 1$, and therefore all states within the Fermi sphere are contained in the first Brillouin zone. On the other hand, when $\ell = 2$ then $1 < k_{\rm F}/(\pi/a) = 2/\sqrt{\pi} < \sqrt{2}$, and the circle penetrates into the second Brillouin zone, but does not reach the third one. Figure 6.51 displays the two Brillouin zones and the two Fermi circles. As seen, in the first case the Fermi circle lies away from the edges of the zone, and therefore the charge carriers are nearly-free electrons. In the second case there are empty places in the first Brillouin zone only near its corners, where electrons in the first band have maximal energy. Thus, the charge carriers in this case are holes. The electrons that "overflow" into the second zone are near the mid points of the edges of the zone, that is, they are near the minimal energy of the second band. It follows that they behave like electrons, but with a different effective mass.
c. The minima of the potential are located on a square lattice of lattice constant a. By writing

$$U(x,y) = -U_0\left[e^{2\pi ix/a} + e^{-2\pi ix/a}\right]\left[e^{2\pi iy/a} + e^{-2\pi iy/a}\right],$$

it is seen that $\widetilde{U}(\mathbf{G}) = -U_0$ only for the four reciprocal-lattice vectors $\mathbf{G} = (\pm 2\pi/a, \pm 2\pi/a)$. The only points in the first Brillouin zone that can be connected by one of these vectors are located at the opposite corners of the zone, e.g., $(\pi/a, \pi/a)$ and $(-\pi/a, -\pi/a)$. The degeneracy of pairs of such corners gives rise to an energy gap in-between the lowest two bands. Extending Eq. (6.91) that has been discussed in problem 6.19, leads to

$$E_{\mathbf{q}_0\pm}^{(1)} = E^{(0)}(\pi/a, \pi/a) \pm |\widetilde{U}(\mathbf{G})| = \frac{(\hbar\pi)^2}{ma^2} \pm |U_0| \ .$$

On the other hand, the energy at points along the edges of the Brillouin zone is not degenerate with any of the points displaced by one of the vectors $G(\pm 2\pi/a, \pm 2\pi/a)$, and therefore the energy gaps created there are small, of the order of $|U_0|^2/[E(\mathbf{q}) - E(\mathbf{q}+\mathbf{G})]$.

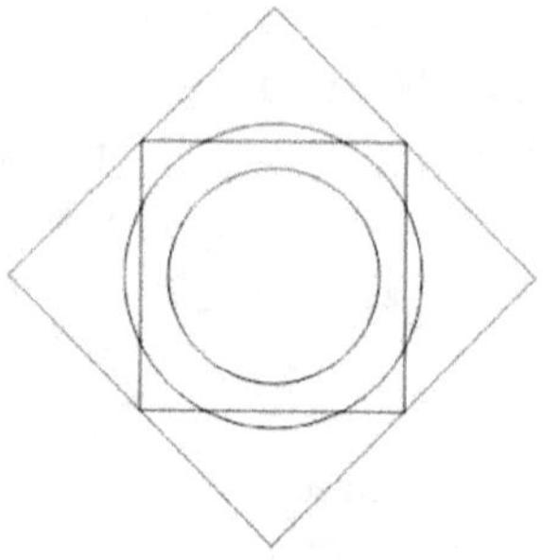

Fig. 6.51

Answer s.6.7.

a. Using a cubic unit cell comprising two atoms, there are eight nearest neighbors, at $(\pm a, \pm a, \pm a)/2$, and six next nearest-neighbors, at $(0, 0, \pm a)$, $(0, \pm a, 0)$, and $(\pm a, 0, 0)$. The band energy is hence

$$\begin{aligned}
E(\mathbf{k}) &= E - \beta - 2\gamma_2[\cos(k_x a) + \cos(k_y a) + \cos(k_z a)] - 2\gamma_1\{\cos[(k_x + k_y + k_z)a/2] \\
&\quad + \cos[(k_x + k_y - k_z)a/2] + \cos[(k_x - k_y + k_z)a/2] + \cos[(-k_x + k_y + k_z)a/2]\} \\
&= E_0 - \beta - 2\gamma_2[\cos(k_x a) + \cos(k_y a) + \cos(k_z a)] \\
&\quad - 8\gamma_1 \cos(k_x a/2)\cos(k_y a/2)\cos(k_z a/2) \ .
\end{aligned}$$

b. Equations (6.154) and (6.155) yield

$$\langle a_y \rangle = -(1/m^*)_{yx} e E_x \ ,$$

$$(1/m^*)_{yx} = 2\gamma_1 a^2 \sin(k_x a/2)\sin(k_y a/2)\cos(k_z a/2)/\hbar^2 \ .$$

c. At the bottom of the band

$$E(\mathbf{k}) \approx E_0 - \beta - \gamma_2(6 - a^2\mathbf{k}^2) - \gamma_1(8 - a^2\mathbf{k}^2) = \widetilde{E}_0 + (\gamma_1 + \gamma_2)a^2\mathbf{k}^2 \ .$$

Hence,

$$\left(\frac{1}{m^*}\right)_{\alpha\beta} = \delta_{\alpha\beta}2(\gamma_1 + \gamma_2)a^2/\hbar^2 \ .$$

d. The Drude formula gives

$$\sigma_{\alpha\beta} = ne^2\tau\left(\frac{1}{m^*}\right)_{\alpha\beta} = \delta_{\alpha\beta}2ne^2\tau(\gamma_1 + \gamma_2)a^2/\hbar^2 \ .$$

Answer s.6.8.

a. The velocity, in the semiclassical approximation, is

$$\langle\mathbf{v}\rangle_{n\mathbf{k}} = \boldsymbol{\nabla}_{\mathbf{k}}E_n(\mathbf{k})/\hbar = [(\hbar/m^*) + \alpha k^2]\mathbf{k} \ .$$

The equations of motion are

$$\hbar\dot{\mathbf{k}} = -e\boldsymbol{\mathcal{E}}_0\cos(\omega t) \ , \quad \Rightarrow \quad \mathbf{k}(t) = -[e/(\hbar\omega)]\boldsymbol{\mathcal{E}}_0\sin(\omega t) \ ,$$

where it has been assumed that $\mathbf{k}(t=0)=0$. Using the expression for the average velocity yields

$$\mathbf{v}(t) = -\frac{e}{m^*\omega}\boldsymbol{\mathcal{E}}_0\sin(\omega t) - \frac{\alpha e^3}{\hbar^3\omega^3}\mathcal{E}_0^2\boldsymbol{\mathcal{E}}_0\sin^3(\omega t) \ .$$

By using the identity $\sin^3 x = [3\sin x - \sin(3x)]/4$, the velocity becomes

$$\mathbf{v}(t) = -\frac{e}{m^*\omega}\boldsymbol{\mathcal{E}}_0\left[\left(1 + \frac{3\alpha e^2 m^*\mathcal{E}_0^2}{4\hbar^3\omega^2}\right)\sin(\omega t) - \frac{\alpha e^2 m^*}{4\hbar^3\omega^2}\mathcal{E}_0^2\sin(3\omega t)\right] \ .$$

Finally, integrating over the time gives, up to a constant,

$$\mathbf{r}(t) = \frac{e}{m^*\omega^2}\boldsymbol{\mathcal{E}}_0\left[\left(1 + \frac{3\alpha e^2 m^*\mathcal{E}_0^2}{4\hbar^3\omega^2}\right)\cos(\omega t) - \frac{\alpha e^2 m^*}{12\hbar^3\omega^2}\mathcal{E}_0^2\cos(3\omega t)\right] \ .$$

b. As seen from the expressions above, the oscillations are at the two frequencies ω and 3ω, with the amplitudes

$$\frac{e}{m^*\omega^2}\left[\left(1 + \frac{3\alpha e^2 m^*\mathcal{E}_0^2}{4\hbar^3\omega^2}\right)\right] \ , \quad \left[-\frac{\alpha e^3\mathcal{E}_0^2}{12\hbar^3\omega^4}\right] \ ,$$

respectively.

c. The contribution of the quartic term in the energy to these results is proportional to α. Thus, both parts of the dispersion relation are of a comparable significance at frequency ω once $3\alpha e^2 m^*\mathcal{E}_0^2/(4\hbar^3\omega^2) = 1$, i.e., $\mathcal{E}_0^2 = 4\hbar^3\omega^2/(3\alpha e^2 m^*)$.

Answer s.6.9.

Assume that the conductivities of the electrons and of the holes are given by the Drude formula Eq. (6.6), with the appropriate concentrations and masses, and with the same relaxation time. The concentration of electrons in an extrinsic semiconductor is x_D; hence

$$\sigma_{\text{int}} = (n_e/m_e + n_h/m_h)e^2\tau = n_e(1/m_e + 1/m_h)e^2\tau \ ,$$

and

$$\sigma_{\text{ext}} = x_D e^2 \tau / m_e \ .$$

By the conditions of the problem,

$$10^4 = \sigma_{\text{ext}}/\sigma_{\text{int}} = (x_D/n_e)m_h/(m_e + m_h) = 3x_D/(5n_e) \ .$$

Equation (6.134) implies that

$$n_e = n_h = \sqrt{C_e C_h}\sqrt{\pi}(k_{\mathrm{B}}T)^{3/2}\exp[-\beta E_g/2]/2 \ ,$$

where (from the explicit expressions in problem 6.29),

$$C_e C_h = 2(m_e m_h)^{3/2}/(\pi^4 \hbar^6) \ .$$

Substituting the value of the energy gap, and in addition $k_{\mathrm{B}}T \approx (1/40)$ eV at room temperature, $m = 0.511$ MeV$/(3 \times 10^8$ m/sec$)^2$, and $\hbar = 6.582$ eV sec, gives $n_e = 3.62 \times 10^{14}$ m^{-3}. Hence, $x_D = 6 \times 10^{18}$ m^{-3}.

Answer s.6.10.

a. The Fermi curve consists of the straight segments given by the equations

$$(|k_x| + |k_y|)a/2 = \arccos(-\alpha) \ .$$

In particular, the equation for the segment in the first quarter of the Brillouin zone is $(k_x + k_y)a/2 = \arccos(-\alpha)$. It follows that the surface is a square whose corners are at $(\pm 2\arccos[-\alpha]/a, 0)$ and $(0, \pm 2\arccos[-\alpha]/a)$.

b. In the absence of the magnetic field, the velocity in the first quarter is

$$\mathbf{v} = \frac{1}{\hbar}\boldsymbol{\nabla}_{\mathbf{k}}E = \frac{a\gamma}{2\hbar}\sin[(k_x + k_y)a/2](1,1,0) \ .$$

On the Fermi curve

$$E_{\mathrm{F}} = -\gamma\cos[(|k_x| + |k_y|)a/2] = \alpha\gamma \ ,$$

and thus the velocity of the electron is constant along the line and is given by

$$\mathbf{v} = \frac{a\gamma}{2\hbar}\sqrt{1-\alpha^2}[\operatorname{sign}(k_{x\mathrm{F}}), \operatorname{sign}(k_{y\mathrm{F}}), 0] \ .$$

In the first quarter of the Brillouin zone the velocity is

$$\mathbf{v} = [a\gamma/(2\hbar)]\sqrt{1-\alpha^2}(1,1,0) \ ,$$

perpendicular to the Fermi curve. The same conclusion pertains for the other three quarters.

c. The equation of motion is

$$\hbar\dot{\mathbf{k}} = -(e/c)\mathbf{v} \times \mathbf{B} \ .$$

Inserting here the velocity from part (b) gives

$$\hbar\dot{\mathbf{k}} = -[2a\gamma\mathrm{B}/(2\hbar c)]\sqrt{1-\alpha^2}(-1,1,0) \ .$$

The force is parallel to the Fermi segment and therefore the lattice momentum varies linearly with time along the segment, with a rate given by

$$\hbar\dot{k} = -[ea\gamma\mathrm{B}/(\sqrt{2}\hbar c)]\sqrt{1-\alpha^2} \ .$$

d. The duration of the motion around the Fermi curve is four times the duration of motion in each quarter, that is

$$\mathbf{T} = 4[2\sqrt{2}]\frac{\arccos(-\alpha)}{ea\gamma\mathrm{B}\sqrt{(1-\alpha^2)/2}/(\hbar^2 c)} = 16\hbar^2 c\frac{\arccos(-\alpha)}{ea\gamma\mathrm{B}\sqrt{1-\alpha^2}} \ .$$

One obtains the same result from Eq. (6.163).

Answer s.6.11.

a. According to problem 6.24, the energies of the electron in the two bands are

$$E = E^{(0)} \pm |\widetilde{\gamma}_{AB}(\mathbf{k})| \ ,$$

$$\widetilde{\gamma}_{AB}(\mathbf{k}) = \sum_{nn} \gamma_{AB} e^{-i\mathbf{k}\cdot\mathbf{r}_{nn}} = \gamma_{AB}(e^{-i\mathbf{k}\cdot\mathbf{a}_1} + e^{-i\mathbf{k}\cdot\mathbf{a}_2} + e^{-i\mathbf{k}\cdot(\mathbf{a}_1-\mathbf{a}_2)}) \ ,$$

that is

$$E = E^{(0)} \pm |\gamma|_{AB}\sqrt{3 + 2\cos(\mathbf{k}\cdot\mathbf{a}_1) + 2\cos(\mathbf{k}\cdot\mathbf{a}_2) + 2\cos(\mathbf{k}\cdot(\mathbf{a}_1-\mathbf{a}_2))}$$

$$= E^{(0)} \pm |\gamma|_{AB}\sqrt{3 + 2\cos(k_x a) + 4\cos(k_x a/2)\cos(k_y a\sqrt{3}/2)} \ .$$

Assume for convenience that $E^{(0)} = 0$ and denote $\epsilon = E/|\gamma_{AB}|$, and $u = 3 + 2\cos(k_x a) + 4\cos(k_x a/2)\cos(k_y a\sqrt{3}/2)$. Then, the derivatives are

$$\epsilon_x = \frac{\partial\epsilon}{\partial k_x} = \frac{u_x}{2\epsilon} \ , \quad \epsilon_y = \frac{\partial\epsilon}{\partial k_y} = \frac{u_y}{2\epsilon} \ ,$$

$$\epsilon_{xx} = \frac{\partial^2\epsilon}{\partial k_x^2} = \frac{u_{xx}}{2\epsilon} - \frac{u_x^2}{4\epsilon^3} \ , \quad \epsilon_{yy} = \frac{\partial^2\epsilon}{\partial k_y^2} = \frac{u_{yy}}{2\epsilon} - \frac{u_y^2}{4\epsilon^3} \ , \quad \epsilon_{xy} = \frac{\partial^2\epsilon}{\partial k_x\partial k_y} = \frac{u_{xy}}{2\epsilon} - \frac{u_x u_y}{4\epsilon^3} \ ,$$

where $u_\ell = \partial u/\partial k_\ell$, $u_{\ell m} = \partial^2 u/(\partial k_\ell \partial k_m)$,

$$u_x = -2a[\sin(k_x a) + \sin(k_x a/2)\cos(k_y a\sqrt{3}/2)] \ ,$$

$$u_y = -2a\sqrt{3}\cos(k_x a)\sin(k_y a\sqrt{3}/2)] \ ,$$

$$u_{xx} = -a^2[2\cos(k_x a) + \cos(k_x a/2)\cos(k_y a\sqrt{3}/2)] \ ,$$

$$u_{yy} = -3a^2\cos(k_x a/2)\cos(k_y a\sqrt{3}/2) \ ,$$

$$u_{xy} == -\sqrt{3}a^2\sin(k_x a/2)\sin(k_y a\sqrt{3}/2) \ .$$

Inserting these expressions in Eq. (6.155) yields

$$\left(\frac{1}{m^*}\right)_{\alpha\beta} = \pm\frac{|\gamma_{AB}|}{\hbar^2}\epsilon_{\alpha\beta} \ .$$

At the Γ point $\mathbf{k} = 0$; the effective-mass tensor is a diagonal matrix, proportional to the unit matrix, i.e.,

$$(m^*)_{\alpha\beta} = \mp 2\hbar^2/(|\gamma_{AB}|a^2)\delta_{\alpha\beta} \ .$$

The effective mass at the bottom of the lower band is positive, and hence corresponds to electrons; the effective mass close to the maximum of the upper band is negative, i.e., it represents holes. At the M point $\mathbf{k} = [\pi/a, -\pi/(a\sqrt{3})]$. It follows that

$$\frac{1}{m^*} = \pm \frac{|\gamma_{AB}|}{\hbar^2} a^2 \begin{bmatrix} 1 & \sqrt{3}/2 \\ \sqrt{3}/2 & 0 \end{bmatrix} , \quad m^* = \pm \frac{\hbar^2}{|\gamma_{AB}|a^2} \begin{bmatrix} 0 & 2/\sqrt{3} \\ 2/\sqrt{3} & -4/3 \end{bmatrix} .$$

The eigenvalues of the effective mass are

$$m^* = \pm \frac{2\hbar^2}{3|\gamma_{AB}|a^2} , \quad \mp \frac{2\hbar^2}{|\gamma_{AB}|a^2} ,$$

and the eigenvectors are

$$(-1, \ \sqrt{3})/2 , \quad (\sqrt{3}, \ 1)/2 .$$

b. Near the Γ point $E = \hbar^2 k^2/(2m^*)$; one may therefore follow the calculation in problem 6.10 to obtain $g(E) = 2m^*/(\pi\hbar^2)$. Near the K point, according to problem 6.24,

$$E = \pm|\widetilde{\gamma}_{AB}(\mathbf{k})| \approx \pm(\gamma_{AB}\sqrt{3}/2)|\mathbf{k}|a .$$

The equal-energy curves are circles in the plane of the lattice momentum. On each such curve $|\boldsymbol{\nabla}_{\mathbf{k}}E| = |\gamma_{AB}|a\sqrt{3}/2$. It then follows from Eq. (6.127) that

$$g(E) = \frac{2\pi|\mathbf{k}|}{2\pi^2|\boldsymbol{\nabla}_{\mathbf{k}}E|} = \frac{4E}{3\pi|\gamma_{AB}|^2 a^2} .$$

Answer s.6.12.

a. The density of the electrons can be obtained from the zero-temperature expression $n = \int_0^{E_F} dE g(E)$. In the units of Fig. 6.50, the density of the electrons that occupy the first band is 10^{28} m^{-3} (this is the area below the density of states of this band). The first step of the second band adds 2×10^{28} m^{-3}, and the second stair of this band can add an additional 3×10^{28} m^{-3}. Since the total density is 5×10^{28} m^{-3}, the electrons fill only 2/3 of the last step. It follows that the Fermi energy is at 5.667×10^{-19} J; as this value is within the band, the material is a metal.

b. At low temperatures the specific heat is given by Eq. (6.129),

$$c_V = \frac{\pi^2}{3} g(E_F) k_B^2 T = 188\, T \ (\text{Jm}^{-3}\,\text{K}^{-1}) ,$$

where the temperature is in K and $k_B = 1.38 \times 10^{-23}$ JK^{-1}.

c. When the total density is 7×10^{28} m^{-3}, the second band is full, and the Fermi energy lies in-between this and the third band. The gap between these two is 1×10^{-19} J, much larger than the room-temperature energy $k_B T \approx 4 \times 10^{-21}$ J, and hence the material is an insulator. The specific heat, for a three-dimensional sample, is determined by Eq. (6.139). This specific heat is rather small, $c_V \propto \beta^{1/2} \exp[-\beta E_g/2]$, with $\beta E_g/2 \approx 25$. The low-temperature specific heat is tiny.

Answer s.6.13.

a. The density of states is finite in the range $E_d - W/2 < E < E_d + W/2$. As there are 5 orbital states in the band, each capable of accommodating two electrons, the density of states (per atom) in this range is $g = 10/W$.

b. At zero temperature the band is filled up to the energy $E_{\max}$, where

$$N = \int_{E_d - W/2}^{E_{\max}} dE g(E) = \frac{10}{W}\left(E_{\max} - E_d + \frac{W}{2}\right) ,$$

from which follows that

$$E_{\max} = E_d + (N - 5)W/10 .$$

The additional energy (to the atomic energy) due to the finite width of the band is

$$\Delta E = \int_{E_d - W/2}^{E_{\max}} dE g(E) E = \frac{5}{W}\left(E_{\max}^2 - (E_d - W/2)^2\right) = N E_d + N W (N - 10)/2 .$$

c. The first term on the right-hand side represents the additional energy in the limit of zero band-width (each electron adds its energy to the atomic energy). The next term is negative, and is the "gain" in energy due to the finite band width, which allows the electron to move among the ions. Hence, the binding energy is

$$-(\Delta E - N E_d) = -W N (N - 10)/2 ,$$

and it is maximal when $N = 5$.

Answer s.6.14.

a. Similar to problem 6.23, the nth unit cell, whose length is a, contains two sites; the wave functions there are ψ_{2n-1} and ψ_{2n}, and the Schrödinger equations in the tight-binding approximation become

$$E\psi_{2n} = -\gamma_- \psi_{2n-1} - \gamma_+ \psi_{2n+1} ,$$
$$E\psi_{2n-1} = -\gamma_- \psi_{2n} - \gamma_+ \psi_{2n-2} .$$

With the Ansatz $\psi_{2n-1} = A \exp[ikna]$ and $\psi_{2n} = B \exp[ikna]$, one obtains

$$EA = -\gamma_+ B e^{-ika} - \gamma_- B = -B(\gamma_- + \gamma_+ e^{-ika}) ,$$
$$EB = -\gamma_- A - \gamma_+ A e^{ika} = -A(\gamma_- + \gamma_+ e^{ika}) .$$

The solution is nontrivial provided that the determinant vanishes, which yields

$$E^2 - (\gamma_- + \gamma_+ e^{-ika})(\gamma_- + \gamma_+ e^{ika}) = 0 ,$$

that is

$$E_\pm = \pm\sqrt{\gamma_-^2 + \gamma_+^2 + 2\gamma_-\gamma_+ \cos(ka)} = \pm 2\gamma_0 \sqrt{\cos^2(ka/2) + \Delta^2 \sin^2(ka/2)} .$$

As in the solution to problem 6.23, there are two energy bands (of positive and of negative energies); the energy gap in-between them vanishes only when $\Delta = 0$.

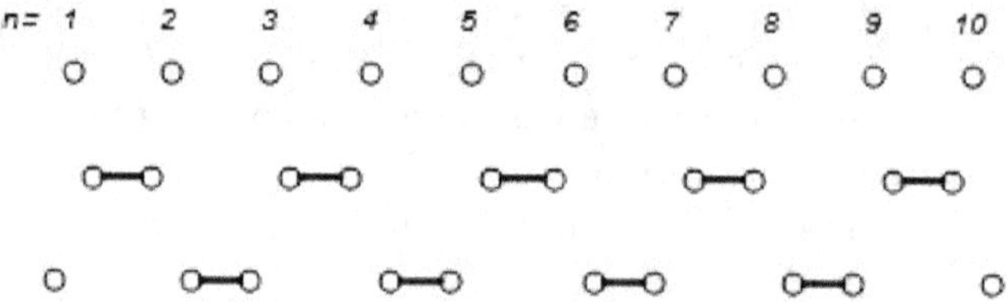

Fig. 6.52

b. Since $E_-(\Delta) < E_-(\Delta = 0)$ electrons in the lower band gain energy when $\Delta \neq 0$. In the "half-filled" case, all levels in the lower band are occupied, and therefore the total electronic energy is

$$E_{\text{tot}}(\Delta) = \sum_k E_-(k) = \frac{Na}{2\pi} \int_{-\pi/(2a)}^{\pi/(2a)} E_-(k) \ .$$

As a result, the system has the energy $E_{\text{tot}}(\Delta) + E_d$, which depends only on Δ^2, i.e., it is even in Δ. Consequently, the states with Δ and with $-\Delta$ are degenerate. The gain in energy is maximal when

$$dE_{\text{tot}}/d(\Delta^2) = -Nx \ . \tag{6.182}$$

One can verify that the integral in the expression for $dE_{\text{tot}}/d(\Delta^2)$ diverges at $\Delta = 0$, rising from $-\infty$ at $\Delta = 0$ and reaching zero as $|\Delta|$ increases. As a result, Eq. (6.182) has always two solutions, $\Delta = \pm\Delta_0$.

c. When $|\Delta| = 1$, the state of the system attains depends on the sign of Δ. When it is negative, $\gamma_+ = 0$; consequently, dimers (i.e., pairs of atoms) are created within each unit cell. Dimers belonging to different unit cells are not coupled. As is the situation with the hydrogen molecule [Eq. (4.14)], the eigenstates of each dimer are the bonding state and the anti-bonding one, $\Psi_\pm = (\psi_{2n-1} \pm \psi_{2n})/\sqrt{2}$, whose energies are $\mp 2\gamma_0$. On the other hand, $\Delta = 1$ implies that $\gamma_- = 0$. The dimers comprise the right atom in a certain cell with the left atom in the cell on its right. Again the dimers are not coupled. In the infinite system the two states are degenerate, but not in the finite system. Assume that the atoms are located in the sites $1,2,3,\ldots,2N$. The upmost row in Fig. 6.52 shows the lattice before the Peierls deformation takes place. The situation for $\Delta = -1$ is illustrated in central row; that of $\Delta = 1$ is shown in the lower row. There is no physical disparity between the two configurations in the bulk of the lattice, but there is a significant difference at the edges. In the first case all electrons belong to the dimers, while in the second there are two "single" electrons, residing in **surface states** (similar to those described in the context of the quantum Hall effect). The energy of each electron in the dimer is $E_-(1) = -2\gamma_0$ (as the energy band is flat), but the energy of an electron at the edge is 0, and hence that electron is in an excited state. Remarkably, this state persists also when $|\Delta|$ deviates from 1. Such edge states appear in **topological insulators** (for instance, in the quantum Hall effect, see Fig. 6.35).

Appendix E

The Schrödinger equation for electrons in a magnetic field

Hamilton's equations in classical mechanics. The classical motion of a particle can be described in terms of **Hamilton's equations of motion**. In this formulation a canonical momentum, $\mathbf{p}$, is conjugated to the coordinate of the particle $\mathbf{r}$. When the spin degree of freedom is ignored, the motion is described by a function of both $\mathbf{p}$ and $\mathbf{r}$, denoted $H(\mathbf{p}, \mathbf{r})$, such that the equations of motion of the particle are

$$\dot{\mathbf{r}} = \boldsymbol{\nabla}_{\mathbf{p}} H , \quad \dot{\mathbf{p}} = -\boldsymbol{\nabla}_{\mathbf{r}} H . \tag{E.1}$$

The function $H(\mathbf{p}, \mathbf{r})$ is called the **Hamiltonian of the particle**, and Eqs. (E.1) are **Hamilton's equations of motion**. In the absence of a magnetic field, the Hamiltonian of a particle in the field of an electric potential that does not vary in time, $\varphi(\mathbf{r})$, is the total energy of the particle,

$$H(\mathbf{p}, \mathbf{r}) = [\mathbf{p}^2/(2m)] - e\varphi(\mathbf{r}) ,$$

and Eqs. (E.1) give

$$\dot{\mathbf{r}} = \mathbf{p}/m , \quad \dot{\mathbf{p}} = e\boldsymbol{\nabla}_{\mathbf{r}}\varphi = -e\boldsymbol{\mathcal{E}} ,$$

where $\boldsymbol{\mathcal{E}} = -\boldsymbol{\nabla}\varphi$ is the electric field. In this case the canonical momentum of the particle coincides with mechanical momentum, $\mathbf{p} = m\mathbf{v} = m\dot{\mathbf{r}}$, and the corresponding equation of motion is just the second law of Newton, $m\ddot{\mathbf{r}} = m\dot{\mathbf{v}} = \dot{\mathbf{p}} = -e\boldsymbol{\mathcal{E}}$. When the particle is also subjected to a magnetic field $\mathbf{B}$, the Lorentz force depends on the velocity; consequently, the canonical momentum does no longer coincide with the mechanical one. The classical equations of motion, $m\dot{\mathbf{v}} = -e(\boldsymbol{\mathcal{E}} + \mathbf{v} \times \mathbf{B}/c)$ [Eq. (6.8)], are reproduced by Hamilton's equations when the Hamiltonian takes the form

$$H(\mathbf{p}, \mathbf{r}) = \frac{1}{2}\left(\mathbf{p} + \frac{e}{c}\mathbf{A}\right)^2 - e\varphi(\mathbf{r}) , \tag{E.2}$$

where $\mathbf{A}$ is the vector potential, which depends on the coordinate and on the time. The electric and the magnetic fields are

$$\boldsymbol{\mathcal{E}} = -\boldsymbol{\nabla}\varphi - \dot{\mathbf{A}}/c , \quad \mathbf{B} = \boldsymbol{\nabla} \times \mathbf{A} .$$

563

Equation (E.1) yields

$$\mathbf{v} = \dot{\mathbf{r}} = \boldsymbol{\nabla}_{\mathbf{p}}H = [\mathbf{p} + (e/c)\mathbf{A}]/m \ , \quad \Rightarrow \quad \mathbf{p} = m\mathbf{v} - (e/c)\mathbf{A} \ .$$

The **canonical momentum** $\mathbf{p}$ differs from the mechanical momentum $m\mathbf{v}$. Differentiating the canonical momentum with respect to time gives

$$\dot{\mathbf{p}} = m\dot{\mathbf{v}} - (e/c)[\partial\mathbf{A}/\partial t + (\mathbf{v}\cdot\boldsymbol{\nabla})\mathbf{A}]$$

(the second term includes both the explicit derivative and the chain derivative due to the the dependence of $\mathbf{r}$ on t). On the other hand, the second of Hamilton's equations becomes

$$\dot{p}_i = -\frac{\partial H}{\partial x_i} = e\frac{\partial\varphi}{\partial x_i} - \left(\frac{e}{mc}\right)\left[\mathbf{p} + \frac{e}{c}\mathbf{A}\right]\cdot\frac{\partial\mathbf{A}}{\partial x_i} = e\frac{\partial\varphi}{\partial x_i} - \mathbf{v}\cdot\left(\frac{e}{c}\right)\frac{\partial\mathbf{A}}{\partial x_i} \ .$$

Comparing the two expressions for p_i, and substituting $\dot{v}_i = \ddot{x}_i$ show that

$$m\ddot{x}_i = \frac{e}{c}\left[\frac{\partial A_i}{\partial t} + \mathbf{v}\cdot\boldsymbol{\nabla}A_i - \mathbf{v}\cdot\frac{\partial\mathbf{A}}{\partial x_i}\right] + e\frac{\partial\varphi}{\partial x_i} \ . \tag{E.3}$$

Problem E.1. demonstrates that this equation is identical to the classical equation of motion (6.8).

Problem E.1.

Use the identity

$$\mathbf{v}\times[\boldsymbol{\nabla}\times\mathbf{A}] = \boldsymbol{\nabla}(\mathbf{v}\cdot\mathbf{A}) - (\mathbf{v}\cdot\boldsymbol{\nabla})\mathbf{A} \ ,$$

(check!) and the expressions for the electric and the magnetic fields to show that Eq. (E.3) reproduces the classical equation of motion, with the Lorentz force.

The Schrödinger equation with a constant magnetic field. The quantum behavior of the electron is obtained from the solution of the Schrödinger equation, $\hat{\mathcal{H}}\psi = E\psi$, where $\hat{\mathcal{H}}$ is the Hamiltonian operator. The latter is derived from the classical Hamiltonian of the electron, replacing the canonical momentum $\mathbf{p}$ by the operator $\hat{\mathbf{p}} = -i\hbar\boldsymbol{\nabla}$; the coordinate operator is the same as the coordinate of the electron, i.e., $\hat{\mathbf{r}} = \mathbf{r}$. Thus, in the absence of the magnetic field $\hat{\mathcal{H}} = \hat{\mathbf{p}}^2/(2m) + \hat{V}(\mathbf{r}) = -\hbar^2\nabla^2/(2m) + V(\mathbf{r})$. In the presence of a magnetic filed, the quantum Hamiltonian is

$$\hat{\mathcal{H}} = \frac{1}{2m}\left(-i\hbar\boldsymbol{\nabla} + \frac{e}{c}\mathbf{A}\right)^2 - e\varphi(\mathbf{r}) \ . \tag{E.4}$$

Choosing the direction of the magnetic field as the $\hat{\mathbf{z}}-$axis, one may use the **Landau gauge**, $\mathbf{A} = -By\hat{\mathbf{x}}$ (check that indeed $\mathbf{B} = \boldsymbol{\nabla}\times\mathbf{A} = B\hat{\mathbf{z}}$). In the absence of the electric potential, the Schrödinger equation takes the form

$$\frac{1}{2m}\left[\left(-i\hbar\frac{\partial}{\partial x} - \frac{e}{c}By\right)^2 - \hbar^2\left(\frac{\partial^2}{\partial y^2} + \frac{\partial^2}{\partial z^2}\right)\right]\psi(\mathbf{r}) = E\psi(\mathbf{r}) \ . \tag{E.5}$$

As the coordinates x and z do not appear in the Hamiltonian, the motion along these two directions is that of a free particle. In other words, substituting $\psi(\mathbf{r}) =$

$\chi(y)\exp[i(k_x x + k_z z)]$, where k_x and k_z are constants (to be determined from the boundary conditions), in Eq. (E.5) yields

$$-\frac{\hbar^2}{2m}\frac{d^2}{dy^2}\chi(y) + \frac{1}{2}m\omega_c^2(y-y_0)^2\chi(y) = \left[E - \frac{\hbar^2}{2m}k_z^2\right]\chi(y)\,, \qquad (E.6)$$

where $\omega_c = e\mathrm{B}/(mc)$ is the cyclotron frequency, Eq. (6.9). This is the Schrödinger equation of a one-dimensional harmonic oscillator, vibrating along the $\hat{\mathbf{y}}$–direction, around the point $y_0 = c\hbar k_x/(e\mathrm{B}) = \ell_B^2 k_x$, where $\ell_B^2 = \hbar c/(e\mathrm{B})$ is the square of the magnetic length. The energy levels of the particle are

$$E(\ell, k_z) = \frac{\hbar^2}{2m}k_z^2 + \hbar\omega_c(\ell + 1/2)\,. \qquad (E.7)$$

The last term, derived from the energy levels of the harmonic oscillator, represents the **Landau levels**, and confirms Eq. (6.165). The first term represents the motion of the electron parallel to the magnetic field (as in the classical case, the motion parallel to the field is not affected by the field, see problem 6.3). For a two-dimensional system, the first term in Eq. (E.7) does not exist. The solution of Eq. (E.6) is

$$\chi_\ell(y) = \frac{1}{\sqrt{2^\ell \ell! \ell_B \sqrt{\pi}}} e^{-[(y-y_0)/\ell_B]^2/2} H_\ell[(y-y_0)/\ell_B]\,,$$

where H_ℓ is the Hermite polynomial of order ℓ. This function is localized within a region of width of the order ℓ_B around the point y_0. Thus, in the two-dimensional system, the wave function is a plane wave along the $\hat{\mathbf{x}}$–axis times a narrow function along $\hat{\mathbf{y}}$.

The degeneracy of the Landau levels in a two-dimensional motion. Assume that the electron is moving in a planar rectangular sample, of dimensions L_x and L_y. Periodic boundary conditions along x imply that $k_x L_x = 2\pi n_x$. where n_x is an integer, and therefore $y_0(n_x) = c\hbar 2\pi n_x/(e\mathrm{B}L_x)$. Since $0 \le y_0 \le L_y$, it follows that

$$n_x \le \frac{L_x L_y \mathrm{B}}{(ch/e)} = \frac{\Phi}{\Phi_0}\,,$$

where $\Phi = L_x L_y \mathrm{B}$ is the magnetic flux through the entire sample, and $\Phi_0 = ch/e$ is the **quantum flux**, precisely the result derived before Fig. 6.34. The degeneracy of each Landau level (i.e., the number of eigenstates of the momentum along $\hat{\mathbf{x}}$) is identical to the number of magnetic-flux quanta contained in the area of the sample. This huge degeneracy is not in contradiction with the Pauli principle, since the wave functions corresponding to different values of the quantum number n_x describe oscillators that vibrate around different centers $y_0(n_x)$.

Problem E.2.

a. *Prove that the expressions for the magnetic filed, $\mathbf{B} = \boldsymbol{\nabla} \times \mathbf{A}$, and for the electric field, $\boldsymbol{\mathcal{E}} = -\boldsymbol{\nabla}\varphi - \dot{\mathbf{A}}/c$, remain unchanged also when the vector potential is displaced by a gradient of an arbitrary scalar function, $\mathbf{A} \to \mathbf{A} + \boldsymbol{\nabla} f(\mathbf{r}, t)$, and* *the scalar potential by the time-derivative of the same function, $\varphi \to \varphi - \dot{f}(\mathbf{r}, t)/c$. This is the* **gauge freedom**.

b. *Prove that a constant magnetic field along $\hat{\mathbf{z}}$ can be also described by the* **symmetric gauge**, $\mathbf{A}_s = \mathbf{B} \times \mathbf{r}/2$. *Prove that the difference between the vector potentials in the symmetric gauge and in the Landau one obeys the result in part (a), and that both gauges obey the* **gauge conditions of Coulomb**, $\nabla \cdot \mathbf{A} = 0$.
Prove that for $\nabla \cdot \mathbf{A} = 0$, Eq. (E.4) yields

$$\hat{\mathcal{H}} = -\frac{\hbar^2}{2m}\nabla^2 - \frac{ie\hbar}{mc}\mathbf{A}\cdot\nabla + \frac{1}{2m}\left(\frac{e}{c}\right)^2 \mathbf{A}^2 - e\varphi .$$

Substitute the symmetric gauge and prove that

$$\hat{\mathcal{H}} = -\frac{\hbar^2}{2m}\nabla^2 + \frac{e\mathrm{B}}{2mc}\hat{L}_z + \frac{m}{8}\omega_c^2(x^2 + y^2) - e\varphi .$$

The second term represents the "normal" Zeeman effect, that is, the coupling of the orbital magnetic moment, $\mu_z = -eL_z/(2mc)$ with the magnetic field. The third term is the potential of a two-dimensional harmonic oscillator.

d. *Using the symmetric gauge, express the Hamiltonian in cylindrical coordinates, $\mathbf{r} = (\rho\cos\theta, \rho\sin\theta, z)$, and show that the eigenstates of the Schrödinger equation for $\varphi = 0$ are of the form*

$$\psi_{nMk_z}(\mathbf{r}) = R_{nM}(\rho)e^{iM\theta}e^{ik_z z} ,$$

where $\hbar M$ is the eigenvalue of the angular-momentum component perpendicular to the plane, $\hat{L}_z = -i\hbar\partial/\partial\theta$, and where $R_{nM}(\rho)$ is the radial wave function of a planar harmonic oscillator. Find the differential equation for $R_{nM}(\rho)$, and show that $R \sim \rho^{|M|}$ near the origin, while $R \sim \exp[-(\rho/\ell_B)^2/4]$ far away from the origin. Explain why this function describes the classical circular motion, as in Fig. 6.35.

The quantum Hall effect. The Hall effect pertains to the geometry portrayed in Fig. 6.1: electrons are moving along the $\hat{\mathbf{x}}$–direction, in the presence of a magnetic field along $\hat{\mathbf{z}}$ and an electric field along $\hat{\mathbf{y}}$ (both fields are constant in space and in time). As $\boldsymbol{\mathcal{E}} = -\nabla\varphi$, it follows that $\varphi = -\mathcal{E}y$; this implies that the Hamiltonian in Eq. (E.4) is augmented by $V(y) = e\mathcal{E}y$. The solution is still that of an harmonic oscillator, because the algebraic manipulation

$$\frac{m\omega_c^2}{2}(y - y_0)^2 + e\mathcal{E}y = \frac{1}{2}m\omega_c^2\left[y - y_0 + \frac{e\mathcal{E}}{m\omega_c^2}\right] + e\mathcal{E}\left[y_0 - \frac{e\mathcal{E}}{2m\omega_c^2}\right]$$

shows that the center of the harmonic oscillator is simply shifted from y_0 to $y_0 - e\mathcal{E}/(m\omega_c^2)$, and the energy is changed into

$$E(\ell, k_z, k_x) = \frac{\hbar^2}{2m}k_z^2 + \hbar\omega_c(\ell + 1/2) - e\mathcal{E}y_0 + \mathcal{O}(\mathcal{E}^2) . \tag{E.8}$$

The Landau levels attain a dependency on $y_0 = c\hbar k_x/(e\mathrm{B})$, and thence on k_x. Since semiclassically

$$v_x = \partial E/\partial p_x = (\partial E/\partial k_x)/\hbar ,$$

[Eq. (6.148)] one finds $v_x = -\mathcal{E}c/B$, and a current density along $\hat{\mathbf{x}}$, $j_x = -nev_x = (nec/B)\mathcal{E} = \sigma_{xy}\mathcal{E}$, in agreement with the classical result (6.15). In a similar fashion, one finds that $v_y = (\partial E/\partial k_y)/\hbar = 0$, which implies that the longitudinal conductance vanishes, $\sigma_{yy} = 0$.

Surface states. Obviously, the sample has a finite width along $\hat{\mathbf{y}}$; the surface there can be described by a potential $U(y)$ which repels the electron (and prevents it from leaking out). As this potential can be ignored in the bulk of the sample, the wave functions corresponding to the Landau levels are $\psi_\ell(k_x)$, that are localized along $\hat{\mathbf{y}}$ around the centers $y_0(k_x)$. The energies of states localized close to the surface, in the lowest order of perturbation theory, are

$$E_\ell \to E(\ell, k_x) \approx E_\ell + \langle \psi_\ell(k_x)|U(y)|\psi_\ell(k_x)\rangle \approx E_\ell + U[y_0(k_x)] \,, \tag{E.9}$$

where it is assumed that the width of $\psi_\ell(k_x)$ (which is of the order of the magnetic length ℓ_B) is small compared to the length over which $U(y)$ varies. (Note that ψ_ℓ is normalized.) The velocity of the electron, along the tangential to the surface, is

$$v_x(\ell, k_x) = \frac{1}{\hbar}\frac{\partial E(\ell, k_x)}{\partial k_x} \approx \frac{1}{\hbar}\frac{\partial U[y_0(k_x)]}{\partial y_0}\frac{\partial y_0}{\partial k_x} = \frac{c}{eB}\frac{\partial U}{\partial y_0} \,. \tag{E.10}$$

This velocity is positive at the vicinity of the right-side edge, where the potential increases as the surface is approached (along the positive $y-$direction), and is negative along the opposite edge, where the potential decreases (again along the positive $y-$direction). When the Fermi energy is in-between two Landau levels the only occupied states are those for which $E_{\rm F} = E(\ell, k_x)$. These states are localized near the edges of the sample, and therefore they correspond to the classical surface states depicted in Fig. 6.35. As in that figure, the electrons in such a state move only along one direction. When the sample is wide enough for the two opposite edges to be far away from one another, the probability of scattering from a state moving near one of the edges into a state moving near the other is small, even when there are foreign atoms in the sample that cause scattering. It follows that the electrical resistance vanishes, as indeed is confirmed in the experiment.

Gauge transformation of the Schrödinger equation, the Aharonov-Bohm effect. The gauge transformation of the potentials is specified in problem E.2. It is shown there that the transformation leaves the electric and magnetic fields intact. The modification caused in the quantum wave functions as a result of this transformation is found as follows. One begins with the time-dependent Schrödinger equation,

$$i\hbar\frac{\partial \psi}{\partial t} = \left[\frac{1}{2m}\left(-i\hbar\boldsymbol{\nabla} + \frac{e}{c}\mathbf{A}\right)^2 - e\varphi\right]\psi \,, \tag{E.11}$$

and applies the transformation

$$\varphi \to \varphi = \varphi' - \dot{f}(\mathbf{r},t)/c \,, \quad \mathbf{A} \to \mathbf{A} = \mathbf{A}' + \boldsymbol{\nabla}f(\mathbf{r},t) \,,$$

to obtain

$$i\hbar\frac{\partial \psi'}{\partial t} = \left[\frac{1}{2m}\left(-i\hbar\boldsymbol{\nabla} + \frac{e}{c}\mathbf{A}'\right)^2 - e\varphi'\right]\psi \,, \tag{E.12}$$

where (as shown in problem E.3.),

$$\psi'(\mathbf{r}, t) = \psi(\mathbf{r}, t) \exp[ief(\mathbf{r}, t)/(c\hbar)] \ . \tag{E.13}$$

Problem E.3.
Show that the wave function in Eq. (E.13) indeed solves Eq. (E.12).

Equation (E.13) demonstrates that the sole effect of the gauge transformation is a change in the phase of the wave function: the transformation does not modify the probability density of the electron's location, $|\psi|^2 = |\psi'|^2$. The modification caused in the phase can be measured in situations in which the wave functions interfere. This can happen when the electron moves along an orbit enclosing a region where a magnetic field is applied. The component of the field normal to the plane where the electron is moving can be described by a vector potential that exists everywhere, also in the regions where the magnetic field vanishes. For instance, the magnetic field created by an infinite solenoid is active only within the cylinder, but the corresponding vector potential is a nonzero constant over circular orbits enclosing the cylinder that extend over the entire space. Though the electron may move in a region where the magnetic field vanishes, it is still affected by it through the vector potential; this is a purely quantum-mechanical phenomenon, with no classical analogue.

Assume that the wave function ψ' corresponds to the situation where $\mathbf{A}' = 0$, that is, the situation in which the magnetic field vanishes. Assume also that the electric potential vanishes as well, and that the system is not subjected to time-dependent forces. The vector potential is then $\mathbf{A} = \boldsymbol{\nabla} f$, where the function f is given by an integration over a path, say path "1", that connects two points,

$$f_1(\mathbf{r}) = \int_{\mathbf{r}_0}^{\mathbf{r}} d\mathbf{r}' \cdot \mathbf{A}(\mathbf{r}') \equiv \int_1 d\mathbf{r}' \cdot \mathbf{A}(\mathbf{r}') \ . \tag{E.14}$$

Why is this useful? consider the two-slit experiment. The electron emerges from a point S, then passes through the two slits via two paths, 1 and 2, and finally impinges on a screen, see Fig. E.1. A magnetic field B is enclosed between the two paths, in the circle marked in the figure; it creates a vector potential in the entire plane which contains the two paths. In particular, the vector potential takes different values on the two paths. The wave functions on these two paths are thus

$$\psi_1(\mathbf{r}) = \psi'(\mathbf{r}) \exp[ief_1(\mathbf{r})/(\hbar c)] \ , \quad \text{and} \quad \psi_2(\mathbf{r}) = \psi'(\mathbf{r}) \exp[ief_2(\mathbf{r})/(\hbar c)] \ , \tag{E.15}$$

where the function f_1 is given in Eq. (E.14), and an analogous equation gives the function f_2. The probability for finding the electron at a point $\mathbf{r}$ on the screen is

$$\begin{aligned}
|\psi(\mathbf{r})|^2 &= |\psi_1(\mathbf{r}) + \psi_2(\mathbf{r})|^2 \\
&= |\psi_1(\mathbf{r})|^2 + |\psi_2(\mathbf{r})|^2 + 2\mathrm{Re}[\psi_1^*(\mathbf{r})\psi_2(\mathbf{r})] = 2|\psi'(\mathbf{r})|^2[1 + \cos(\phi)] \ ,
\end{aligned}$$

where

$$\phi = \frac{e}{\hbar c}\left[\int_2 d\mathbf{r}' \cdot \mathbf{A}(\mathbf{r}') - \int_1 d\mathbf{r}' \cdot \mathbf{A}(\mathbf{r}')\right] = \frac{e}{\hbar c} \oint d\mathbf{r}' \cdot \mathbf{A}(\mathbf{r}') \ . \tag{E.16}$$

The difference between the two path integrals is the integral over a closed path, shown by arrows in Fig. E.1. By the Stokes theorem,

$$\oint d\mathbf{r}' \cdot \mathbf{A}(\mathbf{r}') = \int_S d^2r \,\hat{\mathbf{n}}(\mathbf{r}) \cdot [\boldsymbol{\nabla} \times \mathbf{A}(\mathbf{r})] = \int_S d^2r \, \mathrm{B}_\perp(\mathbf{r}) = \Phi \, ,$$

where $\hat{\mathbf{n}}(\mathbf{r})$ is a unit vector normal to the surface on which the integral is calculated, and $\mathbf{B} = \boldsymbol{\nabla} \times \mathbf{A}$, and thus $\mathrm{B}_\perp$ is the field component normal to the surface. The integral of this component over the area is precisely the total magnetic flux penetrating the surface, Φ. Since the field is active only within the circle in Fig. E.1, the surface integral yields the same result for any path enclosing this circle. Using Eq. (E.16), one obtains

$$\phi = \frac{e}{\hbar c}\Phi = 2\pi\frac{\Phi}{\Phi_0} \, , \tag{E.17}$$

where $\Phi_0 = hc/e$ is the **magnetic flux quantum**, discussed in the context of the quantum Hall effect. The interference pattern in the two-slit experiment displays a periodic dependence of the intensity on the magnetic flux penetrating the closed path, with a full period attained whenever the flux is increased by a flux quantum. These oscillations, as a function of the magnetic flux, are named after Aharonov and Bohm. They serve to detect the coherence of the wave functions in mesoscopic systems, since they disappear when coherence is destroyed (recall that coherence is necessary for constructive interference of the wave functions).

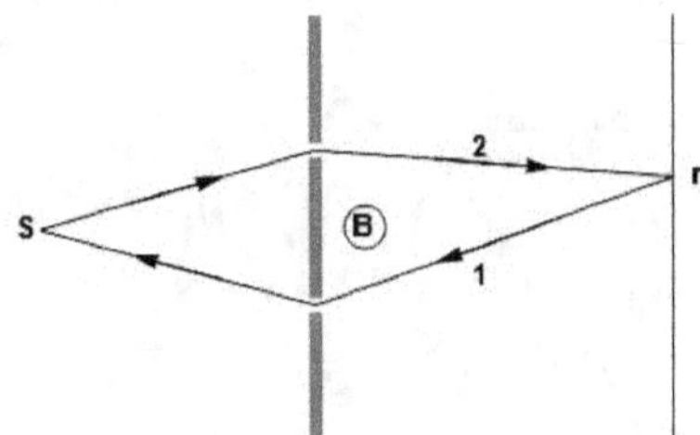

Fig. E.1: The two-slit experiment for an electron in a magnetic field.

Problem E.4.

Figure E.2 illustrates a metallic ring encompassing a solenoid.
a. *Calculate the vector potential along the tangential to the ring; write down the Schrödinger equation for the wave functions of an electron moving on the ring.*
b. *Find the eigenfunctions and the eigenenergies of the Schrödinger equation.*
c. *Show that the average charge-current density around the ring, in an eigenstate with an energy E_n, is $\langle j_\theta \rangle = -c\partial E_n/\partial\Phi$. Explain this result in terms of the relation between the current and the induced magnetic moment.*
d. *Find the total energy of all electrons in the ring at zero temperature. Find the total current around the ring, as a function of the magnetic flux penetrating the*

ring, and show that it is finite and does not depend on time as long as there is a magnetic field within the solenoid. This current is termed **persistent current**. *For which values of the flux does it vanish?*

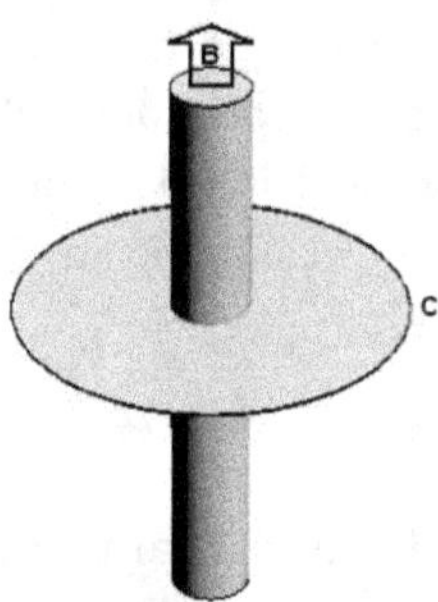

Fig. E.2: The inner cylinder represents a solenoid of radius a, where there is a magnetic field B along the axis. The circle C describes a ring of radius b containing electrons.

E.1   Answers

Answer E.1.

The vector identity in the problem gives

$$(\mathbf{v} \cdot \mathbf{\nabla})A_i = \nabla_i(\mathbf{v} \cdot \mathbf{A}) - (\mathbf{v} \times [\mathbf{\nabla} \times \mathbf{A}])_i = \mathbf{v} \cdot \nabla_i\mathbf{A} - (\mathbf{v} \times \mathbf{B})_i ,$$

using $\nabla_i \mathbf{v} = 0$. Inserting this into Eq. (E.3) gives

$$m\ddot{x}_i = \frac{e}{c}\left[\frac{\partial A_i}{\partial t} - (\mathbf{v} \times \mathbf{B})_i\right] + e\nabla_i\varphi = -e\mathcal{E}_i - \frac{e}{c}\Big([\mathbf{v} \times \mathbf{B}]\Big)_i ,$$

which is Eq. (6.8) (without the scattering term).

Answer E.2.

a. Using the shifted potentials in the expressions for the fields, one finds

$$\mathbf{B} \to \mathbf{\nabla} \times [\mathbf{A} + \mathbf{\nabla}f] = \mathbf{\nabla} \times \mathbf{A} + \mathbf{\nabla} \times [\mathbf{\nabla}f] = \mathbf{B} ,$$

$$\mathcal{E} \to -\mathbf{\nabla}[\varphi - \dot{f}/c] - [\dot{\mathbf{A}} + \mathbf{\nabla}\dot{f}]/c = -\mathbf{\nabla}\varphi - \dot{\mathbf{A}}/c = \mathcal{E} .$$

b. The symmetric gauge gives

$$\mathbf{A}_s = \mathrm{B}[\hat{\mathbf{z}} \times \mathbf{r}]/2 = \mathbf{B} + (-y\hat{\mathbf{x}} + x\hat{\mathbf{y}})/2 ,$$

leading to

$$\mathbf{\nabla} \times \mathbf{A}_s = \mathbf{\nabla} \times (-y\hat{\mathbf{x}} + x\hat{\mathbf{y}})\mathrm{B}/2 = \mathrm{B}\hat{\mathbf{z}} ,$$

$$\mathbf{\nabla} \cdot \mathbf{A}_s = \mathbf{\nabla} \cdot (-y\hat{\mathbf{x}} + x\hat{\mathbf{y}})\mathrm{B}/2 = 0 .$$

In the Landau gauge $\mathbf{A}_L = -By\hat{\mathbf{x}}$, and thus

$$\boldsymbol{\nabla} \cdot \mathbf{A}_L = \boldsymbol{\nabla} \cdot (-y\hat{\mathbf{x}}B) = 0 \ .$$

The difference between the two gauges is

$$\mathbf{A}_s - \mathbf{A}_L = (y\hat{\mathbf{x}} + x\hat{\mathbf{y}})B/2 = \boldsymbol{\nabla} f \ , \quad f = Bxy/2 \ .$$

c. Expanding the square in Eq. (E.4) gives

$$\left[-i\hbar\boldsymbol{\nabla} + \frac{e}{c}\mathbf{A}\right]^2 = -\hbar^2\boldsymbol{\nabla}^2 - \frac{i\hbar e}{c}\left(\boldsymbol{\nabla}\cdot\mathbf{A} + \mathbf{A}\cdot\boldsymbol{\nabla}\right) + \left(\frac{e}{c}\right)^2\mathbf{A}^2 \ .$$

Applying the central term on the wave function ψ results in

$$(\boldsymbol{\nabla}\cdot\mathbf{A} + \mathbf{A}\cdot\boldsymbol{\nabla})\psi = \psi(\boldsymbol{\nabla}\cdot\mathbf{A}) + 2(\mathbf{A}\cdot\boldsymbol{\nabla})\psi = 2(\mathbf{A}\cdot\boldsymbol{\nabla})\psi \ ,$$

where the result of part (b), $\boldsymbol{\nabla}\cdot\mathbf{A} = 0$, is used. The contribution of this central term to the Hamiltonian is

$$-i\frac{\hbar e}{2mc}[\mathbf{B}\times\mathbf{r}]\cdot\boldsymbol{\nabla} = \frac{e\mathrm{B}}{2mc}[\hat{\mathbf{z}}\times\mathbf{r}]\cdot\mathbf{p} = \frac{e\mathrm{B}}{2mc}\hat{\mathbf{z}}\cdot[\mathbf{r}\times\mathbf{p}] = \frac{e\mathrm{B}}{2mc}L_z \ .$$

In the symmetric gauge $\mathbf{A}^2 = \mathrm{B}^2(x^2 + y^2)/4$; thus the required expression is retrieved.

d. In cylindrical coordinates

$$\nabla^2 = \frac{\partial^2}{\partial\rho^2} + \frac{1}{\rho}\frac{\partial}{\partial\rho} + \frac{1}{\rho^2}\frac{\partial^2}{\partial\theta^2} + \frac{\partial^2}{\partial z^2} \ , \quad \hat{L}_z = -i\hbar\frac{\partial}{\partial\theta} \ ,$$

and therefore the Schrödinger equation in these coordinates is

$$\left[-\frac{\hbar^2}{2m}\left(\frac{\partial^2}{\partial\rho^2} + \frac{1}{\rho}\frac{\partial}{\partial\rho} + \frac{1}{\rho^2}\frac{\partial^2}{\partial\theta^2} + \frac{\partial^2}{\partial z^2}\right) - i\mu_\mathrm{B}\mathrm{B}\frac{\partial}{\partial\theta} + \frac{1}{8m}\left(\frac{e\mathrm{B}}{c}\right)^2\rho^2\right]\psi = E\psi \ ,$$

where $\mu_\mathrm{B} = e\hbar/(2mc)$ is the Bohr magneton. Using for the wave function the form

$$\psi(\rho,\theta,z) = R(\rho)\Theta(\theta)Z(z)$$

separates the equation into three equations. First,

$$-\frac{\hbar^2}{2m}\frac{\partial^2 Z}{\partial z^2} = E_z Z \Rightarrow Z = e^{ik_z z} \quad \text{and} \quad E_z = \frac{\hbar^2 k_z^2}{2m} \ .$$

It follows that for $\Theta(\theta) = \exp[iM\theta]$ one finds $\hat{L}_z\Theta = -i\hbar\partial\Theta/\partial\theta = \hbar M\Theta$. The remaining equation is

$$\left[\left(\frac{\partial^2}{\partial\rho^2} + \frac{1}{\rho}\frac{\partial}{\partial\rho} - \frac{M^2}{\rho^2}\right) - \frac{1}{4\ell_B^2}\rho^2\right]R(\rho) = -\frac{2m}{\hbar^2}(E - E_z - \mu_\mathrm{B}M)R(\rho) \ .$$

The Ansatz $R \sim \rho^u$ for small ρ, with $u \geq 0$, yields $(u^2 - M^2)\rho^{u-2} + \mathcal{O}(\rho^u) = 0$, and hence $u = |M|$. For very large ρ, one tries $R \sim \exp[-a\rho^2]$, which gives for the left-hand side the expression

$$[(2a\rho)^2 - 4a - M^2/\rho^2 - \rho^2/(\ell_\mathrm{B}^4)]R \ ,$$

while the right-hand side is proportional to R. Hence, $a = 1/(4\ell_B^2)$, namely, $R \sim \exp[-(\rho/\ell_B)^2]/4$. it follows that the range in which the wave function is significant is of the order of the magnetic length, $\rho \sim \ell_B$. For $M = 0$, the function is finite at the origin and decays at large distances. For $M \neq 0$ the wave function is zero at the origin, then reaches a maximum at a distance of the order of the magnetic length, and finally decays. The result is a circular region around the origin in which the wave function is significant, similar to the classical cyclotron radius.

Answer E.3. Substituting the gauge transformation and Eq. (E.13) in Eq. (E.12) gives

$$i\hbar \frac{\partial}{\partial t}[\psi e^{i\frac{ef}{\hbar c}}] = \left[\frac{1}{2m}\Big(-i\hbar\boldsymbol{\nabla} + \frac{e}{c}(\mathbf{A} - \boldsymbol{\nabla}f)\Big)^2 - e(\varphi + \dot{f}/c)\right][\psi e^{i\frac{ef}{\hbar c}}] . \qquad \text{(E.18)}$$

Using

$$\left[i\hbar\frac{\partial}{\partial t} + e(\varphi + \dot{f}/c)\right][\psi e^{i\frac{ef}{\hbar c}}] = e^{i\frac{ef}{\hbar c}}\left[i\hbar\dot{\psi} - \psi\hbar e\dot{f}/(\hbar c) + \psi e(\varphi + \dot{f}/c)\right]$$

$$= e^{i\frac{ef}{\hbar c}}\left[i\hbar\frac{\partial}{\partial t} + e\varphi\right]\psi ,$$

and

$$\left[-i\hbar\boldsymbol{\nabla} + e(\mathbf{A} - \boldsymbol{\nabla}f)/c\right][\psi e^{i\frac{ef}{\hbar c}}] = e^{i\frac{ef}{\hbar c}}\left[-i\hbar\boldsymbol{\nabla}\psi + \psi\hbar e\boldsymbol{\nabla}f/c + \psi e(\mathbf{A} - \boldsymbol{\nabla}f)/c\right]$$

$$= e^{i\frac{ef}{\hbar c}}\left[-i\hbar\boldsymbol{\nabla} + e\mathbf{A}/c\right]\psi ,$$

one finds that Eq. (E.18) is indeed obeyed when ψ is the solution of Eq. (E.11).

Answer E.4.

a. Due to the rotational symmetry around the solenoid, the vector potential is tangential to the ring and is independent of the location there. The line integral around the ring is thus

$$\oint d\mathbf{r} \cdot \mathbf{A} = 2\pi b A_\theta .$$

On the other hand, by the Stokes theorem this integral is the magnetic flux through the ring, i.e., $2\pi b A_\theta = \Phi = \pi b^2 B$. It follows that $A_\theta = \Phi/(2\pi b) = a^2 B/(2b)$, and then Eq. (E.4) becomes

$$\hat{\mathcal{H}} = \frac{1}{2m}\Big(-i\hbar\nabla_\theta + \frac{e}{c}A_\theta\Big)^2 = \frac{1}{2m}\Big(-i\hbar\frac{1}{b}\frac{\partial}{\partial\theta} + \frac{e\Phi}{2\pi cb}\Big)^2 ,$$

where θ is the angular location of the electron on the ring. The eigenstates are $\psi(\theta)\exp[-iEt/\hbar]$, with

$$\frac{1}{2m}\Big(-i\hbar\frac{1}{b}\frac{\partial}{\partial\theta} + \frac{e\Phi}{2\pi cb}\Big)^2\psi(\theta) = E\psi(\theta) .$$

b. The solution is $\psi_n = \exp[in\theta]/\sqrt{2\pi b}$, as can be shown by substituting this form in the Schrödinger equation; the eigenenergy is

$$E_n = \frac{\hbar^2}{2mb^2}\Big(n + \frac{\Phi}{\Phi_0}\Big)^2 = e_0\Big(n + \frac{\Phi}{\Phi_0}\Big)^2 .$$

The continuity of the function round the ring, $\psi(\theta) = \psi(\theta + 2\pi)$, implies that n is an integer.

c. The classical equations of motion of Hamilton yield that the velocity is $\mathbf{v} = (\mathbf{p} + e\mathbf{A}/c)/m$ [see the discussion following Eq. (E.2)]. Therefore, the charge current carried by each electron is

$$\mathbf{j} = -e\mathbf{v} = -e(\mathbf{p} + e\mathbf{A}/c)/m .$$

In the present case the motion is tangential to the ring, and therefore the current density per electron is

$$j_\theta = -ev_\theta = -e(p_\theta + eA_\theta/c)/m .$$

In the quantum-mechanical description

$$p_\theta \to \hat{p}_\theta = -i\hbar\partial/(b\partial\theta) .$$

The quantum average of the current in the state $\psi_n = \exp[in\theta]/\sqrt{2\pi b}$ is

$$\langle j_\theta \rangle_n = -e\langle v_\theta \rangle = -e\langle \psi_n | \hat{p}_\theta + eA_\theta/c | \psi_n \rangle = -\frac{e\hbar}{2\pi m b^2}\left(n + \frac{\Phi}{\Phi_0}\right) .$$

A straightforward differentiation shows that this expression is identical to $-c\partial E_n/\partial\Phi$. Indeed, the relation between the current around the ring and the induced magnetic moment leads to this identification: the magnetic moment is $M = -\partial E_n/\partial B$, and the current is $\langle j_\theta \rangle = cM/A$.

d. At zero temperature, all energies up to the Fermi energy are filled, and hence the total energy is

$$E_{\text{tot}} = \sum_{E_n \leq E_{\text{F}}} E_n ,$$

while the total current is

$$j = -c\partial E_{\text{tot}}/\partial\Phi = -2e_0 \sum_n (n + \Phi/\Phi_0) ,$$

where each term is multiplied by 2 to account for the two spin states, and the sum runs over all levels for which $E_n \leq E_{\text{F}}$. In the absence of the magnetic flux one may join together terms with $n = \pm|n|$, and thus the sum vanishes. This means that the persistent cutrent flows only in the presence of a magnetic flux. However, it vanishes when this flux equals an integer product or a half integer product of the flux quantum. For instance, when $\Phi = N\Phi_0$ the current is

$$j = -2e_0 \sum_n (n + N) = -2e_0 \sum_n n' , \quad -N_m \leq n' = n + N \leq N_m = \sqrt{E_{\text{F}}/e_0} ,$$

and again the contributions of $n' = \pm|n'|$ cancel each other. In a similar fashion, when $\Phi = (N + 1/2)\Phi_0$ then

$$j = -2e_0 \sum_n (n + N + 1/2) = -2e_0 \sum_{n'=-N_m-1}^{N_m} (n' + 1/2) , \quad E_{\text{F}} = e_0(N_m + 1/2)^2 .$$

The contribution of the term with $n' = |n'|$ cancels that of $n' = -|n'| - 1$, and the current vanishes. Consequently, the persistent current is finite only when Φ/Φ_0 is not an integer or a half integer. Denoting $\Phi/\Phi_0 = N + x$, with $-1/2 \ll x \ll 1/2$, one finds that there are occupied states when $-\sqrt{E_{\mathrm{F}}/e_0} \leq n + N + x \leq \sqrt{E_{\mathrm{F}}/e_0}$. If $N_m - N$ is the integer part of $\sqrt{E_{\mathrm{F}}/e_0}$, then the current is

$$j = -2e_0 \sum_n (n + N + x) = -2e_0 \sum_{n'=-N_m}^{N_m} (n' + x) \ .$$

The terms in the sum can be arranged in pairs, with $n' = \pm|n'|$, and then $j = -2e_0 x (1 + 2N_m)$. It follows that the current is a periodic function of the flux, which decreases linearly between $\Phi/\Phi_0 = N - 1/2$ and $\Phi/\Phi_0 = N + 1/2$, *etc*, around each integer value of N.

Chapter 7

Selected topics

7.1 Preface

This chapter overviews several topics that require certain extensions of the concepts introduced in the previous chapters. It includes two main parts: (i) topics in soft condensed-matter, i.e., materials that are not translationally invariant, and (ii) topics in mesoscopic physics, that is the study of electronic properties in tiny systems for which the macroscopic descriptions presented in Chapter 6 are not sufficient.

Section 7.2 presents aspects of the physics of soft condensed-mater, e.g., liquid crystals, which are periodic along a certain direction in space and behave like liquids along the other directions. Another example are polymers; a ubiquitous simple picture for the latter envisages them as one-dimensional sinuous chains, whose arrangement in space is quite complex. Another subject introduced in Sec. 7.2 are fractals, structures that cannot be mapped onto themselves by translations; they do "return to themselves" when viewed on different length scales, under a transformation called "inflation". An example of fractal structures appears in **percolation theory**, which describes random mixtures of components with different properties.

Section 7.3 is devoted to mesoscopic physics, which describes systems whose typical sizes are in-between the microscopic, atomic scale and the macroscopic one. Though tiny crystals can be represented by finite-size lattices (with the appropriate conditions on their boundaries) their properties, in particular the electric ones, are dominated by quantum mechanics; the semiclassical theory of Drude [even in its quantum-mechanical version, Eq. (6.45)] is not sufficient to capture their behavior. For instance, **quantum dots** behave as artificial atoms, the electrical conductance of a **point contact** attains discrete values, and the electrical conductance of **quantum wires** is specific for each one of them. In contrast with macroscopic samples, whose physical properties are independent of the specific specimen (due to **self averaging** of the properties among the various parts of the sample), measurements taken on mesoscopic samples vary from one to another, though the disparate values, i.e., the fluctuations, follow certain universal properties.

7.2 Soft condensed-matter

An important phase of matter which is not a gas, liquid, or solid, are **liquid crystals**, that are built of long organic molecules. Liquid crystals possess unique optical properties, mostly because the long molecules modify the polarization of the light passing through them; the modification varies with the temperature, or under the effect of electric and magnetic fields. For this reason liquid crystals are used as screens (liquid-crystal displays, LCD), in thermometers, and so on. At low enough temperatures even these linear molecules arrange themselves in periodic solid crystals. Examples of such crystals are displayed in Figs. 4.23, 4.24, and 4.25. In many cases there is no direct phase transition from the solid to the liquid as the temperature is increased. Rather, there appear intermediate phases, with partial periodicity.

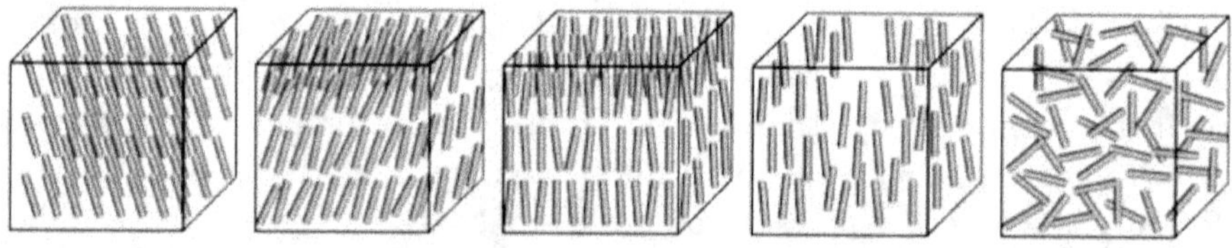

Fig. 7.1: Phases of a liquid crystal; from right to left: liquid, nematic liquid-crystal, smectic liquid-crystal A, smectic liquid-crystal C, solid.

The nematic phase. Intermediate phases between the solid and liquid ones are shown in Fig. 7.1. At high temperatures the molecules are arranged as in a conventional fluid, that is, both the locations of the molecules and the directions of their axes are random. At most there are short-range correlations among the locations and orientations of close-by molecules (see the right panel). When the density is raised, or the temperature is reduced, there appears at times the **nematic phase** (nematic is the Greek word for a thread), in which the centers of mass of the molecules are still like in a normal fluid, but the axes tend to be parallel to each other, as in the second panel of Fig. 7.1, along a vector $\hat{n}$ called the "director". As opposed to the magnetic order, where the magnetic dipole moments are parallel or are antiparallel to a certain vector, there is no such distinction in this case, and the pattern resembles an arrangement of quadrupoles parallel to each other. In other words, whereas the order parameter of a ferromagnet is given by the average (over all magnetic moments) of $\cos(\vartheta)$ (ϑ is the angle between $\hat{n}$ and the magnetic moment), and this average vanishes in the paramagnetic (disordered) phase and increases upon cooling in the ferromagnetic phase, the order parameter of the nematic phase is the average (over all molecules) of

$$[3\cos^2(\vartheta) - 1]/2 \ . \tag{7.1}$$

Here ϑ is the angle between $\hat{n}$ and the molecule's axis. In the fully-ordered nematic phase ϑ vanishes, and the order parameter is 1. In the disordered phase the molecule is free to be oriented along an arbitrary axis, which renders the average of $\cos^2(\vartheta)$ to be 1/3 (check!), namely the order parameter vanishes. As explained in Sec. 4.4, the van der Waals potential energy is lower in configurations in which long molecules are parallel to each other. This energy gain explains the origin of the nematic phase.

The smectic phases. A further reduction in the temperature may lead to the appearance of the **smectic phase** (the Greek word for soap or lather), the third or the fourth panel in Fig. 7.1. In this phase the molecules are located on planes, whose average distance from each other is fixed; the molecules are still free to move within each plane as in a liquid. Denoting the unit vector normal to planes by $\hat{N}$, then the correlation function along $\hat{N}$ is that of a solid, (third panel in Fig. 1.3), while within the plane it is that of a liquid (second panel in that figure). Quantitatively, the order parameter that describes the smectic phase is the average of the local density of the molecules, $\langle n(\mathbf{r}) \rangle$. This average is independent of $\mathbf{r}$ in the nematic phase, but is periodic along the $\hat{N}$ direction in the smectic phases,

$$\langle n(\mathbf{r}) \rangle \approx n_0 + 2n_q \cos(q\hat{N} \cdot \mathbf{r}) + \dots , \tag{7.2}$$

where $q = 2\pi/\ell$ and ℓ is the distance between the planes. n_0 and n_q are constants which represent the mean density of the material and the amplitude of the first harmonic of the oscillations along the normal to the planes, respectively, while $\dots$ represents higher harmonics of the wave vector (integer multiples of q) that appear at lower temperatures. The resulting spatial pattern is a combination of the nematic order parameter, Eq. (7.1), and the spatial one, Eq. (7.2). At sufficiently low temperature the material becomes a periodic solid, the left panel in Fig. 7.1, and then $\langle n(\mathbf{r}) \rangle$ is periodic along any of the basis vectors.

Figure 7.1 exhibits a specific series of transitions among various phases of a liquid crystal. However, other phases are possible in Nature. The third panel (from right) in the figure shows a special case of the smectic phase, with the molecules' axes normal to the plane. This phase is usually marked by A: the director $\hat{n}$ is parallel to $\hat{N}$. The next panel exhibits yet another smectic phase, marked by C. In this phase the axes of the molecules (along $\hat{n}$) make a finite angle with the direction of $\hat{N}$. Other smectic phases are shown in Fig. 7.2 (by a side view, and a single plane viewed from above). The upper row displays phases that are liquid-like in each plane. On the left of the figure the molecules are normal to the plane (the smectic A phase), and on the right they are inclined, as in the smectic C. The two lower rows show solid phases with a periodic order. The second row, on the other hand, illustrates the **hexatic phases**: the pattern of the molecules is not periodic, but the orientations of the lines that join neighboring molecules are correlated; those tend to point along specific directions. The points in the figure denote the periodic lattice, and the circles (or the triangles, or the ellipses) indicate the projections of the molecules on the plane. The centers of the molecules in the solid phases are congruent with the triangular lattice; in the hexatic phase they are not located

on a periodic lattice but the lines that join neighboring molecules are parallel to
the directions of the triangular lattice, with angles that are multiples of 60° (this
is the origin of the term "hexatic"). The hexatic phase appears usually between
the smectic phase and the solid one. This topological phase replaces at times the
periodic one in two dimensions (see Sec. 5.5).

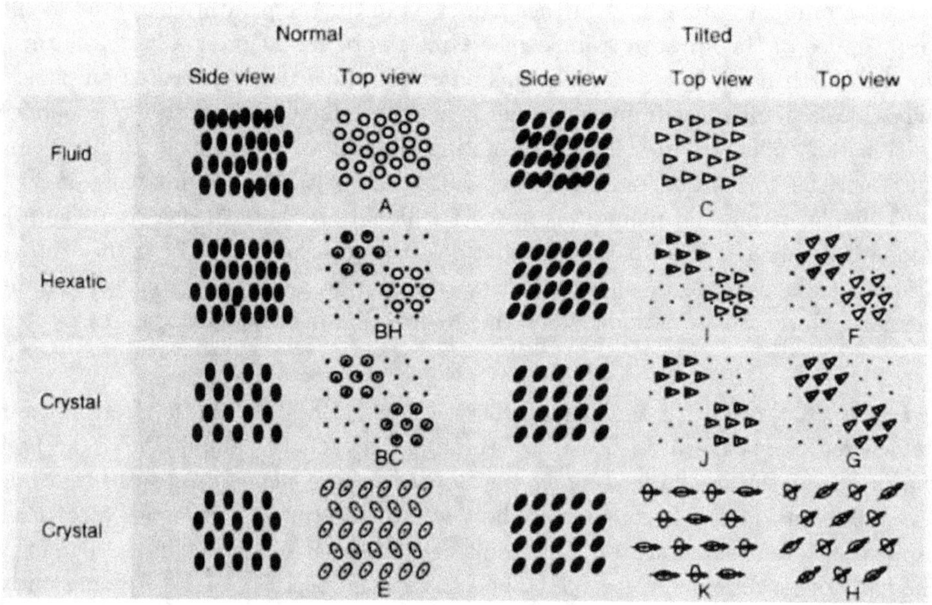

Fig. 7.2: Various phases of liquid crystals. The left column illustrates the order
within the planes (liquid, hexatic phase-with correlations among the bonds, or a
solid). The next two columns show "normal" states, in which the molecules are
oriented normal to the planes, or are inclined with respect to it. [J. D. Brock, R.
J. Birgeneau, J. D. Litster, and A. Aharony, *Liquids, crystals, and liquid crystals,*
Physics Today **42**, 52 (1989).]

The cholesteric phases. The axes of the molecules in the smectic phases are
normal to the plane or are inclined, making a finite angle with the normal to the
plane. Figure 7.3 illustrates the **cholesteric phase**, in which the molecules lie
in the plane, such that their axes are parallel to each other (like in the nematic
phase), but their locations are random, as in a liquid. This is the phase adopted by
molecules derived from cholesterol, and hence the name. The molecules may rotate
from one plane to the other (clockwise or anticlockwise). When the corresponding
angle is a rational fraction times 180°, the structure repeats itself after a finite
number of planes, and thus there is a finite unit cell along the direction normal to
the planes (the structure in the figure includes four planes). Such a pattern, which
depends on the direction of a rotation (of the form of a screw) is called **helical** or

chiral. Along the normal to the planes, the director $\hat{n}$ rotates around $\hat{N}$. That is, for $\hat{N}$ along $\hat{z}$,

$$\hat{n} = \cos(qz)\hat{x} + \sin(qz)\hat{y} \ . \tag{7.3}$$

The distance between successive planes (which determines the wave vector q), and the size of the unit cell along the normal to the planes, depend on the temperature. These lengths dictate the wave length of the light that will be reflected from the liquid crystal. Consequently, the liquid crystal changes its color as the temperature varies; this property serves to make it a thermometer.

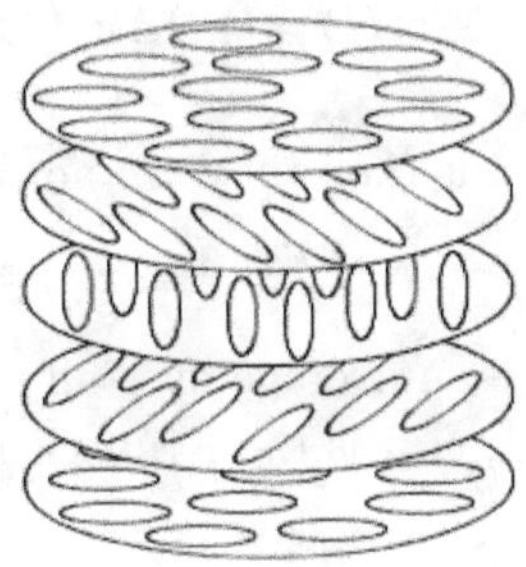

Fig. 7.3: A cholesteric liquid crystal; the molecule are parallel to each other within the planes but are rotated by 45° between neighboring planes.

Problem 7.1.
a. *The symmetries of a liquid crystal are determined by the symmetries of the average density, $\langle n(\mathbf{r}) \rangle$. Determine the symmetries of the phases in Figs. 7.1, 7.2, and 7.3.*
b. *Determine the lattice structures of the solid phases in Fig. 7.2.*

Polymers. Another research field in soft condensed-matter is centered around the properties of polymers. Simple models for those are based on one-dimensional chains. Each chain is built of basic units, called **monomers**; the one-dimensional chain is sometimes augmented by short side-branches. When immersed in a solution, the chain attains various forms, determined by the forces the atoms in the solution exert on it, and also by self-exerted forces: different parts apply forces on other parts. The optimal form of the chain at a given temperature is dictated by the entropy. Polymers, for instance, proteins or the DNA and the RNA molecules, are key objects in biology. Ensembles of polymers create a diversity of flexible materials as rubber and plastic systems. When polymers arrange themselves on a periodic lattice, the diffraction pattern contains information on the structure factor of each unit cell (see Sec. 3.9), and hence on the structure of a single polymer.

Random walks. The simplest model of an isolated polymer is based on **random walks**: beginning with the monomer at one end of the polymer, the successive monomer points at a random direction in space. The next monomer points along another random direction and so on, till the other end of the polymer. A random walk along such a path is called **diffusive motion**; under certain conditions this is the motion of a particle moving in space and colliding with random scatterers. This is also the motion, i.e., the **diffusion**, of an electron in the Drude picture [Sec. 6.2, or Eq. (6.45)], since the direction of the motion is changed randomly after each scattering event.

Let us denote each of the random steps between the successive monomers by $\Delta \mathbf{r}_n$. After N steps the walk reaches the point

$$\mathbf{r}(N) = \sum_{n=1}^{N} \Delta \mathbf{r}_n \ .$$

When the steps are independent (i.e., they are not correlated), then

$$\langle \Delta \mathbf{r}_n \rangle = 0 \ , \quad \langle \Delta \mathbf{r}_n \cdot \Delta \mathbf{r}_m \rangle = a^2 \delta_{nm} \tag{7.4}$$

(the angular brackets indicate an average over all possible configurations, and a is the average length of the step; δ_{nm} is the Kronecker delta function). In this case

$$\langle \mathbf{r}^2(N) \rangle = a^2 N \ , \tag{7.5}$$

(see problem 7.2). The typical distance between the two ends of the polymer, and therefore also the characteristic linear dimension of the region in space covered by the monomers in the chain, is

$$L \propto \sqrt{\langle \mathbf{r}^2(N) \rangle} = a N^{1/2} \ , \quad \text{or} \ \ N = A L^D \ , \tag{7.6}$$

where A is a dimensionless coefficient, and $D = 2$. This power is independent of the spatial dimension of the space in which the random walk is located.

Problem 7.2.

a. *Prove the relation $\langle \mathbf{r}^2(N) \rangle = a^2 N$, given Eq. (7.4).*

b. *Each pair of two consecutive steps, in a random-walk of N monomers where N is an even number, is replaced by a single step,*

$$\Delta \mathbf{r}'_n = \Delta \mathbf{r}_{2n} + \Delta \mathbf{r}_{2n-1} \ .$$

The result is a random walk of $N/2$ steps, that begins at the origin and ends at the same point as the original walk, $\mathbf{r}'(N/2) = \sum_{n=1}^{N/2} \Delta \mathbf{r}'_n = \mathbf{r}(N)$. Prove that

$$\langle \Delta \mathbf{r}'_n \cdot \Delta \mathbf{r}'_m \rangle = (a')^2 \delta_{nm} \ ,$$

and obtain the average length of the new step a'. Use the result to prove that

$$\langle [\mathbf{r}'(N/2)/a']^2 \rangle = N/2 \ .$$

Extend the result to a walk for which each step is a sum of ℓ basic steps. This mapping, where the information pertaining to short distances is averaged over to yield the average behavior at longer distances, is called **"renormalization"**.

Problem 7.3.

*A **biased random walk** is defined by*

$$\langle \Delta \mathbf{r}_n \rangle = \mathbf{r}_0 \ , \qquad \langle (\Delta \mathbf{r}_n - \mathbf{r}_0) \cdot (\Delta \mathbf{r}_m - \mathbf{r}_0) \rangle = a^2 \delta_{nm} \ .$$

The bias, in the case of a particle motion, can be caused by e.g., a non-homogeneous distribution of the scatterers, or an external force applied along a certain direction (for instance, gravity). In the Drude picture the bias is due to an electric field. Assume that the deviation away from $\mathbf{r}_0$ is spatially symmetric, i.e.,

$$\langle (\Delta r_n - r_0)_\alpha (\Delta r_m - r_0)_\beta \rangle = a^2 \delta_{\alpha\beta} \delta_{nm}/3 \ ,$$

$\alpha, \beta = 1, 2, 3$ are the cartesian components.
a. Calculate $\langle \mathbf{r}(N) \rangle$, $\langle [\mathbf{r}(N)]^2 \rangle$, and the mean standard deviation, $\langle [\mathbf{r}(N) - \langle \mathbf{r}(N) \rangle]^2 \rangle$.
b. Prove that the probability per unit volume that the end of a random walk of $N \gg 1$ monomers is at a distance $\mathbf{R}$ from the origin, i.e., the initial point, is

$$P(\mathbf{R}, N) = \frac{\exp\left[\frac{-3[\mathbf{R} - \langle \mathbf{R}(N) \rangle]^2}{2Na^2} \right]}{(2\pi Na^2/3)^{3/2}} \ .$$

*This is a Gaussian (in three dimensions). This formula expresses the **central limit theorem**. Find the probability to return to the origin. [Hint: prove that*

$$P(\mathbf{R}, N) = \int d^3 R' P(\mathbf{R}', N - 1) P(\mathbf{R} - \mathbf{R}', 1) \ .$$

The consequence is that the Fourier transforms obey

$$\widetilde{P}(\mathbf{k}, N) = [\widetilde{P}(\mathbf{k}, 1)]^N \ .$$

Exploit the Taylor expansion of $\widetilde{P}(\mathbf{k}, 1)$, and derive the inverse transform.]
c. Use the results of part (b) to find $\langle [\mathbf{r}(N)]^2 \rangle$.

Fractals. Equation (7.6) is a power law that relates the "mass" of the polymer, which is proportional to N, to the linear length, L, of the region where the polymer resides. Such a relation characterizes **fractals**; the power D is called the **fractal dimension** of this structure. The name originates from the similarity between Eq. (7.6) and the relation for the volume of a $d-$dimensional cube, $V = L^d$, where d is the spatial dimension. The fractal dimension need not be an integer. The density of the fractal is $\rho = N/V = (A/a^d)(L/a)^{D-d}$. When $D < d$ (for instance, a random walk in three dimensions) the average density decreases as the sample is increased. As the number of elementary units increases, the fractal occupies a diminishing part of the volume. On the other hand, for $D > d$ (e.g., a random walk in one dimension), the path "visits" each point numerous times, and the density increases with the number of elementary units.

Consider a sphere centered around the initial point of the random walk (taken as the origin). The number of points within such a sphere of radius r is proportional to $(r/a)^D$, and therefore the radial correlation function in a $d-$dimensional space is

$$g(r) \propto (r/a)^{D-1}/(r/a)^{d-1} = (r/a)^{D-d} \ ,$$

[Eq. (1.1)]. The scattering intensity off such a structure, from Eq. (3.48), is proportional to the Fourier transform of the correlation function, and hence to $\widetilde{\Gamma}(q) \propto (qa)^{-D}$, problem 7.4.

Problem 7.4.

According to problem 3.18 (part (b)), the scattering intensity off a three-dimensional sample, in which the radial correlation function is $\Gamma(r) = g(r)$ is

$$\widetilde{\Gamma}(q) = \frac{4\pi}{q} \int_0^\infty r\,dr\,\Gamma(r)\sin(qr) \ .$$

For a finite fractal, that contains $N = A(L/a)^D$ elementary units in a region whose linear size is L, the correlation function is $g(r) \propto (r/a)^{D-d}$ for $r \leq L$, and $h(x)$ decays to zero for $r \geq L$. Let us hence assume

$$g(r) \propto (r/a)^{D-d}h(r/L) \ ,$$

with $h(x) \approx 1$ for $x < 1$ and decays to zero as $x > 1$.
a. *Prove that when $qL \ll 1$, then*

$$\widetilde{\Gamma}(q) = 1 - (qR_g)^2/3 + \mathcal{O}[(qL)^4] \ ,$$
$$R_g^2 = \frac{\langle r^2 \rangle}{2} = \int d^2r g(r)\frac{r^2}{2} \propto L^2 \ .$$

R_g is the average "radius" of the fractal.
b. *Prove that when $qL \gg 1$, then $\widetilde{\Gamma}(q) \propto q^{-D}$.*

As opposed to solid crystals, which are mapped on themselves under translations, fractals are mapped on themselves when the length scale on which they are examined is varied. For instance, in part (b) of problem 7.2, a reduction of the instantaneous picture of a random walk by a factor of $\sqrt{2}$ produces another picture that appears very much like certain fragments of the original random walk. Figure 7.4 illustrates the same phenomenon: when every four steps out of the 400 of a certain random walk are joined together into a single step, the figure is "squeezed" by a factor of 2. Comparing with the random-walk picture obtained after 100 steps of this random walk, one cannot see a significant disparity. This property is termed **"self-similarity"**. Another example for such a system is displayed in Fig. 2.31: different generations of the Fibonacci lattice appear the same when measured with different "rulers".

Self-avoiding random walks. In one-dimension, a random walker moves forwards and backwards, so that every point is visited many times, and the length covered is far shorter than the number of steps, $(L/a) \propto N^{1/2} \ll N$. A real polymer, on the other hand, is composed of monomers. Each monomer occupies a finite volume in space, and two of them cannot occupy together the same volume in space. (This can be also understood as the result of the repulsion between the electronic clouds at short distances, due to the Pauli principle, see Chapter 4.) Consequently, in one dimension the length of a polymer (which cannot return back,

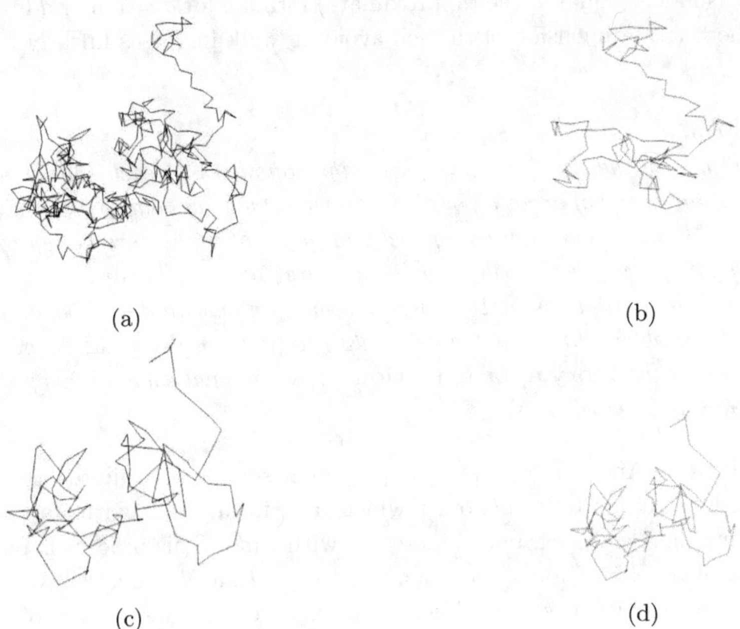

(a)

(b)

(c)

(d)

Fig. 7.4: Random walks with steps of length $a = 1$ and random planar orientations. (a) One hundred steps. (b) Four hundred steps. (c) As in (b), with each step representing four original steps. (d) The pattern of (c), doubly reduced.

in contrast with a random walker) of N monomers must be $L = aN$, and the random-walk model does not describe it faithfully. This is not the case in higher dimensions. There, the probability that a random walker hits the same point more than once is proportional to the product of the probabilities to be in that point, namely to the square of its density, $\rho^2 \propto (L^{D-d})^2$. The number of points visited more than once is proportional to the product of this probability with the volume, that is, $L^d \rho^2 \propto L^{2D-d}$. Hence, for very large $L \gg a$, in dimensions higher than $d_u = 2D = 4$, the path of the random walker does not collide with itself. The dimensionality d_u, above which this simple behavior appears, is called the **upper critical dimension**. Such critical dimensions are ubiquitous in other physical problems; this is the reason why certain problems are investigated as functions of their dimensionality. The random walk model can serve as a faithful model for the polymer in dimensions $d > d_u$. In contrast, in lower dimensions i.e., for $d < d_u$, there is a finite probability (at large L) for the path of the walker to collide with itself and another model is needed. Such a model is the **self-avoiding random walk**. For $1 \le d \le d_u = 4$, this model produces fractal dimensions less than 2, approximately $D_{\mathrm{SAW}} = (d+2)/3$ (problem 7.5). As the dimension of the fractal is smaller than 2, the density of the polymer is smaller than that of the simple random walk. In one dimension, as mentioned, one expects that $N = La$; the dimension

$D_{\text{SAW}} = 1$ is indeed equal to the (approximate) formula for $d = 1$. For d in-between 1 and 4 the fractal dimension of the self-avoiding walk increases linearly from 1 to 2.

Problem 7.5.

a. *Explain why the number of neighboring monomers of each monomer whose volume is of the order of a^d in a polymer described by a random walk, is $a^d(N/L^d)$. This implies that the repulsion energy between pairs of monomers is approximately $E = CN^2/(L/a)^d$, where C is the repulsion energy of a single pair.*
b. *Show that by multiplying the Boltzmann weight factor with the probability derived in part (b) of problem 7.3 gives the probability to find a walker that "avoids" collisions, and prove the* **Flory approximation***: The maximal value of this probability, as a function of L, is when $N \propto L^{(d+2)/3}$.*

    Percolation. Another description of polymers is based on a dilute lattice: every lattice site is occupied by a monomer which may form bonds with its neighbors. The concentration p of such bonds increases with time. This process happens, for instance, while cooking an egg, or among molecules that create a gel. When these bonds are replaced by tubes through which water vapor can flow among coffee grains that fill a container, the process mimics a percolator. For this reason the model is termed **percolation**. Figure 7.5 illustrates examples for clusters of bonds, as a function of p. For $p \ll 1$ there are only small clusters in which each monomer is connected to at least one of its neighbors. The clusters increase as p grows, and at a certain probability p_c, called the **percolation threshold**, there appears a cluster that connects opposite faces of the sample, even when the latter is very large (the mid panel in Fig. 7.5). A further increase of p leads to a further increase of that cluster, which now contains islands with finite clusters that are not attached to it (the left panel). As p increases, the finite clusters gradually diminish in size; the fully-periodic lattice is reached when $p = 1$. When $p < p_c$ the monomers in the gel move as in a viscous liquid. In the example of the percolator, water steam cannot reach from one end of the sample to the other. Once $p > p_c$, the sample behaves like a flexible gel, or a rubber, and is easily deformed. At $p = p_c$ there are clusters of all sizes; the largest cluster has the geometry of a fractal, with a fractal dimension which is 1 at $d = 1$, and reaches 4 for $d > 6$.

    Another example for percolation is a lattice of metallic wires. The complete lattice is capable of conducting electricity between its opposite edges. One may calculate the electrical resistance of this lattice by solving the Kirchhoff equations for a **lattice of resistors**. When the wire bonds between nearest neighbors are cut at random, the conductance decreases. Increasing the concentration of broken bonds leads to the appearance of clusters of bonds that are disconnected from the biggest one, which connects the two edges of the lattice. Once the concentration of broken bonds in a very large sample exceeds $1 - p_c$, that cluster is dismantled and the conductance vanishes. This is the **classical metal-insulator transition**.

Fig. 7.5: Percolation on the square lattice: the probability to form a bond is p. Right - $p = 0.4$, middle - $p_c = 0.5927\ldots$, left - $p = 0.8$. The figures are created by choosing for each possible bond a random number between 0 and 1. If that number is less than p the bond is formed; otherwise, it does not exist.

The percolation model describes also the dynamics of fluids in **porous media**. The sample is viewed as a periodic collection of boxes, some filled with solid material while the others are empty (or filled with a fluid, e.g., water or oil). The permeability of such a system to the flow of the fluid in the voids is large when the concentration of empty boxes is large, and it decreases to zero as that concentration tends to the percolation threshold.

The percolation is also applied to lattices of **magnetic ions**, in which the exchange interaction among neighboring ions is effective only when the bond in-between them is occupied by an additional ion (for instance, the lattice of lanthanum cuprate [Fig. 2.21(b) and problem s.2.17]. The antiferromagnetic interactions between nearest-neighbors coppers is established only through the oxygen that resides on the bond connecting them. Diluting the oxygen ions decreases the magnetic order. In the simplest classical model, the antiferromagnetic order totally disappears once the concentration of the oxygens is less than p_c.

This reduction in the antiferromagnetic moment as a function of the concentration resembles the diminishing of that moment as the temperature is being raised from zero to the transition temperature of the magnetic order in the complete lattice. This similarity leads to investigations of the percolation problem by methods developed for the study of phase transitions.

Fractal growth. The discussion following Fig. 2.3 and the one in Sec. 2.9 describes the growth of periodic crystals. In many cases though, the absorbed atoms are not joined together to form a crystalline cluster. Rather, they create a fractal. For instance, an electrode immersed in a solution collects on its surface ions from the solution that arrive there at random; the electric potential of those prevents additional atoms from moving into the space between the "branches" already grown on the cluster. Figure 7.6 illustrates a cluster grown from a solution on a planar surface; the measured fractal dimension (found by counting the number of atoms in a series of circles of increasing radii, and plotting these numbers as a function of the radii of the circles) is $D \approx 1.66$. A similar growth in three dimensions yields $D \approx 2.5$. In both cases the fractal dimension is smaller than the spatial one, which implies that the cluster density diminishes as its size increases.

A famous numerical procedure to mimic such growths is to consider diffusing particles; once the first one hits randomly the origin, it "sticks" to it. A second particle arriving there is glued to the first, and so on. Since the particles hit the growing cluster at random orientations and then they are not free anymore to reach an equilibrium (as opposed to the description displayed in Fig. 2.3), the growing cluster is not crystalline. The growing branches prevent additional particles to approach the origin, leading thus to a fractal structure, like the one in Fig. 7.6. This numerical procedure is termed **diffusion limited aggregation (DLA)**.

Fig. 7.6: Fractal cluster of zinc atoms, grown from a solution onto a plane. [M. Matsushita, M. Sano, Y. Hayakawa, H. Honjo, and Y. Sawada, *Fractal Structures of Zinc Metal Leaves Grown by Electrodeposition*, Phys. Rev. Lett. **53**, 286 (1984).]

7.3 Mesoscopic physics

Preface. As mentioned in Chapter 1, many recent new technologies are based on small units of materials, of typical sizes of several nanometers up to a few hundreds of them (1 nanometer $= 10^{-9}$ meter $= 10 \ \mathring{A} \ = 10^{-3}$ micron). As this scale is in-between the microscopic (atomistic) one and macroscopic scale, this field of research is termed **mesoscopic physics**, or **nanotechnology**. One of its main features is the fact that the electrons need to be treated quantum-mechanically, as the semiclassical picture of Drude is not applicable there.

Section 1.2 describes properties of systems whose dimensions is less than 3. In the extreme situation, a system is one-dimensional (or is two-dimensional) once its widths along the two directions normal to it (or along the perpendicular direction) comprise a single layer of atoms. Note though, that in problem 5.21 it is found that the specific heat of the phonons can appear as if it belongs to a two-dimensional system even when the width of the system is finite, and that there is a crossover from a two-dimensional behavior to a three-dimensional one as the width increases. The electronic properties of mesoscopic samples depend on their linear dimensions as well: the physical properties of one-dimensional and two-dimensional systems can differ quite significantly from those of the macroscopic three-dimensional ones. For instance, phonon excitations can destroy the solid phase at low dimensions (Sec. 5.5). Likewise, a tiny amount of impurities turns a conductor into an insulator in dimensions less than three due to repeated coherent scattering of the electrons off the impurities, a process which is less detrimental in three dimensions; there only at a finite concentration of impurities a metallic system becomes an insulator (at zero temperature). The density of states is another property that is sensitive to the dimensionality of the sample; this leads to unique quantum phenomena at one and two dimensions (Sec. 6.11).

Consider a system of linear size L (along all directions), in particular when L is larger than all other characteristic lengths of the electron motion. The Drude picture (Sec. 6.2) relates the electric resistance to collisions of electrons with certain imperfections in the material. The electron motion resembles that of a random walker (Sec. 7.2), that is, it moves **diffusively**. The distance between consecutive scatterers is of the order of the mean-free-path, ℓ. At low temperatures, the **collisions** are mostly with impurities (Fig. 6.4), and are **elastic**. The corresponding mean-free-path, ℓ_e (which is determined by the impurities' concentration) is of the order of 1-100 microns.

In the opposite limit, where L is shorter than ℓ_e, the motion is **ballistic**, that is, without collisions; the motion of the electrons has to be treated quantum-mechanically. The wave function of a free electron is either a plane wave, $\propto \exp[i\mathbf{k} \cdot \mathbf{r}]$, with a definite phase determined by the optical path of this wave, or a Bloch function, which also possesses a definite phase (related to the lattice momentum). Upon an **elastic** scattering event, the electron retains its quantum coherence. One has to solve the Schrödinger equation in the presence of the scatterers to obtain the wave function in the entire space. The phase of this wave function is determined by the sum of the optical paths (the products of the wave number with the distances) accumulated between the collisions, augmented by coherent contributions from the scattering); as a result, the motion of the electron is affected by **quantum interference**. In contrast, the collisions with phonons, or with other electrons, are generally **inelastic**, and thus affect also the **surrounding of the electron**, for instance the phonon gas. Consequently, the phase of the wave function is modified, and the electron loses its coherence. The mean-free-path that corresponds to the typical distance between such events is denoted by L_ϕ (the subscript ϕ indicates the phase), it is of the order of the mean-free-path in-between **inelastic collisions**, and is growing as the temperature is reduced. At low temperatures, and in very clean samples, L_ϕ can exceed by several orders of magnitude ℓ_e. In samples larger than L_ϕ the electrons lose the quantum coherence, and then their motion may be described by the semiclassical formalism, like in Sec. 6.10. The term **mesoscopic physics** refers to the intermediate region, $\ell_e \ll L \ll L_\phi$, where there appear deviations away from the semiclassical picture.

When a three-dimensional sample is significantly thicker than L_ϕ along two spatial directions but is in the mesoscopic regime along the third one, the energy levels of Eq. (6.27) are discrete continua along the first two directions, and are discrete along the third one. For a thin enough third dimension, the spacing between the discrete levels, which is proportional to L^{-2} in Eq. (6.27), is larger than $k_{\mathrm{B}}T$; the Boltzmann probability to occupy these levels is quite small. Hence these levels, except the lowest one, can be ignored and the system is described by the two-dimensional spectrum. Slightly larger thicknesses imply that several discrete levels, that represent the perpendicular coordinate, have to be taken into account. Similar considerations apply to a system which is very long along one direction, but finite along the other two. Such a system is described by discrete continuum along the first direction, and a finite number of discrete levels along the perpendicular directions. This is the appropriate description of the energy levels of electrons in the nanowires of Fig. 1.8, or in the nanotubes of Fig. 1.9(a). (In the first case there are free surfaces along the narrow direction; in the second, the nano stripe is rolled over to form a cylinder, and so the states along the direction perpendicular to the axis of the tube obey periodic boundary conditions, see problems 7.6 and s.7.4.)

Problem 7.6.

A two-dimensional electronic system is described by a square lattice as shown in Fig. 7.7(a). The energy bands, derived within the tight-binding approximation, are given in Eq. (6.119).

a. *The planar sheet is folded to form a cylinder, such that the point at the end of the vector $\mathbf{L}_1$ coincides with the origin, and the dashed lines are congruent to one another. The pentagonal prism in Fig. 7.7(b) displays the cylinder formed for $\mathbf{L}_1 = 5\mathbf{a}_1$. Find the unit cell and the energy spectrum when this lattice is viewed as a one-dimensional lattice.*

b. *Find the modifications in the answers to part (a) when $\mathbf{L}_1$ is replaced by $\mathbf{L}_2 = 2\mathbf{a}_1 + \mathbf{a}_2$ [the cylinder in Fig. 7.7(b)] and the dotted lines are congruent to one another.*

Several phenomena based on the coherent motion of electrons are mentioned in Chapter 6. The **Quantum Hall effect**, which determines the quantum of the electrical resistance h/e^2, results from a coherent ballistic motion of electrons along one-dimensional surfaces of a two-dimensional sample. As seen in Sec. 6.11, the absence of scattering is due to the spatial separation between the surfaces, because the width of the sample is longer than the magnetic length, $\ell_{\mathrm{B}} = \sqrt{\hbar c/(eB)}$ (B is the magnitude of the magnetic field). Due to this separation, the scattering of an electron that moves to the right on the lower surface in Fig. 6.35 to a state moving to the left on the opposite surface can be ignored. The fact that both

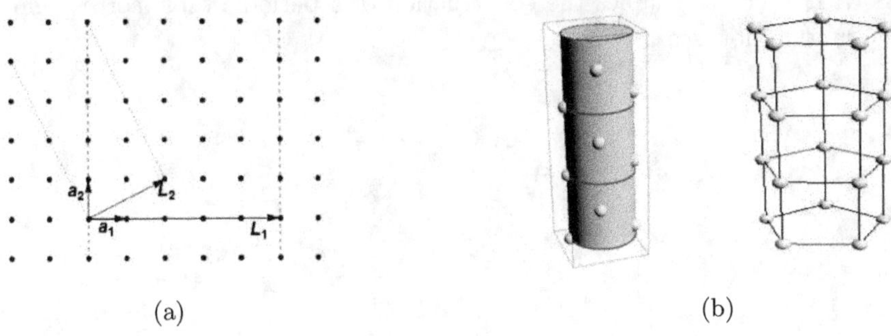

(a)   (b)

Fig. 7.7

expressions, for the quantum of the resistance and for the magnetic length, contain the Planck constant h (or $\hbar$) marks these phenomena as due to quantum origins. The same is true for quite a number of mesoscopic phenomena. For instance, the **Aharonov-Bohm effect**, surveyed in Appendix E and the related phenomenon of **persistent currents** flowing around rings under the effect of magnetic fields. Both effects stem from the quantization of the magnetic flux in units of the flux quantum $\Phi_0 = hc/e$, and both are detected in experiments carried on mesoscopic rings in which the coherence of the wave functions is maintained.

Discrete energy levels, quantum dots. Another length scale in the electronic motion is the wave length of de Broglie at the Fermi energy [see the discussion preceding Eq. (6.28)], $\lambda_F = 2\pi/k_F = h/p_F = h/(m^* v_F)$. In semiconductors like GaAs, the effective mass and the density of the conduction electrons are small; then $\lambda_F \approx 1000$ Å$=0.1$ μm. The energy levels of an electron [Eq. (6.27)] in a sample of size L form a discrete continuum provided that $L \gg \lambda_F$. But when L is small, the spacings between the energy levels increase; once they exceed $k_B T$, the physical properties depend on details of the spectrum, and not only on the density of states (a concept suitable for the description of a continuum or of a discrete continuum of levels). A system whose dimensions are all mesoscopic [for instance, a small cube (or a square in two dimensions), or the Bucky ball in Fig. 1.9(b)] is a **zero-dimensional system**, or a **quantum dot**. Because the energy levels of the electron in such a system are discrete, it is sometimes called an **artificial atom**. The reason for this is related to the Pauli principle according to which each discrete state can accommodate at most two electrons, and to the degeneracy of the energy levels. For instance, the energy levels of a spherically-symmetric potential are characterized by the quantum numbers $\{n\ell m\}$, but the energy is independent of m [the degeneracy is $(2\ell + 1)$]. The energy levels in an atom depend only weakly on ℓ, leading to the structure of the periodic table, where shells of $2n^2$ states are gradually filled. The single-electron energy levels in each shell are close to one another, while the spacing between different shells is larger. Other symmetries of the potential give rise to different degeneracies, but the gradual filling of the discrete levels with electrons enables the construction of a periodic table corresponding to each quantum dot, see Fig. 7.8.

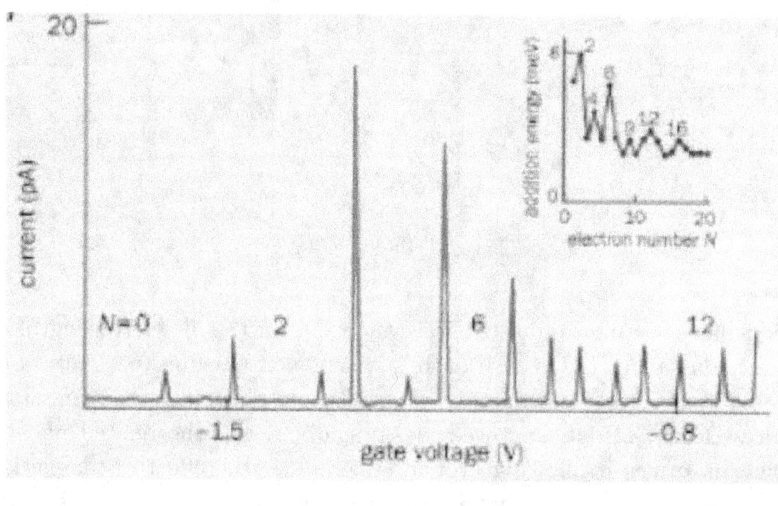

(a)

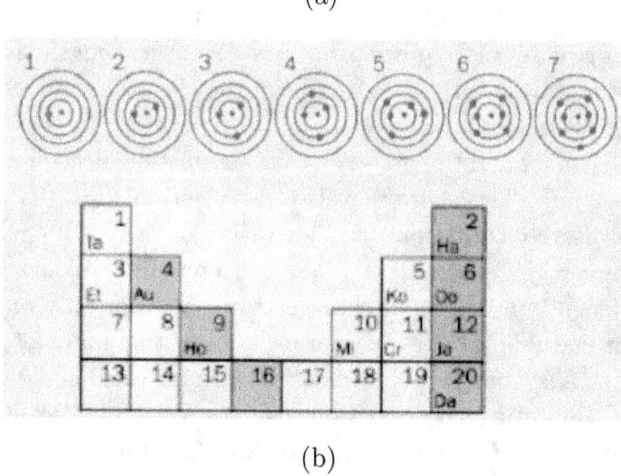

(b)

Fig. 7.8: (a) The current through a two-dimensional circular quantum dot as a function of the gate voltage, and the increment of the electrons' energy as a function of the number of electrons (in the inset). (b) The electronic shells in a two-dimensional harmonic potential (upper part) and the corresponding periodic table (lower part). [L. Kouwenhoven and C. Marcus, *Quantum Dots*, Physics World, June 1998, p. 35.]

Coulomb blockade. What is the energy of N electrons occupying the discrete energy levels $\epsilon_1 \leq \epsilon_2 \leq \epsilon_3 \leq \ldots \leq \epsilon_N$ in a quantum dot? Suppose that an electrode, whose voltage is V_g, is placed near the quantum dot. This voltage is termed **gate**

voltage. It gives the energy $-eV_g$ to each electron on the dot. For $N \gg 1$, the Coulombic repulsion energy of the N electrons is $(Ne)^2/(2C)$ where C is the capacitance of the dot. This capacitance is proportional to the linear size of the quantum dot. The total energy of the N electrons the quantum dot is

$$E(N) = \frac{(Ne)^2}{2C} - NeV_g + \sum_{n=1}^{N} \epsilon_n \ . \tag{7.7}$$

An electron added to the quantum dot increases the total energy by

$$E(N+1) - E(N) = (2N+1)e^2/(2C) - eV_g + \epsilon_{N+1} \ .$$

When the quantum dot is coupled to a bath of electrons whose chemical potential is μ, an extra electron can be added to the dot provided that $\mu = E(N+1) - E(N)$, that is

$$V_g^{(N+1)} = (2N+1)e/(2C) + (\epsilon_{N+1} - \mu)/e \ . \tag{7.8}$$

where μ is the mean chemical potential of the reservoirs. Hence when V_g (or μ) is varied, the number of electrons residing on the dot does not change until one of the discrete values of this equality is reached. The fact that an electron cannot be added to the dot at other values is called **Coulomb blockade**: the Coulomb repulsion prevents the addition of another electron. This device is also referred to as a **single-electron transistor**, since it enables the passage of electrons one by one.

The energy spectrum of a quantum dot can be measured by connecting it to two electronic reservoirs, whose chemical potentials differ from one another only slightly. At low temperatures a current can flow in-between the two reservoirs only when an electron is able to pass through the quantum dot; this happens at certain discrete value of the gate voltage, i.e., when Eq. (7.8) is obeyed. Figure 7.8(a) shows the results of such experiments. The measured system was a circular planar quantum dot. The current displays narrow peaks, at discrete values of the gate voltage, and it almost vanishes everywhere else. Equation (7.7) implies that the spacing between two consecutive peaks is

$$V_g^{(N+1)} - V_g^{(N)} = e/C + (\epsilon_{N+1} - \epsilon_N)/e \ .$$

The figure shows that this spacing is almost a constant for the fourth, fifth, and sixth electrons, i.e., the energy levels of these electrons are quite close to each other. Similarly, the spacings are almost constant for the $7-12$ electrons. Hence, the ground state is doubly degenerate, the second level is four times degenerate, the degeneracy of the third one is 6, and so on. When each such level represents an electronic "shell", the result is the filling of shells and the "periodic table" as described in Fig. 7.3(b). Once a shell is completely full, the spacing between it and the next peak is larger, due to the addition of the difference of energies between the shells.

It has been noted that the "magic" numbers 2, 6, 12, ... are obtained also when the potential confining the motion of the electrons to the quantum dot is harmonic, $m\omega^2\rho^2/2$, rather than spherical. (ρ is the radius of the motion in the plane.) The solution of the Schrödinger equation in problem (E.2.) yields the energy levels $E_n = \hbar\omega(n+1)$, with $(n+1)$ degeneracy (ignoring the spin, the ground-state level is not degenerate, the second level corresponds to the angular momentum $M = \pm 1$ and hence is doubly degenerate, in the third level $M = 0, \pm 2$, and so on). In this potential the nth level can be occupied by $2(n+1)$ electrons (including the spin degree of freedom), and thus the first level is filled after 2 electrons, the second shell is filled after 6 electrons, the third shell is filled after $2 + 4 + 6 = 12$ electrons, and so on, as is indeed found experimentally. Hence, the harmonic potential describes quite faithfully the confining potential of the electrons in the circular quantum dot. Indeed, the harmonic potential, which follows from a Taylor expansion around the minimum of any potential, gives a good approximation for the lower energy levels.

Problem 7.7.

The single-electron levels in a quantum dot are $\epsilon_1 < \epsilon_2 < \ldots < \epsilon_n < \ldots$; the degeneracy of the nth level is g_n. The repulsion energy [see for example Eq. (7.7)] between each pair of electrons residing in the nth level is denoted U_n, and that of electrons in the nth and the mth levels is $U_{n,m}$. The repulsion energies are assumed to be much smaller than the spacings between the single-particle energies.
a. *Find the total energy of N electrons on this dot, and the differences between the values of the gate potential at which an additional electron can enter the dot.*
b. *Discuss the result of part (a) for $g_\ell = 1$ for all ℓ's. What is the result when $U_{\ell,n} = U$? Compare it with Eq. (7.7). Is the approximation in that equation valid?*

Qubits and quantum computation. The mode of operation of a "classical" computer is based on bits which can assume one of two values, e.g., 0 or 1. In recent years there is a vast interest in **quantum computing**, based on **qubits**. The qubit is a superposition of two basis states, with complex coefficients. This superposition can be mapped on others by quantum operations; in this way the two complex coefficients are replaced by another pair of complex coefficients. This is the principle of quantum computing. Since a complex number contains much more information than the classical bit, it is expected that quantum computers will be much faster as compared to the classical one. An example of a qubit is the spinor that describes the state of the spin of an electron. The spinor is a superposition of the two basis states in which a specific component of the spin angular momentum is $\pm\hbar/2$. A possible realization of a quantum computer is built of a network of quantum dots, such that each dot contains a single electron. The application of a magnetic field causes the spin magnetic-moment to point along a desired direction, and thus to determine its spinor, that is, the qubit it represents. Magnetic interactions among the electrons on neighboring dots enable in principle complicated quantum computations. There are other proposals for quantum computers based on quantum systems with two basis states.

Two-dimensional electron gas and quantum wires. A typical mesoscopic system is described in Sec. 6.11. The quantum Hall effect is usually measured on such a system: a thin layer of GaAs, prepared by an epitaxial growth (Sec. 2.9). When the layer is thin enough, the spacings between the energy levels corresponding to the motion along the normal to the film are larger than $k_B T$, and then at low enough temperatures the motion along that direction is at the ground state of that direction, and the electrons move only in the plane of the film. The electron gas there is called a **two-dimensional electron gas**, abbreviated as **2DEG**. This type of thin films is used in many experiments in the mesoscopic regime. The thin film described in Sec. 6.11 is covered by a relatively thick layer of GaAlAs, which is an insulator. Metallic electrodes attached to this layer create in the thin film regions of negative potential, which prevents the electrons from moving there. In this way the rectangular region that serves for the quantum Hall effect measurements is formed.

The energy levels of an electron in the two-dimensional film are

$$E(n, \mathbf{k}) = [\hbar^2/(2m^*)](k_x^2 + k_y^2) + \overline{E}_n ,$$

where $\mathbf{k}$ is the wave vector in the plane (that belongs to the discrete continuum spectrum) and $\overline{E}_n$, $n = 0, 1, 2, 3, \ldots$, are the energy levels of the one-dimensional potential confining the motion of the electrons along the $\hat{\mathbf{z}}$−axis (e.g., a rectangular potential well, or an harmonic oscillator). The first term represents the dense levels; for each value of n their density of states (including the spin degeneracy) is $g^{(2d)} = m^*/(\pi\hbar^2)$ [problem 6.10 and the discussion following Eq. (6.169)]. When the energy $\overline{E}_n$ is not degenerate, the total density of the states is

$$g(E) = \sum_{n=0}^{\infty} g^{(2d)}\Theta(E - \overline{E}_n) . \tag{7.9}$$

Since $\Theta(x)$ is the step function, there appear discrete steps in the density of states as a function of the energy. At low temperatures, an increase in the Fermi energy (for instance, by deepening the gate voltage that attracts the electrons from the nearby bath to the wire, or by changing the Fermi energy in this bath as to add more electrons into the wire) entails jumps in the density of states, and consequently also in the electronic specific heat [Eq. (6.129)].

Quantum point-contacts. Gate voltages also serve to confine the electrons' motion to a narrow stripe (whose width is of the order of the electron's wave length, λ_F). In this case the spacing between the energy levels that correspond to the motion perpendicular to the stripe can be large, making the motion of the electrons one-dimensional, like in a **quantum wire**. Such a wire, that connects two wider regions as in Fig. 7.9(a), is called a **quantum point-contact, QPC**. When its width W is small, $W < \ell_e < L_\phi$, the electronic motion along the normal to the wire is ballistic (with no collisions), and is described in terms of discrete energy levels. The conductance of this device as a function of the gate voltage is depicted in Fig. 7.9(b). When the gate voltage is very negative, it blocks completely the motion of the electrons, and then the effective width vanishes, $W = 0$. As the gate voltage is reduced, the effective width W gradually increases, and current can flow. The measured conductance does not follow Ohm's law, or the Drude's result, Eq. (6.7) with $A \Rightarrow W$: the conductance is not linear in the width. Rather, it has a staircase structure, with steps of $2e^2/h$, the same quantum conductance unit that appears in the quantum Hall effect (the factor 2 is due to the spin as the energy levels are degenerate in the absence of a magnetic field). Similar experiments carried out in the presence of a magnetic field show that each step is split into two. Experimental results like those displayed in Fig. 7.9(b) can serve to deduce the exact value of the quantum of the conductance, and thence the fine-structure constant (see Sec. 6.11).

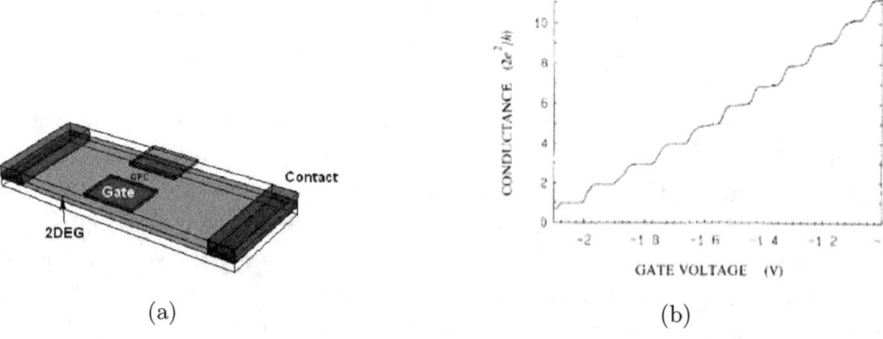

(a)   (b)

Fig. 7.9: (a) A quantum point-contact. The two-dimensional electron gas 2DEG is in the mid layer, in-between two insulating layers. Metallic electrodes, i.e., gates, are placed on the upper layer. These exert negative electric potential on the electron gas and create a narrow passage, i.e., a quantum wire, between its two parts, QPC. A small potential drop between the two contacts at the two edges induces current and enables the measurement of the electrical conductance of this device. (b) The conductance of the device as a function of the gate voltage. [B. J. van Wees, H. van Houten, C. W. J. Beenakker, J. G. Williamson, L. P. Kouwenhoven, D. van der Marel, and C. T. Foxon, *Quantized conductance of point contacts in a two-dimensional electron gas*, Phys. Rev. Lett. **60**, 848 (1988).]

A simple model that describes this phenomenon envisages the wire as a rectangular stripe of width W and length L ($W \ll L < \ell_e < L_\phi$). The spacings between the levels $\overline{E}_{n_y}$ decrease as W is increased. For example, when the potential well is very deep, the transverse wave functions are standing waves, with wave length $\lambda = 2W/n_y$ and energy $\overline{E}_{n_y} \approx \hbar^2(n_y\pi/W)^2/(2m^*)$ (check!). Assuming that the motion normal to the plane is in its ground state (as the first excited level of that motion is significantly larger than $k_B T$) and that the Fermi energy is $E_F \approx \hbar^2 k_F^2/(2m^*)$, all the states for which $n_y\pi/W \leq k_F$ are occupied at low temperatures. Each such transverse state represents a one-dimensional energy band (corresponding to the motion along the passage); $N = W k_F/\pi = 2W/\lambda_F$ such bands are occupied at low temperatures. When a potential drop V is applied between the two wider regions of the electron gas, the respective Fermi energies differ by $\delta\mu = eV$. The density of electrons that participate in the transport (moving from the higher potential to the lower one) is half that of total one, i.e., $\rho^{(1d)}(E)\delta\mu/2 = 2eV/(hv)$ [using the results of part (b) of problem 6.10 for one dimension, $\rho^{(1d)}(E) = n(dk/dE) = 4/(hv)$ where v is the velocity of the electrons; see also Eq. (6.125) for a free particle]. Each electron carries a charge e. Multiplying the density by the velocity, the current is $j_1 = 2e^2 V/h$. The velocity v is cancelled, and hence this contribution to the current does not depend on the band's index. It follows that the current is multiplied by N and the conductance is

$$G = N j_1/V = N(2e^2/h) \ . \tag{7.10}$$

Increasing gradually the width increases in turn N, and adds additional steps to the conductance, as shown in Fig. 7.9(b). The quantum effects become less important when $N \gg 1$ and the conductance is large.

Scattering of electrons and the Landauer formula. Equation (7.9) is based on the assumption that the quantum wire is a perfect conductor, in which the electrons do not undergo collisions. That is, any electron that enters the wire leaves it with the same wave function. When the wire contains scatterers, an electron can traverse it with a probability T ($T \leq 1$), called the **transmission coefficient**, or it can be reflected with a probability R ($R \leq 1$), termed the **reflection coefficient**. Charge conservation dictates that $R + T = 1$. A simple one-dimensional system in which these coefficients are used is presented in Sec. 5.6: the elastic scattering of a one-dimensional wave of a phonon (i.e., acoustic wave). The wave functions on the two sides of the scatterer are displayed in Eqs. (5.87) and (5.88): the impinging wave, $A\exp[i(kna - \omega t)]$, the reflected one, $rA\exp[i(-kna - \omega t)]$, with the complex reflection amplitude r, and the transmitted wave, $tA\exp[i(kna - \omega t)]$, with the transmission amplitude t (all for $n \geq 0$). The explicit calculation yields $R = |r|^2$ and $T = |t|^2$, and confirms the conservation law $R + T = 1$ (check!). A similar calculation for $A\exp[i(-kna - \omega t)]$ for $n \geq 0$ gives the transmitted wave $t'A\exp[i(-kna - \omega t)]$ and the reflected one $r'A\exp[i(kna - \omega t)]$ with $R = |r'|^2 = |r|^2$ and $T = |t'|^2 = |t|^2$. Problem 7.8 presents a similar solution for the scattering of an electron from a point-like scatterer on a one-dimensional lattice, within the tight-binding approximation. Since the probability of the electron to traverse the scatterer, at zero temperature, is T, the ideal conductance $2e^2/h$ is multiplied by T. At zero temperature, when the the Fermi energies of the electrons' reservoirs on the two sides differ only slightly, the only electrons that pass the sample are those at the (common) Fermi energy. The **Landauer formula** states that the conductance of a wire connected to two leads, each with N transversal channels, is

$$G = \frac{2e^2}{h} \sum_{\alpha,\beta=1}^{N} T_{\alpha\beta} \; , \qquad (7.11)$$

where $T_{\alpha\beta} = |t_{\alpha\beta}|^2$ is the transmission coefficient from channel α in one wire to channel β in the other wire, calculated at the Fermi energy. For a perfect transmission, $T_{\alpha\beta} = \delta_{\alpha\beta}$ Eq. (7.9) is retrieved. The **Landauer formula** (and its extension to finite temperatures) is the main tool to analyze the conductance of mesoscopic systems.

Problem 7.8.

An electron of energy $E = E_{\mathrm{F}}$, moves on a one-dimensional chain, modeled in the tight-binding approximation, with the Hamiltonian defined in problem 6.25(b), except for

$$\langle 0|\hat{\mathcal{H}}|0\rangle = \langle 1|\hat{\mathcal{H}}|1\rangle = E' \; ,$$

$$\langle 0|\hat{\mathcal{H}}|1\rangle = \langle 1|\hat{\mathcal{H}}|0\rangle^* = -\gamma' \; .$$

Energies are measured relative to $\overline{E}^{(0)} = 0$.

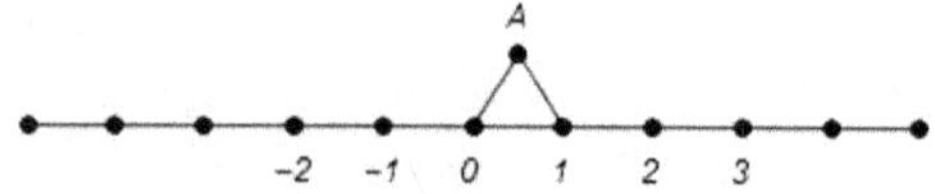

Fig. 7.10

a. *Show that the wave functions that appear in Eqs. (5.87) and (5.88) solve the tight-binding equations of this example, find the transmission and the reflection coefficients, and verify that they obey charge conservation, $T + R = 1$.*
b. *Repeat part (a) for an electron impinging from the opposite direction, and verify that the resulting transmission and reflection coefficients do not change.*
c . *Assume that the site 0 and 1 are also connected to a third atom (or a quantum dot), A (see Fig. 7.10),*

$$\langle A|\hat{\mathcal{H}}|A\rangle = \epsilon_A \; ,$$

$$\langle 0|\hat{\mathcal{H}}|A\rangle = \langle A|\hat{\mathcal{H}}|1\rangle = -\gamma_A e^{-i\phi/2} \; .$$

Assume also that $E' = 0$ and that $\gamma' = \gamma_0'$ is real (it results from the overlap of real wave functions). The phase ϕ represents the magnetic flux, in units of the flux quantum, through the triangle 1A0 due to the Aharonov-Bohm effect, Eq. (E.17). Calculate the transmission coefficient of this system, and show that $T(\phi) = T(-\phi)$ and that the transmission coefficient is periodic in ϕ. Determine the period.

Weak localization. The derivation of the Drude expression for the conductivity, Eqs. (6.6) or (6.45), assumes that successive collisions of the electrons with the scatterers are not correlated. In other words, repeated scattering events by the same scatterer are ignored (as it is assumed that such events are rare). However, this approximation is not always valid. Consider the probability that an electron reaches a point B from a point A; both points are located within the same sample. There are various paths connecting A with B, as illustrated in Fig. 7.11(a). In the quantum-mechanical description, each such path has a finite "width" of the order of the Fermi wave length $\lambda_F = 2\pi/k_F \approx a$. At sufficiently low temperatures one may ignore the inelastic scattering events, and assume that λ_F, $\ell_e \ll L < L_\phi$. To each of the n paths between A and B there corresponds an amplitude, $A_n = |A_n| \exp[i\phi_n]$, where the phase ϕ_n is determined by the optical path-length and the elastic scattering along the path. The total probability P to reach point B from point A is the absolute value squared of the total amplitude, i.e.,

$$P = |A|^2 = |\sum_n A_n|^2 = \sum_n |A_n|^2 + 2 \sum_{n<m} |A_n||A_m| \cos(\phi_n - \phi_m) \ . \qquad (7.12)$$

The first term on the right-hand side represents the probabilities of the independent paths, that is, the classical probability; the second describes quantum interference among the paths. This term is ignored in the Drude's formulation; the argument is that various paths possess different phases (e.g., due to their various lengths). As a result the sum in the second term comprises positive and negative contributions that make the sum far smaller than the one in the first term. There are, though, paths for which this argument breaks down. These are the ones in which the two points A and B coincide, Fig. 7.11(b). The electron can move along the path $A \to C \to D \to E \to B$, (with amplitude $A_1 = |A_1| \exp[i\phi_1]$) or via the reverse path, $A \to E \to D \to C \to B$, (with amplitude $A_2 = |A_2| \exp[i\phi_2]$). The second path is obtained from the first one by the operation of time reversal. Assuming that this operation does not change the Hamiltonian of the system then the electron traverses the same optical path and experiences the same scattering events. It follows that the probability of the electron to come back to its original starting point is $P_0 = |A_1 + A_2|^2 = 4|A_1|^2$, twice the classical probability for the same process, $P_{cl} = |A_1|^2 + |A_2|^2 = 2|A_1|^2$. Hence, the quantum-mechanical diffusion is slower than the classical one, due to constructive interference of paths that return to their origin. This interference causes a reduction in the electrical conductance. The electron tends to remain at the vicinity of the origin point, a phenomenon termed **"weak localization"**. Weak localization, observed at relatively small concentration of scatterers, is a weaker version of the more dramatic phenomenon of strong localization, Sec. 6.12.

According to the (classical) Matthiessen rule [Eq. (6.49) and Fig. (6.4)], the resistance of a metal increases as the temperature is raised. This relation, though, is derived for macroscopic systems by ignoring quantum interference effects. However, as the temperature is decreased, L_ϕ increases. At distances smaller than L_ϕ

the weak localization phenomenon takes place, and the resistance increases as the temperature is lowered, in contrast with the Matthiessen rule.

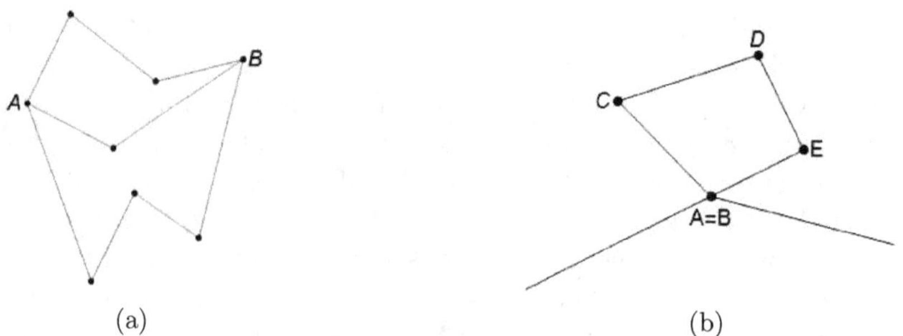

Fig. 7.11: (a) Three different paths between the points A and B. (b) The two points coincide and the path returns to its origin, A=B.

The increase in the probability to return to the origin stems form the constructive interference of the paths, which, in turn, requires that the Hamiltonian is **symmetric with respect to the operation of time reversal**. **Magnetic fields** break this symmetry; indeed, when a magnetic field is applied to the system, say normal to the plane in Fig. 7.11, and when the Zeeman interaction with the electron's spin is ignored, the effect of the field is to induce an additional phase on the amplitude due to the Aharonov-Bohm effect. This phase, ϕ [which is proportional to the magnetic flux through the loop, Eq. (E.17)] is added to the phase of one of the paths, and is subtracted from the phase of the other. The probability to return to the origin is then $P_0 = |A_1 \exp[i\phi] + A_2 \exp[-i\phi]|^2 = 2|A_1|^2[1+\cos(2\phi)]$. Consequently, the electrical conductance is periodic in the magnetic flux. The experimental observation of these periodic oscillations formed the proof for the existence of the phenomenon of weak localization. For a single path, the contribution to the conductance is in the range $\{0,\ 4|A_1|^2\}$, which implies that the oscillations of the conductance are of the same order of magnitude as the conductance itself. This is one of the sources of the **universal conductance fluctuations** discussed below. This is also the order of magnitude of the disparity in the conductances measured in the presence of a magnetic field on different samples with the same mean concentration of impurities (the impurity configurations leading to the closed paths are different in each sample). As the measurement averages over many closed loops that contain various magnetic fluxes, the effect of the weak localization is weakened.

The spin-orbit interaction, which couples the motion of the electron with its spin state, is significant in certain semiconductors. In such materials, the interference of the electronic paths is modified. The reason is that due to the spin-orbit interaction, the angular momentum of the orbital motion of the electron induces an effective magnetic field (somewhat similar to the magnetic field created by an electric current in a loop or in a solenoid), which operates on the spin of the electron. Then, when the electron moves in that effective magnetic field, its spin direction rotates. It turns out that the wave function of the spin reverses its sign when the electron completes a full cycle around a closed path. This implies that the sign of the interference term is different for the two paths in Fig. 7.11(b) when the spins rotate in opposite directions, leading to destructive interference which reduces to probability to return to the origin. As a result, the conductance increases (as compared to its reduced value due to the weak-localization phenomenon). This effect is termed **weak anti-localization**.

Self-averaging. As mentioned, in samples whose linear size is larger than L_ϕ, the electrons lose their quantum coherence, and then their motion can be described within the semiclassical theory. In particular, the conductance of a one-dimensional wire longer than L_ϕ obeys Ohm's law. Consider a wire whose length obeys $L \gg L_\phi$, and divide it into $N = L/L_\phi$ segments, each of length L_ϕ. Because of the scatterers that are randomly located in the wire, the resistances of these segments, $\{r_n,\ n = 1, \dots, N\}$ are distributed at random. The resistance of the wire is the sum of these resistances, $R = \sum_1^N r_n$. Denote by $\bar{r} = \langle r_n \rangle$ the average resistance of a single segment, and by $\Delta r = \sqrt{\langle (r_n - \bar{r})^2 \rangle}$ its standard deviation. Since the electronic wave functions in the different segments lack quantum coherence, the resistances of the various segments are not correlated and therefore $\langle (r_n - \bar{r})(r_m - \bar{r}) \rangle = (\Delta r)^2 \delta_{nm}$. When the total resistance R is averaged over an ensemble of wires, all with the same mean density of scatterers, then

$$\langle R \rangle = N\bar{r}\ , \quad (\Delta R)^2 = \langle (R - \langle R \rangle)^2 \rangle = \sum_{n,m} \langle (r_n - \bar{r})(r_m - \bar{r}) \rangle = N(\Delta r)^2\ ,$$

leading to

$$\frac{\Delta R}{\langle R \rangle} = \frac{\Delta r}{\bar{r}} \frac{1}{\sqrt{N}}\ .$$

This is the statistical law of **large numbers**. Note the similar result obtained for a random walker, Eq. (7.1). In both cases the law of large numbers follows from the addition of many independent, un-correlated, quantities. The mean-square deviation expresses the typical disparity of two samples: though the microscopic compositions are different, the relative difference of the resistances diminishes as the sizes of the samples increase. A large enough sample comprises a large number of the "basis" resistors (of length L_ϕ), and consequently their sum implies an averaging over almost all possible values. This is the meaning of **self averaging**. In fact, other extensive variables, obtained by summations over contributions from various parts of the sample (which are not correlated), are expected to behave similarly.

In dimensions d higher than 1, a sample whose linear size is L is divided into $N = (L/L_\phi)^d$ cubes, each of volume L_ϕ^d. In the absence of any correlations among the cubes, the mean-square deviation of the total resistance is proportional to $\sqrt{N} \sim (L/L_\phi)^{d/2}$, and again the mechanism of self averaging takes place. This argument yields

$$\frac{\Delta G}{\langle G \rangle} = \frac{\Delta G_0}{\langle G_0 \rangle} \left(\frac{L_\phi}{L}\right)^{d/2} ,$$

where $\Delta G_0/\langle G_0 \rangle$ is the relative mean-square deviation of the conductance of a cube of linear size L_ϕ.

Self averaging is realized in any system that possesses a basic length scale L_0, such that there are no correlations between the physical properties of cubes of volume L_0^d. When the linear size of the system is such that $L \gg L_0$, the relative disparities between various samples are proportional to $(L/L_0)^{-d/2}$, and therefore their macroscopic properties are rather similar though microscopically they are quite different. For instance, macroscopic samples of a glass made of the same chemical material have usually the same specific heat, though the microscopic configurations of the atoms in them differ.

Universal conductance fluctuations. The self-averaging feature of the conductance pertains to macroscopic samples, for which $L_\phi \ll L$. Smaller samples, all with the same mean scatterers' concentration, have different conductances, since their scatterers' configurations differ. The differences in the conductances of samples of size L_ϕ are scaled by $\Delta G_0/\langle G_0 \rangle$. At low temperatures, where the conductance of a mesoscopic system is determined by the diffusive paths of the electrons, its measured values fluctuate among samples prepared under identical conditions. Slight modifications in the Fermi energy (which change the de Broglie wave length of the electron, and consequently the lengths of the optical paths), the application of weak magnetic fields (that vary the magnetic fluxes and thus the Aharonov-Bohm phases in closed loops), or displacing the locations of the scatterers, induce significant changes in the conductance, because the interference is sensitive to the phases of the wave functions. Although this disparity affects considerably the quantum interference, and therefore the measured conductance, the standard deviation of the conductance, that is the average deviation from the average conductance, or the amplitude of the fluctuations in the conductance as a function of the Fermi energy or of a magnetic field, are always of the order of the **quantum unit of the conductance,** e^2/h, independent of the size of the samples, of the spatial dimension, or of the impurity concentration.

In contrast to the situation in large samples, where self averaging is active, the properties of various parts of a mesoscopic system of size L_ϕ are correlated. These correlations originate in the interference of wave functions corresponding to paths of the diffusion motion at distances shorter than L_ϕ. According to the Landauer formula, Eq. (7.11), the mean conductance (per spin) is

$$G = \frac{e^2}{h} \langle \sum_{\alpha,\beta=1}^{N} T_{\alpha\beta} \rangle = \frac{e^2}{h} \left(N - \sum_{\alpha,\beta=1}^{N} \langle |r_{\alpha\beta}|^2 \rangle \right) , \qquad (7.13)$$

where the condition of charge conservation

$$\sum_{\alpha,\beta=1}^{N} |t_{\alpha\beta}|^2 = N - \sum_{\alpha,\beta=1}^{N} |r_{\alpha\beta}|^2 \,, \qquad (7.14)$$

is used. Here, $r_{\alpha\beta}$ is the reflection amplitude of a wave impinging in channel α and reflected in channel β. The mean-square deviation of the conductance is found from an estimation of the mean-square deviations of the various terms in Eq. (7.13). It turns out that the correlations among the reflection coefficients are weaker than those among the transmissions, as there are less scattering events which return the electron to the channel where it came from.[1] From Eq. (7.13), the square of the mean deviation is

$$(\Delta G)^2 = (e^2/h)^2 \Big(\Delta \sum_{\alpha,\beta=1}^{N} |r_{\alpha\beta}|^2\Big)^2 = (e^2/h)^2 N^2 \Big(\langle |r_{\alpha\beta}|^4\rangle - \langle |r_{\alpha\beta}|^2\rangle^2\Big) \,. \qquad (7.15)$$

The reflection amplitude can be written as a sum over the amplitudes for returning back to the origin, $r_{\alpha\beta} = \sum_m B_m$. In this representation

$$\langle |r_{\alpha\beta}|^2\rangle = \sum_{n,m} \langle B_n^* B_m\rangle = \sum_{n} \langle |B_n|^2\rangle \,.$$

On the other hand

$$\langle |r_{\alpha\beta}|^4\rangle = \sum_{n,m,i,j} \langle B_n^* B_m B_i^* B_j\rangle = \sum_{n,m,i,j} \langle |B_n|^2|B_i|^2\rangle (\delta_{nm}\delta_{ij} + \delta_{nj}\delta_{mi}) = 2\langle |r_{\alpha\beta}|^2\rangle^2 \,.$$

Correlations between different paths are ignored in the last step, as their number is very small compared to the total number of paths. It follows that

$$\Big(\Delta |r_{\alpha\beta}|^2\Big)^2 = \langle |r_{\alpha\beta}|^4\rangle - \langle |r_{\alpha\beta}|^2\rangle^2 = \langle |r_{\alpha\beta}|^2\rangle^2 \,.$$

It remains to estimate $\langle |r_{\alpha\beta}|^2\rangle$. To this end, consider the Drude formula for the conductivity in two dimensions,

$$\sigma = \frac{e^2 n_e \tau}{m^*} = \frac{e^2}{h} k_F \ell_e \,, \qquad (7.16)$$

where the relations $n_e = k_F^2/(2\pi)$ (from part (a) of problem 6.10) and $\tau = \ell_e/v_F = \ell_e m^*/(\hbar k_F)$ are used. The dimensionless number $k_F \ell_e$ specifies the material. A material with $k_F \ell_e \gg 1$ is a conductor, while when this number is of the order of 1 the conductance is of the order of the quantum unit, and one has to take into account the quantum effects discussed above. When $k_F \ell_e \ll 1$ the material ia an insulator. The condition $k_F \ell_e \gg 1$ for metallic behavior is called the **Ioffe–Regel criterion.** This condition can be understood qualitatively: the semiclassical picture is valid as long as the wave packet that represents the electron is narrow enough, $\Delta k \ll k_F$. By the uncertainty relation, $\Delta k \sim 1/\Delta r \sim 1/\ell_e$: the semiclassical picture is relevant when $k_F \ell_e \gg 1$ and then material is a conductor.

---

[1]P. A. Lee, *Universal conductance fluctuations in disordered metals*, Physica **A140**, 169 (1986).

According to Ohm's law, Eq. (7.16) the states that the conductance of a planar sample of width W and length L is

$$G = \sigma \frac{W}{L} = 2 \frac{e^2}{h} \frac{\pi \ell_e}{2L} N \;, \tag{7.17}$$

where the relation $N = W k_{\rm F}/\pi$ is used [see Eq. (7.10)]. For spinless electrons, Eq. (7.11) yields

$$\langle G \rangle = \frac{e^2}{h} N^2 \langle T_{\alpha\beta} \rangle \;.$$

Comparing it with Eq. (7.17) results in

$$\langle |t_{\alpha\beta}|^2 \rangle = \langle T_{\alpha\beta} \rangle = \frac{\pi \ell_e}{2LN}$$

This result is inserted in Eq. (7.14), and then

$$\langle |r_{\alpha\beta}|^2 \rangle = \frac{1}{N}\left(1 - \frac{\pi \ell_e}{2L} \right) \;.$$

It follows from Eq. (7.15) that when $L \gg \ell_e$,

$$\left(\Delta\left[\sum_{\alpha,\beta=1}^{N} |r_{\alpha\beta}|^2 \right] \right)^2 = N^2 \left(\Delta |r_{\alpha\beta}|^2 \right)^2 = N^2 \langle |r_{\alpha\beta}|^2 \rangle^2 \approx 1 \;.$$

Hence $\Delta G \approx e^2/h$, as indeed observed in experiments. Exact calculations replace the 1 with a number of order 1, which depends on the geometry of the sample.

Thouless energy. The wave function of an electron moving on the nth diffusive path [Fig. 7.11 and Eq. (7.12)] is multiplied by the factor $\exp[-iE_n t/\hbar]$, where E_n is the energy of the moving electron. (This time dependence results from the time-dependent part of the Schrödinger equation.) In case the electron's energies on two paths differ slightly from each other, say by δE, then after time t_1 a phase difference $\delta E t_1/\hbar$ is accumulated between the two paths. Once this phase difference is of the order of 1, the constructive interference is lost. During the time $t_1 \approx \hbar/\delta E$, the electron accomplishes randomly $N = t_1/\tau$ diffusion steps, and therefore according to Eq. (7.5) it traverses the distance $L_1 \sim \ell_e \sqrt{N} \sim \sqrt{\ell_e^2 t_1/\tau} \sim \sqrt{D t_1}$, where the **diffusion coefficient** corresponds to the Fermi energy, $D = D(E_{\rm F}) = v_{\rm F}^2 \tau/d = v_{\rm F}\ell/d$. It follows that once the energy difference is such that $\delta E \approx \hbar D/L_1^2$, phase coherence (and consequently, the constructive interference) is lost. This typical energy, for a sample of length L, is called the **Thouless energy**, $E_{\rm T} = \hbar D/L^2$. The time spent by an electron diffusing from one side of the sample to the other, $\tau_{\rm T} = L^2/D$, is the **Thouless time**; it is related to the Thouless energy by the uncertainty relation, $E_{\rm T}\tau_{\rm T} \sim \hbar$. The destruction of the constructive interference causes a spatial decay of the wave function and consequently a reduction in the conductance. In fact, the ratio of the Thouless energy to the spacing between the energy levels around the Fermi energy conveys information about the character of these wave functions, whether extended (and then the conductance is high) or localized (which corresponds to a poor conductance).

Using the expression for the densities of states (problem 6.10),

$$g(E_{\mathrm{F}}) = \frac{dn_e}{2E_{\mathrm{F}}} = \frac{dn_e}{m^* v_{\mathrm{F}}^2} \ ,$$

in the expression for the Drude conductance [for instance, Eq. (6.161)] results in the **Einstein relation**

$$\sigma = \frac{e^2 n_e \tau}{m^*} = e^2 g(E_{\mathrm{F}}) D(E_{\mathrm{F}}) \ . \tag{7.18}$$

In d dimensions, the conductance is [using Eq. (7.18)]

$$G = \sigma L^{d-2} = \frac{e^2}{\hbar} \frac{E_{\mathrm{T}}}{\delta} \ , \tag{7.19}$$

where $\delta = 1/(g(E_{\mathrm{F}})L^d)$ is the mean spacing between energy levels around the Fermi energy. When $E_{\mathrm{T}} \gg \delta$ the energy levels around the Fermi energy are very dense; then the conductance is much higher than the quantum unit. On the other hand, when $E_{\mathrm{T}} \ll \delta$, the spacing between the levels is large, which usually signals that they are localized at certain regions in the system, and are not extended all over it. Indeed then the conductance is far smaller than the quantum unit.

Problem 7.9.

The wave function of an electron is substantial within a "tube" of width λ_{F}, the wave length at the Fermi momentum (why?), and therefore the volume (in d dimensions) it occupies when the electron moves a time dt is of the order of $\lambda_{\mathrm{F}}^{d-1} v_{\mathrm{F}} dt$. On the other hand, during a time t, a diffusive electron reaches the distance $(Dt)^{1/2}$, and hence it covers a volume of the order of $(Dt)^{d/2}$.
a. *Explain why the probability of the path to return to its original point is*

$$\int_\tau^{\tau_x} dt \frac{\lambda_{\mathrm{F}}^{d-1} v_{\mathrm{F}}}{(Dt)^{d/2}} \ ,$$

where $\tau_x = L_x^2/D$. What is the maximal length L_x, for $L_\phi \gg L$ and for $L_\phi \ll L$? Compute explicitly the integral, and discuss the result as a function of the size of the system, L.
b. *Assume that the probability found in part (a) is small. Explain why in this limit the relative correction to the conductivity, $\Delta\sigma/\sigma$, due to weak localization effects is proportional to this probability. Estimate this correction in one, two, and three dimensions. What can be deduced about the conductivity of metals at low temperatures in arbitrary dimensions?*

Artificial lattices. The technological developments of recents years enable the construction of artificial latices. For example, Fig. 7.12 displays periodic arrays of quantum dots, whose lattice constants are of the order of hundreds of nanometers. Figure 7.13 shows a network of nano-scale transistors; these are manufactured commercially by Intel, for future use in quantum computers.

 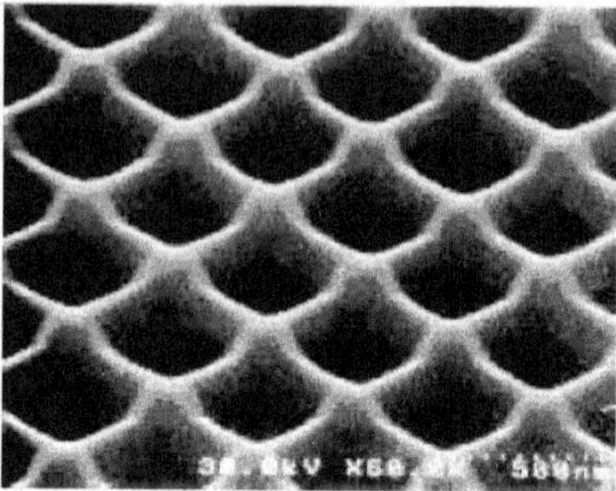

Fig. 7.12: Periodic arrays of quantum dots. Taken from `http://www.rciqe.hokudai.ac.jp/movpe/en/k-naiyou/index.html`.

Fig. 7.13: Wafer of transistors whose size is tens of nanometers. Taken from `http://qbnets.wordpress.com/2011/05/07/intel-waffles`.

Another branch of current research of periodic structures is discussed in problem **s.6.6**, part (c). Figure 6.49 displays an electric potential shaped as an egg tray. It is created by the interference of two electromagnetic waves perpendicular to one another. The interference constructs a two-dimensional standing wave; both the height of the potential barriers in-between the minima and the maxima and their locations can be controlled by a suitable choice of the electromagnetic waves (produced by monochromatic laser radiation). This potential can serve as a substrate that captures atoms from the gas above it. It is termed an **optical lattice**.

7.4 Answers for the problems in the text

Answer 7.1.

a. The right panel in Fig. 7.1 displays long molecules that are arranged in a liquid phase, that is, without any apparent order. The average density of this pattern is homogeneous; this means that on the average the system has continuous symmetry under translations and under rotations. The correlation function $\langle n(0)n(\mathbf{r})\rangle$ is the one shown in Fig. 1.2(b). The next panel in Fig. 7.1, in the nematic phase, the director vector has on the average a preferred direction. This breaks the continuous symmetry with respect to rotations around that direction. In the third panel from the right, that shows the smectic phase A, the continuous symmetry with respect to translations is broken as well. A crystalline pattern is formed along the $\hat{\mathbf{z}}-$direction: the symmetry with respect to translations along this axis is discrete, i.e., $\langle n(\mathbf{r} + \ell\hat{\mathbf{z}})\rangle = \langle n(\mathbf{r})\rangle$, but is continuous in the XY plane, as in the liquid phase. In the fourth panel that displays the smectic phase C the molecules are inclined with respect to $\hat{\mathbf{N}}$, which itself is normal to the planes. As a result, the rotation symmetry around this axis is broken. The left panel shows the solid phase of the inclined smectic crystal. It has discrete translational symmetry as the tetragonal lattice, but lacks the rotational symmetry.

The upper line in Fig. 7.2 shows the smectic phases A (on the left) and C (on the right). The symmetries of these phases are discussed above. The next line displays the hexatic phase, which breaks another continuous rotational symmetry, within the planes. In this phase there is no symmetry with respect to translations in the XY planes, but the symmetry with respect to rotations by 60° around the $\hat{\mathbf{z}}-$axis is maintained.

The cholesteric phase is exhibited in Fig. 7.3. In addition to the broken symmetries of the smectic phases, here the translational symmetry along the direction normal to the planes is by a base vector four times longer than the distance between neighboring planes.

b. The lattice denoted by BC in Fig. 7.2 is a triangular one. A side view reveals that neighboring planes are displaced relative to each other, such that they are ordered in the form ABABAB... (see Fig. 2.24). This is a hexagonal lattice (Fig. 2.28) as opposed to the **HCP** lattice shown in Fig. 2.24. The basis vector, normal to the planes, is longer than the vector that appears in the dense packing of spheres. The unit cell of this lattice comprises a rhombus in the plane, and the normal to the plane, of length 2ℓ. The phase E as well has periodicity of two planes along the $\hat{\mathbf{z}}-$axis, but within each plane the lattice is rectangular, with two molecules in each rectangular unit (at the center and at the corner).

The lattices J, G, K, and H have all periodicity of a single plane. J and G are triangular in the plane and therefore are simple hexagonal lattices. The planes that build the lattices K and H appear as rectangular; the axes of the rectangle are inlined by 60° relative to the horizontal axis in the figure. These lattices are orthorhombic.

Answer 7.2.

a. Using $\mathbf{r}(N) = \sum_{n=1}^{N} \Delta\mathbf{r}_n$ gives

$$\langle [\mathbf{r}(N)]^2 \rangle = \sum_{m=1}^{N} \sum_{n=1}^{N} \langle \Delta\mathbf{r}_m \cdot \Delta\mathbf{r}_n \rangle = \sum_{n=1}^{N} a^2 = Na^2 \ ,$$

since $\langle \Delta\mathbf{r}_n \rangle = 0$ and $\langle \Delta\mathbf{r}_n \cdot \Delta\mathbf{r}_m \rangle = a^2 \delta_{nm}$.

b. In order to describe a "new" step built of two "old" ones, one first examines the correlation of two new steps in terms of that of two old ones,

$$\langle \Delta\mathbf{r}'_n \cdot \Delta\mathbf{r}'_m \rangle = \langle \Delta\mathbf{r}_{2n-1} \cdot \Delta\mathbf{r}_{2m-1} \rangle + \langle \Delta\mathbf{r}_{2n-1} \cdot \Delta\mathbf{r}_{2m} \rangle$$
$$+ \langle \Delta\mathbf{r}_{2n} \cdot \Delta\mathbf{r}_{2m-1} \rangle + \langle \Delta\mathbf{r}_{2n} \cdot \Delta\mathbf{r}_{2m} \rangle \ .$$

This correlation is nonzero only when $n = m$, and then its value is $2a^2$,

$$\langle [\Delta\mathbf{r}'_n]^2 \rangle = 2a^2 \ .$$

The mean value of the new length is $a' = \sqrt{2}a$. Hence

$$\langle [\mathbf{r}'(N')]^2 \rangle = (N/2)2a^2 = N'(a')^2 \ , \quad \text{with} \ \ N' = N/2 \ .$$

For ℓ successive steps one may repeat these arguments to find that

$$a' = \sqrt{\ell}a \ ,$$

and

$$\langle [\mathbf{r}'(N')]^2 \rangle = (N/\ell)\ell a^2 = N'(a')^2 \ , \quad \text{with} \ \ N' = N/\ell \ .$$

The functional dependence of the mean distance on the number of steps is not changed.

Answer 7.3.

a. According to the information given in the problem,

$$\langle \mathbf{r}(N) \rangle = \sum_{n=1}^{N} \langle \Delta\mathbf{r}_n \rangle = N\mathbf{r}_0 \ ,$$

$$\langle [\mathbf{r}(N) - N\mathbf{r}_0]^2 \rangle = \sum_{n,m=1}^{N} \langle (\Delta\mathbf{r}_n - \mathbf{r}_0) \cdot (\Delta\mathbf{r}_m - \mathbf{r}_0) \rangle = Na^2 \ ,$$

and then, with $\mathbf{r}(N) = N\mathbf{r}_0 + [\mathbf{r}(N) - N\mathbf{r}_0]$, one obtains

$$\langle [\mathbf{r}(N)]^2 \rangle = (N\mathbf{r}_0)^2 + Na^2 \ .$$

b. The probability per unit volume to reach from the origin to the point $\mathbf{R} = \mathbf{r}(2)$ after two steps is obtained from the sum of the probabilities to reach the point $\Delta\mathbf{r}_1$

after one step, with the probability per unit volume $P(\Delta\mathbf{r}_1, 1)$; then to do the step $\Delta\mathbf{r}_2$ with the probability per unit volume $P(\Delta\mathbf{r}_2, 1)$,

$$P(\mathbf{R}, 2) = \int d^3(\Delta r_1) P(\Delta\mathbf{r}_1, 1) P(\mathbf{R} - \Delta\mathbf{r}_1, 1) .$$

Similarly, one finds

$$P(\mathbf{R}, N) = \int d^3 R' P(\mathbf{R}', N-1) P(\mathbf{R} - \mathbf{R}', 1) .$$

This is a convolution of two probabilities; the Fourier transform is then, according to Eq. (B.11), the product of the Fourier transforms, and therefore

$$\widetilde{P}(\mathbf{k}, N) = \widetilde{P}(\mathbf{k}, N-1)\widetilde{P}(\mathbf{k}, 1) = \cdots = [\widetilde{P}(\mathbf{k}, 1)]^N .$$

The Fourier transform of the probability of a single step is

$$\widetilde{P}(\mathbf{k}, 1) = \frac{1}{(2\pi)^3} \int d^3 R\, e^{-i\mathbf{k}\cdot\mathbf{R}} P(\mathbf{R}, 1) = \frac{1}{(2\pi)^3} \langle e^{-i\mathbf{k}\cdot\mathbf{R}} \rangle ,$$

where the right-hand side is an average over all possible steps. By using a Taylor expansion, one finds

$$\langle e^{-i\mathbf{k}\cdot\mathbf{R}} \rangle = 1 - i\mathbf{k}\cdot\langle\mathbf{R}\rangle - \langle(\mathbf{k}\cdot\mathbf{R})^2\rangle/2 + \mathcal{O}(k^3)$$
$$= \exp[-i\mathbf{k}\cdot\langle\mathbf{R}\rangle - \langle[\mathbf{k}\cdot(\mathbf{R} - \langle\mathbf{R}\rangle)]^2\rangle + \mathcal{O}(k^3)] .$$

Since the steps are symmetric,

$$\langle[\mathbf{k}\cdot(\mathbf{R} - \langle\mathbf{R}\rangle)]^2\rangle = \langle \sum_{\alpha,\beta=1}^{3} k_\alpha(R - \langle R\rangle)_\alpha k_\beta(R - \langle R\rangle)_\beta \rangle = k^2 a^2/3 .$$

Carrying out the inverse Fourier transform gives the desired probability,

$$P(\mathbf{R}, N) = \frac{1}{(2\pi)^3} \int d^3 k\, e^{i\mathbf{k}\cdot\mathbf{R}} \langle e^{-i\mathbf{k}\cdot\mathbf{R}} \rangle^N = \frac{1}{(2\pi)^3} \int d^3 k\, e^{i\mathbf{k}\cdot(\mathbf{R}-\langle\mathbf{R}\rangle)} e^{-Nk^2 a^2/6 + \mathcal{O}(k^3)}$$

substitution of $\quad \mathbf{k} \Rightarrow \mathbf{q}/\sqrt{N} , \quad \mathbf{R} - \langle\mathbf{R}\rangle \Rightarrow \mathbf{u}\sqrt{N} \quad$ gives

$$P(\mathbf{R}, N) = \frac{1}{(2\pi\sqrt{N})^3} \int d^3 q\, e^{i\mathbf{q}\cdot\mathbf{u}} e^{-q^2 a^2/6 + \mathcal{O}(q^3/\sqrt{N})} .$$

In the large N−limit the last term in the exponent may be ignored, and then one obtains the Fourier transform of the Gaussian distribution,

$$P(\mathbf{R}, N) = \frac{1}{(2\pi\sqrt{N})^3} \int d^3 q\, e^{i\mathbf{q}\cdot\mathbf{u}} e^{-q^2 a^2/6} = (2\pi a^2 N/3)^{-3/2} e^{-3(\mathbf{R}-\langle\mathbf{R}\rangle)^2/(2a^2 N)} .$$

The probability to return to the origin, in three dimensions, is

$$P(0, N) = (a\sqrt{2\pi N/3})^{-3} .$$

b. The mean of the squared distance is

$$\langle[\mathbf{r}(N)]^2\rangle = 2\pi \int R^2 dR\, R^2 P(\mathbf{R}, N) .$$

The outcome of this integration reproduces the results in part (a).

 Introduction to Solid-State Physics

Answer 7.4.

a. In this limit one may use

$$\frac{\sin(qr)}{qr} = 1 - \frac{(qr)^2}{6} + \ldots .$$

One then finds (exploiting the normalization of the correlation function, $\int d^3r g(r) = 1$) the desired result.

b. Denoting $qr = y$, then

$$\widetilde{\Gamma}(q) \propto q^{-D} \int_0^\infty dy\, y^{D-2} \sin(y) h[y/(qL)] .$$

As the integral here converges, one deduces that $\widetilde{\Gamma}(q) \propto q^{-D}$.

Answer 7.5.

a. The number of monomers that exist in a volume a^d around a certain monomer is $a^d \rho = a^d N/L^d$, where L is the linear size of the entire chain. Each of these monomers "collides" with the given monomer, and hence the total energy of the collisions equals this number, times the energy of a single collision event, times the number of monomers,

$$E = CN(a^d N/L^d) = CN^2/(L/a)^d .$$

b. From part (b) of problem 7.3, one finds that the probability to find a random walk whose two edges are at a distance L apart is proportional to $\exp[-3L^2/(2a^2N)]$. The Boltzmann weight of the collision energy is $\exp[-E/(k_\mathrm{B}T)]$. The total probability is hence $\exp[-F(L)/(k_\mathrm{B}T)]$, with the "free energy"

$$F(L) = 3k_\mathrm{B}T(L/a)^2/(2N) + CN^2/(L/a)^d .$$

The probability is maximal when the free energy as a function of L is minimal, which indeed happens when $L \propto N^{3/(d+2)}$.

Answer 7.6.

The energy bands of a square lattice, calculated within the tight-binding approximation that accounts for overlaps of wave functions on nearest-neighbors ions alone, are given in Eq. (6.119),

$$E_n(\mathbf{k}) \approx \overline{E}_n + 4\Gamma_n(a)[\sin^2(k_x a/2) + \sin^2(k_y a/2)] .$$

Wrapping the two-dimensional sheet adds periodic boundary conditions around the circular direction. The cylinder can be described as a one-dimensional lattice along the axis.

a. The cylinder is a one-dimensional lattice, with a lattice vector $\mathbf{a}_2$. The unit cell comprises five lattice sites, $\mathbf{r}_m = m\mathbf{a}_1$, $m = 0,1,2,3,4$. The boundary condition along the lattice vector $\mathbf{a}_1$ is $\psi(\mathbf{r} + \mathbf{L}_1) = \psi(\mathbf{r})$. The component of the lattice momentum k_x can assume the discrete values $k_x = 2\pi n/(5a)$, with $n = 0,1,2,3,4$; the values of k_y are still continuous. Each discrete value of k_x represents an energy band, and thus apparently there are five bands. However, the energies are identical for $n \leftrightarrow 5 - n$, and therefore in reality there are only three energy bands, where the bands of $n = 1, 2$ are doubly degenerate. These bands are illustrated in Fig. 7.14(a).

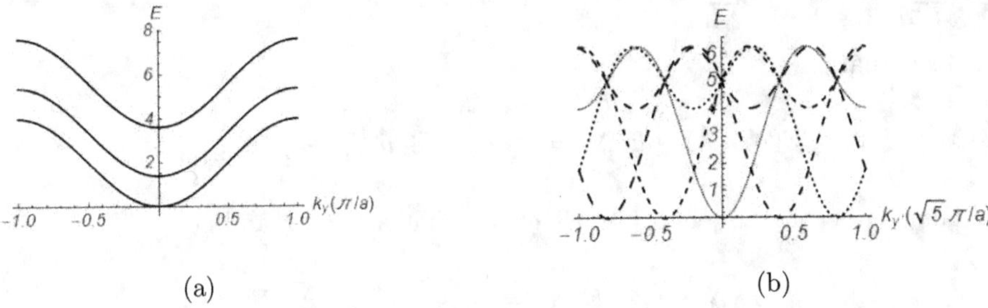

(a)   (b)

Fig. 7.14

b. In this case the unit vector of the one-dimensional lattice is $2\mathbf{a}_2 - \mathbf{a}_1$, perpendicular to the vector $\mathbf{L}_2$. The unit cell is a square whose edge length is $|\mathbf{L}_2| = a\sqrt{5}$; it comprises five sites: $\mathbf{r}_m = 0 , \mathbf{a}_2 , 2\mathbf{a}_2 , \mathbf{a}_1 + \mathbf{a}_2 , \mathbf{a}_1 + 2\mathbf{a}_2$. The periodicity along the normal to the cylinder's axis implies that $\psi(\mathbf{r} + \mathbf{L}_2) = \psi(\mathbf{r})$. In order to use Eq. (6.119) it is convenient to rotate the axes such that they point along the lattice vectors $\mathbf{L}_2$ and $2\mathbf{a}_2 - \mathbf{a}_1$. The rotation is with the angle

$$\phi = \arctan[L_{2y}/L_{2x}] = \arctan[1/2] ,$$

leading to

$$k_x = k_{x'}\cos(\phi) - k_{y'}\sin(\phi) , \quad k_y = k_{x'}\sin(\phi) + k_{y'}\cos(\phi) .$$

Using these values in the expression for the energy, with $k_{x'} = 2\pi n/(\sqrt{5}a)$ and continuous $k_{y'}$ yields five different energies, derived by displacements along $k_{y'}$; they are shown in Fig. 7.14(b).

Answer 7.7.

a. One begins by filling the ground-state energy, to obtain the energy $E(1) = \epsilon_1 - eV_g$. Adding more electrons gives the additional energies $E(2) = 2(\epsilon_1 - eV_g) + U_1$, $E(3) = 3(\epsilon_1 - eV_g) + 3U_1$, *etc.*, until

$$E(g_1) = g_1(\epsilon_1 - eV_g) + g_1(g_1 - 1)U_1/2 .$$

The next electron occupies the level ϵ_2; hence

$$E(g_1 + 1) = g_1(\epsilon_1 - eV_g) + \epsilon_2 - eV_g + g_1(g_1 - 1)U_1/2 + g_1 U_{1,2} ,$$

etc. Finally,

$$E(g_1 + g_2) = g_1(\epsilon_1 - eV_g) + g_2(\epsilon_2 - eV_g) + g_1(g_1 - 1)U_1/2$$
$$+ g_1 g_2 U_{1,2} + g_2(g_2 - 1)U_2/2 .$$

Once the ϵ_n level is filled up, the total energy is

$$E(N) = \sum_{\ell=1}^{n} [g_\ell(\epsilon_n - eV_{\mathrm{g}}) + g_\ell(g_\ell - 1)U_\ell/2] + m(\epsilon_n - eV_{\mathrm{g}}) + m(m-1)U_n/2$$

$$+ \sum_{\ell,\ell'=1}^{n-1} g_\ell g_{\ell'} U_{\ell,\ell'} + \sum_{\ell=1}^{n-1} m g_\ell U_{\ell,n} \; , \quad N = g_1 + g_2 + \ldots + g_{n-1} + m \; , \quad 0 < m \le g_n \; .$$

In this situation,

$$E(N+1) - E(N) = \epsilon_n - eV_{\mathrm{g}} + mU_n + \sum_{\ell=1}^{n-1} g_\ell U_{\ell,n} \; .$$

b. For non-degenerate energy levels $g_\ell = 1$. Then $m = 0$, and hence

$$E(N+1) - E(N) = \epsilon_{N+1} - eV_{\mathrm{g}} + \sum_{\ell=1}^{N} U_{\ell,n} \; .$$

As seen, the assumption $U_{\ell,n} = U$ leads to $E(N+1) - E(N) = \epsilon_{N+1} - eV_{\mathrm{g}} + NU$. The same result is obtained from Eq. (7.7) when the first term there is replaced by $N(N-1)U/2$. The approximation in Eq. (7.7) is valid as long as $N \gg 1$.

Answer 7.8.

a. The tight-binding equations for this situation are

$$E\psi(n) = -\gamma[\psi(n-1) + \psi(n+1)] \; , \quad n \ne 0, 1, \tag{7.20}$$

$$(E - E')\psi(1) = -(\gamma')^*\psi(0) - \gamma\psi(2) \; , \quad (E - E')\psi(0) = -\gamma'\psi(1) - \gamma\psi(-1) \; .$$

For an electron impinging from the left, the solution is of the form

$$\psi(n) = A(e^{ikna} + re^{-ikna}) \; , \quad n \le 0 \; ,$$

$$\psi(n) = Ate^{ikna} \; , \quad n \ge 1 \; .$$

These functions solve the first of Eqs. (7.20) with the dispersion relation

$$E(k) = -2\gamma\cos(ka) \; .$$

The other two equations of (7.20) give

$$t = -\frac{2i\gamma\sin(ka)(\gamma')^*e^{-ika}}{(E - E' + \gamma e^{ika})^2 - |\gamma'|^2} \; , \quad 1 + r = \frac{2i\gamma\sin(ka)(E - E' + \gamma e^{ika})}{(E - E' + \gamma e^{ika})^2 - |\gamma'|^2} \; .$$

When $\gamma = \gamma'$ and $E' = 0$, one finds that $r = 0$ and $t = 1$. In all other cases it can be checked that $T + R = |t|^2 + |r|^2 = 1$.

b. For an electron impinging from the opposite direction, the solution is of the form

$$\psi(n) = A(e^{-ikna} + r'e^{ikna}) \; , \quad n \ge 1 \; , \quad \psi(n) = At'e^{-ikna} \; , \quad n \le 0 \; .$$

Using the equations in part (a) give

$$t' = -\frac{2i\gamma\sin(ka)\gamma'e^{-ika}}{(E - E' + \gamma e^{ika})^2 - |\gamma'|^2} \; , \quad 1 + r'e^{2ika} = \frac{2i\gamma\sin(ka)(E - E' + \gamma e^{ika})}{(E - E' + \gamma e^{ika})^2 - |\gamma'|^2} \; .$$

As seen, $|r'|^2 = |r|^2$ and $|t'|^2 = |t|^2$.

c. The tight-binding equations for the sites 0 and A are

$$(E - E')\psi(0) = -\gamma'\psi(1) - \gamma\psi(-1) - \gamma_A e^{i\phi/2}\psi(A) ,$$

$$(E - E')\psi(1) = -(\gamma')^*\psi(0) - \gamma\psi(2) - \gamma_A e^{-i\phi/2}\psi(A) ,$$

$$(E - \epsilon_A)\psi(A) = -\gamma_A[e^{-i\phi/2}\psi(0) + e^{i\phi/2}\psi(1)] . \tag{7.21}$$

Using the third of Eqs. (7.21) to express $\psi(A)$ in terms of $\psi(0)$ and $\psi(1)$, and substituting it in the other equations there, reproduce the equations in part (a), with the replacements

$$\gamma' \Rightarrow \gamma' - (\gamma_A)^2 e^{i\phi}/(E - \epsilon_A) , \quad E' \Rightarrow E' + (\gamma_A)^2/(E - \epsilon_A) .$$

As seen, these replacements turn the triangle into an effective bond, whose matrix element includes the interference between the (previous) direct bond and the path that passes through the point A. The energies on the sites 0 and 1 are modified as well. One can now use the results obtained in part (a) to find the reflection and transmission coefficients. In particular, the latter depends on the phase ϕ only through the expression

$$|\gamma'|^2 \Rightarrow |\gamma_0' - (\gamma_A)^2 e^{i\phi}/(E - \epsilon_A)|^2$$
$$= (\gamma_0')^2 + (\gamma_A)^4/(E - \epsilon_A)^2 - 2\gamma_0'(\gamma_A)^2 \cos(\phi)/(E - \epsilon_A) ,$$

where it is assumed that the original γ' (denoted here by γ_0' for clarity) is real. Hence, the transmission coefficient is an even periodic function of the magnetic flux, $T(\phi) = T(\phi + 2\pi)$. It follows from Eq. (E.17) that the transmission coefficient returns to the same value when $\Phi \Rightarrow \Phi + \Phi_0$; the periodicity is the quantum unit of the flux, $\Phi_0 = hc/e$.

Answer 7.9.

a. The uncertainty in the momentum of an electron along the direction normal to the direction of the motion is rather small, of the order of the Fermi momentum. This implies that the uncertainty in the location of the electron along that direction is of the order of λ_F, the Fermi wave length. The probability of the electron to "cover" the volume $\lambda_F^{d-1} v_F dt$ at time t is that volume, divided by the total volume at that time, $(Dt)^{d/2}$. This probability is summed over the entire motion of the electron. The minimal relevant time for this sum is τ, i.e., the time for a single collision to occur. On the other hand, the electron's wave function remains coherent as long as the electron does not collide with the surfaces of the sample, or traverses the length L_ϕ. Since the motion is diffusive (a random walk) the maximal allowed time for the coherent motion is

$$\tau_x = [\min(L, L_\phi)]^2/D .$$

Once the integration (i.e., the sum) is carried out, one finds that the probability to return to the point where the motion started is

$$p = \frac{\lambda_F^{d-1} v_F}{D^{d/2}(1 - d/2)} \left(\tau_x^{1-d/2} - \tau^{1-d/2}\right) .$$

In two dimensions, $d = 2$, the result is $(\lambda_{\rm F} v_{\rm F}/D)\ln(\tau_x/\tau)$. When $L_\phi \ll L$, the probability p is independent of the system's size L. In contrast, at low temperatures $L_\phi \gg L$; the probability p is dominated by

$$\tau_x^{1-d/2} \propto L^{2-d} \Rightarrow L \,, \ln[L/\ell] \,, L^{-1} \,, \quad \text{for} \quad d = 1 \,, 2 \,, 3 \,, \quad \text{respectively} \,.$$

b. The probability to return to the origin (the point where the motion began) expresses the loss of conductance. One therefore may argue that $\Delta\sigma/\sigma = -p$, and at low temperatures

$$\Delta\sigma \sim -(e^2/h)L \,, \ \Delta\sigma \sim -(e^2/h)\ln[L/\ell] \,, \ \Delta\sigma \sim -(e^2/h)/L \,,$$

in $d = 1 \,, 2 \,, 3$.

7.5 Problems for self-evaluation

s.7.1.
Figure 7.15 illustrates the molecules of a liquid crystal from a side view. Discuss the order parameter that describes the locations of the centers of the molecules, and the one that corresponds to the directions of their axes. This phase is referred to as "chiral smectic". Explain why, and determine when this term is indeed valid.

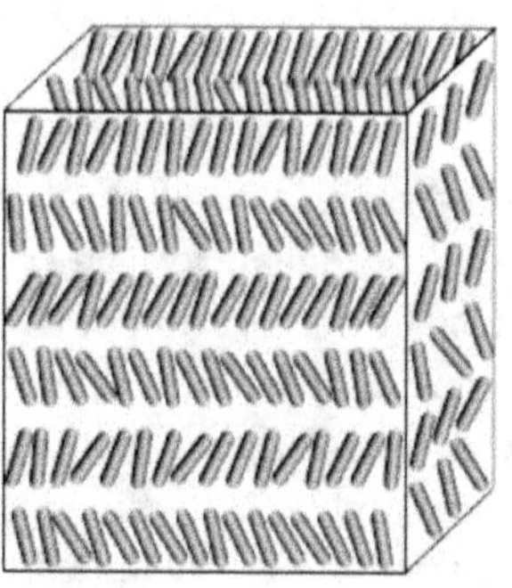

Fig. 7.15

s.7.2.
*Many of the organic molecules are planar, for instance that of benzene, Fig. 4.2. Such molecules form liquid crystals that are called "**discotic**". Describe several examples of discotic liquid crystals, that are built of thin (with width much smaller than the diameter) discs.*

s.7.3.

A particle "walks" randomly on an infinite lattice, beginning the walk at the origin. At each time-step it "jumps" from one lattice site to one of the nearest-neighbor sites, with the probability $w = 1/z$, where z is the coordination number of the lattice (i.e., the number of nearest neighbors).

a. *Express the probability to reach the site $\mathbf{R}$ after N steps, $P(\mathbf{R}, N)$, in terms of the probability to reach one of the nearest neighbors after $(N-1)$ steps. Confining yourselves to one dimension, compare this relation to the ones that appear in previous chapters of the main text.*

b. *Derive the continuum limit of the relation found in part (a) (i.e., the limit where the lattice constants approach zero, while the number of steps is very large). Show that the Gaussian distribution (problem 7.3) solves this relation.*

c. *Obtain the equation for the Fourier transform, $\widetilde{P}(\widetilde{\mathbf{k}}, N)$, show that $\widetilde{P}(\widetilde{\mathbf{k}}, N) = [\widetilde{P}(\mathbf{k}, 1)]^N$, and find $\widetilde{P}(\mathbf{k}, 1)$.*

d. *As in problem 7.3, show that for $N \gg 1$ one may expand $\widetilde{P}(\mathbf{k}, 1)$ in a Taylor series, and use this expansion and the inverse Fourier transform to find an approximate expression for $P(\mathbf{R}, N)$. Compare the result with the one found in problem 7.3.*

s.7.4.

Figure 1.9(a) shows a nanotube. When such a tube is cut along a line parallel to the axis of the cylinder, a rectangular piece of graphene is created. Figure 7.16 displays two vectors on the graphene lattice, $\mathbf{L}_1$ and $\mathbf{L}_2$. This lattice is cut along parallel lines that are normal to each of these vectors at the edges, and is rolled to form nanotubes, such that the parallel lines are congruous to each other.

a. *Determine the unit cell of the one-dimensional lattice that is created in each of the cases. How many lattice sites are contained in this unit cell?*

b. *Find an expression for the energy bands within the tight-binding approximation for the lattice formed by $\mathbf{L}_1$ (assume that the bonds of the tight-binding model are between nearest neighbors).*

c. *In the case where the nanotube is formed by $\mathbf{L}_2$, explain qualitatively how should the band structure be calculated, and how many bands are there.*

s.7.5.

The transmission coefficient of the ring in Fig. 7.17 is $T = 1 - R$. The ring is threaded by a magnetic flux Φ, which is created by a magnetic field perpendicular to the plane of the ring. The reflection amplitude of an electron impinging on the ring comes from the following contributions: r_0 for the path reflected from the impinging point (it is assumed to be real), $r_1 \exp[i(\delta + \phi)]$ from the path that encircles the ring once in the clockwise sense and then returns, (the phase δ is due to elastic scattering by impurities randomly located on the ring, and $\phi = 2\pi\Phi/\Phi_0$ is called the Aharonov-Bohm phase), $r_1 \exp[i(\delta - \phi)]$ from the path that encircles the ring once in the anti-clockwise sense and then returns, and so on.

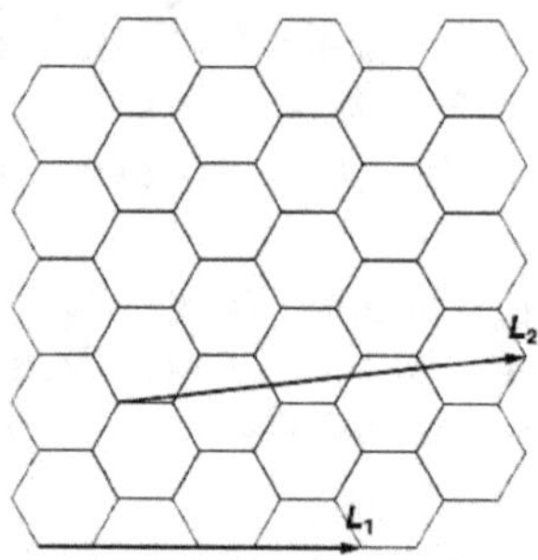

Fig. 7.16

a. *Explain these expressions, and add the expressions for the paths that encircle the ring more times. Use the results to derive the conductance through the ring, and show that it is even and periodic in ϕ. Identify the contributions of the first two harmonics of the conductance.*

b. *A famous experiment measured the conductance of a cylinder between two lines parallel to its axis. Figure 7.17 illustrates the projection of the cylinder along that axis. Assuming that the experiment averaged over the conductances of many parallel rings, what should it measure? Determine the leading harmonic of the measured conductance.*

c. *Part (c) of problem 7.8 solves for the reflection coefficient of a system in which a triangle replaces the ring in Fig. 7.17. Show that an expansion of the reflection coefficient found in problem 7.8 resembles the solution for the reflection from the ring.*

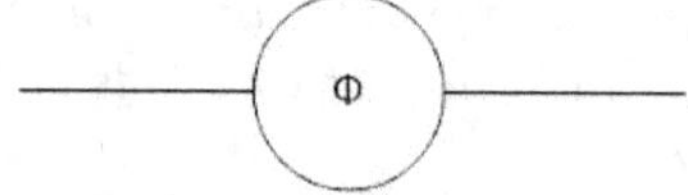

Fig. 7.17

7.6   Answers for the self-evaluation problems

Answer s.7.1. If one accounts only for the centers of mass of the molecules, then Eq. (7.2) for the order parameter holds along the direction normal to the planes, with $q = 2\pi/\ell$. An examination of the directions of the director vectors reveals that they return to themselves only at each second plane; hence the order parameter of Eq. (7.1) behaves in an antiferromagnetic fashion along $\hat{\mathbf{N}}$, with periodicity 2ℓ. The projections of these vectors on the planes, however, behave as in the cholesteric phase, Fig. 7.3, with periodicity 2ℓ. One concludes that this phase is a combination of a smectic phase with a cholesteric (or a chiral) one.

Answer s.7.2.
The right panel in Fig. 7.18 shows a nematic phase, in which the director is perpendicular to the discs. The next panel displays a columnar structure, in which the discs are arranged periodically in separate columns such that each of them is a periodic one-dimensional lattice; the columns themselves are scattered randomly in space. In the third panel the columns are parallel to each other, but arranged at random in the plane perpendicular to their direction. In the left panel the discs are perpendicular to the planes which by themselves are arranged periodically. Within each plane the discs are parallel to each other, and their directions in neighboring planes form a certain finite angle. This is a cholesteric phase.

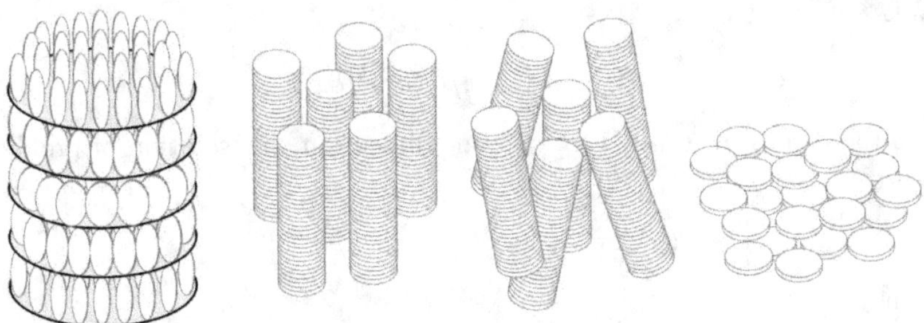

Fig. 7.18

Answer s.7.3.
a. The probability that the particle is at $\mathbf{R}$ in the N-th time step due to the jumps to this point from nearest-neighbor ones is

$$P(\mathbf{R}, N) - P(\mathbf{R}, N-1) = w \sum_{\boldsymbol{\delta}} [P(\mathbf{R} + \boldsymbol{\delta}, N-1) - P(\mathbf{R}, N-1)] , \qquad (7.22)$$

where the sum runs over the z nearest neighbors of $\mathbf{R}$, located at $\mathbf{R}+\boldsymbol{\delta}$. For instance, there are two nearest neighbors, at $\boldsymbol{\delta} = \pm a$, in the one-dimensional lattice, four in the square lattice at $\boldsymbol{\delta} = \pm\mathbf{a}_1, \pm\mathbf{a}_2$, *etc.*

In one dimension, the right hand-side of Eq. (7.22) resembles that of Eq. (5.3), under the mapping $P(na, N) \Rightarrow u_n(t)$. At $N \gg 1$ the left-hand side can be approximated by $P(\mathbf{R}, N) - P(\mathbf{R}, N-1) \approx \partial P/\partial N$. Assuming that each step occurs during the short time t_0, then $t = Nt_0$, and then $\partial P/\partial t = (1/t_0)\partial P/\partial N$, i.e., a first derivative with respect to time in contrast with the second derivative with respect to time that appears in Eq. (5.3). Similar equations appear in Sec. 5.9.

Still confining oneself to one dimension, one notes that the right-hand side of Eq. (7.22) resembles the one that appears in the solution of problem 6.25, for wave functions in the tight-binding approximation. That equation is derived from the Schrödinger equation for the stationary solution $\exp[-iHt/\hbar]$. The energy E replaces the time derivative in the Schrödinger equation, $i\hbar\partial/\partial t$. Hence, the equation for the random walk resembles the tight-binding approximation for the Schrödinger equation, save for the factor i in the latter. This resemblance persists also at dimensions higher than 1.

b. In the continuum limit, the right-hand side of Eq. (7.22) in one dimension is

$$P(x+a) + P(x-a) - 2P(x) \approx a^2 \partial^2 P/\partial x^2 \ .$$

In higher dimensions, the second derivative approaches $\nabla^2 P$, as in the wave equation, the Schrödinger equation, *etc.* Then, replacing the left-hand side

$$\partial P/\partial N \Rightarrow t_0 \partial P/\partial t \ ,$$

one finds

$$\partial P/\partial t = (a^2 w/t_0)\partial^2 P/\partial x^2 \ ,$$

which is the **diffusion equation**.[2] In one dimension, the solution of problem 7.3 becomes

$$P(x, N) = (2\pi a^2 N)^{-1/2} e^{-x^2/(2a^2 N)} \ ,$$

and then

$$\partial P/\partial N = [x^2/(2a^2 N^2) - 1/(2N)]P \ , \quad \partial P/\partial x = -[x/(a^2 N)]P \ ,$$

from which follows that

$$\partial P/\partial N = (a^2/2)\partial^2 P/\partial x^2 \ ,$$

which is the diffusion equation with $w = 1/2$. A similar result can be derived in three dimensions.

c. Applying the Fourier transform

$$\widetilde{P}(\mathbf{k}, N) = \frac{1}{(2\pi)^3} \int d^3 R e^{-i\mathbf{k}\cdot\mathbf{R}} P(\mathbf{R}, N) \ ,$$

on the equation (with a shift of the variables in certain integrals), gives

$$\widetilde{P}(\mathbf{k}, N) = w \sum_{\boldsymbol{\delta}} \widetilde{P}(\mathbf{k}, N-1)e^{i\mathbf{k}\cdot\boldsymbol{\delta}} \quad \Rightarrow \quad \widetilde{P}(\mathbf{k}, N) = \left[w \sum_{\boldsymbol{\delta}} e^{i\mathbf{k}\cdot\boldsymbol{\delta}}\right]^N \ ,$$

---

[2]Fick's law states that the current of matter is proportional to the gradient of its density, i.e., $\mathbf{J} = -D\nabla n$ where D is the diffusion coefficient. The continuity equation, $\partial n/\partial t + \nabla \cdot \mathbf{J} = 0$, then yields $\partial n/\partial t = \nabla \cdot (D\nabla n)$. The usual diffusion equation is obtained when D is a constant.

where the initial condition $P(\mathbf{R}, 0) = \delta(\mathbf{R})$, which implies $\widetilde{P}(\mathbf{k}, 0) = 1$, is used. The expression in the square brackets resembles expressions for the frequencies of lattice vibrations, and for band energies in the tight-binding approximation.

d. The Taylor expansion is

$$\widetilde{P}(\mathbf{k}, 1) = w \sum_{\boldsymbol{\delta}} [1 + i\mathbf{k} \cdot \boldsymbol{\delta} - (\mathbf{k} \cdot \boldsymbol{\delta})^2/2 + \ldots] \ .$$

This is similar to the expansion used in the solution of problem 7.3. In a Bravais lattice $\sum_{\boldsymbol{\delta}} \boldsymbol{\delta} = 0$; $\sum_{\boldsymbol{\delta}} (\mathbf{k} \cdot \boldsymbol{\delta})^2/z$ replaces the expression $\langle (\mathbf{k} \cdot \mathbf{R})^2 \rangle$ that appears there. For hyper-cubic lattices in d dimensions

$$\sum_{\boldsymbol{\delta}} (\mathbf{k} \cdot \boldsymbol{\delta})^2/z = 2 \sum_{\alpha=1}^{d} k_\alpha^2 a^2/(2d) = k^2 a^2/d$$

(check!). For other lattices one obtains different quadratic expressions. For instance,

$$\sum_{\boldsymbol{\delta}} (\mathbf{k} \cdot \boldsymbol{\delta})^2/z = 2[(\mathbf{k} \cdot \mathbf{a}_1)^2 + (\mathbf{k} \cdot \mathbf{a}_2)^2 + (\mathbf{k} \cdot (\mathbf{a}_1 - \mathbf{a}_2))^2]/6 = k^2 a^2/4 \ ,$$

in the triangular lattice. A change of variables in the integral of the inverse Fourier transform, as in problem 7.3, justifies the neglect of the higher-order terms in the expansion, and thus leads again to the Gaussian distribution. Save for the coefficient of k^2, which determines $\langle [\mathbf{R}(N)]^2 \rangle$, this distribution is **universal**.

Answer s.7.4.

a. In order to identify the cylindrical structure which is formed, one has to add lines perpendicular to the vectors $\mathbf{L}_1$ and $\mathbf{L}_2$, at their edges, see Fig. 7.19. These are the edges of the sheet that needs to be wrapped. The perpendicular line ends at a lattice site that hosts the same point as the origin of the line (recall that the hexagonal lattice contains two basis points in each rhombohedral unit cell). The latter lattice site belongs to the next one-dimensional unit cell of the cylinder, and the vector that reaches it from the origin is the basis vector of the one-dimensional lattice. The one-dimensional unit cell contains all points within the rectangle created by this vector and the original vector. These unit cells are displayed in Fig. 7.19 by the dashed lines. The cell built on $\mathbf{L}_1$ contains eight lattice sites (four points of each type of the original hexagonal lattice), and the one built on $\mathbf{L}_2$ contains 76 lattice sites (38 of each type).

b. The energy spectrum of the graphene lattice is calculated in problem 6.24, within the tight-binding model, and accounting only for overlaps between nearest neighbors. The basis vectors used there are marked in Fig. 7.19, $\mathbf{a}_1 = a(3\hat{\mathbf{x}} + \sqrt{3}\hat{\mathbf{y}})/2$ and $\mathbf{a}_2 = a\sqrt{3}\hat{\mathbf{y}}$. The energies are

$$E_\pm(\mathbf{k}) = E^{(0)} \pm |\widetilde{\gamma}_{AB}(\mathbf{k})| \ ,$$

$$\widetilde{\gamma}_{AB}(\mathbf{k}) = \sum_{\text{nn}} \gamma_{AB} e^{-i\mathbf{k} \cdot \mathbf{R}_{\text{nn}}} = \gamma_{AB}(e^{-i\mathbf{k} \cdot \mathbf{a}_1} + e^{-i\mathbf{k} \cdot \mathbf{a}_2} + e^{-i\mathbf{k} \cdot (\mathbf{a}_1 + \mathbf{a}_2)}) \ .$$

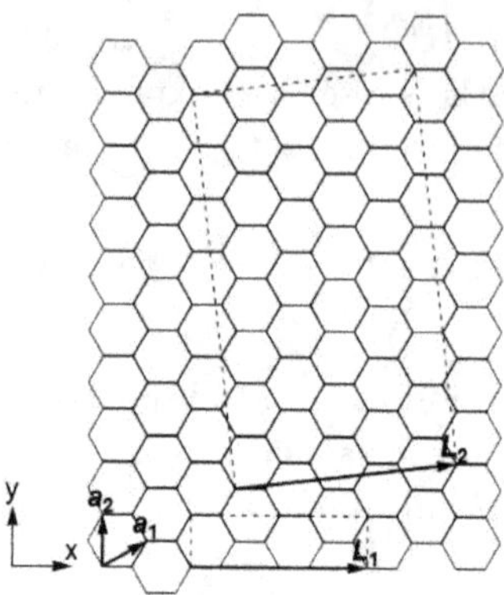

Fig. 7.19

Explicitly,

$$|\tilde{\gamma}_{AB}(\mathbf{k})|^2 = |\gamma_{AB}|^2[3 + 2\cos(\sqrt{3}k_y a) + 4\cos(3k_x a/2)\cos(\sqrt{3}k_y a/2)] \ .$$

In the case of the cylinder built on the vector $\mathbf{L}_1 = 4\mathbf{a}_1 - 2\mathbf{a}_2 = 6a\hat{\mathbf{x}}$, one has to add the periodic boundary condition along the $\hat{\mathbf{x}}$−axis, $k_x L_1 = 6k_x a = 2\pi n$, and therefore there are six discrete bands as a function of k_y (which remains continuous), on the two sides of $E^{(0)}$.

c. In this case one expresses k_x and k_y in terms of k'_x and k'_y, that are obtained from the rotation in the plane that transfers the $\hat{\mathbf{x}}$−axis into the direction of $\mathbf{L}_2 = 5\mathbf{a}_1 - 2\mathbf{a}_2 = (15\hat{\mathbf{x}} + \sqrt{3}\hat{\mathbf{y}})a/2$. The periodic boundary condition yields $k'_x L_2 = \sqrt{57}k'_x a = 2\pi n$.

Answer s.7.5.

a. In the absence of the magnetic flux, the reflection amplitude for traversing the ring is independent of the sense of the motion; i.e., it is $r_1\exp[i\delta]$ for both the clockwise motion and the anti-clockwise one. On the other hand, the magnetic Aharonov-Bohm phase reverses its sign when the direction of motion is reversed. This is the reason for the two different reflection amplitudes for the two directions of motion. When the path encircles the ring m times the two reflection amplitudes are $r_m\exp[im(\delta \pm \phi)]$. It follows that the reflection coefficient is

$$R = |r_0 + \sum_{m=1}^{\infty} r_m e^{im\delta}(e^{im\phi} + e^{-im\phi})|^2$$

$$= |r_0|^2 + 2|r_1|^2 + 4|r_0||r_1|\cos(\delta)\cos(\phi) + 2|r_1|^2\cos(2\phi) + \ldots \ .$$

The reflection coefficient is indeed periodic in Φ, with periodicity Φ_0; the explicit expression contains all harmonics.

b. Different rings contain different realizations of scatterers (but with the same mean concentration), and therefore the phase δ attains many different values. It follows that $\cos(\delta)$ is averaged to zero in an ensemble of many rings. Glancing at the expression obtained in part (a) shows that as a result the odd harmonics (in the flux) of the mean reflection coefficient disappear. The leading harmonics comes from the term with $\cos(2\phi)$, which is a periodic function of periodicity $\Phi_0/2$. This is indeed observed experimentally.

c. The reflection amplitude found in part (c) of problem 7.8 was

$$r = \frac{2i\gamma \sin(ka)(E - E' + \gamma e^{ika})}{(E - E' + \gamma e^{ika})^2 - |\gamma'|^2} - 1 \ ,$$

with

$$\gamma' = \gamma_0 - (\gamma_A)^2 e^{i\phi}/(E - \epsilon_A) \ , \quad E' = E_0 + (\gamma_A)^2/(E - \epsilon_A) \ ,$$

where γ_0 and E_0 are the "bare" values of γ' and of E'. The denominator in the expression for r depends on the parameters of the triangle only via the expression

$$|\gamma'|^2 = (\gamma_0)^2 + (\gamma_A)^4/(E - \epsilon_A)^2 - \gamma_0 \gamma_A^2 (e^{i\phi} + e^{-i\phi})/(E - \epsilon_A) \ .$$

Each of the terms on the RHS represents a "motion" of the electron from site 0 to site 1 and back: the first term represents motion via the lower side of the triangle, the second term represents the motion from 0 to A to 1 and back, and the last term represents the clockwise motion, 0 to A to 1 to 0 or the anticlockwise motion, 0 to 1 to A to 0.

If one ignores the last term in $|\gamma'|^2$ then the resulting expression for r contributes only to r_0. Expanding r in a power series in $\gamma_0 \gamma_A^2 (e^{i\phi} + e^{-i\phi})/(E - \epsilon_A)$, it is easy to see that the first term yields r_1. The second term is proportional to $(e^{2i\phi} + e^{-2i\phi} + 2)$. The first two term contribute to r_2, and the last term (which results from a motion clockwise followed by a motion anticlockwise) contributes to r_0. Similarly, the m's term in the series contributes to all the r_n's with $0 \le n \le m$.

Index

Name	Value		CGS	SI	Notation
Planck's constant	6.6262	$\times$	$10^{-27} erg \cdot sec$	$10^{-34} J \cdot s$	h
in Ev	4.1357	$\times$		$10^{-15} eV \cdot s$	
Reduced Planck's constant	1.0546	$\times$	$10^{-27} erg \cdot sec$	$10^{-34} J \cdot s$	$\hbar$
in Ev	6.5822	$\times$		$10^{-16} eV \cdot s$	
Boltzman's constant	1.3807	$\times$	$10^{-16} erg \cdot K^{-1}$	$10^{-23} J \cdot K^{-1}$	k_B
in Ev	8.6173	$\times$		$10^{-5} eV \cdot K^{-1}$	
Electron charge	1.6021	$\times$	$10^{-12} erg \cdot eV^{-1}$	$10^{-19} coloumb$	e
Electron mass	9.1095	$\times$	$10^{-28} g$	$10^{-31} kg$	m
Bohr magneton	9.2740	$\times$	$10^{-21} erg \cdot G^{-1}$	$10^{-24} J \cdot T^{-1}$	μ_B
in Ev	5.7883	$\times$		$10^{-5} eV \cdot T^{-1}$	
Bohr radius	5.2917	$\times$	$10^{-9} cm$	$10^{-11} m$	a_B
Rydberg's constant	13.605	$\times$		eV	Ry
Speed of light in vacuum	2.9979	$\times$	$10^{10} cm \cdot sec^{-1}$	$10^8 m \cdot s^{-1}$	c
Avogadro's number	6.0221	$\times$		$10^{23} mol^{-1}$	N_A
Klitzing constant	2.5812	$\times$		$10^4 \Omega$	R_K
Permittivity of vacuum	8.8541	$\times$		$10^{-12} F \cdot m^{-1}$	ϵ_0
Permeability of vacuum	4π	$\times$		$10^{-7} H \cdot m^{-1}$	μ_0